A COURSE IN PURE MATHEMATICS

A COURSE IN

PURE MATHEMATICS

MARGARET M. GOW
M.A., Ph.D.
Formerly Head of the Department of Mathematics,
De La Salle College of Education, Manchester

HODDER AND STOUGHTON
LONDON SYDNEY AUCKLAND TORONTO

ISBN 0 340 05217 1 (Hodder International Edition)

First Printed 1960 Fourteenth impression 1982

Printed in Hong Kong for
Hodder and Stoughton Educational,
a division of Hodder and Stoughton Ltd,
Mill Road, Dunton Green, Sevenoaks, Kent,
by Colorcraft Ltd.

FOREWORD

TO THE FIRST EDITION, 1960

The introduction of the new syllabus in Mathematics for Part I of the London B.Sc. General Degree has made desirable a book designed especially to cover its requirements. Dr. Gow, who in many years' teaching has had considerable experience of the needs of students at this level, has now provided a course in Algebra, Co-ordinate Geometry, and the Differential and Integral Calculus, constructed with this new syllabus particularly in mind.

There are greatly increased numbers of students reading Mathematics either as part of a general degree or as an ancillary subject. For them Mathematics should be presented in as straightforward a manner as is consistent with a moderate standard of rigour. In this country the output of books catering for these students has been surprisingly small. For this reason alone the book should be welcome and prove very useful to first-year (and some second-year) students not only in London but also in other universities, and to students reading Mathematics as an ancillary subject in universities, colleges of advanced technology, and some of the technical colleges.

The book assumes a knowledge of Mathematics up to Advanced Level of the General Certificate of Education and builds up a comprehensive First Year Course which should be well suited to the needs of students who are not actually specialising in Mathematics. Each chapter contains worked examples illustrating the text. Excellent and lengthy collections of miscellaneous examples, taken from examinations set by London University and several of the Northern Universities, are to be found at the end of each chapter, and groups of selected examples are inserted at points within the chapters to help the student to acquire practice with new ideas and techniques as they are introduced. These features should be of great value to students who may have to work without much supervision.

D. C. PACK, C.B.E., M.A., D.Sc.,
Professor of Pure Mathematics
in the University of Strathclyde

PREFACE

TO THE FIRST EDITION, 1960

Recent changes in the examination requirements of various Universities and professional institutions have prompted me to write a new book on Pure Mathematics which I hope will prove suitable for a wide range of scientists and engineers.

This book contains a progression of integrated courses in Algebra, Co-ordinate Geometry, and Differential and Integral Calculus suitable for candidates who wish to read for Part I of the London B.Sc. (General) Degree under the revised regulations, also for those who wish to offer Mathematics ancillary to Honours courses in Physics, Chemistry, Physiology, Psychology, Geology and Geography in Universities and Colleges of Technology. A knowledge of Pure Mathematics up to G.C.E. Advanced Level is assumed, but wherever possible some revision of the subject at that level is provided. To facilitate private study brief hints as well as answers are given to practice examples.

The book contains about 350 examples which are fully worked out in the text, and over 1500 exercises drawn from recent examination papers of the Universities of Durham, Leeds, Liverpool, London and Sheffield. I am grateful to the Senates of these Universities for permission to use their questions.

In writing this book I have received advice and encouragement from many friends and colleagues, but I am particularly indebted to Mr. A. D. D. McKay, former Head of the Department of Mathematics, Dundee Technical College, for his helpful suggestions when the manuscript was in its first draft; also to Professor D. C. Pack, to Dr. James Cossar, Mathematics Department, University of Edinburgh, and to Professor Raymond Smart, Professor of Mathematics, Heriot-Watt College, Edinburgh, who all read the manuscript in its final form and offered constructive criticism.

My thanks are also due to Mr. T. L. Sime, B.Sc., and Miss M. Yeaman, D.A., who prepared the diagrams, to Mr. A. F. Ewan, M.A., B.Sc., and Miss J. E. R. Thomson, B.Sc., who read the proofs, and to the staff of Messrs. T. and A. Constable Ltd. for their unfailing cooperation.

MARGARET M. GOW

CONTENTS

1. POLYNOMIALS; THE REMAINDER AND FACTOR THEOREMS; UNDETERMINED COEFFICIENTS; PARTIAL FRACTIONS . 1

Polynomials. The remainder and factor theorems. Further properties of polynomials. The method of undetermined coefficients. Factorisation of cyclic homogeneous polynomials. Rational functions. Partial fractions.

2. THEORY OF EQUATIONS; INEQUALITIES 14

Algebraic equations. Relations between the coefficients and the roots of an algebraic equation. Transformation of equations. Reciprocal equations. Condition for two quadratic equations to have a common root. Condition for an equation to have a multiple root. Note on inequalities.

3. DETERMINANTS 26

Definitions. Properties of determinants. Note on determinants of the fourth and higher orders. Solution of linear simultaneous equations. Homogeneous equations. Consistency of equations.

4. LIMITS AND INFINITE SERIES 53

Definitions. The idea of infinity. The idea of a limit. Convergent series. Series which do not converge. Behaviour of x^n. The geometrical progression. Theorems on limits. General theorems on convergence. A necessary condition for convergence. The harmonic series. Series of positive terms. The comparison tests. Two standard series. The ratio test for series of positive terms. Series of positive and negative terms. Absolute convergence. Tests for absolute convergence. Properties of power series. Functions of a continuous variable. Limit of a function of a continuous variable. Continuous functions. A fundamental property of a continuous function.

5. THE BINOMIAL, EXPONENTIAL AND LOGARITHMIC SERIES . 74

The binomial series. Vandermonde's theorem. The binomial theorem for any rational index. The number e. The exponential theorem. Properties of e^x. Series for a^x, $a>0$. Hyperbolic functions. Inverse hyperbolic functions. The logarithmic series. Series related to the logarithmic series. Limits.

6. COMPLEX NUMBERS 102

Complex numbers. Operations with complex numbers. Geometrical representation of complex numbers. Modulus and argument of a complex number. Vectorial representation of a complex number. Geometrical representation of addition and subtraction of two complex numbers. Multiplication and division of complex numbers. Geometrical construction for the product of two complex numbers. Geometrical application

7. COMPLEX NUMBERS (*continued*) 119

Demoivre's theorem. Roots of a complex number. The nth roots of unity. Complex roots of an equation. Real quadratic factors. Expression of powers of $\sin\theta$ and $\cos\theta$ in terms of multiple angles. Expansions of circular functions of multiple angles. Series of complex terms. The exponential series. Exponential values of circular functions. Generalised circular and hyperbolic functions. Connection between the circular and hyperbolic functions. Logarithms of a complex number. The logarithmic series.

8. SUMMATION OF SERIES 149

Standard results. Methods of summing series. Series whose rth term is a polynomial in r. The method of induction. Series reducible to exponential, binomial or logarithmic series. The method of differences. Summation of trigonometrical series using complex numbers.

9. DIFFERENTIATION AND APPLICATIONS 170

Differentiation. General rules. Two important limits. Differentiation of x^n, $\sin x$, $\sin^{-1} x$, e^x and $\log x$. Standard results (algebraic, logarithmic, exponential, trigonometrical, hyperbolic, inverse trigonometrical and inverse hyperbolic functions). Extensions of standard results. Differentiation of logarithmic functions. Successive differentiation. The theorem of Leibniz. Formulae for dy/dx and d^2y/dx^2 when x and y are given in terms of a parameter. The gradient of a curve. The positive tangent. The tangent and normal to a curve. The mean value theorem. The significance of the sign of $f'(x)$. Maximum and minimum points. Concavity and convexity. Tests for maxima and minima.

10. INTEGRATION 200

Integration as the inverse of differentiation. The indefinite integral. The definite integral. Standard integrals. Integration by substitution. Extensions of standard forms. Integration of rational functions. Irrational functions. Trigonometrical integrals evaluated by substitution. Products of sines and cosines of multiple angles. Powers of $\sin x$ and $\cos x$. Powers of $\tan x$ and $\cot x$. Hyperbolic functions. Integration by parts. Improper or infinite integrals. Reduction formulae for $\int \sin^n x\,dx$ and $\int \cos^n x\,dx$. Reduction formula for $\int \sin^m x \cos^n x\,dx$, where m and n are positive integers. Wallis's formulae. Reduction formulae for $\int \tan^n x\,dx$ and $\int \sec^n x\,dx$.

11. EXPANSIONS IN SERIES 244

Power series. Maclaurin's series. Taylor's series. Series for $\sin x$, $\cos x$, $(1+x)^n$ and $\log(1+x)$. Other methods of expanding a function $f(x)$. Use of Leibniz's theorem in the expansion of $f(x)$. Indeterminate forms. De l'Hospital's rule. Newton's approximation to a root of an equation. Modification of Newton's formula.

12. COORDINATE GEOMETRY OF THE STRAIGHT LINE AND CIRCLE 269

Useful formulae (revision). Parametric equations of a straight line. Change of axes. Pairs of straight lines. The homogeneous equation of the second degree in x and y. The angle between two lines. The bisectors of the angles between two lines. Condition that the general equation of the second degree in x and y should represent two straight lines. The circle. Useful formulae (revision). The equation of the tangent to a circle. Tangents of gradient m to the circle $x^2+y^2=a^2$. The chord of contact of tangents to a circle. The power of a point. The radical axis of two circles. Coaxal circles. Intersecting and non-intersecting systems of coaxal circles. Limiting points.

13. THE PARABOLA 300

Conic sections. The parabola (revision). Chord of contact. The chord $[t_1, t_2]$. Focal chords. The tangent and normal at $[t]$. The point of intersection of tangents. The tangent of gradient m. The locus of the mid-points of parallel chords. Conormal points. Concyclic points.

14. THE ELLIPSE AND HYPERBOLA 313

The ellipse (revision). The auxiliary circle and eccentric angle. The chord $[\theta, \phi]$. Tangent and normal at $[\theta]$. The point of intersection of tangents. Tangents which are parallel to the diameter $y=mx$. The director circle. The locus of the mid-points of parallel chords. Conjugate diameters. The hyperbola (revision). Parametric representation. Standard results. Conjugate diameters. The conjugate hyperbola. The rectangular hyperbola. The rectangular hyperbola referred to its asymptotes as axes. Parametric representation. The tangent and normal. Conormal points. Concyclic points.

15. THE STRAIGHT LINE, CIRCLE AND CONIC IN POLAR COORDINATES 337

Polar coordinates. Relations between cartesian and polar coordinates. The straight line, circle and conic in polar coordinates. The chord joining two points on the conic $l/r=1+e\cos\theta$. The tangent to the conic.

16. COORDINATE GEOMETRY OF THREE DIMENSIONS: THE PLANE AND STRAIGHT LINE 346

Coordinates of a point in space. Section formula. Direction cosines of a straight line. Length, direction and equations of the line joining two points. Direction ratios. Note on projection. The angle between two straight lines. The equation of a plane. The angle between two planes. The perpendicular distance of a point from a plane. The condition for coplanar lines. The perpendicular distance of a point from a line. The shortest distance between two skew lines. Simplest form of the equations of two skew lines.

17. THE SPHERE 378

The equation of a sphere. The diametral form of the equation of a sphere. Tangent plane to a sphere. Condition that a plane should touch a sphere. The intersection of a plane and a sphere. The power of a point. The radical plane. Orthogonal spheres. Pencils of spheres. Polar plane. Polar lines.

18. THE QUADRIC 388

The equation of a surface. Quadric surfaces. Central quadrics. The intersections of a line and a quadric. Tangent plane and normal to a quadric. The condition for a plane to touch a quadric. The plane containing all chords bisected at a given point. Polar planes and polar lines. Note on ruled surfaces.

19. PARTIAL DIFFERENTIATION 403

Continuous functions of several variables. Partial derivatives. Geometrical interpretation. Partial derivatives of higher orders. Total variation. Differentials. Total differential of a function of two variables. Total derivative. Applications of differentials. Euler's theorems on homogeneous functions. Note on envelopes. The tangent to a space curve. Tangent lines to a surface. The tangent plane and normal to a surface.

20. APPLICATIONS OF INTEGRATION—CARTESIAN COORDINATES . 441

The definite integral as the limit of a sum. Areas. Mean value. Volumes. Curve sketching from cartesian equations. Length of arc and surface area. Differential relations. Curves given in parametric form. Centres of mass. Centroid of a uniform plane lamina, and solid of revolution. Theorems of Pappus. Moments of inertia. Moments of inertia about perpendicular axes. The principle of parallel axes. Calculation of moments of inertia.

21. APPLICATIONS OF INTEGRATION—POLAR COORDINATES . 477

Length of arc of a polar curve. Tangents. The perpendicular from the pole to a tangent. Curve sketching from polar equations. Some well-known curves. Areas in polar coordinates. Areas of closed curves given in polar and in parametric forms. Surface area in polar coordinates. Centroid of a plane area. Roulettes: the cycloid, epicycloid and hypocycloid.

22. CURVATURE 499

Curvature. Curvature of a circle. Radius of curvature. Formulae for ρ applicable to curves given in cartesian, parametric or polar forms. Formula for ρ in terms of p and r. The circle of curvature. The evolute. Contact of two curves.

23. DIFFERENTIAL EQUATIONS OF THE FIRST ORDER . . . 512

Definitions. Formation and solution of differential equations. Equations of the first order and first degree. Variables separable. The homogenous equation. Equations reducible to homogeneous form. Exact equations. Integrating factors. The linear equation. Bernoulli's equation. Change of variable. Clairaut's equation. Geometrical applications. Tangents and normals in cartesian and polar coordinates. Orthogonal trajectories.

24. LINEAR DIFFERENTIAL EQUATIONS WITH CONSTANT COEFFICIENTS 537

The operator D. Applications of the operator D. Linear equation with constant coefficients. Solution of the reduced equation $\phi(D)y=0$. Solution of the equation $\phi(D)y=f(x)$ when $f(x)\neq 0$. Inverse operators. Use of operators in the evaluation of integrals. Simple harmonic motion; damped harmonicmotion; forced oscillations. The homogeneous linear equation. Simultaneous linear equations with constant coefficients. Change of variable.

25. SPHERICAL TRIGONOMETRY 562

Spherical triangle. Area of a lune. Area of a spherical triangle. The cosine formula. The sine formula. The cotangent formula. The polar triangle. Sides and angles of polar triangles. Supplemental formulae. Right-angled triangles. Napier's rules. Quadrantal triangles. Measurements on the earth's surface.

HINTS AND ANSWERS 577

INDEX 613

CHAPTER 1

POLYNOMIALS; THE REMAINDER AND FACTOR THEOREMS; UNDETERMINED COEFFICIENTS; PARTIAL FRACTIONS

1.1. Polynomials in one variable

If x is a variable, n is a positive integer and $p_0, p_1, p_2, \ldots, p_n$ are given constants of which p_0 is not zero, then

$$p_0x^n+p_1x^{n-1}+ \ldots +p_{n-1}x+p_n$$

is a polynomial of degree n in x. We shall denote this polynomial by $P(x)$.

1.2. The remainder and factor theorems

Let $P(x)$, a polynomial of degree n in x, be divided by $x-a$, where a is a constant. Then the quotient $Q(x)$ is a polynomial of degree $n-1$, the remainder R is a constant and

$$P(x)\equiv(x-a)Q(x)+R \qquad \text{(i)}$$

The identity sign $\equiv$ indicates that the equality is true for *all* values of x. An equation such as (i) which is true for all values of x is called an identity.

Putting $x=a$ in (i), we get $P(a)=R$.

Hence when $P(x)$ is divided by $x-a$, the remainder is $P(a)$. This result is known as the *remainder theorem.* The *factor theorem* follows immediately: if $P(x)$ is a polynomial, $x-a$ is a factor of $P(x)$ if, and only if, $P(a)=0$.

1.3. Further properties of polynomials

1. *If the polynomial $P(x)\equiv p_0x^n+p_1x^{n-1}+\ldots+p_{n-1}x+p_n$ $(p_0\neq 0)$ is equal to zero when x has any one of the n distinct values $a_1, a_2, \ldots, a_n$, then*

$$p_0x^n+p_1x^{n-1}+\ldots+p_{n-1}x+p_n\equiv p_0(x-a_1)(x-a_2)\ldots(x-a_n).$$

By the factor theorem, since $P(a_1)=0$, $x-a_1$ is a factor of $P(x)$ and the quotient when we divide $P(x)$ by $x-\alpha_1$ is a polynomial of degree $n-1$ whose first term is p_0x^{n-1}. Hence we write

$$P(x)\equiv(x-\alpha_1)Q_{n-1}(x), \qquad \text{(i)}$$

where $\quad Q_{n-1}(x)\equiv p_0x^{n-1}+\ldots.$

Since $\quad P(\alpha_2)=0$, we have from (i)

$$0=P(\alpha_2)=(\alpha_2-\alpha_1)Q_{n-1}(\alpha_2)$$

i.e. $\quad Q_{n-1}(\alpha_2)=0$, since $\alpha_2\neq\alpha_1$.

Hence $x-\alpha_2$ is a factor of $Q_{n-1}(x)$, and we write

$$Q_{n-1}(x)\equiv(x-\alpha_2)Q_{n-2}(x)$$

and, by (i),

$$P(x)\equiv(x-\alpha_1)(x-\alpha_2)Q_{n-2}(x),$$

where $\quad Q_{n-2}(x)\equiv p_0x^{n-2}+\ldots.$

Proceeding in this way, we see that

$$P(x)\equiv(x-\alpha_1)(x-\alpha_2)\ldots(x-\alpha_n)Q_0(x),$$

where $\quad Q_0(x)\equiv p_0.$

$$\therefore\ P(x)\equiv p_0(x-\alpha_1)(x-\alpha_2)\ldots(x-\alpha_n).$$

II. *If the polynomial* $P(x)\equiv p_0x^n+p_1x^{n-1}+\ldots+p_{n-1}x+p_n$ *is equal to zero for more than n distinct values of x,* $P(x)$ *is equal to zero for all values of x and each of the coefficients* $p_0, p_1, \ldots, p_n$ *is zero. In this case* $P(x)$ *is identically zero, that is*

$$p_0x^n+p_1x^{n-1}+\ldots+p_{n-1}x+p_n\equiv0.$$

We suppose that $P(x)=0$ when $x=\alpha_1, \alpha_2, \ldots, \alpha_n$.

Then, by I, $P(x)\equiv p_0(x-\alpha_1)(x-\alpha_2)\ldots(x-\alpha_n)$.

Now suppose that $P(\alpha)=0$, where α is different from any of $\alpha_1, \alpha_2, \ldots, \alpha_n$.

Then $\quad p_0(\alpha-\alpha_1)(\alpha-\alpha_2)\ldots(\alpha-\alpha_n)=0,$

and since none of the factors $(\alpha-\alpha_1)$, $(\alpha-\alpha_2)$, $\ldots$, $(\alpha-\alpha_n)$ is zero, p_0 must be zero. Hence

$$P(x)\equiv p_1x^{n-1}+p_2x^{n-2}+\ldots+p_n.$$

This polynomial of degree $n-1$ vanishes for more than $(n-1)$ values of x and so, applying the same argument, we see that $p_1=0$. Continuing in this way, we prove that

$$p_1=p_2=\ldots=p_n=0.$$

As a corollary to II, we have the important property that if, for all values of x,

$$p_0x^n+p_1x^{n-1}+\ \ldots\ +p_{n-1}x+p_n=q_0x^n+q_1x^{n-1}+\ \ldots\ +q_{n-1}x+q_n, \qquad \text{(i)}$$

then $p_0=q_0$, $p_1=q_1$, $\ldots$, $p_n=q_n$.

To see this we consider the polynomial

$$(p_0-q_0)x^n+(p_1-q_1)x^{n-1}+ \ldots +(p_{n-1}-q_{n-1})x+(p_n-q_n)$$

which, by hypothesis, is zero for all values of x. It follows that all its coefficients are zero, and so

$$p_0=q_0,\ p_1=q_1,\ \ldots,\ p_n=q_n \quad . \quad . \quad . \quad \text{(ii)}$$

The process of deducing from the identity (i) the results (ii) is called *equating coefficients.*

1.4. Polynomials in several variables

A polynomial in several variables $x, y, z, \ldots$ is a sum of terms of the form $kx^py^qz^r \ldots$, where the indices $p, q, r \ldots$ are positive integers or zero and k is a constant.

The degree of any term is the sum of its degrees with respect to the variables, so that a term $7xy^3z^2$ is of the sixth degree.

The degree of a polynomial is that of the term of the highest degree in the polynomial. For example $x^2y^2+y^2+2x-5$ is a polynomial of the fourth degree in x and y; $xyz-2x^2+3z+4$ is a polynomial of the third degree in x, y and z.

1.5. The method of undetermined coefficients

The method of undetermined coefficients is based on the principle that if two polynomials in x are identically equal, coefficients of like powers of x must be equal. The principle is valid for polynomials in several variables $x, y, z, \ldots$ and the method of undetermined coefficients may be applied to such polynomials.

Example 1

Find the values of the constants a, b, c and d such that

$$r^3 \equiv ar(r-1)(r-2)+br(r-1)+cr+d \quad . \quad . \quad \text{(i)}$$

Multiplying out, we have

$$r^3 \equiv a(r^3-3r^2+2r)+b(r^2-r)+cr+d,$$

$$r^3 \equiv ar^3+r^2(b-3a)+r(c+2a-b)+d \quad . \quad . \quad . \quad \text{(ii)}$$

We shall use the symbol $((r^n))$ to denote the coefficient(s) of r^n.

Equating $((r^3))$ on each side of (ii), we have $1=a$;
equating $((r^2))$, $\quad 0=b-3a, \quad \therefore\ b=3$;
equating $((r))$, $\quad 0=c+2a-b, \quad \therefore\ c=1$;
equating $((r^0))$, $\quad 0=d.$

$$\therefore\ r^3 \equiv r(r-1)(r-2)+3r(r-1)+r.$$

Alternatively, we may use the fact that identity (i) is true for all values of r. Substituting in turn the values $r=0, 1, 2$ and 3, we obtain as before $d=0$, $c=1$, $b=3$, $a=1$.

Example 2

Factorise the expression

$$2x^2-3xy-2y^2+x+13y-15.$$

Since $$2x^2-3xy-2y^2\equiv(2x+y)(x-2y)$$

we try

$$2x^2-3xy-2y^2+x+13y-15\equiv\{(2x+y)+A\}\{(x-2y)+B\} \quad . \quad . \quad \text{(i)}$$
$$\equiv 2x^2-3xy-2y^2+A(x-2y)+B(2x+y)+AB,$$

A and B being constants.

Then $$x+13y-15\equiv x(A+2B)+y(B-2A)+AB \quad . \quad . \quad \text{(ii)}$$

Equating $((x))$ on each side, we have $1=A+2B$;

equating $((y))$ on each side, we have $13=B-2A$;

$$\therefore A=-5,\ B=3.$$

By equating the constants on each side of (ii) we obtain $AB=-15$, so that our original conjecture is justified. Hence

$$2x^2-3xy-2y^2+x+13y-15\equiv(2x+y-5)(x-2y+3).$$

Example 3

Prove that if $(a+b+c)=0$ *and* $(bc+ca+ab)+3m=0$, *then the expression* E *where* $E=(x^2+ax+m)(x^2+bx+m)(x^2+cx+m)$ *will contain no powers of* x *except those whose index is a multiple of three.*

Given that the expression x^6+16x^3+64 *has a factor of the form* x^2-2x+m, *resolve it into three quadratic factors of the form similar to* E. [L.U.]

We have $$E=(y+ax)(y+bx)(y+cx),$$

where $$y=x^2+m \quad . \quad . \quad . \quad . \quad . \quad . \quad . \quad \text{(i)}$$

$$\therefore E=y^3+(a+b+c)y^2x+(bc+ca+ab)yx^2+abcx^3.$$

If $a+b+c=0$ and $(bc+ca+ab)+3m=0$,

$$E=y^3-3myx^2+abcx^3$$
$$=y(y^2-3mx^2)+abcx^3$$
$$=(x^2+m)(x^4-mx^2+m^2)+abcx^3 \text{ by (i)}$$
$$=x^6+abcx^3+m^3 \quad . \quad . \quad . \quad . \quad . \quad . \quad \text{(ii)}$$

Thus E contains no powers of x except those whose index is a multiple of three.

If x^6+16x^3+64 is identically equal to E

$$64=m^3, \qquad \therefore m=4.$$

Hence if x^6+16x^3+64 has a factor of the form x^2-2x+m and two other similar factors, as in E, we may assume that

$$x^6+16x^3+64\equiv(x^2-2x+4)(x^2+bx+4)(x^2+cx+4)$$
$$\equiv x^6-2bcx^3+64, \text{ by (ii)}.$$
$$\therefore -2bc=16, \text{ i.e. } bc=-8.$$

But $$-2+b+c=0, \quad \therefore b+c=2.$$

Hence $$b=4,\ c=-2 \text{ or } b=-2,\ c=4$$

and $$x^6+16x^3+64\equiv(x^2-2x+4)^2(x^2+4x+4).$$

The method of undetermined coefficients may be used to establish identities between functions other than polynomials.

Example 4

Show that when a and b are positive constants, a positive constant R and a constant acute angle α may be found such that

$$a \sin \theta + b \cos \theta \equiv R \sin(\theta + \alpha).$$

We have, if possible, to choose R, α so that

$$\begin{aligned} a \sin \theta + b \cos \theta &\equiv R \sin(\theta+\alpha) \\ &\equiv R \sin \theta \cos \alpha + R \cos \theta \sin \alpha \\ &\equiv (R \cos \alpha) \sin \theta + (R \sin \alpha) \cos \theta. \end{aligned}$$

The identity is valid if

$$a = R \cos \alpha \quad . \quad . \quad . \quad . \quad . \quad \text{(i)}$$

$$b = R \sin \alpha \quad . \quad . \quad . \quad . \quad . \quad \text{(ii)}$$

Sqnaring and adding corresponding sides of these equations we get

$$a^2 + b^2 = R^2,$$

$$\therefore R = \sqrt{(a^2+b^2)} \quad (R > 0).$$

When a, b, R are positive, we see from (i) and (ii) that $\sin \alpha$ and $\cos \alpha$ are positive. Hence α is an acute angle such that

$$\sin \alpha : \cos \alpha : 1 = b : a : R.$$

With this value of α,

$$a \sin \theta + b \cos \theta \equiv \sqrt{(a^2+b^2)} \sin(\theta+\alpha).$$

Similarly, when a and b are positive,

$$a \sin \theta - b \cos \theta \equiv \sqrt{(a^2+b^2)} \sin(\theta - \alpha),$$

$$\text{and } a \cos \theta - b \sin \theta \equiv \sqrt{(a^2+b^2)} \cos(\theta+\alpha),$$

where α is the same acute angle as above.

Exercises 1 (*a*)

1. Express $n(n+1)(2n+1)$ in the form
$$An + Bn(n-1) + Cn(n-1)(n-2),$$
where A, B and C are constants independent of n.

2. Express $(2n-1)(2n+1)(2n+3)$ in the form
$$A + B(2n) + C(2n)(2n-1) + D(2n)(2n-1)(2n-2),$$
where A, B, C and D are constants independent of n.

3. Factorise the expressions
(i) $2x^2 - 3xy - 2y^2 + 2x + 11y - 12$,
(ii) $6x^2 - 5xy - 6y^2 - 5x + 14y - 4$,
(iii) $6x^2 + xy - y^2 - 3x + y$.

4. Express the following functions in the form indicated :
(i) $\sin \theta + \cos \theta \equiv R \sin(\theta + \alpha)$,
(ii) $\sin \theta - \sqrt{3} \cos \theta \equiv R \sin(\theta - \alpha)$,
(iii) $3 \cos \theta - 4 \sin \theta \equiv R \cos(\theta + \alpha)$,
where in all cases R is a positive constant and α is a constant acute angle measured in degrees.

5. Determine the coefficients a, b, c in the polynomial $f(x)$, where $f(x)=ax^4+bx^3+cx^2$, if $f\{n(n+1)\}-f\{n(n-1)\}\equiv n^7$.
   Hence, or otherwise, find the sum of the seventh powers of the first n integers. [L.U.]

1.6. Factorisation of cyclic homogeneous polynomials

A function of several variables is said to be *homogeneous* if all its terms are of the same degree.

A function of two or more variables is said to be *symmetrical* in these variables when its value is unaltered by the interchange of any two of the variables ; it is said to be an *alternating* (or a *skew*) function if the interchange multiplies the value of the function by -1.

For example, $x+y+2$ is a symmetrical function of the first degree in x and y, x^2-y^2 is an alternating homogeneous function of the second degree in x and y.

A function of x, y and z which is unaltered when we write y for x, z for y and x for z is said to be *cyclic*, or to have cyclic symmetry. For example, $x+y+z$ is a cyclic homogeneous function of the first degree ; $x^2+y^2+z^2$ and $yz+zx+xy$ are cyclic homogeneous functions of the second degree.

Every symmetrical function has cyclic symmetry.

The following examples illustrate methods of factorising cyclic homogeneous polynomials.

Example 5

Factorise $x^4(y^2-z^2)+y^4(z^2-x^2)+z^4(x^2-y^2)$.

Let $E=x^4(y^2-z^2)+y^4(z^2-x^2)+z^4(x^2-y^2)$ . . . . . (i)

When $x=y$, $E=0$, $\therefore$ $x-y$ is a factor of E.

When $x=-y$, $E=0$, $\therefore$ $x+y$ is a factor of E.

Similarly, $y-z$, $y+z$, $z-x$ and $z+x$ are factors of E.

Now E is of the sixth degree, so that in addition to the six factors already found there can be only a numerical factor k (say).

Hence $E=k(y-z)(z-x)(x-y)(y+z)(z+x)(x+y)$ . . . (ii)

and comparing the coefficients of x^4y^2 in (i) and (ii) we obtain $k=-1$.

$$\therefore E=-(y-z)(z-x)(x-y)(y+z)(z+x)(x+y).$$

Example 6

Factorise $x(y-z)^3+y(z-x)^3+z(x-y)^3$.

Let $E=x(y-z)^3+y(z-x)^3+z(x-y)^3$ . . . . . (i)

As in Example 5, $(y-z)(z-x)(x-y)$ is a factor of E. But E and this factor are both cyclic and E is a homogeneous function of the fourth degree ; hence the remaining factor must be a cyclic homogeneous expression of the first degree, and the only such expression is $k(x+y+z)$, where k is a numerical constant.

$$\therefore E=k(x+y+z)(y-z)(z-x)(x-y) \quad . \quad . \quad \text{(ii)}$$

Comparing coefficients of xy^3 in (i) and (ii), we obtain

$$E=(x+y+z)(y-z)(z-x)(x-y).$$

Example 7

Factorise $x(y^4-z^4)+y(z^4-x^4)+z(x^4-y^4)$.

Let $E=x(y^4-z^4)+y(z^4-x^4)+z(x^4-y^4)$ . . . . . . (i)

As in the previous examples, $(y-z)(z-x)(x-y)$ is a factor of E. Since E and this factor are both cyclic and E is a homogeneous function of the fifth degree, the remaining factor is a cyclic homogeneous expression of the second degree, and the most general expression of this type is $k_1(x^2+y^2+z^2)+k_2(yz+zx+xy)$, where k_1 and k_2 are numerical constants.

$$\therefore\ E=(y-z)(z-x)(x-y)\{k_1(x^2+y^2+z^2)+k_2(yz+zx+xy)\} \quad . \quad \text{(ii)}$$

Comparing coefficients of x^4y in (i) and (ii) we obtain $k_1=1$; comparing coefficients of x^3y^2 we have $k_2=k_1$.

$$\therefore\ E=(y-z)(z-x)(x-y)(x^2+y^2+z^2+yz+zx+xy).$$

Exercises 1 (*b*)

Factorise the following expressions:

1. $(b+c)^3(b-c)+(c+a)^3(c-a)+(a+b)^3(a-b)$.
2. $a^3(b^2-c^2)+b^3(c^2-a^2)+c^3(a^2-b^2)$.
3. $a^3+b^3+c^3-3abc$.
4. $a^4(b-c)+b^4(c-a)+c^4(a-b)$.
5. $a^2b^2(a-b)+b^2c^2(b-c)+c^2a^2(c-a)$.
6. $(b-c)(b+c)^4+(c-a)(c+a)^4+(a-b)(a+b)^4$.
7. $a(b-c)(b+c)^3+b(c-a)(c+a)^3+c(a-b)(a+b)^3$.
8. $(b-c)^5+(c-a)^5+(a-b)^5$.
9. $(bc+ca+ab)^3-b^3c^3-c^3a^3-a^3b^3$.
10. $a^5(b-c)+b^5(c-a)+c^5(a-b)$.

1.7. Rational functions

A rational function is the ratio of two polynomials.

If $$P(x)\equiv p_0x^n+p_1x^{n-1}+\ \dots\ +p_{n-1}x+p_n$$

and $$Q(x)\equiv q_0x^m+q_1x^{m-1}+\ \dots\ +q_{m-1}x+q_m,$$

where m and n are positive integers or zero, the function $P(x)/Q(x)$ is rational except when x takes any value which makes the divisor $Q(x)$ zero. The term "rational function" includes a polynomial, for if $m=0$, $Q(x)$ reduces to the constant q_0 and $P(x)/Q(x)$ is a polynomial in x.

If $n<m$, the rational function $P(x)/Q(x)$ is said to be a *proper* fraction, but if $n\geqslant m$, it is an *improper* fraction. An improper fraction may be expressed, by division, as the sum of a polynomial and a proper fraction.

For example, $$\frac{2x^2+6x+1}{x+2}=2x+2-\frac{3}{x+2}.$$

1.8. Partial fractions

The sum or difference of a number of proper fractions is itself a proper fraction. For example,

$$\frac{2}{x+1}+\frac{1}{x+2}=\frac{3x+5}{x^2+3x+2}$$

and

$$\frac{1}{x-1}-\frac{x+2}{x^2+x+1}=\frac{3}{x^3-1}.$$

The converse result is also true: a proper fraction $P(x)/Q(x)$ whose denominator $Q(x)$ breaks up into real factors may be expressed as the sum or difference of simpler fractions, known as *partial fractions*, each with one of the factors of $Q(x)$ as denominator. For example,

$$\frac{2x}{x^2-9}=\frac{1}{x-3}+\frac{1}{x+3}$$

and

$$\frac{x^3+x^2-x+3}{x^4-1}=\frac{x-1}{x^2+1}+\frac{1}{x-1}-\frac{1}{x+1}.$$

We now apply the method of undetermined coefficients to the problem of expressing as a sum or difference of partial fractions a rational function $N(x)/D(x)$ given in its lowest terms. Detailed discussion of this problem is to be found in text-books on algebra; we merely outline the rules by which the partial fractions may be found.

We assume that $N(x)$ is of lower degree than $D(x)$, i.e. that the fraction is proper. Should this not be the case the fraction must be expressed, by division, as the sum of a polynomial and a proper fraction (see § 1.7).

Rule I

To each non-repeated linear factor $(x-a)$ of $D(x)$, there corresponds a fraction of the form $\dfrac{A}{x-a}$, where A is a constant not equal to zero.

Example 8

Express $(x+5)/(x-3)(x+1)$ *in partial fractions.*

Since the given fraction is proper, the partial fractions must be proper, and so we assume that

$$\frac{x+5}{(x-3)(x+1)}\equiv\frac{A}{x-3}+\frac{B}{x+1}.$$

This assumption is justified because it is equivalent to

$$x+5\equiv A(x+1)+B(x-3) \quad \quad \text{(i)}$$

or

$$x+5\equiv x(A+B)+A-3B;$$

and this identity is valid since unique values of A and B can be found to satisfy the equations $A+B=1$, $A-3B=5$, obtained by equating coefficients.

In practice, after verifying that the original assumption is valid, it is shorter to use the method given below.

Substitute $x=3$ in (i); then $8=4A$, i.e. $A=2$;

substitute $x=-1$ in (i); then $4=-4B$, i.e. $B=-1$.

$$\therefore \frac{x+5}{(x-3)(x+1)} \equiv \frac{2}{x-3} - \frac{1}{x+1}.$$

From (i) we see that A can be immediately obtained by deleting the factor $(x-3)$ in $(x+5)/(x-3)(x+1)$ and substituting $x=3$ in the resulting fraction. Similarly B can be immediately determined.

Partial fractions corresponding to all non-repeated linear factors of $D(x)$ can be found in this way.

Example 9

Express $(3x^3-x^2+2)/x(x^2-1)$ *in partial fractions.*

By division the improper fraction

$$\frac{3x^3-x^2+2}{x(x^2-1)} \equiv 3+\frac{P(x)}{x(x+1)(x-1)},$$

where $P(x)$ is of degree less than three. We therefore assume that

$$\frac{3x^3-x^2+2}{x(x^2-1)} \equiv 3+\frac{A}{x}+\frac{B}{x+1}+\frac{C}{x-1}$$

and, clearing fractions, obtain

$$3x^3-x^2+2 \equiv 3x(x^2-1)+A(x^2-1)+Bx(x-1)+Cx(x+1). \qquad \text{(i)}$$

Having verified that our first assumption is justified (since by equating coefficients in identity (i) we obtain three equations from which A, B and C may be uniquely determined), we substitute in turn the values $x=0$, -1 and 1 in (i) and obtain $A=-2$, $B=-1$, $C=2$.

$$\therefore \frac{3x^3-x^2+2}{x(x^2-1)} = 3-\frac{2}{x}-\frac{1}{x+1}+\frac{2}{x-1}.$$

From (i) we see that the value of A can be immediately determined by deleting the factor x in $(3x^3-x^2+2)/x(x-1)(x+1)$ and substituting $x=0$ in the resulting fraction. Similarly B and C may be immediately found.

It is useful to note that when $N(x)/D(x)$ is a proper fraction, the number of constants to be determined is equal to the degree of $D(x)$.

Rule II

The quadratic factor ax^2+bx+c is said to be irreducible if it has no real linear factors. To such a factor in $D(x)$ there corresponds a partial fraction of the form $\dfrac{Ax+B}{ax^2+bx+c}$.

Example 10

Express $2/(x^4+x^2+1)$ *in partial fractions.*

Since $x^4+x^2+1=(x^2+x+1)(x^2-x+1)$, the given proper fraction must be expressed as the sum of two proper fractions with irreducible quadric denominators. The numerators of such fractions may be of the first degree, and so we assume that

$$\frac{2}{x^4+x^2+1}\equiv\frac{Ax+B}{x^2+x+1}+\frac{Cx+D}{x^2-x+1}$$

i.e.
$$2\equiv(Ax+B)(x^2-x+1)+(Cx+D)(x^2+x+1).$$

Equating coefficients we have

$$\begin{aligned}((x^3)):&\quad A+C=0,\\ ((x^2)):&\quad B+D-A+C=0,\\ ((x)):&\quad A-B+C+D=0,\\ ((x^0)):&\quad B+D=2.\end{aligned}$$

Solving these equations, we get $A=B=D=1$, $C=-1$.

$$\therefore\ \frac{2}{x^4+x^2+1}\equiv\frac{x+1}{x^2+x+1}+\frac{1-x}{x^2-x+1}.$$

Example 11

Express $(5x-12)/(x+2)(x^2-2x+3)$ *in partial fractions.*

Assume that

$$\frac{5x-12}{(x+2)(x^2-2x+3)}\equiv\frac{A}{x+2}+\frac{Bx+C}{x^2-2x+3}$$

so that
$$5x-12\equiv A(x^2-2x+3)+(Bx+C)(x+2)\ . \qquad \text{(i)}$$

This assumption is justified since there are three coefficients to equate, and so the three unknowns can be found uniquely.

By substituting $x=-2$ we find that $A=-2$.

Equating coefficients we have

$$\begin{aligned}((x^2)):&\quad 0=A+B, &\therefore\ B=2.\\ ((x^0)):&\quad -12=3A+2C, &\therefore\ C=-3.\end{aligned}$$

Hence
$$\frac{5x-12}{(x+2)(x^2-2x+3)}\equiv\frac{2x-3}{x^2-2x+3}-\frac{2}{x+2}.$$

Rule III

If $D(x)$ contains a repeated linear factor, $(x-a)^2$ say, it would be correct to assume that the corresponding partial fraction is of the form $\dfrac{A_1x+B_1}{(x-a)^2}$, for the denominator is of the second degree and hence two undetermined coefficients are required.

But we may write

$$A_1x+B_1 \equiv A_1(x-a)+B_1+aA_1$$

i.e. $$A_1x+B_1 \equiv A_1(x-a)+A_2, \text{ where } A_2=B_1+aA_1.$$

Thus $\dfrac{A_1x+B_1}{(x-a)^2}$ is equivalent to $\dfrac{A_1(x-a)+A_2}{(x-a)^2}$ or $\dfrac{A_1}{x-a}+\dfrac{A_2}{(x-a)^2}$, and in general it is more convenient to have the partial fractions in this form.

Similarly, if $(x-a)^r$, but not $(x-a)^{r+1}$, is a factor of $D(x)$, we have corresponding to this factor r partial fractions of the form

$$\frac{A_1}{x-a}+\frac{A_2}{(x-a)^2}+\ldots+\frac{A_r}{(x-a)^r}.$$

Example 12

Express $(2x^2+x-2)/x^3(x-1)$ *in partial fractions.*

Assume that $$\frac{2x^2+x-2}{x^3(x-1)} \equiv \frac{A}{x}+\frac{B}{x^2}+\frac{C}{x^3}+\frac{D}{x-1},$$

so that $$2x^2+x-2 \equiv Ax^2(x-1)+Bx(x-1)+C(x-1)+Dx^3 \quad . \quad \text{(i)}$$

This assumption is justified since there are four coefficients to equate and so the four unknowns A, B, C and D can be uniquely determined.

Substituting in turn the values $x=0$ and $x=1$ in (i), we find that $C=2$ and $D=1$.

Equating coefficients, we have

$$((x^3)): \quad 0=A+D, \qquad \therefore A=-1.$$

$$((x)): \quad 1=-B+C, \qquad \therefore B=1.$$

$$\therefore \frac{2x^2+x-2}{x^3(x-1)} \equiv \frac{1}{x-1}-\frac{1}{x}+\frac{1}{x^2}+\frac{2}{x^3}.$$

It is useful to note that the value of C can be immediately determined by deleting x^3 in the denominator of the given fraction and substituting $x=0$ in the resulting fraction. Similarly, D can be immediately found. We may then use the method given below.

From (i),

$$2x^2+x-2 \equiv Ax^2(x-1)+Bx(x-1)+2(x-1)+x^3$$

i.e. $$-(x^3-2x^2+x) \equiv x(x-1)(Ax+B)$$

$$-(x-1) \equiv Ax+B.$$

$$\therefore A \equiv -1, \quad B=1.$$

Example 13

Express $(x^3+5x^2+4x+5)/(x-1)(x^3-1)$ *in partial fractions.*

The factors of the denominator are $(x-1)^2(x^2+x+1)$ and so we assume that

$$\frac{x^3+5x^2+4x+5}{(x-1)^2(x^2+x+1)} \equiv \frac{A}{x-1}+\frac{B}{(x-1)^2}+\frac{Cx+D}{x^2+x+1}$$

i.e. $x^3+5x^2+4x+5 \equiv A(x-1)(x^2+x+1)+B(x^2+x+1)+(Cx+D)(x-1)^2$.

By putting $x=1$ we get $B=5$, and so

$$x^3-x \equiv (x-1)\{A(x^2+x+1)+(Cx+D)(x-1)\}$$

$$\therefore\ x(x+1) \equiv A(x^2+x+1)+(Cx+D)(x-1).$$

By putting $x=1$ we get $A=\frac{2}{3}$ and so

$$\tfrac{1}{3}(x^2+x-2) \equiv (Cx+D)(x-1)$$

$$\therefore\ \tfrac{1}{3}(x+2) \equiv Cx+D.$$

Hence $$\frac{x^3+5x^2+4x+5}{(x-1)(x^3-1)} \equiv \frac{1}{3}\left\{\frac{2}{x-1}+\frac{15}{(x-1)^2}+\frac{x+2}{x^2+x+1}\right\}.$$

Rule IV

If $D(x)$ contains a repeated factor $(ax^2+bx+c)^r$, where ax^2+bx+c is irreducible, the corresponding partial fractions are

$$\frac{A_1x+B_1}{ax^2+bx+c}+\frac{A_2x+B_2}{(ax^2+bx+c)^2}+\frac{A_3x+B_3}{(ax^2+bx+c)^3}+\ldots+\frac{A_rx+B_r}{(ax^2+bx+c)^r}.$$

Example 14

Express $(2x^2+x+4)/x(x^2+2)^2$ *in partial fractions.*

$$\frac{2x^2+x+4}{x(x^2+2)^2} \equiv \frac{A}{x}+\frac{B_1x+C_1}{x^2+2}+\frac{B_2x+C_2}{(x^2+2)^2}$$

$$\therefore\ 2x^2+x+4 \equiv A(x^2+2)^2+x(B_1x+C_1)(x^2+2)+x(B_2x+C_2).$$

By substituting $x=0$, we find that $A=1$.

Equating coefficients, we have

$$\begin{array}{lll} ((x^4)): & 0=A+B_1, & \therefore\ B_1=-1. \\ ((x^3)): & 0=C_1. & \\ ((x^2)): & 2=4A+2B_1+B_2, & \therefore\ B_2=0. \\ ((x)): & 1=2C_1+C_2, & \therefore\ C_2=1. \end{array}$$

$$\therefore\ \frac{2x^2+x+4}{x(x^2+2)^2} \equiv \frac{1}{x}-\frac{x}{x^2+2}+\frac{1}{(x^2+2)^2}.$$

Exercises 1 (*c*)

Express in terms of partial fractions :

1. $(x+1)/(x^2+7x+12)$.
2. $3x/(x-4)(x+2)$.
3. $(2x-1)/(x^3-x^2-2x)$.
4. $(3x^2-7x)/(1-x)(2-x)(3-x)$.
5. $x^3/(16-x^2)$.
6. $(20x-5)/(x-1)(2x+1)(2x+3)$.
7. $x^2/(x-2)^3$.
8. $(3x^2+5x-4)/(x-1)(x+1)^2$.
9. $(x^2-2)/(x^3+2x^2)$.
10. $(13-x^2)/(x-2)(x+1)^2$.
11. $(x^2+6)/(2x-1)(x+2)^2$.
12. $(17x+53)/(x+4)^2(2x-7)$.
13. $(1-x)/(1+x+x^2+x^3)$.
14. $(9-x)/x(x^2+9)$.
15. $(2x^2-x)/(x^3+1)$.
16. $32/x(4-x^2)^2$.
17. $81/x(9+x^2)^2$.
18. $(x^5-1)/x^2(x^3+1)$.
19. $(x^3-1)/(x^2-2x+3)(x+1)^2$.
20. $(x^2+1)/(x-1)^3$.
21. $(2x^3+x+3)/(x^2+1)^2$.
22. $x(1+x+5x^2+x^3+x^4)/(1-x^3)^2$.
23. $x^3(x^2+4)/(x^2+2)^3$.
24. $16x/(1-x^2)^3$.

CHAPTER 2

THEORY OF EQUATIONS; INEQUALITIES

2.1. Algebraic equations

If $f(x)$ denotes the polynomial

$$p_0x^n+p_1x^{n-1}+p_2x^{n-2}+\ldots+p_rx^{n-r}+\ldots+p_{n-1}x+p_n$$

where the coefficients $p_0, p_1, \ldots, p_n$ are all real and $p_0 \neq 0$, the equation $f(x)=0$ is the general algebraic equation of degree n, and we shall assume that every such equation has at least one root.

This is the fundamental theorem of algebra. From it we can deduce the following theorem :

An equation of the nth degree has n and only n roots.

Let α_1 be a root of the equation $f(x)=0$. Then $f(\alpha_1)=0$ and by the factor theorem $x-\alpha_1$ is a factor of $f(x)$

$$\therefore f(x)=(x-\alpha_1)g_{n-1}(x), \quad \ldots \quad \text{(i)}$$

where $\quad g_{n-1}(x)=p_0x^{n-1}+\ldots,\quad$ cf. § 1.3, I.

Again $g_{n-1}(x)=0$ must have at least one root α_2 (say). Hence

$$g_{n-1}(x)=(x-\alpha_2)g_{n-2}(x)$$

and, by (i), $\quad f(x)=(x-\alpha_1)(x-\alpha_2)g_{n-2}(x),$

where $\quad g_{n-2}(x)=p_0x^{n-2}+\ldots.$

Proceeding in this way we prove that

$$f(x)=p_0(x-\alpha_1)(x-\alpha_2)\ldots(x-\alpha_n) \quad . \quad \text{(ii)}$$

Thus $f(x)$ will vanish when x has any of the values $\alpha_1, \alpha_2, \ldots, \alpha_n$, and for no other value of x. Hence the equation $f(x)=0$ has exactly n roots.

The roots of the equation $f(x)=0$ may not be real and they need not all be distinct. Thus if

$$f(x)=(x-\alpha)^rF(x),$$

where $F(\alpha)\neq 0$, the equation $f(x)=0$ is said to have an r-fold root α (or a root α of multiplicity r). *Multiple* roots are sometimes referred to as *repeated* roots. In particular, if $r=2$, α is called a *double* root.

In the application of the above theorem any r-fold root is reckoned r times.

2.2. Relations between the coefficients and the roots of an algebraic equation

If α, β, γ are the roots of the cubic equation $x^3+px^2+qx+r=0$,

$$x^3+px^2+qx+r\equiv(x-\alpha)(x-\beta)(x-\gamma)$$
$$\equiv x^3-x^2(\alpha+\beta+\gamma)+x(\beta\gamma+\gamma\alpha+\alpha\beta)-\alpha\beta\gamma.$$

By equating coefficients of corresponding powers of x, we obtain the relations

$$\alpha+\beta+\gamma=-p,$$
$$\beta\gamma+\gamma\alpha+\alpha\beta=q,$$
$$\alpha\beta\gamma=-r.$$

In the same way, if α, β, γ, δ are the roots of the quartic equation $x^4+px^3+qx^2+rx+s=0$,

$$\alpha+\beta+\gamma+\delta=-p,$$
$$\alpha\beta+\alpha\gamma+\alpha\delta+\beta\gamma+\beta\delta+\gamma\delta=q,$$
$$\beta\gamma\delta+\alpha\gamma\delta+\alpha\beta\delta+\alpha\beta\gamma=-r,$$
$$\alpha\beta\gamma\delta=s.$$

Finally, if $\alpha_1, \alpha_2, \alpha_3, \ldots, \alpha_n$ are the roots of the equation

$$p_0x^n+p_1x^{n-1}+p_2x^{n-2}+\ldots+p_{n-1}x+p_n=0,$$

then
$$p_0x^n+p_1x^{n-1}+p_2x^{n-2}+\ldots+p_{n-1}x+p_n$$
$$\equiv p_0(x-\alpha_1)(x-\alpha_2)(x-\alpha_3)\ldots(x-\alpha_n)$$
$$\equiv p_0\{x^n-(\Sigma\alpha_1)x^{n-1}+(\Sigma\alpha_1\alpha_2)x^{n-2}-(\Sigma\alpha_1\alpha_2\alpha_3)x^{n-3}$$
$$+\ldots+(-1)^n\alpha_1\alpha_2\alpha_3\ldots\alpha_n\},$$

where $\Sigma\alpha_1=$ the sum of the roots,

$\Sigma\alpha_1\alpha_2=$ the sum of the products of the roots in pairs,

$\Sigma\alpha_1\alpha_2\alpha_3=$ the sum of the products of the roots in threes,

. . . .

Equating coefficients on each side of the above identity, we have

$$\Sigma\alpha_1=-p_1/p_0,$$
$$\Sigma\alpha_1\alpha_2=p_2/p_0,$$
$$\Sigma\alpha_1\alpha_2\alpha_3=-p_3/p_0,$$

. . . .

$$\alpha_1\alpha_2\alpha_3\ldots\alpha_n=(-1)^np_n/p_0.$$

By means of these relations any symmetrical function of the roots of an equation may be expressed in terms of the coefficients.

Example 1

If α, β, γ are the roots of the equation $x^3+px^2+qx+r=0$, express $\alpha^3+\beta^3+\gamma^3$ in terms of p, q and r.

We have
$$\left.\begin{aligned}\alpha+\beta+\gamma&=-p\\ \beta\gamma+\gamma\alpha+\alpha\beta&=\ \ q\\ \alpha\beta\gamma&=-r\end{aligned}\right\} \quad . \quad . \quad . \quad . \quad . \quad . \qquad \text{(i)}$$

and $\alpha^3+\beta^3+\gamma^3-3\alpha\beta\gamma=(\alpha+\beta+\gamma)(\alpha^2+\beta^2+\gamma^2-\beta\gamma-\gamma\alpha-\alpha\beta)$.

$$\begin{aligned}\therefore\ \alpha^3+\beta^3+\gamma^3&=3\alpha\beta\gamma+(\alpha+\beta+\gamma)\{(\alpha+\beta+\gamma)^2-3(\beta\gamma+\gamma\alpha+\alpha\beta)\}\\ &=-\{3r+p(p^2-3q)\}\text{ by (i).}\end{aligned}$$

2.3. The equation whose roots are the reciprocals of those of a given equation

Let $\alpha_1, \alpha_2, \ldots, \alpha_n$ be the roots of the equation

$$p_0x^n+p_1x^{n-1}+p_2x^{n-2}+\ \ldots\ +p_{n-1}x+p_n=0 \quad . \quad . \qquad \text{(i)}$$

and suppose that $\alpha_1, \alpha_2, \ldots, \alpha_n$ are all different from zero.

Writing $x=1/y$, we obtain the equation

$$p_ny^n+p_{n-1}y^{n-1}+\ \ldots +p_1y+p_0=0 \quad . \quad . \qquad \text{(ii)}$$

which is satisfied by the values $1/y=\alpha_1, \alpha_2, \ldots, \alpha_n$,

i.e. by $y=1/\alpha_1, 1/\alpha_2, 1/\alpha_3, \ldots, 1/\alpha_n$.

Hence (ii) is the equation whose roots are the reciprocals of the roots of (i).

Example 2

If α, β, γ are the roots of the equation $x^3+px+q=0$, find the equation whose roots are (a) $1/\alpha^2, 1/\beta^2, 1/\gamma^2$; (b) $1/\beta\gamma, 1/\gamma\alpha, 1/\alpha\beta$.

We have
$$\left.\begin{aligned}\alpha+\beta+\gamma&=\ \ 0\\ \beta\gamma+\gamma\alpha+\alpha\beta&=\ \ p\\ \alpha\beta\gamma&=-q\end{aligned}\right\} \quad . \quad . \quad . \quad . \quad . \qquad \text{(i)}$$

(*a*) We shall find the equation whose roots are α^2, β^2, γ^2 and use the result of § 2.3 to deduce the required equation.

By (i),
$$\begin{aligned}\alpha^2+\beta^2+\gamma^2&=(\alpha+\beta+\gamma)^2-2(\beta\gamma+\gamma\alpha+\alpha\beta)=-2p,\\ \beta^2\gamma^2+\gamma^2\alpha^2+\alpha^2\beta^2&=(\beta\gamma+\gamma\alpha+\alpha\beta)^2-2\alpha\beta\gamma(\alpha+\beta+\gamma)=p^2,\end{aligned}$$

and
$$\alpha^2\beta^2\gamma^2=q^2.$$

Hence the equation whose roots are α^2, β^2, γ^2 is

$$x^3+2px^2+p^2x-q^2=0 \quad . \quad . \quad . \quad . \qquad \text{(ii)}$$

and so the equation whose roots are $1/\alpha^2, 1/\beta^2, 1/\gamma^2$ is

$$q^2x^3-p^2x^2-2px-1=0.$$

Equation (ii) may be obtained in another way.

The given equation $x^3+px+q=0$ may be written in the form

$$x(x^2+p)=-q,$$

and so the roots α, β, γ satisfy the equation

$$x^2(x^2+p)^2=q^2.$$

Writing $y=x^2$, we have

$$y(y+p)^2=q^2$$

i.e.

$$y^3+2py^2+p^2y-q^2=0$$

This is the cubic equation which is satisfied by α^2, β^2, γ^2.

(*b*) We shall first find the equation whose roots are $\beta\gamma$, $\gamma\alpha$, $\alpha\beta$.

Since by (i)

$$\beta\gamma+\gamma\alpha+\alpha\beta=p,$$

$$\alpha^2\beta\gamma+\beta^2\gamma\alpha+\gamma^2\alpha\beta=\alpha\beta\gamma(\alpha+\beta+\gamma)=0,$$

and

$$\alpha^2\beta^2\gamma^2=q^2,$$

the equation whose roots are $\beta\gamma$, $\gamma\alpha$, $\alpha\beta$ is $x^3-px^2-q^2=0$ and so the required equation is $\qquad q^2x^3+px-1=0.$

Example 3

Find the conditions that the roots of the equation

$$x^3+px^2+qx+r=0$$

should be

(a) *in geometric progression (i.e. of the form* ξ, $\xi\eta$, $\xi\eta^2$),

(b) *in arithmetic progression (i.e. of the form* $\xi-\eta$, ξ, $\xi+\eta$),

(c) *in harmonic progression* (α, β, γ *are in harmonic progression if* $1/\alpha$, $1/\beta$, $1/\gamma$ *are in arithmetic progression*). [Sheffield.]

(*a*) Suppose the roots of the equation $x^3+px^2+qx+r=0$ are ξ, $\xi\eta$, $\xi\eta^2$. Then by § 2.2,

$$\xi(1+\eta+\eta^2)=-p \quad . \quad . \quad . \quad . \quad \text{(i)}$$

$$\xi^2(\eta+\eta^2+\eta^3)=q \quad . \quad . \quad . \quad . \quad \text{(ii)}$$

and

$$\xi^3\eta^3=-r \quad . \quad . \quad . \quad . \quad \text{(iii)}$$

The required condition is obtained by eliminating ξ and η between the above equations.

From (i) and (ii), $\xi\eta=-q/p$ and so, by (iii),

$$q^3=p^3r.$$

(*b*) Suppose the roots of $x^3+px^2+qx+r=0$ are $\xi-\eta$, ξ, $\xi+\eta$. Then by § 2.2,

$$3\xi=-p \quad . \quad . \quad . \quad . \quad . \quad \text{(i)}$$

$$\xi(\xi-\eta)+\xi^2-\eta^2+\xi(\xi+\eta)=q$$

i.e.

$$3\xi^2-\eta^2=q \quad . \quad . \quad . \quad . \quad . \quad \text{(ii)}$$

and

$$\xi(\xi^2-\eta^2)=-r \quad . \quad . \quad . \quad . \quad . \quad \text{(iii)}$$

From (ii) and (iii), $\qquad \xi(q-2\xi^2)=-r$

whence, by (i), $\qquad p(9q-2p^2)=27r.$

(c) The roots of $x^3+px^2+qx+r=0$ are in harmonic progression if the roots of $rx^3+qx^2+px+1=0$ are in arithmetic progression. Hence the required condition is obtained by writing q/r, p/r and $1/r$ for p, q and r respectively in the result of (b). The condition is

$$q(9pr-2q^2)=27r^2.$$

2.4. The equation whose roots are those of a given equation each increased (or decreased) by the same amount

Let $\alpha_1, \alpha_2, \alpha_3, \ldots, \alpha_n$ be the roots of the equation

$$f(x)\equiv p_0x^n+p_1x^{n-1}+\ \ldots\ +p_{n-1}x+p_n=0 \quad . \quad \text{(i)}$$

The substitution $x=y-k$ transforms (i) into the equation $f(y-k)=0$, which is satisfied by $\alpha_1+k, \alpha_2+k, \ldots, \alpha_n+k$.

Hence the roots of $f(y-k)=0$ are those of $f(x)=0$ each increased by k.

Similarly, the roots of $f(y+k)=0$ are those of $f(x)=0$ each decreased by k.

Example 4

Find the equation whose roots are those of the equation

$$x^4+4x^3-2x^2-12x-3=0$$

each increased by 1. *Hence, or otherwise, solve the given equation.*

The required equation is

$$(y-1)^4+4(y-1)^3-2(y-1)^2-12(y-1)-3=0,$$

which reduces to $y^4-8y^2+4=0$,

$$\therefore\ y^2=4\pm2\sqrt{3}$$

$$\text{or } y=\pm\sqrt{(4\pm2\sqrt{3})}.$$

Let $\sqrt{(4\pm2\sqrt{3})}=\sqrt{a}\pm\sqrt{b}.$

Then $4\pm2\sqrt{3}=a+b\pm2\sqrt{(ab)}$

so that $a+b=4$ and $ab=3$.

$$\therefore\ a=3 \text{ or } 1,\ b=1 \text{ or } 3.$$

Hence $y=\pm(\sqrt{3}\pm1)$,

and so $x=y-1=\pm\sqrt{3},\ -2\pm\sqrt{3}.$

Example 5

If α, β, γ are the roots of the equation $x^3+px+q=0$, find the equation whose roots are $\beta^2+\gamma^2, \gamma^2+\alpha^2, \alpha^2+\beta^2$.

From Example 2(a) we have

$$\alpha^2+\beta^2+\gamma^2=-2p.$$

$$\therefore\ \beta^2+\gamma^2=-(\alpha^2+2p),$$

$$\gamma^2+\alpha^2=-(\beta^2+2p),$$

$$\alpha^2+\beta^2=-(\gamma^2+2p).$$

Also, from Example 2(a), the equation whose roots are $\alpha^2, \beta^2, \gamma^2$ is

$$x^3+2px^2+p^2x-q^2=0.$$

The required equation, found by writing $y=-(x+2p)$ in this equation, is

$$-(y+2p)^3+2p(y+2p)^2-p^2(y+2p)-q^2=0,$$

i.e.

$$y^3+4py^2+5p^2y+(2p^3+q^2)=0.$$

2.5. Reciprocal equations

A reciprocal equation is one which is unaltered when the unknown is replaced by its reciprocal.

For example,

$$ax^3+bx^2+bx+a=0$$

and

$$ax^4+bx^3+cx^2+bx+a=0$$

are reciprocal equations.

If α is a root of a reciprocal equation, it follows by definition that $1/\alpha$ is also a root of the equation and that the roots of such an equation are reciprocal in pairs. If a reciprocal equation is of odd degree, one root must be its own reciprocal and hence $+1$ or -1 is a root of the equation.

To solve a reciprocal equation of even degree in x, we use the substitution $x+1/x=t$; then $x^2+1/x^2=t^2-2$ and $x^3+1/x^3=t^3-3t$. The method of solving a reciprocal equation of odd degree is illustrated in the following example.

Example 6

Solve the equation

$$8x^5+62x^4+155x^3+155x^2+62x+8=0.$$

An obvious solution is $x=-1$ and we factorise the left-hand side of the equation as follows:

$$8x^4(x+1)+54x^3(x+1)+101x^2(x+1)+54x(x+1)+8(x+1)=0,$$

$$\therefore\ x=-1 \quad \text{or} \quad 8x^2+54x+101+54/x+8/x^2=0.$$

Putting $x+1/x=t$, we get

$$8t^2+54t+85=0$$

$$(2t+5)(4t+17)=0.$$

$$\therefore\ t=-2\tfrac{1}{2} \text{ and } x=-2,\ -\tfrac{1}{2},$$

$$\text{or } t=-4\tfrac{1}{4} \text{ and } x=-4,\ -\tfrac{1}{4}.$$

2.6. Condition for two quadratic equations to have a common root

If the equations

$$a_1x^2+b_1x+c_1=0,\ a_2x^2+b_2x+c_2=0$$

have a common root α,

$$a_1\alpha^2+b_1\alpha+c_1=0$$

and

$$a_2\alpha^2+b_2\alpha+c_2=0.$$

Solving these equations simultaneously for α^2 and α, we have

$$\frac{\alpha^2}{b_1c_2-b_2c_1}=\frac{\alpha}{c_1a_2-c_2a_1}=\frac{1}{a_1b_2-a_2b_1},$$

whence, eliminating α, we have

$$(b_1c_2-b_2c_1)(a_1b_2-a_2b_1)=(c_1a_2-c_2a_1)^2.$$

This is a necessary condition for the existence of a common root.

2.7. Condition for an equation to have a multiple root

We prove two theorems :

I. *If $f(x)$ is a polynomial in x and if the equation $f(x)=0$ has a root α of multiplicity r, $(r>1)$, then the equation $f'(x)=0$ has a root α of multiplicity $r-1$.*

We may write $f(x)=(x-\alpha)^rF(x)$, where $F(x)$ is a polynomial in x, and $F(\alpha)\neq 0$.

Then
$$\begin{aligned} f'(x)&=r(x-\alpha)^{r-1}F(x)+(x-\alpha)^rF'(x)\\ &=(x-\alpha)^{r-1}\{rF(x)+(x-\alpha)F'(x)\}. \end{aligned}$$

The second factor on the right does not vanish when $x=\alpha$ since $F(\alpha)\neq 0$. Hence $f'(x)=0$ has a root α of multiplicity $r-1$.

II. *If the equation $f'(x)=0$ has a root α of multiplicity $r-1$ then, provided that $f(\alpha)=0$, the equation $f(x)=0$ has a root α of multiplicity r.*

Since $f(\alpha)=0$, $f(x)$ contains a factor $(x-\alpha)$ and by supposing this factor to be an $n-$ fold one we can use theorem I to prove that $n=r$.

From the above theorems it follows that $f(x)=0$ has a multiple root if, and only if, the equations $f(x)=0$ and $f'(x)=0$ have a common root.

Example 7

If the equation $x^3+3ax^2+3bx+c=0$ has a repeated root, show that this root also satisfies the quadratic equation $x^2+2ax+b=0$; hence show that the value of the repeated root is $(c-ab)/2(a^2-b)$.

Solve the equation $4x^3-12x^2-15x-4=0$. [L.U.]

By § 2.7, a repeated root of the equation

$$x^3+3ax^2+3bx+c=0 \quad . \quad . \quad . \quad . \quad \text{(i)}$$

satisfies the equation
$$x^2+2ax+b=0 \quad . \quad . \quad . \quad . \quad \text{(ii)}$$

Multiplying (ii) by x and subtracting from (i) we get

$$ax^2+2bx+c=0 \quad . \quad . \quad . \quad . \quad \text{(iii)}$$

By eliminating x^2 between (ii) and (iii) we find the value of the repeated root of (i)

$$x=(c-ab)/2(a^2-b).$$

The equation
$$4x^3-12x^2-15x-4=0 \quad . \quad . \quad . \quad . \quad \text{(iv)}$$

has a repeated root if it is satisfied by a root of the equation

$$4x^2-8x-5=0$$

i.e. by $$x=-\tfrac{1}{2} \text{ or } x=\tfrac{5}{2}.$$

$x=-\frac{1}{2}$ satisfies (iv) and so $-\frac{1}{2}$ is a repeated root of (iv). We readily find that $4x^3-12x^2-15x-4\equiv(2x+1)(2x+1)(x-4)$ since we know that $2x+1$ is a repeated factor of the left-hand side.

Hence 4 is the remaining root of (iv).

2.8. Note on inequalities

Many properties are common to equations and inequalities: an inequality remains valid if each side is increased (or decreased) by the same amount. This means that we can transpose terms from one side to the other provided we change their signs; an inequality remains valid if both sides are multiplied or divided by the same positive number; if, however, we multiply or divide both sides by a negative number, the inequality sign must be reversed.

Three useful inequalities (valid for real numbers) are given below. We use $|x|$ to denote the numerical value of x. If $x\geqslant 0$, $|x|=x$; if $x<0$, $|x|=-x$.

I. The product $(x-\alpha)(x-\beta)\geqslant 0$ unless x lies between α and β.

II. An expression of the form $a^2+b^2+\ldots+k^2$ is positive unless $a=b=\ldots=k=0$.

III. The arithmetic mean of two positive numbers is greater than or equal to their geometric mean. The equality occurs when the two numbers are equal.

The proof is as follows: since $(x-y)^2\geqslant 0$,

$$x^2+y^2\geqslant 2xy \quad . \quad . \quad . \quad . \qquad \text{(i)}$$

Substitution of $x=\sqrt{a}$, $y=\sqrt{b}$, where a and b are positive numbers, gives

$$a+b\geqslant 2\sqrt{(ab)}$$

$$\therefore\ \tfrac{1}{2}(a+b)\geqslant\sqrt{(ab)}.$$

If x and y are of opposite signs, substitution of $|x|$ for x and $|y|$ for y in (i) shows that

$$x^2+y^2\geqslant 2|xy| \quad . \quad . \quad . \quad . \qquad \text{(ii)}$$

Example 8

Prove the following inequalities, in which a, b stand for unequal positive numbers:

(i) $a^3-a^2>a^{-2}-a^{-3}$, $(a\neq 1)$;

(ii) $a^{m+n}+b^{m+n}>a^mb^n+a^nb^m$ (*m, n positive integers*). [L.U.]

(i) $$a^3+a^{-3}-a^2-a^{-2}=(a^3-a^2)(1-a^{-5}).$$

If $a>1$, a^3-a^2 and $1-a^{-5}$ are both positive; if $a<1$, a^3-a^2 and $1-a^{-5}$ are both negative.

It follows that when $a \neq 1$

$$a^3+a^{-3}-a^2-a^{-2}>0$$

$$\therefore\ a^3-a^2>a^{-2}-a^{-3}.$$

(ii) $$a^{m+n}+b^{m+n}-a^mb^n-a^nb^m=(a^m-b^m)(a^n-b^n).$$

a^m-b^m and a^n-b^n are both positive when $a>b$ and both negative when $a<b$.

$$\therefore\ a^{m+n}+b^{m+n}>a^mb^n+a^nb^m.$$

Example 9

If a, b, c, d are any real numbers, prove that

$$a^4+b^4 \geqslant 2a^2b^2, \quad a^4+b^4+c^4+d^4 \geqslant 4abcd.$$

Prove also that $(a^2+b^2)^2+(c^2+d^2)^2 \geqslant 2(ab+cd)^2$.

Show that, if $a^4+b^4+c^4+d^4 \leqslant 1$, *then*

$$1/a^4+1/b^4+1/c^4+1/d^4 \geqslant 16. \qquad \text{[L.U.]}$$

By III (i), $a^4+b^4 \geqslant 2a^2b^2$ and $c^4+d^4 \geqslant 2c^2d^2$

$$\therefore\ a^4+b^4+c^4+d^4 \geqslant 2(a^2b^2+c^2d^2) \quad . \quad . \quad . \quad \text{(i)}$$

and so, by III (ii), $$a^4+b^4+c^4+d^4 \geqslant 4|abcd| \geqslant 4abcd \quad . \quad . \quad \text{(ii)}$$

Also, $$(a^2+b^2)^2+(c^2+d^2)^2=a^4+b^4+c^4+d^4+2(a^2b^2+c^2d^2)$$
$$\geqslant 4abcd+2(a^2b^2+c^2d^2), \text{ by (ii)}$$

$$\therefore\ (a^2+b^2)^2+(c^2+d^2)^2 \geqslant 2(ab+cd)^2.$$

Finally, by III (i), $1/a^4+1/b^4 \geqslant 2/a^2b^2$ and $1/c^4+1/d^4 \geqslant 2/c^2d^2$.

$$\therefore\ 1/a^4+1/b^4+1/c^4+1/d^4 \geqslant 2(1/a^2b^2+1/c^2d^2)$$
$$\geqslant 4/|abcd|$$
$$\geqslant 16/(a^4+b^4+c^4+d^4) \text{ by (ii)}$$
$$\geqslant 16 \text{ if } a^4+b^4+c^4+d^4 \leqslant 1.$$

Exercises 2

1. Find the equation whose roots are those of the equation
$$x^4-4x^3-7x^2+22x+24=0$$
each diminished by 1.
Hence, or otherwise, solve the given equation.

2. Find the equation whose roots are those of the equation
$$4x^4+32x^3+83x^2+76x+21=0$$
each increased by 2.
Hence, or otherwise, solve the given equation.

3. Solve the equation
$$6x^4-35x^3+62x^2-35x+6=0.$$

4. Solve the equation
$$2x^4-x^3-6x^2-x+2=0.$$

5. Solve the equation
$$x^5-9x^4+16x^3+16x^2-9x+1=0.$$

6. (i) By substituting $t=x^2+x$, or otherwise, solve the equation
$$(x^2+x-10)(x^2+x-26)=80.$$
(ii) If $x^2-yz=a$, $y^2-zx=b$, $z^2-xy=c$, then prove that $cx+ay+bz=0$. Find another relation of this type and hence, or otherwise, solve the equations when $a=1$, $b=2$, $c=3$. [Durham.]

7. If α, β, γ are the roots of $x^3+px^2+qx+r=0$, find the equation whose roots are $1/\alpha^2$, $1/\beta^2$, $1/\gamma^2$.
Find an expression in terms of p, q, r for $\Sigma(\alpha\beta/\gamma)$. [Leeds.]

8. Find the equation whose roots are the squares of the roots of the equation
$$2x^3-3x^2-x+7=0.$$ [Sheffield.]

9. Given that α, β, γ are the roots of the equation
$$x^3+px^2+qx+r=0,$$
express $\alpha^3+\beta^3+\gamma^3$ in terms of p, q, r, and show that
$$1/\alpha^3+1/\beta^3+1/\gamma^3=(3pqr-q^3-3r^2)/r^3.$$ [L.U.]

10. Find the condition for the roots of the equation
$$x^3+px^2+qx+r=0$$
to be in geometric progression, and solve the equation
$$3x^3-26x^2+52x-24=0.$$ [L.U.]

11. The roots of $x^2+px+q=0$ are α and β. If $y(p-x)=p+x$, prove that
$$(2p^2+q)y^2+2y(q-p^2)+q=0.$$
Show that the roots of this equation are $\beta/(2\alpha+\beta)$ and $\alpha/(2\beta+\alpha)$.
Express $$\{\alpha/(2\beta+\alpha)\}^2+\{\beta/(2\alpha+\beta)\}^2$$
in terms of p and q only. [Durham.]

12. If the roots of the equation
$$x^4-ax^3+bx^2-abx+1=0$$
are α, β, γ, δ, show that
$$(\alpha+\beta+\gamma)(\alpha+\beta+\delta)(\alpha+\gamma+\delta)(\beta+\gamma+\delta)=1.$$ [Sheffield.]

13. Given that the equation whose roots are the squares of the roots of the cubic
$$x^3-px^2+qx-1=0$$
is identical with this cubic, prove that either (i) $p=q=0$, or (ii) $p=q=3$, or (iii) p and q are the roots of $t^2+t+2=0$.
[Sheffield.]

14. Given that α, β, γ are the roots of the equation
$$x^3+px^2+r=0,$$
express $\alpha^2+\beta^2+\gamma^2$ and $\beta^2\gamma^2+\gamma^2\alpha^2+\alpha^2\beta^2$ in terms of p and r. Find also, by means of the relations $\beta^2+\gamma^2=\alpha^2+\beta^2+\gamma^2-\alpha^2$, etc., or otherwise, the equation with roots $\beta^2+\gamma^2$, $\gamma^2+\alpha^2$, $\alpha^2+\beta^2$. [Sheffield.]

15. The equation

$$x^3+px^2+qx+r=0$$

has roots α, β, γ. By means of the expressions

$$\beta+\gamma=(\alpha+\beta+\gamma)-\alpha,$$

etc., or otherwise, find the equation with roots

$$\alpha/(\beta+\gamma),\ \beta/(\gamma+\alpha),\ \gamma/(\alpha+\beta).$$ [Sheffield.]

16. By the substitution $x+1/x=t$, or otherwise, solve the equation

$$2x^4-13x^3+24x^2-13x+2=0.$$ [Durham.]

17. (i) If α, β, γ are the roots of the equation

$$x^3+px+q=0,$$

express $\alpha^4+\beta^4+\gamma^4$ in terms of p and q.

(ii) By the substitution $y=x+1/x$, or otherwise, solve the equation

$$6x^4-25x^3+37x^2-25x+6=0.$$ [L.U.]

18. If one root of $x^3+ax+b=0$ is twice the difference of the other two, show that the roots are

$$-13b/12a,\ 13b/3a,\ -13b/4a,$$

and that

$$144a^3+2197b^2=0.$$ [L.U.]

19. If α and β are the roots of the quadratic equation $ax^2+2hx+b=0$, find the quartic equation whose roots are $\pm 1/\alpha$, $\pm 1/\beta$.

Find the equation whose roots are the roots of the following quartic equation, each augmented by 2 :

$$x^4+8x^3+12x^2-16x-28=0.$$

Hence, or otherwise, solve the given equation. [L.U.]

20. If a, b and c are the roots of the equation

$$x^3+px^2+qx+r=0,$$

express the quantities $a+b+c$, $bc+ca+ab$ and abc in terms of p, q and r.

Prove that

(i) $a^2+b^2+c^2=p^2-2q$,

(ii) $a^3+b^3+c^3=3pq-3r-p^3$,

and, if $abc\neq 0$,

(iii) $b/c+c/a+a/b+c/b+a/c+b/a=(pq/r)-3.$ [L.U.]

21. (i) Obtain the equation whose roots exceed by 3 the roots of the equation

$$x^4+12x^3+49x^2+78x+42=0.$$

Hence, or otherwise, solve the original equation.

(ii) If the reciprocals of the roots of the cubic equation

$$x^3+3px^2+3qx+r=0$$

are in arithmetical progression, prove that

$$2q^3=r(3pq-r).$$ [L.U.]

22. If none of the quantities $(a+1)$, $(b+1)$ and $(c+1)$ is zero, and if

$$y+z=ax+xyz,$$
$$z+x=by+xyz,$$
$$x+y=cz+xyz,$$

prove that $(a+1)x=(b+1)y=(c+1)z$, and hence solve the first three equations for x, y and z. [L.U.]

23. (i) If $x^2=y^2+z^2+2ayz$, $y^2=z^2+x^2+2bzx$, $z^2=x^2+y^2+2cxy$, where none of the quantities a^2, b^2, c^2 is equal to unity, prove that

$$x^2/(1-a^2)=y^2/(1-b^2)=z^2/(1-c^2).$$

(ii) Prove that, if x, y, z are any positive numbers,

$$x^2+y^2 \geqslant 2xy, \quad (y+z)(z+x)(x+y) \geqslant 8xyz.$$

Prove also that, if a, b, c are any three positive numbers such that each is less than the sum of the other two,

$$(b+c-a)(c+a-b)(a+b-c) \leqslant abc.$$ [L.U.]

24. (i) If a, b, c, d and x, y, z, u are all real, prove that

$$(a^2+b^2+c^2+d^2)(x^2+y^2+z^2+u^2) \geqslant (ax+by+cz+du)^2.$$

(ii) If $x^3y=aZ^2+bZ^4$, where $Z \geqslant x$ and a, b, Z, x and y are real and positive, prove that $y \geqslant 2(ab)^{1/2}$. [L.U.]

25. If a, b, c are positive real numbers, prove that

$$(a^3+b^3+c^3)/(a+b+c) \geqslant \tfrac{1}{3}(a^2+b^2+c^2).$$

If y and x are positive real numbers such that $y-x>0$, prove that

$$\{y^3+3yx^2+4(y^2-x^2)^{3/2}\}/(3y^2-x^2) > \tfrac{2}{3}\{y+(y^2-x^2)^{1/2}\}.$$ [L.U.]

26. Prove that the equation $ax^3+bx^2+cx+d=0$ has a repeated root if, and only if,

$$(bc-9ad)^2-4(b^2-3ac)(c^2-3bd)=0.$$ [L.U.]

27. If a, b, c are positive, prove that

(i) $(b+c)(c+a)(a+b) \geqslant 8abc$,

(ii) $(a^7+b^7)(a^2+b^2) \geqslant (a^5+b^5)(a^4+b^4)$.

If $y+z>x$, $z+x>y$, $x+y>z$, use (i) to prove that

$$(y+z-x)(z+x-y)(x+y-z) \leqslant xyz.$$ [L.U.]

28. If $x/y+y/x=a$, $y/z+z/y=b$, $x/z+z/x=c$ and x, y, z are all real, prove that

(i) a^2, b^2, c^2 are each not less than 4,

(ii) if two of a, b, c are equal to -2, the other must be equal to $+2$,

(iii) $\{a \pm \sqrt{(a^2-4)}\}\{b \pm \sqrt{(b^2-4)}\}\{c \pm \sqrt{(c^2-4)}\}=8$, where one of the ambiguous signs is opposite to the other two. [L.U.]

CHAPTER 3

DETERMINANTS

3.1. Definitions

The equations
$$a_2x+b_2y+c_2=0,$$
$$a_3x+b_3y+c_3=0$$
are satisfied by the values
$$x=(b_2c_3-b_3c_2)/(a_2b_3-a_3b_2),\quad y=(a_3c_2-a_2c_3)/(a_2b_3-a_3b_2),$$
provided that $a_2b_3 \neq a_3b_2$.

These results may be written in the form
$$\frac{x}{b_2c_3-b_3c_2}=\frac{-y}{a_2c_3-a_3c_2}=\frac{1}{a_2b_3-a_3b_2}$$
or
$$\frac{x}{\begin{vmatrix} b_2 & c_2 \\ b_3 & c_3 \end{vmatrix}}=\frac{-y}{\begin{vmatrix} a_2 & c_2 \\ a_3 & c_3 \end{vmatrix}}=\frac{1}{\begin{vmatrix} a_2 & b_2 \\ a_3 & b_3 \end{vmatrix}} \tag{3.1}$$
where
$$\begin{vmatrix} b_2 & c_2 \\ b_3 & c_3 \end{vmatrix}=b_2c_3-b_3c_2,\quad \begin{vmatrix} a_2 & c_2 \\ a_3 & c_3 \end{vmatrix}=a_2c_3-a_3c_2$$
and
$$\begin{vmatrix} a_2 & b_2 \\ a_3 & b_3 \end{vmatrix}=a_2b_3-a_3b_2.$$

The expression
$$\begin{vmatrix} a_2 & b_2 \\ a_3 & b_3 \end{vmatrix}$$
is known as a *determinant of the second order*, and the quantities a_2, a_3, b_2, b_3 are called its *elements*. They are arranged in two rows, a_2, b_2 and a_3, b_3, and two columns, a_2, a_3 and b_2, b_3. The diagonal consisting of the elements a_2 and b_3 is called the *leading* (or *principal*) *diagonal*, and a_2 is known as the *leading element.*

The condition that the values of x and y given by (3.1) should satisfy the equation
$$a_1x+b_1y+c_1=0$$
is
$$a_1(b_2c_3-b_3c_2)-b_1(a_2c_3-a_3c_2)+c_1(a_2b_3-a_3b_2)=0,$$
i.e.
$$a_1\begin{vmatrix} b_2 & c_2 \\ b_3 & c_3 \end{vmatrix}-b_1\begin{vmatrix} a_2 & c_2 \\ a_3 & c_3 \end{vmatrix}+c_1\begin{vmatrix} a_2 & b_2 \\ a_3 & b_3 \end{vmatrix}=0.$$

This condition is written compactly as

$$\begin{vmatrix} a_1 & b_1 & c_1 \\ a_2 & b_2 & c_2 \\ a_3 & b_3 & c_3 \end{vmatrix} = 0.$$

The expression on the left-hand side of this equation is a *determinant of the third order* with a_1 as leading element and with the elements a_1, b_2, c_3 along the leading diagonal. Denoting this determinant by Δ, we have, by definition,

$$\Delta = a_1 \begin{vmatrix} b_2 & c_2 \\ b_3 & c_3 \end{vmatrix} - b_1 \begin{vmatrix} a_2 & c_2 \\ a_3 & c_3 \end{vmatrix} + c_1 \begin{vmatrix} a_2 & b_2 \\ a_3 & b_3 \end{vmatrix}. \tag{3.2}$$

If, in Δ, we delete the row and column containing a given element, we obtain a second-order determinant which is known as the minor of that element. Denoting by α_r, β_r, γ_r the minors of a_r, b_r, c_r respectively, we have from (3.2)

$$\Delta = a_1\alpha_1 - b_1\beta_1 + c_1\gamma_1. \tag{3.3}$$

Each term on the right-hand side contains an element of the first row of Δ, and so $a_1\alpha_1 - b_1\beta_1 + c_1\gamma_1$ is the expansion of Δ from the first row.

Again, from (3.2),

$$\begin{aligned} \Delta &= a_1(b_2c_3 - b_3c_2) - b_1(a_2c_3 - a_3c_2) + c_1(a_2b_3 - a_3b_2) \qquad (3.4) \\ &= a_1(b_2c_3 - b_3c_2) - a_2(b_1c_3 - b_3c_1) + a_3(b_1c_2 - b_2c_1) \\ &= a_1 \begin{vmatrix} b_2 & c_2 \\ b_3 & c_3 \end{vmatrix} - a_2 \begin{vmatrix} b_1 & c_1 \\ b_3 & c_3 \end{vmatrix} + a_3 \begin{vmatrix} b_1 & c_1 \\ b_2 & c_2 \end{vmatrix} \\ &= a_1\alpha_1 - a_2\alpha_2 + a_3\alpha_3. \end{aligned}$$

This formula gives the expansion of Δ from the first column. In a similar way, by re-arrangement of the terms in (3.4) a formula may be found for expanding Δ from any row or column, the expansion consisting in each case of the three elements of the row (or column) each multiplied by its minor with the appropriate sign attached. The sign may be determined by setting up a chessboard pattern with positive signs along the leading diagonal, as shown below.

$$\begin{vmatrix} + & - & + \\ - & + & - \\ + & - & + \end{vmatrix}$$

Thus, for example, $\Delta = -b_1\beta_1 + b_2\beta_2 - b_3\beta_3$,

$$= a_3\alpha_3 - b_3\beta_3 + c_3\gamma_3.$$

The term $a_1b_2c_3$ obtained by multiplying together the elements of the leading diagonal is known as the *leading term* in the expansion of Δ.

We shall denote the ith row and the jth column of Δ by R_i and C_j respectively.

Example 1

Evaluate the determinants

$$\text{(i)}\quad \Delta_1=\begin{vmatrix} 3 & -4 & 5 \\ 6 & -5 & -3 \\ -2 & 1 & 2 \end{vmatrix}, \qquad \text{(ii)}\quad \Delta_2=\begin{vmatrix} 2 & 7 & 6 \\ 4 & -1 & 3 \\ 3 & 0 & 1 \end{vmatrix}.$$

Expanding Δ_1 from R_1 we have

$$=3(-10+3)+4(12-6)+5(6-10)$$
$$=-17.$$

If Δ_2 is expanded from C_2 or from R_3 the expansion consists of two terms only. Expanding from R_3, we obtain

$$\Delta_2=3(21+6)+1(-2-28)$$
$$=51.$$

3.2. Properties of determinants

The process of expanding determinants in the absence of zero elements is sometimes lengthy. It is greatly simplified if use is made of the following fundamental properties which apply to determinants of any order and may be easily proved in the case of third order determinants from the definition (3.2).

(i) If two rows (or columns) of a determinant are interchanged, the determinant retains the same numerical value but changes sign.

(ii) If a determinant has two identical rows (or columns), its value is zero.

(iii) If the elements of any row (or column) of a determinant are all multiplied by the same factor, the value of the determinant is also multiplied by that factor.

(iv) If each element of any row (or column) of a determinant consists of the sum of two terms, the determinant may be expressed as the sum of two other determinants, in each of which the elements are single terms. For example,

$$\begin{vmatrix} a_1+x & b_1 & c_1 \\ a_2+y & b_2 & c_2 \\ a_3+z & b_3 & c_3 \end{vmatrix} = \begin{vmatrix} a_1 & b_1 & c_1 \\ a_2 & b_2 & c_2 \\ a_3 & b_3 & c_3 \end{vmatrix} + \begin{vmatrix} x & b_1 & c_1 \\ y & b_2 & c_2 \\ z & b_3 & c_3 \end{vmatrix}.$$

(v) The value of a determinant is unaltered if the elements of any row (or column) are multiplied by a constant and the results added to the corresponding elements of any other row (or column). Thus, by several applications of the rule,

$$\begin{vmatrix} a_1 & b_1 & c_1 \\ a_2 & b_2 & c_2 \\ a_3 & b_3 & c_3 \end{vmatrix} = \begin{vmatrix} a_1+pb_1+qc_1 & b_1+kc_1 & c_1 \\ a_2+pb_2+qc_2 & b_2+kc_2 & c_2 \\ a_3+pb_3+qc_3 & b_3+kc_3 & c_3 \end{vmatrix}.$$

When the elements of a third-order determinant are numerical, there is often little to be gained by simplifying it instead of expanding it directly as in § 3.1. If, however, the elements are very large, it is sometimes profitable to simplify the determinant before expansion as in the following case.

Example 2

Evaluate the determinant

$$\Delta = \begin{vmatrix} 100 & 101 & 102 \\ 101 & 102 & 103 \\ 102 & 103 & 104 \end{vmatrix}.$$

By (v), if from R_2 we take R_1, we obtain

$$\Delta = \begin{vmatrix} 100 & 101 & 102 \\ 1 & 1 & 1 \\ 102 & 103 & 104 \end{vmatrix}.$$

Again, taking R_1 from R_3, we have

$$\Delta = \begin{vmatrix} 100 & 101 & 102 \\ 1 & 1 & 1 \\ 2 & 2 & 2 \end{vmatrix}$$

$$= 2\begin{vmatrix} 100 & 101 & 102 \\ 1 & 1 & 1 \\ 1 & 1 & 1 \end{vmatrix}, \text{ by property (iii)},$$

$$= 0 \quad \text{by property (ii).}$$

The first two steps of the above simplification may be effected simultaneously, but until the processes of manipulation have been thoroughly mastered, it is advisable to do one simplification at a time.

Further applications of the properties of determinants are illustrated below.

Example 3

Factorise the determinant

$$\Delta = \begin{vmatrix} x & a & a \\ a & x & a \\ a & a & x \end{vmatrix}.$$

$$\Delta = \begin{vmatrix} x+2a & x+2a & x+2a \\ a & x & a \\ a & a & x \end{vmatrix} \qquad R_1+(R_2+R_3);$$

$$= (x+2a)\begin{vmatrix} 1 & 1 & 1 \\ a & x & a \\ a & a & x \end{vmatrix}$$

$$= (x+2a)\begin{vmatrix} 1 & 0 & 0 \\ a & x-a & 0 \\ a & 0 & x-a \end{vmatrix} \qquad (C_2-C_1);\ (C_3-C_1);$$

$$= (x+2a)(x-a)^2, \text{ expanding from } R_1.$$

Example 4

Factorise the determinant

$$\Delta = \begin{vmatrix} 1 & 1 & 1 \\ a & b & c \\ a^3 & b^3 & c^3 \end{vmatrix}.$$

We note that, if $a=b$, Δ has two identical columns, and so $(a-b)$ is a factor of Δ (see § 1.2). Similarly, $(b-c)$ and $(c-a)$ are factors. Now Δ is a cyclic homogeneous polynomial of the fourth degree in the variables a, b and c, as can be seen by expanding Δ from R_1. Hence (see § 1.6) we may write

$$\Delta = k(b-c)(c-a)(a-b)(a+b+c),$$

where k is a numerical factor.

The leading term in the expansion of Δ is bc^3, and this is the only term in bc^3 in the expansion. From the factorised form of Δ we see that the term in bc^3 is kbc^3. Hence $k=1$ and $\Delta=(b-c)(c-a)(a-b)(a+b+c)$.

Example 5

Prove that

$$\begin{vmatrix} 1 & a^2 & a^3 \\ 1 & b^2 & b^3 \\ 1 & c^2 & c^3 \end{vmatrix} = (bc+ca+ab)\begin{vmatrix} 1 & a & a^2 \\ 1 & b & b^2 \\ 1 & c & c^2 \end{vmatrix}.$$

This result may be proved by working out or factorising both deter minants. Alternatively we may proceed as follows:

$$\Delta=(bc+ca+ab)\begin{vmatrix} 1 & a & a^2 \\ 1 & b & b^2 \\ 1 & c & c^2 \end{vmatrix} = \begin{vmatrix} bc+ca+ab & a & a^2 \\ bc+ca+ab & b & b^2 \\ bc+ca+ab & c & c^2 \end{vmatrix}, \text{ by property (iii)},$$

$$= \begin{vmatrix} bc+a(b+c) & a & a^2 \\ ca+b(c+a) & b & b^2 \\ ab+c(a+b) & c & c^2 \end{vmatrix}$$

$$= \begin{vmatrix} bc & a & a^2 \\ ca & b & b^2 \\ ab & c & c^2 \end{vmatrix} + \begin{vmatrix} a(b+c) & a & a^2 \\ b(c+a) & b & b^2 \\ c(a+b) & c & c^2 \end{vmatrix}, \text{ by property (iv)}$$

$=\Delta_1+\Delta_2$, say.

Multiplying R_1 by a, R_2 by b, and R_3 by c, we have

$$abc\,\Delta_1= \begin{vmatrix} abc & a^2 & a^3 \\ abc & b^2 & b^3 \\ abc & c^2 & c^3 \end{vmatrix} = abc \begin{vmatrix} 1 & a^2 & a^3 \\ 1 & b^2 & b^3 \\ 1 & c^2 & c^3 \end{vmatrix}, \text{ by (iii)},$$

$$\therefore\ \Delta_1= \begin{vmatrix} 1 & a^2 & a^3 \\ 1 & b^2 & b^3 \\ 1 & c^2 & c^3 \end{vmatrix}.$$

Also,

$$\Delta_2=abc \begin{vmatrix} b+c & 1 & a \\ c+a & 1 & b \\ a+b & 1 & c \end{vmatrix}, \text{ by (iii)},$$

$$=abc \begin{vmatrix} a+b+c & 1 & a \\ a+b+c & 1 & b \\ a+b+c & 1 & c \end{vmatrix} \qquad (C_1+C_3);$$

$$=abc(a+b+c) \begin{vmatrix} 1 & 1 & a \\ 1 & 1 & b \\ 1 & 1 & c \end{vmatrix}, \text{ by (iii)},$$

$=\ 0$, by (ii).

Hence $\Delta=\Delta_1$ and the required result is established.

Example 6

If a, b, c all have different values, and if

$$\Delta \equiv \begin{vmatrix} a & a^2 & a^3-1 \\ b & b^2 & b^3-1 \\ c & c^2 & c^3-1 \end{vmatrix} = 0,$$

prove that $abc=1$.

$$\Delta = \begin{vmatrix} a & a^2 & a^3 \\ b & b^2 & b^3 \\ c & c^2 & c^3 \end{vmatrix} - \begin{vmatrix} a & a^2 & 1 \\ b & b^2 & 1 \\ c & c^2 & 1 \end{vmatrix}, \text{ by (iv)},$$

$$= abc \begin{vmatrix} 1 & a & a^2 \\ 1 & b & b^2 \\ 1 & c & c^2 \end{vmatrix} - \begin{vmatrix} 1 & a & a^2 \\ 1 & b & b^2 \\ 1 & c & c^2 \end{vmatrix},$$

making two transpositions of columns in the second determinant.

$$\therefore \Delta = (abc-1) \begin{vmatrix} 1 & a & a^2 \\ 1 & b & b^2 \\ 1 & c & c^2 \end{vmatrix} = (abc-1)\,\Delta', \text{ say.}$$

Thus, if $\Delta=0$, either $abc=1$, or $\Delta'=0$.

But Δ' is a cyclic homogeneous polynomial of the third degree which vanishes when $a=b$, when $b=c$ and when $c=a$. Hence

$$\Delta' \propto (b-c)(c-a)(a-b),$$

and so, if a, b, c are all different, $\Delta' \neq 0$.

$$\therefore abc=1.$$

Example 7

Express the determinant

$$\begin{vmatrix} 2 & a+\alpha+b+\beta & ab+\alpha\beta \\ a+\alpha+b+\beta & 2(a+b)(\alpha+\beta) & ab(\alpha+\beta)+\alpha\beta(a+b) \\ ab+\alpha\beta & ab(\alpha+\beta)+\alpha\beta(a+b) & 2a\alpha b\beta \end{vmatrix}$$

as the sum of eight determinants, and hence, or otherwise, show that its value is zero. [L.U.]

We rewrite Δ, the given determinant, expressing each element as a sum of two terms:

$$\Delta = \begin{vmatrix} 1+1 & (a+b)+(\alpha+\beta) & ab+\alpha\beta \\ (a+b)+(\alpha+\beta) & (\alpha+\beta)(a+b)+(\alpha+\beta)(a+b) & ab(\alpha+\beta)+\alpha\beta(a+b) \\ ab+\alpha\beta & \alpha\beta(a+b)+ab(\alpha+\beta) & ab\alpha\beta+\alpha\beta ab \end{vmatrix}$$

$$\quad (1)\ (2) \qquad\qquad (3)\ (4) \qquad\qquad (5)\ (6)$$

Thus Δ may be regarded as being subdivided into six "subcolumns" numbered as indicated. The scheme for expressing Δ as the sum of eight determinants using property (iv) is outlined below. We use [(1), (3), (5)] to denote the determinant whose columns are the subcolumns (1), (3), (5) of Δ.

$$\Delta=[(1)+(2),\ (3)+(4),\ (5)+(6)],$$
$$=[(1)+(2),\ (3)+(4),\ (5)]\ +\ [(1)+(2),\ (3)+(4),\ (6)],$$
$$=[(1)+(2),\ (3),\ (5)]\ +\ [(1)+(2),\ (4),\ (5)]\ +\ [(1)+(2),\ (3),\ (6)]\ +\ [(1)+(2),\ (4),\ (6)],$$
$$=[(1),\ (3),\ (5)]\ +\ [(2),\ (3),\ (5)]\ +\ [(1),\ (4),\ (5)]\ +\ [(2),\ (4),\ (5)]\ +\ [(1),\ (3),\ (6)]\ +\ [(2),\ (3),\ (6)]\ +\ [(1),\ (4),\ (6)]\ +\ [(2),\ (4),\ (6)].$$

$$\text{Now } [(1),(3),(5)] = \begin{vmatrix} 1 & a+b & ab \\ a+b & (\alpha+\beta)(a+b) & ab(\alpha+\beta) \\ ab & \alpha\beta(a+b) & ab\alpha\beta \end{vmatrix}$$

$$=ab(a+b)\begin{vmatrix} 1 & 1 & 1 \\ a+b & \alpha+\beta & \alpha+\beta \\ ab & \alpha\beta & \alpha\beta \end{vmatrix}=0, \text{ by (ii).}$$

Similarly, the other seven determinants vanish; and so $\Delta=0$.

3.3. Note on determinants of the fourth and higher orders

By an extension of (3.3), a determinant of any order may be defined in terms of the minors of its elements. For example, the fourth-order determinant

$$\Delta=\begin{vmatrix} a_1 & b_1 & c_1 & d_1 \\ a_2 & b_2 & c_2 & d_2 \\ a_3 & b_3 & c_3 & d_3 \\ a_4 & b_4 & c_4 & d_4 \end{vmatrix}$$

$$= a_1\alpha_1-b_1\beta_1+c_1\gamma_1-d_1\delta_1,$$

where α_1 is the minor of a_1, i.e. the third-order determinant which is obtained by deleting from Δ the row and column containing a_1.

$$\text{Thus} \qquad \alpha_1=\begin{vmatrix} b_2 & c_2 & d_2 \\ b_3 & c_3 & d_3 \\ b_4 & c_4 & d_4 \end{vmatrix}$$

and β_1, γ_1 and δ_1 are similarly defined.

Δ may be expanded from any row or column, the expansion consisting of the four elements of the row (or column) each multiplied by its minor with the appropriate sign attached. The sign may be determined from the sketch which shows a chessboard pattern with positive signs along the leading diagonal.

$$\begin{vmatrix} + & - & + & - \\ - & + & - & + \\ + & - & + & - \\ - & + & - & + \end{vmatrix}$$

Thus, for example, $\Delta = -b_1\beta_1+b_2\beta_2-b_3\beta_3+b_4\beta_4,$
$= a_3\alpha_3-b_3\beta_3+c_3\gamma_3-d_3\delta_3.$

In a similar manner determinants of higher order may be defined.

The properties listed in § 3.2 should be used to simplify determinants of order higher than the third before expansion. We give examples to show how they may be used systematically to reduce the order of a given determinant.

Example 8

Evaluate the determinant

$$\Delta = \begin{vmatrix} 1 & 1 & 1 & 1 \\ 1 & 1+a & 1 & 1 \\ 1 & 1 & 1+b & 1 \\ 1 & 1 & 1 & 1+c \end{vmatrix},$$ [Sheffield.]

We have

$$\Delta = \begin{vmatrix} 1 & 0 & 0 & 0 \\ 1 & a & 0 & 0 \\ 1 & 0 & b & 0 \\ 1 & 0 & 0 & c \end{vmatrix} \qquad (C_2-C_1);\ (C_3-C_1);\ (C_4-C_1).$$

$$= \begin{vmatrix} a & 0 & 0 \\ 0 & b & 0 \\ 0 & 0 & c \end{vmatrix}, \text{ expanding from } R_1,$$

$$= abc.$$

Example 9

Evaluate the determinant

$$\Delta = \begin{vmatrix} 1 & -2 & 3 & -4 \\ -2 & 3 & -4 & 1 \\ 3 & -4 & 1 & -2 \\ -4 & 1 & -2 & 3 \end{vmatrix}.$$

We have

$$\Delta = \begin{vmatrix} 1 & -2 & 3 & -4 \\ 0 & -1 & 2 & -7 \\ 0 & 2 & -8 & 10 \\ 0 & -7 & 10 & -13 \end{vmatrix} (R_2+2R_1);\ (R_3-3R_1);\ (R_4+4R_1);$$

$$= \begin{vmatrix} -1 & 2 & -7 \\ 2 & -8 & 10 \\ -7 & 10 & -13 \end{vmatrix}, \text{ expanding from } C_1.$$

$$\therefore\ \Delta = -(104-100)-2(70-26)-7(20-56), \text{ expanding from } R_1$$
$$= 160.$$

Exercises 3 (*a*)

1. By expanding directly, show that

(i) $$\begin{vmatrix} 0 & c & b \\ c & 0 & a \\ b & a & 0 \end{vmatrix} = 2abc,$$

(ii) $$\begin{vmatrix} a & b & c \\ c & a & b \\ b & c & a \end{vmatrix} = a^3+b^3+c^3-3abc,$$

(iii) $$\begin{vmatrix} 1 & a & -b \\ -a & 1 & c \\ b & -c & 1 \end{vmatrix} = 1+a^2+b^2+c^2,$$

(iv) $$\begin{vmatrix} 1 & 1 & 1 \\ z & x & y \\ y & z & x \end{vmatrix} = x^2+y^2+z^2-yz-zx-xy,$$

(v) $$\begin{vmatrix} \cos(x+y) & \sin(x+y) & -\cos(x+y) \\ \sin(x-y) & \cos(x-y) & \sin(x-y) \\ \sin 2x & 0 & \sin 2y \end{vmatrix} = \sin 2(x+y).$$

2. If

$$\Delta = \begin{vmatrix} 1 & 1 & 1 \\ \sin A & \sin B & \sin C \\ \cos A & \cos B & \cos C \end{vmatrix},$$

prove, by expanding from the first row, that

$$\Delta = \sin(A-B) + \sin(B-C) + \sin(C-A).$$

By subtracting columns before expansion and by converting differences into products, show that

$$\Delta = -4 \sin \tfrac{1}{2}(A-B) \sin \tfrac{1}{2}(B-C) \sin \tfrac{1}{2}(C-A).$$

3. Solve the equation

$$\begin{vmatrix} x^2+x+2 & x^2 & 0 \\ x+4 & 2 & x^2 \\ 1 & 1 & 1 \end{vmatrix} = 0.$$

[Liverpool.]

4. Evaluate

$$\text{(i)} \begin{vmatrix} 2 & 4 & 16 \\ 3 & 9 & 81 \\ 5 & 25 & 625 \end{vmatrix}, \quad \text{(ii)} \begin{vmatrix} 1! & 2! & 3! \\ 2! & 3! & 4! \\ 3! & 4! & 5! \end{vmatrix}.$$

[Liverpool.]

5. Solve the equation

$$\begin{vmatrix} x & a & b \\ a & x & b \\ a & b & x \end{vmatrix} = 0.$$

[Sheffield.]

6. Find the roots of the equation

$$\begin{vmatrix} x-3 & 1 & -1 \\ 1 & x-5 & 1 \\ -1 & 1 & x-3 \end{vmatrix} = 0.$$

[Sheffield.]

7. Show that

$$\begin{vmatrix} x & a & b \\ x^2 & a^2 & b^2 \\ a+b & x+b & x+a \end{vmatrix} = (b-a)(x-a)(x-b)(x+a+b).$$

[Sheffield.]

8. If

$$\Delta = \begin{vmatrix} a & h & g \\ h & b & f \\ g & f & c \end{vmatrix},$$

prove that $\Delta = abc + 2fgh - af^2 - bg^2 - ch^2$.

If B, C and F are the minors of b, c and f respectively in Δ, show, by expressing the determinant $\begin{vmatrix} B & F \\ F & C \end{vmatrix}$ as the sum of four second-order determinants, that

$$BC - F^2 = a\Delta.$$

9. Factorise the determinants

$$\text{(i)} \begin{vmatrix} a & 1 & 1 \\ 1 & a & 1 \\ 1 & 1 & a \end{vmatrix}, \quad \text{(ii)} \begin{vmatrix} 1 & ab & a+b \\ 1 & bc & b+c \\ 1 & ca & c+a \end{vmatrix},$$

$$\text{(iii)} \begin{vmatrix} a & b & c \\ b+c & c+a & a+b \\ bc & ca & ab \end{vmatrix}.$$

10. Show that

$$\begin{vmatrix} 1 & 1 & 1 \\ 1 & 1+a & 1 \\ 1 & 1 & 1+b \end{vmatrix} = ab,$$

and that

$$\begin{vmatrix} 1+a & 1 & 1 \\ 1 & 1+b & 1 \\ 1 & 1 & 1+c \end{vmatrix} = abc(1+1/a+1/b+1/c).$$

11. Show that

$$\begin{vmatrix} \log x & \log y & \log z \\ \log 2x & \log 2y & \log 2z \\ \log 3x & \log 3y & \log 3z \end{vmatrix} = 0.$$

12. Show, without expanding the determinants, that

$$\begin{vmatrix} 1 & 1 & 1 \\ ab & bc & ca \\ a^2b^2 & b^2c^2 & c^2a^2 \end{vmatrix} = -abc \begin{vmatrix} c^2 & b^2 & a^2 \\ c & b & a \\ 1 & 1 & 1 \end{vmatrix}$$

$$= -abc(b-c)(c-a)(a-b).$$

13. Show, without expanding the determinants, that

$$\begin{vmatrix} x & y & z \\ x^2 & y^2 & z^2 \\ yz & zx & xy \end{vmatrix} = \begin{vmatrix} x^2 & y^2 & z^2 \\ x^3 & y^3 & z^3 \\ 1 & 1 & 1 \end{vmatrix}$$

$$= (y-z)(z-x)(x-y)(yz+zx+xy).$$

14. Prove that

$$\begin{vmatrix} 1 & 1 & 1 \\ a-b & b-c & c-a \\ (a-b)^2 & (b-c)^2 & (c-a)^2 \end{vmatrix} = -(2a-b-c)(2b-c-a)(2c-a-b).$$

15. By expressing the determinant

$$\begin{vmatrix} x+y & y+z & z+x \\ y+z & z+x & x+y \\ z+x & x+y & y+z \end{vmatrix}$$

as the sum of eight third-order determinants, show that its value is

$$2\begin{vmatrix} x & y & z \\ y & z & x \\ z & x & y \end{vmatrix}.$$

16. (i) Find the values for p and q such that the determinant

$$\begin{vmatrix} 1 & 1 & 1 \\ a & pb & c \\ a & b & qc \end{vmatrix}$$

is a constant multiple of $bc+ca+ab$.

(ii) Factorise the determinant

$$\begin{vmatrix} a & b+c & a^3 \\ b & c+a & b^3 \\ c & a+b & c^3 \end{vmatrix}.$$

[Liverpool.]

17. Factorise the determinants

(i) $$\begin{vmatrix} 2a & b+c & b-c \\ 2b & a-c & a+c \\ a+b & a & b \end{vmatrix},$$

(ii) $$\begin{vmatrix} 1 & a & b \\ a & 1 & b \\ a & b & 1 \end{vmatrix}.$$

18. Evaluate the determinant

$$\begin{vmatrix} b+c & b-c & c-b \\ a-c & c+a & c-a \\ a-b & b-a & a+b \end{vmatrix}.$$

19. Solve the following equation in x:

$$\begin{vmatrix} -1 & 2 & 2 \\ x^2+ax & -ax & ax+a^2 \\ x^2+ax & ax+a^2 & -ax \end{vmatrix} = 0.$$

[L.U.]

20. Express the determinant

$$\begin{vmatrix} 1 & a^2-bc & a^4 \\ 1 & b^2-ca & b^4 \\ 1 & c^2-ab & c^4 \end{vmatrix}$$

as the product of one quadratic and four linear factors. [Sheffield.]

21. Evaluate the determinant

$$\begin{vmatrix} 4a & 3a-b-c & 3a-b-c \\ 3b-c-a & 4b & 3b-c-a \\ 3c-a-b & 3c-a-b & 4c \end{vmatrix}.$$

22. By multiplying the first row by abc, and taking factors out of the resulting columns, prove that

$$\begin{vmatrix} 1 & 1 & 1 \\ bc(c-b) & ca(a-c) & ab(b-a) \\ b^2c & c^2a & a^2b \end{vmatrix} = abc(a^3+b^3+c^3-3abc).$$

23. Factorise the determinant

$$\begin{vmatrix} 1 & 1 & 1 \\ bc(b+c) & ca(c+a) & ab(a+b) \\ b^2c^2 & c^2a^2 & a^2b^2 \end{vmatrix}.$$

24. (i) Find the value of the determinant

$$\begin{vmatrix} a_1+\lambda a_2 & b_1+\lambda b_2 & c_1+\lambda c_2 \\ a_2+\mu a_3 & b_2+\mu b_3 & c_2+\mu c_3 \\ a_3+\nu a_1 & b_3+\nu b_1 & c_3+\nu c_1 \end{vmatrix}$$

in terms of λ, μ, ν and D, where D is the value of the determinant when $\lambda=\mu=\nu=0$.

(ii) Find the roots of the equation

$$\begin{vmatrix} x^3 & 3 & 8 \\ x & 2 & 2 \\ 1 & 3 & 1 \end{vmatrix} = 0.$$

[Durham.]

25. (i) Evaluate

$$\begin{vmatrix} 7 & 13 & 10 & 6 \\ 5 & 9 & 7 & 4 \\ 8 & 12 & 11 & 7 \\ 4 & 10 & 6 & 3 \end{vmatrix}.$$

(ii) Prove that, if $a^2+b^2+c^2+d^2=1$, then

$$\begin{vmatrix} a^2-1 & ab & ac & ad \\ ba & b^2-1 & bc & bd \\ ca & cb & c^2-1 & cd \\ da & db & dc & d^2-1 \end{vmatrix} = 0.$$

[Sheffield.]

26. Evaluate the determinant

$$\begin{vmatrix} 8 & 3 & 4 & 7 \\ 4 & 1 & 2 & 3 \\ 11 & 3 & 7 & 6 \\ 9 & 2 & 3 & 5 \end{vmatrix}.$$

[Leeds.]

27. Evaluate the determinant

$$\begin{vmatrix} 3 & 13 & 11 & 9 \\ 1 & 4 & 17 & 8 \\ 1 & 8 & 0 & -2 \\ 2 & 6 & 24 & 11 \end{vmatrix}.$$

[Leeds.]

28. (i) Show that

$$\begin{vmatrix} a+1 & 1 & 1 & 1 \\ 1 & b+1 & 1 & 1 \\ 1 & 1 & c+1 & 1 \\ 1 & 1 & 1 & d+1 \end{vmatrix} = abcd(1/a+1/b+1/c+1/d+1).$$

(ii) Show that

$$\begin{vmatrix} a & 1 & 1 & 1 \\ 1 & a & 1 & 1 \\ 1 & 1 & a & 1 \\ 1 & 1 & 1 & a \end{vmatrix} = (a+3)(a-1)^3.$$

29. Show that $ax+by+cz$ is a factor of

$$\begin{vmatrix} 0 & x & y & z \\ x & 0 & -c & b \\ y & c & 0 & -a \\ z & -b & a & 0 \end{vmatrix},$$

and evaluate the determinant. [Leeds.]

3.4. Solution of linear simultaneous equations

The solution of the equations

$$a_1x+b_1y=k_1,$$
$$a_2x+b_2y=k_2,$$

may be written in the form

$$\frac{x}{\begin{vmatrix} k_1 & b_1 \\ k_2 & b_2 \end{vmatrix}} = \frac{y}{\begin{vmatrix} a_1 & k_1 \\ a_2 & k_2 \end{vmatrix}} = \frac{1}{\begin{vmatrix} a_1 & b_1 \\ a_2 & b_2 \end{vmatrix}} \quad \text{if } a_1b_2 \neq a_2b_1. \tag{3.5}$$

This result may be extended to the equations

$$\left.\begin{aligned} a_1x+b_1y+c_1z=k_1 \\ a_2x+b_2y+c_2z=k_2 \\ a_3x+b_3y+c_3z=k_3 \end{aligned}\right\} \tag{3.6}$$

where we get

$$\frac{x}{\begin{vmatrix} k_1 & b_1 & c_1 \\ k_2 & b_2 & c_2 \\ k_3 & b_3 & c_3 \end{vmatrix}} = \frac{y}{\begin{vmatrix} a_1 & k_1 & c_1 \\ a_2 & k_2 & c_2 \\ a_3 & k_3 & c_3 \end{vmatrix}} = \frac{z}{\begin{vmatrix} a_1 & b_1 & k_1 \\ a_2 & b_2 & k_2 \\ a_3 & b_3 & k_3 \end{vmatrix}} = \frac{1}{\begin{vmatrix} a_1 & b_1 & c_1 \\ a_2 & b_2 & c_2 \\ a_3 & b_3 & c_3 \end{vmatrix}} \tag{3.7}$$

provided that

$$\Delta \equiv \begin{vmatrix} a_1 & b_1 & c_1 \\ a_2 & b_2 & c_2 \\ a_3 & b_3 & c_3 \end{vmatrix},$$

the determinant of the coefficients of the unknowns, does not have the value zero.

This result, known as *Cramer's rule,* may be written in the form

$$\frac{x}{\Delta_x} = \frac{y}{\Delta_y} = \frac{z}{\Delta_z} = \frac{1}{\Delta},$$

where Δ_x, Δ_y, Δ_z are obtained from Δ by substituting the k's in place of the $x-$, $y-$ and $z-$ coefficients respectively. Cramer's rule may be extended to the case of n simultaneous equations in n unknowns.

3.5. Homogeneous equations

If in equations (3.6) we write $k_1 = k_2 = k_3 = 0$, we obtain a set of three homogeneous equations

$$\left.\begin{aligned} a_1x + b_1y + c_1z = 0 \\ a_2x + b_2y + c_2z = 0 \\ a_3x + b_3y + c_3z = 0 \end{aligned}\right\} \tag{3.8}$$

which are satisfied by the values $x = y = z = 0$. If they have any other solution,

$$\Delta \equiv \begin{vmatrix} a_1 & b_1 & c_1 \\ a_2 & b_2 & c_2 \\ a_3 & b_3 & c_3 \end{vmatrix}$$

must have the value zero.

To see this, we suppose that the equations have a solution in which x, y and z are not all zero, and that $\Delta \neq 0$. Then by Cramer's rule

$$x = \frac{\Delta_x}{\Delta} = \frac{0}{\Delta} = 0,$$

and, similarly, $y = z = 0$. This is a contradiction. Hence $\Delta = 0$ is a necessary condition for equations (3.8) to have a solution other than $x = y = z = 0$. We state without proof that this condition is also sufficient. It follows that equations (3.8) will have such a solution if, and only if,

$$\begin{vmatrix} a_1 & b_1 & c_1 \\ a_2 & b_2 & c_2 \\ a_3 & b_3 & c_3 \end{vmatrix} = 0. \tag{3.9}$$

3.6. Consistency of equations

A system of n linear equations in n unknowns will, if the determinant of the coefficients of the unknowns is not zero, determine these n quantities uniquely. If, however, the number of equations exceeds the number of unknowns, it is not usually possible to find values of the n unknowns which will simultaneously satisfy all the equations. If such values can be found, the system is said to be *consistent*, and, at most, n of the given equations are independent.

If the equations

$$\left.\begin{aligned} a_1x+b_1y+c_1=0 \\ a_2x+b_2y+c_2=0 \\ a_3x+b_3y+c_3=0 \end{aligned}\right\} \tag{3.10}$$

are consistent, they are satisfied by (say) $x=x_0$, $y=y_0$. This is equivalent to saying that equations (3.8) have a solution $x=x_0, y=y_0, z=1$ and so, by (3.9),

$$\begin{vmatrix} a_1 & b_1 & c_1 \\ a_2 & b_2 & c_2 \\ a_3 & b_3 & c_3 \end{vmatrix} = 0. \tag{3.11}$$

This is a necessary condition that equations (3.10) should be consistent. That this condition is not sufficient is clear from consideration of the equations

$$x+y+1=0, \quad x+y+2=0, \quad x+y+3=0,$$

where condition (3.11) is satisfied but the equations are obviously inconsistent.

It has been shown in § 3.1 that when $a_2b_3-a_3b_2\neq 0$, equations (3.10) are consistent if condition (3.11) is satisfied. Similarly, it can be shown that if the minor of at least one of c_1, c_2, c_3 is not zero, (3.11) is sufficient condition for equations (3.10) to be consistent.

Example 10

Solve the equations

$$\begin{aligned} ax+by+cz=1, \\ a^2x+b^2y+c^2z=1, \\ a^4x+b^4y+c^4z=1, \end{aligned}$$

simplifying the results as far as possible. [L.U.]

By Cramer's rule

$$\frac{x}{\begin{vmatrix} 1 & b & c \\ 1 & b^2 & c^2 \\ 1 & b^4 & c^4 \end{vmatrix}} = \frac{y}{\begin{vmatrix} a & 1 & c \\ a^2 & 1 & c^2 \\ a^4 & 1 & c^4 \end{vmatrix}} = \frac{z}{\begin{vmatrix} a & b & 1 \\ a^2 & b^2 & 1 \\ a^4 & b^4 & 1 \end{vmatrix}} = \frac{1}{\begin{vmatrix} a & b & c \\ a^2 & b^2 & c^2 \\ a^4 & b^4 & c^4 \end{vmatrix}}$$

Now

$$\begin{vmatrix} a & b & c \\ a^2 & b^2 & c^2 \\ a^4 & b^4 & c^4 \end{vmatrix} = abc \begin{vmatrix} 1 & 1 & 1 \\ a & b & c \\ a^3 & b^3 & c^3 \end{vmatrix}$$

$$= abc(b-c)(c-a)(a-b)(a+b+c), \text{ see § 3.2, Example 4.}$$

Substituting the values $a=1$, $b=1$, $c=1$ in turn in this result, we can evaluate the other determinants. After simplification we obtain

$$x=(1-b)(1-c)(1+b+c)/a(a-b)(a-c)(a+b+c),$$
$$y=(1-c)(1-a)(1+c+a)/b(b-c)(b-a)(b+c+a),$$
$$z=(1-a)(1-b)(1+a+b)/c(c-a)(c-b)(c+a+b).$$

If any of a, b, c has the value zero, if $a+b+c=0$, or if $a=b$, $b=c$ or $c=a$, the solution is not valid.

Example 11

Express as a determinant equated to zero the condition that the equations

$$2\lambda x-3y+\lambda-3=0,$$
$$3x-2y+1=0,$$
$$4x-\lambda y+2=0,$$

should be satisfied by the same values of x and y. Find the two values of λ for which the equations are consistent and the corresponding solutions.

[L.U. Anc.]

By (3.11) the required condition is

$$\begin{vmatrix} 2\lambda & -3 & \lambda-3 \\ 3 & -2 & 1 \\ 4 & -\lambda & 2 \end{vmatrix} = 0.$$

If we take $2R_2$ from R_3 we obtain

$$\begin{vmatrix} 2\lambda & -3 & \lambda-3 \\ 3 & -2 & 1 \\ -2 & 4-\lambda & 0 \end{vmatrix} = 0.$$

Expanding from C_3 we have

$$(\lambda-3)(8-3\lambda)+2\lambda^2-8\lambda+6=0,$$

which reduces to

$$(\lambda-3)(6-\lambda)=0,$$
$$\lambda=3 \text{ or } 6.$$

When $\lambda=3$, the given equations become

$$6x-3y=0,$$
$$3x-2y+1=0,$$
$$4x-3y+2=0,$$

which have the solution $x=1$, $y=2$.

When $\lambda=6$, the given equations become

$$12x-3y+3=0,$$
$$3x-2y+1=0,$$
$$4x-6y+2=0,$$

and these are satisfied by the values $x=-\frac{1}{5}$, $y=\frac{1}{5}$.

Exercises 3 (*b*)

1. Solve by determinants the equations

$$5x+3y+3z=48,$$
$$2x+6y-3z=18,$$
$$8x-3y+2z=21.$$

2. Solve by determinants the equations

$$x+y+z=5,$$
$$x+2y+3z=11,$$
$$3x+y+4z=13.$$

3. Eliminate x, y and z from the equations

$$ax+hy+gz=0,$$
$$hx+by+fz=0,$$
$$gx+fy+cz=0.$$

4. Find the value of λ for which the following equations are consistent:

$$4x+\lambda y=10,$$
$$x-2y=8,$$
$$5x+7y=6.$$

Find the values of x and y corresponding to this value of λ.

5. Find the values of λ for which the equations

$$(2-\lambda)x+2y+3=0,$$
$$2x+(4-\lambda)y+7=0,$$
$$2x+5y+6-\lambda=0,$$

are consistent, and find the values of x and y corresponding to each of these values of λ.

6. Find all the values of t for which the equations

$$(t-1)x+(3t+1)y+2tz=0,$$
$$(t-1)x+(4t-2)y+(t+3)z=0,$$
$$2x+(3t+1)y+3(t-1)z=0,$$

are compatible and find the ratios of $x:y:z$ when t has the smallest of these values. What happens when t has the greatest of these values? [L.U.]

7. Solve by determinants the equations

(i)
$$x-y+z=1,$$
$$x-2y+4z=8,$$
$$x+3y+9z=27.$$

(ii)
$$x+y-z=1,$$
$$8x+3y-6z=1,$$
$$-4x-y+3z=1.$$

8. Solve the equations
$$\tfrac{1}{2}x+\tfrac{1}{3}y+\tfrac{1}{4}z=1,$$
$$\tfrac{1}{3}x+\tfrac{1}{4}y+\tfrac{1}{5}z=1,$$
$$\tfrac{1}{4}x+\tfrac{1}{5}y+\tfrac{1}{6}z=1$$
by means of determinants.

If
$$y-z=ax,$$
$$z-x=by,$$
$$x-y=cz,$$
and x, y, z are not all zero, show, without solving any equations, that
$$abc+a+b+c=0.$$

9. Prove that
$$\begin{vmatrix} a & b & c \\ b+c & c+a & a+b \\ a^2 & b^2 & c^2 \end{vmatrix} = -(a-b)(b-c)(c-a)(a+b+c).$$

Solve completely the equations
$$ax+by+cz=a+b+c,$$
$$(b+c)x+(c+a)y+(a+b)z=2(a+b+c),$$
$$a^2x+b^2y+c^2z=a^2+b^2+c^2,$$
where a, b, c are non-zero and distinct. [Durham.]

Miscellaneous Exercises 3

1. (i) Prove that the determinant
$$\begin{vmatrix} 1 & b+c & (b+c)\,(b^2+c^2) \\ 1 & c+a & (c+a)\,(c^2+a^2) \\ 1 & a+b & (a+b)\,(a^2+b^2) \end{vmatrix}$$
vanishes.

(ii) Evaluate the determinant
$$\begin{vmatrix} 1 & b+c & (b-c)\,(b^2-c^2) \\ 1 & c+a & (c-a)\,(c^2-a^2) \\ 1 & a+b & (a-b)\,(a^2-b^2) \end{vmatrix}.$$
[L.U.]

2. (i) Show that $(b-c)$ is a factor of

$$\begin{vmatrix} a^2 & bc & b+c \\ b^2 & ac & c+a \\ c^2 & ab & a+b \end{vmatrix}$$

and find the other factors.

(ii) Prove that

$$\begin{vmatrix} 1 & \cos 2\alpha & \sin \alpha \\ 1 & \cos 2\beta & \sin \beta \\ 1 & \cos 2\gamma & \sin \gamma \end{vmatrix} = 2(\sin \beta - \sin \gamma)(\sin \gamma - \sin \alpha)(\sin \alpha - \sin \beta).$$

[L.U.]

3. (i) Express the determinant

$$\begin{vmatrix} 1 & b+c-a & a^3 \\ 1 & c+a-b & b^3 \\ 1 & a+b-c & c^3 \end{vmatrix}$$

as a product of factors.

(ii) Show that

$$\begin{vmatrix} \cos\theta & \cos\alpha\cos\theta & \cos(\alpha+\theta) \\ \cos(\alpha+\theta) & \cos\theta & \cos\alpha\cos(\alpha+\theta) \\ \cos(\alpha+\theta) & \cos\theta\sin^2\alpha & -\cos\alpha\sin\alpha\sin\theta \end{vmatrix}$$

is equal to $-\cos^2\alpha \sin^2\alpha \cos\theta$. [L.U.]

4. (i) Show that $a+b+c$ is a factor of the determinant

$$\begin{vmatrix} b+c-a & b & c \\ a & a+c-b & c \\ a & b & a+b-c \end{vmatrix}$$

and deduce that its value is $3abc-a^3-b^3-c^3$.

(ii) If $A+B+C=180°$, prove that

$$\begin{vmatrix} \cos(B-C) & \cos(C+A) & \cos(A+B) \\ \cos(B+C) & \cos(C-A) & \cos(A+B) \\ \cos(B+C) & \cos(C+A) & \cos(A-B) \end{vmatrix} = 0.$$

[L.U.]

5. (i) Prove that

$$\begin{vmatrix} 3 & 3x & 3x^2+2a^2 \\ 3x & 3x^2+2a^2 & 3x^3+6a^2x \\ 3x^2+2a^2 & 3x^3+6a^2x & 3x^4+12a^2x^2+2a^4 \end{vmatrix}$$

is independent of x

(ii) Find all the roots of the equation

$$\begin{vmatrix} 1 & 1 & 1 \\ x & 2 & 1 \\ x^3 & 8 & 1 \end{vmatrix} = 0.$$

[L.U.]

6. Prove that

(i) $$\begin{vmatrix} b+c & c+a & a+b \\ a^2 & b^2 & c^2 \\ a^3 & b^3 & c^3 \end{vmatrix} = (b^2-c^2)(c^2-a^2)(a^2-b^2);$$

(ii) $$\begin{vmatrix} 1 & 1 & 1 \\ a^2 & b^2 & c^2 \\ a^3 & b^3 & c^3 \end{vmatrix} = (b-c)(c-a)(a-b)(ab+bc+ca).$$

[L.U.]

7. (i) Show that $\sin \alpha$, $\cos \alpha$, $(\sin \alpha - \cos \alpha)$ are factors of the determinant

$$\begin{vmatrix} \cos \alpha & \sin 2\alpha & \cos^2 \alpha \\ \sin \alpha & \sin 2\alpha & \sin^2 \alpha \\ \sin \alpha & \sin^2 \alpha & \cos^2 \alpha \end{vmatrix}$$

and find the remaining factor.

(ii) Find the values of λ for which the following equations are consistent:

$$3x + \lambda y = 5,$$
$$\lambda x - 3y = -4,$$
$$3x - y = -1.$$

Solve the equations for these values of λ. [L.U.]

8. (i) Show that

$$\begin{vmatrix} a-b-c & 2a & 2a \\ 2b & b-c-a & 2b \\ 2c & 2c & c-a-b \end{vmatrix} = (a+b+c)^3.$$

(ii) By multiplying the second column by b, the third column by c, and subtracting the elements of one column from the other (or by any other method) show that a is a factor of the determinant

$$\begin{vmatrix} b^2+c^2 & ab & ac \\ ab & c^2+a^2 & bc \\ ac & bc & a^2+b^2 \end{vmatrix}.$$

Evaluate the determinant. [L.U.]

9. (i) Factorise the expression

$$\begin{vmatrix} 1 & a & a^3 \\ 1 & b & b^3 \\ 1 & c & c^3 \end{vmatrix}.$$

(ii) Prove that

$$\begin{vmatrix} a-1 & a^2-1 & a^3-1 \\ a^2-1 & a^4-1 & a^6-1 \\ a^3-1 & a^6-1 & a^9-1 \end{vmatrix} = a^4(a-1)^6(a+1)^2(a^2+a+1).$$

[L.U.]

10. (i) Solve the following equation in x:

$$\begin{vmatrix} -1 & 1 & 1 \\ x^2+2ax & -ax & ax+2a^2 \\ x^2+2ax & ax+2a^2 & -ax \end{vmatrix} = 0.$$

(ii) Show that $(a+b+c)$ and $(a^2+b^2+c^2)$ are factors of the determinant

$$\begin{vmatrix} a^2 & (b+c)^2 & bc \\ b^2 & (c+a)^2 & ca \\ c^2 & (a+b)^2 & ab \end{vmatrix}$$

and find all the factors. [L.U.]

11. Prove that

$$\Delta \equiv \begin{vmatrix} a & b & c \\ b & c & a \\ c & a & b \end{vmatrix} = -(a+b+c)(a^2+b^2+c^2-bc-ca-ab).$$

Show that there are three real values of λ for which the equations

$$(a-\lambda)x+by+cz=0,$$
$$bx+(c-\lambda)y+az=0,$$
$$cx+ay+(b-\lambda)z=0,$$

are simultaneously true, and that the product of these values of λ is Δ.

[L.U.]

12. (i) Solve the equation

$$\begin{vmatrix} x+1 & x+2 & 3 \\ 2 & x+3 & x+1 \\ x+3 & 1 & x+2 \end{vmatrix} = 0.$$

(ii) Prove that

$$\begin{vmatrix} (x+1)(x+2) & (x+2) & 1 \\ (x+2)(x+3) & (x+3) & 1 \\ (x+3)(x+4) & (x+4) & 1 \end{vmatrix} = -2.$$

[L.U.]

13. (i) Show that $x=3$ is a root of the equation

$$\begin{vmatrix} x & -6 & -1 \\ 3 & -2x & x-4 \\ -2 & 3x & x-2 \end{vmatrix} = 0,$$

and solve it completely.

(ii) Prove that

$$\begin{vmatrix} (y-z)^3 & (z-x)^3 & (x-y)^3 \\ (y-z)^2 & (z-x)^2 & (x-y)^2 \\ 1 & 1 & 1 \end{vmatrix}$$

$$=(-2x+y+z)(x-2y+z)(x+y-2z)(x^2+y^2+z^2-xy-yz-zx).$$

[L.U.]

14. (i) Show that $x+y+z$ is a factor of the determinant

$$\begin{vmatrix} y+z & -y & 2z \\ -x & z+x & -z \\ 2x & 2y & x+y \end{vmatrix},$$

and hence, or otherwise, evaluate the determinant as a product of linear factors.

(ii) Solve completely the equation

$$\begin{vmatrix} x & 2 & x-4 \\ 2x-2 & 3x-2 & 4 \\ 2x+3 & 3x & 5 \end{vmatrix} = 0.$$

[L.U.]

15. (i) Solve completely the equation :

$$\begin{vmatrix} 3x & 2 & 3x-4 \\ 6x-2 & 9x-2 & 4 \\ 6x+3 & 9x & 5 \end{vmatrix} = 0.$$

(ii) Show that the equations

$$2x+3y=4,$$
$$3x+\lambda y=-1,$$
$$\lambda x-2y=c,$$

are consistent for real values of λ if

$$4c^2-156c-439 \geqslant 0.$$

[L.U.]

16. If the corresponding elements of two rows of a three-rowed determinant are equal, prove that the value of the determinant is zero.

Prove that

$$\begin{vmatrix} \cos\theta & (1+\frac{1}{2}\sec\theta)\sec^2\frac{1}{2}\theta & \sin\theta \\ \cos^2\theta & 1 & \sin\theta\cos\theta \\ 2\sin\theta\cos^2\frac{1}{2}\theta & c & \sin^2\theta-\cos\theta \end{vmatrix}$$

$$=\tfrac{1}{2}\cos\theta\sec^2\tfrac{1}{2}\theta.$$

[L.U.]

17. (i) Express as a product of linear factors

$$\begin{vmatrix} x^2 & y^2 & z^2 \\ yz & zx & xy \\ 1 & 1 & 1 \end{vmatrix}.$$

(ii) If no two of the numbers a, b, c are equal, find the condition that the equations

$$x+y+z=0,$$
$$x/a+y/b+z/c=0,$$
$$a^2x+b^2y+c^2z=0,$$

may have a consistent set of non-zero solutions. Find the ratios $x:y:z$ when $a=1$, $b=-3$, $c=2$. [L.U.]

18. (i) Factorise

$$\begin{vmatrix} x+y+nz & (n-1)x & (n-1)y \\ (n-1)z & y+z+nx & (n-1)y \\ (n-1)z & (n-1)x & z+x+ny \end{vmatrix}.$$

(ii) Prove that

$$\begin{vmatrix} \cos(x+\alpha) & \sin(x+\alpha) & 1 \\ \cos(x+\beta) & \sin(x+\beta) & 1 \\ \cos(x+\gamma) & \sin(x+\gamma) & 1 \end{vmatrix}$$

is independent of x. [L.U.]

19. Prove that if, in a third-order determinant, the corresponding elements of any two columns are identical, the determinant is zero.

(i) Factorise the determinant

$$\begin{vmatrix} 1 & 1 & 1 \\ a^2 & b^2 & c^2 \\ (b+c)^2 & (c+a)^2 & (a+b)^2 \end{vmatrix};$$

(ii) Solve the equation

$$\begin{vmatrix} x+1 & 2x & 1 \\ x & 3x-2 & 2x \\ 1 & x & x \end{vmatrix} = 0.$$

[L.U.]

20. (i) If Δ is a determinant of the third order, whose elements are polynomials in x, and if all the rows of Δ become identical when x takes the value α, prove that $(x-\alpha)^2$ is a factor of Δ.

(ii) Factorise the determinants

$$\text{(a)} \begin{vmatrix} 1 & x & x^2 \\ 1 & x^2 & x^4 \\ 1 & x^3 & x^6 \end{vmatrix}, \quad \text{(b)} \begin{vmatrix} \beta^2\gamma^2+\alpha^2\delta^2 & \beta\gamma+\alpha\delta & 1 \\ \gamma^2\alpha^2+\beta^2\delta^2 & \gamma\alpha+\beta\delta & 1 \\ \alpha^2\beta^2+\gamma^2\delta^2 & \alpha\beta+\gamma\delta & 1 \end{vmatrix}.$$

[L.U.]

21. Factorise the determinants

$$\begin{vmatrix} 1 & 1 & 1 \\ a & b & c \\ a^2 & b^2 & c^2 \end{vmatrix}, \quad \begin{vmatrix} 2 & a+b & a^2+b^2 \\ a+b & a^2+b^2 & a^3+b^3 \\ 1 & c & c^2 \end{vmatrix}.$$

Hence, or otherwise, show that

$$\begin{vmatrix} 2 & a+b & a^2+b^2 \\ a+b & a^2+b^2 & a^3+b^3 \\ 1 & c & c^2 \end{vmatrix} = \begin{vmatrix} 1 & 1 & 0 \\ a & b & 0 \\ 0 & 0 & 1 \end{vmatrix} \times \begin{vmatrix} 1 & 1 & 1 \\ a & b & c \\ a^2 & b^2 & c^2 \end{vmatrix}.$$

[L.U.]

22. Find the values of λ for which the equations

$$\lambda x+y+\sqrt{2}z=0,$$
$$x+\lambda y+\sqrt{2}z=0,$$
$$\sqrt{2}x+\sqrt{2}y+(\lambda-2)z=0,$$

have a solution other than $x=y=z=0$. Find also the ratios $x:y:z$ which correspond to each of these values of λ. [L.U.]

23. (i) Show that $x+y+z$ is a factor of the determinant

$$\begin{vmatrix} y+z & x & x^3 \\ z+x & y & y^3 \\ x+y & z & z^3 \end{vmatrix},$$

and express it as a product of five linear factors.

(ii) Show that $x=0$ satisfies the equation

$$\begin{vmatrix} 2x+7 & x+6 & 2x+10 \\ 2x+14 & 2x+12 & 3x+20 \\ x+6 & x+9 & 2x+13 \end{vmatrix} = 0,$$

[L.U. Anc.]

and solve it completely.

24. (i) Show that

$$\begin{vmatrix} 1 & a & a^2 \\ \cos(n-1)x & \cos nx & \cos(n+1)x \\ \sin(n-1)x & \sin nx & \sin(n+1)x \end{vmatrix} = (1-2a\cos x+a^2)\sin x.$$

(ii) If a, b, c are three distinct, non-zero numbers, solve the equation

$$\begin{vmatrix} x & x^2 & a^3-x^3 \\ b & b^2 & a^3-b^3 \\ c & c^2 & a^3-\;^3 \end{vmatrix} = 0.$$

[L.U.]

25. (i) Determine which, if any, of the following two sets of equations are consistent and, when possible, solve them.

(a)
$$x+y+z=1,$$
$$2x+4y-3z=9,$$
$$3x+5y-2z=11.$$

(b)
$$2x-y+z=7,$$
$$3x+y-5z=13,$$
$$x+y+z=5.$$

(ii) Factorise

$$\begin{vmatrix} x+y+2z & x & y \\ z & y+z+2x & y \\ z & x & z+x+2y \end{vmatrix}.$$

[L.U. Anc.]

26. Show that x and $x+y+z$ are factors of the determinant

$$\begin{vmatrix} (y+z)^2 & x^2 & x^2 \\ y^2 & (z+x)^2 & y^2 \\ z^2 & z^2 & (x+y)^2 \end{vmatrix}.$$

Hence, or otherwise, evaluate the determinant as a product of linear factors. [L.U.]

27. (i) Solve the equations

$$4x+3y+5z=11,$$
$$9x+4y+15z=13,$$
$$12x+10y-3z=4.$$

(ii) Show that $a+b+c$ is a factor of

$$\begin{vmatrix} a & b-c & c+b \\ a+c & b & c-a \\ a-b & b+a & c \end{vmatrix},$$

and factorise the determinant. [L.U.]

28. (i) Factorise

$$\begin{vmatrix} (a-x)^2 & (a-y)^2 & (a-z)^2 \\ (b-x)^2 & (b-y)^2 & (b-z)^2 \\ (c-x)^2 & (c-y)^2 & (c-z)^2 \end{vmatrix}.$$

(ii) Find the condition for the equations

$$(a-b-c)x+2ay+2a=0,$$
$$2bx+(b-c-a)y+2b=0,$$
$$2cx+2cy+(c-a-b)=0,$$

to have a common solution, and show that when this condition is satisfied the equations have infinitely many common solutions.

[L.U.]

CHAPTER 4

LIMITS AND INFINITE SERIES

4.1. Definitions

The indicated sum $\quad u_1+u_2+u_3+\ldots+u_n$

of n terms each formed according to a definite law $u_r=f(r)$, say, is called a finite series of n terms.* We use the Greek letter Σ (sigma) to denote summation and write the series briefly as $\sum_{r=1}^{n} u_r$. The arithmetic series

$$a+(a+d)+(a+2d)+\ldots+(a+\overline{n-1}d)$$

in which each of the n terms is formed by adding a constant d to the preceding term is an example of a finite series.

When every term of a series is followed by another term as in

$$u_1+u_2+u_3+\ldots+u_n+\ldots$$

the series is said to be infinite and is denoted by $\sum_{r=1}^{\infty} u_r$ or, where there is no ambiguity, by Σu_r. For example, the geometrical progression

$$a+ax+ax^2+\ldots+ax^{n-1}+\ldots$$

in which each term bears the ratio x to the preceding term is an infinite series.

The sum of the first n terms of an infinite series is usually denoted by S_n. Throughout this chapter the letter n denotes a positive integer.

4.2. The idea of infinity

Suppose that n takes successively the positive integral values 1, 2, 3,... Then there is no limit to the values which n can assume. However large a number we may think of, n will ultimately exceed it. When n increases in this way we say that n tends to infinity (through positive integral values) or, in symbols, $n\to\infty$.

Again, if $f(n)\equiv n^2$, as n increases without limit, $f(n)$ also increases without limit and will ultimately exceed and remain greater than any pre-assigned positive number however large. In this case we say that $f(n)$ tends to infinity as n tends to infinity (through positive integral values) i.e. $f(n)\to\infty$ as $n\to\infty$.

The statement that $f(n)\to\infty$ as $n\to\infty$ implies that there is an integer

* The series could be written $f(1)+f(2)+f(3)+\ldots+f(n)$ but the notation u_r is more convenient, it being understood that r can take only positive integral values.

N_1 such that $f(n) > 1{,}000$, say, provided that $n \geqslant N_1$, and an integer N_2 such that $f(n) > 100{,}000$, say, provided that $n \geqslant N_2$; and corresponding to *each* positive number G however large, there is an integer N_0 such that $f(n) > G$ for all integers $n \geqslant N_0$.

Although in the above example $f(n) \equiv n^2$ tends to infinity through certain positive *integral* values, any function $F(n)$ which increases without limit through positive values as $n \to \infty$ is said to tend to infinity as $n \to \infty$. For example $F(n) \equiv n + 1/n \to \infty$ as $n \to \infty$.

The function $\phi(n) \equiv 1/n - n$ which increases without limit through negative values as $n \to \infty$, is said to tend to negative infinity as $n \to \infty$, and we write $\phi(n) \to -\infty$ as $n \to \infty$.

4.3. The idea of a limit

If n takes in succession the values 1, 10, 100, 1,000, 10,000,... the values of the function $f(n) \equiv 1 + 1/n$ are respectively

$$2,\ 1{\cdot}1,\ 1{\cdot}01,\ 1{\cdot}001,\ 1{\cdot}0001, \ldots$$

As n increases, $f(n)$ steadily decreases, and the larger n becomes the more closely $f(n)$ approaches the value 1. There is no value of n for which $f(n)$ assumes the value 1; but $f(n)$ differs from 1 by an amount less than any pre-assigned positive quantity however small if n is sufficiently large. For example, $f(n)$ differs from 1 by less than 0·00001 for all values of $n > 100{,}000$.

This behaviour is described by the statement: $f(n)$ tends to 1 as n tends to infinity ($f(n) \to 1$ as $n \to \infty$), or the limit of $f(n)$ as n tends to infinity is 1 ($\lim_{n\to\infty} f(n) = 1$).

More generally, any function S_n of n is said to tend to a limit S as n tends to infinity if there exists a definite number S independent of n such that as n increases $|S_n - S|$ ultimately becomes and remains less than any pre-assigned positive number however small.

This definition implies that there is an integer n_1, such that $|S_n - S| < 10^{-3}$, say, provided $n \geqslant n_1$, and an integer n_2 such that $|S_n - S| < 10^{-6}$, say, provided $n \geqslant n_2$, and so on; and corresponding to *each* positive number ϵ, there is an integer n_0 such that

$$|S_n - S| < \epsilon \text{ provided } n \geqslant n_0.$$

The value of n_0 will (in general) depend on the number ϵ.

We state without proof four useful theorems:

I. If, as $n \to \infty$, $u_n \to S$ and $v_n \to S$, and if $u_n \leqslant S_n \leqslant v_n$ for each n, then $S_n \to S$.

II. If $S_n > 0$ for each n, and $S_n \to 0$ as $n \to \infty$, then $1/S_n \to \infty$.

III. If $S_n = a$ for each n, then $\lim_{n\to\infty} S_n = a$.

IV. If, as $n\to\infty$, $S_{2n}\to S$ and $S_{2n+1}\to S$, then $S_n\to S$.

4.4. Convergent series

The sum S_n of n terms of a finite series has an obvious meaning, but the sum of an infinite series must be defined.

If $S_n \equiv u_1+u_2+u_3+\ldots+u_n$ is the sum of the first n terms of the infinite series Σu_r, and if, as n tends to infinity, S_n tends to a finite limit S, the infinite series Σu_r is said to be *convergent* and S is called its *sum to infinity* (S_∞) or, more briefly, its sum.

For example, if $u_r = \dfrac{1}{r(r+1)}$,

$$S_n = \frac{1}{1.2}+\frac{1}{2.3}+\frac{1}{3.4}+\ldots+\frac{1}{n(n+1)}$$

$$=1-\frac{1}{2}+\frac{1}{2}-\frac{1}{3}+\frac{1}{3}-\frac{1}{4}+\ldots+\frac{1}{n}-\frac{1}{n+1}$$

$$=1-\frac{1}{n+1}.$$

Now as $n\to\infty$, $S_n\to 1$; hence the series $\Sigma\dfrac{1}{r(r+1)}$ converges and its sum is 1.

4.5. Series which do not converge

If S_n does not tend to a limit as defined in § 4.3, the series Σu_r does not converge and does not possess a sum to infinity.

For example, if $u_r = r$, $S_n = 1+2+3+\ldots+n = \frac{1}{2}n(n+1)$. Hence as $n\to\infty$, $S_n\to\infty$ and the series Σr is said to be *divergent* or to diverge to $+\infty$.

If $S_n\to-\infty$ as $n\to\infty$, the series is said to diverge to $-\infty$.

Again, for the series $1-1+1-1+\ldots$, $S_n = 1$ if n is odd and $S_n = 0$ if n is even. Hence the series is not convergent. It is said to *oscillate finitely*.

In the case of the series $1-1+3-3+5-5+\ldots$, $S_{2n}=0$ and $S_{2n+1} = 2n+1$. Hence when $n\to\infty$, $S_{2n}\to 0$ and $S_{2n+1}\to\infty$. The series does not converge and is said to *oscillate infinitely*.

The reader should note that some writers use the word "divergent" to describe any series which does not converge. With this terminology a series is either convergent or divergent, and a divergent series may diverge to $+\infty$ or to $-\infty$, oscillate finitely or oscillate infinitely.

4.6. Behaviour of x^n when $n\to\infty$, where n is a positive integer

(i) When $0<x<1$, $1/x>1$ and so we write $1/x=1+y$ where $y>0$. Then by the binomial theorem for a positive integral index

$$\begin{aligned}1/x^n&=(1+y)^n\\&=1+ny+\frac{n(n-1)}{2!}y^2+\ldots+y^n\\&>ny, \text{ since each term on the right is positive.}\end{aligned}$$

Hence $0<x^n<1/ny$, and as $n\to\infty$, $1/ny\to0$ (since, when x is given, y is fixed).

$$\therefore\ x^n\to0 \text{ by theorem I of § 4.3.}$$

(ii) If $x>1$, $0<1/x<1$, and so by (i), $1/x^n\to0$ as $n\to\infty$. Hence $x^n\to\infty$ as $n\to\infty$ by theorem II of § 4.3.

(iii) If $-1<x<0$, let $z=-x$ so that $0<z<1$. Then, by (i),

$$z^n=(-1)^nx^n\to0 \text{ as } n\to\infty$$

$$\therefore\ x^n\to0 \text{ as } n\to\infty.$$

(iv) If $x<-1$, $-x>1$ and $(-x)^n=(-1)^nx^n\to\infty$ as $n\to\infty$ by (ii). Now $x^n>0$ when n is even and $x^n<0$ when n is odd. Hence as n increases through integral values, the numerical value of x^n increases without limit and the sign of x^n oscillates between positive and negative. In this case x^n is said to oscillate infinitely (cf. § 4.5).

(v) By theorem III of § 4.3, if $x=1$, $\lim x^n=1$ and if $x=0$, $\lim x^n=0$. If $x=-1$, $x^n=(-1)^n$ and so as $n\to\infty$, x^n oscillates between the values $+1$ and -1, i.e. x^n oscillates finitely.

4.7. The geometrical progression

The geometrical progression

$$a+ax+ax^2+\ldots+ax^{n-1}+\ldots$$

where $a\neq0$, provides an important example of the behaviour of an infinite series.

When $x\neq1$, S_n, the sum of the first n terms of the series is given by

$$S_n=\frac{a(1-x^n)}{1-x}.$$

We apply the results of § 4.6.

If $|x|<1$, $x^n\to0$ as $n\to\infty$; hence S_n tends to the finite limit $a/(1-x)$ and the series converges.

If $x>1$, $x^n\to\infty$ as $n\to\infty$; hence $S_n\to+\infty$ if $a>0$, $S_n\to-\infty$ if $a<0$, and so the series diverges.

If $x<-1$, x^n oscillates infinitely; hence S_n oscillates infinitely and the series cannot converge.

If $x=1$, the series is $a+a+a+\ldots$ and $S_n=na$; hence $S_n \to +\infty$ if $a>0$, $S_n \to -\infty$ if $a<0$; and so the series diverges.

If $x=-1$, the series is $a-a+a-a+\ldots$; hence $S_n=0$ or a according as n is even or odd, and the series oscillates finitely.

It is important to note that the geometrical progression converges only when the common ratio x is numerically less than unity.

4.8. Theorems on limits

The following theorems on limits follow from the definition given in § 4.3:

1. If $\lim_{n\to\infty} S_n=S$ and k is any constant, then $\lim_{n\to\infty} kS_n=kS$; if however $S_n\to\infty$ as $n\to\infty$, $kS_n\to+\infty$ if $k>0$ and $kS_n\to-\infty$ if $k<0$.

2. If $S_n\to S$ and $T_n\to T$ as $n\to\infty$, then

 (i) $pS_n+qT_n\to pS+qT$, where p and q are constants,

 (ii) $S_nT_n\to ST$,

 (iii) $S_n/T_n\to S/T$ if $T\neq 0$.

(ii) and (iii) may be deduced from the identities

$$S_nT_n-ST\equiv S_n(T_n-T)+T(S_n-S)$$

and $S_n/T_n-S/T\equiv\{T(S_n-S)-S(T_n-T)\}/T_nT$ respectively.

The use of (iii) is demonstrated in the following example:

Example 1

Find the limits of the following expressions when $n\to\infty$:

$$\text{(i)}\ \frac{n^2+4}{2n^2+1},\quad \text{(ii)}\ \frac{n+1}{n^2+2},\quad \text{(iii)}\ \frac{n^3+3}{n^2+1}.$$

In each case we divide numerator and denominator by the highest power of n which occurs in the denominator.

(i) $$\frac{n^2+4}{2n^2+1}=\frac{1+4/n^2}{2+1/n^2}\qquad \therefore \lim_{n\to\infty}\frac{n^2+4}{2n^2+1}=\frac{1}{2};$$

(ii) $$\frac{n+1}{n^2+2}=\frac{1/n+1/n^2}{1+2/n^2}\qquad \therefore \lim_{n\to\infty}\frac{n+1}{n^2+2}=0;$$

(iii) $$\frac{n^3+3}{n^2+1}=\frac{n+3/n^2}{1+1/n^2}\qquad \therefore \frac{n^3+3}{n^2+1}\to\infty \text{ as } n\to\infty.$$

4.9. General theorems on convergence

The theorems on limits given in § 4.8 lead to some general theorems on convergence of infinite series:

(i) If Σu_r is convergent with sum S, then Σku_r, where k is any constant, is convergent with sum kS. Similarly, Σu_r and Σku_r diverge or oscillate together, unless $k=0$.

(ii) If $\sum_{r=1}^{\infty} u_r$ is convergent with sum S, the series $\sum_{r=m+1}^{\infty} u_r$ is convergent with sum $S-\sum_{r=1}^{m} u_r$, m being any given positive integer. Similarly, if $\sum_{=1}^{\infty} u_r$ diverges or oscillates, so also does $\sum_{r=m+1}^{\infty} u_r$. It follows that in discussing the convergence of a series, any finite number of terms at the beginning of the series may be ignored.

(iii) If Σu_r and Σv_r are two convergent series with sums S and T respectively, the series $\Sigma(pu_r \pm qv_r)$ is also convergent with sum $pS \pm qT$, p and q being any constants.

4.10. A necessary condition for convergence

If $u_1+u_2+u_3+\dots$ is a convergent series, then $u_n \to 0$ as $n \to \infty$.

We have
$$u_n = S_n - S_{n-1}$$
$$\therefore \lim_{n\to\infty} u_n = \lim_{n\to\infty} S_n - \lim_{n\to\infty} S_{n-1}$$
$$=0,$$
since S_n and S_{n-1} have the same limit.

This condition, although necessary, is not sufficient to guarantee the convergence of Σu_r. For example, if

$$S_n = \frac{1}{\sqrt{1}} + \frac{1}{\sqrt{2}} + \frac{1}{\sqrt{3}} + \dots + \frac{1}{\sqrt{n}}$$

$$S_n > n \cdot \frac{1}{\sqrt{n}} = \sqrt{n}.$$

$$\therefore S_n \to \infty \text{ as } n \to \infty \text{ although } u_n \to 0 \text{ as } n \to \infty.$$

4.11. The harmonic series

The series $\Sigma(1/r) \equiv 1+\frac{1}{2}+\frac{1}{3}+\frac{1}{4}+\frac{1}{5}+\dots$ is known as the *harmonic series*.

Now $\frac{1}{3}+\frac{1}{4} > \frac{1}{4}+\frac{1}{4} = \frac{1}{2}$, $\frac{1}{5}+\frac{1}{6}+\frac{1}{7}+\frac{1}{8} > \frac{4}{8} = \frac{1}{2}$, and so on.

Hence, if S_n is the sum of n terms of the harmonic series $S_1=1$ $S_2=1+\frac{1}{2}$, $S_4>1+\frac{1}{2}+\frac{1}{2}$, $S_8>1+\frac{1}{2}+\frac{1}{2}+\frac{1}{2}$, and if $n=2^p$, $S_n>1+\frac{1}{2}p$.

When $n \to \infty$, $p \to \infty$, and so $S_n \to \infty$.

Hence $\Sigma(1/r)$ diverges to $+\infty$, although $u_n = 1/n \to 0$ as $n \to \infty$.

4.12. Series of positive terms

The investigation of the convergence of a series is simplified if the terms are all positive, for we can then apply the following theorem which we do not prove but which we illustrate graphically:

If S_n is the sum of the first n terms of a series of positive terms and if, for all values of n, S_n remains less than a fixed number k, independent of n, then the series converges, and its sum to infinity is less than or equal to k.

On a straight line Ox (fig. 1) following the usual sign conventions

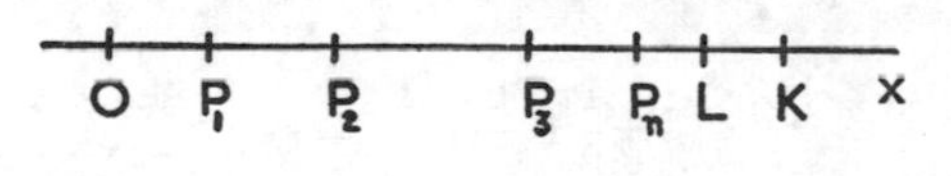

Fig. 1

we mark the points $P_1, P_2, P_3, \ldots$ such that

$$OP_1 = S_1, OP_2 = S_2, OP_3 = S_3, \ldots$$

Let $OK = k$. Then since the series contains only positive terms, S_n increases with n, and so $P_1, P_2, P_3 \ldots$ advance steadily from left to right along Ox. For all values of n, $S_n < k$ and so every point P_n lies to the left of K. The theorem states that there exists a point L either to the left of K or coincident with K beyond which P_n never passes and such that $P_n L \to 0$ as $n \to \infty$. Thus if $OL = l$, $\lim_{n \to \infty} S_n = l$, where $l \leqslant k$, and so the series converges.

If no such number k can be found, S_n increases without limit and the series diverges to $+\infty$. A series of positive terms cannot oscillate.

A series of negative terms will either converge, or diverge to $-\infty$.

A series whose terms are all positive only after the mth term (say) is said to be *ultimately positive*. The convergence of such a series is unaffected by the removal of the first m terms (see § 4.9 (ii)) and we can apply to the series

$$u_{m+1} + u_{m+2} + u_{m+3} + \ldots$$

tests for a series of positive terms.

4.13. The comparison tests

Although the question of convergence of an infinite series depends on the behaviour as $n \to \infty$ of S_n, the sum to n terms of the series, there are comparatively few series for which an expression for S_n in terms of n can be obtained. Hence other tests, applicable to the individual terms of the series, are used.

TEST 1. *A series of positive terms converges if its terms are less than (or equal to) the corresponding terms of some convergent series.*

Let Σu_r and Σv_r be two series of positive terms. Let Σv_r be convergent with sum V, and let $u_r \leqslant v_r$ for all values of r.

Then $u_1+u_2+\ldots+u_r \leqslant v_1+v_2+\ldots+v_r < V$ for all values of r; and so by § 4.12, Σu_r is convergent.

For example, the terms of the series

$$\frac{1}{1!}+\frac{1}{2!}+\frac{1}{3!}+\frac{1}{4!}+\ldots$$

are less than or equal to the corresponding terms of the convergent series

$$1+\frac{1}{2}+\frac{1}{2^2}+\frac{1}{2^3}+\ldots$$

Hence the given series converges.

Corollary to Test 1: If $0 < u_r \leqslant kv_r$, where k is a positive constant and Σv_r converges, then Σu_r is also convergent, for Σkv_r converges (see § 4.9 (i)).

TEST 2. *A series of positive terms diverges if its terms are greater than (or equal to) the corresponding terms of a divergent series of positive terms.*

Let Σu_r and Σv_r be two series of positive terms; let Σv_r be divergent and let $u_r \geqslant v_r$ for all values of r.

Then $u_1+u_2+\ldots+u_r \geqslant v_1+v_2+\ldots+v_r$ for all values of r.

But $\qquad v_1+v_2+\ldots+v_r \to \infty$ when $r \to \infty$

$$\therefore\ u_1+u_2+\ldots+u_r \to \infty \text{ when } r \to \infty.$$

Hence the series Σu_r diverges.

For example, if $r>0$, $\qquad r(r+1)<(r+1)^2$,

and so

$$\frac{1}{\sqrt{\{r(r+1)\}}} > \frac{1}{r+1}.$$

But $\Sigma \dfrac{1}{r+1}$ is a divergent series (see § 4.11);

$$\therefore\ \Sigma \frac{1}{\sqrt{\{r(r+1)\}}} \text{ is divergent.}$$

Corollary to Test 2: If $u_r \geqslant kv_r > 0$, where k is a positive constant and Σv_r diverges, then Σu_r is also divergent, for Σkv_r is divergent (see § 4.9 (i)).

TEST 3. *If Σu_r and Σv_r are series of positive terms and if* $\lim\limits_{n\to\infty} \dfrac{u_n}{v_n} = L$, *where L is a constant other than zero, then Σu_r and Σv_r are series of the same type, that is both are convergent or both are divergent.*

Since $\dfrac{u_n}{v_n} \to L$ as $n \to \infty$, then corresponding to any positive number ϵ there exists a value N of n such that

$$\left| \frac{u_n}{v_n} - L \right| < \epsilon \quad \text{for } n \geqslant N.$$

Choose ϵ so that $0 < \epsilon < L$; then

$$0 < L - \epsilon < \frac{u_n}{v_n} < L + \epsilon$$

i.e.

$$0 < (L - \epsilon)v_n < u_n < (L + \epsilon)v_n.$$

Since $u_n < (L + \epsilon)v_n$ for all values of $n \geqslant N$, it follows from Test 1 (corollary) and § 4.9 (ii) that if Σv_r is convergent, Σu_r converges.

Since $u_n > (L - \epsilon)v_n$ for all values of $n \geqslant N$, it follows from Test 2 (corollary) and § 4.9 (ii) that if Σv_r is divergent, Σu_r diverges.

Example 2

Test for convergence the series Σu_r where

(i) $u_r = (3r+1)/(3r^2-2)$, (ii) $u_r = (3r-1)/(3r^2+2)$.

(i) $$\frac{3r+1}{3r^2-2} > \frac{3r}{3r^2} = \frac{1}{r}, \text{ and } \Sigma \frac{1}{r} \text{ is a divergent series.}$$

Hence, by comparison test 2, Σu_r is divergent.

(ii) $$\frac{3r-1}{3r^2+2} = \frac{(3-1/r)}{r(3+2/r^2)} \simeq \frac{1}{r} \text{ when } r \text{ is large.}$$

Hence we compare the given series Σu_r with Σv_r where $v_r = 1/r$.

We have $$\frac{u_r}{v_r} = \frac{r(3r-1)}{3r^2+2} = \frac{3-1/r}{3+2/r^2} \to 1 \text{ as } r \to \infty.$$

But Σv_r is divergent. Hence by comparison test 3, Σu_r is divergent.

4.14. Two standard series

Two series are regarded as standard and are frequently used in applications of the comparison tests :

(1) The geometrical series $\sum\limits_{n=1}^{\infty} x^{n-1}$, which (as has been shown in § 4.7) converges only when $|x| < 1$.

(2) The series $\sum_{r=1}^{\infty}(1/r^p)$, which we now show to converge if $p>1$ and diverge if $p\leqslant 1$.

(i) Let $p>1$.

$$S_n=\frac{1}{1^p}+\frac{1}{2^p}+\frac{1}{3^p}+\ldots+\frac{1}{n^p}.$$

Now
$$\frac{1}{2^p}+\frac{1}{3^p}<\frac{2}{2^p}=2^{1-p},$$

$$\frac{1}{4^p}+\frac{1}{5^p}+\frac{1}{6^p}+\frac{1}{7^p}<\frac{4}{4^p}=4^{1-p},$$

$$\frac{1}{8^p}+\frac{1}{9^p}+\ldots+\frac{1}{15^p}<\frac{8}{8^p}=8^{1-p},$$

and so on.

$$\therefore\ S_n<1+\frac{1}{2^{p-1}}+\frac{1}{4^{p-1}}+\frac{1}{8^{p-1}}+\ldots$$

The right-hand side of this inequality is a geometrical series whose common ratio $(\frac{1}{2})^{p-1}$ is less than unity when $p>1$

$$\therefore\ S_n<\frac{1}{1-(\frac{1}{2})^{p-1}}\text{ for all values of }n$$

and so by § 4.12 the series converges.

(ii) Let $p=1$. The series is then $\Sigma(1/r)$, which has been shown in § 4.11 to be divergent.

(iii) Let $p<1$.

When $p<1$, $1/r^p>1/r$ if $r>1$.

Hence each term of $\Sigma(1/r^p)$ after the first exceeds the corresponding term of the divergent series $\Sigma(1/r)$ and so, by comparison test 2, $\Sigma(1/r^p)$ is divergent when $p<1$.

Thus the series $\Sigma(1/r^p)$ converges when $p>1$ and diverges when $p\leqslant 1$.

4.15. The ratio test for a series of positive terms

If Σu_r is a series of positive terms such that $\lim\limits_{n\to\infty}\dfrac{u_{n+1}}{u_n}=p$, Σu_r converges if $p<1$ and diverges if $p>1$.

(i) Let $p<1$ and suppose that q is a number such that $p<q<1$. Then since u_{n+1}/u_n can be made to differ from p by as small a quantity as we please by making n sufficiently large, we can find a number N such that when $n \geqslant N$, $u_{n+1}/u_n<q<1$.

Then $u_{N+1}<qu_N$, $u_{N+2}<qu_{N+1}<q^2u_N$, and so on.

$$\therefore\ u_N+u_{N+1}+u_{N+2}+\ldots+u_{N+K}<u_N(1+q+q^2+\ldots+q^K)$$

$$<\frac{u_N}{1-q}$$

for all values of K however large, since $0<q<1$.

$$\therefore\ u_1+u_2+\ldots+u_{N-1}+u_N+\ldots+u_{N+K}<u_1+u_2+\ldots+u_{N-1}+\frac{u_N}{1-q}$$

for all values of K.

But the right-hand side is a positive number independent of K.

$$\therefore\ \Sigma u_r \text{ is convergent.}$$

(ii) Let $p>1$. Then since $\lim\limits_{n\to\infty}\dfrac{u_{n+1}}{u_n}=p$, $\dfrac{u_{n+1}}{u_n}>1$ ultimately, and so u_n does not tend to zero. Hence Σu_r is divergent (see § 4.10).

When $p=1$, the ratio test gives no conclusive result (see Example 4 below).

Example 3

Test for convergence the series

$$\text{(i) } \Sigma\frac{a^r}{r!} \text{ when } a>0, \quad \text{(ii) } \Sigma\frac{1.3.5\ldots(2r-1)}{3.6.9\ldots(3r)}.$$

(i)
$$\frac{u_{n+1}}{u_n}=\frac{a}{n+1}\to 0 \text{ as } n\to\infty.$$

Hence $\Sigma\dfrac{a^r}{r!}$ converges.

(ii)
$$\frac{u_{n+1}}{u_n}=\frac{2n+1}{3(n+1)}=\frac{2+1/n}{3(1+1/n)}\to\frac{2}{3} \text{ as } n\to\infty.$$

Hence the series converges.

Example 4

Test for convergence the series $\Sigma(1/r^p)$.

$$\frac{u_{n+1}}{u_n}=\left(\frac{n}{n+1}\right)^p\to 1 \text{ as } n\to\infty.$$

Now we have shown in § 4.14 that $\Sigma(1/r^p)$ is convergent when $p>1$ and divergent when $p\leqslant 1$. It is therefore seen that if $\lim_{n\to\infty}\frac{u_{n+1}}{u_n}=1$ for a given series we cannot, from this fact, draw any conclusion as to the convergence of the series Σu_r.

4.16. Series of positive and negative terms

A simple type of series in which the terms are not all of one sign is the *alternating series* in which the terms are alternately positive and negative. For example, the series

$$u_1-u_2+u_3-u_4+\dots \qquad \text{(i)}$$

where each u is positive is an alternating series.

We shall show that if $u_n>u_{n+1}$ for all values of n, and if $u_n\to 0$ as $n\to\infty$, then series (i) converges.

We consider separately the sum of an even number of terms and the sum of an odd number of terms. We have

$$S_{2n}=(u_1-u_2)+(u_3-u_4)+\dots+(u_{2n-1}-u_{2n}).$$

Each bracket is positive and so S_{2n} increases as n increases.

But $$S_{2n}=u_1-(u_2-u_3)-\dots-(u_{2n-2}-u_{2n-1})-u_{2n}$$

$$\therefore\ S_{2n}\leqslant u_1.$$

Hence (see § 4.12) S_{2n} tends to a limit which is less than or equal to u_1.

Now $$S_{2n+1}=S_{2n}+u_{2n+1}$$

$$\therefore\ \lim_{n\to\infty} S_{2n+1}=\lim_{n\to\infty} S_{2n}+\lim_{n\to\infty} u_{2n+1}.$$

But since $\lim_{n\to\infty} u_{2n+1}=0$, $\lim_{n\to\infty} S_{2n+1}$ exists and is equal to $\lim_{n\to\infty} S_{2n}$. Hence by § 4.3, theorem IV the given series is convergent.

Example 5

The alternating series $1-\frac{1}{2}+\frac{1}{3}-\frac{1}{4}+\dots$ is convergent, since $u_n>u_{n+1}$ and $u_n\to 0$ as $n\to\infty$.

Example 6

The alternating series $1{\cdot}1-1{\cdot}01+1{\cdot}001-1{\cdot}0001+\dots$ does not converge because $\lim_{n\to\infty} u_n\neq 0$ (see § 4.10). In this case

$$S_{2n}=\tfrac{1}{11}[1-(0{\cdot}1)^{2n}]\to\tfrac{1}{11} \text{ as } n\to\infty$$

$$S_{2n+1}=1+\tfrac{1}{11}[1+(0{\cdot}1)^{2n+1}]\to\tfrac{12}{11} \text{ as } n\to\infty.$$

Hence the series oscillates finitely.

4.17. Absolute convergence

If Σu_r is a series which contains positive and negative terms, Σu_r i said to be *absolutely convergent* if $\Sigma \mid u_r \mid$ is convergent.

A series which is absolutely convergent is also convergent.

To prove this result for a series Σu_r we construct two series of *positive* terms Σv_r and Σw_r by taking

$$v_r = u_r \text{ when } u_r \geqslant 0, \qquad v_r = 0 \text{ when } u_r < 0;$$
$$w_r = -u_r \text{ when } u_r < 0, \quad w_r = 0 \text{ when } u_r > 0.$$

Then
$$\mid u_r \mid = v_r + w_r \quad . \quad . \quad . \quad \text{(i)}$$
and
$$u_r = v_r - w_r \quad . \quad . \quad . \quad \text{(ii)}$$

Since $v_r \geqslant 0$ and $w_r \geqslant 0$ we have from (i)

$$v_r \leqslant \mid u_r \mid \text{ and } w_r \leqslant \mid u_r \mid.$$

But, by hypothesis, $\Sigma \mid u_r \mid$ is convergent; hence by comparison test 1 Σv_r and Σw_r are both convergent. It follows from (ii) that Σu_r is convergent (see § 4.9 (iii)).

From this important theorem it follows that if a series of positive terms is convergent and a new series is obtained by changing the signs of any of the terms of the given series, then the new series is convergent.

If Σu_r is a convergent series of positive and negative terms such that $\Sigma \mid u_r \mid$ is divergent, we say that Σu_r is *conditionally convergent.*

4.18. Tests for absolute convergence

Any test applicable to series of positive terms can be used as a test for absolute convergence of the series Σu_r since $\Sigma \mid u_r \mid$ is a series of positive terms. For example, applying the ratio test we see that if $\left| \dfrac{u_{n+1}}{u_n} \right| \to p$ as $n \to \infty$, Σu_n is absolutely convergent if $p < 1$ and is not convergent if $p > 1$. For if $p < 1$, $\Sigma \mid u_r \mid$ is convergent, while if $p > 1$, the terms of Σu_r ultimately increase numerically and so Σu_r cannot converge (cf. § 4.10). The ratio test is inconclusive when $p = 1$.

Example 7

The series $1 - \dfrac{1}{2^2} - \dfrac{1}{3^2} + \dfrac{1}{4^2} - \dfrac{1}{5^2} - \dfrac{1}{6^2} + \ldots$ is absolutely convergent since $\Sigma \dfrac{1}{r^2}$ is convergent.

Example 8

The series $1 - \frac{1}{2} + \frac{1}{3} - \frac{1}{4} + \ldots$ is convergent (see Example 5), but since $\Sigma(1/r)$ is divergent (see § 4.11) the given series is only conditionally convergent.

Example 9

Examine for absolute convergence the series (i) $\Sigma(-1)^{n+1}\, n(n+1)/3^n$ *and* (ii) $\Sigma(-1)^{n+1}/\sqrt{(n^2+1)}$.

(i) Here $\quad |u_n|=n(n+1)/3^n,\ |u_{n+1}|=(n+1)(n+2)/3^{n+1}$

and $$\left|\frac{u_{n+1}}{u_n}\right|=\tfrac{1}{3}(1+2/n)\to\tfrac{1}{3} \text{ when } n\to\infty.$$

Hence by the ratio test, Σu_n is absolutely convergent.

(ii) $$\left|\frac{u_{n+1}}{u_n}\right|=\frac{\sqrt{(n^2+1)}}{\sqrt{(n^2+2n+2)}}=\frac{\sqrt{(1+1/n^2)}}{\sqrt{(1+2/n+2/n^2)}}\to 1 \text{ when } n\to\infty.$$

Hence the ratio test fails.

Now Σu_n is an alternating series in which each term is numerically less than the preceding one and $u_n\to 0$ as $n\to\infty$. Hence Σu_n is convergent (see § 4.16). Also $|u_n|\simeq\dfrac{1}{n}$ when n is large and so we compare $\Sigma|u_n|$ with Σv_n where $v_n=\dfrac{1}{n}$. Then the comparison test in limit form (see § 4.13) gives

$$\frac{|u_n|}{v_n}=\frac{n}{\sqrt{(n^2+1)}}\to 1 \text{ as } n\to\infty.$$

$\therefore\ \Sigma|u_n|$ is divergent since Σv_n is divergent,

$\therefore\ \Sigma u_n$ is conditionally convergent.

4.19. Power series

A series of the form $a_0+a_1x+a_2x^2+a_3x^3+\ldots$, where the coefficients $a_0, a_1, a_2, \ldots$ are independent of x, is called a power series in x. Such a series cannot diverge for all values of x since, when $x=0$, it converges to the value a_0.

Associated with any power series there is a number R such that the series converges absolutely if $|x|<R$ but does not converge if $|x|>R$. R is called the *radius of convergence* of the series and the interval containing all the values of x for which the series converges is called the *interval of convergence* of the series. Except for its end-points this interval may often be found by the ratio test. Other methods must be used to determine the behaviour of the series at the end-points of the interval.

Example 10

Find the values of x for which the exponential series

$$1+\frac{x}{1!}+\frac{x^2}{2!}+\frac{x^3}{3!}+\ldots$$

is convergent.

The test ratio is $\left|\dfrac{u_{n+1}}{u_n}\right|=\dfrac{|x|}{n}$.

$$\therefore \lim_{n\to\infty}\left|\frac{u_{n+1}}{u_n}\right|=0 \text{ for any finite value of } x,$$

and so the exponential series is convergent for any finite value of x. The interval of convergence is $(-\infty, \infty)$.

Example 11

Find the radius and interval of convergence of the logarithmic series

$$x-\frac{x^2}{2}+\frac{x^3}{3}-\frac{x^4}{4}+\ldots$$

$$\left|\frac{u_{n+1}}{u_n}\right|=\left(\frac{n}{n+1}\right)|x|=\left(1-\frac{1}{n+1}\right)|x|$$

$$\therefore \lim_{n\to\infty}\left|\frac{u_{n+1}}{u_n}\right|=|x|.$$

Hence the logarithmic series converges when $|x|<1$ and does not converge when $|x|>1$. The radius of convergence is 1 and values of x in the range $-1<x<1$ belong to the interval of convergence. The end-points of this interval, i.e. $x=\pm 1$, must be tested individually.

$x=1$ gives the series $1-\frac{1}{2}+\frac{1}{3}-\frac{1}{4}+\ldots$ which converges, see § 4.16, Example 5.

$x=-1$ gives the series $-(1+\frac{1}{2}+\frac{1}{3}+\frac{1}{4}+\ldots)$ which diverges, see § 4.11.

Hence the interval of convergence of the series is $-1<x\leqslant 1$.

Example 12

Prove that, when m is not a positive integer, the binomial series

$$1+mx+\frac{m(m-1)}{2!}x^2+\frac{m(m-1)(m-2)}{3!}x^3+\ldots$$

converges absolutely if $-1<x<1$.

The test ratio of this series for $n>1$ has the form

$$\frac{u_{n+1}}{u_n}=\frac{m(m-1)(m-2)\ldots(m-n+1)}{n!}x^n\div\frac{m(m-1)(m-2)\ldots(m-n+2)}{(n-1)!}x^{n-1}$$

i.e.

$$\frac{u_{n+1}}{u_n}=\frac{m-n+1}{n}x.$$

$$\therefore \lim_{n\to\infty}\left|\frac{u_{n+1}}{u_n}\right|=|x|$$

and so by the ratio test the binomial series converges absolutely when $|x|<1$.

4.20. Properties of power series

(i) From § 4.9 (iii) it follows that two power series may be added or subtracted term by term for all values of x for which both series are convergent. Hence if $\Sigma a_n x^n = f(x)$ when $|x| < R_1$ and $\Sigma b_n x^n = g(x)$ when $|x| < R_2$, $R_1 < R_2$, then

$$\Sigma(a_n + b_n)x^n = f(x) + g(x) \text{ when } |x| < R_1.$$

(ii) It can also be shown that the above series may be multiplied together like polynomials for any value of x for which both series are absolutely convergent (and thus for every value of x within the smaller of their intervals of convergence), i.e.

$$f(x) . g(x) = a_0 b_0 + (a_0 b_1 + a_1 b_0)x + (a_0 b_2 + a_1 b_1 + a_2 b_0)x^2 + \ldots$$

We state without proof further properties of power series:

(iii) If two power series have the same sum over some interval $-l < x < l$, the coefficients of corresponding powers of x in these series are identical.

(iv) The sum of a power series is a continuous function (cf. § 4.23) in the interval of convergence of the series. It follows that, if the radius of convergence is not zero,

$$\lim_{x \to 0} (a_0 + a_1 x + a_2 x^2 + \ldots) = a_0.$$

The corresponding result for series other than power series may not be true.

(v) A power series which converges to $f(x)$ in the interval $-l < x < l$ may be differentiated term by term and the resultant series will converge to $\frac{d}{dx}\{f(x)\}$ in the interval $-l < x < l$.

(vi) A power series which converges may be integrated term by term between any limits lying within the interval of convergence of the series.

Exercises 4 (*a*)

1. Discuss the convergence of the series whose nth term is given by

(i) $\dfrac{2^n}{(n+1)(n+3)}$,

(ii) $\dfrac{1}{n(n+2)3^n}$,

(iii) $\dfrac{1}{2n(2n+1)}$,

(iv) $\dfrac{2n+1}{2.4\ldots(2n+2)}$,

(v) $\dfrac{1+3n^2}{1+n^2}$,

(vi) $\dfrac{1-n+n^2}{n!}$,

(vii) $\dfrac{n}{\sqrt{(4n^3+1)}}$,

(viii) $\sqrt{\left(\dfrac{n}{4n^4+7}\right)}$.

(ix) $\dfrac{n^p}{n!}$ (p constant), (xi) $\dfrac{\sqrt{(n^2+n+1)}+\sqrt{(n^2-n+1)}}{n}$,

(x) $\dfrac{1}{\sqrt{(n+2)}-\sqrt{n}}$, (xii) $\dfrac{\sqrt{(n+1)}-\sqrt{n}}{n^2}$.

2. Determine for what values of x the following series are (a) absolutely convergent, (b) conditionally convergent:

(i) $\Sigma \dfrac{r}{(r+1)(r+2)} x^r$, (iv) $\Sigma \dfrac{r^{10}x^r}{r!}$,

(ii) $\Sigma \dfrac{r!}{(2r)!} x^r$, (v) $\Sigma \dfrac{(-1)^r x^{2r+1}}{2r+1}$.

(iii) $\Sigma \dfrac{x^r}{(r^2+1)^{1/3}}$,

3. Prove that an absolutely convergent series with real terms is convergent.

Show that $\sum_{r=1}^{\infty} \dfrac{\cos rx}{r^2}$ is convergent. [L.U.]

4. Prove that, when $|x|<1$, $nx^n \to 0$ as $n \to \infty$.

[Assume first that $0<x<1$ and write $\dfrac{1}{x}=1+y$ where $y>0$; then use the inequality $(1+y)^n > \frac{1}{2}n(n-1)y^2$ to show that $0<nx^n<2/(n-1)y^2$].

5. Prove that the following alternating series are convergent:

(i) $1-\frac{1}{2}+\frac{1}{4}-\frac{1}{8}+\ldots$,

(ii) $\dfrac{1}{3^2}-\dfrac{2}{3^3}+\dfrac{3}{3^4}-\dfrac{4}{3^5}+\ldots$,

(iii) $\dfrac{3}{2}-\dfrac{4}{3}\left(\dfrac{1}{2}\right)+\dfrac{5}{4}\left(\dfrac{1}{2^2}\right)-\dfrac{6}{5}\left(\dfrac{1}{2^3}\right)+\ldots$,

(iv) $\dfrac{1}{3}-\dfrac{1.2}{3.5}+\dfrac{1.2.3}{3.5.7}-\dfrac{1.2.3.4}{3.5.7.9}+\ldots$,

(v) $\dfrac{1}{3}-\dfrac{1.3}{3.6}+\dfrac{1.3.5}{3.6.9}-\dfrac{1.3.5.7}{3.6.9.12}+\ldots$.

6. Show that the following series are conditionally convergent:

(i) $\dfrac{2}{1.3}-\dfrac{3}{2.4}+\dfrac{4}{3.5}-\dfrac{5}{4.6}+\ldots$,

(ii) $1-\dfrac{1}{\sqrt{2}}+\dfrac{1}{\sqrt{3}}-\dfrac{1}{\sqrt{4}}+\ldots$,

(iii) $\dfrac{1}{2}-\dfrac{2}{5}+\dfrac{3}{10}-\dfrac{4}{17}+\ldots$

7. Explain the meaning of the statement

"the infinite series $t_1+t_2+\ldots+t_n+\ldots$ is convergent".

Prove that the series $1+\dfrac{1}{\sqrt{2}}+\dfrac{1}{\sqrt{3}}+\ldots+\dfrac{1}{\sqrt{n}}+\ldots$ is not convergent.

State the comparison test for the convergence of series with positive terms, and use it to show that if $\sum_{n=1}^{\infty} t_n$ is absolutely convergent then so also is $\sum_{n=1}^{\infty} t_n^2$. [L.U.]

8. Define *absolute convergence* of an infinite series $\sum_{r=1}^{\infty} u_r$ and prove *from your definition* that $\sum_{r=1}^{\infty} x^{r-1}$ is absolutely convergent when $-1<x<1$.

Determine for which real values of x the series $\sum_{r=1}^{\infty} \dfrac{x^r}{r(r+1)}$ is convergent and for which it is divergent. [L.U.]

4.21. Functions of a continuous variable

We may represent real numbers by points on a straight line $x'Ox$ (fig. 2), the origin O representing the number zero and points to the right and left of O representing positive and negative numbers respectively, the distances of the points from O being proportional to the magnitude of the numbers they represent.

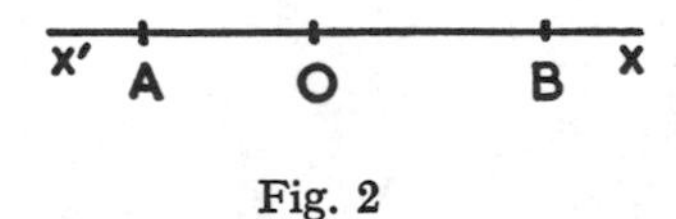

Fig. 2

If the points A and B represent the numbers a and b respectively, a number x which takes successively *all* values between a and b may be represented by a point P which moves along the line from A to B. x is then said to vary continuously from a to b. If P progresses always to the right, starting from A, x is said to tend *continuously* to infinity.

Suppose that n is a positive integer represented by a point Q moving along $x'Ox$. Then if $n \to \infty$, Q moves progressively to the right along Ox by a series of equal jumps whereas, when $x \to \infty$, P moves smoothly along $x'Ox$.

If for each value of x in some interval the value of another number y is uniquely determined, y is said to be a single-valued function of the continuous variable x. In what follows we shall assume that the functions with which we have to deal are single-valued.

4.22. Limit of a function of a continuous variable

Corresponding to the definitions given in § 4.2 and § 4.3 of the behaviour of $f(n)$ when $n\to\infty$ (n being a positive integer), we have definitions for $f(x)$ where x is a continuous variable. For example,

(i) A function $f(x)$ is said to tend to a finite limit l as x tends to $+\infty$ if, corresponding to *each* positive number ϵ there is a number X such that $|f(x)-l|<\epsilon$, provided that $x\geqslant X$.
We then write $f(x)\to l$ as $x\to\infty$, or $\lim\limits_{x\to\infty} f(x)=l$.

For example, $\lim\limits_{x\to\infty}\dfrac{1}{x}=0$.

(ii) A function $f(x)$ is said to tend to a finite limit l as x approaches a if $|f(x)-l|$ can be made as small as we please by taking values of x sufficiently near to a.

This may be stated more formally: a function $f(x)$ is said to tend to a finite limit l as x tends to a, if, corresponding to *each* positive number ϵ, we can find a positive number η (which generally depends on ϵ) such that $|f(x)-l|<\epsilon$ when $0<|x-a|\leqslant\eta$. In this case we write $\lim\limits_{x\to a} f(x)=l$, or $f(x)\to l$ as $x\to a$.

Now x can approach the value a through values greater than a (denoted by $x\to a+$) or through values less than a ($x\to a-$). When we write $\lim\limits_{x\to a} f(x)=l$, we imply that x may approach the value a from either side, and the statement $\lim\limits_{x\to a} f(x)=l$ is equivalent to the two statements $\lim\limits_{x\to a+} f(x)=l$, and $\lim\limits_{x\to a-} f(x)=l$. If $\lim\limits_{x\to a+} f(x)\neq\lim\limits_{x\to a-} f(x)$, $\lim\limits_{x\to a} f(x)$ does not exist.

(iii) The function $f(x)$ is said to tend to infinity as x tends to $+\infty$ if, corresponding to *each* positive number Δ, however large, there is a number X_0 (which generally depends on Δ) such that $f(x)\geqslant\Delta$ for all values of $x\geqslant X_0$. We then write $f(x)\to\infty$ as $x\to\infty$.

Similar definitions describe the behaviour of $f(x)$ as $x\to-\infty$.

If $f(x)$ neither tends to a finite limit nor to $+\infty$ or $-\infty$ as $x\to\infty$, $f(x)$ is said to oscillate.

The theorems on limits given in § 4.8 are valid for limits of functions of a continuous variable.

Example 13

Find (i) $\lim\limits_{x\to0}\,(ax^2+bx+c)/(a'x^2+b'x+c')$, $c'\neq0$,

(ii) $\lim\limits_{x\to0}\,x/\{\sqrt{(a+x)}-\sqrt{(a-x)}\}$, $a>0$.

(i) As $x\to0$, $ax^2+bx+c\to c$ and $a'x^2+b'x+c'\to c'$

$$\therefore\ \lim_{x\to0}\frac{ax^2+bx+c}{a'x^2+b'x+c'}=\frac{c}{c'},\text{ since } c'\neq0.$$

In this case the limit is found by substituting $x=0$ in the given function, i.e. $\lim_{x\to 0} f(x)=f(0)$.

(ii) If we substitute $x=0$ in $\dfrac{x}{\sqrt{(a+x)}-\sqrt{(a-x)}}$, we obtain $\dfrac{0}{0}$ which is meaningless.

But if $x\neq 0$, $$\frac{x}{\sqrt{(a+x)}-\sqrt{(a-x)}}=\tfrac{1}{2}\{\sqrt{(a+x)}+\sqrt{(a-x)}\}$$

and hence as $x\to 0$, $$\frac{x}{\sqrt{(a+x)}-\sqrt{(a-x)}}\to\sqrt{a}.$$

In this case $\lim_{x\to 0} f(x)\neq f(0)$ since $f(0)$ is not defined.

4.23. Continuous functions

It is natural to call a function *continuous* if its graph is a continuous curve. For example, the curve in fig. 25 is continuous; the curve in fig. 42 is continuous except at $x=0$, where it is said to be *discontinuous*.

Before we give a definition of continuity for all values of x we must first define continuity for a particular value of x, $x=a$ (say). The simplest properties of a function $f(x)$ which is continuous when $x=a$ are

(i) $f(x)$ must be defined for $x=a$, otherwise there would be a point missing from the curve.

(ii) $f(x)$ must be defined for all values of x near $x=a$.

(iii) $f(x)\to f(a)$ as $x\to a$ from either side.

These properties lead to the following definition:

A function $f(x)$ is said to be continuous for $x=a$ if $f(x)$ tends to a limit as $x\to a$ from either side and each of these limits is equal to $f(a)$. More formally:

A function $f(x)$ is continuous at $x=a$ if corresponding to *each* positive number ϵ, however small, there is a positive number η (depending on ϵ) such that $|f(x)-f(a)|<\epsilon$ whenever $0<|x-a|\leqslant\eta$.

A function $f(x)$ is said to be continuous throughout a certain interval of values of x if it is continuous for all values of x in that interval; it is continuous everywhere if it is continuous for all values of x.

Using extensions of the limit theorems given in § 4.8, we can show that the sum (or difference) and the product of two continuous functions is a continuous function; the quotient of two continuous functions is continuous except for values of x where the denominator takes the value zero.

In particular, a polynomial $P(x)$ is continuous for all finite values of x and the rational function $N(x)/D(x)$ is continuous except at values of x where $D(x)=0$.

Example 14

$f(x)=\frac{1}{x}$ is continuous for all values of x except $x=0$.

As $x\to 0+$, $\frac{1}{x}\to+\infty$; as $x\to 0-$, $\frac{1}{x}\to-\infty$; when $x=0$, $f(x)$ is not defined.

The function $\frac{1}{x}$ is said to have an infinite discontinuity at $x=0$.

Its graph is shown in fig. 42.

Example 15

$f(x)=\frac{x^2-4}{x-2}$ has the value $x+2$ except when $x=2$.

When $x=2$, $f(x)$ is not defined, hence the function is discontinuous at $x=2$ and its graph is that of the straight line $y=x+2$ except that there is no point which corresponds to $x=2$.

4.24. A fundamental property of a continuous function

Suppose that A and B are the points corresponding to the values $x=\alpha$, $x=\beta$ respectively on the graph of a continuous function $y=f(x)$. Then the ordinates of A and B are $f(\alpha)$ and $f(\beta)$ respectively. We shall suppose that $f(\beta)>f(\alpha)$.

The fact that the curve is continuous between A and B suggests that the line $y=k$ where $f(\alpha)<k<f(\beta)$ will cut the curve at least once, i.e. there is a value of x between α and β for which $f(x)=k$. In other words, as x varies from α to β, y must assume at least once every value between $f(\alpha)$ and $f(\beta)$. This is a fundamental property of a continuous function.

In particular, if $f(\alpha)<0$ and $f(\beta)>0$, $f(x)$ takes the value zero at least once as x varies from α to β.

Exercises 4 (*b*)

Evaluate the following limits:

1. $\lim\limits_{a\to 0} \frac{x^2-a^2}{x^2-2ax+a^2}$.
2. $\lim\limits_{x\to 0} \frac{1-\sqrt{(1-x)}}{x}$.
3. $\lim\limits_{x\to\infty} \frac{x^2+3x+2}{x^2+4x+3}$.
4. $\lim\limits_{x\to 1} \frac{x^2-4x+3}{x^2-3x+2}$.
5. $\lim\limits_{x\to\infty} \frac{5+x}{x^2+x+1}$.
6. $\lim\limits_{h\to 0} \frac{\sqrt{(x+h)}-\sqrt{x}}{h}$.

Find the values of x for which the following functions are discontinuous:

7. $\frac{x+4}{x-2}$.
8. $\frac{x^4-1}{x^2}$.
9. $\frac{x^2-4}{x^2-x-2}$.
10. $\tan 2x$.
11. $\frac{\sin^2 x}{1-\cos x}$.
12. $\operatorname{cosec} 2x$.

CHAPTER 5

THE BINOMIAL, EXPONENTIAL AND LOGARITHMIC SERIES

5.1. The binomial series

The series

$$1+\frac{nx}{1!}+\frac{n(n-1)}{2!}x^2+\frac{n(n-1)(n-2)}{3!}x^3+\quad .$$
$$+\frac{n(n-1)\dots(n-r+1)}{r!}x^r+\dots \qquad (5.1)$$

is known as the *binomial series* and is denoted by $\sum_0^\infty \binom{n}{r} x^r$, where $\binom{n}{r}=\frac{n(n-1)\dots(n-r+1)}{r!}$ and $\binom{n}{0}$ is interpreted as unity.

If n is a positive integer, the series ends after a finite number of terms and is in fact the expression for $(1+x)^n$ given by the binomial theorem for a positive integral index n. In this case $\binom{n}{r}$ is equal to ${}^nC_r$, the number of ways of choosing a group of r things from n different things. When n is not a positive integer the above series is infinite and converges absolutely for all values of n when $|x|<1$ (see § 4.19, Example 12). We shall use Vandermonde's theorem to find its sum when $|x|<1$.

5.2. Vandermonde's theorem

If m and n are positive integers, we have, by the binomial theorem for a positive integral index,

$$(1+x)^{m+n}=1+{}^{m+n}C_1x+{}^{m+n}C_2x^2+\dots+{}^{m+n}C_rx^r+\dots+x^{m+n} \qquad \text{(i)}$$

But $(1+x)^{m+n}\equiv(1+x)^m(1+x)^n$

$$\equiv(1+{}^mC_1x+{}^mC_2x^2+\dots+{}^mC_rx^r+\dots+x^m)$$
$$\times(1+{}^nC_1x+{}^nC_2x^2+\dots+{}^nC_rx^r+\dots+x^n)\quad . \qquad \text{(ii)}$$

Equating coefficients of x^r in (i) and (ii) we have

$${}^{m+n}C_r={}^mC_r+{}^mC_{r-1}{}^nC_1+{}^mC_{r-2}{}^nC_2+\dots+{}^mC_{r-s}{}^nC_s+\dots+{}^nC_r \qquad \text{(iii)}$$

The method of proof of this result assumes that m and n are positive integers such that $m+n\geqslant r$; but if we write

$${}^mC_r=\{m(m-1)\dots(m-r+1)\}/r!, \quad {}^nC_r=\{n(n-1)\dots(n-r+1)\}/r!$$

we see that equation (iii) is of the rth degree in m and n and since it is satisfied by all integers m and n such that $m+n \geqslant r$ (iii) is an identity, true for *all* values of m and n and for all positive integral values of r. Hence we may write (iii) in the form

$$\binom{m+n}{r}=\binom{m}{r}+\binom{m}{r-1}\binom{n}{1}+\binom{m}{r-2}\binom{n}{2}+\dots$$
$$+\binom{m}{r-s}\binom{n}{s}+\dots+\binom{n}{r}.$$

This is Vandermonde's theorem.

5.3. The binomial theorem for any rational index

Let us consider the product of the two series $\sum_0^\infty \binom{m}{r}x^r$ and $\sum_0^\infty \binom{n}{r}x^r$, where $|x|<1$ so that both series are absolutely convergent. Denoting their respective sums by $f(m)$ and $f(n)$ respectively, we have by § 4.20 (ii)

$$f(m)\times f(n)=\sum_0^\infty a_r x^r,$$

where $a_r=\binom{m}{r}+\binom{m}{r-1}\binom{n}{1}+\binom{m}{r-2}\binom{n}{2}+\dots$
$$+\binom{m}{r-s}\binom{n}{s}+\dots+\binom{n}{r}$$
$=\binom{m+n}{r}$ by Vandermonde's theorem, since r is a positive integer.

$$\therefore f(m)\times f(n)=1+\binom{m+n}{1}x+\binom{m+n}{2}x^2+\dots+\binom{m+n}{r}x^r+\dots$$

$$\text{i.e. } f(m)\times f(n)=f(m+n) \quad . \quad . \quad . \quad \text{(i)}$$

This result may be extended to any number of factors; for

$$f(m)\times f(n)\times f(k)=f(m+n)\times f(k)=f(m+n+k),$$

and so on. In general,

$$f(n_1)\times f(n_2)\times \dots \times f(n_p)=f(n_1+n_2+\dots+n_p) \quad . \quad \text{(i)}$$

If $n_1=n_2=\dots=n_p=n$ we have from (i)

$$\{f(n)\}^p=f(pn) \quad . \quad . \quad . \quad . \quad \text{(ii)}$$

where p is a positive integer and n can have any value.

Putting $n=1$ in (ii), we have

$$f(p)=\{f(1)\}^p=(1+x)^p \quad . \quad . \quad . \quad \text{(iii)}$$

This is the binomial theorem for a positive integral index

Next let n be any positive rational number, i.e. let $n=q/p$, where p and q are positive integers. Then by (ii)

$$\{f(q/p)\}^p = f(q) = (1+x)^q \quad . \quad . \quad . \quad \text{(iv)}$$

so that

$$f(q/p) = (1+x)^{q/p}.$$

Hence $f(q/p)$ is a pth root of $(1+x)^q$. Now by § 4.20 (iv), $f(q/p)$ is a continuous function of x when $|x|<1$, and for values of x in this interval $f(q/p)$ does not take the value zero. It follows from § 4.24 that $f(q/p)$ does not change sign in the interval. But when $x=0$, $f(q/p)=1$; hence when $|x|<1$, $f(q/p)>0$ and so $f(q/p)$ is the *positive* pth root of $(1+x)^q$.

We have now proved that when n is a positive rational number

$$f(n) = (1+x)^n \quad . \quad . \quad . \quad . \quad \text{(v)}$$

Finally, if m is a negative rational number we put $m=-n$, where n is a positive rational number. Then from (i),

$$f(m)f(n+1) = f(1)$$

i.e.

$$f(m)(1+x)^{n+1} = 1+x \quad \text{by (v)}$$

i.e.

$$f(m) = (1+x)^m.$$

It follows that for all rational values of n, positive or negative,

$$1+x)^n = 1+\binom{n}{1}x+\binom{n}{2}x^2+\binom{n}{3}x^3+\ldots+\binom{n}{r}x^r+\ldots \text{when} |x|<1.$$

5.4. Particular cases of the binomial theorem

When $|x|<1$, we have by (5.1)

$$(1-x)^{-1} = 1+x+x^2+x^3+\ldots.$$

When $|b|<|a|$,

$$(a-b)^{-2} = \{a(1-b/a)\}^{-2} = \frac{1}{a^2}\{1+2b/a+3b^2/a^2+4b^3/a^3+\ldots\}.$$

When $|a|<|b|$,

$$(a+b)^{-3} = \{b(1+a/b)\}^{-3} = \frac{1}{b^3}\left\{1-3\frac{a}{b}+\frac{3.4}{2}\frac{a^2}{b^2}-\frac{4.5}{2}\frac{a^3}{b^3}+\ldots\right.$$

$$\left.+(-1)^r\frac{(r+1)(r+2)}{2}\left(\frac{a}{b}\right)^r+\ldots\right\}.$$

When $|x|<1$,

$$(1+x)^{\frac{1}{2}} = 1+\frac{1}{2}x-\frac{1}{2.4}x^2+\frac{1.3}{2.4.6}x^3-\ldots$$

$$(1-x)^{-\frac{1}{2}} = 1+\frac{1}{2}x+\frac{1.3}{2.4}x^2+\frac{1.3.5}{2.4.6}x^3+\ldots$$

and, more generally,

$$(1+x)^{-n}=1-nx+\frac{n(n+1)}{2!}x^2-\frac{n(n+1)(n+2)}{3!}x^3+\ldots$$

$$(1-x)^{-n}=1+nx+\frac{n(n+1)}{2!}x^2+\frac{n(n+1)(n+2)}{3!}x^3+\ldots.$$

Replacing n by p/q in these last results, we have

$$(1+x)^{-p/q}=1-p\left(\frac{x}{q}\right)+\frac{p(p+q)}{2!}\left(\frac{x}{q}\right)^2-\frac{p(p+q)(p+2q)}{3!}\left(\frac{x}{q}\right)^3+\ldots.$$

$$(1-x)^{-p/q}=1+p\left(\frac{x}{q}\right)+\frac{p(p+q)}{2!}\left(\frac{x}{q}\right)^2+\frac{p(p+q)(p+2q)}{3!}\left(\frac{x}{q}\right)^3+\ldots.$$

Note that when p/q is positive, all the coefficients in the expansion $(1-x)^{-p/q}$ are positive.

5.5. Miscellaneous examples

Example 1

Express in partial fractions

$$f(x)=25x/(1-x)^2(1-6x)$$

and find the coefficient of x^n in the expansion of $f(x)$ in a series of ascending powers of x.

Deduce, or prove otherwise, that an integer of the form $6^{n+1}-5(n+1)-1$ is divisible by 25. [L.U.]

$$\frac{25x}{(1-x)^2(1-6x)}=\frac{6}{1-6x}-\frac{1}{1-x}-\frac{5}{(1-x)^2}.$$

When $|x|<\frac{1}{6}$, $\frac{1}{1-6x}=1+6x+(6x)^2+\ldots+(6x)^r+\ldots;$

when $|x|<1$, $\frac{1}{1-x}=1+x+x^2+\ldots+x^r+\ldots;$

when $|x|<1$, $\frac{1}{(1-x)^2}=1+2x+3x^2+\ldots+(r+1)x^r+\ldots.$

Hence, when $|x|<\frac{1}{6}$, $f(x)=6\sum_0^\infty(6x)^r-\sum_0^\infty x^r-5\sum_0^\infty(r+1)x^r$

$$=\sum_0^\infty\{6^{r+1}-5(r+1)-1\}x^r.$$

The coefficient of x^n in the expansion is the integer $6^{n+1}-5(n+1)-1$.

But $f(x)=25x\{(1-x)^2(1-6x)\}^{-1}=25x\{1-(8x-13x^2+6x^3)\}^{-1}$ and so if we expand the function in this form, the coefficient of every power of x will be an integer of which 25 is a factor, and the expansion will be valid when $|x|<\frac{1}{6}$.

Equating the coefficients of x^n (see § 4.20 (iii)) we establish the required result.

Example 2

Sum the series $\frac{1}{2}t+\frac{1}{2.4}t^3+\frac{1.3}{2.4.6}t^5+\ldots$ *when convergent.*

If $t=2x/(1+x^2)$ *show that for a given value of t there are two values of x which are reciprocals and hence prove that the series*

$$\frac{1}{2}\left(\frac{2x}{1+x^2}\right)+\frac{1}{2.4}\left(\frac{2x}{1+x^2}\right)^3+\frac{1.3}{2.4.6}\left(\frac{2x}{1+x^2}\right)^5+\ldots$$

converges to x or $1/x$ *according as* $|x|$ *is less or greater than unity.* [L.U.]

When $|t|<1$,

$$\frac{1}{2}t+\frac{1}{2.4}t^3+\frac{1.3}{2.4.6}t^5+\ldots \qquad \text{(i)}$$

$$=\{1-(1-t^2)^{\frac{1}{2}}\}/t \qquad \text{(ii)}$$

see § 5.4.

If
$$t=\frac{2x}{1+x^2} \qquad \text{(iii)}$$

and $x\neq 1$, it follows from the inequality $1+x^2>2x$ that $|t|<1$. Hence substituting from (iii) in (i) we obtain the convergent series

$$\frac{1}{2}\left(\frac{2x}{1+x^2}\right)+\frac{1}{2.4}\left(\frac{2x}{1+x^2}\right)^3+\frac{1.3}{2.4.6}\left(\frac{2x}{1+x^2}\right)^5+\ldots \qquad \text{(iv)}$$

whose sum by (ii) is

$$\{1-\sqrt{(1-t^2)}\}/t, \qquad \text{(v)}$$

where t is given by (iii).

But from (iii)
$$tx^2-2x+t=0, \qquad \text{(vi)}$$

so that to each value of t there are two values of x

$$x_1=\{1+\sqrt{(1-t^2)}\}/t \text{ and } x_2=\{1-\sqrt{(1-t^2)}\}/t \qquad \text{(vii)}$$

which are reciprocals.

Now when $|t|<1$, $|x_1|>\left|\frac{1}{t}\right|$, i.e. $|x_1|>1$ and so $|x_2|<1$ since $x_1x_2=1$. It follows from (v) and (vii) that series (iv) converges to x_2 that root of (vi) which is numerically less than unity. Hence the series converges to x or $1/x$ according as $|x|$ is less than or greater than unity.

The reader should verify this result by taking a particular value of t, e.g. $t=\frac{3}{5}$.

Exercises 5 (*a*)

1. Find the coefficients of x^{2r} and x^{2r+1} in the expansion of $\dfrac{1-x}{(1-2x^2)(1-2x)}$ in ascending powers of x, and state for what range of values of x the expansion is valid. [L.U.]

2. Express $\dfrac{11x-2}{(x-2)^2(x^2+1)}$ in terms of partial fractions.

Show that the coefficient of x^{2n} in the expansion of this expression in positive powers of x is $(4n+3)2^{-2n-1}+2(-1)^{n-1}$, and find the coefficient of x^{2n+1}. State the range of values of x for which the expansion is valid. [L.U.]

3. Prove that

$$\frac{(a-b)^3}{(1-ax)^2(1-bx)^2}=\frac{a^2(a-b)}{(1-ax)^2}+\frac{b^2(a-b)}{(1-bx)^2}-\frac{2a^2b}{(1-ax)}+\frac{2ab^2}{(1-bx)}.$$

Find the coefficient of x^n in the expansion of

$$\frac{(a-b)^3}{(1-ax)^2(1-bx)^2}$$

in ascending powers of x, and state the range of values of x for which the expansion is valid. [L.U.]

4. Find a, b, c, d so that the coefficient of x^n in the expansion of

$$\frac{a+bx+cx^2+dx^3}{(1-x)^4}$$

is $(n+1)^3$. Deduce that $\sum_{n=0}^{\infty}(n+1)^3(\sqrt{3}-2)^n=0$. [Sheffield.]

5. Find coefficients a, b and c such that

$$27+32(1-4x)(1-x)^2=a(1-x)^2+b(1-4x)^2+c(1-x)(1-4x)^2.$$

Express $\dfrac{27}{(1-5x+4x^2)^2}$ in terms of partial fractions, and obtain its expansion in positive integral powers of x, stating the range in which the expansion is valid.

Deduce that $4^{n+2}(3n+1)+3n+11$ has 27 as a factor for all positive integral values of n. [L.U.]

6. Express $\dfrac{(a-b)^2}{(1-ax)^2(1-bx)}$ in terms of partial fractions. If the function is expanded in positive integral powers of x, find the coefficient of x^n, and state the range of values of x for which the expansion is valid.

Deduce, or prove by any other means, that the expression

$$(n+1)a^{n+2}-(n+2)a^{n+1}b+b^{n+2},$$

where n is a positive integer, contains $(a-b)^2$ as a factor. [L.U.]

7. Express the function $\dfrac{1}{(x^2+1)(x-2)}$ as a sum of partial fractions.

For what range of values of x can the function be expanded as a series of ascending powers of x?

Find the coefficients of x^{2n} and x^{2n+1} in the expansion. [L.U.]

8. If $$\Sigma_1=1.2+3.4x^2+5.6x^4+7.8x^6+\ldots$$

and $$\Sigma_2=2.3+4.5x^2+6.7x^4+8.9x^6+\ldots$$

where x is positive, prove that $\Sigma_1/\Sigma_2=(1+3x^2)/(x^2+3)$. [L.U.]

9. If $\phi(x)=A_0+A_1x+A_2x^2+\ldots+A_nx^n+\ldots$, show that

$$(A_0+A_1+\ldots+A_n)$$

is equal to the coefficient of x^n in the expansion of $\dfrac{\phi(x)}{1-x}$ in ascending powers of x.

Obtain the expansion of $(1-x)^{-\frac{1}{2}}$ in the form

$$1+\tfrac{1}{2}x+\frac{1.3}{2.4}x^2+\frac{1.3.5}{2.4.6}x^3+\ldots$$

and prove that the sum of the series

$$1+\tfrac{1}{2}+\frac{1.3}{2.4}+\frac{1.3.5}{2.4.6}+\ldots+\frac{1.3.5\ldots(2n-1)}{2.4.6\ldots 2n}$$

is equal to $$\frac{3.5.7\ldots(2n+1)}{2^n n!}.$$ [L.U.]

5.6. The number *e*

If S_n is the sum of the first n terms of the infinite series

$$1+\frac{1}{1!}+\frac{1}{2!}+\frac{1}{3!}+\ldots+\frac{1}{r!}+\ldots$$

$$S_n=1+\frac{1}{1}+\frac{1}{1.2}+\frac{1}{1.2.3}+\ldots+\frac{1}{1.2.3\ldots(n-1)}$$

$$S_n\leqslant 1+1+\frac{1}{2}+\frac{1}{2^2}+\ldots+\frac{1}{2^{n-2}}$$

<3 summing the geometrical progression.

But S_n increases steadily as n increases, and so by § 4.12, as $n\to\infty$, S_n tends to a limit which is less than or equal to 3 and is denoted by e; hence

$$e=1+\frac{1}{1!}+\frac{1}{2!}+\frac{1}{3!}+\ldots$$

The value of e may be calculated to any required degree of accuracy from the above series. Correct to 4 significant figures, $e=2{\cdot}718$.

5.7. The exponential theorem

It has been shown (see Example 10 of § 4.19) that the series

$$1+\frac{x}{1!}+\frac{x^2}{2!}+\frac{x^3}{3!}+\frac{x^4}{4!}+\ldots+\frac{x^r}{r!}+\ldots$$

converges absolutely for any finite value of x.

Let us now consider the product $f(x).f(y)$, where

$$f(x)=1+\frac{x}{1!}+\frac{x^2}{2!}+\frac{x^3}{3!}+\ldots+\frac{x^r}{r!}+\ldots \quad . \quad . \quad \text{(i)}$$

$$f(y)=1+\frac{y}{1!}+\frac{y^2}{2!}+\frac{y^3}{3!}+\ldots+\frac{y^r}{r!}+\ldots.$$

Since the series are absolutely convergent, we have

$$f(x).f(y)=1+\left(\frac{x}{1!}+\frac{y}{1!}\right)+\left(\frac{x^2}{2!}+\frac{xy}{1!\,1!}+\frac{y^2}{2!}\right)$$

$$+\left(\frac{x^3}{3!}+\frac{x^2y}{2!\,1!}+\frac{xy^2}{1!\,2!}+\frac{y^3}{3!}\right)+\ldots$$

$$=1+\frac{(x+y)}{1!}+\frac{(x+y)^2}{2!}+\frac{(x+y)^3}{3!}+\ldots$$

the general term being

$$\frac{x^r}{r!}+\frac{x^{r-1}y}{(r-1)!\,1!}+\frac{x^{r-2}y^2}{(r-2)!\,2!}+\ldots+\frac{x^{r-s}y^s}{(r-s)!\,s!}+\ldots+\frac{y^r}{r!}$$

$$=(x+y)^r/r!$$

$$\therefore f(x).f(y)=f(x+y).$$

Extending this result, we have

$$f(x).f(y).f(z)=f(x+y+z)$$

and, more generally,

$$f(x_1).f(x_2).f(x_3)\ldots f(x_n)=f(x_1+x_2+x_3+\ldots+x_n), \quad . \quad \text{(ii)}$$

where n is any positive integer.

From (i), putting $x=1$, we have $f(1)=e$, and from (ii), putting $x_1=x_2=\ldots=x_n=1$,

$$\{f(1)\}^n=f(n),$$

i.e. $\qquad f(n)=e^n$, when n is a positive integer.

In (ii) put $x_1=x_2=\ldots=x_n=m/n$, where m and n are positive integers.

Then $f(m/n).f(m/n)\ldots$to n factors$=f(m/n+m/n+\ldots$to n terms),

i.e. $$\{f(m/n)\}^n=f(m)$$

$$=e^m, \text{ since } m \text{ is a positive integer.}$$

Taking the nth root of each side, we see that $f(m/n)$ is one of the values of $e^{m/n}$ and since, by (i), $f(m/n)>0$ when m and n are positive, (m/n) is given by the positive value of $e^{m/n}$.

Hence if x is a positive rational number,

$$f(x)=e^x \quad . \quad . \quad . \quad . \qquad \text{(iii)}$$

the positive value of e^x being understood.

Finally, if x is a negative rational number, we write $x=-y$, where y is a positive rational number.

Then
$$f(x).f(y)=f(x+y) \text{ by (ii)}$$
$$=f(0)=1.$$
$$\therefore f(x)=\frac{1}{f(y)}=\frac{1}{e^y} \text{ by (iii)}$$
$=e^x$, the positive value being understood.

Thus, if x is any rational number, $f(x)=e^x$ and so

$$e^x=1+x+\frac{x^2}{2!}+\frac{x^3}{3!}+\frac{x^4}{4!}+\ldots \qquad (5.2)$$

for all rational values of x. This is the *exponential theorem.*

Writing $-x$ for x in (5.2), we have

$$e^{-x}=1-x+\frac{x^2}{2!}-\frac{x^3}{3!}+\frac{x^4}{4!}-\ldots \qquad (5.3)$$

When x is irrational, we define e^x by (5.2). It then follows from (ii) that

$$e^{x_1}.e^{x_2}\ldots e^{x_n}=e^{x_1+x_2+\ldots+x_n} \quad . \quad . \quad . \qquad \text{(iv)}$$

for all *real* values of $x_1, x_2 \ldots x_n$.

5.8. Properties of e^x

1. e^x is a continuous function of x, being the sum of a power series (cf. § 4.20 (iv)).
2. e^x increases as x increases. This is clear from (5.2) when $x \geqslant 0$ and from $e^x=1/e^{-x}$ when $x<0$.
3. $e^x \to \infty$ as $x \to \infty$. This is clear from (5.2).
4. $e^x \to 0$ as $x \to -\infty$ since $e^{-x}=1/e^x$.

It is important to note that e^x is always positive.

Since the function e^x can take any value between 0 and $+\infty$ it follows that corresponding to any real positive number y there is a number x such that $y=e^x$, i.e. $x=\log_e y$. Logarithms to the base e are called *natural* or *Napierian logarithms.* In subsequent sections when the base of a logarithm is not stated it may be assumed that the base is e.

The elementary laws of logarithms such as $\log m+\log n=\log mn$, will be assumed.

5.9. Series for a^x, $a > 0$

If a is a positive number and x is a positive rational number p/q, p and q being positive integers, the positive value y of $a^{p/q}$ is given by $y^q = a^p$. From this equation we have

$$q \log y = p \log a \text{ so that } \log y = (p/q) \log a = x \log a.$$

Hence $$y = e^{x \log a}$$

i.e. $$a^x = e^{x \log a} \quad . \quad . \quad . \quad . \quad \text{(i)}$$

Again, when x is a negative rational number, by writing $x = -t$ where t is a positive rational number we obtain

$$a^x = \frac{1}{a^t} = \frac{1}{e^{t \log a}} = e^{-t \log a} = e^{x \log a}.$$

Hence (i) is true for all rational values of x.

Except in the case when $a = e$, a^x has not been defined for irrational values of x. We take (i) as our definition of a^x when x is irrational. For such a value of x, a^x is defined only for positive values of a.

From this definition we can show that the laws of indices hold for irrational as well as for rational indices. For example, by (iv) of § 5.7,

$$a^x \times a^y = e^{x \log a} \times e^{y \log a} = e^{(x+y) \log a} = a^{x+y}$$

and $$(a^x)^y = e^{y \log a^x} = e^{xy \log a} = a^{xy}.$$

Again, by definition,

$$a^x = e^{x \log a} = 1 + \frac{x \log a}{1\,!} + \frac{x^2 (\log a)^2}{2\,!} + \frac{x^3 (\log a)^3}{3\,!} + \ldots, \text{ by (5.2).}$$

The expansion is valid for any finite value of x.

5.10. Hyperbolic functions

We define the hyperbolic sine of x (written $\sinh x$) and the hyperbolic cosine of x (written $\cosh x$) by the relations

$$\sinh x = \tfrac{1}{2}(e^x - e^{-x}), \ \cosh x = \tfrac{1}{2}(e^x + e^{-x}) \quad (5.4)$$

Writing $-x$ for x in these equations, we have

$$\sinh(-x) = -\sinh x, \ \cosh(-x) = \cosh x. \quad (5.5)$$

Also, $$\cosh x + \sinh x = e^x \quad (5.6)$$

and $$\cosh x - \sinh x = e^{-x} \quad (5.7)$$

whence $$\cosh^2 x - \sinh^2 x = 1. \quad (5.8)$$

From (5.6) and (5.7) by squaring and adding corresponding sides we have

$$\cosh^2 x + \sinh^2 x = \tfrac{1}{2}(e^{2x} + e^{-2x}) = \cosh 2x, \text{ by definition,}$$

i.e. $$\cosh 2x = 2\cosh^2 x - 1 = 2\sinh^2 x + 1 \text{ by (5.8).}$$

Again from (5.6) and (5.7) by squaring and subtracting corresponding sides we get

$$\sinh 2x = 2 \sinh x \cosh x.$$

By writing $\sinh (x+y) = \frac{1}{2}(e^x . e^y - e^{-x} . e^{-y})$ and using (5.6) and (5.7) we have

$$\sinh (x+y) = \tfrac{1}{2}\{(\cosh x + \sinh x)(\cosh y + \sinh y) - (\cosh x - \sinh x)(\text{ccsh } y - \sinh y)\}$$

$$\therefore \sinh (x+y) = \sinh x \cosh y + \cosh x \sinh y.$$

Similarly,

$$\cosh (x+y) = \cosh x \cosh y + \sinh x \sinh y.$$

Substituting $-y$ for y in these results and using (5.5), we obtain

$$\sinh (x-y) = \sinh x \cosh y - \cosh x \sinh y,$$

$$\cosh (x-y) = \cosh x \cosh y - \sinh x \sinh y.$$

It should now be clear that to every trigonometrical relationship involving sines and cosines there corresponds a hyperbolic identity involving hyperbolic sines and cosines.

The analogy is carried still further by the introduction of hyperbolic tangents, cosecants, secants and cotangents defined by the relations

$$\tanh x = \sinh x/\cosh x, \quad \text{cosech } x = 1/\sinh x, \quad \text{sech } x = 1/\cosh x,$$
$$\coth x = 1/\tanh x.$$

Dividing throughout (5.8) by $\cosh^2 x$ and $\sinh^2 x$ in turn we have

$$\text{sech}^2 x = 1 - \tanh^2 x \text{ and } \text{cosech}^2 x = \coth^2 x - 1;$$

from the formulae for $\sinh (x+y)$ and $\cosh (x+y)$ we deduce that

$$\tanh (x \pm y) = \frac{\tanh x \pm \tanh y}{1 \pm \tanh x \tanh y}.$$

The following rule (known as Osborn's rule) enables us to write down any relationship connecting hyperbolic functions if we know the one which connects the corresponding circular functions: in any identity connecting circular functions of general angles replace each circular function by the corresponding hyperbolic function but change the sign in front of a product or an implied product of *two* sines; for example, in front of $\sin^2 x$, $\tan^2 x$, $\cot^2 x$, $\sin x \sin y$.

The identities $\cos 3\theta = 4 \cos^3 \theta - 3 \cos \theta$, $\quad \sin 3\theta = 3 \sin \theta - 4 \sin^3 \theta$

give $\quad \cosh 3x = 4 \cosh^3 x - 3 \cosh x$, $\sinh 3x = 3 \sinh x + 4 \sinh^3 x$;

and from the formulae

$$\sin \theta \cos \phi = \tfrac{1}{2}\{\sin (\theta + \phi) + \sin (\theta - \phi)\},$$
$$\cos \theta \cos \phi = \tfrac{1}{2}\{\cos (\theta + \phi) + \cos (\theta - \phi)\},$$
$$\cos \theta \sin \phi = \tfrac{1}{2}\{\sin (\theta + \phi) - \sin (\theta - \phi)\},$$
$$\sin \theta \sin \phi = \tfrac{1}{2}\{\cos (\theta - \phi) - \cos (\theta + \phi)\}.$$

we obtain the relations

$$\sinh x \cosh y = \tfrac{1}{2}\{\sinh (x+y) + \sinh (x-y)\},$$
$$\cosh x \cosh y = \tfrac{1}{2}\{\cosh (x+y) + \cosh (x-y)\},$$
$$\cosh x \sinh y = \tfrac{1}{2}\{\sinh (x+y) - \sinh (x-y)\},$$
$$\sinh x \sinh y = \tfrac{1}{2}\{\cosh (x+y) - \cosh (x-y)\}.$$

All the above hyperbolic identities may be proved from the definitions of the hyperbolic functions.

5.11. Series for sinh *x* and cosh *x*

By addition and subtraction of (5.2) and (5.3) we have (see § 4.20 (i))

$$\sinh x = \tfrac{1}{2}(e^x - e^{-x}) = x + \frac{x^3}{3!} + \frac{x^5}{5!} + \frac{x^7}{7!} + \ldots$$

$$\cosh x = \tfrac{1}{2}(e^x + e^{-x}) = 1 + \frac{x^2}{2!} + \frac{x^4}{4!} + \frac{x^6}{6!} + \ldots.$$

Both series are convergent for any finite value of x.

5.12. Note on the graphs of hyperbolic functions

The graphs of the functions $y = \sinh x$ and $y = \cosh x$ are shown in fig. 3 (*a*); the graph of $y = \tanh x$ is shown in fig. 3 (*b*).

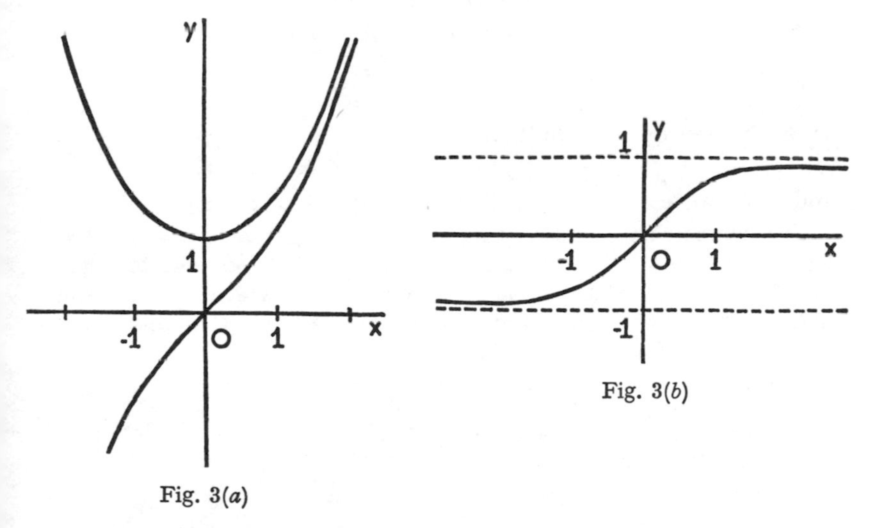

Fig. 3(*a*)

Fig. 3(*b*)

We note that:

(i) If x is real and $y = \sinh x$, y can take all values, positive and negative.

(ii) If x is real and $y = \cosh x$, $y \nless 1$; also $\cosh x = \cosh (-x)$ and so the graph of $y = \cosh x$ is symmetrical about Oy.

(iii) If $y=\tanh x=\dfrac{e^x-e^{-x}}{e^x+e^{-x}}$, it is clear that $y=0$ when $x=0$ and that $|y|<1$. By a rearrangement we can deduce at once the limiting values of y as $x\to\pm\infty$; for by writing $y=\dfrac{1-e^{-2x}}{1+e^{-2x}}$ we see that $y\to 1$ as $x\to+\infty$ and by writing $y=\dfrac{e^{2x}-1}{e^{2x}+1}$ we see that $y\to-1$ as $x\to-\infty$.

5.13. Inverse hyperbolic functions

If x and y are numbers connected by the relation $x=\sinh y$, we write $y=\sinh^{-1} x$, and call y the inverse hyperbolic sine of x. If $x=\cosh y$ we write $y=\cosh^{-1} x$ and so on.

Inverse hyperbolic functions may be expressed in logarithmic form. For example, if $y=\sinh^{-1} x$,

$$\sinh y=x,\ \cosh y=+\sqrt{(x^2+1)},\ \text{since } \cosh y\geqslant 1,$$

and by (5.6)

$$e^y=x+\sqrt{(x^2+1)}$$

$$\therefore\ y=\sinh^{-1} x=\log\{x+\sqrt{(x^2+1)}\}.$$

If

$$y=\cosh^{-1} x,$$

$$\cosh y=x,\ \sinh y=\pm\sqrt{(x^2-1)}$$

$$\therefore\ e^y=x\pm\sqrt{(x^2-1)},$$

and so

$$y=\cosh^{-1} x=\log\{x\pm\sqrt{(x^2-1)}\}.$$

But

$$x-\sqrt{(x^2-1)}=\frac{1}{x+\sqrt{(x^2-1)}}$$

as can be seen by cross-multiplication

$$\therefore\ \log\{x-\sqrt{(x^2-1)}\}=-\log\{x+\sqrt{(x^2-1)}\}$$

and so we write

$$\cosh^{-1} x=\pm\log\{x+\sqrt{(x^2-1)}\}.$$

The double sign is due to the fact that the graph of $y=\cosh^{-1} x$, i.e. of $x=\cosh y$, is symmetrical about the x-axis so that to every value of $x\geqslant 1$ there correspond two values of y which are equal in magnitude and opposite in sign. The principal value* of $\cosh^{-1} x$ is conventionally taken as $+\log\{x+\sqrt{(x^2-1)}\}$.

When

$$y=\tanh^{-1} x,\quad (|x|<1)$$

$$x=\tanh y=\frac{e^{2y}-1}{e^{2y}+1},$$

i.e.

$$e^{2y}=\frac{1+x}{1-x},$$

$$\therefore\ y=\tanh^{-1} x=\tfrac{1}{2}\log\frac{1+x}{1-x}\qquad (x^2<1).$$

Similarly,

$$\coth^{-1} x=\tfrac{1}{2}\log\frac{x+1}{x-1}\qquad (x^2>1).$$

* See also § 9.6

5.14. Miscellaneous examples

Example 3

Evaluate $\lim_{h\to 0}\dfrac{e^h-1}{h}$.

By (5.2)
$$e^h-1=h+\frac{h^2}{2\,!}+\frac{h^3}{3\,!}+\dots$$
$$\frac{e^h-1}{h}=1+\frac{h}{2\,!}+\frac{h^2}{3\,!}+\dots$$
and so, by § 4.20 (iv), $\lim_{h\to 0}\dfrac{e^h-1}{h}=1.$

Example 4

Solve for real values of x the equations

(a) $2\cosh 2x-\sinh 2x=2,$

(b) $\cosh(\log x)=\sinh(\log \tfrac{1}{4}x)+\tfrac{19}{8}.$

(a) If
$$2\cosh 2x-\sinh 2x=2,$$
$$e^{2x}+e^{-2x}-\tfrac{1}{2}(e^{2x}-e^{-2x})=2$$
$$e^{2x}-4+3e^{-2x}=0$$
$$(e^{x}-e^{-x})(e^{x}-3e^{-x})=0.$$
Hence $\quad e^{x}-e^{-x}=0,\quad e^{2x}=1\quad \therefore x=0;$

or $\quad e^{x}-3e^{-x}=0,\quad e^{2x}=3\quad \therefore x=\tfrac{1}{2}\log 3.$

(b)
$$\cosh(\log x)=\tfrac{1}{2}(e^{\log x}+e^{-\log x})$$
$$=\tfrac{1}{2}(x+1/x)$$
$$\sinh(\log \tfrac{1}{4}x)=\tfrac{1}{2}(e^{\log \frac{1}{4}x}-e^{-\log \frac{1}{4}x})$$
$$=\tfrac{1}{2}(x/4-4/x).$$
Hence if
$$\cosh(\log x)=\sinh(\log \tfrac{1}{4}x)+\tfrac{19}{8}$$
$$\tfrac{1}{2}(x+1/x)=\tfrac{1}{2}(x/4-4/x)+\tfrac{19}{8},$$
$$3x^2-19x+20=0,$$
$$x=\tfrac{4}{3}\text{ or }5.$$

Example 5

Prove that $\tanh^{-1}x+\tanh^{-1}y=\tanh^{-1}\dfrac{x+y}{1+xy}$.

Let $\tanh^{-1}x=A$ and $\tanh^{-1}y=B$ so that $x=\tanh A$ and $y=\tanh B$. Then $\tanh^{-1}x+\tanh^{-1}y=A+B$
$$=\tanh^{-1}\{\tanh(A+B)\}$$
$$=\tanh^{-1}\left(\frac{\tanh A+\tanh B}{1+\tanh A\tanh B}\right),\text{ see § 5.10,}$$
$$=\tanh^{-1}\left(\frac{x+y}{1+xy}\right).$$

Exercises 5 (*b*)

1. Write down the first four terms of series whose sums to infinity are respectively

$$\text{(i) } e-1/e, \qquad \text{(ii) } e^2+1/e^2, \qquad \text{(iii) } e^3.$$

Show that $\frac{1}{2}(1+e)^2/e=2+\dfrac{1}{2\,!}+\dfrac{1}{4\,!}+\dfrac{1}{6\,!}+\ldots.$

2. Express as power series in x giving the first four terms and the general term

$$\text{(i) } e^{2x}, \qquad \text{(ii) } (1+x)e^x, \qquad \text{(iii) } (1-x^2)/e^x.$$

3. Expand e^{x+x^2} as a power series in x as far as the term in x^4.

4. Find the values of a, b, c so that the coefficient of x^n in the expansion of $(a+bx+cx^2)e^x$ in ascending powers of x shall be $(n+1)^2/n\,!$. [L.U.]

5. Prove that

(i) $(\cosh 2x-1)/(\cosh 2x+1)=\tanh^2 x$,

(ii) $\sinh x=2 \tanh \frac{1}{2}x/(1-\tanh^2 \frac{1}{2}x)$,

(iii) $\cosh x=(1+\tanh^2 \frac{1}{2}x)/(1-\tanh^2 \frac{1}{2}x)$.

6. Prove that

(i) $\sinh (x-y)=\sinh x \cosh y-\cosh x \sinh y$,

(ii) $\cosh 3x \cosh^3 x+\sinh 3x \sinh^3 x=\cosh^3 2x$,

(iii) $\sinh^{-1}(\frac{4}{3})=\log 3$. [L.U.]

7. Find real numbers r and α such that $5 \sinh x+3 \cosh x=r \sinh (x+\alpha)$ for all values of x. Hence, or otherwise, find the real value of x which satisfies the equation $5 \sinh x+3 \cosh x=-3$. [L.U.]

8. Solve the equation $5 \cosh x+\sinh x=7$. [L.U.]

9. (i) Write down the series for $\sinh x$, $\cosh x$ and show that when $x>0$, $x(2+\cosh x)>3 \sinh x$.

(ii) Solve the equation $\cosh (\log x)=\sinh (\log \frac{1}{2}x)+7/4$.

(iii) If $\tan x=\tan \lambda \tanh \mu$, and $\tan y=\cot \lambda \tanh \mu$, prove that $\tan (x+y)=\operatorname{cosec} 2\lambda \sinh 2\mu$. [L.U.]

10. If $y=\log \tan (\frac{1}{4}\pi+\frac{1}{2}x)$, prove that

(i) $\sinh y=\tan x$; (ii) $\cosh y=\sec x$. [L.U.]

11. If $\cot x=\sinh u$ and $0<x<\pi$, prove that

(i) $\cos x=\tanh u$; (ii) $e^u=\cot \frac{1}{2}x$.

Solve the equation $4 \sinh u+3e^u+3=0$ for real values of u. [L.U.]

12. If $\tan \frac{1}{2}x = \tan \alpha \tanh \beta$, prove that

$$\tan x = \sin 2\alpha \sinh 2\beta/(1+\cos 2\alpha \cosh 2\beta),$$

and that $\sin x = \sin 2\alpha \sinh 2\beta/(\cos 2\alpha + \cosh 2\beta)$. [L.U.]

13. (i) Define the hyperbolic sine, cosine and tangent and show that

$$\cosh^{-1} x = \pm \log \{x+\sqrt{(x^2-1)}\}, \quad \tanh^{-1} x = \tfrac{1}{2} \log \frac{1+x}{1-x}$$

and deduce that $\tanh^{-1}(\sin \theta) = \cosh^{-1}(\sec \theta)$.

(ii) Draw rough graphs of the functions $\cot x$ and $\coth x$, and show that there is a root of the equation $\coth x = \cot x$ in every interval $\{n\pi, (n+1)\pi\}$, $n \geqslant 1$, and $\{n\pi, (n-1)\pi\}$, $n \leqslant 1$, n being an integer. Give an approximate formula for the numerically large roots of the equation. [L.U.]

14. Show that $\cosh x + \sinh x = e^x$ and deduce that

$$\cosh nx + \sinh nx = \cosh^n x + \binom{n}{1} \cosh^{n-1} x \sinh x$$

$$+ \binom{n}{2} \cosh^{n-2} x \sinh^2 x + \ldots + \sinh^n x.$$

Obtain a similar expansion for $\cosh nx - \sinh nx$.

Prove that

$\cosh 7x = 64 \cosh^7 x - 112 \cosh^5 x + 56 \cosh^3 x - 7 \cosh x$. [Durham.]

15. Given that

$$\frac{\sinh x}{5} = \frac{\sinh y}{9} = \frac{\sinh (x+y)}{28},$$

show that either $\sinh x = 0$ or $5 \cosh y = 28 - 9 \cosh x$. Hence eliminate y and prove that either $x=0$, $y=0$, or $x = \log 3$, $y = \log 5$ or $x = -\log 3$, $y = -\log 5$. [L.U. Anc.]

16. Express $\tanh 2x$ and $\tanh 3x$ in terms of $\tanh x$, and show that, if $-1 < k < 1$, then $u = \tanh(\frac{1}{3} \tanh^{-1} k)$ is a root of the cubic equation $u^3 - 3ku^2 + 3u - k = 0$. [Sheffield.]

17. Prove the formulae

$$\cosh 3x = 4 \cosh^3 x - 3 \cosh x; \quad \sinh 3x = 4 \sinh^3 x + 3 \sinh x.$$

Show that the curves

$$y = \cosh x, \quad y = \tfrac{1}{3} \cosh 3x$$

intersect at an angle $\cos^{-1}(7/9)$. [Sheffield.]

18. If $\tan x = \tan \lambda \tanh \mu$ and $\tan y = \cot \lambda \tanh \mu$, prove that

$$\tan (x-y) = -\cot 2\lambda \tanh 2\mu.$$ [L.U.]

19. Prove that $\sinh 3u = 4 \sinh^3 u + 3 \sinh u$.

Deduce that the real root of the equation $4x^3 + 3x = 3$ is

$$\tfrac{1}{2}\{(3+\sqrt{10})^{1/3} - (3+\sqrt{10})^{-1/3}\}.$$ [L.U.]

20. Define the functions $\sinh x$ and $\cosh x$, and sketch their graphs roughly in the same figure.

Defining $\cosh^{-1} x$ to be always positive or zero, show that

$$\cosh^{-1} x = \log \{x+\sqrt{(x^2-1)}\} \text{ when } x \geqslant 1.$$

Prove that, in the range $0<\theta<\frac{1}{2}\pi$, the equation

$$\cosh^{-1} (\sec \theta)+\log (\sin 2\theta)=0$$

has just one solution, viz. $\theta=\sin^{-1} \frac{1}{2}(\sqrt{3}-1)$. [L.U.]

5.15. The logarithmic series

In § 5.14, Example 3, we have shown that

$$\lim_{h\to 0} \frac{e^h-1}{h}=1.$$

If we substitute $h=\pm n \log (1-x)$, where $0<x<1$ and n is a positive number, then as $h\to 0$, $n\to 0+$ and we obtain

$$\lim_{n\to 0+} \frac{(1-x)^n-1}{n \log (1-x)}=1=\lim_{n\to 0+} \frac{(1-x)^{-n}-1}{-n \log (1-x)}.$$

Replacing $\log (1-x)$, which is negative, by $-\log \dfrac{1}{1-x}$, we obtain

$$\log \frac{1}{1-x}=\lim_{n\to 0+} \frac{1-(1-x)^n}{n}=\lim_{n\to 0+} \frac{(1-x)^{-n}-1}{n} \quad . \qquad \text{(i)}$$

Now, by the binomial theorem,

$$\frac{1-(1-x)^n}{n}=x-\frac{(n-1)}{1.2}x^2+\frac{(n-1)(n-2)}{1.2.3}x^3-\ldots$$

$$=x+\frac{(1-n)}{1}\frac{x^2}{2}+\frac{(1-n)(2-n)}{1.2}\frac{x^3}{3}+\ldots$$

$$<x+\frac{x^2}{2}+\frac{x^3}{3}+\ldots$$

since x is positive, and n may be taken less than 1 since n ultimately tends to zero.

Also

$$\frac{(1-x)^{-n}-1}{n}=x+\frac{(n+1)}{1}\frac{x^2}{2}+\frac{(n+1)(n+2)}{1.2}\frac{x^3}{3}+\ldots$$

$$>x+\frac{x^2}{2}+\frac{x^3}{3}+\ldots$$

since x and n are positive.

Hence, when $0<n<1$,

$$\frac{1-(1-x)^n}{n}<x+\frac{x^2}{2}+\frac{x^3}{3}+\ldots<\frac{(1-x)^{-n}-1}{n}$$

and since, by (i), $\lim\limits_{n\to 0+}\dfrac{1-(1-x)^n}{n}=\lim\limits_{n\to 0+}\dfrac{(1-x)^{-n}-1}{n}=\log\dfrac{1}{1-x}$

it follows that

$$\log\frac{1}{1-x}=x+\frac{x^2}{2}+\frac{x^3}{3}+\ldots \text{when } 0<x<1 \quad . \quad . \qquad \text{(ii)}$$

Again, by putting x^2 for x in (ii), we have

$$\log\frac{1}{1-x^2}=x^2+\frac{x^4}{2}+\frac{x^6}{3}+\ldots \text{ when } 0<x<1.$$

$$\therefore \log\frac{1}{1-x^2}-\log\frac{1}{1-x}=-x+\frac{x^2}{2}-\frac{x^3}{3}+\frac{x^4}{4}-\ldots \text{ when } 0<x<1,$$

i.e.

$$\log\frac{1}{1+x}=-x+\frac{x^2}{2}-\frac{x^3}{3}+\frac{x^4}{4}-\ldots \text{ when } 0<x<1 \quad . \qquad \text{(iii)}$$

We have thus proved that

$$\log(1+x)=x-\frac{x^2}{2}+\frac{x^3}{3}-\frac{x^4}{4}+\ldots \text{ when } 0<x<1$$

and

$$\log(1-x)=-x-\frac{x^2}{2}-\frac{x^3}{3}-\frac{x^4}{4}-\ldots \text{ when } 0<x<1$$

These results can be expressed by writing

$$\log(1+x)=x-\frac{x^2}{2}+\frac{x^3}{3}-\frac{x^4}{4}+\ldots \text{ when } -1<x<1$$

The series on the right-hand side is known as the *logarithmic series* and has been shown in § 4.19, Example 11 to converge when $x=1$. Hence by § 4.20 (iv) it converges to log 2.

$$\therefore \log(1+x)=x-\frac{x^2}{2}+\frac{x^3}{3}-\frac{x^4}{4}+\ldots \text{ when} -1<x\leqslant 1 \qquad (5.9)$$

Replacing x by $-x$, we have

$$\log(1-x)=-\left(x+\frac{x^2}{2}+\frac{x^3}{3}+\frac{x^4}{4}+\ldots\right) \text{ when } -1\leqslant x<1. \quad (5.10)$$

5.16. Series related to the logarithmic series

By subtraction we obtain from (5.9) and (5.10), cf. § 4.20 (i)

$$\tfrac{1}{2}\log\frac{1+x}{1-x}=x+\frac{x^3}{3}+\frac{x^5}{5}+\ldots \text{ when } -1<x<1. \tag{5.11}$$

This series, which converges more rapidly than (5.9) or (5.10), is useful in numerical computations.

If $x=\dfrac{1}{2n+1}$, (5.11) gives

$$\tfrac{1}{2}\log(1+1/n)=\frac{1}{2n+1}+\frac{1}{3(2n+1)^3}+\frac{1}{5(2n+1)^5}+\ldots \tag{5.12}$$

This series is valid when $|2n+1|>1$,

i.e. when $(2n+1)^2>1$,

$n(n+1)>0$.

This condition is fulfilled when $n<-1$ and when $n>0$.

Again, if $x=\dfrac{m-n}{m+n}$, (5.11) gives

$$\tfrac{1}{2}\log\frac{m}{n}=\frac{m-n}{m+n}+\frac{1}{3}\left(\frac{m-n}{m+n}\right)^3+\frac{1}{5}\left(\frac{m-n}{m+n}\right)^5+\ldots \tag{5.13}$$

This series is valid when $\left|\dfrac{m-n}{m+n}\right|<1$

i.e. when $(m-n)^2<(m+n)^2$.

This condition is fulfilled when m and n have the same sign, i.e. when $m/n>0$, and so if we write $y=m/n$ in (5.13) we obtain when $y>0$

$$\tfrac{1}{2}\log y=\frac{y-1}{y+1}+\frac{1}{3}\left(\frac{y-1}{y+1}\right)^3+\frac{1}{5}\left(\frac{y-1}{y+1}\right)^5+\ldots \tag{5.14}$$

Example 6

Show that, when $|x|>1$,

$$\frac{1}{x^2}+\frac{1}{2x^4}+\frac{1}{3x^6}+\ldots=2\left\{\frac{1}{2x^2-1}+\frac{1}{3(2x^2-1)^3}+\frac{1}{5(2x^2-1)^5}+\ldots\right\}.$$

By (5.10) when $|x|>1$,

$$\frac{1}{x^2}+\frac{1}{2x^4}+\frac{1}{3x^6}+\ldots=-\log\left(1-\frac{1}{x^2}\right)$$

$$=\log\frac{x^2}{x^2-1}.$$

By (5.11) when $|2x^2-1|>1$, i.e. when $|x|>1$,

$$2\left\{\frac{1}{2x^2-1}+\frac{1}{3(2x^2-1)^3}+\frac{1}{5(2x^2-1)^5}+\ldots\right\}=\log\left(\frac{1+\dfrac{1}{2x^2-1}}{1-\dfrac{1}{2x^2-1}}\right)$$

$$=\log\frac{x^2}{x^2-1}.$$

The required result follows.

Example 7

If $0<x<1$, *prove that* $(1+x)\log(1+x)+(1-x)\log(1-x)>0$. *Deduce, or prove otherwise that, if* $n>1$, $(n+1)^{1+1/n}.(n-1)^{1-1/n}>n^2$. [L.U.]

When $0<x<1$,

$$(1+x)\log(1+x)=(1+x)(x-\tfrac{1}{2}x^2+\tfrac{1}{3}x^3-\tfrac{1}{4}x^4+\ldots)$$

$$(1-x)\log(1-x)=-(1-x)(x+\tfrac{1}{2}x^2+\tfrac{1}{3}x^3+\tfrac{1}{4}x^4+\ldots)$$

$$\therefore (1+x)\log(1+x)+(1-x)\log(1-x)=2\{(1-\tfrac{1}{2})x^2+(\tfrac{1}{3}-\tfrac{1}{4})x^4+(\tfrac{1}{5}-\tfrac{1}{6})x^6+\ldots\}$$

$$>0.$$

When $x=1/n$, where $n>1$, this gives

$$(1+1/n)\log(1+1/n)+(1-1/n)\log(1-1/n)>0$$

i.e.
$$\log\{(1+1/n)^{1+1/n}.(1-1/n)^{1-1/n}\}>0$$

$$\log\left\{\frac{(n+1)^{1+1/n}.(n-1)^{1-1/n}}{n^2}\right\}>0$$

$$\therefore (n+1)^{1+1/n}.(n-1)^{1-1/n}>n^2.$$

Exercises 5 (*c*)

1. Expand the following expressions in ascending powers of x as far as the fourth term, giving the coefficient of x^r and stating the values of x for which the expansions are valid:

   (i) $\log(1+2x)$,
   (ii) $\log(1-\tfrac{1}{4}x)$,
   (iii) $\log(3+x)$,
   (iv) $\log(2-3x)$,
   (v) $\log(4+x)^2$,
   (vi) $\log\sqrt{(1-x-2x^2)}$,
   (vii) $\log(1+x+x^2)$,
   (viii) $\log\dfrac{1+2x}{1-x}$.

2. If $\log(1+x+x^2+x^3)$ is expanded in a series of ascending powers of x, and $|x|<1$, show that the coefficient of x^n is $1/n$ unless n is a multiple of 4, in which case the coefficient is $-3/n$.

3. If $|x|<1$, expand in powers of x as far as the fourth term
$$(1-x)\log(1+x)+(1+x)\log(1-x).$$

4. Find the coefficient of x^r in the following expansions :

$$\text{(i) } (1+3x)\log(1+3x)\;;\quad \text{(ii) } (1+2x)^2\log(1+2x)$$

the expansions being assumed valid.

5. Prove that, if $-\frac{1}{2}<x\leqslant\frac{1}{2}$, then

$$\log(1+x-2x^2)=x-\tfrac{5}{2}x^2+\tfrac{7}{3}x^3-\tfrac{17}{4}x^4+\ldots$$

6. Sum to infinity the series

$$\frac{1}{3}+\frac{1}{3^2.2}+\frac{1}{3^3.3}+\frac{1}{3^4.4}+\ldots,$$

and
$$\frac{1}{2}+\frac{1}{3.2^3}+\frac{1}{5.2^5}+\frac{1}{7.2^7}+\ldots.$$

7. Find the expansion of $\log(1+x)$ in ascending powers of x and deduce that

$$\log\frac{1+x}{1-x}=2(x+\tfrac{1}{3}x^3+\tfrac{1}{5}x^5+\ldots),\text{ when } |x|<1.$$

Find the range of values of x for which the series

$$\tfrac{2}{3}x^3+\tfrac{4}{5}x^5+\tfrac{6}{7}x^7+\ldots$$

converges, and prove that its sum, when $x=\frac{1}{2}$, is $\frac{2}{3}-\frac{1}{2}\log 3$.

[Sheffield.]

8. Show that, if θ is not a multiple of π,

$$\log\operatorname{cosec}\theta=\tfrac{1}{2}\cos^2\theta+\tfrac{1}{4}\cos^4\theta+\tfrac{1}{6}\cos^6\theta+\ldots.$$

9. Show that, if $p>1$,

$$\frac{1}{p+1}+\frac{1}{2(p+1)^2}+\frac{1}{3(p+1)^3}+\ldots=\frac{1}{p}-\frac{1}{2p^2}+\frac{1}{3p^3}-\ldots.$$

10. Show that, if

$$S_1=\tfrac{1}{2}-\tfrac{1}{2}(\tfrac{1}{2})^2+\tfrac{1}{3}(\tfrac{1}{2})^3-\tfrac{1}{4}(\tfrac{1}{2})^4+\ldots,$$
$$S_2=\tfrac{1}{5}+\tfrac{1}{3}(\tfrac{1}{5})^3+\tfrac{1}{5}(\tfrac{1}{5})^5+\tfrac{1}{7}(\tfrac{1}{5})^7+\ldots,$$

then
$$S_1=2S_2.$$

11. Show that if $y>2$ or $y<-2$,

$$\frac{2}{y^3-3y}+\frac{1}{3}\left(\frac{2}{y^3-3y}\right)^3+\frac{1}{5}\left(\frac{2}{y^3-3y}\right)^5+\ldots$$
$$=\log\frac{y-1}{y+1}+\tfrac{1}{2}\log\frac{y+2}{y-2}.$$

[L.U.]

12. Prove that, if m and n have the same sign,

$$\log\frac{m}{n}=2\left\{\frac{m-n}{m+n}+\frac{1}{3}\left(\frac{m-n}{m+n}\right)^3+\frac{1}{5}\left(\frac{m-n}{m+n}\right)^5+\ldots\right\}$$

and state why m and n must have the same sign for the series to converge.

Deduce, or prove by any other means, that

$$\theta = \tanh\theta + \tfrac{1}{3}\tanh^3\theta + \tfrac{1}{5}\tanh^5\theta + \dots \qquad \text{[L.U.]}$$

13. Prove that
$$\tfrac{1}{2}\sin^2\theta + \tfrac{1}{4}\sin^4\theta + \tfrac{1}{6}\sin^6\theta + \dots$$
$$= 2\{\tan^2(\tfrac{1}{2}\theta) + \tfrac{1}{3}\tan^6(\tfrac{1}{2}\theta) + \tfrac{1}{5}\tan^{10}(\tfrac{1}{2}\theta) + \dots\}$$
and state the range of values of θ for which this result is true. [L.U.]

14. Expand $\log y$ as a series in ascending powers of $\dfrac{y-1}{y+1}$ and find the range of values of y for which the expansion is valid.

If $-\tfrac{1}{3}\pi < \theta < \tfrac{1}{3}\pi$, prove that $\log\cos\theta$ differs from $-2\tan^2(\tfrac{1}{2}\theta)$ by less than $\tfrac{1}{36}$. [L.U.]

15. Show that if p, q, r are three consecutive positive integers
$$\log q = \tfrac{1}{2}\log(pr) + \frac{1}{2pr+1} + \frac{1}{3}\left(\frac{1}{2pr+1}\right)^3 + \dots$$

Use this series to calculate the value of log 19 correct to four decimal places, given that

$$\log 2 = 0{\cdot}693147,$$
$$\log 3 = 1{\cdot}098612,$$
$$\log 10 = 2{\cdot}302585. \qquad \text{[L.U.]}$$

16. Prove that, if p, q and x are positive,
$$2(p-q)\left\{1 + \frac{1}{3}\left(\frac{p-q}{p+q+2x}\right)^2 + \frac{1}{5}\left(\frac{p-q}{p+q+2x}\right)^4 + \dots\right\}$$
$$= (p+q+2x)\log\frac{p+x}{q+x}.$$

Deduce that, if $0 < p < q$, the effect on the expression $(p/q)^{p+q}$ of increasing p and q by the *same* positive numbers is to increase its value. [L.U.]

17. If p, q, r are consecutive integers, prove that
$$\log q - \tfrac{1}{2}(\log p + \log r) = \sum_{n=1}^{\infty} \frac{x^{2n-1}}{2n-1},$$
where $x = 1/(2pr+1)$.

Deduce that $\log 2$ exceeds $\dfrac{4}{7} + \dfrac{2}{17} + \dfrac{4}{1029}$ by less than $\cdot 0003$. [L.U.]

5.17. Limits

The following examples illustrate the use of series in the evaluation of limits. We assume that the method is valid if the functions involved can be expanded in convergent power series (see § 4.20 (iv)).

Example 8

Evaluate $\lim_{x\to\infty}\{x\sqrt{(x^2+a^2)}-\sqrt{(x^4+a^4)}\}$.

$$x\sqrt{(x^2+a^2)}=x^2\left(1+\frac{a^2}{x^2}\right)^{\frac{1}{2}}$$

$$=x^2\left(1+\frac{1}{2}\frac{a^2}{x^2}-\frac{1}{8}\frac{a^4}{x^4}+\ldots\right) \text{ when } x^2>a^2$$

$$\sqrt{(x^4+a^4)}=x^2\left(1+\frac{a^4}{x^4}\right)^{\frac{1}{2}}$$

$$=x^2\left(1+\frac{1}{2}\frac{a^4}{x^4}+\ldots\right) \text{ when } x^2>a^2$$

$$\therefore\ x\sqrt{(x^2+a^2)}-\sqrt{(x^4+a^4)}=\tfrac{1}{2}a^2-\frac{5}{8}\frac{a^4}{x^2}+\text{terms involving higher powers of } (1/x^2).$$

$$\therefore\ \lim_{x\to\infty}\{x\sqrt{(x^2+a^2)}-\sqrt{(x^4+a^4)}\}=\tfrac{1}{2}a^2.$$

Example 9

Evaluate $\lim_{x\to 0}\dfrac{1-e^{-2x}}{\log(1+x)}$.

The given function takes the indeterminate form $\frac{0}{0}$ when $x=0$.

But if $0<|x|<1$,

$$\frac{1-e^{-2x}}{\log(1+x)}=\frac{2x-\frac{4x^2}{2!}+\ldots}{x-\frac{1}{2}x^2+\frac{1}{3}x^3-\ldots}$$

$$=\frac{2-2x+\ldots}{1-\frac{1}{2}x+\ldots} \text{ since } x\neq 0$$

$$\therefore\ \lim_{x\to 0}\frac{1-e^{-2x}}{\log(1+x)}=2.$$

5.18. Miscellaneous examples

Example 10

Prove that, if n is large,

$$(n-1/3n)\log\frac{n+1}{n-1}=2+\frac{8}{45n^4}+\ldots$$

and that $\left(\dfrac{n+1}{n-1}\right)^{\frac{3n^2-1}{3n}}=e^2\left(1+\dfrac{8}{45n^4}+\ldots\right)$.

$$\log\frac{n+1}{n-1}=\log\frac{1+1/n}{1-1/n}$$

$$=2\left\{\frac{1}{n}+\frac{1}{3n^3}+\frac{1}{5n^5}+\ldots\right\} \text{ when } |n|>1, \text{ by (5.11), page 92}$$

$$\therefore (n-1/3n)\log\frac{n+1}{n-1}=2\left(1-\frac{1}{3n^2}\right)\left(1+\frac{1}{3n^2}+\frac{1}{5n^4}+\ldots\right)$$

$$=2\left\{1+\frac{4}{45n^4}+\ldots\right\}$$

$$\therefore \left(\frac{n+1}{n-1}\right)^{n-1/3n}=e^{2+\frac{8}{45n^4}+\ldots}=e^2\left\{e^{\frac{8}{45n^4}+\ldots}\right\}$$

i.e. $$\left(\frac{n+1}{n-1}\right)^{n-1/3n}=e^2\left\{1+\frac{8}{45n^4}+\ldots\right\} \text{ by (5.2), page 82.}$$

Example 11

Expand in powers of x as far as the term in x^4 the logarithms of both sides of the identity $(1-ax)(1-bx)\equiv 1-px+qx^2$, where $p=a+b$ and $q=ab$. Hence find values of a^3+b^3 and a^4+b^4 in terms of p and q.

We have $\log(1-ax)+\log(1-bx)\equiv\log\{1-x(p-qx)\}$. Expanding both sides and multiplying throughout by -1 we obtain

$$(a+b)x+\tfrac{1}{2}(a^2+b^2)x^2+\tfrac{1}{3}(a^3+b^3)x^3+\tfrac{1}{4}(a^4+b^4)x^4+\ldots$$
$$=x(p-qx)+\tfrac{1}{2}x^2(p-qx)^2+\tfrac{1}{3}x^3(p-qx)^3+\tfrac{1}{4}x^4(p-qx)^4+\ldots$$

Equating in turn $((x^3))$ and $((x^4))$ on each side, we have

$$a^3+b^3=p^3-3pq$$
$$a^4+b^4=p^4-4p^2q+2q^2.$$

Example 12

By expanding $(e^x-1)^n$ in two ways, n being a positive integer, and comparing suitable coefficients in the two expansions, show that

$$n^n-\frac{n}{1}(n-1)^n+\frac{n(n-1)}{1.2}(n-2)^n-\frac{n(n-1)(n-2)}{1.2.3}(n-3)^n+\ldots=n!.$$

By using a similar method with the expression $(e^x+1)^n-(e^x-1)^n$, or otherwise, show that, if $n>3$,

$$c_1(n-1)^3+c_3(n-3)^3+c_5(n-5)^3+\ldots=n^2(n+3)2^{n-4},$$

where c_1, $c_2,\ldots$ are the usual binomial coefficients. [L.U.]

By the binomial theorem

$$(e^x-1)^n=e^{nx}-{}^nC_1e^{(n-1)x}+{}^nC_2\,e^{(n-2)x}-\ldots$$

$$=\sum_{r=0}^{\infty}\frac{(nx)^r}{r!}-\frac{n}{1!}\sum_{r=0}^{\infty}\frac{(n-1)^rx^r}{r!}+\frac{n(n-1)}{2!}\sum_{r=0}^{\infty}\frac{(n-2)^rx^r}{r!}-\ldots. \quad \text{(i)}$$

By (5.2), page 82, $$(e^x-1)^n=\left(x+\frac{x^2}{2!}+\frac{x^3}{3!}+\ldots\right)^n$$

$$=x^n\left(1+\frac{x}{2!}+\frac{x^2}{3!}+\ldots\right)^n \quad . \quad . \quad \text{(ii)}$$

Equating $((x^n))$ in (i) and (ii), we have

$$\frac{n^n}{n!}-\frac{n}{1!}\frac{(n-1)^n}{n!}+\frac{n(n-1)}{2!}\frac{(n-2)^n}{n!}-\frac{n(n-1)(n-2)}{3!}\frac{(n-3)^n}{n!}+\ldots=1,$$

$$n^n-n(n-1)^n+\frac{n(n-1)}{2!}(n-2)^n-\frac{n(n-1)(n-2)}{3!}(n-3)^n+\ldots=n!.$$

If c_r represents the binomial coefficient ${}^nC_r$,

$$(e^x+1)^n-(e^x-1)^n=2\{c_1e^{(n-1)x}+c_3e^{(n-3)x}+c_5e^{(n-5)x}+\ldots\}$$

$$=2\left\{c_1\sum_{r=0}^{\infty}\frac{(n-1)^rx^r}{r!}+c_3\sum_{r=0}^{\infty}\frac{(n-3)^rx^r}{r!}+c_5\sum_{r=0}^{\infty}\frac{(n-5)^rx^r}{r!}+\ldots\right\}. \quad \text{(iii)}$$

Also $(e^x+1)^n-(e^x-1)^n$

$$=\left\{2+x\left(1+\frac{x}{2!}+\frac{x^2}{3!}+\ldots\right)\right\}^n-x^n\left\{1+\frac{x}{2!}+\frac{x^2}{3!}+\ldots\right\}^n$$

$$=2^n+2^{n-1}c_1x\left(1+\frac{x}{2}+\frac{x^2}{6}+\ldots\right)+2^{n-2}c_2x^2\left(1+\frac{x}{2}+\ldots\right)^2$$

$$+2^{n-3}c_3x^3\left(1+\frac{x}{2}+\ldots\right)^3+\ldots-x^n\left\{1+\frac{x}{2!}+\frac{x^2}{3!}+\ldots\right\}^n \quad \text{(iv)}$$

Equating $((x^3))$ in (iii) and (iv), we have, if $n>3$,

$$\frac{2}{3!}\{c_1(n-1)^3+c_3(n-3)^3+c_5(n-5)^3+\ldots\}=\tfrac{1}{6}.2^{n-1}c_1+2^{n-2}c_2+2^{n-3}c_3$$

$$=\tfrac{1}{6}(2^{n-3})\{4n+6n(n-1)+n(n-1)(n-2)\}$$

$$\therefore\ \{c_1(n-1)^3+c_3(n-3)^3+c_5(n-5)^3+\ldots\}=n^2(n+3)2^{n-4}.$$

Exercises 5 (*d*)

1. Evaluate the following limits:

(i) $\displaystyle\lim_{x\to 0}\frac{(2-x)\log(1-x)+(2+x)\log(1+x)}{x^4}$,

(ii) $\displaystyle\lim_{x\to 1+}\frac{\log x}{x-1}$,

(iii) $\displaystyle\lim_{x\to 0}\frac{\log(1+\frac{1}{2}x)-(1+x)^{\frac{1}{2}}+1}{x^3}$,

(iv) $\lim\limits_{x\to 0} \dfrac{\log(1-\frac{1}{3}x)-(1-x)^{\frac{1}{3}}+1}{x^2}$,

(v) $\lim\limits_{x\to 0} \dfrac{e^{-x}-1+x}{x-\log(1+x)}$.

2. If x and y are small, show that

$$\frac{(1+y)^x}{(1+x)^y}=1-\tfrac{1}{2}xy(y-x)$$

provided that the ratio x/y is finite and that terms of the fourth and higher orders are neglected. [L.U.]

3. If $|x|>1$, show that $\left(1+\dfrac{1}{x}\right)^{x+1}=e\left(1+\dfrac{1}{2x}-\dfrac{1}{24x^2}+\dots\right)$ and that, if x is small, $(1+x)^{1+x}\simeq 1+x+x^2+\frac{1}{2}x^3$.

4. If $n>1$, show that $(n+\frac{1}{2})\log\left(1+\dfrac{1}{n}\right)=\left(1+\dfrac{1}{12n^2}-\dfrac{1}{12n^3}+\dots\right)$ and hence that $\left(1+\dfrac{1}{n}\right)^{n+\frac{1}{2}}=e\left(1+\dfrac{1}{12n^2}-\dfrac{1}{12n^3}+\dots\right)$.

5. By using the fact that $(1+x/n)^n=e^{n\log(1+x/n)}$ prove that

$$(1+x/n)^n+(1-x/n)^{-n}=2e^x\left\{1+\frac{1}{n^2}\left(\frac{x^3}{3}+\frac{x^4}{8}\right)\right\}$$

if $1/n^4$ and higher powers of $1/n$ are neglected. [L.U.]

6. If $y=(1+x)^{1+x}(1-x)^{1-x}$ and $0<x<1$, prove that

$$\log y=2\left\{\frac{x^2}{1.2}+\frac{x^4}{3.4}+\frac{x^6}{5.6}+\dots\right\}.$$

Show also that $\dfrac{1}{1-x^2}>y>1+x^2$. [L.U.]

7. Find the coefficient of x^r in the expansion of

$$\log(1+ax)+\log(1+bx)+\log(1+cx).$$

Deduce that, if $a+b+c=0$, then

$$2(bc+ca+ab)^2=a^4+b^4+c^4.$$

8. Write down the expansion of $\log\{1+(a+b)x+abx^2\}$ in ascending powers of x, and state the range of values of x for which the expansion is valid.

If a and b are both positive, show that for positive integral n

$$a^n+b^n=(a+b)^n-nab(a+b)^{n-2}$$
$$+\frac{n(n-3)}{2!}a^2b^2(a+b)^{n-4}-\frac{n(n-4)(n-5)}{3!}a^3b^3(a+b)^{n-6}+\dots$$ [L.U.]

9. Prove that the sum of the nth powers of the roots of the equation $x^3+px+q=0$ is $-na_n$ where a_n is the coefficient of t^n in the expansion of $\log (1+pt^2+qt^3)$.

   Find the sum of the ninth powers of the roots in terms of p and q. [Sheffield.]

10. (i) Find the first four terms and the coefficient of x^n in the expansion of $\log (1+x-2x^2)$ in ascending powers of x. For what values of x is the expansion valid ?

    (ii) If $\log y=1+2x-2x^2$ where x is small, show that an approximate value of y is given by $y=e(1+2x-\frac{8}{3}x^3)$. [Durham.]

11. If $|x|<1$, show, by using the logarithmic series, or otherwise, that $(1+x)^{1+x}.(1-x)^{1-x} \geqslant 1$.

    If a, b are positive and unequal, deduce by putting $x=\dfrac{a-b}{a+b}$ in the above inequality that $a^a b^b > \left(\dfrac{a+b}{2}\right)^{a+b}$. [L.U.]

12. Show that, if $-1<x<+1$,
$$(1-x)^{-x}=1-x \log (1-x)+\frac{x^2}{2\,!}[\log (1-x)]^2-\frac{x^3}{3\,!}[\log (1-x)]^3+ \dots$$
and hence expand $(1-x)^{-x}$ in powers of x up to the term in x^6. [L.U.]

13. Show that if $n>0$,
$$\log (1+1/n)=2\left\{\frac{1}{2n+1}+\frac{1}{3}\frac{1}{(2n+1)^3}+\frac{1}{5}\frac{1}{(2n+1)^5}+\dots\right\}.$$
Hence, or otherwise, show that if $n>0$, $\log (1+1/n)$ lies between $\dfrac{1}{n+\frac{1}{2}}$ and $\dfrac{1}{2}\left(\dfrac{1}{n}+\dfrac{1}{n+1}\right)$. [L.U.]

14. Show that when $n>1$,
$$(1)\ \log (n+1)-\log n<1/n,$$
$$(2)\ \log n-\log (n-1)>1/n.$$
Deduce that $0<1+\frac{1}{2}+\frac{1}{3}+\frac{1}{4}+\dots+\dfrac{1}{n}-\log n<1$. [L.U.]

15. Prove that, if $p>1$,
$$\frac{2}{p}+\frac{2}{3p^3}<\log \frac{p+1}{p-1}<\frac{2}{p}+\frac{2}{3p^3}+\frac{2}{5p^3(p^2-1)}.$$
By giving p the values 26, 31, 49, calculate to three places of decimals the value of $\log 5$. [L.U.]

16. Write down the series for the expansions of $\log (1+x)$ and $-\log (1-x)$, when x is numerically less than unity. Deduce, or prove otherwise, that if n is a positive integer
$$\log \frac{n}{n-1}>\frac{1}{n}>\log \frac{n+1}{n}.$$

Prove also that for any integer k,

$$\log\frac{n+k}{n-1} > \left(\frac{1}{n}+\frac{1}{n+1}+\ldots+\frac{1}{n+k}\right) > \log\frac{n+k+1}{n}.$$

By putting $k=2n$ in this result prove that, when n tends to infinity the sum

$$\frac{1}{n}+\frac{1}{n+1}+\ldots+\frac{1}{3n}$$

tends to the value log 3. [L.U.]

17. (i) By writing $\log(1+1/n)$ as $-\log\dfrac{n}{n+1}$, or otherwise, prove that, when $n>1$,

$$\log(1+1/n)^n = 1-\sum_{r=1}^{\infty}\frac{1}{r(r+1)(n+1)^r}.$$

(ii) Prove that, when n is large,

$$(1+1/n)^n = e\left(1-\frac{1}{2n}+\frac{11}{24n^2}+\ldots\right).$$ [L.U.]

18. Show that, for all positive values of n,

$$\tfrac{1}{2}\log\frac{n+1}{n} = \frac{1}{2n+1}+\frac{1}{3(2n+1)^3}+\frac{1}{5(2n+1)^5}+\ldots$$

and hence, or otherwise, show that the ratio of two consecutive terms of the sequence $u_1, u_2, u_3, \ldots, u_n, \ldots$, where

$$u_n = \frac{n!\,e^n}{n^{(2n+1)/2}}$$

approaches unity as n increases. [L.U.]

19. Show that the coefficient of x^n in the expansion of $e^{(e^x)}$ is

$$\frac{1}{n!}\left(\frac{1^n}{1!}+\frac{2^n}{2!}+\frac{3^n}{3!}+\ldots+\frac{r^n}{r!}+\ldots\right).$$

Hence find the sums of the infinite series

$$\frac{1^3}{1!}+\frac{2^3}{2!}+\frac{3^3}{3!}+\ldots,$$

and

$$\frac{1^4}{1!}+\frac{2^4}{2!}+\frac{3^4}{3!}+\ldots.$$ [L.U.]

20. (i) Expand $\left(x-\dfrac{1}{x}\right)\log\dfrac{1+x}{1-x}$ as a series in ascending powers of x, giving the general term and stating the condition under which the expansion is valid.

(ii) Express as an infinite series the coefficient of x^n in the power series expansion of e^{e^x}, and by summing this series for the case $n=2$, or otherwise, find the coefficient of x^2 in the expansion. [L.U.]

CHAPTER 6

COMPLEX NUMBERS

6.1. Complex numbers

The quadratic equation

$$x^2+1=0 \quad . \quad . \quad . \quad . \qquad \text{(i)}$$

has no real roots since there is no real number whose square is -1. If, however, we assume the existence of a number i such that $i^2=-1$, equation (i) is satisfied by the values $x=\pm i$. If, further, we suppose that i obeys the rules of algebra of real numbers, we find that the roots of the equation

$$x^2-2x+5=0$$

are $x=1\pm 2i$.

If a and b are real, the number $c=a+ib$ is called a *complex number* and a and b are known respectively as its *real* and *imaginary* parts. When $a=0$, the complex number is said to be *purely imaginary*; when $b=0$ the complex number is *real*.

The number $a-ib$ is said to be *conjugate* to c and is denoted by $\bar{c}$.

6.2. Operations with complex numbers

We operate with complex numbers in exactly the same way as we operate with real numbers ; for example :

$$(a+ib)+(c+id)=(a+c)+i(b+d), \quad . \quad . \quad . \quad . \qquad \text{(i)}$$

$$(a+ib)-(c+id)=(a-c)+i(b-d), \quad . \quad . \quad . \quad . \qquad \text{(ii)}$$

$$(a+ib)(c+id)=(ac-bd)+i(bc+ad) \quad . \quad . \quad . \qquad \text{(iii)}$$

and $$(a+ib)(a-ib)=a^2+b^2, \text{ since } i^2=-1\,;$$

also $$\frac{a+ib}{c+id}=\frac{(a+ib)(c-id)}{c^2+d^2}=\frac{(ac+bd)+i(bc-ad)}{c^2+d^2} \quad . \qquad \text{(iv)}$$

unless $c=d=0$.

Hence, the sum, difference, product and quotient of two complex numbers is a complex number ; the product of two conjugate complex numbers is real and so is their sum.

6.3. Geometrical representation of complex numbers

Complex numbers may be represented geometrically in what is known as an *Argand diagram*, as follows :

Let $x'Ox$, $y'Oy$ be rectangular axes with the usual sign conventions. Then the point P whose cartesian coordinates are (x, y) may be taken to represent the complex number $x+iy$. Real numbers are represented by points on the x-axis which is called the *real axis*, imaginary numbers are represented by points on the y-axis which is known as the *imaginary axis*. The origin O represents the number zero.

We generally write z for the complex number $x+iy$ and refer to the plane of the Argand diagram as the complex or z-plane. If P represents the number $z(=x+iy)$, the complex number z is said to be the affix of P (see fig. 4 overleaf).

6.4. Definition of complex numbers

In the preceding sections we have based the study of complex numbers on the hypothesis that there is a number i whose square is -1 and which obeys the fundamental laws of algebra—associative, commutative and distributive.

To place the theory of complex numbers on a sound logical basis we may define a complex number $x+iy$ as the point in the Argand diagram with coordinates (x, y). From this definition it immediately follows that two complex numbers x_1+iy_1, x_2+iy_2 are equal if and only if they are identical, i.e. if they are the same point. This is so if, and only if, $x_1=x_2$ and $y_1=y_2$. Hence *if two complex numbers are equal their real parts are equal and their imaginary parts are equal.*

Relations (i), (ii), (iii) and (iv) of § 6.2 are adopted as the definitions of addition, subtraction, multiplication and division of complex numbers. From these it is readily proved that complex numbers satisfy the fundamental laws of algebra—associative, commutative and distributive. For example, we establish the commutative law of multiplication as follows :

If $$z_1=x_1+iy_1 \text{ and } z_2=x_2+iy_2,$$

$$z_1z_2=(x_1+iy_1)(x_2+iy_2)=(x_1x_2-y_1y_2)+i(x_1y_2+x_2y_1) \text{ by (iii) of § 6.2}$$
$$=(x_2+iy_2)(x_1+iy_1)=z_2z_1.$$

The complex number $x+i0$ behaves like a real number and so we identify $x+i0$ with x ; in the same way we write iy for $0+iy$. With this notation

$$i^2=(0+1i)(0+1i)=-1+0i \text{ by (iii) of § 6.2}$$

i.e. $$i^2=-1.$$

6.5. Modulus and argument of a complex number

If the length of OP is r and $\angle xOP=\theta$, $r=+\sqrt{(x^2+y^2)}$ and $\tan\theta=y/x$ (fig. 4). r is called the *modulus* of z and written $|z|$; θ is called the *argument* or *amplitude* of z and written arg z or am z. We shall measure θ in radians unless otherwise stated.

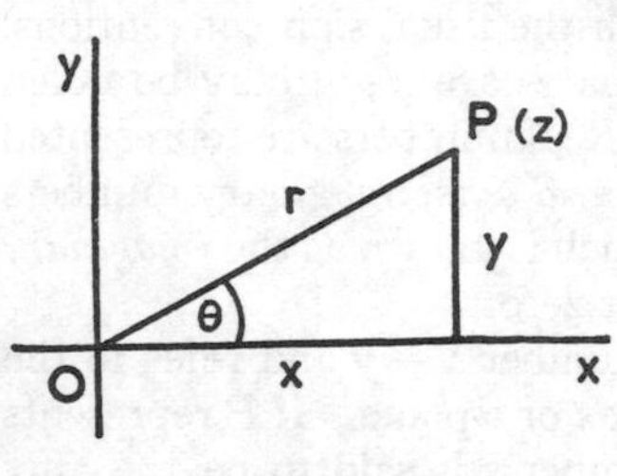

Fig. 4

Since r is by definition positive, and $\cos\theta : \sin\theta : 1 = x : y : \sqrt{(x^2+y^2)}$, for a given value of $z(=x+iy)$ there is a unique value of θ in the range $-\pi<\theta\leqslant\pi$. This is known as the *principal value* of arg z, other values being given by the formula $\theta+2k\pi$ where k is any integer, not zero. In subsequent work, arg z will denote the principal value unless otherwise stated.

Since $x=r\cos\theta$ and $y=r\sin\theta$, z may be written in the cartesian form $x+iy$ or in the polar form $r(\cos\theta+i\sin\theta)$ which is often abbreviated as $r\angle\theta$. The expression $\cos\theta+i\sin\theta$ is sometimes denoted by cis θ and $\cos\theta-i\sin\theta\equiv\cos(-\theta)+i\sin(-\theta)$ by cis $(-\theta)$. Alternatively, since $(\cos\theta+i\sin\theta)(\cos\theta-i\sin\theta)=1$, we may denote $\cos\theta-i\sin\theta$ by $(\text{cis}\,\theta)^{-1}$.

If two numbers z and z' are equal, then $|z|=|z'|$ and arg $z=$ arg z'.

Example 1

Represent the following complex numbers in the Argand diagram and express them in polar form :

(i) $3-3i$, (ii) -4, (iii) $2i$, (iv) $-3+4i$.

In fig. 5, A, B, C and D represent the complex numbers $3-3i$, -4, $2i$ and $-3+4i$ respectively

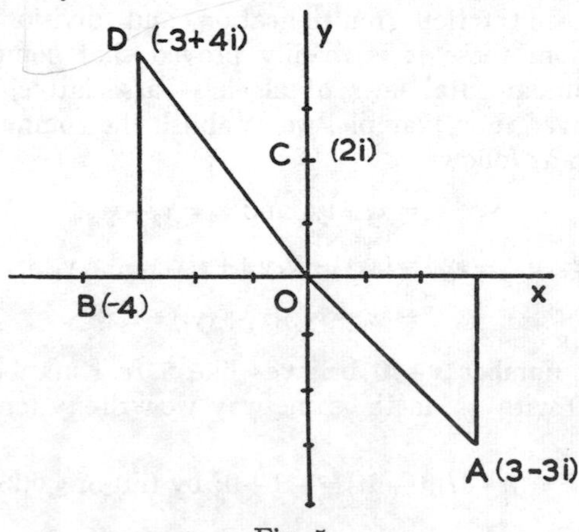

Fig. 5

(i) $OA=3\sqrt{2}$ and $\angle xOA=-\frac{1}{4}\pi$

$\therefore\ 3-3i=3\sqrt{2}\ \text{cis}\ (-\frac{1}{4}\pi)$.

(ii) $OB=4$, $\angle xOB=\pi$

$\therefore\ -4=4\ \text{cis}\ \pi$

(iii) $OC=2$, $\angle xOC=\frac{1}{2}\pi$

$\therefore\ 2i=2\ \text{cis}\ \frac{1}{2}\pi$.

(iv) $OD=5$, $\angle xOD=180°-\tan^{-1} 4/3=126°\ 52'=2{\cdot}214$ radians

$\therefore\ -3+4i=5\ \text{cis}\ (2{\cdot}214)$.

Alternatively, let $-3+4i=r(\cos\theta+i\sin\theta)$.

Then, equating real and imaginary parts on each side of this equation, we have

$$-3=r\cos\theta$$

and $$4=r\sin\theta$$

whence $$r^2=25,\ r=5.$$

Also, $$\sin\theta=\tfrac{4}{5},\ \cos\theta=-\tfrac{3}{5}$$

$$\therefore\ \theta=126°\ 52'=2{\cdot}214\ \text{radians.}$$

Hence $$-3+4i=5\ \text{cis}\ (2{\cdot}214).$$

Example 2

Find the modulus and argument of $(1-\cos\theta-i\sin\theta)/(1+\cos\theta+i\sin\theta)$ *when* $0<\theta<\pi$. *What are the modulus and argument when* $\pi<\theta<2\pi$?

$$1-\cos\theta=2\sin^2\tfrac{1}{2}\theta,\ 1+\cos\theta=2\cos^2\tfrac{1}{2}\theta \text{ and } \sin\theta=2\sin\tfrac{1}{2}\theta\cos\tfrac{1}{2}\theta.$$

Hence
$$\frac{1-\cos\theta-i\sin\theta}{1+\cos\theta+i\sin\theta}=\frac{\sin\frac{1}{2}\theta(\sin\frac{1}{2}\theta-i\cos\frac{1}{2}\theta)}{\cos\frac{1}{2}\theta(\cos\frac{1}{2}\theta+i\sin\frac{1}{2}\theta)}$$
$$=-i\tan\tfrac{1}{2}\theta.$$
$$=\tan\tfrac{1}{2}\theta\ \text{cis}\ (-\tfrac{1}{2}\pi).$$

When $0<\theta<\pi$, $\tan\frac{1}{2}\theta>0$ and so the modulus and argument of the given complex number are $\tan\frac{1}{2}\theta$ and $-\frac{1}{2}\pi$ respectively.

When $\pi<\theta<2\pi$, $\tan\frac{1}{2}\theta<0$ and we write

$$\frac{1-\cos\theta-i\sin\theta}{1+\cos\theta+i\sin\theta}=(-\tan\tfrac{1}{2}\theta)i$$
$$=(-\tan\tfrac{1}{2}\theta)\ \text{cis}\ (\tfrac{1}{2}\pi).$$

Hence the modulus and argument are $(-\tan\frac{1}{2}\theta)$ and $\frac{1}{2}\pi$ respectively.

Example 3

(i) *Find the modulus and argument of the complex number*

$$(1+i)(2+i)/(3-i).$$

(ii) *If* $x+iy=a+b(1+it)/(1-it)$, *where* a *and* b *are real constants and* x, y, t *are real variables, show that the locus of the point* (x, y) *as* t *varies is a circle.*

(i) $$z=\frac{(1+i)(2+i)}{3-i}=\frac{1+3i}{3-i}=i=\text{cis}\ \tfrac{1}{2}\pi$$

$$\therefore\ |z|=1 \text{ and } \arg z=\tfrac{1}{2}\pi.$$

(ii) If
$$x+iy=a+b\left(\frac{1+it}{1-it}\right)$$
$$=a+b\frac{(1+it)^2}{1+t^2}$$
$$=a+b\left(\frac{1-t^2}{1+t^2}\right)+\frac{2ibt}{1+t^2}$$
by equating the real parts and the imaginary parts on each side of this equation we have
$$x=a+b\left(\frac{1-t^2}{1+t^2}\right),\ y=\frac{2bt}{1+t^2}$$
so that $(x-a)^2+y^2=b^2$.

Hence the locus of the point (x, y) is a circle centre $(a. 0)$, radius b.

Example 4

If z is a variable complex number subject to the condition $|z|=1$, and, if $w=2z+1/z$, show that the point of the complex plane corresponding to w describes an ellipse. [Sheffield.]

Since $|z|=1$, we may let $z=\cos\theta+i\sin\theta$; then $1/z=\cos\theta-i\sin\theta$, and if $w=u+iv$ where u and v are real,
$$u+iv=2z+1/z=3\cos\theta+i\sin\theta.$$
By equating real and imaginary parts we obtain
$$u=3\cos\theta,\ v=\sin\theta$$
so that $\frac{1}{9}u^2+v^2=1$.

Hence w moves on an ellipse with centre at $(0, 0)$ and semi-axes of length 3 and 1 (see § 14.1).

6.6. Vectorial representation of a complex number

Since a complex number z is determined by its modulus r and its argument θ we may represent z by a vector of length r drawn in a direction which makes an angle θ with the positive direction of the real axis. For example, the complex number $x+iy$ which in fig. 6 is represented by the point $P(x, y)$ may also be represented by the vector OP (which we shall denote by $\overline{OP}$) or by any equivalent vector $\overline{AB}$, that is by any line equal to, parallel to and drawn in the same sense as $\overline{OP}$.

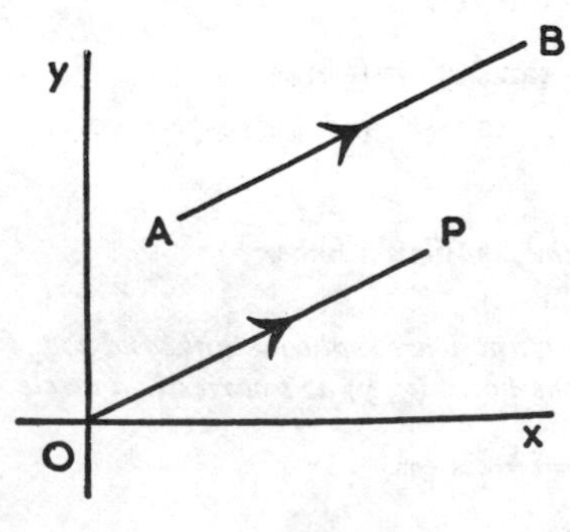

Fig. 6

A real number x is represented by a vector of length $|x|$, drawn along or parallel to the real axis in the positive or negative direction according as x is positive or negative.

In the same way the purely imaginary number iy is represented by a vector of length $|y|$ drawn along or parallel to the imaginary axis in the positive or negative direction according as y is positive or negative.

We shall find it convenient to use both the point and vector methods of representing a complex number.

The length of a vector $\overline{AB}$ will be denoted by AB.

6.7. Geometrical representation of addition or subtraction of two complex numbers

Let A and B represent the complex numbers $z_1(=x_1+iy_1)$ and $z_2(=x_2+iy_2)$ respectively in an Argand diagram (fig. 7). Complete parallelograms $OACB$ and $ODAB$.

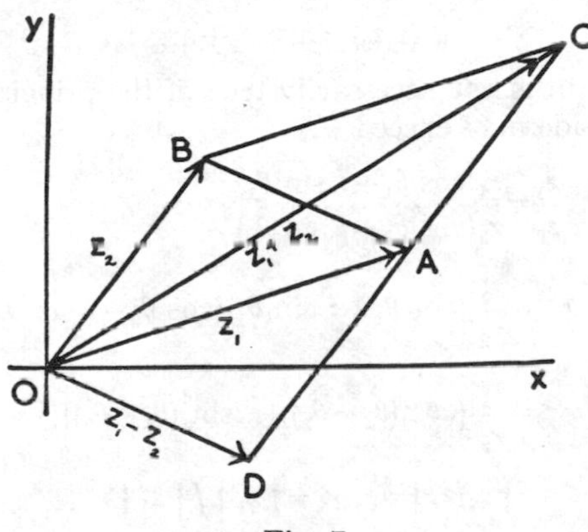

Fig. 7

Then since the mid-point of AB has affix $\frac{1}{2}\{(x_1+x_2)+i(y_1+y_2)\}$, C has affix $\{(x_1+x_2)+i(y_1+y_2)\}$, i.e. z_1+z_2.

In the same way, D has affix z_1-z_2.

In the vector representation z_1, z_2, z_1+z_2 are represented by $\overline{OA}$, $\overline{OB}$, $\overline{OC}$ respectively. We thus see that this representation is in conformity with the usual parallelogram law of vector addition:

$$\overline{OA}+\overline{OB}=\overline{OC}.$$

z_1-z_2 is represented by $\overline{OD}$ and also by BA.

In triangle OAC

$$OC<OA+AC$$

$$\therefore |z_1+z_2|<|z_1|+|z_2|.$$

This result is true for any two complex numbers except when O, A and C lie in order on a straight line. In this case

$$|z_1+z_2|=|z_1|+|z_2|.$$

Hence

$$|z_1+z_2|\leqslant|z_1|+|z_2|.$$

6.8. Multiplication and division of complex numbers

Let $z_1=r_1\angle\theta_1$ and $z_2=r_2\angle\theta_2$. Then

$$\begin{aligned}z_1z_2&=r_1r_2(\cos\theta_1+i\sin\theta_1)(\cos\theta_2+i\sin\theta_2)\\&=r_1r_2\{(\cos\theta_1\cos\theta_2-\sin\theta_1\sin\theta_2)+i(\sin\theta_1\cos\theta_2+\cos\theta_1\sin\theta_2)\}\\&=r_1r_2\{\cos(\theta_1+\theta_2)+i\sin(\theta_1+\theta_2)\}.\end{aligned} \tag{6.1}$$

From this result we see that

$$|z_1z_2|=|z_1|.|z_2|$$

and

$$\arg(z_1z_2)=\arg z_1+\arg z_2.$$

The latter result is not necessarily true of the principal values since the right-hand side may exceed π.

Again,

$$\begin{aligned}\frac{z_1}{z_2}&=\frac{r_1(\cos\theta_1+i\sin\theta_1)}{r_2(\cos\theta_2+i\sin\theta_2)}\\&=\frac{r_1}{r_2}(\cos\theta_1+i\sin\theta_1)(\cos\theta_2-i\sin\theta_2)\\&=\frac{r_1}{r_2}\{\cos(\theta_1-\theta_2)+i\sin(\theta_1-\theta_2)\}.\end{aligned} \tag{6.2}$$

Hence

$$|z_1/z_2|=r_1/r_2=|z_1|/|z_2|$$

and

$$\arg(z_1/z_2)=\arg z_1-\arg z_2.$$

The latter result is not necessarily true of the principal values.

If $z_1=r\angle\theta$ and $z_2=1\angle\phi$, we have by (6.1)

$$z_1z_2=r\angle(\theta+\phi).$$

Thus the effect of multiplying a complex number z_1 by a complex number with unit modulus and argument ϕ is to rotate the vector which represents z_1 counter-clockwise through an angle ϕ.

When $\phi=90°$, $z_2=i$, and so the vector which represents iz_1 is obtained by rotating counter-clockwise through 90° the vector which represents z_1.

More generally, if z_1 is represented by the vector $\overline{OP}$ and $z_2=r_2\angle\theta_2$, the product z_1z_2 is represented by a vector $\overline{OQ}$ such that $\angle POQ=\theta_2$ and $OQ=r_2.OP$.

6.9. Geometrical construction for the product of two complex numbers

Let P_1, P_2 and A represent the numbers z_1, z_2 and 1 respectively. Construct the triangle OP_2P directly similar to triangle OAP_1 (fig. 8). Then $\quad OP : OP_2 = OP_1 : OA \qquad \therefore OP = |z_1|.|z_2| = |z_1z_2|.$
Also

$$\angle xOP = \angle xOP_2 + \angle P_2OP = \angle xOP_2 + \angle xOP_1 = \arg z_2 + \arg z_1 = \arg (z_1z_2).$$

Hence P represents the number z_1z_2.

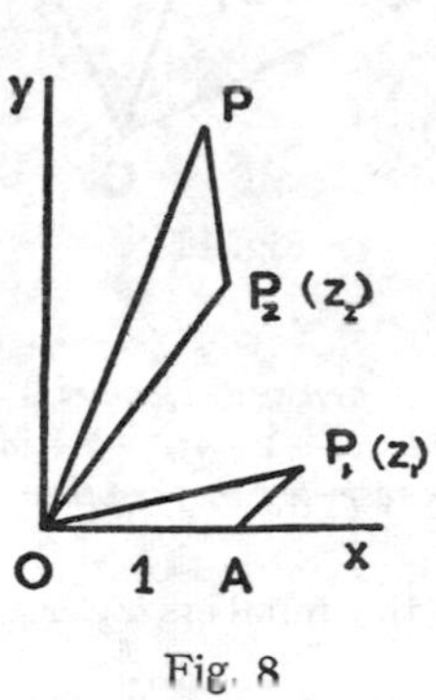

Fig. 8

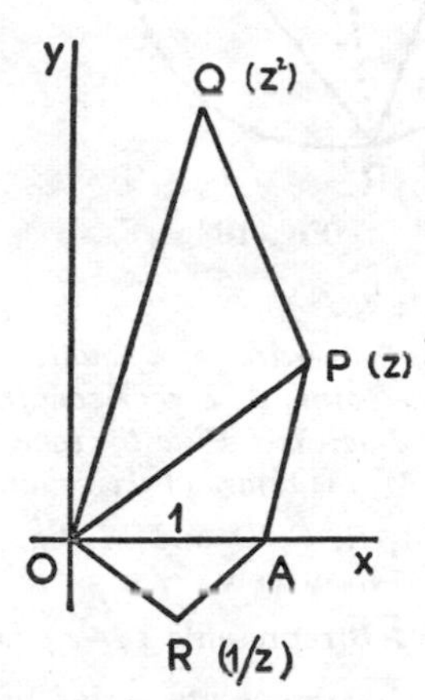

Fig. 9

Example 5

If P represents the number z in the Argand diagram, show how to represent z^2 and $1/z$.

Join P to A, the point which represents the number 1 (fig. 9).

Construct triangles ORA, OPQ directly similar to triangle OAP. Then, as above, it may be shown that Q represents the number z^2 and R represents $1/z$.

6.10. Miscellaneous examples

Example 6

In the Argand diagram, PQR is an equilateral triangle of which the circumcentre is at the origin. If P represents the number $2+i$, find the numbers represented by Q and R. [L.U.]

In fig. 10 $\angle POQ = \angle QOR = \angle ROP = \frac{2}{3}\pi$ and $OP = OQ = OR$,

$$\therefore \overline{OQ} = \overline{OP} \times \text{cis}\,(\tfrac{2}{3}\pi) \quad \text{and} \quad \overline{OR} = \overline{OP} \times \text{cis}\,(-\tfrac{2}{3}\pi)$$

i.e. $\overline{OQ} = (2+i) \times \frac{1}{2}(-1+i\sqrt{3})$

$\qquad = -(1+\frac{1}{2}\sqrt{3}) + i(\sqrt{3}-\frac{1}{2}).$

Similarly, $\qquad \overline{OR} = -(1-\frac{1}{2}\sqrt{3}) - i(\sqrt{3}+\frac{1}{2}).$

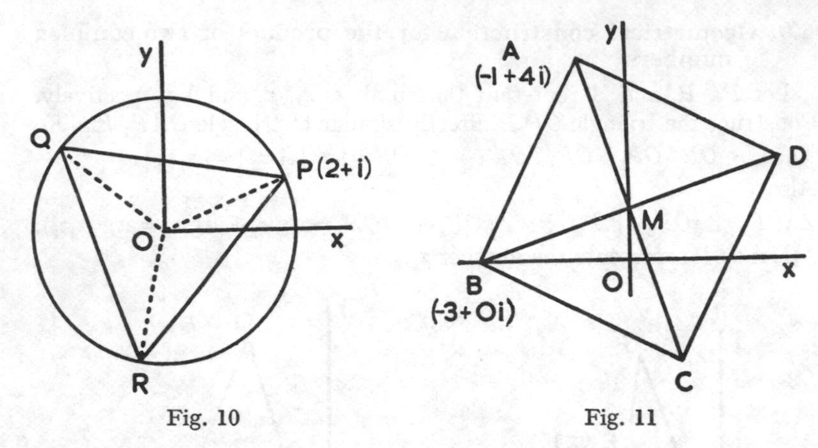

Fig. 10 Fig. 11

Example 7

When the vertices of a square, A, B, C, D are taken anti-clockwise in that order, the points A, B represent the complex numbers $-1+4i$, $-3+0i$ *in the Argand diagram. Find the complex numbers represented by the other vertices and by M, the centre of the square.* [L.U.]

Let A, B, C, D and M (fig. 11) represent the numbers z_A, z_B, z_C, z_D and z_M respectively.

Then $\overline{AB}$ represents z_B-z_A, i.e. $-2-4i$.

$\overline{BC}$ represents $-i\overline{BA}$ or $i\overline{AB}$, since $AB=BC$ and $\angle CBA=\frac{1}{2}\pi$.

$\therefore$ $\overline{BC}$ represents $4-2i$, i.e. $z_C-z_B=4-2i$.

$$\therefore z_C=1-2i.$$

Now
$$z_A+z_C=z_B+z_D=2z_M,$$
i.e.
$$0+2i=(-3+0i)+z_D=2z_M$$
$$\therefore z_D=3+2i \text{ and } z_M=0+i.$$

Example 8

Define the modulus of a complex number and prove that

$$|z_1+z_2|\leqslant|z_1|+|z_2|.$$

By considering the modulus of the left-hand side of the following equation in z, or otherwise, prove that all the roots of

$$z^n \cos n\alpha+z^{n-1}\cos(n-1)\alpha+\ldots+z\cos\alpha=1,$$

where α is real, lie outside the circle $|z|=\frac{1}{2}$. [L.U.]

The modulus of a complex number is defined in § 6.5 and the inequality is proved in § 6.7.

If
$$z^n \cos n\alpha+z^{n-1}\cos(n-1)\alpha+\ldots+z\cos\alpha=1 \quad . \quad \text{(i)}$$
$$|z^n \cos n\alpha+z^{n-1}\cos(n-1)\alpha+\ldots+z\cos\alpha|=1 \quad . \quad \text{(ii)}$$

But from $|z_1+z_2| \leqslant |z_1|+|z_2|$

we have $|z_1+z_2+z_3| \leqslant |z_1|+|z_2|+|z_3|$, and so on.

Hence $|z^n \cos n\alpha + z^{n-1}\cos(n-1)\alpha + \ldots + z\cos\alpha|$

$$\leqslant |z^n \cos n\alpha| + |z^{n-1}\cos(n-1)\alpha| + \ldots + |z\cos\alpha|$$
$$\leqslant |z^n| + |z^{n-1}| + \ldots + |z|,$$

since, if α is real, $|\cos n\alpha| \leqslant 1$ for all real values of n.

Now if $z=r(\cos\theta + i\sin\theta)$ so that $|z|=r$ and $|z^n|=r^n$,

$$|z^n \cos n\alpha + z^{n-1}\cos(n-1)\alpha + \ldots + z\cos\alpha| \leqslant r^n + r^{n-1} + \ldots + r,$$

and so, from (ii)

$$r^n + r^{n-1} + \ldots + r \geqslant 1 \quad . \quad . \quad . \quad . \qquad \text{(iii)}$$

This condition is satisfied if $r \geqslant 1$.

If $r<1$, $r^n + r^{n-1} + \ldots + r = \dfrac{r(1-r^n)}{1-r}$

$$< \frac{r}{1-r}, \text{ since } 0<r<1$$

and so from (iii) $\dfrac{r}{1-r} > 1$

$$r > \tfrac{1}{2}$$

i.e. $|z| > \frac{1}{2}$.

Hence the roots of equation (i) all lie outside the circle $|z| = \frac{1}{2}$.

6.11. Geometrical applications

If the points A, B, C represent the complex numbers z_1, z_2, z_3 respectively, then as in § 6.7, $\overline{CA}$ and $\overline{BA}$ represent the complex numbers z_1-z_3 and z_1-z_2. Hence $|z_1-z_3|$ and $|z_1-z_2|$ are the respective lengths of these vectors.

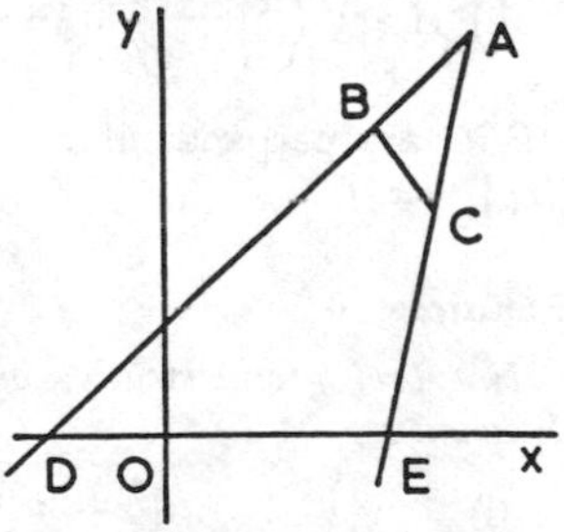

Also, if AB and AC (fig. 12) meet Ox at D and E respectively,

$$\arg(z_1-z_2) = \angle xDA$$

and $\arg(z_1-z_3) = \angle xEA$.

But (see § 6.8)

$$\arg\left(\frac{z_1-z_3}{z_1-z_2}\right) = \arg(z_1-z_3) - \arg(z_1-z_2)$$
$$= \angle xEA - \angle xDA$$
$$= \angle BAC.$$

Fig. 12

Hence the argument of the quotient of two complex numbers is the angle between the vectors which represent the complex numbers.

The above results enable us to interpret loci in the Argand diagram.

Example 9

Prove that

(i) *if* $|z_1+z_2|=|z_1-z_2|$, *the difference of the arguments of* z_1 *and* z_2 *is* $\frac{1}{2}\pi$;

(ii) *if* $\arg\left(\frac{z_1+z_2}{z_1-z_2}\right)=\frac{1}{2}\pi$, *then* $|z_1|=|z_2|$. [L.U.]

Suppose that the points P and Q (fig. 13) represent the complex numbers z_1 and z_2 respectively in an Argand diagram, and complete parallelogram $OPRQ$.

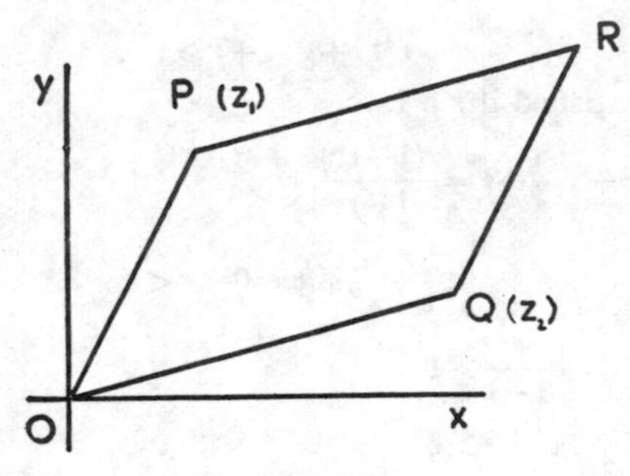

Fig. 13

Then the vectors $\overline{OR}$ and $\overline{QP}$ represent the complex numbers z_1+z_2 and z_1-z_2 respectively.

(i) If $|z_1+z_2|=|z_1-z_2|$, $OR=QP$ and so parallelogram $OPRQ$ has equal diagonals. Hence $OPRQ$ must be a rectangle, OP is perpendicular to OQ, and so $\arg z_1-\arg z_2=\pm\frac{1}{2}\pi$.

(ii) If $\arg\left(\frac{z_1+z_2}{z_1-z_2}\right)=\frac{1}{2}\pi$, the diagonals Ok and QP of parallelogram $OPRQ$ are perpendicular. Hence $OPRQ$ is a rhombus, $OP=OQ$ and so $|z_1|=|z_2|$.

Example 10

Interpret geometrically, or otherwise, the following loci in the Argand diagram :

(i) $$|z+3i|^2-|z-3i|^2=12,$$

(ii) $$|z+ik|^2+|z-ik|^2=10k^2, \quad k>0.$$

(i) Let P, A, B represent the numbers $z(=x+iy)$, $-3i$ and $3i$ respectively in an Argand diagram (fig. 14).

Draw PM perpendicular to Oy so that $PM=x$ and $OM=y$.

Then $$|z+3i|=AP,\ |z-3i|=BP$$

and if $$|z+3i|^2-|z-3i|^2=12$$

$$AP^2-BP^2=12$$

i.e. $$AM^2-BM^2=12$$

$$2OM.AB=12,$$

i.e. $$OM=1.$$

Hence P lies on the straight line $y=1$ which is parallel to the real axis.

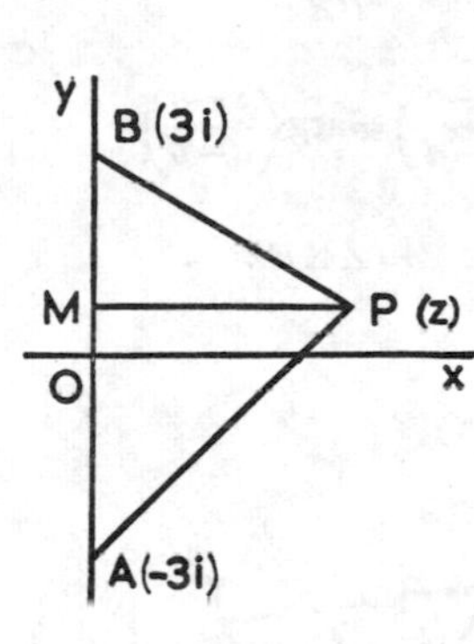

Fig. 14

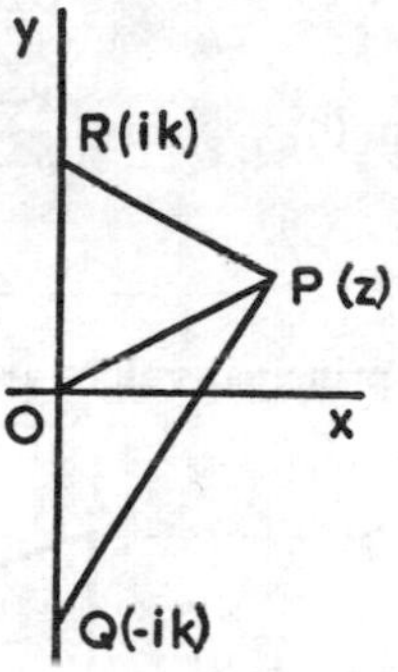

Fig. 15

(ii) Let P, Q, R represent the points $z(=x+iy)$, $-ik$, and ik respectively in an Argand diagram (fig. 15).

Then if $|z+ik|^2+|z-ik|^2=10k^2$

$$PQ^2+PR^2=10k^2$$

i.e. $$2OP^2+2OR^2=10k^2, \text{ since } O \text{ is mid-point of } QR$$

$$\therefore OP^2=4k^2$$

and so $$OP=2k.$$

Hence the locus of P is a circle with centre O and radius $2k$.

Example 11

Triangles BCX, CAY, ABZ are described on the sides of a triangle ABC. If the points A, B, C, X, Y, Z in the Argand diagram represent the complex numbers a, b, c, x, y, z respectively, and

$$\frac{x-c}{b-c}=\frac{y-a}{c-a}=\frac{z-b}{a-b}$$

show that the triangles BCX, CAY, ABZ are similar.

Prove also that the centroids of ABC, XYZ are coincident. [L.U.]

The vectors $\overline{CX}$, $\overline{AY}$ and $\overline{BZ}$ (fig. 16) represent the complex numbers $x-c$, $y-a$ and $z-b$ respectively and so if

$$\frac{x-c}{b-c} = \frac{y-a}{c-a} = \frac{z-b}{a-b} \quad . \quad . \quad . \quad \text{(i)}$$

$$\left|\frac{x-c}{b-c}\right| = \left|\frac{y-a}{c-a}\right| = \left|\frac{z-b}{a-b}\right|$$

i.e.

$$\frac{CX}{CB} = \frac{AY}{AC} = \frac{BZ}{BA} \quad . \quad . \quad . \quad \text{(ii)}$$

Also from (i), $\arg\left(\frac{x-c}{b-c}\right) = \arg\left(\frac{y-a}{c-a}\right) = \arg\left(\frac{z-b}{a-b}\right)$

$$\therefore \angle BCX = \angle CAY = \angle ABZ \quad . \quad . \quad . \quad \text{(iii)}$$

because principal values are taken.

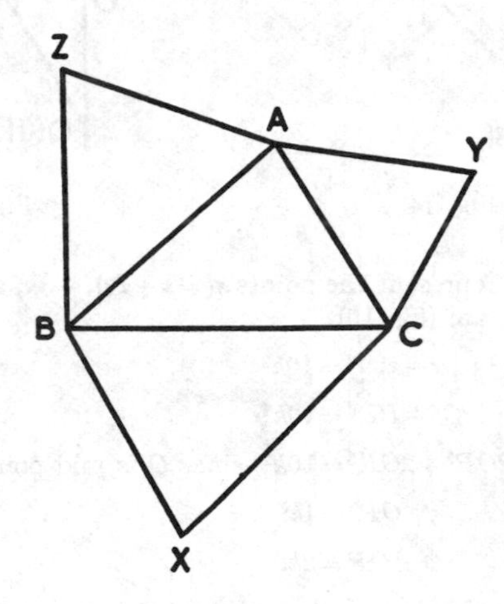

Fig. 16

From (ii) and (iii) triangles BCX, CAY, ABZ are similar.

If G and G' are the centroids of ΔABC and ΔXYZ respectively, G is the point of affix $\frac{1}{3}(a+b+c)$ and G' is the point of affix $\frac{1}{3}(x+y+z)$.

But if in (i) k is the common value of the given ratios we have

$$x-c+y-a+z-b = k(b-c+c-a+a-b) = 0$$

$$\therefore x+y+z = a+b+c$$

Hence G and G' represent the same number and so coincide.

Example 12

If a and b are complex constants, interpret geometrically in an Argand diagram the following loci :

$$\text{(i)}\ arg\left(\frac{z-a}{z-b}\right)=constant.$$

$$\text{(ii)}\ \left|\frac{z-a}{z-b}\right|=constant.$$

Let P, A and B represent the complex numbers z, a and b respectively in an Argand diagram.

(i) If $$\arg\left(\frac{z-a}{z-b}\right)=\text{constant},$$

$$\angle BPA=\text{constant}$$

and so P moves on *either* the major arc AB *or* the minor arc AB of a circle through A and B.

(ii) If $$\left|\frac{z-a}{z-b}\right|=\text{constant},$$

$$\frac{AP}{BP}=\text{constant}=k\ \text{(say)}.$$

When $k=1$, P lies on the perpendicular bisector of AB.

When $k\neq 1$, we divide AB internally at C and externally at D in the ratio $k:1$ (fig. 17).

Then $$\frac{AP}{PB}=\frac{AC}{CB}=\frac{AD}{DB}.$$

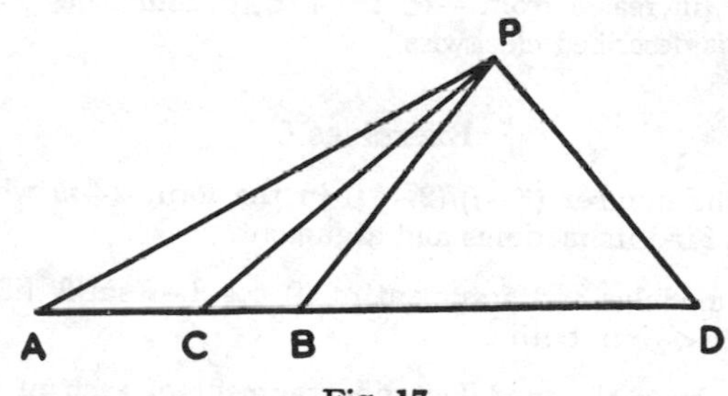

Fig. 17

It follows that PC and PD are the internal and external bisectors of $\angle APB$ and so $\angle CPD$ is a right angle. But C and D are fixed; hence P lies on the circle whose diameter is CD.

Example 13

The complex numbers z_1 and z_2 are represented by points P_1 and P_2 in an Argand diagram. If $z_1(1-z_2)=z_2$ and P_1 describes the line $2x+1=0$, prove that P_2 describes a circle whose centre is at the origin. Find the radius of this *circle and the sense in which it is described as P_1 moves along the line in the direction which makes the ordinate increase.* [L.U.]

Let $z_1=x_1+iy_1$ and $z_2=x_2+iy_2$.

Then since $$z_1=\frac{z_2}{1-z_2} \quad . \quad . \quad . \quad . \quad . \quad . \quad . \quad \text{(i)}$$

$$x_1+iy_1=\frac{x_2+iy_2}{1-(x_2+iy_2)}=\frac{(x_2+iy_2)(1-x_2+iy_2)}{(1-x_2)^2+y_2^2}.$$

Separating the real and imaginary parts we have

$$x_1=\frac{x_2(1-x_2)-y_2^2}{(1-x_2)^2+y_2^2} \quad . \quad . \quad . \quad . \quad \text{(i)}$$

$$y_1=\frac{y_2}{(1-x_2)^2+y_2^2} \quad . \quad . \quad . \quad . \quad \text{(ii)}$$

But P_1 describes the line $2x+1=0$, i.e. $x_1=-\frac{1}{2}$, and substituting this value in (i) we obtain

$$x_2^2+y_2^2=1.$$

Hence z_2 describes the circle

$$x^2+y^2=1 \quad . \quad . \quad . \quad . \quad . \quad \text{(iii)}$$

which has unit radius and centre at the origin.

The point $(\cos\theta, \sin\theta)$ lies on this circle for all values of θ, and so we may put $x_2=\cos\theta$, $y_2=\sin\theta$ in (ii). This gives

$$y_1=\frac{\sin\theta}{2(1-\cos\theta)}=\tfrac{1}{2}\cot\tfrac{1}{2}\theta.$$

Hence as y_1 increases from $-\infty$ to $+\infty$, θ diminishes from 2π to 0 i.e. circle (iii) is described clockwise.

Exercises 6

1. Express the number $(5-i)/(2-3i)$ in the form $a+ib$ where a and b are real. Find its modulus and argument.

2. Find the modulus and argument of (i) $\cos\theta-i\sin\theta$, (ii) $1+i\tan\theta$, where $0<\theta<\frac{1}{2}\pi$ in both cases. [Liverpool.]

3. (*a*) Write down the modulus and argument of each of the complex numbers $1+i$, $-1+i$, $1-i$.

   (*b*) The points A, B, C, D in the Argand diagram correspond to the complex numbers $9+i$, $4+13i$, $-8+8i$ and $-3-4i$. Prove that $ABCD$ is a square. [L.U.]

4. Find the modulus and argument of $(1+i)/(1-i)$, and of $\sqrt{2}/(1-i)$. Show, without using tables, that the argument of $(1+\sqrt{2}+i)/(1-i)$ is $3\pi/8$. [Sheffield.]

5. If $z=(2+i)/(1-i)$, find the real and imaginary parts of $z+1/z$. [L.U.]

6. If the real part of $(z+1)/(z+i)$ is equal to 1, prove that the point z lies on a certain straight line in the Argand plane. [L.U.]

7. The point of affix z in the Argand diagram moves in such a way that $|(2z+1)/(iz+1)|=2$. Show that it describes a straight line. [Leeds.]

8. If the ratio $(z-i)/(z-1)$ is purely imaginary, show that the point z lies on a circle whose centre is $\frac{1}{2}(1+i)$ and radius $1/\sqrt{2}$. [L.U.]

9. If the argument of $(z-1)/(z+1)$ is $\frac{1}{4}\pi$, show that z lies on a fixed circle of radius $\sqrt{2}$ and centre $(0, 1)$. [L.U.]

10. The point P representing $z(=x+iy)$ in the Argand diagram lies on the line $6x+8y=R$ where R is real. Q is the point representing R^2/z. Prove that the locus of Q is a circle and find its centre and radius. [L.U.]

11. P is the point in the Argand diagram representing the complex number z, and Q is the point representing $1/(z-3)+17/3$. Find the locus of Q as P describes the circle $|z-3|=3$. [L.U.]

12. Interpret the relation $|z-a|=|z-b|$ in the Argand diagram, where a, b, and z are complex numbers and $a\neq b$. Show that, when this relation holds, $\arg\{(2z-a-b)/(a-b)\}=\pm\frac{1}{2}\pi$.

13. (i) In the Argand diagram, find the locus of the point z if

$$\arg\{(z-2)/(z+1)\}=\tfrac{1}{2}\pi.$$

(ii) If the point z describes the circle $|z-1|=1$, show that the point z^2 describes the curve $r=2+2\cos\theta$. [L.U. Anc.]

14. (i) If p is real, and the complex number $(1+i)/(2+pi)+(2+3i)/(3+i)$ is represented in the Argand diagram by a point on the line $x=y$, show that $p=-5\pm\sqrt{21}$.

(ii) The complex numbers $z_1=x_1+iy_1$ and $z_2=x_2+iy_2$ are connected by the relation $z_1=z_2+1/z_2$. If the point representing z_2 in the Argand diagram describes a circle of radius a with centre at the origin, show that the point representing z_1 describes the ellipse $x^2/(1+a^2)^2+y^2/(1-a^2)^2=1/a^2$. [L.U.]

15. Show that if z_1, z_2 are complex numbers,

$$|z_1z_2|=|z_1|.|z_2| \text{ and } \arg(z_1z_2)=\arg z_1+\arg z_2.$$

If the complex number $z=r(\cos\theta+i\sin\theta)$ is represented by the point P in an Argand diagram, find the complex number represented by the point P_1 which is the reflexion of P in the y-axis. If P_2 represents the complex number $-4/z$, show that $OP_2.OP_1=4$, where O is the origin, and that OP_1P_2 is a straight line.

By taking $z=x+iy$, and $-4/z=u+iv$, where x, y, u, v are real, or otherwise, show that if P describes the line $x=c(c>0)$, in the Argand diagram, the point P_2 describes a circle whose centre is on the real axis and which passes through the origin. [L.U. Anc.]

16. (i) If z, z_0 are two complex numbers, $\bar{z}_0$ the conjugate of z_0 and $|z|=1$, and if the numbers z, z_0, $z\bar{z}_0$, 1, 0 are represented in an Argand diagram by the points P, P_0, Q, A and the origin O respectively, show that the triangles POP_0, AOQ are congruent, and hence, or otherwise, prove that $|z-z_0|=|z\bar{z}_0-1|$.

(ii) Show that, if the points representing two complex numbers z_1 and z_2 form with the origin an equilateral triangle, $z_1^2-z_1z_2+z_2^2=0$. [L.U. Anc.]

17. Prove that if z and a are complex numbers

$$|z+a|^2+|z-a|^2=2\{|z|^2+|a|^2\}$$

The complex number z is represented by a point on the circle whose centre is at the point $1+0i$, and whose radius is unity. Show in a diagram how $z-2$ may be represented, and prove that

$$(z-2)/z=i\tan(\arg z).$$ [L.U.]

18. If $z\equiv\rho(\cos\theta+i\sin\theta)$ and $a\equiv r(\cos\alpha+i\sin\alpha)$ are two complex numbers so that $|z|=\rho$ and $|a|=r$, find the value of $|z-a|^2$ in terms of the real quantities r, ρ, θ, α. Deduce that if $\bar{a}$ is the conjugate of the number a, $|1-\bar{a}z|^2-|z-a|^2=(1-|z|^2)(1-|a|^2)$. [L.U.]

19. (i) If A, B, P represent the numbers z_1, z_2, z respectively and $z=\lambda z_1+\mu z_2$, where λ and μ are real numbers such that $\lambda+\mu=k$, a constant, prove that the locus of P is a straight line parallel to AB.

(ii) If P represents the number z and $\arg\{z/(z-i)\}=\frac{1}{3}\pi$, prove that P lies on a circle. [L.U.]

20. $ABCD$ is a parallelogram whose diagonals AC and BD intersect at E. The angle AED is 45°, the length of AC is to the length of BD in the ratio 3 : 2, and the sense of the description of $ABCD$ is counter-clockwise. If the points A and C represent the complex numbers $-2-3i$ and $4+i$ respectively in the Argand diagram, determine the numbers which are represented by the points B and D. [L.U.]

21. If the complex numbers z_1, z_2, z_3 are connected by the relation

$$2/z_1=1/z_2+1/z_3,$$

show that the points Z_1, Z_2, Z_3 representing them in an Argand diagram lie on a circle passing through the origin. [L.U.]

CHAPTER 7

COMPLEX NUMBERS (*continued*) DEMOIVRE'S THEOREM; REAL QUADRATIC FACTORS; GENERALISED CIRCULAR AND HYPERBOLIC FUNCTIONS

7.1. Demoivre's theorem

By (6.1), page 108,

$$(\cos\theta_1+i\sin\theta_1)(\cos\theta_2+i\sin\theta_2)=\cos(\theta_1+\theta_2)+i\sin(\theta_1+\theta_2),$$

and by repeated application of this result we obtain

$$(\text{cis }\theta_1)(\text{cis }\theta_2)\ldots(\text{cis }\theta_n)=\text{cis }(\theta_1+\theta_2+\ldots+\theta_n).$$

Putting $\theta_1=\theta_2=\ldots=\theta_n=\theta$, we have

$$(\cos\theta+i\sin\theta)^n=\cos n\theta+i\sin n\theta \quad . \quad . \quad \text{(i)}$$

or

$$\text{cis}^n\,\theta=\text{cis }n\theta.$$

This result is known as Demoivre's theorem for a positive integral index. We shall now investigate the value of $\text{cis}^n\,\theta$ when n is a negative integer and when n is a rational fraction, positive or negative.

When n is a negative integer, we let $n=-m$, where m is a positive integer.

$$\text{Then } (\cos\theta+i\sin\theta)^n=\frac{1}{(\cos\theta+i\sin\theta)^m}$$

$$=\frac{1}{\cos m\theta+i\sin m\theta} \quad \text{by (i)}$$

$$=\cos m\theta-i\sin m\theta$$

$$=\cos(-m\theta)+i\sin(-m\theta)$$

$$=\cos n\theta+i\sin n\theta.$$

Hence (i) is true when n is a negative integer.

Finally, when n is a rational fraction, we put $n=p/q$, where p and q are integers, and no loss of generality results by taking q as positive.

Since up to now no meaning has been assigned to the symbol $z^{p/q}$ when z is not real, we adopt the definition used for real values of z and define $z^{p/q}$ as a number w which satisfies the equation $w^q=z^p$. Then w is a qth root of z^p.

By Demoivre's theorem for a positive index q

$$\left(\cos\frac{p\theta}{q}+i\sin\frac{p\theta}{q}\right)^q=\cos p\theta+i\sin p\theta$$
$$=(\cos\theta+i\sin\theta)^p, \text{ since } p \text{ is an integer.}$$

It follows that $\cos\frac{p\theta}{q}+i\sin\frac{p\theta}{q}$ is a qth root of $(\cos\theta+i\sin\theta)^p$, i.e. $\cos\frac{p\theta}{q}+i\sin\frac{p\theta}{q}$ is a value of $(\cos\theta+i\sin\theta)^{p/q}$.

Hence Demoivre's theorem has been proved for all rational values of n.

To find the other values of $(\cos\theta+i\sin\theta)^{p/q}$ we suppose that

$$(\cos\theta+i\sin\theta)^{p/q}=\rho(\cos\phi+i\sin\phi).$$

Then
$$(\cos\theta+i\sin\theta)^p=\rho^q(\cos\phi+i\sin\phi)^q,$$
i.e.
$$\cos p\theta+i\sin p\theta=\rho^q(\cos q\phi+i\sin q\phi).$$

Equating real and imaginary parts we have

$$\cos p\theta=\rho^q\cos q\phi, \quad \sin p\theta=\rho^q\sin q\phi \quad . \quad . \quad \text{(ii)}$$

Squaring and adding, we obtain $\rho^{2q}=1$; and since ρ, the modulus of a complex number is positive, $\rho=1$.

Equations (ii) then become

$$\cos p\theta=\cos q\phi, \ \sin p\theta=\sin q\phi$$

and these equations are satisfied when

$$q\phi=p\theta+2k\pi,$$
$$\phi=\frac{p\theta+2k\pi}{q} \quad . \quad . \quad . \quad . \quad \text{(iii)}$$

where k is an integer or zero.

When ϕ is given by (iii), cis ϕ is a value of $(\cos\theta+i\sin\theta)^{p/q}$. Now corresponding to the values $k=0, 1, 2, \ldots, (q-1)$, the function cis $\frac{p\theta+2k\pi}{q}$ takes q values which are distinct, and there are no further values, for any other value of k gives a repetition of one of the values already found. Hence

$$(\cos\theta+i\sin\theta)^{p/q}=\text{cis}\,\frac{p\theta+2k\pi}{q}, \ k=0, 1, 2, \ldots, (q-1).$$

7.2. Roots and fractional powers of a complex number

When n is a positive integer, the nth roots of a complex number are by definition the values of w which satisfy the equation

$$w^n=z \quad . \quad . \quad . \quad . \quad . \quad \text{(i)}$$

If $w=\rho(\cos \phi+i \sin \phi)$ and $z=r(\cos \theta+i \sin \theta)$,

$$\rho^n(\cos n\phi+i \sin n\phi)=r(\cos \theta+i \sin \theta),$$

whence $\rho^n=r$ and $n\phi=\theta+2k\pi$, where k is an integer or zero. Now by definition ρ and r are positive so that $\rho=\sqrt[n]{(r)}$, the unique positive nth root of r; also, $\phi=\dfrac{\theta+2k\pi}{n}$, where k is an integer or zero; and taking, in succession, the values $k=0, 1, 2, \ldots, (n-1)$ we find as in § 7.1 that cis $\dfrac{\theta+2k\pi}{n}$ has n, and only n distinct values. Hence there are n distinct nth roots of z given by the formula

$$w_k=\sqrt[n]{(r)} \text{ cis } \frac{\theta+2k\pi}{n}, \quad k=0, 1, 2, \ldots, (n-1) \tag{7.1}$$

In the case where n is a rational number, $n=p/q$, say, where p and q are integers and q is positive, the values of z^n are the values of W which satisfy the equation

$$W^q=z^p$$

Hence if $z=r(\cos \theta+i \sin \theta)$ there are q values of $z^{p/q}$ given by the formula

$$W_m=\sqrt[q]{(r^p)} \text{ cis } \frac{p\theta+2m\pi}{q}, \quad m=0, 1, 2, \ldots, (q-1)$$

where $\sqrt[q]{(r^p)}$ is the unique positive qth root of r^p.

7.3. The nth roots of unity

If in (7.1) we put $r=1$, $\theta=0$, we obtain the n roots of the equation $w^n=1$. They are cis $(2k\pi/n)$, $k=0, 1, 2, \ldots, (n-1)$. If ω denotes the root cis $(2\pi/n)$ the nth roots of unity may be written in the form $1, \omega, \omega^2, \omega^3, \ldots, \omega^{n-1}$, whence we see that they form a geometric progression whose sum $(1-\omega^n)/(1-\omega)$ is zero since $\omega^n=1$. We note that the nth roots of unity are represented in the Argand diagram by points which are vertices of a regular polygon of n sides inscribed in the circle $|z|=1$, one of the vertices being $z=1$.

If we write $z^n=1=\cos 2k\pi \pm i \sin 2k\pi$, where k is zero or any integer, we see that the nth roots of unity are given by

$$z=\cos (2k\pi/n) \pm i \sin (2k\pi/n),$$

where $k=0, 1, 2, \ldots, \frac{1}{2}(n-1)$ when n is odd, and $k=0, 1, 2, \ldots, \frac{1}{2}n$ when n is even. (Note that $k=0$ and $k=\frac{1}{2}n$ give but one root each—a real root.) Hence when n is odd, $z=1$ and $z=\text{cis } (\pm 2k\pi/n)$, $k=1, 2, 3, \ldots, \frac{1}{2}(n-1)$. When n is even, $z=\pm 1$ and $z=\text{cis } (\pm 2k\pi/n)$, $k=1, 2, 3, \ldots, \frac{1}{2}(n-2)$.

From these formulae we see that the roots of unity which are not real occur in conjugate pairs, and, in the same way, the non-real roots of any real number occur in conjugate pairs.

Example 1

Find the fifth roots of -1.

$$-1=\text{cis}\,\pi$$

Hence if $z^5=-1=\text{cis}\,(\pi+2k\pi)$, where k is zero or any integer,

$$z=\text{cis}\,\{(2k+1)\pi/5\},\ k=0,\ 1,\ 2,\ 3,\ 4$$

i.e. $z=\text{cis}\,(\pi/5),\ \text{cis}\,(3\pi/5),\ -1,\ \text{cis}\,(7\pi/5),\ \text{cis}\,(9\pi/5)$

or $z=-1,\ \text{cis}\,(\pm\pi/5),\ \text{cis}\,(\pm 3\pi/5)$.

Example 2

Find the square roots of i.

$$i=\text{cis}\,(\tfrac{1}{2}\pi).$$

Hence if $z^2=i=\text{cis}\,(2k\pi+\frac{1}{2}\pi)$, where k is zero or any integer,

$$z=\text{cis}\,(4k+1)\pi/4,\ k=0,\ 1$$

i.e. $z=\text{cis}\,(\pi/4)=(1+i)/\sqrt{2}$

and $z=\text{cis}\,(5\pi/4)=-(1+i)/\sqrt{2}$.

Example 3

Find the three cube roots of $(1-\cos\phi-i\sin\phi)$ *where* $0<\phi<2\pi$ *and state the argument and modulus of each.*

$$\begin{aligned}1-\cos\phi-i\sin\phi&=2\sin\tfrac{1}{2}\phi(\sin\tfrac{1}{2}\phi-i\cos\tfrac{1}{2}\phi)\\&=2\sin\tfrac{1}{2}\phi\times\text{cis}\,\tfrac{1}{2}\phi\times(-i)\\&=2\sin\tfrac{1}{2}\phi\times\text{cis}\,\tfrac{1}{2}\phi\times\text{cis}\,(-\tfrac{1}{2}\pi)\\&=2\sin\tfrac{1}{2}\phi\times\text{cis}\,(\tfrac{1}{2}\phi-\tfrac{1}{2}\pi).\end{aligned}$$

Thus, if

$$\begin{aligned}z^3&=1-\cos\phi-i\sin\phi\\&=2\sin\tfrac{1}{2}\phi\times\text{cis}\,(\tfrac{1}{2}\phi-\tfrac{1}{2}\pi+2k\pi),\end{aligned}$$

where k is zero or any integer,

$$z=\sqrt[3]{(2\sin\tfrac{1}{2}\phi)}\times\text{cis}\,\tfrac{1}{6}\{(4k-1)\pi+\phi\},$$

where $k=0,\ 1,\ 2$.

Since $\sin\frac{1}{2}\phi>0$, the modulus of each cube root is $\sqrt[3]{(2\sin\frac{1}{2}\phi)}$; the arguments are $\frac{1}{6}(\phi-\pi)$, $\frac{1}{6}(\phi+3\pi)$, $\frac{1}{6}(\phi+7\pi)$.

7.4. Solution of equations

Demoivre's theorem may be used as in § 7.2 to solve an equation of the form $az^n+b=0$. We give examples of the solution of equations of other types.

Example 4

Obtain the roots of the equation $3z^2-(2+11i)z+3-5i=0$ *in the form* $a+ib$, *where* a *and* b *are real.* [Leeds.]

Solving this quadratic equation for z we get

$$z=\tfrac{1}{6}\{2+11i\pm\sqrt{(104i-153)}\},$$

where $\sqrt{(104i-153)}$ denotes either of the two solutions of the equation

$$w^2=104i-153.$$

Now if $\quad w=p+iq$, where p and q are real,

$$w^2=p^2-q^2+2ipq$$

$$\therefore\ p^2-q^2=-153$$

and $\quad pq=52,$

so that $p=\pm 4$, $q=\pm 13$, like signs being taken together.

It follows that $\pm\sqrt{(104i-153)}=\pm(4+13i)$ and so

$$z=1+4i \text{ or } -\tfrac{1}{3}(1+i).$$

Example 5

Solve the equation $z^6+z^5+z^4+z^3+z^2+z+1=0$, *and deduce that*

$$\cos\frac{2\pi}{7}+\cos\frac{4\pi}{7}+\cos\frac{6\pi}{7}=-\tfrac{1}{2}.$$

$$z^6+z^5+z^4+z^3+z^2+z+1=(z^7-1)/(z-1),\ z\neq 1$$

and so we consider the equation $z^7-1=0$, which is satisfied by $z=1$ and by $z=\text{cis}\,(\pm 2k\pi/7)$, where $k=1, 2, 3$.

Hence the roots of the given equation are

$$z=\text{cis}\,(\pm 2k\pi/7),\ k=1, 2, 3.$$

The sum of these roots is $2\left(\cos\frac{2\pi}{7}+\cos\frac{4\pi}{7}+\cos\frac{6\pi}{7}\right)$; but from the given equation the sum of the roots is also -1 (see § 2.2).

Hence $\quad \cos\frac{2\pi}{7}+\cos\frac{4\pi}{7}+\cos\frac{6\pi}{7}=-\tfrac{1}{2}.$

Example 6

Show that every root of the equation $(z+1)^{2n}+(z-1)^{2n}=0$, *where* n *is a positive integer, is purely imaginary.*

If the roots are represented in the Argand diagram by points $P_1, P_2, \ldots, P_{2n}$, *prove that, if* O *is the origin,* $OP_1^2+OP_2^2+\ldots+OP_{2n}^2=2n(2n-1)$. [L.U.]

If $\quad (z+1)^{2n}+(z-1)^{2n}=0 \quad . \quad . \quad . \quad . \quad . \quad$ (i)

$$\left(\frac{z+1}{z-1}\right)^{2n}=-1$$

$$=\text{cis}\,(2k-1)\pi,$$

where k is zero or any integer.

$$\therefore\ \frac{z+1}{z-1}=\text{cis}\,\{(2k-1)\pi/2n\},\ k=1, 2, 3, \ldots, 2n.$$

Writing $2\alpha_k$ for $(2k-1)\pi/2n$ and solving for z we have

$$z=\frac{1+\cos 2\alpha_k+i\sin 2\alpha_k}{\cos 2\alpha_k+i\sin 2\alpha_k-1}$$

$$=\frac{2\cos\alpha_k(\text{cis }\alpha_k)}{2\sin\alpha_k(i\cos\alpha_k-\sin\alpha_k)}$$

$$=-i\cot\alpha_k$$

$$\therefore\ z=-i\cot\{(2k-1)\pi/4n\},\ k=1,2,3,\ldots,2n,$$

and so the roots of (i) are purely imaginary.

If P_k represents the number $-i\cot\alpha_k$ in the Argand diagram,

$$OP_k=|\ -i\cot\alpha_k\ |=\cot\alpha_k$$

and so $\sum_{k=1}^{2n} OP_k{}^2=\cot^2\alpha_1+\cot^2\alpha_2+\ldots+\cot^2\alpha_{2n}$

$$=(\cot\alpha_1+\cot\alpha_2+\ldots+\cot\alpha_{2n})^2$$

$$-2(\cot\alpha_1\cot\alpha_2+\cot\alpha_1\cot\alpha_3+\ldots)\qquad\text{(ii)}$$

Now writing (i) in the form

$$z^{2n}+{}^{2n}C_2\,z^{2n-2}+{}^{2n}C_4\,z^{2n-4}+\ldots+1=0$$

we see by § 2.2 that the sum of the roots is zero, and the sum of the products of the roots in pairs is ${}^{2n}C_2=n(2n-1)$.

Hence $$\sum_{k=1}^{2n}\cot\alpha_k=0$$

and $$-(\cot\alpha_1\cot\alpha_2+\cot\alpha_1\cot\alpha_3+\ldots)=n(2n-1).$$

Substituting these values in (ii) we get

$$\sum_{k=1}^{2n} OP_k{}^2=2n(2n-1).$$

Exercises 7 (*a*)

1. (i) Find the cube roots of $(1+i)$.

(ii) Find all the roots of the equation $x^6-2x^3+2=0$. [Leeds.]

2. (i) Find the modulus and argument of $1+\cos\theta-i\sin\theta$, $-\pi<\theta\leqslant\pi$.

(ii) Find all the cube roots of $2i-2$.

3. Find the roots of the equation $z^2-(3+5i)z+8i-4=0$. [L.U.]

4. Solve the equation $z^2+(4-6i)z=9+15i$. [Leeds.]

5. Write down the solutions of the equation $w^4=16$ and deduce the solutions of the equation $(z+1)^4=16(z-1)^4$. [Liverpool.]

6. Find all solutions of the equations

(i) $z^6+z^4+z^2+1=0$, (ii) $z^4=(z+1)^4$. [Liverpool.]

7. Solve the equation $z^3=i(z-1)^3$, and show that the points in the Argand diagram which represent the roots are collinear.

8. Express the four roots of the equation $(z-2)^4+(z+1)^4=0$ in the form $a_\nu+ib_\nu (\nu=1, 2, 3, 4)$, where a_ν, b_ν are real numbers. [Sheffield.]

9. Write down the five roots of $z^5-1=0$. Show that the roots of the equation $(5+z)^5-(5-z)^5=0$ can be written in the form $5i \tan (r\pi/5)$, where $r=0, \pm 1, \pm 2$. [L.U. Anc.]

10. Find in the form $p+iq$, where p and q are real, all the solutions of the equations

$$\text{(i) } z^2-4iz-4-2i=0, \qquad \text{(ii) } z^6+8i=0.$$

11. Determine the roots of the equation $z^5=1$ and describe their positions in the Argand diagram.

Let ω be the root, other than 1, which lies in the first quadrant. If $u=\omega+\omega^4$ and $v=\omega^2+\omega^3$ prove that

$$u+v=uv=-1 \text{ and } u-v=+\sqrt{5}.$$

Deduce that $\cos 72° = (\sqrt{5}-1)/4$. [Durham.]

12. Prove that, with the exception of one zero root, the roots of the equation $(1+z)^n=(1-z)^n$ are all imaginary. [Sheffield.]

13. Solve completely the equation $z^6+z^3+1=0$, expressing the solutions in terms of trigonometric functions of acute angles. Make a rough sketch exhibiting the position of the solutions in the complex plane. [Sheffield.]

14. Prove that $$\frac{(1+\sin\theta+i\cos\theta)}{(1+\sin\theta-i\cos\theta)}=\sin\theta+i\cos\theta,$$

and hence show that

$$\{1+\sin(\pi/5)+i\cos(\pi/5)\}^5+i\{1+\sin(\pi/5)-i\cos(\pi/5)\}^5=0.$$

15. Find the nth roots of unity and prove that their sum is zero. If ω is a complex fifth root of unity, prove that $\omega+\dfrac{1}{\omega}$ is real and satisfies the equation $x^2+x-1=0$.

Hence show that $\cos(2\pi/5)=\frac{1}{4}(-1+\sqrt{5})$, $\cos(\pi/5)=\frac{1}{4}(1+\sqrt{5})$. [Sheffield.]

16. Indicate on the Argand diagram the positions of the points

$$z=1+\sin\theta\pm i\cos\theta$$

for a given value of the angle θ.

Prove that one of the values of $\left(\dfrac{1+\sin\theta+i\cos\theta}{1+\sin\theta-i\cos\theta}\right)^n$ is equal to $\cos n(\frac{1}{2}\pi-\theta)+i\sin n(\frac{1}{2}\pi-\theta)$.

Obtain all the values of $\left(\dfrac{\sqrt{2}+1+i}{\sqrt{2}+1-i}\right)^{1/4}$ in the form $a+ib$, where a and b are real. [L.U.]

17. Express the complex number $1+i$ in the form $r(\cos\theta + i\sin\theta)$. Hence, or otherwise, prove that, n being any positive integer,

$$(1+i)^n+(1-i)^n=2(2^{n/2}\cos\tfrac{1}{4}n\pi).$$

If $(1+x)^n=p_0+p_1x+p_2x^2+\ldots+p_nx^n$, prove that

$$p_0-p_2+p_4-\ldots=2^{n/2}\cos\tfrac{1}{4}n\pi,$$

and $$p_1-p_3+p_5-\ldots=2^{n/2}\sin\tfrac{1}{4}n\pi.$$ [L.U.]

18. Show that the equation $z^3=1$ has one real root and two other roots which are not real, and that, if one of the non-real roots is denoted by ω, the other is then ω^2. Mark on the Argand diagram the points which represent the three roots, and show that they are the vertices of an equilateral triangle.

Prove that $$1+\omega+\omega^2=0$$

and

$$(a+b+c)(a+\omega b+\omega^2c)(a+\omega^2b+\omega c)=a^3+b^3+c^3-3abc.$$ [L.U.]

19. What conditions have to be satisfied by the complex number z in order that the points representing all integral powers of z should (i) lie on a circle with centre at the origin, (ii) be finite in number? Mark on the diagram the points which represent a number z such that there are only three distinct points in the sequence given by $z, z^2, \ldots$ [L.U.]

20. Solve the equation $(x+1)^8+x^8=0$. [L.U.]

7.5. Complex roots of an equation

If $z=x+iy$ and $f(z)$ is a polynomial in z with real coefficients, $f(z)$ may be expressed in the form $X+iY$ where X and Y are real. Since even powers of (iy) are real and odd powers are purely imaginary, X will contain only even powers of y, while Y will contain only odd powers of y. It follows that if we change the sign of y, X will be unaltered but Y will change sign.

Hence, if $f(x+iy)=X+iY$, $f(x-iy)=X-iY$.

If $x+iy$ is a root of $f(z)=0$, $f(x+iy)=0$, i.e. $X+iY=0$ and so, equating real and imaginary parts, $X=0$ and $Y=0$. Hence $X-iY=0$, i.e. $f(x-iy)=0$ so that $x-iy$ is a root of $f(z)=0$.

Hence, in an equation with real coefficients, roots which are not real occur in conjugate pairs.

7.6. Real quadratic factors

The problem of factorising a given expression is closely related to that of solving an equation; for if $z=z_1$ is a solution of the equation $f(z)=0$, $z-z_1$ is a factor of $f(z)$.

Suppose that the coefficients which occur in the polynomial $f(z)$ are real, and suppose that the complex number $a(\cos\theta+i\sin\theta)$ is a root of the equation $f(z)=0$. Then, by § 7.5, $a(\cos\theta-i\sin\theta)$ is also a root of the equation. Hence $z-a(\cos\theta+i\sin\theta)$ and $z-a(\cos\theta-i\sin\theta)$ are factors of $f(z)$, and multiplied together they give the real quadratic factor $z^2-2az\cos\theta+a^2$.

For example, the factors of $F(z)\equiv z^6+z^5+z^4+z^3+z^2+z+1$ can be deduced from the solution of the equation $F(z)=0$ (see Example **5**, p. 123). They are $z-\cos(2k\pi/7)\pm i\sin(2k\pi/7)$, $k=1, 2, 3$. Grouped in pairs they give three real quadratic factors

$$\{z^2-2z\cos(2\pi/7)+1\}\{z^2-2z\cos(4\pi/7)+1\}\{z^2-2z\cos(6\pi/7)+1\}$$

and we write

$$F(z)=\prod_{k=1}^{3}\{z^2-2z\cos(2k\pi/7)+1\}.$$

The symbol Π for products corresponds to the symbol Σ for sums. Thus we denote the product $(z-z_1)(z-z_2)\dots(z-z_n)$ by $\prod_{k=1}^{n}(z-z_k)$.

Similarly, from the result of Example 1, p. 122, we deduce that

$$z^5+1=(z+1)\prod_{k=0}^{1}\left\{z^2-2z\cos\frac{(2k+1)\pi}{5}+1\right\}.$$

Example 7

Find all the roots of the equation

$$x^{2n}-2a^nx^n\cos n\theta+a^{2n}=0$$

where n is a positive integer, and a is a real constant. Show that

$$x^2-2ax\cos(\theta+2r\pi/n)+a^2,\ r=0, 1, 2, \dots, (n-1)$$

is a factor of $x^{2n}-2a^nx^n\cos n\theta+a^{2n}$, *and deduce, or prove by any other means, that*

$$\cos n\phi-\cos n\theta=2^{n-1}\prod_{r=0}^{n-1}\{\cos\phi-\cos(\theta+2r\pi/n)\}. \qquad \text{[L.U.]}$$

Solving the equation $x^{2n}-2a^nx^n\cos n\theta+a^{2n}=0$ as a quadratic in x^n, we have

$$\begin{aligned} x^n &= a^n\cos n\theta\pm\sqrt{(a^{2n}\cos^2 n\theta-a^{2n})}\\ &= a^n(\cos n\theta\pm i\sin n\theta)\\ &= a^n\{\cos(n\theta+2r\pi)\pm i\sin(n\theta+2r\pi)\},\ \text{where } r \text{ is zero or any integer.}\end{aligned}$$

Hence by Demoivre's theorem,

$$x=a\{\cos(\theta+2r\pi/n)\pm i\sin(\theta+2r\pi/n)\},\ r=0, 1, 2, \dots, (n-1),$$

and, as in § 7.6,

$$x^{2n}-2a^nx^n\cos n\theta+a^{2n}=\prod_{r=0}^{n-1}\{x^2-2ax\cos(\theta+2r\pi/n)+a^2\} \qquad \text{(i)}$$

Hence $x^2-2ax\cos(\theta+2r\pi/n)+a^2$, $r=0, 1, 2, \dots, (n-1)$, is a factor of $x^{2n}-2a^nx^n\cos n\theta+a^2$.

Dividing throughout (i) by x^n and putting $a=1$, we get

$$x^n+x^{-n}-2\cos n\theta=\prod_{r=0}^{n-1}\{x+x^{-1}-2\cos(\theta+2r\pi/n)\} \qquad \text{(ii)}$$

If now we let $x=\cos\phi+i\sin\phi$ so that $x^{-1}=\cos\phi-i\sin\phi$ $x^n=\cos n\phi+i\sin n\phi$ and $x^{-n}=\cos n\phi-i\sin n\phi$, we have from (ii)

$$2(\cos n\phi-\cos n\theta)=\prod_{r=0}^{n-1}\{2\cos\phi-2\cos(\theta+2r\pi/n)\}$$

$$\therefore\ \cos n\phi-\cos n\theta=2^{n-1}\prod_{r=0}^{n-1}\{\cos\phi-\cos(\theta+2r\pi/n)\}.$$

Example 8

Solve the equation $(z+1)^8-z^8=0$, *and prove that*

$$(z+1)^8-z^8=\tfrac{1}{16}(2z+1)\prod_{s=1}^{3}\{4z^2+4z+\operatorname{cosec}^2(s\pi/8)\}.$$

Hence show that

$$16(\cos^{16}\theta-\sin^{16}\theta)=\cos 2\theta\prod_{s=1}^{3}\{\cos^2 2\theta+\cot^2(s\pi/8)\}.$$ [L.U.]

If $$(z+1)^8-z^8=0 \quad . \quad . \quad . \quad . \quad \text{(i)}$$

$$\left(\frac{z+1}{z}\right)^8=\operatorname{cis}2s\pi,\text{ where } s \text{ is zero or any integer,}$$

$$\frac{z+1}{z}=\operatorname{cis}(s\pi/4),\ s=0,1,2,\ldots,7.$$

$$\therefore\ z\{1-\cos(s\pi/4)-i\sin(s\pi/4)\}=-1,$$
$$2iz\sin(s\pi/8)\operatorname{cis}(s\pi/8)=1.$$

Hence ignoring the infinite root given by $s=0$ we have

$$z=-\tfrac{1}{2}i\operatorname{cis}(-s\pi/8)\operatorname{cosec}(s\pi/8),\ s=1,2,\ldots,7.$$
$$=-\tfrac{1}{2}\{1+i\cot(s\pi/8)\}.$$

When $s=4$, $z=-\frac{1}{2}$; the other roots are $z=-\frac{1}{2}\{1\pm i\cot(s\pi/8)\}$ where $s=1, 2, 3$, since $\cot(5\pi/8)=-\cot(3\pi/8)$ and so on.

Hence $$(z+1)^8-z^8=8(z+\tfrac{1}{2})\prod_{s=1}^{3}[z+\tfrac{1}{2}\{1\pm i\cot(s\pi/8)\}]$$

the numerical factor 8 being determined by comparing coefficients of z^7 on the two sides,

$$=4(2z+1)\prod_{s=1}^{3}\{(z+\tfrac{1}{2})^2+\tfrac{1}{4}\cot^2(s\pi/8)\}$$

$$=\tfrac{1}{16}(2z+1)\prod_{s=1}^{3}\{4z^2+4z+\operatorname{cosec}^2(s\pi/8)\}.$$

The substitution $z=-\sin^2\theta$ gives

$$\cos^{16}\theta-\sin^{16}\theta=\tfrac{1}{16}\cos 2\theta\prod_{s=1}^{3}\{(1-2\sin^2\theta)^2+\cot^2(s\pi/8)\}$$

$$\therefore\ 16(\cos^{16}\theta-\sin^{16}\theta)=\cos 2\theta\prod_{s=1}^{3}\{\cos^2 2\theta+\cot^2(s\pi/8)\}.$$

Exercises 7 (*b*)

1. Prove that, when n is a positive integer,

$$x^{2n}-2x^n\cos n\theta+1=\prod_{k=0}^{n-1}\{x^2-2x\cos(\theta+2k\pi/n)+1\}.$$

Deduce the following results:

(i) $\cos n\phi-\cos n\theta=2^{n-1}\prod_{k=0}^{n-1}\{\cos\phi-\cos(\theta+2k\pi/n)\}$,

$$1-\sin^2\tfrac{1}{2}n\phi\operatorname{cosec}^2\tfrac{1}{2}n\theta=\prod_{k=0}^{n-1}\{1-\sin^2\tfrac{1}{2}\phi\operatorname{cosec}^2(\tfrac{1}{2}\theta+k\pi/n)\}.$$

(ii) $\sin^2 n\beta=2^{2n-2}\prod_{k=0}^{n-1}\sin^2(\beta+k\pi/n)$,

$$\prod_{k=0}^{n-1}\sin\{(6k+1)\pi/6n\}=2^{-n}.$$

(iii) $\prod_{k=0}^{n-1}\{1+2\cos(\theta+2k\pi/n)\}$

$=(-1)^{n-1}(1+2\cos n\theta)$ if n is not a multiple of 3,

$=(-1)^n 2(1-\cos n\theta)$ if n is a multiple of 3.

(Put $\phi=2\pi/3$ in (i).)

(iv) $\sin\theta=2^{n-1}\prod_{k=0}^{n-1}\sin\dfrac{\theta+k\pi}{n}$ and $\cos\theta=2^{n-1}\prod_{k=0}^{n-1}\sin\dfrac{2\theta+(2k+1)\pi}{2n}$.

(v) $\sin n\theta=2^{n-1}\prod_{k=0}^{n-1}\sin(\theta+k\pi/n)$.

If r is a positive integer,

$$2^{2r-1}\sin\theta\sin(\theta+\pi/r)\sin(\theta+2\pi/r)\ldots\sin\{\theta+(2r-1)\pi/r\}=(-1)^r(1-\cos 2r\theta)$$

2. If n is a positive integer, prove that

$$x^{2n}-1=(x^2-1)\prod_{k=1}^{n-1}\{x^2-2x\cos(k\pi/n)+1\}$$

and deduce that

(i) $(\sin n\theta)/\sin\theta=2^{n-1}\prod_{k=1}^{n-1}\{\cos\theta-\cos(k\pi/n)\}$,

(ii) $\sqrt{n}=2^{n-1}\prod_{k=1}^{n-1}\sin(k\pi/2n)$,

(iii) $(\sinh n\theta)/\sinh\theta=2^{n-1}\prod_{k=1}^{n-1}\{\cosh\theta-\cos(k\pi/n)\}$.

3. By finding the real quadratic factors of $x^{2n}+x^{2n-1}+\ldots+x+1$, where n is a positive integer, show that

$$2^n\sin\frac{\pi}{2n+1}\sin\frac{2\pi}{2n+1}\ldots\sin\frac{n\pi}{2n+1}=\sqrt{(2n+1)}.$$ [L.U.]

4. Prove that, if n is a positive integer,

$$x^n+x^{-n}=\prod_{r=1}^{n}[x+x^{-1}-2\cos\{(2r-1)\pi/2n\}].$$

Prove that $$2^{2n-1}\prod_{r=1}^{n}\sin^2\{(2r-1)\pi/4n\}=1,$$

and that $$\cos n\theta=\prod_{r=1}^{n}[1-\sin^2\tfrac{1}{2}\theta\operatorname{cosec}^2\{(2r-1)\pi/4n\}].$$ [L.U.]

5. Prove that if n is a positive integer

$$(1+x)^{2n}-(1-x)^{2n}=4nx\prod_{r=1}^{n-1}\{x^2+\tan^2(r\pi/2n)\}.$$

Hence, or otherwise, prove that

$$\prod_{r=1}^{n-1}\cos(r\pi/2n)=2^{1-n}\sqrt{n}.$$ [L.U.]

6. If n is a positive integer, prove that

$$a^{2n}+b^{2n}=\prod_{r=1}^{n}[a^2-2ab\cos\{(2r-1)\pi/2n\}+b^2].$$

Deduce that

(i) $\cos n\theta=2^{n-1}\prod_{r=1}^{n}[\cos\theta-\cos\{(2r-1)\pi/2n\}]$,

(ii) $\cos 2n\theta=2^{2n-1}\prod_{r=1}^{n}[\cos^2\theta-\cos^2\{(2r-1)\pi/4n\}]$,

(iii) $\prod_{r=1}^{n}\sin\{(2r-1)\pi/4n\}=2^{\frac{1}{2}-n}$. [L.U.]

7. Prove that, if a and b are any given complex numbers, the locus of a point z in the Argand diagram such that $\left|\dfrac{z-a}{z-b}\right|$ is constant is in general a circle.

Show that the roots of the equation $(z-1)^5=32(z+1)^5$ are represented in the Argand diagram by points lying on a circle of radius 4/3, and that the values of z are

$$\{-3+4i\sin(2r\pi/5)\}/\{5-4\cos(2r\pi/5)\}\ (r=0, 1, \ldots, 4).$$

Deduce that

$$31\prod_{r=1}^{5}\{-3+4i\sin(2r\pi/5)\}=-33\prod_{r=1}^{5}\{5-4\cos(2r\pi)/5\}.$$ [L.U.]

8. $ABCDEF$ is a regular hexagon inscribed in the circle $|z|=a$ in the Argand diagram, A being the point $(a, 0)$. If P, representing the complex number z, is any point on the circle, write down the complex numbers represented by the six points obtained by drawing lines from the origin equal and parallel to the directed lines

$$\overrightarrow{AP}, \overrightarrow{BP}, \overrightarrow{CP}, \overrightarrow{DP}, \overrightarrow{EP}, \overrightarrow{FP},$$

and prove that their product is z^6-a^6. Hence prove that

$$AP.BP.CP.DP.EP.FP\leqslant 2a^6.$$ [L.U.]

9. Resolve x^6-x^3+1 into real quadratic factors and deduce that

$$\text{(i)}\quad \cos\frac{\pi}{9}+\cos\frac{5\pi}{9}+\cos\frac{7\pi}{9}=0,$$

$$\text{(ii)}\quad \cos\frac{\pi}{9}\cos\frac{5\pi}{9}\cos\frac{7\pi}{9}=\frac{1}{8},$$

$$\text{(iii)}\quad \sin\frac{\pi}{9}\sin\frac{5\pi}{9}\sin\frac{7\pi}{9}=\frac{1}{8}\sqrt{3}.$$ [L.U.]

10. (i) Prove that the points which represent the roots of the equation $(1-z)^n=z^n$ in the Argand diagram are collinear.

(ii) State Demoivre's theorem, and prove it for integral indices, positive or negative.

Express x^9+1 as a product of one linear and four quadratic real factors. [L.U.]

7.7. Expression of powers of cos θ and sin θ in terms of multiple angles

Let $z=\cos\theta+i\sin\theta$; then $z^n=\cos n\theta+i\sin n\theta$

$$\frac{1}{z}=\cos\theta-i\sin\theta \qquad \text{and} \qquad \frac{1}{z^n}=\cos n\theta-i\sin n\theta$$

$$\therefore\ \left.\begin{aligned} z+\frac{1}{z}&=2\cos\theta\\ z-\frac{1}{z}&=2i\sin\theta\end{aligned}\right\}\ (7.2) \qquad \left.\begin{aligned} z^n+\frac{1}{z^n}&=2\cos n\theta\\ z^n-\frac{1}{z^n}&=2i\sin n\theta\end{aligned}\right\}\ (7.3)$$

The relations (7.2) and (7.3) enable us to express powers of cos θ and sin θ in terms of sines and cosines of multiples of θ.

Example 9

Express $\cos^3\theta\sin^4\theta$ as a sum of cosines of multiples of θ.

If $z=\cos\theta+i\sin\theta$

$$\begin{aligned}(2\cos\theta)^3(2i\sin\theta)^4&=\left(z+\frac{1}{z}\right)^3\left(z-\frac{1}{z}\right)^4, \text{ by (7.2),}\\ &=\left(z^2-\frac{1}{z^2}\right)^3\left(z-\frac{1}{z}\right)\\ &=\left(z^6-3z^2+\frac{3}{z^2}-\frac{1}{z^6}\right)\left(z-\frac{1}{z}\right)\\ &=\left(z^7+\frac{1}{z^7}\right)-\left(z^5+\frac{1}{z^5}\right)-3\left(z^3+\frac{1}{z^3}\right)+3\left(z+\frac{1}{z}\right)\\ &=2(\cos 7\theta-\cos 5\theta-3\cos 3\theta+3\cos\theta), \text{ by (7.3).}\end{aligned}$$

Hence $\quad\cos^3\theta\sin^4\theta=\frac{1}{64}(\cos 7\theta-\cos 5\theta-3\cos 3\theta+3\cos\theta).$

Example 10

Express $\sin^5\theta$ as a sum of sines of multiples of θ and hence find all the solutions of the equation $16 \sin^5\theta = \sin 5\theta$.

If $z = \cos\theta + i \sin\theta$

$$(2i \sin\theta)^5 = \left(z - \frac{1}{z}\right)^5$$

$$= \left(z^5 - \frac{1}{z^5}\right) - 5\left(z^3 - \frac{1}{z^3}\right) + 10\left(z - \frac{1}{z}\right)$$

$$= 2i(\sin 5\theta - 5 \sin 3\theta + 10 \sin\theta), \text{ by (7.3).}$$

$$\therefore \sin^5\theta = \tfrac{1}{16}(\sin 5\theta - 5\sin 3\theta + 10 \sin\theta).$$

If $$16 \sin^5\theta = \sin 5\theta,$$

$$2 \sin\theta - \sin 3\theta = 0,$$

i.e. $$2 \sin\theta - (3 \sin\theta - 4 \sin^3\theta) = 0,$$

$$\sin\theta\,(4 \sin^2\theta - 1) = 0.$$

$$\therefore \sin\theta = 0, \pm\tfrac{1}{2}.$$

$$\therefore \theta = k\pi,\ k\pi \pm \tfrac{1}{6}\pi, \text{ where } k \text{ is zero or any integer,}$$

i.e. $$\theta = k\pi,\ (6k \pm 1)\pi/6.$$

7.8. Expansions of circular functions of multiple angles

By Demoivre's theorem, when n is a positive integer,

$$\cos n\theta + i \sin n\theta = (\cos\theta + i \sin\theta)^n.$$

If we expand the right-hand side using the binomial theorem and equate real parts and imaginary parts in the resultant equation, we obtain expressions for $\cos n\theta$ and $\sin n\theta$ in terms of powers of $\cos\theta$ and $\sin\theta$.

Example 11

Prove that $\cos 6\theta = 32 \cos^6\theta - 48 \cos^4\theta + 18 \cos^2\theta - 1$. *By putting* $x = \cos^2\theta$, *or otherwise, show that the roots of the equation* $64x^3 - 96x^2 + 36x - 3 = 0$ *are* $\cos^2(\pi/18)$, $\cos^2(5\pi/18)$, $\cos^2(7\pi/18)$, *and deduce that*

$$\sec^2(\pi/18) + \sec^2(5\pi/18) + \sec^2(7\pi/18) = 12. \qquad \text{[L.U.]}$$

By Demoivre's theorem,

$$\cos 6\theta + i \sin 6\theta = (\cos\theta + i \sin\theta)^6$$

$$= \cos^6\theta + 6i \cos^5\theta \sin\theta - 15 \cos^4\theta \sin^2\theta - 20i \cos^3\theta \sin^3\theta + 15 \cos^2\theta \sin^4\theta + 6i \cos\theta \sin^5\theta - \sin^6\theta.$$

Equating real terms on each side of this equation we have

$$\cos 6\theta = \cos^6\theta - 15 \cos^4\theta \sin^2\theta + 15 \cos^2\theta \sin^4\theta - \sin^6\theta.$$

$$= \cos^6\theta - 15 \cos^4\theta(1 - \cos^2\theta) + 15 \cos^2\theta(1 - \cos^2\theta)^2 - (1 - \cos^2\theta)^3$$

$$= 32 \cos^6\theta - 48 \cos^4\theta + 18 \cos^2\theta - 1 \quad . \quad . \quad . \quad . \quad \text{(i)}$$

If, in the equation

$$64x^3-96x^2+36x-3=0 \quad . \quad . \quad . \quad \text{(ii)}$$

we substitute $x=\cos^2\theta$, we have

$$64\cos^6\theta-96\cos^4\theta+36\cos^2\theta-2=1,$$

i.e.

$$\cos 6\theta=\tfrac{1}{2}, \text{ by (i)}$$

$$6\theta=2k\pi\pm\tfrac{1}{3}\pi$$

where k is zero or any integer,

$$\therefore\ \theta=(6k\pm1)\pi/18 \quad . \quad . \quad . \quad \text{(iii)}$$

For values of θ given by (iii), $x=\cos^2\theta$ is a root of (ii), and since (iii) gives x three distinct values: $\cos^2(\pi/18)$, $\cos^2(5\pi/18)$ and $\cos^2(7\pi/18)$, these are the roots of (ii).

The equation whose roots are the reciprocals of the roots of (ii) is (see § 2.3)

$$3x^3-36x^2+96x-64=0,$$

and by § 2.2 the sum of the roots of this equation is 12

$$\therefore\ \sec^2(\pi/18)+\sec^2(5\pi/18)+\sec^2(7\pi/18)=12.$$

Exercises 7 (*c*)

1. Show that

   (i) $\sin 7\theta=7\sin\theta-56\sin^3\theta+112\sin^5\theta-64\sin^7\theta$,

   (ii) $64\sin^7\theta=35\sin\theta-21\sin 3\theta+7\sin 5\theta-\sin 7\theta$.

2. By writing $2\cos\theta=z+z^{-1}$ and $2i\sin\theta=z-z^{-1}$, where $z=\cos\theta+i\sin\theta$, or otherwise, show that $2^6\sin^5\theta\cos^2\theta=\sin 7\theta-3\sin 5\theta+\sin 3\theta+5\sin\theta$. [L.U.]

3. By writing $z=\cos\theta+i\sin\theta$, $z^{-1}=\cos\theta-i\sin\theta$ express $32i\sin\theta.\cos^4\theta$ in terms of z, and hence prove that

   $$16\cos^4\theta\sin\theta=\sin 5\theta+3\sin 3\theta+2\sin\theta. \quad \text{[L.U. Anc.]}$$

4. Prove that, if $\cos\theta+i\sin\theta=t$, then $2\cos n\theta=t^n+t^{-n}$, $2i\sin n\theta=t^n-t^{-n}$, where n is any integer. Hence, or otherwise, establish the formulae

   $$16\sin^5\theta=\sin 5\theta-5\sin 3\theta+10\sin\theta,$$

   $$32\cos^6\theta=\cos 6\theta+6\cos 4\theta+15\cos 2\theta+10.$$

   Solve completely the equation $\cos 5\theta+5\cos 3\theta+10\cos\theta=\tfrac{1}{2}$, where θ is real. [L.U.]

5. Express $(\sin 6\theta)/(\sin\theta)$ as a polynomial in $\cos\theta$. [Sheffield.]

6. Prove that

   $$\cos 8A=\cos^8 A(1-28\tan^2 A+70\tan^4 A-28\tan^6 A+\tan^8 A)$$

   and that $\tan(\pi/16)\tan(3\pi/16)\tan(5\pi/16)\tan(7\pi/16)=1$.

7. Express $\sin 9\theta/\sin\theta$ as a polynomial in $\cos\theta$ and deduce, or prove otherwise, that

   (i) $\sec^2(\pi/9)+\sec^2(2\pi/9)+\sec^2(4\pi/9)=36$,

   (ii) $\sec(\pi/9)\sec(2\pi/9)\sec(4\pi/9)=8$. [L.U.]

8. Express the left-hand side of the equation

$$\cos 6\phi+6\cos 4\phi-9\cos 3\phi+15\cos 2\phi-27\cos\phi+14=0$$

as a polynomial in $\cos\phi$, and hence, or otherwise, find all angles ϕ between 0° and 360° inclusive satisfying the equation. [L.U.]

9. By considering the real and imaginary parts of $(\cos\theta+i\sin\theta)^n$, where n is a positive integer, obtain formulae expressing $\cos n\theta$ and $\sin n\theta$ in terms of $\cos\theta$ and $\sin\theta$, and deduce that

$$\tan n\theta=\frac{{}^nC_1\cot^{n-1}\theta-{}^nC_3\cot^{n-3}\theta+\dots}{\cot^n\theta-{}^nC_2\cot^{n-2}\theta+\dots}.$$

Prove that the roots of the equation

$$x^n+{}^nC_1x^{n-1}-{}^nC_2x^{n-2}-{}^nC_3x^{n-3}+\,+\,-\,-\dots=0$$

are $\cot(3\pi/4n)$, $\cot(7\pi/4n)$, $\cot(11\pi/4n),\dots,\cot\{(4n-1)\pi/4n\}$. [Sheffield.]

10. By means of Demoivre's theorem, or otherwise, show that

$$\frac{\sin 2k\theta}{\sin\theta\cos\theta},$$

where k is a positive integer, can always be expressed as a polynomial in $\sin^2\theta$. Obtain this polynomial for $k=4$ and hence solve the equation

$$x^6-6x^4+10x^2-4=0.$$ [Durham.]

11. By first solving the equation $\cos 3\theta+\sin 3\theta=0$, or otherwise, show that the roots of the equation $t^2+4t+1=0$ are $t=-\tan(\pi/12)$ and $t=-\tan(5\pi/12)$. [L.U.]

7.9. Series of complex terms

The series Σz_r where $z_r=x_r+iy_r$ and x_r, y_r are real is said to be convergent if the series of real terms Σx_r and Σy_r separately converge.

If, as $n\to\infty$, $\sum_1^n x_r\to x$ and $\sum_1^n y_r\to y$, then we say that $\sum_1^n z_r\to x+iy$, and $x+iy$ is called the sum to infinity of Σz_r.

The series of positive real terms $\Sigma|z_r|$ is known as the *series of moduli*. If $\Sigma|z_r|$ is convergent, Σz_r is convergent; for since x_r and y_r are real, $|x_r|\leqslant\sqrt{(x_r^2+y_r^2)}$, i.e. $|x_r|\leqslant|z_r|$ and so, by comparison test 1 (see § 4.13) if $\Sigma|z_r|$ is convergent, $\Sigma|x_r|$ is convergent and hence Σx_r is absolutely convergent.

Similarly Σy_r is absolutely convergent if $\Sigma|z_r|$ is convergent; and so, by definition Σz_r is convergent if $\Sigma|z_r|$ is convergent.

When the series of moduli $\Sigma|z_r|$ converges, the series Σz_r is said to be *absolutely convergent.*

7.10. The exponential series

Consider the series

$$1+z+\frac{z^2}{2!}+\frac{z^3}{3!}+\frac{z^4}{4!}+\dots \qquad \text{(i)}$$

where

$$z=\rho(\cos\theta+i\sin\theta).$$

The series of moduli is

$$1+\rho+\frac{\rho^2}{2!}+\frac{\rho^3}{3!}+\frac{\rho^4}{4!}+\dots$$

and this series converges for all finite values of ρ. Hence the given series converges absolutely for all values of z.

Now when z is real, the sum of the given series is e^z and so we define e^z when z is complex as the sum of series (i) i.e.

$$e^z=1+z+\frac{z^2}{2!}+\frac{z^3}{3!}+\frac{z^4}{4!}+\dots=\sum_0^\infty\frac{z^r}{r!}. \qquad (7.4)$$

The sum of this series is sometimes denoted by exp z.

7.11. Exponential values of circular functions

When $z=i\theta$, where θ is real, we have from (7.4)

$$e^{i\theta}=1+i\theta-\frac{\theta^2}{2!}-i\frac{\theta^3}{3!}+\frac{\theta^4}{4!}+\dots$$

$$=\left(1-\frac{\theta^2}{2!}+\frac{\theta^4}{4!}-\frac{\theta^6}{6!}+\dots\right)+i\left(\theta-\frac{\theta^3}{3!}+\frac{\theta^5}{5!}-\dots\right)$$

The real and imaginary parts of this series are the Maclaurin expansions of $\cos\theta$ and $\sin\theta$ respectively (See § 11.5)

$$\therefore\ e^{i\theta}=\cos\theta+i\sin\theta \qquad (7.5)$$

and, writing $-\theta$ for θ,

$$e^{-i\theta}=\cos\theta-i\sin\theta. \qquad (7.6)$$

From these results,

$$\cos\theta=\tfrac{1}{2}(e^{i\theta}+e^{-i\theta}) \text{ and } \sin\theta=\frac{1}{2i}(e^{i\theta}-e^{-i\theta}). \qquad (7.7)$$

By means of (7.5) and (7.6) we can express the complex numbers $r(\cos\theta+i\sin\theta)$ and $r(\cos\theta-i\sin\theta)$ in the compact form $re^{i\theta}$, $re^{-i\theta}$ or $r\exp(i\theta)$, $r\exp(-i\theta)$ respectively.

It is beyond the scope of this book to prove that when z is complex the function e^z as defined by (7.4) may be treated in the same way as we could treat it if z were real; but we have shown in § 6.8 that,

$$(r_1e^{i\theta_1})(r_2e^{i\theta_2})=r_1r_2e^{i(\theta_1+\theta_2)}$$

$$(r_1e^{i\theta_1})/(r_2e^{i\theta_2})=(r_1/r_2)e^{i(\theta_1-\theta_2)}$$

and by Demoivre's theorem, $(e^{i\theta})^n = e^{in\theta}$ when n is rational. Also, the result $e^{z_1}e^{z_2} = e^{z_1+z_2}$ can be proved by multiplication of series (see § 5.7) and so we shall assume that e^z when z is complex obeys the index laws. With this assumption

$$e^z = e^{x+iy} = e^x . e^{iy} = e^x(\cos y + i \sin y).$$

Also, since $e^{2i\pi} = 1$, $e^{2ni\pi} = 1$, n being any integer, and so

$$e^{z+2ni\pi} = e^z . e^{2ni\pi} = e^z.$$

Hence e^z is a periodic function, its period being $2\pi i$.

The following examples illustrate the use of the exponential form of a complex number.

Example 12

If the complex numbers z_1 and z_2 are represented in the Argand diagram by the points P and Q respectively, interpret geometrically the modulus and amplitude (argument) of $z_2 - z_1$.

If a third complex number z_3 is represented by the point R, and the angles of the triangle PQR at Q and R are each $\frac{1}{2}(\pi - \alpha)$, prove that

$$(z_3 - z_2)^2 = 4(z_3 - z_1)(z_1 - z_2) \sin^2 \tfrac{1}{2}\alpha. \qquad \text{[L.U.]}$$

Let QP (fig. 18) meet the real axis at S. Then $|z_2 - z_1| = PQ$ and arg $(z_2 - z_1) = \angle xSQ$. Also $PQ = PR$ and $\angle QPR = \alpha$.

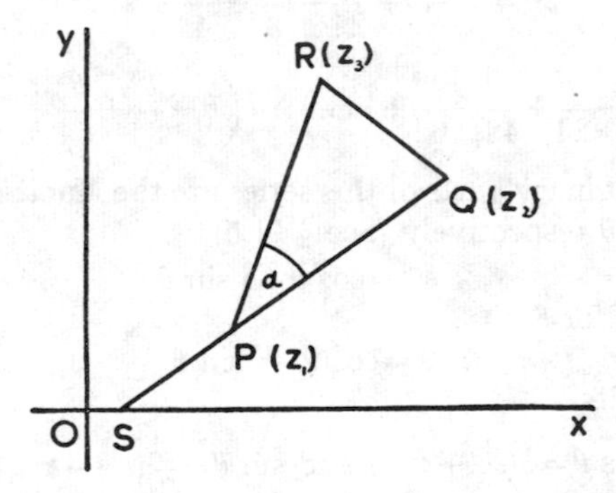

Fig. 18

Let $\overline{PQ}$ represent the complex number $ae^{i\theta}$. Then since PR is the vector $\overline{PQ}$ turned counter-clockwise through an angle α, $\overrightarrow{PR}$ represents the number $(ae^{i\theta}).(e^{i\alpha}) = ae^{i(\theta+\alpha)}$ (see § 6.8).

The vector $\overline{QR}$ is obtained by turning the vector QP clockwise through an angle $(\frac{1}{2}\pi - \frac{1}{2}\alpha)$ and multiplying its length by 2 sin $\frac{1}{2}\alpha$. Hence $\overline{QR}$ represents the number $2 \sin \frac{1}{2}\alpha \, (-ae^{i\theta})\{e^{-i(\frac{1}{2}\pi - \frac{1}{2}\alpha)}\}$, i.e. $2aie^{i(\theta+\frac{1}{2}\alpha)} \sin \frac{1}{2}\alpha$.

But $\overline{PQ}$, $\overline{PR}$, $\overline{QR}$ represent the numbers z_2-z_1, z_3-z_1, z_3-z_2 respectively and so $z_1-z_2=-ae^{i\theta}$, $z_3-z_1=ae^{i(\theta+\alpha)}$, $z_3-z_2=2aie^{i(\theta+\frac{1}{2}\alpha)}\sin\frac{1}{2}\alpha$.

$$\therefore \quad \frac{(z_3-z_2)^2}{(z_3-z_1)(z_1-z_2)}=\frac{4a^2e^{i(2\theta+\alpha)}\sin^2\frac{1}{2}\alpha}{a^2e^{i(2\theta+\alpha)}}$$

$$(z_3-z_2)^2=4(z_3-z_1)(z_1-z_2)\sin^2\tfrac{1}{2}\alpha.$$

Example 13

If a is a complex number, and r and θ are real, show that the point representing z, where r is a constant and $z=a+re^{i\theta}$ lies on a fixed circle, whose centre is a, for all values of θ.

Let T be the length of the tangent to this circle from the point representing Z. If $Z=a+Re^{i\phi}$ where R and ϕ are real and $R>r$, show that

$$\sqrt{(T^2+r^2)}=\text{mod}(Z-a).$$

Explain why the last result is independent of ϕ. [L.U.]

If
$$z=a+re^{i\theta} \quad . \quad . \quad . \quad . \quad . \quad \text{(i)}$$

$$z-a=r(\cos\theta+i\sin\theta)$$

$$\therefore \quad |z-a|=r=\text{constant}.$$

Hence, if P and C represent z and a respectively, PC is constant and equal to r and so P describes a circle with centre C and radius r. Since $\theta=\arg(z-a)$, θ is the angle which the radius CP makes with the positive direction of the real axis (see fig. 19).

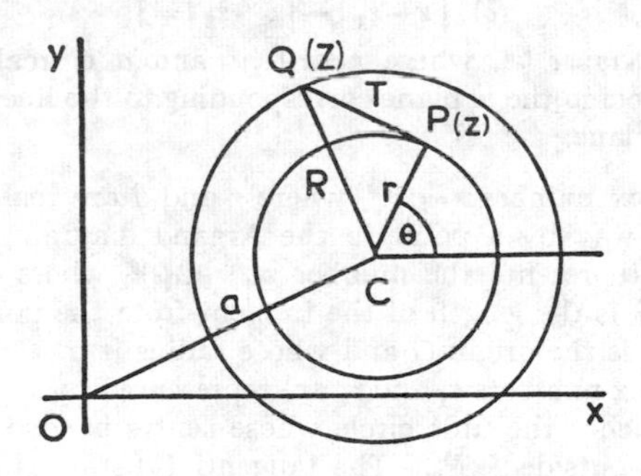

Fig. 19

If
$$Z=a+Re^{i\phi} \quad . \quad . \quad . \quad . \quad . \quad \text{(ii)}$$

the point Q which represents Z lies on a circle with centre C and radius R, and if T is the length of the tangent drawn from Q to circle (i),

$$PQ^2=CQ^2-CP^2$$

$$T^2=|Z-a|^2-r^2$$

$$\therefore \sqrt{(T^2+r^2)}=|Z-a|=R, \text{ by (ii)}.$$

R is constant for all positions of Q and hence $\sqrt{(T^2+r^2)}$ is independent of ϕ.

Exercises 7 (*d*)

1. Find u and v, the real and imaginary parts of
$$u+iv=(z-1)e^{-i\alpha}+e^{i\alpha}/(z-1),$$
where $z=x+iy$ and α is real.

Prove that the locus of the points on the Argand diagram repre senting the complex number z such that $v=0$ is a circle of unit radius with centre at the point (1, 0) and a straight line through the centre of the circle. [L.U.]

2. Show that, if z lies on the circle $|z|=1$ and $2W=z/k+k/z$, where $k=ae^{i\alpha}\neq 0$, then W lies on an ellipse whose foci are the points $W=(\pm 1, 0)$. [Sheffield.]

3. Z and z are two complex numbers represented in the Argand diagram by the points P and Q respectively, and connected by the relation $Zz+Z-z+1=0$. Show that if $z=e^{i\theta}$, then $|Z|=\tan \frac{1}{2}\theta$. Show also that if Q describes the circle of unit radius, centre the origin, then P describes the y-axis. [L.U.]

4. (i) If z_1 and z_2 are two given complex numbers, describe the locus in the Argand diagram represented by
$$(a) \arg\left(\frac{z-z_1}{z-z_2}\right)=\tfrac{1}{8}\pi.$$
$$(b) \; |z-z_1|-|z-z_2|=1.$$
(ii) If $w=u+iv=e^{\pi z/a}$, where $z=x+iy$, and a is real and positive, find the loci in the w plane corresponding to the lines $y=0$ and $y=a$ in the z plane. [L.U.]

5. If the complex number $z=re^{i\theta}$, where r and θ are real, is represented in the usual way by a point in the Argand diagram, show that the point which represents the number $ae^{i\theta}+\lambda ie^{i\theta}$, where a, λ, θ are real, is such that λ is the length of the tangent from the point to the circle whose centre is the origin O and whose radius is a.

The complex numbers z_1 and z_2 are represented by P_1 and P_2 and the line P_1P_2 touches the unit circle whose centre is O so that the point of contact is outside P_1P_2. The tangential distances of these points from the circle are λ_1 and $\lambda_2 (\lambda_2<\lambda_1)$ respectively. If $\lambda_2 z_1-\lambda_1 z_2$ is represented by the point Q, show that the magnitude of OQ is $\lambda_1-\lambda_2$ and that its direction is perpendicular to P_1P_2. [L.U.]

6. If the complex numbers z_1 and z_2 are represented in the Argand diagram by the points P and Q respectively, interpret geometrically the modulus and argument (amplitude) of z_2-z_1.

If a third complex number z_3 is represented by the point R, show that the triangle PQR is similar to a given triangle ABC if
$$(z_3-z_1)/(z_2-z_1)=(b/c)e^{\pm iA},$$
the positive or negative sign being taken according as the similitude is direct or inverse. [L.U.]

7. The complex numbers z_1, z_2, z_3 are represented in the Argand diagram by the points A, B, C in a counter-clockwise order. Prove that a necessary and sufficient condition that the triangle ABC may be equilateral is $z_1-z_2=e^{\frac{1}{3}\pi i}(z_3-z_2)$.

   On the side BC of the above triangle ABC, an equilateral triangle BCA' is drawn externally to the triangle ABC. Find the complex number represented by the point A'. [L.U.]

8. The vertices A_1, A_2, A_3 of an isosceles triangle whose equal sides are A_1A_2 and A_1A_3 represent the complex numbers z_1, z_2, z_3 respectively and the angle $A_2A_1A_3$ is α.

   Show that $(z_1-z_2)^2+(z_1-z_3)^2=2(z_1-z_2)(z_1-z_3)\cos\alpha$. [L.U.]

9. $ABCD$ is a parallelogram. The angle ABC is equal to α, the ratio of the length of BC to the length of AB is $k:1$, and the sense of description is counter-clockwise. If A and B represent two complex numbers z_1 and z_2, determine the complex numbers represented by C and D. [L.U.]

10. (i) $OABC$ is a rectangle in which $OC=kOA$ and the vertices O, A, B, C occur counter-clockwise. If the vector $\overline{OA}$ represents the complex number z, write down the complex numbers represented by $\overline{OC}$, $\overline{OB}$, and $\overline{CA}$.

    (ii) In an Argand diagram L, Z and W are the points representing the complex numbers l, z and w respectively, the triangle ZLW being described counter-clockwise. If $LW=ZW$ and angle $ZLW=\theta$, prove that $2(w-l)\cos\theta=e^{-i\theta}(z-l)$. Hence show that, if Z describes a circle centre the origin, l and $\theta(<\frac{1}{2}\pi)$ remaining constant, the locus of W is a circle with its centre at the point representing $\frac{1}{2}le^{i\theta}/\cos\theta$ and of radius $\frac{1}{2}|z|/\cos\theta$. [L.U.]

11. State briefly how complex numbers may be represented by points in an Argand diagram. Show that, if the point A represents the number $ae^{i\theta}$, any point on the perpendicular through A to OA is defined by the number $ae^{i\theta}+i\lambda e^{i\theta}$, where a, θ, λ are real.

    The vertex A_1 of the regular polygon $OA_1\,A_2\ldots A_{n-1}$ of n sides represents the number $ae_i{}^{\theta}$ in an Argand diagram with origin at O. Prove that the vertex A_{r+1} represents $ae^{i(\theta+r\alpha)}\sin(r+1)\alpha\operatorname{cosec}\alpha$ where $\alpha=\pi/n$.

    (The vertices of the polygon are taken in the anti-clockwise sense.) [L.U.]

12. Express $1+i$ in the form $re^{i\theta}$ and hence prove that $\left(\dfrac{1+i}{\sqrt{2}}\right)^n$, where n is an integer, has eight distinct values, and indicate their positions in the Argand diagram. Show that all but one of these are roots of the equation $1+z+z^2\ldots+z^7=0$, and deduce the roots of the equation $1-z+z^2-\ldots-z^7=0$. [L.U.]

13. Prove that, if n is a positive integer,

$$x^{2n}-2x^n \cos n\theta+1=\prod_{s=0}^{n-1} [x^2-2x \cos (\theta+2s\pi/n)+1].$$

By writing $x=e^{i\phi}$, deduce the result that

$$\cos n\phi-\cos n\theta=2^{n-1} \prod_{s=0}^{n-1} [\cos \phi-\cos (\theta+2s\pi/n)].$$

By writing $\theta=\pi/2n$ and $\phi=0$ in the last result, deduce further that

$$2^{\frac{1}{2}-n}=\prod_{s=0}^{n-1} \sin \{(4s+1)\pi/4n\}.$$ [L.U.]

14. If θ is real and n is a positive integer, prove that

$$(\cos \theta+i \sin \theta)^n(\sin \theta+i \cos \theta)^n=e^{in\pi/2}.$$

Express the three values of $(1+\sqrt{3}i)^{1/3}(\sqrt{3}+i)^{1/3}$ in the form $a+ib$, where a and b are real. (The positive value of $\sqrt{3}$ is to be used.) [L.U.]

15. Defining $e^{ix}=\cos x+i \sin x$, prove that $e^{ix}.e^{iy}=e^{i(x+y)}$, and that, for integral n, $e^{inx}=(\cos x+i \sin x)^n$.

Hence, or otherwise, verify that

$$\cos 7x=64 \cos^7 x-112 \cos^5 x+56 \cos^3 x-7 \cos x,$$

$$\sin 7x/\sin x=64 \cos^6 x-80 \cos^4 x+24 \cos^2 x-1.$$ [Durham.]

16. Show that, for all integral values of r, $x=\cos (2r\pi/5)$ satisfies the equation $16x^5-20x^3+5x-1=0$. [Leeds.]

17. Find all the roots of the equation $\omega^n=a^n$ when a is a given complex number.

By writing $\omega=(z+i)/(z-i)$, $a=e^{2i\theta}$, or otherwise, find the roots of the equation $e^{-ni\theta}(z+i)^n-e^{ni\theta}(z-i)^n=0$, and prove that, if $n\theta$ is not a multiple of π, $\sum_{r=1}^{n} \cot (\theta+r\pi/n)=n \cot n\theta$. [Durham.]

18. Prove that, if $z=e^{i\theta}$ and n is a positive integer,

$$z^{2n}+z^{2n-2}+z^{2n-4}+\ldots+z^{-2n}=\sin (2n+1)\theta/\sin \theta.$$

Deduce that $8 \cos^3 2\theta+4 \cos^2 2\theta-4 \cos 2\theta-1=\sin 7\theta/\sin \theta$.

Prove that $\cos (2\pi/7)$ is one root of the equation $8x^3+4x^2-4x-1=0$, and find the other two roots. [Sheffield.]

19. If $0\leqslant\theta\leqslant\pi$, find the modulus and amplitude of $(1+e^{i\theta})^n$. Prove that

$$\sum_{r=0}^{n} \binom{n}{r} \cos r\theta=2^n \cos^n \tfrac{1}{2}\theta \cos \tfrac{1}{2}n\theta,$$

where $\binom{n}{r}$ denotes the usual binomial coefficient. [Sheffield.]

20. Express in partial fractions the function

$$\frac{1-t^2}{(1-at)(1-bt)},$$

where the non-zero constants a and b are (i) unequal, (ii) equal.

By taking $a=1/b=e^{i\theta}$ and a suitable value of t deduce, or otherwise prove that, if $0<\phi<\frac{1}{2}\pi$,

$$\frac{\cos\phi}{1-\sin\phi\cos\theta}=1+\sum_{n=1}^{\infty}2\tan^n\tfrac{1}{2}\phi\cos n\theta. \qquad \text{[L.U.]}$$

7.12. Generalised circular and hyperbolic functions

The circular functions of any complex number z are defined by the relations

$$\sin z=\frac{1}{2i}(e^{iz}-e^{-iz}) \quad . \quad . \quad . \quad \text{(i)}$$

$$\cos z=\tfrac{1}{2}(e^{iz}+e^{-iz}) \quad . \quad . \quad . \quad \text{(ii)}$$

$$\tan z=\sin z/\cos z,\ \operatorname{cosec} z=1/\sin z,\ \sec z=1/\cos z,$$

$$\cot z=1/\tan z.$$

From (i) and (ii),

$$(\cos z+i\sin z)(\cos z-i\sin z)=e^{iz}\times e^{-iz}$$

$$\therefore \cos^2 z+\sin^2 z=1.$$

Similarly it may be shown that the circular functions as defined above for complex z satisfy all the fundamental identities established for real values of z.

$\sin z$ and $\cos z$ are periodic functions with period 2π; $\tan z$ is periodic with period π.

We define the generalised hyperbolic functions by the relations

$$\sinh z=\tfrac{1}{2}(e^z-e^{-z}) \quad . \quad . \quad . \quad . \quad . \quad . \quad \text{(iii)}$$

$$=z+\frac{z^3}{3!}+\frac{z^5}{5!}+\ldots \text{ for all values of } z;$$

$$\cosh z=\tfrac{1}{2}(e^z+e^{-z}) \quad . \quad . \quad . \quad . \quad . \quad . \quad \text{(iv)}$$

$$=1+\frac{z^2}{2!}+\frac{z^4}{4!}+\ldots \text{ for all values of } z;$$

$$\tanh z=\sinh z/\cosh z,\ \operatorname{cosech} z=1/\sinh z,\ \operatorname{sech} z=1/\cosh z,$$

$$\coth z=1/\tanh z.$$

From (iii) and (iv)

$$(\cosh z+\sinh z)(\cosh z-\sinh z)=e^z\times e^{-z}$$

$$\cosh^2 z-\sinh^2 z=1,$$

and in the same way it may be shown that the hyperbolic functions defined in this way satisfy all the fundamental identities established in Chapter 5 for real values of z.

$\sinh z$ and $\cosh z$ are periodic functions with period $2\pi i$; $\tanh z$ is periodic with period πi.

7.13. Connection between the circular and hyperbolic functions

There is a simple connection between the generalised circular and hyperbolic functions.

$$\sin(iz)=\frac{1}{2i}(e^{-z}-e^z)=i\sinh z \tag{7.8}$$

and

$$\cos(iz)=\tfrac{1}{2}(e^{-z}+e^z)=\cosh z. \tag{7.9}$$

Again,

$$\sinh(iz)=\tfrac{1}{2}(e^{iz}-e^{-iz})=i\sin z \tag{7.10}$$

$$\cosh(iz)=\tfrac{1}{2}(e^{iz}+e^{-iz})=\cos z. \tag{7.11}$$

(7.8) and (7.9) justify Osborn's rule (given in § 5.10) for deducing formulae connecting hyperbolic functions from the corresponding formulae for the circular functions.

Also, from the above results we can express $\sin(x+iy)$ and $\cos(x+iy)$ where x and y are real in the form $a+ib$.

$$\sin(x+iy)=\sin x\cos(iy)+\cos x\sin(iy)=\sin x\cosh y+i\cos x\sinh y,$$

$$\cos(x+iy)=\cos x\cos(iy)-\sin x\sin(iy)=\cos x\cosh y-i\sin x\sinh y.$$

7.14. Logarithms of a complex number

If z is any complex number such that $z=e^w$, then w is defined as a natural logarithm of z and we write $w=\text{Log } z$.

If $z=r(\cos\theta+i\sin\theta)$ and $w=u+iv$, the relation $z=e^w$ becomes

$$r(\cos\theta+i\sin\theta)=e^{u+iv}$$

$$=e^u(\cos v+i\sin v)$$

$$\therefore\ r\cos\theta=e^u\cos v,$$

$$r\sin\theta=e^u\sin v,$$

and so

$$r=e^u, \text{ that is } u=\log r.$$

Now r is real and positive and u is real, so that u is the ordinary real natural logarithm of r which is uniquely defined. On the other hand, $v=\theta+2k\pi$, where k is zero or any integer.

Now

$$\text{Log } z=w=u+iv$$

$$\therefore\ \text{Log } z=\log r+i(\theta+2k\pi) \tag{7.12}$$

i.e.
$$\text{Log } z = \log |z| + i \arg z.$$
Hence there are infinitely many logarithms of a complex number z, each pair differing by a multiple of $2\pi i$.

We define the principal value, $\log z$, of $\text{Log } z$ by the relation
$$\log z = \log r + i\theta',$$
where θ' is the principal value of $\arg z$, i.e. $-\pi < \theta' \leqslant \pi$.

7.15. The logarithmic series

When x is real and $-1 < x \leqslant 1$,
$$\log(1+x) = x - \tfrac{1}{2}x^2 + \tfrac{1}{3}x^3 - \tfrac{1}{4}x^4 + \dots$$
If z is complex, it can be shown that when $|z| < 1$ the series
$$z - \tfrac{1}{2}z^2 + \tfrac{1}{3}z^3 - \tfrac{1}{4}z^4 + \dots$$
converges to $\log(1+z)$, the principal value of $\text{Log}(1+z)$. We therefore write
$$\log(1+z) = z - \tfrac{1}{2}z^2 + \tfrac{1}{3}z^3 - \tfrac{1}{4}z^4 + \dots, \text{ if } |z| < 1 \qquad (7.13)$$
It can also be shown that (7.13) is valid for all values of z for which $|z| = 1$ with the exception of $z = -1$.

7.16. A useful principle

In § 7.5 we showed that if $f(z)$ is a polynomial in z with real coefficients and $f(x+iy) = X + iY$, then $f(x-iy) = X - iY$.

We now assume that we can extend this result to functions which can be represented by convergent series in ascending powers of z with real coefficients (see § 11.2) and we illustrate its use in the following examples.

Example 14

If $x+iy = c \tanh(u+iv)$ where x, y, c, u and v are all real, determine x and y in terms of u, v and c.

Prove that this relationship implies both
$$x^2+y^2+c^2-2cx \coth 2u = 0$$
and
$$x^2+y^2-c^2+2cy \cot 2v = 0. \qquad \text{[L.U.]}$$

If $\quad x+iy = c \tanh(u+iv)$,

then $\quad x-iy = c \tanh(u-iv)$

$$\therefore\ 2x/c = \tanh(u+iv) + \tanh(u-iv)$$
$$= \frac{\sinh(u+iv)\cosh(u-iv) + \cosh(u+iv)\sinh(u-iv)}{\cosh(u+iv)\cosh(u-iv)}$$
$$= \frac{2\sinh 2u}{\cosh 2u + \cosh 2iv}, \text{ see § 5.10,}$$
$$x = \frac{c \sinh 2u}{\cosh 2u + \cos 2v}.$$

Similarly $2iy=\dfrac{2c\sinh 2iv}{\cosh 2u+\cos 2v}$,

$$\therefore\ y=\frac{c\sin 2v}{\cosh 2u+\cos 2v},\ \text{by (7.10).}$$

Now since $c\tanh(u+iv)=x+iy$ and $c\tanh(u-iv)=x-iy$,

$$c\tanh\{(u+iv)+(u-iv)\}=\frac{c\{\tanh(u+iv)+\tanh(u-iv)\}}{1+\tanh(u+iv)\tanh(u-iv)},$$

i.e.
$$c\tanh 2u=\frac{2x}{1+(x^2+y^2)/c^2}$$

$$\tanh 2u=\frac{2cx}{x^2+y^2+c^2},$$

and so
$$x^2+y^2+c^2-2cx\coth 2u=0.$$

Similarly,
$$c\tanh\{(u+iv)-(u-iv)\}=\frac{2iy}{1-(x^2+y^2)/c^2}.$$

Hence
$$\tanh 2iv=i\tan 2v=\frac{-2ciy}{x^2+y^2-c^2}$$

and so
$$x^2+y^2-c^2+2cy\cot 2v=0.$$

Example 15

(i) *Find the general solution of the equation* $\sin z=3i\cos z$.

(ii) *If* $u+iv=\coth(x+iy)$, *show that* $v=-\sin 2y/(\cosh 2x-\cos 2y)$.

Hence show that
$$iv=\frac{1}{1-e^{2(x-iy)}}-\frac{1}{1-e^{2(x+iy)}}$$

and deduce that if $x<0$, v *can be expressed as the infinite series*

$$-2\sum_{r=1}^{\infty}e^{2rx}\sin 2ry.$$ [L.U.]

(i) If
$$\sin z=3i\cos z$$
$$e^{iz}-e^{-iz}=-3(e^{iz}+e^{-iz})$$

i.e.
$$e^{2iz}=-\tfrac{1}{2}=+\tfrac{1}{2}e^{(2k+1)i\pi},\ \text{where } k \text{ is zero or any integer,}$$
$$2iz=\log\tfrac{1}{2}+(2k+1)\pi i$$
$$z=\tfrac{1}{2}\{(2k+1)\pi+i\log 2\}.$$

(ii) If $u+iv=\coth(x+iy)$,

$$u-iv=\coth(x-iy)$$

$$\therefore\ 2iv=\coth(x+iy)-\coth(x-iy)$$

$$=\frac{\sinh(x-iy)\cosh(x+iy)-\cosh(x-iy)\sinh(x+iy)}{\sinh(x+iy)\sinh(x-iy)}$$

$$=\frac{-2\sinh 2iy}{\cosh 2x-\cosh 2iy}$$

$$\therefore\ v=\frac{-\sin 2y}{\cosh 2x-\cos 2y}.$$

$$\text{Now } iv=\frac{-i \sin 2y}{\cosh 2x-\cos 2y}$$

$$=\frac{e^{-2iy}-e^{2iy}}{e^{2x}+e^{-2x}-e^{2iy}-e^{-2iy}}$$

$$=\frac{e^{2(x-iy)}-e^{2(x+iy)}}{1-e^{2(x+iy)}-e^{2(x-iy)}+e^{4x}},$$

multiplying top and bottom by e^{2x}

$$=\frac{1}{1-e^{2(x-iy)}}-\frac{1}{1-e^{2(x+iy}}.$$

Hence

$$iv=\{1-e^{2(x-iy)}\}^{-1}-\{1-e^{2(x+iy)}\}^{-1},$$

and expanding by the binomial theorem we have

$$iv=-\{e^{2x}(e^{2iy}-e^{-2iy})+e^{4x}(e^{4iy}-e^{-4iy})+e^{6x}(e^{6iy}-e^{-6iy})+\ldots\}$$

$$=-2i\{e^{2x}\sin 2y+e^{4x}\sin 4y+e^{6x}\sin 6y+\ldots\},$$

$$v=-2\sum_{r=1}^{\infty} e^{2rx}\sin 2ry.$$

The binomial expansion is valid only if $|e^{2(x-iy)}|$ and $|e^{2(x+iy)}|$ are both less than unity. But $|e^{2(x\pm iy)}|=e^{2x}$ and so x must be negative for the expansion to be valid.

Example 16

If $u+iv=\log\dfrac{(x+iy+a)}{(x+iy-a)}$, *show that*

$$x^2+y^2-2ax\coth u+a^2=0$$

and

$$x^2+y^2+2ay\cot v-a^2=0.$$

Verify that the circles given by u=constant, v=constant intersect at right angles. [L.U.]

Let $x+iy+a=re^{i\theta}$; then $r=\sqrt{\{(x+a)^2+y^2\}}$, $\tan\theta=y/(x+a)$ (i)

Similarly if

$x+iy-a=Re^{i\phi}$, $R=\sqrt{\{(x-a)^2+y^2\}}$, $\tan\phi=y/(x-a)$ (ii)

Then $u+iv=\log(re^{i\theta}/Re^{i\phi})=\log(r/R)+i(\theta-\phi)$ . . (iii)

and $u-iv=\log(r/R)-i(\theta-\phi)$ . . . . . (iv)

$$\therefore\ 2u=\log(r^2/R^2)$$

$$e^{2u}=\frac{(x+a)^2+y^2}{(x-a)^2+y^2}, \text{ by (i) and (ii),}$$

and

$$-\coth u=\frac{1+e^{2u}}{1-e^{2u}}=\frac{x^2+y^2+a^2}{-2ax}$$

$$\therefore\ x^2+y^2-2ax\coth u+a^2=0 \quad . \quad . \quad . \quad \text{(v)}$$

Again, from (iii) and (iv), $v=\theta-\phi$

and
$$\tan v=\frac{\tan\theta-\tan\phi}{1+\tan\theta\tan\phi}$$
$$=\frac{-2ay}{x^2+y^2-a^2}, \text{ by (i) and (ii),}$$
$$\therefore\ x^2+y^2+2ay\cot v-a^2=0 \quad . \quad . \quad . \quad \text{(vi)}$$

When u is constant, $\coth u=\text{constant}=\lambda$ (say) and (v) becomes
$$x^2+y^2-2a\lambda x+a^2=0.$$

Similarly, if v is constant, $\cot v=\text{constant}=\mu$ (say) and (vi) becomes
$$x^2+y^2+2a\mu y-a^2=0.$$

The last two equations represent two systems of coaxal circles such that any two members of opposite systems cut each other orthogonally (see § 12.9 (iv)).

Exercises 7 (*e*)

1. If $z=x+iy=\tanh(u+\frac{1}{4}i\pi)$, where u is real, find x and y in terms of u, and show that for all values of u, the point z lies on the circle $x^2+y^2=1$.

2. If $\sin(u+iv)=x+iy$, where u, v, x, y are all real, find x and y in terms of u and v.

   Show that, if v is constant and u varies, the point whose coordinates are (x, y) describes an ellipse, while, if u is constant and v varies, the point (x, y) describes a hyperbola.

   By considering the intersections of the graphs of $\tan u$ and $\tanh u$, show also that $x=y$ when $u=v$ once in every interval $\{\frac{1}{2}(2n-1)\pi, \frac{1}{2}(2n+1)\pi\}$ of values of u, where n is an integer.

3. If u, v, x, y are real numbers such that $u+iv=e^{x+iy}$, prove that $u^2+v^2=e^{2x}$ and $v/u=\tan y$.

   Draw a sketch of the path of the point (u, v) when the point (x, y) describes the rectangle formed by the axes of reference and the lines $x=1$, $y=\frac{1}{2}\pi$.

4. If $x+iy=c\cosh(u+iv)$, where u, v, x, y and c are all real, prove that $x^2\sinh^2 u+y^2\cosh^2 u=c^2\sinh^2 u\cosh^2 u$.

5. If $\tan^{-1}(x+iy)=u+iv$, where x, y, u, v are all real, show that $u=\frac{1}{2}\tan^{-1}\left\{\frac{2x}{1-x^2-y^2}\right\}$ and $v=\frac{1}{2}\tanh^{-1}\left\{\frac{2y}{1+x^2+y^2}\right\}$. Show that, if the real part of $\tan^{-1}(x+iy)=\frac{1}{8}\pi$, the representative point of the complex number $(x+iy)$ on an Argand diagram lies on the circle of radius $\sqrt{2}$, with its centre at $-1+0i$. [L.U.]

   (Note: if $w=\tan^{-1} z$, $z=\tan w$).

6. Define the functions sinh x, cosh x and show that $\sin ix = i \sinh x$, $\cos ix = \cosh x$. If $x+iy = a \tan(u+iv)$, show that

$$\frac{x}{\sin 2u} = \frac{y}{\sinh 2v} = \frac{a}{\cos 2u + \cosh 2v}.$$

If x and y are the rectangular cartesian coordinates of a point P in a plane, show that, for any given value of v, P lies on the circle

$$x^2+y^2-2ay \coth 2v + a^2 = 0.$$ [L.U.]

7. (i) Starting from the exponential values of $\sin z$ and $\cos z$, prove that $\sin iz = i \sinh z$, $\cos iz = \cosh z$ and use these relations to find the hyperbolic identities corresponding to $\sin 3z = 3 \sin z - 4 \sin^3 z$ and $\cos 2z = (1-\tan^2 z)/(1+\tan^2 z)$.

(ii) If $$\frac{\sinh x}{a} = \frac{\sinh y}{b} = \frac{\sinh(x+y)}{c},$$

prove that $\cosh x = (b^2+c^2-a^2)/2bc$. Hence show that, if $a = 8\sqrt{2}$, $b = 3$, $c = 17$, the value of x is $\pm \log 3$ and find the corresponding values of y. [L.U.]

8. (i) If $\tanh x = 0{\cdot}5$, find the value of $\sinh 3x$.

(ii) Express as complex numbers $\sin^{-1} 3$ and $\tan^{-1}(1+i)$. [L.U.]
($\sin^{-1} 3$ is defined to be any value of z which satisfies the equation $\sin z = 3$).

9. (i) Define $\sin z$ and $\cos z$, where z is complex, and show that $\sin iz = i \sinh z$, $\cos iz = \cosh z$.
Show that the general value of $\cos^{-1} 2$ is $2n\pi \pm i \log(2+\sqrt{3})$, where n is any integer.

(ii) Prove that, if $\tan z = \cos \alpha + i \sin \alpha$ where α is real and acute, then $z = (n+\frac{1}{4})\pi + \frac{1}{2} i \log \tan(\frac{1}{4}\pi + \frac{1}{2}\alpha)$, where n is any integer. [L.U.]

10. Determine the general values of the complex number z for which (i) e^z and (ii) $\cos z$ have real values.
If $w = z^2$, sketch in an Argand diagram the path traced out by the point w as the point z describes the rectangle whose vertices are at the points $\pm a$, $\pm a + ia$, where a is real. [L.U.]

11. Express the modulus and argument (amplitude) of e^z, where z is a complex number, in terms of the modulus and argument of z.
Find the polar equation of the curve in the Argand diagram described by the point z when it varies so that ze^z is real.
Sketch that part of the curve which is such that $\arg(ze^z) = 0$ and $-\pi < \arg z < \pi$, indicating the asymptotes. [L.U.]

12. Show that all the points in the Argand diagram which represent the values of $\log(1+i)$, lie on a straight line parallel to the imaginary axis. What is the distance between consecutive points? [L.U.]

13. (i) Given that θ is real, find the real and imaginary parts of

$$\frac{(1+i)^4(\sqrt{3}+i)}{1+\sqrt{3}i}, \quad \frac{\cosh(\theta+i\pi/3)}{(1+\sqrt{3}i)^{\frac{1}{2}}}.$$

(ii) Locate on the Argand diagram all the points which satisfy the relations (*a*) $z^2+2=2z$, (*b*) $\left|\frac{z-1}{z+1}\right| \leqslant 1$. [Sheffield.]

14. Find the cartesian equation of the curve in the Argand diagram described by the point $z \equiv x+iy$ when it varies in such a way that $\sinh z - z$ is real. Sketch roughly the shape of that part of the curve between the lines $y = \pm\pi$ and show that it asymptotically approaches these lines. [L.U.]

15. If $\sin^{-1}(\cos x + i \sin x) = u + iv$, where x, u and v are real and $i=\sqrt{-1}$, and $0<x<\frac{1}{2}\pi$ whilst v is positive, prove that $u=\sin^{-1}\sqrt{(1-\sin x)}$ and $v=\log\{\sqrt{(\sin x)}+\sqrt{(1+\sin x)}\}$. [L.U.]

16. Find the real and imaginary parts of $\cos z$, where $z=x+iy$. Solve completely the equation $\cos z=(\frac{1}{4}\sqrt{2})\{e(1-i)+e^{-1}(1+i)\}$. [L.U.]

17. (*a*) If z_1 and z_2 are two given complex numbers such that $|z_1-z_2| \leqslant \frac{1}{2}|z_2|$, prove that (i) $|z_1| \geqslant \frac{1}{2}|z_2|$; (ii) $|z_1+z_2| \geqslant \frac{3}{2}|z_2|$.

(*b*) If $z=x+iy$ and $y>0$, prove that $\tanh y \leqslant |\tan z| \leqslant \coth y$. [L.U.]

18. If $\tan(x+iy)=\sin(p+iq)$, where x, y, p, q are real, prove that

$$\tan p \sinh 2y = \tanh q \sin 2x. \qquad \text{[L.U.]}$$

19. Write down the expressions for $\sin x$ and $\cos x$ in terms of e^{ix} and e^{-ix}, and for e^{ix} in terms of $\sin x$ and $\cos x$.

Show that, if $\tan\theta=m$ and $\tan\theta'=m'$, then

$$\frac{m-i}{m'-i} \div \frac{m+i}{m'+i} = e^{2i(\theta-\theta')}.$$

Deduce that, if $\theta-\theta'=\frac{1}{2}\pi$, then $mm'=-1$. [L.U.]

20. Find the coefficient of x^n in the expansion of $\log\{1-(a+b)x+abx^2\}$, and assuming this expansion is valid for complex values of a and b, find the coefficient of x^n in the real part of $\log(1+xe^{i\theta})$ where x and θ are real. [L.U.]

CHAPTER 8

SUMMATION OF SERIES

8.1. Standard results

We assume that the student is familiar with the following elementary results :

(i) For the arithmetical progression (A.P.)

$$a,\ (a+d),\ (a+2d),\ldots,\ (a+\overline{n-1}d),$$

S_n, the sum to n terms, is given by

$$S_n=\tfrac{1}{2}n\{2a+(n-1)d\},\text{ or }\tfrac{1}{2}n(a+l),$$

where l is the nth term.

(ii) For the geometrical progression (G.P.)

$$a,\ ax,\ ax^2,\ldots,\ ax^{n-1},\text{ where } x\neq 1,$$

the sum $S_n=\dfrac{a(1-x^n)}{1-x}$.

When $|x|<1$, $S_\infty=\dfrac{a}{1-x}$.

(iii)

$$\sum_{r=1}^{n} r=1+2+3+\ldots+n=\tfrac{1}{2}n(n+1),$$

$$\sum_{r=1}^{n} r^2=1^2+2^2+3^2+\ldots+n^2=\tfrac{1}{6}n(n+1)(2n+1),$$

$$\sum_{r=1}^{n} r^3=1^3+2^3+3^3+\ldots+n^3=\{\tfrac{1}{2}n(n+1)\}^2.$$

(iv) The arithmetico-geometrical series is of the form

$$a,\ (a+d)x,\ (a+2d)x^2,\ (a+3d)x^3,\ldots$$

The nth term is $\{a+(n-1)d\}x^{n-1}$, and the sum to n terms, S_n, is found as follows :

$$S_n=a+(a+d)x+(a+2d)x^2+\ldots+\{a+(n-1)d\}x^{n-1},$$

$$xS_n=\quad ax+\ (a+d)x^2+\ldots+\{a+(n-2)d\}x^{n-1}+\{a+(n-1)d\}x^n.$$

Subtracting, we obtain

$$(1-x)S_n=a+\{dx+dx^2+dx^3+\ldots+dx^{n-1}\}-\{a+(n-1)d\}x^n.$$

The terms within the square brackets form a G.P. of $(n-1)$ terms and common ratio x. Summing these, we have

$$(1-x)S_n=a+\frac{dx(1-x^{n-1})}{1-x}-\{a+(n-1)d\}x^n, \quad x\neq 1,$$

i.e.
$$S_n=\frac{a-x^n\{a+(n-1)d\}}{1-x}+\frac{dx(1-x^{n-1})}{(1-x)^2}.$$

8.2. Methods of summing series

The types of series whose summation is considered here may be classified as follows:

(i) series summed by using the results of § 8.1 (iii),
(ii) series summed by the method of induction,
(iii) series summed by using the exponential, binomial or logarithmic series,
(iv) series summed by the method of differences,
(v) series summed by using complex numbers.

8.3. Series whose rth term is a polynomial in r

Example 1

Sum to n terms the series $2.3+3.4+4.5+\ldots$

In this case $u_r=(r+1)(r+2)$.

Hence
$$\sum_1^n u_r=\sum_1^n(r^2+3r+2)$$
$$=\sum_1^n r^2+3\sum_1^n r+2n$$
$$=\tfrac{1}{6}n(n+1)(2n+1)+\tfrac{3}{2}n(n+1)+2n, \text{ by § 8.1 (iii)}$$
$$=\tfrac{1}{3}n(n^2+6n+11).$$

Example 2

Sum to n terms the series $1+(1+2)+(1+2+3)+(1+2+3+4)+\ldots.$

In this case $u_r=(1+2+3+\ldots+r)$
$$=\tfrac{1}{2}r(r+1), \text{ by § 8.1 (iii)}.$$

Thus
$$\sum_1^n u_r=\tfrac{1}{2}\sum_1^n r^2+\tfrac{1}{2}\sum_1^n r$$
$$=\tfrac{1}{12}n(n+1)(2n+1)+\tfrac{1}{4}n(n+1), \text{ by § 8.1 (iii)}$$
$$=\tfrac{1}{6}n(n+1)(n+2).$$

If u_r is a polynomial in r of degree higher than the third, this method of solution is not applicable without first calculating sums of higher powers of r. In some cases, however, it is possible to sum such a series by the method of differences (see § 8.6) or, when the result to be proved is given, by the method of induction.

8.4. The method of induction

This method is best illustrated by an example.

Example 3

Prove that the sum of n terms of the series $1.3.2^2+2.4\ 3^2+3.5.4^2+\ldots$ *is* $\frac{1}{10}n(n+1)(n+2)(n+3)(2n+3)$.

First we assume that the result is true when $n=p$ and prove that on this assumption it is true when $n=p+1$.

Suppose, then, that

$$\sum_1^p u_r=\tfrac{1}{10}p(p+1)(p+2)(p+3)(2p+3) \quad . \quad . \quad . \qquad \text{(i)}$$

Since $\quad u_r=r(r+2)(r+1)^2$

$$u_{p+1}=(p+1)(p+3)(p+2)^2$$

$$\therefore \sum_1^{p+1} u_r=\sum_1^p u_r+u_{p+1}$$

$$=\tfrac{1}{10}p(p+1)(p+2)(p+3)(2p+3)+(p+1)(p+3)(p+2)^2 \quad \text{by (i)}$$

$$=\tfrac{1}{10}(p+1)(p+2)(p+3)\{(2p^2+3p)+10(p+2)\}$$

$$=\tfrac{1}{10}(p+1)(p+2)(p+3)(p+4)(2p+5).$$

But this is the result we should obtain by substituting $n=(p+1)$ in the given result. Hence if the given formula is true when $n=p$, it is true when $n=p+1$. But the formula is true when $n=1$ since $1.3.2^2=\frac{1}{10}(1.2.3.4.5)$; hence it is true for $n=2$. Since it is true for $n=2$, it is true for $n=3$, and proceeding in this way we may show that the given result is true for all positive integral values of n.

Exercises 8 (*a*)

Sum to n terms the following series:

1. $2.5+3.7+4.9+\ldots.$
2. $1^2+4^2+7^2+\ldots.$
3. $1.2.3+3.4.5+5.6.7+\ldots.$
4. $1.3.4+2.5.7+3.7.10+\ldots.$
5. $1^2.3+2^2.5+3^2.7+\ldots.$
6. $1+2x+3x^2+4x^3+\ldots.$
7. $1+4x+7x^2+10x^3+\ldots.$

Use mathematical induction to establish the following results:

8. $1^2+2^2+3^2+\ldots+n^2=\frac{1}{6}n(n+1)(2n+1)$.
9. $1.2.4+2.3.5+\ldots+n(n+1)(n+3)=\frac{1}{12}n(n+1)(n+2)(3n+13)$.

8.5. Series reducible to exponential, binomial or logarithmic series

Example 4

Sum to infinity the series whose rth term is $(3r^2+4r+1)/r\,!$.

The presence of $r\,!$ in the denominator suggests an exponential series. We proceed as follows :

$$u_r=\frac{3r}{(r-1)\,!}+\frac{4}{(r-1)\,!}+\frac{1}{r\,!}$$

$$=\frac{3\{(r-1)+1\}}{(r-1)\,!}+\frac{4}{(r-1)\,!}+\frac{1}{r\,!}$$

$$=\frac{3}{(r-2)\,!}+\frac{7}{(r-1)\,!}+\frac{1}{r\,!},\; r\geqslant 2.$$

By inspection, $u_1=\frac{7}{0!}+\frac{1}{1!}$, $0\,!$ being defined as 1.

$$\text{Hence}\qquad \sum_{r=1}^{\infty}u_r=3\sum_{r=2}^{\infty}\frac{1}{(r-2)\,!}+7\sum_{r=1}^{\infty}\frac{1}{(r-1)\,!}+\sum_{r=1}^{\infty}\frac{1}{r\,!}\quad . \quad . \quad \text{(i)}$$

provided that all the series on the R.H.S. are convergent. But by § 5.7, the first and second of these series converge to the value e; the third converges to $e-1$. Hence by (i) the sum of the given series is $11e-1$.

Example 5

Sum to infinity the series

$$\frac{1}{9}+\frac{1.3}{9.12}+\frac{1.3.5}{9.12.15}+\frac{1.3.5.7}{9.12.15.18}+\ldots$$

Here we have a set of factors in A.P. in each numerator, and the same number of factors in A.P. in each denominator. This suggests a binomial series. Suppose its sum is S.

The general term in the expansion of $(1+x)^n$ is

$$\frac{n(n-1)(n-2)\ldots(n-r+1)}{1.2.3\ldots r}x^r$$

and so we divide each of the factors in the numerators of the given series by 2, the common difference of the A.P. which they form. In the same way we divide each of the factors of the denominators by 3. Then

$$S=\frac{\frac{1}{2}}{3}\left(\frac{2}{3}\right)+\frac{(\frac{1}{2})(\frac{3}{2})}{3.4}\left(\frac{2}{3}\right)^2+\frac{(\frac{1}{2})(\frac{3}{2})(\frac{5}{2})}{3.4.5}\left(\frac{2}{3}\right)^3+\ldots.$$

The next step is to introduce factorials in the denominators of each term and to change the sign of each factor in the numerators so that they form a decreasing A.P.

$$S=2\left\{-\frac{(-\frac{1}{2})}{3\,!}\left(\frac{2}{3}\right)+\frac{(-\frac{1}{2})(-\frac{3}{2})}{4\,!}\left(\frac{2}{3}\right)^2-\frac{(-\frac{1}{2})(-\frac{3}{2})(-\frac{5}{2})}{5\,!}\left(\frac{2}{3}\right)^3+\ldots\right\}$$

In a binomial series, the number of factors in the numerator of each coefficient is equal to the number of factors in the denominator. Hence we introduce two factors $\frac{3}{2}$ and $\frac{1}{2}$ in each numerator

$$S=\frac{2}{(\frac{3}{2})(\frac{1}{2})}\left\{-\frac{(\frac{3}{2})(\frac{1}{2})(-\frac{1}{2})}{3!}\left(\frac{2}{3}\right)+\frac{(\frac{3}{2})(\frac{1}{2})(-\frac{1}{2})(-\frac{3}{2})}{4!}\left(\frac{2}{3}\right)^2-\ldots\right\}$$

$$=\frac{8}{3}\left(\frac{3}{2}\right)^2\left\{-\frac{(\frac{3}{2})(\frac{1}{2})(-\frac{1}{2})}{3!}\left(\frac{2}{3}\right)^3+\frac{(\frac{3}{2})(\frac{1}{2})(-\frac{1}{2})(-\frac{3}{2})}{4!}\left(\frac{2}{3}\right)^4-\ldots\right\}.$$

The series in the large brackets is the expansion of $(1-\frac{2}{3})^{3/2}$ with the first three terms missing; and so

$$S=6\left[\left(1-\frac{2}{3}\right)^{3/2}-\left\{1-\frac{3}{2}\left(\frac{2}{3}\right)+\frac{(\frac{3}{2})(\frac{1}{2})}{2!}\left(\frac{2}{3}\right)^2\right\}\right]$$

$$=\frac{2}{3}\sqrt{3}-1.$$

Example 6

Find the sum to infinity of the series whose nth term is

$$(-1)^{n-1}\frac{2.5.8\ldots(3n-1)}{6.9.12\ldots(3n+3)}\frac{1}{3^{3n+3}}.$$ [L.U.]

The series is

$$\frac{2}{6}\left(\frac{1}{3}\right)^6-\frac{2.5}{6.9}\left(\frac{1}{3}\right)^9+\frac{2.5.8}{6.9.12}\left(\frac{1}{3}\right)^{12}-\ldots.$$

Proceeding as in Example 5 we have

$$S=-\left\{\frac{(-\frac{2}{3})}{2!}\left(\frac{1}{3^3}\right)^2+\frac{(-\frac{2}{3})(-\frac{5}{3})}{3!}\left(\frac{1}{3^3}\right)^3+\frac{(-\frac{2}{3})(-\frac{5}{3})(-\frac{8}{3})}{4!}\left(\frac{1}{3^3}\right)^4+\ldots\right\}$$

$$=-3\left\{\frac{(\frac{1}{3})(-\frac{2}{3})}{2!}\left(\frac{1}{27}\right)^2+\frac{(\frac{1}{3})(-\frac{2}{3})(-\frac{5}{3})}{3!}\left(\frac{1}{27}\right)^3-\ldots\right\}$$

$$=-3\left\{\left(1+\frac{1}{27}\right)^{1/3}-\left(1+\frac{1}{81}\right)\right\}$$

$$=82/27-28^{1/3}.$$

Example 7

Show that $$\sum_{r=1}^{\infty}\frac{r+2}{r(r+1)}\left(\frac{1}{3}\right)^r=1+\log\frac{2}{3}.$$ [L.U.]

By partial fractions, $u_r=\left\{\frac{2}{r}-\frac{1}{r+1}\right\}\left(\frac{1}{3}\right)^r.$

We therefore consider the two series

$$\sum_{r=1}^{\infty}\frac{1}{r}\left(\frac{1}{3}\right)^r \text{ and } \sum_{r=1}^{\infty}\frac{1}{r+1}\left(\frac{1}{3}\right)^r.$$

The first by (5.10), page 91, converges to $-\log(1-\frac{1}{3})$, i.e. to $-\log\frac{2}{3}$; the second is $\frac{1}{2}(\frac{1}{3})+\frac{1}{3}(\frac{1}{3})^2+\frac{1}{4}(\frac{1}{3})^3+\ldots$

$$=3\{\tfrac{1}{2}(\tfrac{1}{3})^2+\tfrac{1}{3}(\tfrac{1}{3})^3+\tfrac{1}{4}(\tfrac{1}{3})^4+\ldots\}$$
$$=3\{-\log\tfrac{2}{3}-\tfrac{1}{3}\}\text{ by (5.10).}$$

Hence the sum of the given series is $1+\log\frac{2}{3}$.

Exercises 8 (*b*)

Sum to infinity the following series:

1. $1+\frac{1}{10}+\frac{1.4}{10.20}+\frac{1.4.7}{10.20.30}+\ldots.$

2. $1-\frac{3}{4}+\frac{3.5}{4.8}-\frac{3.5.7}{4.8.12}+\ldots.$

3. $1+\frac{2}{1!}+\frac{3}{2!}+\frac{4}{3!}+\ldots.$

4. $1+\frac{2^2}{1!}+\frac{3^2}{2!}+\frac{4^2}{3!}+\ldots.$

5. $\frac{1}{1.3}+\frac{1}{2.3^2}+\frac{1}{3.3^3}+\ldots.$

6. $\frac{1}{1.2}+\frac{1}{3.2^3}+\frac{1}{5.2^5}+\ldots.$

7. $\frac{1}{1!}+\frac{1+2}{2!}+\frac{1+2+3}{3!}+\frac{1+2+3+4}{4!}+\ldots.$

8. $1+\frac{3}{8}+\frac{3.9}{8.16}+\frac{3.9.15}{8.16.24}+\ldots.$

9. $\frac{1}{2.3}+\frac{1}{4.5}+\frac{1}{6.7}+\ldots.$

10. $1-\frac{1}{5}+\frac{1.4}{5.10}-\frac{1.4.7}{5.10.15}+\ldots.$

11. $\frac{3}{1.2}\left(\frac{1}{2}\right)+\frac{4}{2.3}\left(\frac{1}{2}\right)^2+\frac{5}{3.4}\left(\frac{1}{2}\right)^3+\ldots.$

12. $\frac{2}{3!}+\frac{4}{5!}+\frac{6}{7!}+\ldots.$

8.6. The method of differences

Consider the series $\sum_1^n u_r$.

If u_r can be expressed in the form $v_{r+1}-v_r$, where v_r is a known function of r, then

$$\sum_1^n u_r=(v_2-v_1)+(v_3-v_2)+(v_4-v_3)+\ldots+(v_n-v_{n-1})+(v_{n+1}-v_n)$$
$$=v_{n+1}-v_1.$$

The method of differences is particularly useful for summing a series each of whose terms is the reciprocal of a product of a constant number of factors in A.P., the first factors of successive terms being in the same A.P.

Example 8

Sum to n terms the series

$$\frac{1}{1.3.5}+\frac{1}{3.5.7}+\frac{1}{5.7.9}+\dots,$$

and deduce the sum to infinity.

Here $u_r=\dfrac{1}{(2r-1)(2r+1)(2r+3)}=\dfrac{1}{4}\left\{\dfrac{1}{(2r-1)(2r+1)}-\dfrac{1}{(2r+1)(2r+3)}\right\}.$

Summing as above, we obtain

$$S_n=\frac{1}{4}\left\{\frac{1}{3}-\frac{1}{(2n+1)(2n+3)}\right\}$$

and $$S_\infty=\frac{1}{12}.$$

Example 9

Sum to n terms the series whose rth term is $1/r(r+1)(r+3)$ *and deduce the sum to infinity.*

We write

$$u_r=\frac{r+2}{r(r+1)(r+2)(r+3)}=\frac{1}{(r+1)(r+2)(r+3)}+\frac{2}{r(r+1)(r+2)(r+3)}$$

$$=\frac{1}{2}\left\{\frac{1}{(r+1)(r+2)}-\frac{1}{(r+2)(r+3)}\right\}+\frac{2}{3}\left\{\frac{1}{r(r+1)(r+2)}-\frac{1}{(r+1)(r+2)(r+3)}\right\}$$

and summing we obtain

$$S_n=\frac{1}{2}\left\{\frac{1}{6}-\frac{1}{(n+2)(n+3)}\right\}+\frac{2}{3}\left\{\frac{1}{6}-\frac{1}{(n+1)(n+2)(n+3)}\right\}$$

and $S_\infty=\dfrac{7}{36}.$

The method of differences may also be used to sum a series each of whose terms is the product of a constant number of factors in A.P., the first factors of successive terms being in the same A.P.

Example 10

Sum to n terms the series $3.5.7.9+5.7.9.11+\dots.$

Here $u_r=(2r+1)(2r+3)(2r+5)(2r+7).$

Let $v_{r+1}=(2r+1)(2r+3)(2r+5)(2r+7)(2r+9).$

Then $v_r=(2r-1)(2r+1)(2r+3)(2r+5)(2r+7).$

and $v_{r+1}-v_r=10(2r+1)(2r+3)(2r+5)(2r+7)=10u_r.$

Hence $u_r=\frac{1}{10}(v_{r+1}-v_r)$

and $\sum_1^n u_r=\frac{1}{10}(v_{n+1}-v_1)$

$$=\tfrac{1}{10}\{(2n+1)(2n+3)(2n+5)(2n+7)(2n+9)-1.3.5.7.9\}.$$

The method of differences may be used to sum certain trigonometric series. For example, to sum the series

$$S_n = \sin A + \sin(A+B) + \sin(A+2B) + \dots + \sin\{A+(n-1)B\},$$

when B is not a multiple of 2π, we multiply each term by $2\sin\frac{1}{2}B$. Then

$$2\sin A\sin\tfrac{1}{2}B = \cos(A-\tfrac{1}{2}B) - \cos(A+\tfrac{1}{2}B),$$
$$2\sin(A+B)\sin\tfrac{1}{2}B = \cos(A+\tfrac{1}{2}B) - \cos(A+\tfrac{3}{2}B),$$
$$2\sin(A+2B)\sin\tfrac{1}{2}B = \cos(A+\tfrac{3}{2}B) - \cos(A+\tfrac{5}{2}B),$$
$$\cdot \quad \cdot \quad \cdot \quad \cdot \quad \cdot \quad \cdot \quad \cdot \quad \cdot$$
$$2\sin\{A+(n-1)B\}\sin\tfrac{1}{2}B = \cos\{A+(n-\tfrac{3}{2})B\} - \cos\{A+(n-\tfrac{1}{2})B\},$$

and by addition,

$$(2\sin\tfrac{1}{2}B)S_n = \cos(A-\tfrac{1}{2}B) - \cos\{A+(n-\tfrac{1}{2})B\}$$

$$\therefore S_n = \frac{\sin\{A+\frac{1}{2}(n-1)B\}\sin\frac{1}{2}nB}{\sin\frac{1}{2}B} \quad . \quad . \quad \text{(i)}$$

The same method may be used to sum the series

$$C_n = \cos A + \cos(A+B) + \cos(A+2B) + \dots + \cos\{A+(n-1)B\}$$

but we deduce its sum from (i) by substituting $(A+\frac{1}{2}\pi)$ for A. This gives

$$C_n = \frac{\cos\{A+\frac{1}{2}(n-1)B\}\sin\frac{1}{2}nB}{\sin\frac{1}{2}B}.$$

If $\pi+B$ is written for B in the above two series, we obtain the series

$$\sin A - \sin(A+B) + \sin(A+2B) + \dots$$
$$\cos A - \cos(A+B) + \cos(A+2B) + \dots$$

whose sum to n terms can also be found directly by using the multiplier $2\sin\frac{1}{2}(\pi+B) = 2\cos\frac{1}{2}B$.

8.7. Summation of trigonometrical series using complex numbers

If we use the identities

$$\cos\theta + i\sin\theta = e^{i\theta} \text{ and } \cos n\theta + i\sin n\theta = e^{in\theta}$$

it will be seen that it is possible to reduce certain trigonometrical series to algebraic series of the types already considered.

Example 11

Use complex numbers to sum to n terms the series whose rth term is $\sin\{A+(r-1)B\}$, *where B is not a multiple of* 2π.

Let $C_n = \cos A + \cos(A+B) + \cos(A+2B) + \dots + \cos\{A+(n-1)B\}$

and $S_n = \sin A + \sin(A+B) + \sin(A+2B) + \dots + \sin\{A+(n-1)B\}$.

Then

$$C_n+iS_n=e^{iA}+e^{i(A+B)}+e^{i(A+2B)}+\dots+e^{i\{A+(n-1)B\}}$$

$$=\frac{e^{iA}(1-e^{niB})}{1-e^{iB}}, \text{ summing the G.P., since } B \text{ is not a multiple of } 2\pi$$

$$=\frac{e^{iA}.e^{\frac{1}{2}inB}(e^{-\frac{1}{2}inB}-e^{\frac{1}{2}inB})}{e^{\frac{1}{2}iB}(e^{-\frac{1}{2}iB}-e^{\frac{1}{2}iB})}$$

$$=e^{i\{A+\frac{1}{2}(n-1)B\}}\frac{\sin\frac{1}{2}nB}{\sin\frac{1}{2}B}.$$

Equating the imaginary parts on each side of the equation we have

$$S_n=\frac{\sin\{A+\frac{1}{2}(n-1)B\}\sin\frac{1}{2}nB}{\sin\frac{1}{2}B}.$$

Example 12

Find the sum to infinity of the series whose rth term is $(\cos r\theta)/r!$

Let
$$C=\frac{\cos\theta}{1!}+\frac{\cos 2\theta}{2!}+\frac{\cos 3\theta}{3!}+\dots$$

and
$$S=\frac{\sin\theta}{1!}+\frac{\sin 2\theta}{2!}+\frac{\sin 3\theta}{3!}+\dots.$$

Then
$$C+iS=\frac{e^{i\theta}}{1!}+\frac{e^{2i\theta}}{2!}+\frac{e^{3i\theta}}{3!}+\dots$$

$$=e^{e^{i\theta}}-1 \text{ by § 7.10}$$

$$=e^{(\cos\theta+i\sin\theta)}-1$$

$$=e^{\cos\theta}\{\cos(\sin\theta)+i\sin(\sin\theta)\}-1.$$

Equating the real parts on each side of this equation, we have

$$C=e^{\cos\theta}\cos(\sin\theta)-1.$$

Exercises 8 (*c*)

1. Sum to n terms and to infinity the series:

   (i) $\frac{1}{1.2.3}+\frac{1}{2.3.4}+\frac{1}{3.4.5}+\dots.$

   (ii) $\frac{2}{1.3.5}+\frac{4}{3.5.7}+\frac{6}{5.7.9}+\dots.$

2. Sum to n terms the series:

   (i) $1.3.5+3.5.7+5.7.9+\dots.$

   (ii) $\cos A\cos 2A+\cos 2A\cos 3A+\cos 3A\cos 4A+\dots.$

3. Prove that if x is real and numerically less than 1,
$$1+2x\cos\theta+2x^2\cos 2\theta+2x^3\cos 3\theta+\ldots=(1-x^2)/(1-2x\cos\theta+x^2).$$

4. By expressing r^2+2r+3 in the form $A+B(r+3)+C(r+2)(r+3)$ find the sum to infinity of the series whose rth term is
$$(r^2+2r+3)/r(r+1)(r+2)(r+3).$$

8.8. Miscellaneous examples

Example 13

Sum to n terms the series of which the rth term is $(2r-1)x^{r-1}$.
If $\alpha=2\pi/n$, *where n is a positive integer, show that*
$$1+3\cos\alpha+5\cos 2\alpha+\ldots+(2n-1)\cos(n-1)\alpha=-n$$
and $$3\sin\alpha+5\sin 2\alpha+\ldots+(2n-1)\sin(n-1)\alpha=-n\cot\tfrac{1}{2}\alpha.$$

[L.U.]

Let $$s_n=1+3x+5x^2+\ldots+(2n-3)x^{n-2}+(2n-1)x^{n-1}.$$

The series is arithmetico-geometrical, and summing as in § 8.1 (iv) we obtain
$$s_n=\frac{1-(2n-1)x^n}{1-x}+\frac{2x(1-x^{n-1})}{(1-x)^2},\ x\neq 1. \qquad \text{(i)}$$

If $x=1$, the series is the A.P. $1+3+5+\ldots+(2n-1)$ and $s_n=n^2$.

Now consider the series
$$C_n=1+3\cos\alpha+5\cos 2\alpha+\ldots+(2n-1)\cos(n-1)\alpha,$$
and $$S_n=\quad 3\sin\alpha+5\sin 2\alpha+\ldots+(2n-1)\sin(n-1)\alpha.$$
Then $C_n+iS_n=1+3e^{i\alpha}+5e^{2i\alpha}+\ldots+(2n-1)e^{(n-1)i\alpha}$
$$=\frac{1-(2n-1)e^{ni\alpha}}{1-e^{i\alpha}}+\frac{2e^{i\alpha}(1-e^{(n-1)i\alpha})}{(1-e^{i\alpha})^2}\text{ by (i)},$$
since α is not a multiple of 2π.

But $\alpha=2\pi/n$, $\therefore e^{ni\alpha}=\cos 2\pi+i\sin 2\pi=1$.

Hence
$$C_n+iS_n=\frac{2-2n}{1-e^{i\alpha}}+\frac{2(e^{i\alpha}-1)}{(1-e^{i\alpha})^2}$$
$$=\frac{-2n}{1-e^{i\alpha}}$$
$$=\frac{-2ne^{-\frac{1}{2}i\alpha}}{e^{-\frac{1}{2}i\alpha}-e^{\frac{1}{2}i\alpha}}$$
$$=\frac{-in(\cos\frac{1}{2}\alpha-i\sin\frac{1}{2}\alpha)}{\sin\frac{1}{2}\alpha}.$$

Separating the real and imaginary parts we obtain
$$C_n=-n$$
and $$S_n=-n\cot\tfrac{1}{2}\alpha.$$

Example 14

Prove that the infinite series whose rth term is $\dfrac{x^r}{r(r+1)}$ *converges when* $-1 \leqslant x \leqslant 1$, *and that its sum when* $-1 \leqslant x < 1$ *is* $(1/x-1) \log(1-x)+1$. *Find the sum when* $x=1$.

Here
$$\left|\frac{u_{r+1}}{u_r}\right| = \frac{r}{r+2} \mid x \mid \to \mid x \mid \text{ as } r \to \infty.$$

Hence by the ratio test (§ 4.18) the given series is convergent when $|x| < 1$ and divergent when $|x| > 1$.

When $x=1$, the series is $\Sigma \dfrac{1}{r(r+1)}$, which is seen to be convergent if we compare it with the convergent series $\Sigma \dfrac{1}{r^2}$ (see § 4.13).

When $x=-1$, the series is $\Sigma \dfrac{(-1)^r}{r(r+1)}$, which converges since it is obtained from the convergent series $\Sigma \dfrac{1}{r(r+1)}$ by changing the signs of alternate terms (see § 4.17).

Hence the given series is convergent when $-1 \leqslant x \leqslant 1$.

To sum the series we write $u_r = x^r \left(\dfrac{1}{r} - \dfrac{1}{r+1}\right)$ and consider the series $\Sigma \dfrac{x^r}{r}$ and $\Sigma \dfrac{x^r}{r+1}$.

By (5.10) the first converges to $-\log(1-x)$ when $-1 \leqslant x < 1$; the second is

$$\frac{x}{2}+\frac{x^2}{3}+\frac{x^3}{4}+\dots$$
$$=\frac{1}{x}\left(\frac{x^2}{2}+\frac{x^3}{3}+\frac{x^4}{4}+\dots\right)$$
$$=\frac{1}{x}\{-\log(1-x)-x\} \text{ when } -1 \leqslant x < 1 \text{ by (5.10).}$$

Hence
$$\Sigma u_r = \left(\frac{1}{x}-1\right) \log(1-x)+1 \text{ when } -1 \leqslant x < 1.$$

When $x=1$, $u_r = \dfrac{1}{r} - \dfrac{1}{r+1}$.

Hence
$$\sum_1^n u_r = 1 - \frac{1}{n+1} \text{ and } \sum_1^\infty u_r = 1.$$

Example 15

(i) *Sum the infinite series*

$$x \sin \theta + \frac{x^3 \sin 3\theta}{3\,!} + \frac{x^5 \sin 5\theta}{5\,!} + \dots$$

where x *and* θ *are real.*

(ii) *In the triangle ABC, show that, with the usual notation and with $b<c$,*

$$\frac{a^n}{c^n}\left\{\sin A+\frac{nb}{c}\sin 2A+\frac{n(n+1)}{2!}\frac{b^2}{c^2}\sin 3A+\ldots\right\}=\sin(A+nB) \text{ [L.U.]}$$

(i) Let $\qquad C=x\cos\theta+\dfrac{x^3\cos 3\theta}{3!}+\dfrac{x^5\cos 5\theta}{5!}+\ldots$

and $\qquad S=x\sin\theta+\dfrac{x^3\sin 3\theta}{3!}+\dfrac{x^5\sin 5\theta}{5!}+\ldots.$

Then $C+iS=xe^{i\theta}+\dfrac{x^3e^{3i\theta}}{3!}+\dfrac{x^5e^{5i\theta}}{5!}+\ldots=\sinh(xe^{i\theta})$

(see § 7.12, the summation being valid for all values of x and θ)

$$=\sinh\{x(\cos\theta+i\sin\theta)\},$$

$$=\sinh(x\cos\theta)\cosh(ix\sin\theta)+\cosh(x\cos\theta)\sinh(ix\sin\theta), \text{ see § 5.10}$$

$$=\sinh(x\cos\theta)\cos(x\sin\theta)+i\cosh(x\cos\theta)\sin(x\sin\theta)$$

see § 7.13.

Equating imaginary parts on each side of this equation we have

$$S=\cosh(x\cos\theta)\sin(x\sin\theta).$$

(ii) Let $C=\dfrac{a^n}{c^n}\left\{\cos A+\dfrac{nb}{c}\cos 2A+\dfrac{n(n+1)}{2!}\dfrac{b^2}{c^2}\cos 3A+\ldots\right\}$

and $S=\dfrac{a^n}{c^n}\left\{\sin A+\dfrac{nb}{c}\sin 2A+\dfrac{n(n+1)}{2!}\dfrac{b^2}{c^2}\sin 3A+\ldots\right\}.$

Then

$$C+iS=\frac{a^n}{c^n}\left\{e^{iA}+\frac{nb}{c}e^{2iA}+\frac{n(n+1)}{2!}\frac{b^2}{c^2}e^{3iA}+\ldots\right\}$$

$$=\frac{a^n}{c^n}e^{iA}\left\{1+(-n)\left(-\frac{b}{c}e^{iA}\right)+\frac{(-n)(-n-1)}{2!}\left(-\frac{b}{c}e^{iA}\right)^2+\ldots\right\}$$

$$=\frac{a^n}{c^n}e^{iA}\left\{1-\frac{b}{c}e^{iA}\right\}^{-n} \text{ the series being convergent when } b<c,$$

$$=\frac{a^ne^{iA}}{(c-be^{iA})^n},$$

$$=\frac{a^ne^{iA}}{(a\cos B-ib\sin A)^n}$$

$$=\frac{e^{iA}}{(\cos B-i\sin B)^n}$$

$$=e^{i(A+nB)}.$$

Equating the imaginary parts on each side of this equation we obtain $S=\sin(A+nB)$.

Example 16

(i) *Express sin 3x in terms of sin x, and find the sum to n terms and to infinity of the series whose rth term is* $3^{r-1}\sin^3(\theta/3^r)$.

(ii) *If* $0<b<a$, *sum to infinity the series*

$$(b/a)\sin C+(b^2/2a^2)\sin 2C+(b^3/3a^3)\sin 3C+\ldots$$

If a, b, c, A, B, C are the elements of a plane triangle, prove that this sum is equal to B [L.U.]

(i) $$\sin 3x=3\sin x-4\sin^3 x$$

$$\therefore\ \sin^3 x=\tfrac{1}{4}(3\sin x-\sin 3x).$$

Now consider the series $\sum\limits_1^{\infty} u_r$ where

$$u_r=3^{r-1}\sin^3\frac{\theta}{3^r}=\frac{3^{r-1}}{4}\left(3\sin\frac{\theta}{3^r}-\sin\frac{\theta}{3^{r-1}}\right)$$

Then $$u_1=\frac{1}{4}\left(3\sin\frac{\theta}{3}-\sin\theta\right),$$

$$u_2=\frac{3}{4}\left(3\sin\frac{\theta}{9}-\sin\frac{\theta}{3}\right).$$

$$u_3=\frac{9}{4}\left(3\sin\frac{\theta}{27}-\sin\frac{\theta}{9}\right).$$

$$\cdot\quad\cdot\quad\cdot\quad\cdot\quad\cdot\quad\cdot\quad\cdot$$

$$u_n=\frac{3^{n-1}}{4}\left(3\sin\frac{\theta}{3^n}-\sin\frac{\theta}{3^{n-1}}\right).$$

Adding : $$S_n=\frac{1}{4}\left\{3^n\sin\frac{\theta}{3^n}-\sin\theta\right\}$$

$$=\frac{1}{4}\left[\theta.\left\{\frac{\sin(\theta/3^n)}{\theta/3^n}\right\}-\sin\theta\right]$$

As $n\to\infty$, $\theta/3^n\to 0$ and $\dfrac{\sin(\theta/3^n)}{\theta/3^n}\to 1$ see § 9.3, (2)

$$\therefore\ S_\infty=\tfrac{1}{4}\{\theta-\sin\theta\}.$$

(ii) Let $$X=\frac{b}{a}\cos C+\frac{b^2}{2a^2}\cos 2C+\frac{b^3}{3a^3}\cos 3C+\ldots$$

and $$Y=\frac{b}{a}\sin C+\frac{b^2}{2a^2}\sin 2C+\frac{b^3}{3a^3}\sin 3C+\ldots$$

Then $$X+iY=\frac{b}{a}e^{iC}+\frac{b^2}{2a^2}e^{2iC}+\frac{b^3}{3a^3}e^{3iC}+\ldots$$

$$=-\log\{1-(b/a)e^{iC}\}$$ since $b/a<1$, see § 7.15.

In fig. 20 $\overline{OP}=1$

and $\overline{OQ}=\overline{RP}=(b/a)e^{iC}$

$\therefore\ \overline{OR}=1-(b/a)e^{iC}$

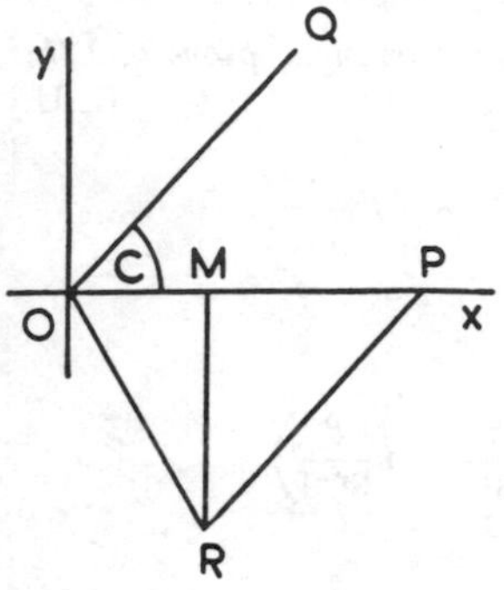

The imaginary part of

$$-\log\{1-(b/a)e^{iC}\}=-\angle POR \text{ (see § 7.14)}$$

i.e. $$Y=\tan^{-1}\left\{\frac{(b/a)\sin C}{1-(b/a)\cos C}\right\}$$

$$=\tan^{-1}\left\{\frac{b\sin C}{a-b\cos C}\right\}.$$

If a, b, c, A, B, C are the elements of a plane triangle

$b\sin C=c\sin B$ and $a-b\cos C=c\cos B$

$$\therefore \tan^{-1}\left(\frac{b\sin C}{a-b\cos C}\right)=B.$$

Fig. 20

Miscellaneous Exercises 8

1. Find $\sum_{r=1}^{n}(r^3+3r^2-r+1)$. [Durham.]

2. Sum the series

(a) $\frac{1}{1.2}+\frac{1}{2.3}+\frac{1}{3.4}+\ldots+\frac{1}{n(n+1)}$.

(b) $1.3+2.4+3.5+\ldots+100.102$. [Durham.]

3. Sum the series :

(i) $\frac{1}{1.3}+\frac{1}{3.5}+\frac{1}{5.7}+\ldots$ to n terms.

(ii) $1.2.3+2.3.4+3.4.5+\ldots$ to 15 terms.

(iii) $x+2x^2+3x^3+\ldots$ to n terms. [Durham.]

4. Find the sum of the infinite series $\frac{2\frac{1}{2}}{1!}+\frac{3\frac{1}{3}}{2!}+\frac{4\frac{1}{4}}{3!}+\frac{5\frac{1}{5}}{4!}+\ldots$. [Durham.]

5. By expressing $(2r-1)(2r+1)$ in the form $A+B(2r)+C(2r)(2r-1)$, where A, B, C are independent of r, show that the sum to infinity of the series

$$\frac{1.3}{2!}+\frac{3.5}{4!}+\frac{5.7}{6!}+\ldots$$

is $(e^2+2e-1)/2e$. [L.U. Anc.]

6. Show that, provided $|r|<1$, the sum to infinity of the series

$$r \sin \theta + r^2 \sin 2\theta + r^3 \sin 3\theta + \ldots \text{ is } (r \sin \theta)/(1-2r \cos \theta + r^2).$$

[L.U. Anc.]

7. Find the sum to infinity of each of the following series :

(a) $\dfrac{1}{1.3}+\dfrac{1}{3.5}+\dfrac{1}{5.7}+\ldots.$

(b) $\dfrac{2^2}{1!}+\dfrac{3^2}{2!}+\dfrac{4^2}{3!}+\ldots.$ [L.U.]

8. Sum to infinity the series

$$1+2x+\frac{3x^2}{2!}+\frac{4x^3}{3!}+\frac{5x^4}{4!}+\ldots.$$

and

$$1^2+2^2x+\frac{3^2x^2}{2!}+\frac{4^2x^3}{3!}+\frac{5^2x^4}{4!}+\ldots.$$ [L.U.]

9. Sum to infinity the series

(i) $1+\dfrac{1}{8}+\dfrac{1.3}{2!\,8^2}+\dfrac{1.3.5}{3!\,8^3}+\ldots.$

(ii) $\dfrac{1}{1.2}-\dfrac{x}{2.3}+\dfrac{x^2}{3.4}-\dfrac{x^3}{4.5}+\ldots,$

x being chosen so that the series is convergent. [L.U.]

10. Show that, if $-1<x<+1$,

$$\sum_{r=1}^{\infty} \frac{x^r}{r(r+2)}=\frac{1}{2}\left\{\frac{1-x^2}{x^2}\log(1-x)+\frac{x+2}{2x}\right\}.$$

[L.U.]

11. (i) Sum to infinity the series having for its nth term $(n^2+2)/n\,!$.

(ii) Prove that $\tan\theta=\cot\theta-2\cot 2\theta$, and sum the series

$$\tan\theta+\frac{1}{2}\tan\frac{\theta}{2}+\frac{1}{2^2}\tan\frac{\theta}{2^2}+\ldots+\frac{1}{2^n}\tan\frac{\theta}{2^n}.$$ [L.U.]

12. (i) Sum the infinite series

$$\frac{1.3}{2!}+\frac{2.4}{3!}+\frac{3.5}{4!}+\ldots+\frac{(n-1)(n+1)}{n!}+\ldots.$$

(ii) Prove that $\tan^{-1}(n+1)-\tan^{-1}n=\cot^{-1}(1+n+n^2)$.

Hence, or otherwise, sum the finite series

$$\cot^{-1}3+\cot^{-1}7+\cot^{-1}13+\ldots+\cot^{-1}(1+n+n^2).$$ [L.U.]

13. Prove that, if x is not an integral multiple of π,

$$\tfrac{1}{2}+\cos 2x+\cos 4x+\ldots+\cos 2nx=\tfrac{1}{2}\sin(2n+1)x/\sin x$$

and

$$\sin x+\sin 3x+\ldots+\sin(2n-1)x=\sin^2 nx/\sin x.$$

Hence show that

$$\int_0^{\frac{1}{2}\pi}\left(\frac{\sin nx}{\sin x}\right)^2 dx=\tfrac{1}{2}n\pi.$$ [L.U.]

14. (i) Find the sum to infinity of the series

(a) $$1+\frac{1}{2}\left(\frac{9}{25}\right)+\frac{1.3}{2.4}\left(\frac{9}{25}\right)^2+\frac{1.3.5}{2.4.6}\left(\frac{9}{25}\right)^3+\ldots$$

(b) $$\frac{1^2.2}{1!}+\frac{2^2.2^2}{2!}+\frac{3^2.2^3}{3!}+\frac{4^2.2^4}{4!}+\frac{5^2.2^5}{5!}+\ldots.$$

(ii) If $$(1-x+x^2)^{3n}=a_0+a_1x+a_2x^2+\ldots,$$

and $$(x+1)^{3n}=c_0x^{3n}+c_1x^{3n-1}+c_2x^{3n-2}+\ldots,$$

prove that $$a_0+a_1+a_2+\ldots=1$$

and $$a_0c_0+a_1c_1+a_2c_2+\ldots=(3n)!/(n!)(2n)!.$$ [L.U.]

15. Show that, if $-1\leqslant x<1$, where x is a real variable, the infinite series whose rth term is $(4r-1)x^{r-1}/r(r+1)$ is convergent. Assuming that x has a value in this range, sum the series. [L.U.]

16. (i) If $0<(n+1)\phi<\tfrac{1}{2}\pi$, show that

$$\sin\phi\sec n\phi\sec(n+1)\phi=\tan(n+1)\phi-\tan n\phi,$$

and sum the finite series

$$\sum_{r=1}^{n}\sec r\phi\sec(r+1)\phi.$$

(ii) Find the sum of the infinite series

$$\frac{1}{2}+\frac{1}{3}.\frac{x^2}{1!}+\frac{1}{4}.\frac{x^4}{2!}+\frac{1}{5}.\frac{x^6}{3!}+\ldots.$$ [L.U.]

17. Find the sums to infinity of the series

(i) $$\frac{x^2}{2!}+\frac{2^2x^3}{3!}+\frac{3^2x^4}{4!}+\ldots.$$

(ii) $$1+\frac{1}{2^3}+\frac{1.5}{2!}\,\frac{1}{2^6}+\frac{1.5.9}{3!}\,\frac{1}{2^9}+\ldots.$$ [L.U.]

(i) By expressing $1/x(x+4)$ in partial fractions, or otherwise, prove that the sum of the infinite series

$$\frac{1}{1.5}+\frac{1}{2.6}+\frac{1}{3.7}+\ldots+\frac{1}{n(n+4)}+\ldots$$

is 25/48.

(ii) Find the value of $\sum_{n=1}^{\infty} \frac{(\cos n\theta)}{2^n}$. [L.U.]

19. Show that, if $1>x>-1$, the sum to infinity of the series

$$\frac{1}{1.3}+\frac{x^2}{3.5}+\frac{x^4}{5.7}+\frac{x^6}{7.9}+\ldots$$

is

$$\frac{1}{2x^2}+\frac{(x^2-1)}{4x^3}\log\frac{1+x}{1-x}.$$

What is the sum of this series when $x=1$? [L.U. Anc.]

20. Show that the sum of the infinite series whose rth term is

$\frac{10r+1}{2r(2r-1)(2r+1)}\cdot\frac{1}{2^{2r}}$ is $2-\log 2-\frac{3}{4}\log 3$. [L.U.]

21. (i) By expressing $(2n-1)(2n+1)(2n+3)$ in the form

$$A+B(2n)+C(2n)(2n-1)+D(2n)(2n-1)(2n-2),$$

where A, B, C, D are constants independent of n, show that the sum to infinity of the series

$$\frac{1.3.5}{2!}+\frac{3.5.7}{4!}+\frac{5.7.9}{6!}+\ldots$$

is $3+\frac{1}{2}(7e-e^{-1})$.

(ii) Show that

$$\sum_{n=1}^{\infty}\frac{r^{2n}\sin(2n\theta)}{(2n)!}=\sin(r\sin\theta)\sinh(r\cos\theta).$$ [L.U.]

22. Find the sums to infinity of the series

(i) $1+\frac{1}{2}x+\frac{1.3}{2.4}x^2+\frac{1.3.5}{2.4.6}x^3+\ldots,$

(ii) $\frac{1}{2!}\frac{x^4}{4}+\frac{1}{3!}\frac{x^5}{5}+\frac{1}{4!}\frac{x^6}{6}+\frac{1}{5!}\frac{x^7}{7}+\ldots,$

(iii) $\frac{2}{1.4}x^2+\frac{3}{2.5}x^3+\frac{4}{3.6}x^4+\frac{5}{4.7}x^5+\ldots,$

stating for what range of values of x each result is valid. [L.U.]

23. Resolve $1/(2n-1)(2n+1)(2n+3)$ into partial fractions and show that the sum to infinity of the series of which this is the nth term is $1/12$.

Show that the sum to infinity of the series whose nth term is $(-1)^{n-1}/(2n-1)(2n+1)(2n+3)$ is $\frac{1}{8}\pi-\frac{1}{3}$. [L.U.]

24. Sum to infinity the series whose nth terms are

(i) $1/n(n+1)(n+2)$; (ii) $(n^2 \cos n\theta)/(n+1)!$.

Deduce from (i), or otherwise prove that

$$\frac{1}{2^3}+\frac{1}{3^3}+\ldots+\frac{1}{(n+1)^3}<\frac{1}{4}.$$ [L.U.]

25. (i) Sum to n terms the series whose rth term is $1/(2r-1)(2r+1)(2r+3)$, and find the sum to infinity.

(ii) Find the sum to infinity of the series whose rth terms are $(4r^2+3r+1)/r!$, $(\sin r\theta)/r!$. [L.U.]

26. (i) By expressing $4n^2-1$ in the form $A+Bn+Cn(n-1)$, where A, B, C are constants independent of n, or otherwise, show that

$$\sum_{n=1}^{\infty}\frac{(4n^2-1)}{n!}=7e+1.$$

(ii) Find the value of λ for which the expansion of

$$f(x)\equiv(1+\lambda x)(1+\tfrac{1}{8}x)^{-1}\log(1+x)$$

contains no term in x^2. [L.U. Anc.]

27. (i) By expressing $n(n+1)(2n+1)$ in the form

$$An+Bn(n-1)+Cn(n-1)(n-2),$$

where A, B, C are constants independent of n, or otherwise, show that

$$\sum_{n=1}^{\infty}\frac{1^2+2^2+3^2+\ldots+n^2}{2^n n!}=\frac{(11\sqrt{e})}{12}$$

(ii) By writing the nth term of the series

$$\frac{3}{1.2.4}+\frac{5}{2.3.5}+\frac{7}{3.4.6}\ldots$$

as a sum of partial fractions, or otherwise, prove that the sum to infinity of the series is 37/36. [L.U.]

28. (i) Prove that

$$\sum_{r=1}^{n}\binom{n}{r}\sin 2r\theta=2^n\cos^n\theta\sin n\theta.$$

(ii) Sum the infinite series

$$\sum_{n=1}^{\infty}\frac{1}{n(n+2)(n+4)}.$$ [L.U.]

29. Find the sum to n terms of the series

$$1+5x+9x^2+\ldots+(4n-3)x^{n-1}+\ldots$$

If $\alpha=2\pi/n$, prove that

$1+5\cos\alpha+9\cos 2\alpha+\ldots+(4n-3)\cos(n-1)\alpha=-2n$, and

$1+5\sin\alpha+9\sin 2\alpha+\ldots+(4n-3)\sin(n-1)\alpha=-2n\cot\tfrac{1}{2}\alpha$. [L.U.]

30. Show that $\sum_{n=1}^{\infty}\{(1/n2^n)\cos\tfrac{1}{3}n\pi\}=\log 2-\tfrac{1}{2}\log 3$.

Find the sum of the series $1+\sum_{r=1}^{\infty}\frac{(\cos 2r\theta)}{(2r)!}$. [L.U.]

31. If
$$u_n=\frac{n^2+9n+5}{(n+1)(2n+3)(2n+5)(n+4)},$$
prove that
$$\sum_{n=1}^{\infty} u_n=5/36.$$
[L.U.]

32. (i) By expressing $r/(2r-1)(2r+1)(2r+3)$ in partial fractions, or otherwise, sum to n terms the series
$$\frac{1}{1.3.5}+\frac{2}{3.5.7}+\frac{3}{5.7.9}+\ldots$$
and show that the sum to infinity is 1/8.

(ii) Show that the series
$$\frac{1^3}{3!}+\frac{2^3}{5!}+\frac{3^3}{7!}+\frac{4^3}{9!}+\ldots$$
converges to the value $(e^2+3)/16e$. [L.U.

33. (i) Find the sum of the series
$$\cos(x+y)+\cos(x+2y)+\cos(x+3y)+\ldots+\cos(x+ny).$$

(ii) Show that the sum to infinity of the series whose rth term is
$$(r^2+r+1)/r! \text{ is } 4e-1.$$

(iii) If $u_n=\dfrac{\sinh x}{\sinh nx \sinh (n+1)x}$, $(x\neq 0)$

show that $u_n=\coth nx-\coth(n+1)x$, and find the sum to infinity of the series whose nth term is u_n. Distinguish between the cases when x is positive and when x is negative. [L.U.]

34. Find the sum to n terms of the series whose rth term is
$$1/(3r-2)(3r+1)(3r+4)$$
and find the sum to infinity.

Find also the sum to infinity of the series whose rth terms are

(i) $(r+1)^2/r!$; (ii) $\cos r\theta/(2^r+2^{r+1})$. [L.U.]

35. Evaluate

(i) $\displaystyle\sum_{1}^{n}\frac{p-2}{p(p+1)(p+3)}$, (ii) $\displaystyle\sum_{p=1}^{n} p\sin p\theta$. [L.U.]

36. (i) Prove that the sum to infinity of the series
$$\frac{2^2+1}{2!}+\frac{2(3^2+1)}{3!}+\frac{3(4^2+1)}{4!}+\frac{4(5^2+1)}{5!}+\ldots$$
is $3e+1$.

(ii) By expanding the expression $(e^x-1)^n$ in two ways and equating coefficients of x^n, show that, if n is a positive integer,
$$n^n-n(n-1)^n+\frac{n(n-1)}{1.2}(n-2)^n-\frac{n(n-1)(n-2)}{1.2.3}(n-3)^n+\ldots$$
to n terms is equal to $n!$. [L.U.]

37. If $|x|<1$, find the sum to infinity of the series of which the rth terms are (i) $x^r/r(r+1)$; (ii) $x^r \cos rx$. [L.U.]

38. (i) If $r>0$, show that $\tan^{-1}(1/2r^2)=\tan^{-1}(2r+1)-\tan^{-1}(2r-1)$.

Hence find the value of $\sum_{r=1}^{\infty} \tan^{-1}(1/2r^2)$.

(ii) Sum the series $\sum_{r=1}^{n} \frac{r}{(r+4)(r+5)(r+6)}$. [L.U.]

39. (i) Show that $4\sin^3\phi=3\sin\phi-\sin 3\phi$.

Sum to n terms and to infinity the series whose mth term is $\{\sin^3(3^{m-1}\theta)\}/3^{m-1}$.

(ii) Sum the infinite series

$$\cos\theta-\frac{1}{2}\cos 3\theta+\frac{1.3}{2.4}\cos 5\theta-\frac{1.3.5}{2.4.6}\cos 7\theta+\ldots$$

where $-\frac{1}{2}\pi\leqslant\theta\leqslant\frac{1}{2}\pi$. [L.U.]

40. Sum to infinity the series whose nth terms are

$$1/n(n+1) \text{ and } (n\cos n\theta)/(n+1)!.$$

Also sum to n terms the series whose rth term is $\cos\{a+(r-1)\beta\}$. [L.U.]

41. Sketch the graph of the curve $y=\cot^{-1}x$, for values of y lying between $-\pi$ and $+\pi$, the inverse function being regarded as many-valued.

If $a>0$, and the inverse functions are interpreted as acute angles, prove that $\tan^{-1}(1+a)-\tan^{-1}a=\cot^{-1}(1+a+a^2)$.

Hence prove that, if n is a positive integer,

$$\cot^{-1}3+\cot^{-1}7+\cot^{-1}13+\ldots+\cot^{-1}(1+n+n^2)=\cot^{-1}(1+2/n),$$

and deduce that the infinite series $\sum_{n=1}^{\infty}\cot^{-1}(n^2)$ converges. [L.U.]

42. Find the sum of the first n terms of the series whose rth term is $1/r(r+4)$ and show that the sum to infinity is 25/48.

Show that $\tanh n\theta-\tanh(n-1)\theta=\sinh\theta \operatorname{sech}(n-1)\theta \operatorname{sech} n\theta$.

Find the sum of the first n terms of the series whose rth term is $\operatorname{sech}(r-1)\theta \operatorname{sech} r\theta$ and deduce the sum to infinity.

Find the sum of n terms of the series whose rth term is $\sin r\theta$. [L.U.]

43. (*a*) Show how to evaluate the sum $\sum_{r=1}^{n} u_r$ given that $u_r=v_r-v_{r-1}$, where v_r is a known function of r.

Hence sum to n terms the series whose rth term is

(i) $1/r(r+1)(r+2)(r+3)$; (ii) $\operatorname{cosec} 2^r\theta$.

(*b*) Sum to infinity the series whose rth term is $(\sin^2 r\theta)/r\,!$. [L.U.]

44. Prove that

$$\sin \alpha - \sin(\alpha+\beta) + \sin(\alpha+2\beta) - \ldots - \sin\{\alpha+(2n-1)\beta\} = -\sin n\beta \cos\{\alpha+(n-\tfrac{1}{2})\beta\} \sec \tfrac{1}{2}\beta,$$

where α and β are real constants.

$A_1, A_2, \ldots, A_{2n}$ are the vertices of a regular polygon of $2n$ sides. Prove that the sum of the lengths $A_1A_3, A_1A_5, \ldots, A_1A_{2n-1}$, differs from the sum of the lengths $A_1A_2, A_1A_4, \ldots, A_1A_{2n}$ by $a \sec^2(\pi/4n)$, where $2a$ is the length of the side of the polygon. [L.U.]

45. Find the sum to infinity of the series whose rth terms are

(i) $1/r(r+1)(r+3)$; (ii) $(2r^2-4r+3)/r!$; (iii) $(\cos r\theta)/(r+1)!$. [L.U.]

46. Prove that the infinite series whose nth term is $x^n/n(n+1)(n+2)$ converges when $-1 \leqslant x \leqslant 1$ and that its sum, when $-1 \leqslant x < 1$, is

$$\frac{3}{4} - \frac{1}{2x} - \frac{(1-x)^2}{2x^2} \log(1-x).$$

Find the sum when $x=1$. [L.U.]

47. (i) Find the sum to n terms of the series

$$\frac{3^2}{1.2.3.4} + \frac{4^2}{2.3.4.5} + \frac{5^2}{3.4.5.6} + \ldots$$

and show that the sum to infinity is 29/36.

(ii) Evaluate $\displaystyle\sum_{n=1}^{\infty} \frac{(-1)^n}{n!} \frac{(n+1)}{n+2}$. [L.U.]

48. Express $(x^2+5x+2)/x^2(x+1)^2$ as a sum of partial fractions.

Find the sum of the first n terms of the series whose rth term is $(r^2+5r+2)/r^2(r+1)^2$. Deduce the sum to infinity.

What is the smallest value of n for which these two sums differ by less than 0·1 ? [L.U.]

49. Find the sum of the infinite series whose nth term is $\cos n\theta \sin^n \theta$, where $0 < \theta < \frac{1}{2}\pi$. [L.U.]

50. (i) Sum the geometric progression $1+x+\ldots+x^n$, and, hence or otherwise, evaluate the sum $1+2x+\ldots+nx^{n-1}$ and the sum to infinity

$$1+\frac{2}{3}+\frac{3}{3^2}+\frac{4}{3^3}+\ldots+\frac{n}{3^{n-1}}+\ldots.$$

(ii) Prove that ${}^nC_1+2({}^nC_2)+\ldots+n({}^nC_n)=n2^{n-1}$. [L.U.]

51. Find the sums to infinity of

(i) $$\frac{5}{3.6}+\frac{5.7}{3.6.9}+\frac{5.7.9}{3.6.9.12}+\ldots.$$

(ii) $$\frac{5}{1.2.3}+\frac{8}{2.3.5}+\frac{11}{3.4.7}+\ldots.$$ [L.U.]

CHAPTER 9

DIFFERENTIATION AND APPLICATIONS

9.1. Differentiation

Let y be a single-valued continuous function of x defined by the equation

$$y=f(x). \quad \quad \quad \quad \text{(i)}$$

Then an increase in the value of x will produce an increment (positive, negative or zero) in the value of y. Assuming a fixed initial value for x, let δy be the increment in y corresponding to an increment δx in x. Then

$$y+\delta y=f(x+\delta x) \quad \quad \quad \quad \text{(ii)}$$

so that from (i) and (ii)

$$\delta y=f(x+\delta x)-f(x),$$

and

$$\frac{\delta y}{\delta x}=\frac{f(x+\delta x)-f(x)}{\delta x}, \quad \delta x\neq 0.$$

$\dfrac{\delta y}{\delta x}$ measures the average change in y per unit change in x, i.e. the average rate of change of y with respect to x in the interval δx.

If, as $\delta x\to 0$, $\dfrac{f(x+\delta x)-f(x)}{\delta x}$ tends to a finite limit, this limit may be interpreted as the rate of change of y with respect to x for the initial value of x. It is called the *differential coefficient* (or the *derivative*) of $f(x)$ with respect to x and is denoted by $f'(x)$, by $\dfrac{d}{dx}f(x)$, by Df, by $\dfrac{dy}{dx}$, by y_1 or by y'.

In order that

$$\lim_{\delta x\to 0}\frac{f(x+\delta x)-f(x)}{\delta x} \quad \quad \quad \text{(iii)}$$

should exist it is necessary that $f(x+\delta x)-f(x)$ should tend to zero as $\delta x\to 0$, i.e. that $f(x)$ should be continuous for the value of x under consideration. In subsequent chapters it may be assumed that all functions discussed are differentiable, i.e. their derivatives exist except possibly at isolated values of x.

9.2. General Rules

From definition (iii) the following rules (with which it is assumed that the student is already familiar) may be established. We use $u, v, w, \ldots$ to denote functions of x, and $a, b, c, \ldots$ to denote constants.

I. $$\frac{d}{dx}(u+v-w)=\frac{du}{dx}+\frac{dv}{dx}-\frac{dw}{dx}.$$

II. $$\frac{d}{dx}(ay)=a\frac{dy}{dx}.$$

III. $$\frac{d}{dx}(uv)=u\frac{dv}{dx}+v\frac{du}{dx}. \quad \text{(Product rule.)}$$

IV. $$\frac{d}{dx}\left(\frac{u}{v}\right)=\frac{v\dfrac{du}{dx}-u\dfrac{dv}{dx}}{v^2}. \quad \text{(Quotient rule.)}$$

V. If y is a function of u, where u is a function of x,

$$\frac{dy}{dx}=\frac{dy}{du}\times\frac{du}{dx}.$$

VI. $$\frac{dx}{dy}=1\bigg/\frac{dy}{dx} \text{ if } \frac{dy}{dx}\neq 0.$$

By way of revision we establish several fundamental results. Other standard formulae are listed below.

9.3. Two important limits

(1) For all rational values of n

$$\lim_{x\to a}\frac{x^n-a^n}{x-a}=na^{n-1} \text{ and } \lim_{h\to 0}\frac{(x+h)^n-x^n}{h}=nx^{n-1}.$$

(i) When n is a positive integer, we obtain by division

$$\frac{x^n-a^n}{x-a}=x^{n-1}+ax^{n-2}+\ldots+a^{n-2}x+a^{n-1}.$$

As $x\to a$ each of the n terms on the right tends to a^{n-1}

$$\therefore \lim_{x\to a}\frac{x^n-a^n}{x-a}=na^{n-1}.$$

(ii) When n is a positive rational number, we write $n=p/q$, where p and q are positive integers. We also suppose a to be positive. Then if $y=x^{1/q}$ and $b=a^{1/q}$

$$\frac{x^n-a^n}{x-a}=\frac{y^p-b^p}{y^q-b^q}=\frac{(y^p-b^p)/(y-b)}{(y^q-b^q)/(y-b)}.$$

As $x\to a$, $y\to b$, and by (i)

$$(y^p-b^p)/(y-b)\to pb^{p-1} \text{ and } (y^q-b^q)/(y-b)\to qb^{q-1}.$$

$$\therefore \lim_{x\to a}\frac{x^n-a^n}{x-a}=(p/q)b^{p-q}=(p/q)a^{p/q-1}=na^{n-1}.$$

(iii) When n is a negative rational number, we write $n=-m$, where m is a positive rational number. Then

$$\frac{x^n-a^n}{x-a}=\frac{x^{-m}-a^{-m}}{x-a}=\frac{-1}{x^m a^m}\cdot\frac{x^m-a^m}{x-a}$$

$$\therefore \lim_{x\to a}\frac{x^n-a^n}{x-a}=-\frac{1}{a^{2m}}\cdot m a^{m-1} \text{ by (ii)}$$

$$=-ma^{-m-1}=na^{n-1}.$$

Hence, for all rational values of n,

$$\lim_{x\to a}\frac{x^n-a^n}{x-a}=na^{n-1}.$$

If in this result we replace x by $x+h$ and a by x, we obtain

$$\lim_{h\to 0}\frac{(x+h)^n-x^n}{h}=nx^{n-1}.$$

(2) When θ is measured in radians, $\lim\limits_{\theta\to 0}\dfrac{\sin\theta}{\theta}=1.$

In fig. 21, PQ is an arc of a circle with centre O and radius r. $\angle QOP=\theta$ radians and since θ must ultimately tend to zero we shall assume that $0<\theta<\frac{1}{2}\pi$. The tangent at P meets OQ produced at T.

Then since the area of sector POQ lies between the areas of the triangles POQ and POT.

$$\tfrac{1}{2}r^2\sin\theta<\tfrac{1}{2}r^2\theta<\tfrac{1}{2}r^2\tan\theta,$$

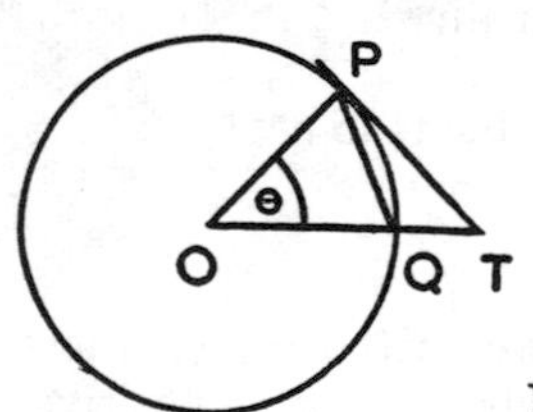

Fig. 21

$$1<\frac{\theta}{\sin\theta}<\sec\theta,$$

$$1>\frac{\sin\theta}{\theta}>\cos\theta.$$

But $\cos\theta\to 1$ as $\theta\to 0+$, hence also

$$\frac{\sin\theta}{\theta}\to 1 \text{ as } \theta\to 0+.$$

Since $\dfrac{\sin\theta}{\theta}=\dfrac{\sin(-\theta)}{-\theta}$, it follows that $\lim\limits_{\theta\to 0-}\dfrac{\sin\theta}{\theta}=1,$

$$\therefore \lim_{\theta\to 0}\frac{\sin\theta}{\theta}=1.$$

9.4. Differentiation of x^n

If $y=x^n$, where n is any rational number,

$$\frac{dy}{dx}=\lim_{\delta x\to 0}\frac{(x+\delta x)^n-x^n}{\delta x}$$
$$=nx^{n-1} \text{ from (1) of § 9.3.}$$

9.5. Differentiation of sin x

If $$y=\sin x$$

$$\frac{dy}{dx}=\lim_{\delta x\to 0}\frac{\sin(x+\delta x)-\sin x}{\delta x}$$
$$=\lim_{\delta x\to 0}\frac{2\cos(x+\frac{1}{2}\delta x)\sin\frac{1}{2}\delta x}{\delta x}$$
$$=\lim_{\theta\to 0}\cos(x+\theta).\frac{\sin\theta}{\theta}, \text{ putting } \delta x=2\theta$$
$$=\cos x \text{ from (2) of § 9.3.}$$

Similarly we prove that if $y=\cos x$, $\dfrac{dy}{dx}=-\sin x$.

9.6. Differentiation of $\sin^{-1} x$

If x and y are numbers connected by the relation $x=\sin y$, we write $y=\sin^{-1}x$ and call y the inverse sine of x. To each value of x in the range $-1\leqslant x\leqslant 1$ there is an infinite number of values of y, but there is one, and only one, value between $-\frac{1}{2}\pi$ and $+\frac{1}{2}\pi$. The angle between $-\frac{1}{2}\pi$ and $+\frac{1}{2}\pi$ whose sine is x is called the *principal value* of $\sin^{-1}x$. The principal value of $\tan^{-1} x$ is defined similarly.

If $y=\sin^{-1}x$, where $-\frac{1}{2}\pi<y<\frac{1}{2}\pi$,

$$x=\sin y, \quad \frac{dx}{dy}=\cos y,$$

and so, by rule VI, $$\frac{dy}{dx}=\frac{1}{\cos y}.$$

Now when $-\frac{1}{2}\pi<y<\frac{1}{2}\pi$, $\cos y>0$; hence $\cos y=+\sqrt{(1-x^2)}$

and $$\frac{dy}{dx}=\frac{1}{+\sqrt{(1-x^2)}}.$$

The angle between 0 and π whose cosine is x is taken as the principal value of $\cos^{-1}x$. The method given above may be used to show that if $y=\cos^{-1}x$, where $0<y<\pi$,

$$\frac{dy}{dx}=-\frac{1}{\sqrt{1-x^2}}.$$

9.7. Differentiation of e^x

If $y=e^x$,

$$\frac{dy}{dx}=\lim_{h\to 0}\frac{e^{x+h}-e^x}{h}$$

$$=\lim_{h\to 0} e^x\frac{(e^h-1)}{h}.$$

But (see § 5.14, Example 3), $\lim_{h\to 0}\frac{e^h-1}{h}=1$;

$$\therefore\ \frac{dy}{dx}=e^x.$$

9.8. Differentiation of log x

If $y=\log x$

$$x=e^y,$$

$$\frac{dx}{dy}=e^y \quad \text{by § 9.7}$$

$$=x$$

$$\therefore\ \frac{dy}{dx}=\frac{1}{x}.$$

9.9. Standard results (algebraic, logarithmic and exponential functions)

Unless otherwise stated, the base of logarithms is e.

	y	$\frac{dy}{dx}$	*Notes as to Method*
(i)	x^n	nx^{n-1}	See § 9.4.
(ii)	u^n	$nu^{n-1}\frac{du}{dx}$	By rule V, $\frac{d}{dx}(u^n)=\frac{d}{du}(u^n)\times\frac{du}{dx}$.
(iii)	e^{ax}	ae^{ax}	See § 9.7 and § 9.13.
(iv)	$\log x$	$\frac{1}{x}$	See § 9.8.
(v)	$\log u$	$\frac{1}{u}\frac{du}{dx}$	By rule V, $\frac{d}{dx}(\log u)=\frac{d}{du}(\log u)\times\frac{du}{dx}$
(vi)	a^x	$a^x \log a$	See § 9.15, Example 5.
(vii)	$\log_a x$	$\frac{1}{x}\log_a e$	See § 9.15, Example 6.

9.10. Standard results (trigonometrical functions): x is in radians

	y	$\frac{dy}{dx}$	*Notes as to Method*
(i)	$\sin x$	$\cos x$	See § 9.5.
(ii)	$\cos x$	$-\sin x$	
(iii)	$\tan x$	$\sec^2 x$	Write $\tan x=\sin x/\cos x$ and use rule IV.
(iv)	$\cot x$	$-\operatorname{cosec}^2 x$	Write $\cot x=\cos x/\sin x$ and use rule IV.
(v)	$\sec x$	$\sec x \tan x$	Write $\sec x=1/\cos x$ and use rule IV.
(vi)	$\operatorname{cosec} x$	$-\operatorname{cosec} x \cot x$	Write $\operatorname{cosec} x=1/\sin x$ and use rule IV.
(vii)	$\sin (ax+b)$	$a \cos (ax+b)$	See § 9.13.
(viii)	$\cos (ax+b)$	$-a \sin (ax+b)$	See § 9.13.

9.11. Standard results (hyperbolic functions)

	y	$\frac{dy}{dx}$	*Notes as to Method*
(i)	$\sinh x$	$\cosh x$	Write $\sinh x=\frac{1}{2}(e^x-e^{-x})$ and use § 9.9 (iii).
(ii)	$\cosh x$	$\sinh x$	Write $\cosh x=\frac{1}{2}(e^x+e^{-x})$ and use § 9.9 (iii).
(iii)	$\tanh x$	$\operatorname{sech}^2 x$	Write $\tanh x=\sinh x/\cosh x$ and use rule IV.
(iv)	$\coth x$	$-\operatorname{cosech}^2 x$	Write $\coth x=\cosh x/\sinh x$ and use rule IV.
(v)	$\operatorname{sech} x$	$-\operatorname{sech} x \tanh x$	Write $\operatorname{sech} x=1/\cosh x$ and use rule IV.
(vi)	$\operatorname{cosech} x$	$-\operatorname{cosech} x \coth x$	Write $\operatorname{cosech} x=1/\sinh x$ and use rule IV.

9.12. Standard results (inverse trigonometrical and hyperbolic functions)

y	$\dfrac{dy}{dx}$	*Notes as to Method*
(i) $\sin^{-1}\dfrac{x}{a}$	$\dfrac{1}{\sqrt{(a^2-x^2)}}$	See § 9.6; $-\frac{1}{2}\pi<y<\frac{1}{2}\pi$.
(ii) $\cos^{-1}\dfrac{x}{a}$	$\dfrac{-1}{\sqrt{(a^2-x^2)}}$	See § 9.6; $0<y<\pi$.
(iii) $\tan^{-1}\dfrac{x}{a}$	$\dfrac{a}{a^2+x^2}$	$x=a\tan y$; find $\dfrac{dx}{dy}$ in terms of x and use rule VI; $-\frac{1}{2}\pi<y<\frac{1}{2}\pi$.
(iv) $\sinh^{-1}\dfrac{x}{a}$	$\dfrac{1}{\sqrt{(x^2+a^2)}}$	$x=a\sinh y$; find $\dfrac{dx}{dy}$ in terms of x and use rule VI.
(v) $\cosh^{-1}\dfrac{x}{a}$	$\dfrac{1}{\sqrt{(x^2-a^2)}}$	$x=a\cosh y$; find $\dfrac{dx}{dy}$ in terms of x and use rule VI; $0<a<x$ and $y>0$.
(vi) $\tanh^{-1}\dfrac{x}{a}$	$\dfrac{a}{a^2-x^2}$	$x=a\tanh y$; find $\dfrac{dx}{dy}$ in terms of x and use rule VI; $x^2<a^2$.
(vii) $\coth^{-1}\dfrac{x}{a}$	$\dfrac{-a}{x^2-a^2}$	$x=a\coth y$; find $\dfrac{dx}{dy}$ in terms of x and use rule VI; $x^2>a^2$.

9.13. Extensions of standard results

The above standard results may be used in conjunction with the six general rules to find the derivative of a given function.

For example, if $y=f(ax+b)$, we write $y=f(u)$, where $u=ax+b$, and by rule V, $\dfrac{dy}{dx}=af'(ax+b)$.

Applying this result, we have

$$\frac{d}{dx}(ax+b)^n=na(ax+b)^{n-1}, \text{ using § 9.4,}$$

$$\frac{d}{dx}(e^{ax})=ae^{ax}, \text{ using § 9.7,}$$

$$\frac{d}{dx}\log(3x+4)=\frac{3}{3x+4}, \text{ using § 9.8,}$$

$$\frac{d}{dx}\sin(2x-5)=2\cos(2x-5) \text{ using § 9.5.}$$

9.14. Differentiation of logarithmic functions

If $y=\log f(x)$, we write $y=\log u$, where $u=f(x)$, and obtain by rule V

$$\frac{dy}{dx}=\frac{f'(x)}{f(x)}.$$

For example, $\frac{d}{dx}\log(x^3+x-9)=\frac{3x^2+1}{x^3+x-9}$; $\frac{d}{dx}\log\sin 2x=2\cot 2x$.

Some logarithmic functions are more easily differentiated if the laws of logarithms are used to simplify them.

Example 1

If
$$y=\log\left\{e^{-x}\sqrt{\left(\frac{1+2x}{1-2x}\right)}\right\}$$
$$=-x+\tfrac{1}{2}\{\log(1+2x)-\log(1-2x)\},$$
$$\frac{dy}{dx}=-1+\tfrac{1}{2}\left\{\frac{2}{1+2x}+\frac{2}{1-2x}\right\}$$
$$=\frac{1+4x^2}{1-4x^2}.$$

9.15. Logarithmic differentiation

In the case of complicated products or quotients or functions of the form u^v, where u and v are both variable, it is advisable to take logarithms before differentiating. This process is called *logarithmic differentiation.*

Example 2

If
$$y=\frac{x(1+x^2)^3}{\sqrt[3]{(1+x^3)}},$$
$$\log y=\log x+3\log(1+x^2)-\tfrac{1}{3}\log(1+x^3).$$

Differentiating with respect to x we have, by § 9.9 (v),

$$\frac{1}{y}\frac{dy}{dx}=\frac{1}{x}+\frac{6x}{1+x^2}-\frac{x^2}{1+x^3}$$
$$=\frac{1+7x^2+6x^5}{x(1+x^2)(1+x^3)}$$
$$\therefore\ \frac{dy}{dx}=\frac{(1+7x^2+6x^5)(1+x^2)^2}{(1+x^3)^{4/3}}.$$

Example 3

Differentiate with respect to x,

$$\text{(i)} \sqrt{\left(\frac{a^2+ax+x^2}{a^2-ax+x^2}\right)}, \qquad \text{(ii)} (\cosh x)^x. \qquad \text{[L.U.]}$$

(i) Let
$$y=\sqrt{\left(\frac{a^2+ax+x^2}{a^2-ax+x^2}\right)}.$$

Then
$$\log y=\tfrac{1}{2}\{\log (a^2+ax+x^2)-\log (a^2-ax+x^2)\}$$

$$\therefore \frac{1}{y}\frac{dy}{dx}=\frac{1}{2}\left\{\frac{a+2x}{a^2+ax+x^2}-\frac{(-a+2x)}{a^2-ax+x^2}\right\}$$

$$=\frac{1}{2}\left\{\frac{2a(a^2+x^2)-4ax^2}{(a^2+ax+x^2)(a^2-ax+x^2)}\right\}$$

$$\frac{dy}{dx}=\frac{a(a^2-x^2)}{\sqrt{\{(a^2+ax+x^2)(a^2-ax+x^2)^3\}}}.$$

(ii) Let
$$u=(\cosh x)^x.$$

Then
$$\log u=x \log (\cosh x)$$

$$\therefore \frac{1}{u}\frac{du}{dx}=\log (\cosh x)+x \tanh x,$$

$$\frac{du}{dx}=\{\log (\cosh x)+x \tanh x\}(\cosh x)^x.$$

Example 4

If $y=uvw$, where u, v and w are all functions of x, then

$$\log y=\log u+\log v+\log w,$$

$$\therefore \frac{1}{y}\frac{dy}{dx}=\frac{1}{u}\frac{du}{dx}+\frac{1}{v}\frac{dv}{dx}+\frac{1}{w}\frac{dw}{dx} \quad \text{by § 9.9 (v)},$$

$$\frac{dy}{dx}=vw\frac{du}{dx}+uw\frac{dv}{dx}+uv\frac{dw}{dx}.$$

This is an extension of the product rule.

Example 5

If $y=a^x$, where a is a positive constant, then

$$\log y=x \log a$$

$$\therefore \frac{1}{y}\frac{dy}{dx}=\log a$$

$$\text{i.e. } \frac{dy}{dx}=a^x \log a.$$

Example 6

If $y=\log_a x$, then

$$a^y=x$$

and so

$$y \log a=\log x.$$

Hence

$$\frac{dy}{dx}.\log a=\frac{1}{x}$$

$$\frac{dy}{dx}=\frac{1}{x \log a}$$

$$=\frac{1}{x} \log_a e.$$

9.16. Successive differentiation

If y is a function of x, then, in general, $\frac{dy}{dx}$ is also a function of x. The derivative of $\frac{dy}{dx}$ i.e. $\frac{d}{dx}\left(\frac{dy}{dx}\right)$ is denoted by $\frac{d^2y}{dx^2}$, by y_2 or by y''; the derivative of $\frac{d^2y}{dx^2}$ is denoted by $\frac{d^3y}{dx^3}$, by y_3 or by y''' and so on.

Below are some examples of functions for which a general formula can be found for the nth derivative; such formulae cannot usually be obtained.

(i) If
$$y=(ax+b)^m,$$
$$y_1=am(ax+b)^{m-1},$$
$$y_2=a^2m(m-1)(ax+b)^{m-2},$$
$$\cdot \quad \cdot \quad \cdot \quad \cdot \quad \cdot \quad \cdot$$

and, in general,

$$y_n=a^nm(m-1)(m-2)\ldots(m-n+1)(ax+b)^{m-n}.$$

(ii) If
$$y=\log(ax+b),$$
$$y_n=\frac{(-1)^{n-1}(n-1)!\,a^n}{(ax+b)^n}.$$

(iii) If
$$y=\sin(ax+b),$$
$$y_1=a\cos(ax+b)=a\sin(ax+b+\tfrac{1}{2}\pi),$$
$$y_2=-a^2\sin(ax+b)=a^2\sin(ax+b+\pi),$$
$$y_3=a^3\sin(ax+b+\tfrac{3}{2}\pi)$$

and, in general,

$$y_n=a^n\sin(ax+b+\tfrac{1}{2}n\pi).$$

Similarly, if $y=\cos(ax+b)$,

$$y_n=a^n\cos(ax+b+\tfrac{1}{2}n\pi).$$

(iv) If $y = e^{ax} \sin bx$,

$$y_1 = e^{ax}(a \sin bx + b \cos bx)$$
$$= Re^{ax} \sin (bx + \alpha)$$

where (see § 1.5, Example 4)

$$R = \sqrt{(a^2+b^2)} \text{ and } \cos \alpha : \sin \alpha : 1 = a : b : R.$$

Repeating the process, we have

$$y_2 = R^2 e^{ax} \sin (bx + 2\alpha),$$

and, in general,

$$y_n = R^n e^{ax} \sin (bx + n\alpha).$$

Similarly, if $y = e^{ax} \cos bx$,

$$y_n = R^n e^{ax} \cos (bx + n\alpha),$$

where R and α are as defined above.

Applications of these results are given in the following examples:

Example 7

Find the fourth derivative of $\dfrac{1+3x}{2-3x-2x^2}$ *with respect to* x.

If
$$y = \frac{1+3x}{2-3x-2x^2}$$
$$= \frac{1}{1-2x} - \frac{1}{2+x},$$

by (i),
$$y_4 = \frac{2^4(4\,!)}{(1-2x)^5} - \frac{4\,!}{(2+x)^5}$$
$$= 24\left\{\frac{16}{(1-2x)^5} - \frac{1}{(2+x)^5}\right\}.$$

Example 8

Find the sixth derivative of $e^{3x} \cos 3x$ *with respect to* x.

By (iv), if
$$y = e^{3x} \cos 3x,$$
$$y_1 = e^{3x}(3 \cos 3x - 3 \sin 3x)$$
$$= 3\sqrt{2}e^{3x} \cos (3x + \tfrac{1}{4}\pi)$$

and
$$y_6 = (3\sqrt{2})^6 e^{3x} \cos (3x + \tfrac{3}{2}\pi)$$
$$= 5832 e^{3x} \sin 3x.$$

9.17. The theorem of Leibniz

Let u and v be functions of x and let $u_r = \dfrac{d^r u}{dx^r}$, $v_r = \dfrac{d^r v}{dx^r}$.

Then, if
$$y = uv,$$
$$y_1 = u_1 v + v_1 u,$$
$$y_2 = u_2 v + 2u_1 v_1 + u v_2,$$
$$y_3 = u_3 v + 3u_2 v_1 + 3u_1 v_2 + u v_3.$$

These results suggest that when n is a positive integer

$$y_n = u_n v + {}^nC_1 u_{n-1} v_1 + {}^nC_2 u_{n-2} v_2 + \ldots + {}^nC_r u_{n-r} v_r + \ldots$$
$$+ {}^nC_{n-1} u_1 v_{n-1} + u v_n \quad . \quad \text{(i)}$$

where the coefficients ${}^nC_1$, ${}^nC_2, \ldots$, ${}^nC_r, \ldots$ are those which occur in the binomial expansion of $(1+x)^n$.

This is Leibniz's theorem, and we prove it by induction.

Assuming that (i) is true for $n=k$, we obtain by differentiation

$$y_{k+1} = u_{k+1} v + (1 + {}^kC_1) u_k v_1 + ({}^kC_1 + {}^kC_2) u_{k-1} v_2 + \ldots$$
$$+ ({}^kC_{r-1} + {}^kC_r) u_{k-r+1} v_r + \ldots + ({}^kC_{k-1} + 1) u_1 v_k + u v_{k+1}$$
$$= u_{k+1} v + {}^{k+1}C_1 u_k v_1 + {}^{k+1}C_2 u_{k-1} v_2 + \ldots + {}^{k+1}C_r u_{k-r+1} v_r + \ldots$$
$$+ {}^{k+1}C_k u_1 v_k + u v_{k+1}$$

since ${}^kC_{r-1} + {}^kC_r = {}^{k+1}C_r$ by Vandermonde's theorem (p. 74). This result for y_{k+1} is exactly that which we would obtain by substituting $n=k+1$ in (i) and so, if (i) is true for $n=k$, it is true for $n=k+1$. But the theorem is true for $n=1, 2, 3$ and so it is true for all positive integral values of n.

Example 9

Find the nth derivative of x^3e^x with respect to x.

Using Leibniz's theorem with $u=e^x$ and $v=x^3$, we have $u_n=e^x$, and $v_1=3x^2$, $v_2=6x$, $v_3=6$, $v_n=0$, when $n>3$.

$$\frac{d^n}{dx^n}(x^3e^x) = e^x\{x^3 + 3nx^2 + 3n(n-1)x + n(n-1)(n-2)\}.$$

Example 10

If $y=\{x+\sqrt{(x^2+a^2)}\}^n$, prove that $(x^2+a^2)\dfrac{d^2y}{dx^2}+x\dfrac{dy}{dx}-n^2y=0$, and by differentiating this result k times show that

$$(x^2+a^2)y_{k+2} + (2k+1)xy_{k+1} + (k^2-n^2)y_k = 0, \text{ where } y_k = \frac{d^ky}{dx^k}.$$

$$y = \{x + \sqrt{(x^2+a^2)}\}^n$$

$$\frac{dy}{dx} = n\{x + \sqrt{(x^2+a^2)}\}^{n-1} \cdot \left\{1 + \frac{x}{\sqrt{(x^2+a^2)}}\right\}$$

$$= \frac{ny}{\sqrt{(x^2+a^2)}}$$

$$\therefore \sqrt{(x^2+a^2)}\frac{dy}{dx} = ny. \quad . \quad . \quad . \quad \text{(i)}$$

Differentiating again, we have

$$\sqrt{(x^2+a^2)}\,\frac{d^2y}{dx^2} + \frac{x}{\sqrt{(x^2+a^2)}}\,\frac{dy}{dx} - n\,\frac{dy}{dx} = 0,$$

$$\therefore (x^2+a^2)\,\frac{d^2y}{dx^2} + x\,\frac{dy}{dx} - n^2y = 0 \quad \text{by (i)}.$$

We now differentiate this result k times using Leibniz's theorem for each product :

$$\{(x^2+a^2)y_{k+2}+2kxy_{k+1}+k(k-1)y_k\}+\{xy_{k+1}+ky_k\}-n^2y_k=0,$$

i.e. $$(x^2+a^2)y_{k+2}+(2k+1)xy_{k+1}+(k^2-n^2)y_k=0.$$

Further examples of this kind will be found in Chapter 11.

9.18. To find $\dfrac{dy}{dx}$ and $\dfrac{d^2y}{dx^2}$

(*a*) when x and y are given in terms of a parameter ;

(*b*) in the case of an implicit function (i.e. one in which neither variable can be conveniently expressed in terms of the other).

(*a*) Suppose x and y are given in terms of a parameter t by the equations $x=x(t)$, $y=y(t)$.

Then denoting $\dfrac{dx}{dt}$ by $\dot{x}$, $\dfrac{dy}{dt}$ by $\dot{y}$, $\dfrac{d^2x}{dt^2}$ by $\ddot{x}$ and $\dfrac{d^2y}{dt^2}$ by $\ddot{y}$, we have

$$\frac{dy}{dx}=\frac{dy}{dt}\times\frac{dt}{dx}=\frac{dy}{dt}\div\frac{dx}{dt}=\dot{y}/\dot{x}$$

and $$\frac{d^2y}{dx^2}=\frac{d}{dx}\left(\frac{dy}{dx}\right)=\frac{d}{dt}\left(\frac{\dot{y}}{\dot{x}}\right)\frac{dt}{dx}=(\dot{x}\ddot{y}-\ddot{x}\dot{y})/\dot{x}^3.$$

Example 11

If $x=\sinh t$, $y=\sinh pt$, prove that

$$(1+x^2)\frac{d^2y}{dx^2}+x\frac{dy}{dx}=p^2y.$$

$$x=\sinh t,\qquad y=\sinh pt,$$
$$\dot{x}=\cosh t\qquad \dot{y}=p\cosh pt$$
$$\ddot{x}=\sinh t\qquad \ddot{y}=p^2\sinh pt.$$

$$\therefore\ \frac{dy}{dx}=\frac{\dot{y}}{\dot{x}}=\frac{p\cosh pt}{\cosh t}\quad . \quad . \quad . \quad . \qquad \text{(i)}$$

and $$\frac{d^2y}{dx^2}=(\dot{x}\ddot{y}-\ddot{x}\dot{y})/\dot{x}^3=(p^2\cosh t\sinh pt-p\sinh t\cosh pt)/\cosh^3 t\ . \qquad \text{(ii)}$$

From (i) and (ii) after simplification we obtain

$$(1+x^2)\frac{d^2y}{dx^2}+x\frac{dy}{dx}=p^2\sinh pt=p^2y.$$

(*b*) The method of dealing with implicit functions is best demonstrated by an example.

Example 12

Find $\dfrac{d^2y}{dx^2}$ *when* $ax^2+2hxy+by^2=1$ *(a, b and h being constants).*

$$ax^2+2hxy+by^2=1 \quad . \quad . \quad . \quad . \quad . \quad \text{(i)}$$

$$\therefore\ 2ax+2h\left(y+x\frac{dy}{dx}\right)+2by\frac{dy}{dx}=0$$

$$\therefore\ \frac{dy}{dx}=-\frac{ax+hy}{hx+by} \quad . \quad . \quad . \quad . \quad . \quad . \quad \text{(ii)}$$

$$\frac{d^2y}{dx^2}=-\frac{(hx+by)\left(a+h\dfrac{dy}{dx}\right)-(ax+hy)\left(h+b\dfrac{dy}{dx}\right)}{(hx+by)^2}$$

$$=\frac{(h^2-ab)\left(y-x\dfrac{dy}{dx}\right)}{(hx+by)^2}$$

$$=\frac{(h^2-ab)(ax^2+2hxy+by^2)}{(hx+by)^3} \quad \text{by (ii)}$$

$$=\frac{h^2-ab}{(hx+by)^3} \quad \text{by (i).}$$

Exercises 9 (*a*)

For brevity, y_r is written for d^ry/dx^r.

1. If $y=\tan^{-1}x$, prove that $(1+x^2)y_2+2xy_1=0$ and deduce that
$$(1+x^2)y_{n+2}+2(n+1)xy_{n+1}+n(n+1)y_n=0.$$

2. If $y=\log\{\sqrt{(x+1)}+\sqrt{(x-1)}\}$, prove that $(x^2-1)y_2+xy_1=0$ and that $(x^2-1)y_{n+2}+(2n+1)xy_{n+1}+n^2y_n=0$.

3. If $y=\sin\log(1+x)$, prove that $(1+x)^2y_2+(1+x)y_1+y=0$ and that $(1+x)^2y_{n+2}+(2n+1)(1+x)y_{n+1}+(n^2+1)y_n=0$.

4. If $y=\sqrt{(1-x^2)}\sin^{-1}x$, show that $(1-x^2)y_1+xy=1-x^2$ and that, when $n\geqslant 2$, $(1-x^2)y_{n+2}-(2n+1)xy_{n+1}-(n^2-1)y_n=0$.

5. If $y=(\sinh^{-1}x)/\sqrt{(1+x^2)}$, prove that $(1+x^2)y_1+xy=1$ and that $(1+x^2)y_{n+2}+(2n+3)xy_{n+1}+(n+1)^2y_n=0$.

6. If $y=\sin(m\sinh^{-1}x)$, prove that $(1+x^2)y_2+xy_1+m^2y=0$ and that $(1+x^2)y_{n+2}+(2n+1)xy_{n+1}+(n^2+m^2)y_n=0$.

7. If $y=(x^2-1)^n$, where n is a constant, prove that $(1-x^2)y_1+2nxy=0$. Deduce that $(1-x^2)y_{n+2}-2xy_{n+1}+n(n+1)y_n=0$. [Leeds.]

8. If $y=x^m\log x$, show that $xy_1=my+x^m$. Differentiate this equation n times, where $n>m$. [L.U.]

9. Find y_2 when (i) $x^2/a^2+y^2/b^2=1$; (ii) $x^{2/3}+y^{2/3}=a^{2/3}$; (iii) $x^3+y^3=a^3$.

10. Find y_2 when $x=3\cos\theta-\cos^3\theta$, $y=3\sin\theta-\sin^3\theta$.

11. Find y_2 when $x=a\cos^3\theta$, $y=a\sin^3\theta$.

12. If $y=(2x-\pi)^5\sin(x/3)$, find the value of y_7 when $x=\frac{1}{2}\pi$. [L.U.]

13. By first proving that the nth derivative of $\cos\pi x$ with respect to x is $\pi^n\cos(\pi x+\frac{1}{2}n\pi)$, show that the $2m$th derivative of $x^2\cos\pi x$ when $x=1$ has the value $(-1)^{m+1}\pi^{2m-2}(\pi^2+2m-4m^2)$. [L.U.]

14. If $y_0=e^{x^2}$ and $y_n=\dfrac{d^ny_0}{dx^n}$ for all positive integers n, prove that

$$y_{n+1}-2xy_n-2ny_{n-1}=0.$$

If $u_n=e^{-x^2}y_n$, prove by induction with respect to r that, for $r\leqslant n$,

$$\frac{d^ru_n}{dx^r}=2^rn(n-1)\ldots(n-r+1)u_{n-r},$$

and hence evaluate $\dfrac{d^nu_n}{dx^n}$.

Miscellaneous Exercises 9

For brevity, y_r is written for d^ry/dx^r.

1. (i) Differentiate $\{x(1-x)\}^{\frac{1}{2}}$ and $\cot\{1/(x^2+1)\}$.
   (ii) If $y=\sqrt{(a^2-x^2)}$, prove that $x(a^2-x^2)y_2=a^2y_1$. [Durham.]

2. (i) Differentiate $(x^2+2x+7)/(3x-1)^{\frac{1}{2}}$ and $1/\{\sin(x^3)\}$.
   (ii) If $y=x\sin(1/x)$, show that $x^4y_2+y=0$. [Durham.]

3. (i) Differentiate $(x-1/x^4)^{\frac{1}{2}}$ and $(1-\sin x)/(1-\cos x)$.
   (ii) Prove that, if $\sin\theta=2\sin\phi$ and $x=\cos\theta-2\cos\phi$, then
$$dx/d\theta=x\tan\phi.$$ [Durham.]

4. Differentiate with respect to x:
   (a) e^{-1/x^2}; (b) $\sqrt{\{\cot(1/x)\}}$; (c) $e^{-x}(3x+5)/(7x-1)$. [Sheffield.]

5. Differentiate $y=\sin^{-1}\{2ax\sqrt{(1-a^2x^2)}\}$ and $y=(e^{\cos x}-1)/(e^{\cos x}+1)$. [Durham.]

6. Find the derivatives with respect to x of
   (a) $\sqrt{\{(2+\sin^2 x)/(1-\sin x)\}}$; (b) $\tan^{-1}\{1/(1-x^2)\}$. [Sheffield.]

7. Differentiate with respect to x:
   (i) $(x^2-2)\sin^{-1}(\frac{1}{2}x)+\frac{1}{2}x\sqrt{(4-x^2)}$; (ii) $\log\sec^2(x/a)$;
   (iii) $(1+e^{2x})/(1-e^{2x})$. [Sheffield.]

8. (i) Differentiate with respect to x:

$$\frac{x \sin x}{(1+\cos x)}, \quad \tan^{-1}\{(a+bx)/(a-bx)\}, \quad \log(\sec x+\tan x).$$

(ii) If $y^2=\sec 2x$, prove that $y_2+y=3y^5$.

9. (i) If $x=a(2\cos t+\cos 2t)$, $y=a(2\sin t-\sin 2t)$, find dy/dx in its simplest form in terms of t and prove that

$$8a\frac{d^2y}{dx^2}=\operatorname{cosec}\tfrac{3}{2}t\sec^3\tfrac{1}{2}t.$$

(ii) Prove that $\dfrac{d}{d\theta}\left(\dfrac{1}{\sin\theta\cos\theta}\right)=\dfrac{2}{\cos^2\theta}-\dfrac{1}{\sin^2\theta\cos^2\theta}$

and hence find $\displaystyle\int\frac{d\theta}{\sin^2\theta\cos^2\theta}$. [L.U.]

10. (i) If $y=(\log x)^x$, find dy/dx.

(ii) If $y=\tan(m\tan^{-1}x)$, prove that $(1+x^2)y_2=2(my-x)y_1$.

(iii) Given that $x=4b\cos\theta-b\cos 4\theta$, $y=4b\sin\theta-b\sin 4\theta$, find dy/dx in terms of θ, and prove that $d^2y/dx^2=(5/16b)\sec^3\frac{5}{2}\theta\operatorname{cosec}\frac{3}{2}\theta$. [L.U.]

11. (i) If $y=\sin(m\sin^{-1}x)$, prove that $(1-x^2)y_2-xy_1+m^2y=0$.

(ii) If $y=\tan^{-1}(\sinh x)$, prove that $y_2+(\tan y)y_1{}^2=0$. [L.U.]

12. (i) Evaluate $\dfrac{d}{dx}(x^2e^{2x}\log 2x)$ when $x=\frac{1}{2}$.

(ii) If $y=3x/(x-2)(x+1)$, show that dy/dx is negative for all real values of x.

(iii) If $y=(n+1+x)^{n+1}/(n+x)^n$, n is a fixed positive integer and x is positive, find dy/dx by logarithmic differentiation, and show that y increases with x.

Hence, or otherwise, show that $\{1+(x/n)\}^n<\{1+x/(n+1)\}^{n+1}$. [L.U.]

13. (i) Differentiate with respect to x

$$\sec(x^{3/4}) \text{ and } \log(\tan e^{\sqrt{x}}).$$

(ii) If $y=a\cos(\log bx)+b\sin(\log ax)$, show that

$$x^2y_2+xy_1+y=0.$$ [L.U.]

14. (i) Obtain in their simplest forms the derivatives of

(a) $\log\log x$, (b) $\tanh^{-1}\{(2\sqrt{x})/(1+x)\}+\tan^{-1}\{(2\sqrt{x})/(1-x)\}$.

(ii) If $x=\lambda^3+1$, $y=\lambda^2+1$, where λ is a variable, show that

$$\frac{d^2y}{dx^2}\Big/\left(\frac{dy}{dx}\right)^4$$

is constant. [L.U.]

15. If $y=\dfrac{d^n}{dx^n}(\tan^{-1} x)$, show that $(x^2+1)y_2+2(n+1)xy_1+n(n+1)y=0$.

16. Differentiate with respect to x the following functions:

(i) $(1-x)/(1-x^3)^{\frac{1}{2}}$; (ii) $\tan^{-1}(m \tan x)$; (iii) $\sec\{\frac{1}{2}\log(a^2+x^2)\}$. [L.U.]

17. Differentiate with respect to x

$$\log(1+\sin 2x)+2\log\{\sec(\tfrac{1}{4}\pi-x)\},$$

and express the result in its simplest form. Explain why the result is of such simple form. [L.U.]

18. Find from first principles the derivative of $x \sin x$ with respect to x.

Express in their simplest forms the derivatives with respect to x of

(i) $\tan^{-1}\{2\sqrt{x}/(1-x)\}$; (ii) $\log[e^x\{(x-1)/(x+1)\}^{\frac{1}{2}}]$.

If $x=\tan t$ and $y=\tan pt$, where p is a constant, show that

$$(1+x^2)d^2y/dx^2=2(py-x)dy/dx.$$ [L.U.]

19. Find the nth differential coefficients with respect to x of

(i) $\cos x$; (ii) $\log\{(1-x)/(1+x)\}$.

Find the value of $\dfrac{d^5}{dx^5}(16\sin^4 x\cos x)$ when $x=\pi/10$. [L.U.]

20. (*a*) Differentiate $(1/2a)\log\{(x-a)/(x+a)\}+(1/a)\coth^{-1}(x/a)$. What deductions can be made from the result?

(*b*) If $y=e^{ax}\cos bx$, show that $y_n=r^n e^{ax}\cos(bx+n\theta)$, where $r^2=a^2+b^2$, and $\tan\theta=b/a$. [L.U.]

21. (*a*) Differentiate

(i) $\tan^{-1}\{(1-\sqrt{x})/(1+\sqrt{x})\}$; (ii) x^x.

(*b*) If $x=\cos t$, $y=\cos 2pt$, prove that $(1-x^2)d^2y/dx^2-xdy/dx+4p^2y=0$, and deduce that $(1-x^2)y_{n+2}-(2n+1)xy_{n+1}+(4p^2-n^2)y_n=0$. [L.U.]

22. (*a*) Define the derivative of a function and from your definition, assuming the exponential theorem, prove that the derivative of a^x is $a^x \log a$.

(*b*) If $y=t^m+t^{-m}$ and $x=t+t^{-1}$, prove that

(i) $(x^2-4)(dy/dx)^2=m^2(y^2-4)$.

(ii) $(x^2-4)d^2y/dx^2+xdy/dx-m^2y=0$. [L.U.]

23. Differentiate with respect to x, expressing each differential coefficient in its simplest form:—

(*a*) $(3x+1)^5/(2-x)^{10}$; (*b*) $\{x+\sqrt{(1+x^2)}\}^n$; (*c*) $\sin^{-1}\{2x\sqrt{(1-x^2)}\}$. [L.U.]

24. Differentiate $\tan^{-1}\{\frac{1}{2}x^2/\sqrt{(1+x^2)}\}$; also show that

$$\frac{d^2}{dx^2}(x\tan^{-1}x)=2\sin^4\theta, \text{ where } x=\cot\theta.$$ [L.U.]

25. (i) Differentiate with respect to x, the functions $\log\tanh\frac{1}{2}x$ and $(1-x^2)^{3/2}\sin^{-1}x$.

(ii) Find and simplify the differential coefficient of $\log x-(x-1)/\sqrt{x}$ with respect to x, and hence show that, if $x>1$, $\log x<(x-1)/\sqrt{x}$. [L.U.]

26. Find from first principles the derivative of $x\cos x$ with respect to x. Express in their simplest forms the derivatives with respect to x of

(i) $\tan^{-1}\{(3-x)\sqrt{x}/(1-3x)\}$, (ii) $x^x\log x$. [L.U.]

27. If $y=\sin n\theta\operatorname{cosec}\theta$ and $x=\cos\theta$, prove that

$$(1-x^2)dy/dx-xy+n\cos n\theta=0$$

and

$$(1-x^2)d^2y/dx^2-3xdy/dx+(n^2-1)y=0.$$

Show that, if $n=7$, the latter equation is satisfied by a polynomial of the form $x^6+bx^4+cx^2+d$, and find the values of b, c and d. [L.U.]

28. Define the derivative of a function, and from the definition find the derivative of $\sin x^2$.

Find (i) dy/dx if $y=\sin(x+y)^2$,

(ii) d^2y/dx^2 if $x=3\cos\theta-\cos 3\theta$, $y=3\sin\theta-\sin 3\theta$. [L.U.]

29. (i) Differentiate $\sin^{-1}\{x^2/(a^2+x^2)\}$ with respect to x.

(ii) If $y=x^n\log x$, prove that $xy_{n+1}=n!$ [L.U.]

30. Find from first principles the differential coefficient of $\tan^{-1}x$ with respect to x. Hence prove that

$$\frac{d}{dx}\{\tan^{-1}(ax+b)\}=a/\{1+(ax+b)^2\},$$

stating any general theorems on differentiation used in your proof.

If $y=2x-\tan^{-1}x$, prove that

$$\frac{d^2x}{dy^2}=-2x(1+x^2)/(1+2x^2)^3.$$ [L.U.]

31. Obtain and simplify the first derivatives of the two functions

$\cos^{-1}\{(a\cos x+b)/(a+b\cos x)\}$ and $\tan^{-1}[\sqrt{\{(a-b)/(a+b)\}}\tan\frac{1}{2}x]$

and explain the significance of your results. [L.U.]

32. (i) Prove that $\dfrac{1}{n!}\left(\dfrac{d}{dx}\right)^n\dfrac{1}{x(1-x)}=\dfrac{(-1)^n}{x^{n+1}}+\dfrac{1}{(1-x)^{n+1}}$.

(ii) If $y=A\tan\frac{1}{2}\theta+B(2+\theta\tan\frac{1}{2}\theta)$, where A and B are any constants, prove that $(1+\cos\theta)\dfrac{d^2y}{d\theta^2}=y$. [L.U.]

33. Differentiate with respect to x

(i) $\cos^{-1} 2x\sqrt{(1-x^2)}$; (ii) $\tan^{-1}\{(\cos x-\sin x)/(\cos x+\sin x)\}$

reducing each result to its simplest form.

If $y=ax \sin (b/x)$, prove that $x^4y_2+b^2y=0$. [L.U.]

34. Find the functions $p(x)$ and $q(x)$ such that $y=(\sin^{-1} x)^2$ satisfies the differential equation $p(x)y_{n+2}+q(x)y_{n+1}+n^2y_n=0$ provided $n \geqslant 1$. [Durham.]

35. (i) Find the derivatives with respect to x of the functions

$$\sin^{-1} \{\sqrt{(x-1)}\}, \quad \tanh^{-1} \{\sqrt{(2-x)}\}$$

for $1<x<2$.

(ii) If $(x-a)^2+(y-b)^2=r^2$, where a, b, r are independent of x and y, find a relation between y_1, y_2, y_3 which is independent of a, b, r. [L.U.]

36. (i) Differentiate with respect to x

(*a*) $\sin^{-1} \sqrt{(1-x^2)}$; (*b*) $\tan^{-1} (x/e^{x^2})+\tan^{-1} (e^{x^2}/x)$.

(ii) Find the nth differential coefficient of $1/(x-1)^2(x-2)$. [L.U.]

37. (i) Differentiate with respect to x

(*a*) $\tan^{-1}\{4\sqrt{x}/(1-4x)\}$; (*b*) $\log [e^x\{(x-2)/(x+2)\}^{3/4}]$.

(ii) If

$$x=a \sin t-b \sin (at/b),$$
$$y=a \cos t-b \cos (at/b),$$

where a, b are independent of t, obtain expressions for dy/dx, d^2y/dx^2 in terms of t.

38. (*a*) Differentiate the following functions with respect to x:

$$\log (\sin x),\ \tan^{-1} (\log x),\ (x+1/x)^x.$$

(*b*) Prove that the differential coefficient of the function

$$(x+1)(x+2)\dots(x+n)$$

has the value $$n!\left\{1+\frac{1}{2}+\frac{1}{3}+\dots+\frac{1}{n}\right\}$$

when $x=0$. [L.U.]

39. Find the following derivatives:

(i) $\dfrac{d}{dx}\{\sin^{-1} (\cos x)\}$ $(0<x<\pi)$,

(ii) $\dfrac{dy}{dx}$ where $x=\cos 2\theta$ and $y=\theta+\sin 2\theta$,

(iii) $\dfrac{d^3}{dx^3}\{e^{2x} \tan^{-1} x\}$ when $x=0$. [L.U.]

9.19. The gradient of a curve. The positive tangent

If P and Q are neighbouring points on a continuous curve, the tangent to the curve at P is defined as the limiting position of the chord PQ as Q moves along the curve towards P. The gradient of the curve at P is defined as the gradient of the tangent at P.

A line PT drawn along the tangent at P in the direction of x increasing is said to be drawn in the positive direction of the tangent at P and is called the *positive tangent* at P. The angle between the positive x-axis and the positive tangent at P is generally denoted by ψ, and fig. 22 shows that ψ is either a positive or negative acute angle.

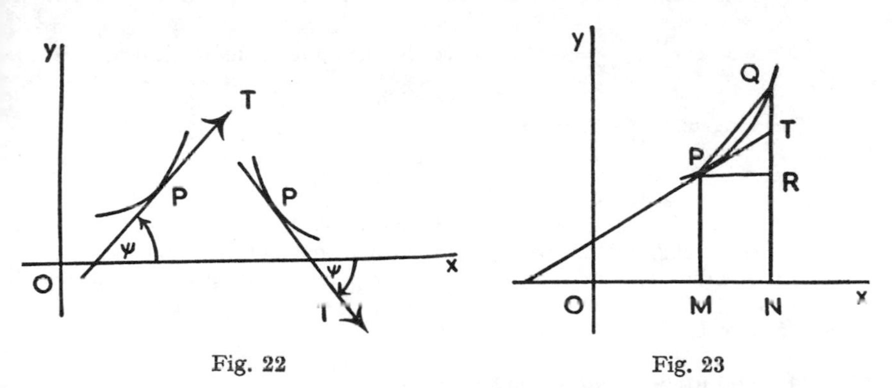

Fig. 22 Fig. 23

In fig. 23, $P(x, y)$ and $Q(x+\delta x, y+\delta y)$ are neighbouring points on the continuous curve $y=f(x)$. PM, QN are perpendicular to Ox, PR is perpendicular to QN and the positive tangent PT at P makes with Ox an angle ψ where $-\frac{1}{2}\pi<\psi<\frac{1}{2}\pi$. We shall assume that the curve has a unique tangent at P.

We have $PR=\delta x$, $QR=\delta y=f(x+\delta x)-f(x)$ and $\tan RPQ=\delta y/\delta x$ As Q moves along the curve towards coincidence with P, $\delta x \to 0$ $\angle RPQ \to \psi$ and so

$$\lim_{\delta x \to 0} \frac{\delta y}{\delta x} = \lim_{\delta x \to 0} \frac{f(x+\delta x)-f(x)}{\delta x} = \tan \psi.$$

i.e.

$$\frac{dy}{dx} = f'(x) = \tan \psi.$$

Hence $\dfrac{dy}{dx}$ measures the gradient of the curve at $P(x, y)$.

Note that the fact that the tangent at P is not parallel to Oy implies that $\delta y/\delta x$ tends to a finite limit as $\delta x \to 0$. If the tangent at P is parallel to Oy, $\delta y/\delta x$ does not tend to a finite limit,

9.20. The tangent and normal to a curve

The equation of the tangent at the point (x_1, y_1) to the curve $y=f(x)$ is

$$y-y_1=f'(x_1)(x-x_1)$$

where $f'(x_1)$ denotes the value of $f'(x)$ when $x=x_1$.

The equation of the normal at the same point is

$$f'(x_1)(y-y_1)+(x-x_1)=0.$$

Example 13

The tangent to the curve $y(1+x^2)=2$ at the point P $(2, \frac{2}{5})$ meets the curve again at Q. Find the coordinates of Q.

If $y(1+x^2)=2$, $(1+x^2)\dfrac{dy}{dx}+2xy=0.$

Hence at $P(2, \frac{2}{5})$, $\dfrac{dy}{dx}=-8/25$ and so the equation of the tangent at P is

$$8x+25y=26.$$

This tangent meets the curve again where

$$\frac{2}{1+x^2}=\frac{26-8x}{25},$$

$$4x^3-13x^2+4x+12=0.$$

This equation gives the abscissae of the three points in which the tangent at P meets the curve, and since two of these points coincide at P we know that $(x-2)^2$ is a factor of $4x^3-13x^2+4x+12$. The remaining factor is $4x+3$, whence Q is the point $(-\frac{3}{4}, \frac{32}{25})$.

9.21. The mean value theorem

The mean value theorem states that if $f(x)$ is differentiable in the interval $a \leqslant x \leqslant b$, there is a number ξ between a and b such that

$$\frac{f(b)-f(a)}{b-a}=f'(\xi).$$

Geometrically this means that if A and B are the points on the curve $y=f(x)$ (fig. 24) at which $x=a$, $x=b$ respectively, there is at least one

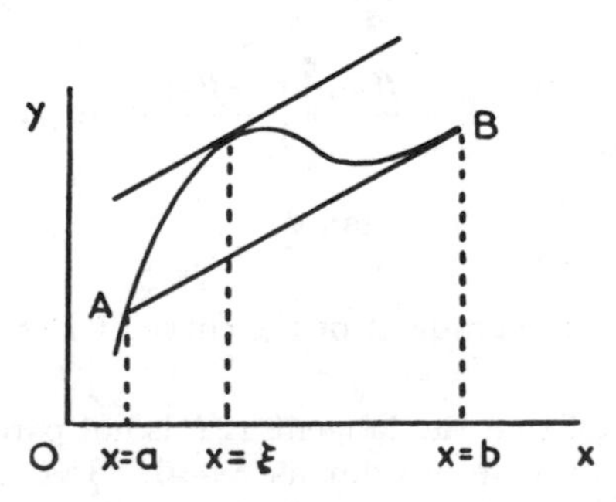

Fig. 24

point on the arc AB at which the gradient is equal to that of the chord AB.

The analytical proof of the theorem is beyond the scope of this book.

9.22. The significance of the sign of $f'(x)$

A function $f(x)$ is *increasing* in the interval $a \leqslant x \leqslant b$ if $f(x)$ increases as x increases from a to b inclusive. A function $f(x)$ is *decreasing* in the interval $a \leqslant x \leqslant b$ if $f(x)$ decreases as x increases from a to b inclusive.

If $f'(x) > 0$ throughout an interval, or even if $f'(x) > 0$ except at a finite number of points at which $f'(x) = 0$, then $f(x)$ is increasing in the interval. This appears from consideration of the curve $y = f(x)$; for $f'(x)$ measures the gradient of the curve at any point (x, y).

There is a similar test for a decreasing function.

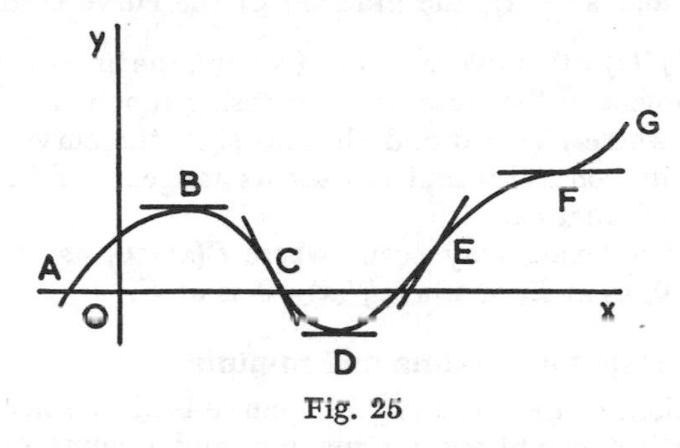

Fig. 25

Fig. 25 shows the graph of a continuous function $f(x)$ which is increasing between the points A and B, and between the points D and G. Between the points B and D, $f(x)$ is decreasing.

9.23. Maximum and minimum points

If $f'(x)$ changes sign from + to − as x increases through the value h, $f(x)$ has a *maximum* value when $x = h$, i.e. the ordinate $f(h)$ of the curve $y = f(x)$ exceeds (algebraically) neighbouring ordinates on either side.

This is clear from the fact that $f(x)$ is increasing immediately to the left of $x = h$ since $f'(x) > 0$, and decreasing to the right of $x = h$ since $f'(x) < 0$.

Since $f'(h) = 0$, it follows that the tangent to the curve $y = f(x)$ at a maximum point is parallel to the x-axis. In fig. 25, B is a maximum point.

There is a similar test for a minimum point: if $f'(x)$ changes sign from − to + as x increases through the value k, then $f(x)$ has a *minimum* value at $x = k$.

Since $f'(k) = 0$, the tangent to the curve $y = f(x)$ at a minimum point is parallel to the x-axis. In fig. 25, D is a minimum point.

Points on the curve $y = f(x)$ at which $f'(x) = 0$, i.e. at which the tangent to the curve is parallel to the x-axis are called *stationary* points. In fig. 25, B, D and F are stationary points. The points B and D at which $f'(x)=0$ and changes sign are called *turning points.*

9.24. Concavity and convexity

If the gradient of an arc of a curve increases as x increases, the arc bends upwards and it is said to be *concave up* or *convex down.* The arc lies above the tangent at any point of it.

If the gradient of an arc of a curve decreases as x increases, the arc bends downwards and it is said to be *concave down* or *convex up.* The arc lies below the tangent at any point of it.

If $f''(x)>0$ at every point of an arc, the arc is concave up, for $\dfrac{d}{dx}\{f'(x)\}>0$ and so $f'(x)$, the gradient of the curve is an increasing function. If $f''(x)<0$ at every point of an arc, the arc is concave down since the gradient of the curve is a decreasing function.

At points where $f''(x)=0$ and changes sign, the curve changes the direction of its concavity and crosses its tangent. Such points are called *points of inflexion.*

A point of inflexion may occur where $f'(x)<0$, as at C (fig. 25), where $f'(x)>0$, as at E, or where $f'(x)=0$ as at F.

9.25. Other tests for maxima and minima

Considerations of the concavity of a curve lead to other tests which provide sufficient conditions for maxima and minima of a function with continuous second derivatives :

if $f'(a)=0$ and $f''(a)<0$, $f(x)$ has a maximum value at $x=a$;
if $f'(b)=0$ and $f''(b)>0$, $f(x)$ has a minimum value at $x=b$.

Example 14

In the triangle ABC, the side BC is of length a. P is a point between B and C such that $BP=x$. The perpendicular from P to AB meets AB in D, and the perpendicular to AC through P meets AC at E. Find the position of P when $PD.PE$ is a maximum.

Show that DE is a minimum when $x=AB \cos B$, and show that the greatest value of DE is $a \sin B$ or $a \sin C$. [L.U.]

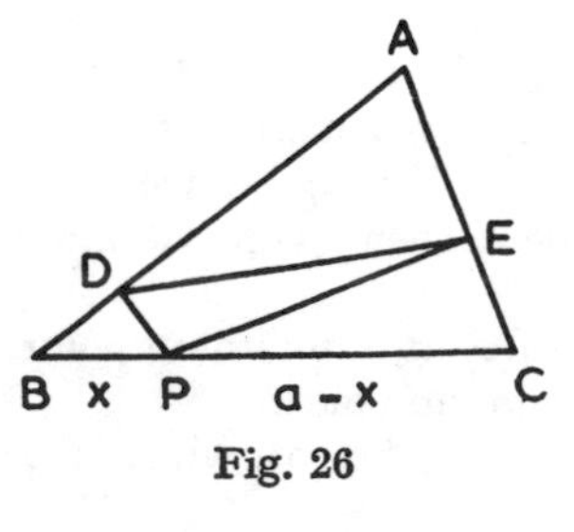

Fig. 26

If $\quad y=PD.PE$ (fig. 26)

$$=x(a-x)\sin B \sin C$$

$$\frac{dy}{dx}=(a-2x)\sin B \sin C$$

and $\quad \dfrac{d^2y}{dx^2}=-2\sin B \sin C.$

Hence $\dfrac{dy}{dx}=0$ when $x=\frac{1}{2}a$ and for this value of x $\dfrac{d^2y}{dx^2}<0$, since $\angle B$ and $\angle C$ are both less than π. It follows that the product $PD.PE$ is a maximum when P is mid-point of BC.

Now quadrilateral $ADPE$ is cyclic and AP is a diameter of its circumcircle. Hence, applying the sine rule to $\triangle DAE$ we get $DE=AP \sin A$.

Thus DE is a minimum when AP is a minimum, i.e. when AP is perpendicular to BC and $x=AB \cos B$.

DE has its greatest value when P coincides with B or C depending on whether AB or AC is greater. This value is either $AB \sin A$ or $AC \sin A$, i.e. $a \sin C$ or $a \sin B$.

Example 15

The brightness of an illuminated surface varies inversely as the square of the distance from the source and directly as the cosine of the angle which the rays make with the normal to the surface. Find at what height on the wall of a room a source of light must be placed to produce the greatest brightness at a point on the floor at a given distance a from the wall.

If, owing to the wall being insufficiently high, it is impossible to place the source of light at the point which gives maximum mathematical brightness, where should the source be placed to give the best results? Give reasons. [L.U.]

Let I be the brightness of the illumination at P, a point distant r from the source O, and let θ be the angle which the rays make with the normal to the surface at P (fig. 27).

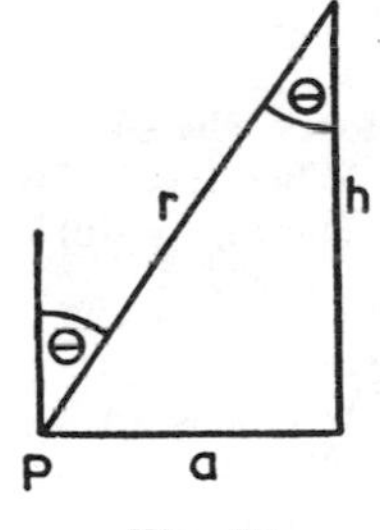

Fig. 27

Then $I=(k \cos \theta)/r^2$ where k is a positive constant

$$=(k/a^2) \sin^2 \theta \cos \theta \quad . \quad . \quad . \quad \text{(i)}$$

$$\frac{dI}{d\theta}=(k/a^2) \sin \theta \,(2 \cos^2 \theta-\sin^2 \theta),$$

and $\quad \dfrac{dI}{d\theta}=0$ when $\theta=0$, or $\tan^{-1} \pm\sqrt{2}$.

From (i), $\theta=0$ gives zero brightness.

If θ is slightly less than $\tan^{-1}\sqrt{2}$, $\dfrac{dI}{d\theta}>0$; if θ is slightly more than $\tan^{-1}\sqrt{2}$, $\dfrac{dI}{d\theta}<0$.

Thus $\tan^{-1}\sqrt{2}$ gives I_{max} and in this case O must be placed at a height $\frac{1}{2}a\sqrt{2}$ up the wall.

If the height of the wall is less than $\frac{1}{2}a\sqrt{2}$, the source of light O should be placed at the top of the wall, for I increases from zero (when $\theta=\frac{1}{2}\pi$ and O is at the foot of the wall) to its maximum value (when O is at a height $\frac{1}{2}a\sqrt{2}$) and then decreases as θ tends to zero and O moves further up the wall.

Example 16

The ends of a right prism are regular polygons of n sides and the area of the whole surface of the prism is S. When the volume of the prism is a maximum, prove that the area of an end is independent of the value of n. Deduce that, when n=4, the right prism of maximum volume is a cube [L.U.]

Let the polygonal ends of the prism have sides of length $2a$. Then their area A is made up of n isosceles triangles of semi-vertical angle π/n and base $2a$.

$$\therefore A = na^2 \cot(\pi/n) \quad . \quad . \quad . \quad . \quad \text{(i)}$$

If l is the length of the prism, S its total surface area and V its volume

$$S = 2A + 2nal \quad . \quad . \quad . \quad . \quad . \quad . \quad \text{(ii)}$$

and

$$V = Al = A(S-2A)/2na, \quad \text{from (ii)}$$

$$= \tfrac{1}{2}a \cot(\pi/n)\{S - 2na^2 \cot(\pi/n)\} \text{ by (i)},$$

$$\frac{dV}{da} = \tfrac{1}{2}\cot(\pi/n)\{S - 6na^2 \cot(\pi/n)\},$$

and

$$\frac{d^2V}{da^2} = -6na \cot^2(\pi/n), \text{ since } S \text{ is constant.}$$

$$\frac{dV}{da} = 0 \text{ when } S = 6na^2 \cot(\pi/n)$$

$$\text{and } a = \sqrt{\{(\tfrac{1}{6}S/n)\tan(\pi/n)\}}.$$

When a has this value, $\dfrac{d^2V}{da^2} < 0$, and from (i)

$$A = \tfrac{1}{6}S \quad . \quad . \quad . \quad . \quad . \quad \text{(iii)}$$

Hence when V is a maximum, A is independent of n.

When $n=4$, $A = 4a^2$ by (i) and $S = 8a^2 + 8al$ by (ii).

Hence, by (iii), for maximum volume

$$4a^2 = \tfrac{1}{6}(8a^2 + 8al),$$

i.e.

$$l = 2a.$$

It follows that the prism of maximum volume is a cube.

Exercises 9 (*b*)

1. A curve is traced out by the point whose co-ordinates are $x = 2\cos\theta + \cos 2\theta$, $y = 2\sin\theta - \sin 2\theta$, where θ varies from $-\pi$ to π.

   Find $dx/d\theta$ and $dy/d\theta$, and deduce that the gradient at the point θ is $-\tan\tfrac{1}{2}\theta$. Show also that the equation of the tangent to the curve at this point is $x\sin\tfrac{1}{2}\theta + y\cos\tfrac{1}{2}\theta = \sin\tfrac{3}{2}\theta$. [L.U.]

2. Find the equation of the tangent to the curve $y = x^3 + x^2$ at the point $(-1, 0)$. Find also the maximum and minimum ordinates. [L.U.]

3. Show that the tangent to the curve $x^3 + y^3 = 3axy$ at the point (x_1, y_1) is $x(x_1^2 - ay_1) + y(y_1^2 - ax_1) = ax_1y_1$. Write down the equation of the tangent at the point $(6a/7, -12a/7)$, and verify that it meets the curve again in the point $(-16a/21, 4a/21)$. [L.U.]

4. If the tangent at the point $P(at^2, at^3)$ on the curve $ay^2=x^3$ meets the curve again at Q, find the coordinates of Q.

If N is the foot of the perpendicular from P to the x-axis, R is the point where the tangent at P cuts the y-axis, and O is the origin, prove that OQ and RN are equally inclined to the x-axis. [L.U.]

5. Find the equation of the normal to the curve $y=2x/(x^2+1)$ at the point $(3, \frac{3}{5})$. What is the equation of the tangent at the origin ?

Sketch the curve roughly, giving the coordinates of the maximum and minimum turning-points. [Leeds.]

6. Show that the equation of the tangent at the point t on the curve $x=a\cos^3 t$, $y=a\sin^3 t$ is $x\sin t+y\cos t-a\sin t\cos t=0$.

Prove that the locus of intersection of tangents at right angles to one another is the curve whose equation in polar coordinates can be expressed in the form $2r^2=a^2\cos^2 2\theta$. [L.U.]

7. Sketch the curve $y=x^3$.

A is the point (1, 1) on this curve. Find the equation of the tangent to the curve at A. If this tangent meets the curve again at B, show that B is the point $(-2, -8)$.

8. Sketch the graphs of the functions $1+2x$ and e^x with the same axes, and prove that the greatest value assumed by the function $1+2x-e^x$ is $\log(4/e)$.

Find the greatest value assumed by the function $\sinh(2x-e^x)$. [L.U.]

9. Given that $f(0)=0$ and $f'(x)\geqslant 0$ when $x\geqslant 0$, prove that $f(x)\geqslant 0$ when $x>0$.

Prove that (i) $\sqrt{(1+4x)}\geqslant 1+2x-2x^2$ when $x>0$,

(ii) $\log(1+x)\leqslant x-\frac{1}{2}x^2+\frac{1}{3}x^3$ when $x>0$. [Sheffield.]

10. Show that if $f'(0)$ is positive, and $f(0)=0$, then $f(x)$ has the same sign as x for values of x near zero.

If $$f(x)=x-\tfrac{1}{2}x^2+\ldots+\frac{(-1)^{n-1}}{n}x^n$$

and $$F(x)=x-\tfrac{1}{2}x^2+\ldots+\frac{(-1)^{n-2}}{n-1}x^{n-1}+\frac{(-1)^{n-1}}{n}\frac{x^n}{1+x},$$

find and simplify the differential coefficients of $f(x)-\log(1+x)$ and of $F(x)-\log(1+x)$ and prove that, for all positive values of x, $\log(1+x)$ lies between $f(x)$ and $F(x)$. [L.U.]

11. By considering the derivative of the function

$$f(x)=\sin x\tan x-2\log\sec x$$

prove that $f(x)$ steadily increases as x increases from 0 to $\frac{1}{2}\pi$. Show also that the graph of the function has no inflexion between these limits.

Show that the function $2\sin x\tan x-5\log\sec x$ has one minimum value in the range $0<x<\frac{1}{2}\pi$. [L.U.]

12. If $f(x)=e^x(x^2-6x+12)-(x^2+6x+12)$, show that $f'''(x)>0$ when $x\neq 0$.
    Deduce that $f(x)>0$ when $x>0$ and that $\tanh x>3x/(x^2+3)$ when $x>0$. [L.U.]

13. Discuss the stationary values of the function $6\log(x/7)+(x-7)(x-1)$ for positive values of x.
    Deduce that the equation $6\log(x/7)+(x-7)(x-1)=0$ has only one real root and state its value. [Sheffield.]

14. Find the equations of the inflexional tangents of the curve $y=x^3-3/x$ and the coordinates of the points where they meet the curve again. [Sheffield.]

15. Find all the maxima and minima of the function $(3x^2-5)/(2x^2-x-6)$. [Durham.]

16. (i) Discuss the stationary values of x^4-2x^3+2x.
    (ii) Prove that the maximum value of $e^{tx}/(e^x+1)$ is $t^t(1-t)^{1-t}$, where t is a constant such that $0<t<1$. [Sheffield.]

17. Determine the gradients of the inflexional tangents of the curve
$$y=e^{2x}-12e^x+4x^2.$$
    Prove that $(0, b)$, $(b, 0)$ are the only points of inflexion of the curve whose equation is $x^3+3axy+y^3=b^3$ $(b\neq a)$. [Sheffield.]

18. From a fixed point A on the circumference of a circle of radius a the perpendicular AQ is drawn on to the tangent at a variable point P. If AP makes an angle θ with the diameter through A, prove that the area of the triangle APQ is $2a^2\sin\theta\cos^3\theta$.
    Find the maximum area of the triangle. [L.U.]

19. Prove that the weight of the heaviest right circular cylinder that can be cut from a given sphere of uniform material is $\frac{1}{3}\sqrt{3}$ times the weight of the sphere. [L.U.]

20. Show that the points of inflexion on the curve $y^2=a^2(a-x)/x$ are $(\frac{3}{4}a, \pm\frac{1}{3}a\sqrt{3})$.

21. A curve is given by the parametric equations $x=1/(1+t^2)$, $y=t^3/(1+t^2)$. Find the equation of the tangent to the curve at the point whose parameter is t, and show that the area of the triangle formed by the tangent and the coordinate axes is not greater than $(3\sqrt{3})/8$ units.
    Sketch the curve. [L.U.]

22. (i) Obtain the values of x for which $x^3(x-1)^2$ is stationary, determining which give maxima and which minima. Sketch the graph of the function.
    (ii) A canister, of total length l, is made of metal which is thin compared with the dimensions of the canister. It consists of a cylinder of radius r closed at its ends by cones of vertical angle 2θ. The weight of metal per unit area for the cones is n times that for the cylinder. Prove that if only θ varies, the weight of the canister

cannot have a true minimum unless n is greater than 2 and r is less than $\frac{1}{4}l\sqrt{(n^2-4)}$. [L.U.]

23. If $y=2+2x-x^2$ for $-1 \leqslant x \leqslant 2$, and $y=16/x+x-8$ for $2<x\leqslant 6$, find (i) the maximum and minimum values of y, (ii) the greatest and least values of y for values of x between -1 and 6. [L.U.]

24. The equation of a plane curve is $y^3+x^3-9xy+1=0$, and (x_1, y_1) is a point on the curve at which the tangent is parallel to the x-axis. Prove that, at (x_1, y_1), $d^2y/dx^2=18/(27-x_1^3)$.

Prove also that the stationary values of y occur at the points for which $x=(27\pm 3\sqrt{78})^{1/3}$ and determine which of these gives a maximum value of y and which a minimum. [L.U.]

25. Show that $\cos x/(1+\cos^2 x)^{3/2}$ has $2/3\sqrt{3}$ and $-1/2\sqrt{2}$ as maximum values, and $1/2\sqrt{2}$ and $-2/3\sqrt{3}$ as minimum. Illustrate by drawing a rough graph of the function. [L.U.]

26. A and B are two points on either side of a straight line which separates two different types of country. M and N are the feet of the perpendiculars from A and B respectively on this line. $MA=a$, $NB=b$, $MN=c$, and P is a point on the line between M and N distant x from M. If, in the type of country containing A, a man can walk with velocity u and, in the type containing B, with velocity v, find the time taken along the path APB, and show that when this time is a minimum $\sin \angle MAP/\sin \angle NBP=u/v$ [L.U.]

27. Show that the function $e^{ax}/(1+x^2)$, where a is real, has a maximum or a minimum value if $|a|<1$, but that there are no turning points if $|a|\geqslant 1$. Draw rough graphs of the function for the cases $a=\frac{1}{2}$, $a=1$ $a=2$, for values of x from $-\infty$ to $+\infty$ showing clearly how they differ. [L.U.]

28. (i) A right circular cylinder is inscribed in a given right circular cone so that one circular end is on the base of the cone and the circumference of the other end on the surface of the cone. Prove that the maximum volume of such a cylinder is 4/9 that of the cone.

(ii) A chord which cuts off a given segment of a certain parabola is perpendicular to the axis of the parabola. A rectangle is inscribed in the segment, with one side lying along the chord and the vertices of the parallel side lying on the parabola. Prove that the maximum area of such a rectangle is $1/\sqrt{3}$ that of the segment. [L.U.]

29. A right circular cylinder is inscribed in a right circular cone of height h and with base of radius a, one plane end of the cylinder being in contact with the base of the cone. Show that there is always a cylinder of maximum volume, but that there is no proper cylinder of maximum total superficial area (that is, the sum of the areas of the curved surface and the two plane ends) unless $2a$ is less than h. [L.U.]

30. Show that $y=\tan^3 x \tan^4 (\frac{1}{4}\pi-x)$ has a maximum value when $x=\tan^{-1} (\frac{1}{3})$. [L.U.]

31. A right circular cone is circumscribed to a sphere of radius a, with the base of the cone touching the sphere. Find an expression for the volume of the cone in terms of a and the semi-vertical angle of the cone, and show that when the volume of the cone is a minimum it is double the volume of the sphere.
    Find also the smallest volume of the cone if its semi-vertical angle is restricted to lie between $\sin^{-1}(\frac{1}{5})$ and $\sin^{-1}(\frac{1}{4})$ inclusive. [L.U.]

32. A tree trunk, in the form of a frustum of a cone is h feet long, and the greatest and least diameters are a and b feet respectively. A beam of square cross-section is cut from the tree. Show that if $2a > 3b$, then the beam has maximum volume when its length is $\frac{1}{3}ha/(a-b)$. What is the length of the beam for maximum volume when $2a < 3b$? [L.U.]

33. The brightness of a small surface varies inversely as the square of its distance r from the source of light and directly as the cosine of the angle between r and the normal to the surface. Two equal light sources are situated at the points A and B, not in the same vertical line, at heights a and $2a$ above a horizontal plane. The verticals through A and B meet the plane at M and N, where $MN=3a$. If a small surface parallel to the plane is moved along the line MN, show that its brightness is a minimum when it is situated at a point of trisection of MN. [L.U.]

34. Sketch the graph of the curve $y=4\cos x-3\cos 2x$, from $x=0$ to $x=\pi$.
    Find the values of x giving maximum and minimum values of the function $y=4\alpha\cos x-\beta\cos 2x$, where α and β are positive numbers. Distinguish between the cases (i) $\alpha>\beta$, (ii) $\alpha=\beta$ and (iii) $\alpha<\beta$. [L.U.]

35. The illumination of an area by a source of light is proportional to

$$\frac{x}{(x^2+a^2)^{\frac{3}{2}}}-\frac{x}{(x^2+b^2)^{\frac{3}{2}}},$$

where a and $b(>a)$ are constants, and x can be varied. Find the value of x which gives maximum illumination. [L.U.]

36. The vertices of a quadrilateral are the centres of the circles

$$x^2+y^2+2\lambda x=0, \quad x^2+y^2+2y/\lambda=0,$$

and the points of intersection of these circles. Prove that the area of the quadrilateral is the same for all values of λ.
    Find the length of the common chord of the circles, and show that it has a stationary value when the circles are of equal radius. [L.U.]

37. If r_1, r_2 be the focal distances of a point on an ellipse whose major axis is $2a$, find the maximum and minimum values of $r_1r_2(r_1-r_2)$, $(r_1>r_2)$, distinguishing between the cases where the eccentricity is greater than or less than $1/\sqrt{3}$.
    Illustrate by sketching the graph of the function $x(x-a)(2a-x)$. [L.U.]

38. P is a variable point on a parabola of latus rectum $4a$ and vertex O. The ordinate at P meets the axis at M, and Q is the foot of the perpendicular from M to OP. Find the length OM when $QP-OQ$ is a maximum. [L.U.]

39. (i) Find all maxima and minima of $(x^4+5x^2+8x+8)e^{-x}$.
(ii) Find the minimum value of the function $a+b+c+x-4(abcx)^{1/4}$, where a, b, c are positive constants, and sketch the graph. Hence, or otherwise, show that

$$a+b+c+d-4(abcd)^{1/4} \geqslant a+b+c-3(abc)^{1/3},$$

for any positive numbers a, b, c, d. [L.U.]

40. Show that the distance r between a point P on the curve $x=2a\cos^3 t$, $y=2a\sin^3 t$ and the point $(a, 0)$ is least when the parameter of the point P is given by $8\cos t=\sqrt{33}+1$.

Draw a rough graph showing the variation of r with t as the point P completely describes the curve. [L.U.]

41. Prove that the length of the tangent to the ellipse $b^2x^2+a^2y^2=a^2b^2$ intercepted between the axes has one finite stationary value. Prove analytically that it is a minimum and find this value. [L.U.]

42. (i) Find the least value of each of the following expressions, x and θ being real:

$$\sqrt{(x^2+4x+6)}, \quad \cos^2\theta+4\cos\theta+6.$$

(ii) Find the greatest value assumed by the function $x^2e^{-x^2}$. Sketch the graph of the function and indicate where it has inflexions.

[L.U.]

CHAPTER 10

INTEGRATION

10.1. Integration as the inverse of differentiation

In Chapter 9 we dealt with the process of differentiation. The inverse process, that of finding a function whose derivative is a given continuous function, is known as *integration*. It will be assumed that every continuous function is the derivative of some function.

10.2. The indefinite integral

A function $F(x)$ is an integral of $f(x)$ if

$$\frac{d}{dx}\{F(x)\}=f(x) \qquad \text{(i)}$$

and we write $$F(x)=\int f(x)\,dx\,;$$

$f(x)$ is called the integrand of $\int f(x)\,dx$.

If $G(x)$ is any other integral of $f(x)$, then

$$\frac{d}{dx}\{G(x)\}=f(x) \qquad \text{(ii)}$$

and so, from (i) and (ii),

$$\frac{d}{dx}\{G(x)-F(x)\}=0,$$

so that $$G(x)=F(x)+C,$$

where C is a constant. Hence every integral of $f(x)$ is of the form

$$F(x)+C \quad (C \text{ constant}) \qquad \text{(iii)}$$

Conversely, since $\frac{d}{dx}\{F(x)+C\}=\frac{d}{dx}\{F(x)\}=f(x)$, we see that *every* function of the form (iii), whatever be the value of the constant C, is an integral of $f(x)$. Thus the number of integrals of $f(x)$ is infinite. They are obtained by giving C all values in (iii). The constant C is known as an *arbitrary constant* or *constant of integration*. Also, $F(x)+C$ is called the *indefinite integral* of $f(x)$ with respect to x.

Throughout this chapter, for convenience, the constant of integration will be frequently omitted.

10.3. The definite integral

If $f(x)$ is continuous when $a \leqslant x \leqslant b$, and if $\dfrac{d}{dx}\{F(x)\}=f(x)$, we use the symbol $\int_a^b f(x)\,dx$ to denote $F(b)-F(a)$, a and b being assumed finite. $\int_a^b f(x)\,dx$ is called the *definite integral* from a to b of $f(x)$ and is so called because its value does not involve an arbitrary constant but depends on the values of b and a which are known respectively as the upper and lower limits of the integral.

In evaluating the integral it is convenient to write

$$\int_a^b f(x)\,dx = \Big[F(x)\Big]_a^b = F(b)-F(a).$$

Three elementary properties of the definite integral should be noted:

1. $\int_b^a f(x)\,dx = F(a)-F(b) = -\int_a^b f(x)\,dx.$
2. $\int_a^b f(x)\,dx = F(b)-F(a) = \int_a^b f(t)\,dt.$
3. $\int_a^c f(x)\,dx + \int_c^b f(x)\,dx = F(c)-F(a)+F(b)-F(c)$

$$= F(b)-F(a) = \int_a^b f(x)\,dx.$$

The significance of the definite integral is explained in Chapter 20.

10.4. Standard integrals

We give three lists of results—a set of general rules and two lists of standard forms. In these, u, v, w,... denote functions of x, and a, b, c,... denote constants. Unless otherwise stated, the base of logarithms is e. To avoid repeated use of the modulus sign in this and subsequent chapters it is to be understood whenever the logarithm of a function occurs that only positive values of the function are considered. Thus we write $\int \dfrac{dx}{x} = \log x$ instead of $\int \dfrac{dx}{x} = \log |x|$ which is established in § 10.5. It is also implied that x is confined to values for which the integrand exists.

General Rules

I. $\int (u+v-w)\,dx = \int u\,dx + \int v\,dx - \int w\,dx.$

II. $\int af(x)\,dx = a\int f(x)\,dx.$

III. $\displaystyle\int \frac{f'(x)}{f(x)}\,dx = \log f(x)$. See § 10.7.

IV. $\displaystyle\int \frac{\frac{1}{2}f'(x)}{\sqrt{\{f(x)\}}}\,dx = \sqrt{\{f(x)\}}$. See § 10.7.

V. $\displaystyle\int f(x)\,dx = \int \phi(u)\,\frac{dx}{du}\,du$. (Rule for change of variable.) See § 10.5.

VI. $\displaystyle\int u\,dv = uv - \int v\,du$. (Rule for integration by parts.) See § 10.17.

STANDARD FORMS

(i) (*a*) $\displaystyle\int x^n\,dx = \frac{x^{n+1}}{n+1},\ n \neq -1.$

(*b*) $\displaystyle\int (ax+b)^n\,dx = \frac{1}{a}\,\frac{(ax+b)^{n+1}}{n+1},\ n \neq -1.$ See § 10.6.

(ii) (*a*) $\displaystyle\int \frac{dx}{x} = \log x.$ See § 10.5.

(*b*) $\displaystyle\int \frac{dx}{ax+b} = \frac{1}{a}\log(ax+b).$ See § 10.6.

(iii) $\displaystyle\int e^{ax}\,dx = \frac{1}{a}\,e^{ax}.$

(iv) $\displaystyle\int \sin x\,dx = -\cos x.$

(v) $\displaystyle\int \cos x\,dx = \sin x.$

(vi) $\displaystyle\int \frac{dx}{a^2+x^2} = \frac{1}{a}\tan^{-1}\frac{x}{a}.$

(vii) When $x^2 < a^2$, $\displaystyle\int \frac{dx}{a^2-x^2} = \frac{1}{2a}\log\frac{a+x}{a-x} = \frac{1}{a}\tanh^{-1}\frac{x}{a}.$

(viii) When $x^2 > a^2$, $\displaystyle\int \frac{dx}{x^2-a^2} = \frac{1}{2a}\log\frac{x-a}{x+a} = -\frac{1}{a}\coth^{-1}\frac{x}{a}.$

In (ix), (x) and (xi) a is supposed positive.

(ix) When $x^2 < a^2$, $\displaystyle\int \frac{dx}{\sqrt{(a^2-x^2)}} = \sin^{-1}\frac{x}{a}.$

(x) $\displaystyle\int \frac{dx}{\sqrt{(a^2+x^2)}} = \log\left\{\frac{x+\sqrt{(a^2+x^2)}}{a}\right\} = \sinh^{-1}\frac{x}{a}.$

(xi) When $x > a$, $\displaystyle\int \frac{dx}{\sqrt{(x^2-a^2)}} = \log\left\{\frac{x+\sqrt{(x^2-a^2)}}{a}\right\} = \cosh^{-1}\frac{x}{a}.$

OTHER STANDARD RESULTS

(xii) When $a>0$, $\int a^x dx = \dfrac{a^x}{\log a}$.

(xiii) $\int \sec^2 x\, dx = \tan x$.

(xiv) $\int \operatorname{cosec}^2 x\, dx = -\cot x$.

(xv) $\int \sec x \tan x\, dx = \sec x$.

(xvi) $\int \operatorname{cosec} x \cot x\, dx = -\operatorname{cosec} x$.

(xvii) $\int \tan x\, dx = \log \sec x$. See § 10.7.

(xviii) $\int \cot x\, dx = \log \sin x$. See § 10.7.

(xix) $\int \operatorname{cosec} x\, dx = \begin{cases} \log \tan \frac{1}{2}x. & \text{See § 10.10 (b).} \\ \log(\operatorname{cosec} x - \cot x). & \text{See § 10.7.} \end{cases}$

(xx) $\int \sec x\, dx = \begin{cases} \log \tan(\frac{1}{4}\pi + \frac{1}{2}x). & \text{See § 10.10 (b).} \\ \log(\sec x + \tan x). & \text{See § 10.7.} \end{cases}$

(xxi) $\int \sin^2 x\, dx = \frac{1}{2}(x - \frac{1}{2}\sin 2x)$. See § 10.11.

(xxii) $\int \cos^2 x\, dx = \frac{1}{2}(x + \frac{1}{2}\sin 2x)$. See § 10.11.

(xxiii) $\int \sinh x\, dx = \cosh x$.

(xxiv) $\int \cosh x\, dx = \sinh x$.

(xxv) $\int \tanh x\, dx = \log \cosh x$. See § 10.7.

(xxvi) $\int \coth x\, dx = \log \sinh x$. See § 10.7.

(xxvii) $\int \operatorname{sech} x\, dx = \begin{cases} 2 \tan^{-1}(\tanh \frac{1}{2}x). \\ 2 \tan^{-1}(e^x). & \text{See § 10.14.} \end{cases}$

(xxviii) $\int \operatorname{cosech} x\, dx = \log \tanh \frac{1}{2}x$. See § 10.14.

(xxix) $\int \operatorname{sech}^2 x\, dx = \tanh x$.

(xxx) $\int \operatorname{cosech}^2 x\, dx = -\coth x$.

(xxxi) $\int \operatorname{sech} x \tanh x\, dx = -\operatorname{sech} x$.

(xxxii) $\int \operatorname{cosech} x \coth x\, dx = -\operatorname{cosech} x$.

10.5. Integration by substitution

Suppose that, in the indefinite integral,

$$I=\int f(x)\,dx \qquad \text{(i)}$$

we wish to change the variable from x to u by means of the substitution $x=x(u)$ which transforms $f(x)$ into $\phi(u)$.

By definition, $$\frac{dI}{dx}=f(x)$$

and $$\frac{dI}{du}=\frac{dI}{dx}\cdot\frac{dx}{du}=\phi(u)\,\frac{dx}{du},$$

$$\therefore\ I=\int\phi(u)\,\frac{dx}{du}\,du \qquad \text{(ii)}$$

Hence, when changing the variable from x to u we replace $f(x)$ by $\phi(u)$ and dx by $\frac{dx}{du}\,du$.

Example 1

When $x>0$, $\int\frac{dx}{x}=\log x$.

If $x<0$, we put $x=-u$ where $u>0$. Then dx is replaced by $-du$ and

$$\int\frac{dx}{x}=-\int\frac{du}{(-u)}=\int\frac{du}{u}=\log u.$$

Hence when $x<0$, $\int\frac{dx}{x}=\log(-x)$.

We can summarise the result for both positive and negative values of x by writing

$$\int\frac{dx}{x}=\log|x|.$$

Example 2

$I=\int\frac{dx}{\sqrt{(9-x^2)}}$. Put $x=3\sin u$, so that dx is replaced by $3\cos u\,du$.

Then $$I=\int du=u=\sin^{-1}\tfrac{1}{3}x.$$

Example 3

$I=\int\frac{8x}{(3+2x)^4}\,dx$. Put $3+2x=u$ so that dx is replaced by $\frac{1}{2}du$.

Then
$$\begin{aligned}I&=\int\frac{2u-6}{u^4}\,du\\&=-\frac{1}{u^2}+\frac{2}{u^3}\\&=-\frac{(1+2x)}{(3+2x)^3}.\end{aligned}$$

From the result $\int f(x)\,dx = \int \phi(u)\frac{dx}{du}\,du$, it follows that if an integral is recognised to be of the form $\int \phi(u)\,\frac{dx}{du}\,du$ it may be replaced by $\int f(x)\,dx$, where $f(x)=\phi(u)$.

Example 4

$I=\int \sin^6 u \cos u\,du$. Let $x=\sin u$, $dx=\cos u\,du$.

$$\therefore\ I=\int x^6\,dx=\tfrac{1}{7}x^7=\tfrac{1}{7}\sin^7 u.$$

With practice it will be found to be more direct to write

$$I=\int \sin^6 u \cos u\,du=\int \sin^6 u\,d(\sin u)=\tfrac{1}{7}\sin^7 u.$$

If a definite integral is to be evaluated and a change of variable is required to perform the integration, we may first find the corresponding indefinite integral in terms of the original variable and then insert the limits of integration. Alternatively we may, if convenient, change the given limits of integration to the corresponding values of the new variable as shown in Examples 5 and 6. This makes it unnecessary to restore the original variable after integration.

Example 5

$I=\int_{-2}^{2} \frac{x^2}{64+x^6}\,dx$. Let $u=x^3$ so that $\tfrac{1}{3}du=x^2\,dx$.

When $x=-2$, $u=-8$ and when $x=2$, $u=8$.

$$\therefore\ I=\tfrac{1}{3}\int_{-8}^{8}\frac{du}{64+u^2}=\frac{1}{24}\left[\tan^{-1}\frac{u}{8}\right]_{-8}^{8} \text{ by standard form (vi).}$$

$$=\tfrac{1}{24}[\tan^{-1}1-\tan^{-1}(-1)]$$

$$=\tfrac{1}{24}[\tfrac{1}{4}\pi-(-\tfrac{1}{4}\pi)]$$

$$=\pi/48.$$

Example 6

$I=\int_0 xe^{-x^2}\,dx$. Let $u=x^2$ so that $\tfrac{1}{2}du=x\,dx$.

When $x=0$, $u=0$; when $x=2$, $u=4$.

$$\therefore\ I=\tfrac{1}{2}\int_0^4 e^{-u}\,du=\tfrac{1}{2}\Big[-e^{-u}\Big]_0^4=\tfrac{1}{2}(1-e^{-4})=0{\cdot}4909.$$

10.6. Extensions of standard forms

By substituting $ax+b=u$ and applying standard forms we establish the results

$$\int (ax+b)^n\,dx = (ax+b)^{n+1}/a(n+1),\ n \neq -1,$$

$$\int \frac{dx}{ax+b} = (1/a)\log(ax+b),$$

$$\int e^{ax+b}\,dx = (1/a)e^{ax+b},$$

$$\int \cos(ax+b)\,dx = (1/a)\sin(ax+b),$$

and, if $p>0$,

$$\int \frac{dx}{\sqrt{\{p^2-(ax+b)^2\}}} = (1/a)\sin^{-1}\{(ax+b)/p\}.$$

10.7. Two useful results

(*a*) If in $I=\int \frac{f'(x)}{f(x)}dx$, we put $u=f(x)$, $du=f'(x)\,dx$

$$I=\int \frac{du}{u} = \log u = \log f(x)$$

i.e.

$$\int \frac{f'(x)}{f(x)}dx = \log f(x). \qquad (10.1)$$

Hence the integral of a fraction whose numerator is the derivative cf the denominator is the logarithm of the denominator.

(*b*) The substitution $u=f(x)$ shows that

$$\int \frac{\frac{1}{2}f'(x)\,dx}{\sqrt{\{f(x)\}}} = \sqrt{\{f(x)\}}. \qquad (10.2)$$

Example 7

If k is a constant, $\int \cot kx\,dx = (1/k)\int \frac{k\cos kx}{\sin kx}dx$

$$= (1/k)\log\sin kx \quad \text{by (10.1).}$$

Similarly, $\int \tan kx\,dx = -(1/k)\log\cos kx = (1/k)\log\sec kx$,

$\int \tanh kx\,dx = (1/k)\log\cosh kx$ and $\int \coth kx\,dx = (1/k)\log\sinh kx$.

Example 8

When k is a constant,

$$\int \sec kx\,dx = \int \frac{\sec^2 kx + \sec kx\tan kx}{\sec kx + \tan kx}dx$$

$$= (1/k)\log(\sec kx + \tan kx) \quad \text{by (10.1).}$$

Similarly, $\int \operatorname{cosec} kx\,dx = (1/k)\log(\operatorname{cosec} kx - \cot kx)$.

Example 9

$$\int \frac{x-1}{\sqrt{(x^2-2x+3)}}dx = \sqrt{(x^2-2x+3)} \quad \text{by (10.2).}$$

Example 10

$$\int\sqrt{\left(\frac{4-x}{4+x}\right)}\,dx=\int\frac{4-x}{\sqrt{(16-x^2)}}\,dx$$

$$=4\int\frac{dx}{\sqrt{(16-x^2)}}-\int\frac{x}{\sqrt{(16-x^2)}}\,dx$$

$=4\sin^{-1}\frac{1}{4}x+\sqrt{(16-x^2)}$ by standard form (ix) and (10.2).

Exercises 10 (*a*)

Use the principle given in § 10.6 and the tables of standard forms to integrate with respect to x the functions in Nos. 1-8:

1. $(2x-3)^2$.
2. $\dfrac{1}{3x-4}$.
3. $\dfrac{1}{(5-3x)^3}$.
4. $\dfrac{1}{(5+2x)^2+9}$.
5. $\dfrac{1}{\sqrt{\{9-(3+4x)^2\}}}$.
6. $\dfrac{1}{\sqrt{\{(4+3x)^2+4\}}}$.
7. $\dfrac{1}{\sqrt{\{(6x-1)^2-4\}}}$.
8. $\dfrac{1}{(7x-2)^2}$.

By a suitable substitution evaluate the integrals in Nos. 9-22:

9. $\int x\sqrt{(3+x)}\,dx$.
10. $\int\dfrac{x}{\sqrt{(x^2-5)}}\,dx$.
11. $\int\dfrac{x^2}{\sqrt{(5+x^3)}}\,dx$.
12. $\int\dfrac{x^2}{5+x^3}\,dx$.
13. $\int\dfrac{x}{9+x^4}\,dx$.
14. $\int x\sqrt{(9+x^2)}\,dx$.
15. $\int\sin^7 x\cos x\,dx$.
16. $\int\cos^6 x\sin x\,dx$.
17. $\int\sin^2 x\cos^3 x\,dx$.
18. $\int\dfrac{\cos x}{\sin^3 x}\,dx$.
19. $\int\dfrac{\sec^2 x}{3-4\tan x}\,dx$.
20. $\int\dfrac{e^{\sqrt{x}}}{\sqrt{x}}\,dx$.
21. $\int\dfrac{(\sin^{-1}x)^2}{\sqrt{(1-x^2)}}\,dx$.
22. $\int\dfrac{(\log x)^3}{x}\,dx$.

Integrate with respect to x the functions in Nos. 23-32:

23. $\dfrac{x}{1+x^2}$.
24. $\dfrac{x}{\sqrt{(4+x^2)}}$.
25. $\dfrac{x+2}{x^2+4x-5}$.
26. $\dfrac{x+2}{\sqrt{(x^2+4x-5)}}$.
27. $\dfrac{\cos x}{\sin x}$.
28. $\cot 3x$.
29. $\tan 4x$.
30. $\dfrac{\sin x}{2+3\cos x}$.
31. $\dfrac{\sec^2 x}{1+2\tan x}$.
32. $\dfrac{e^{2x}}{1+e^{2x}}$.

10.8. Integration of rational functions

The fractions are assumed to be proper, i.e. the degree of the numerator is less than the degree of the denominator. In cases where this is not so, the fraction should be converted by division into the sum of a polynomial and a proper fraction.

CASE 1. *Denominator of first degree*

Example 11

$$\frac{x^2-1}{x+2}=\frac{x(x+2)-2(x+2)+3}{x+2}$$
$$=x-2+3/(x+2).$$
$$\int\frac{x^2-1}{x+2}\,dx=\tfrac{1}{2}x^2-2x+3\log(x+2).$$

CASE 2. *Denominator of the second degree, i.e. functions of the form*

$$\frac{Px+Q}{ax^2+bx+c}$$

(a) *Denominator expressible as a product of linear factors with rational coefficients.*

Here we resolve the integrand into partial fractions.

Example 12

$$\int\frac{18-x}{12x^2-7x-12}\,dx=2\int\frac{dx}{3x-4}-3\int\frac{dx}{4x+3}$$
$$=\tfrac{2}{3}\log(3x-4)-\tfrac{3}{4}\log(4x+3),$$

by standard form (ii).

Example 13

$$\int\frac{6x-10}{(2x+1)^2}\,dx=\int\frac{3(2x+1)-13}{(2x+1)^2}\,dx$$
$$=3\int\frac{dx}{2x+1}-13\int\frac{dx}{(2x+1)^2}$$
$$=\tfrac{1}{2}\{3\log(2x+1)+13/(2x+1)\}.$$ See § 10.6.

(b) *Denominator which does not resolve into linear factors with rational coefficients.*

Here the denominator is a product of two linear factors with irrational coefficients or it can be expressed in the form $(x+a)^2+b^2$.

Example 14

$$\int \frac{dx}{x^2+4x+13} = \int \frac{dx}{(x+2)^2+9}$$

$$= \int \frac{du}{u^2+9}, \text{ where } u = x+2,$$

$$= \tfrac{1}{3} \tan^{-1} \tfrac{1}{3}u, \text{ by standard form (vi).}$$

$$= \tfrac{1}{3} \tan^{-1} \tfrac{1}{3}(x+2).$$

Example 15

$$I = \int \frac{4x-3}{2x^2-4x-1}\,dx.$$

Since $\dfrac{d}{dx}(2x^2-4x-1) = 4x-4$, we express the numerator in the form $\lambda(4x-4)+\mu$, where λ and μ are constants. I can then be divided into two parts: an integral of the form $\displaystyle\int \frac{f'(x)}{f(x)}\,dx$ and another reducible to one of the standard forms (vi), (vii) or (viii). We have, by inspection,

$$I = \int \frac{4x-3}{2x^2-4x-1}\,dx = \int \frac{4x-4}{2x^2-4x-1}\,dx + \int \frac{dx}{2x^2-4x-1}$$

$$= \log(2x^2-4x-1) + I_1 \text{ by (10.1),}$$

where
$$I_1 = \tfrac{1}{2} \int \frac{dx}{x^2-2x-\frac{1}{2}}$$

$$= \tfrac{1}{2} \int \frac{dx}{(x-1)^2-\frac{3}{2}}$$

$$= \frac{1}{2\sqrt{6}} \log \left(\frac{x-1-\sqrt{(\frac{3}{2})}}{x-1+\sqrt{(\frac{3}{2})}} \right) \text{ by standard form (viii).}$$

$$\therefore\ I = \log(2x^2-4x-1) + \frac{1}{2\sqrt{6}} \log \left\{ \frac{(x-1)\sqrt{2}-\sqrt{3}}{(x-1)\sqrt{2}+\sqrt{3}} \right\}.$$

Example 16

$$I = \int \frac{3+4x}{1+4x-4x^2}\,dx = \int \frac{5-\frac{1}{2}(4-8x)}{1+4x-4x^2}\,dx$$

$$\therefore\ I = \tfrac{5}{4} \int \frac{dx}{\frac{1}{4}-(x^2-x)} - \tfrac{1}{2} \int \frac{4-8x}{1+4x-4x^2}\,dx$$

$$= \tfrac{5}{4} I_1 - \tfrac{1}{2} \log(1+4x-4x^2)$$

where
$$I_1 = \int \frac{dx}{\frac{1}{2}-(x-\frac{1}{2})^2}$$

$$= \frac{1}{\sqrt{2}} \log \left(\frac{\sqrt{2}+2x-1}{\sqrt{2}-2x+1} \right) \text{ by standard form (vii).}$$

$$\therefore\ I = \frac{5}{4\sqrt{2}} \log \left(\frac{\sqrt{2}+2x-1}{\sqrt{2}-2x+1} \right) - \tfrac{1}{2} \log(1+4x-4x^2).$$

Note that if I_1 is written in the form $\int \frac{dx}{\frac{1}{2}-(\frac{1}{2}-x)^2}$

$$I_1=-\int \frac{du}{(1/\sqrt{2})^2-u^2}, \quad u=\tfrac{1}{2}-x,$$

$$=-\frac{1}{\sqrt{2}}\log\left(\frac{1/\sqrt{2}+u}{1/\sqrt{2}-u}\right)$$

$$=\frac{1}{\sqrt{2}}\log\left(\frac{\sqrt{2}-1+2x}{\sqrt{2}+1-2x}\right) \text{ as before.}$$

CASE 3. *Denominator of degree higher than the second*

The denominator should be factorised and the function expressed in terms of partial fractions. In general, the integral will then break up into integrals of the types already considered.

Example 17

$$I=\int \frac{6x-2x^2-8}{(x-1)(x^4-1)}dx=\int\left\{\frac{2}{x-1}-\frac{1}{(x-1)^2}-\frac{2}{x+1}-\frac{3}{x^2+1}\right\}dx$$

$$\therefore\ I=2\log(x-1)+1/(x-1)-2\log(x+1)-3\tan^{-1}x.$$

Exercises 10 (*b*)

Integrate with respect to x:

1. $\frac{x}{2x^2-x-3}$.
2. $\frac{x-3}{3x^2+2x-5}$.
3. $\frac{2-3x}{3x^2-4x+1}$.
4. $\frac{x+3}{x^3-6x^2+8x}$.
5. $\frac{4x^2-x+12}{x(x^2+4)}$.
6. $\frac{3x^2+3x+18}{(3-x)(3+x)^2}$.
7. $\frac{x^2+1}{(x+2)^2}$.
8. $\frac{x^2+3x-2}{x^4-2x^3+x^2}$.
9. $\frac{1}{x^2+6x-4}$.
10. $\frac{1}{x^2+6x+25}$.
11. $\frac{4x-2}{12-6x-x^2}$.
12. $\frac{2x+1}{x^2+2x+26}$.
13. $\frac{4x+1}{4x^2+16x+25}$.
14. $\frac{5x+1}{3x^2+6x-2}$.
15. $\frac{6x-1}{1+6x-2x^2}$.

10.9. Irrational functions

(a) *Any algebraic expression containing only a single irrational expression of the form* $\sqrt{(ax+b)}$.

Such a function may be reduced to one of the forms considered in § 10.8 by means of the substitution $u=\sqrt{(ax+b)}$.

Example 18

$$I=\int\frac{x-2}{x\sqrt{(x+1)}}\,dx.$$

Let $u=\sqrt{(x+1)}$ so that $du=\frac{1}{2}dx/\sqrt{(x+1)}$.

Then $I=2\int\frac{u^2-3}{u^2-1}\,du-2\int\left(1-\frac{2}{u^2-1}\right)du$

$$=2\left(u-\log\frac{u-1}{u+1}\right)$$

$$=2\left\{\sqrt{(x+1)}-\log\frac{\sqrt{(x+1)}-1}{\sqrt{(x+1)}+1}\right\}.$$

(b) *Irrational fractions of the form* $\dfrac{Px+Q}{\sqrt{(ax^2+bx+c)}}$.

Example 19

$$\int\frac{dx}{\sqrt{(4x^2-4x+5)}}=\frac{1}{2}\int\frac{dx}{\sqrt{(x^2-x+\frac{5}{4})}}$$

$$=\frac{1}{2}\int\frac{dx}{\sqrt{\{(x-\frac{1}{2})^2+1\}}}$$

$$=\tfrac{1}{2}\sinh^{-1}(x-\tfrac{1}{2})\text{ by standard form (x)}$$

$$=\tfrac{1}{2}\log\{(x-\tfrac{1}{2})+\sqrt{(x^2-x+\tfrac{5}{4})}\}.$$

Example 20

$$I=\int\frac{x-1}{\sqrt{(x^2-4x+3)}}\,dx.$$

The derivative of x^2-4x+3 is $2(x-2)$, so we write

$$I=\int\frac{x-2}{\sqrt{(x^2-4x+3)}}\,dx+\int\frac{dx}{\sqrt{\{(x-2)^2-1\}}}$$

$=\sqrt{(x^2-4x+3)}+\cosh^{-1}(x-2)$ by (10.2) and standard form (xi).

$$=\sqrt{(x^2-4x+3)}+\log\{(x-2)+\sqrt{(x^2-4x+3)}\}.$$

(c) *Irrational functions of the form* $\dfrac{1}{(x-p)\sqrt{(ax^2+2bx+c)}}$.

These may be integrated by means of the substitution $x-p=\dfrac{1}{u}$.

Example 21

$$I=\int\frac{dx}{(1-x)\sqrt{(16-26x+9x^2)}},\ (1-x>0).$$

Put $1-x=1/u$; then $\log(1-x)=-\log u$ and so $dx/(1-x)=du/u$.

$$\therefore\ I=\int\frac{du}{u\sqrt{(8/u-1+9/u^2)}}$$

$$=\int\frac{du}{\sqrt{(8u-u^2+9)}},\ \text{since } u>0$$

$$=\int\frac{du}{\sqrt{\{25-(u-4)^2\}}}$$

$$=\sin^{-1}\tfrac{1}{5}(u-4) \text{ by standard form (ix)}$$

$$=\sin^{-1}\frac{4x-3}{5(1-x)}.$$

Exercises 10 (*c*)

Integrate with respect to x:

1. $\sqrt{(9-2x)}$.
2. $\dfrac{1}{\sqrt{(9-4x^2)}}$.
3. $\dfrac{1}{\sqrt{(x^2+2x+26)}}$.
4. $\dfrac{1}{\sqrt{(x^2-4x-21)}}$.
5. $\dfrac{1}{\sqrt{(7-6x-x^2)}}$.
6. $\dfrac{x+3}{\sqrt{(x^2+2x+10)}}$.
7. $\dfrac{1}{\sqrt{\{x(4-x)\}}}$.
8. $\dfrac{x+3}{\sqrt{(x^2-7x+12)}}$.
9. $\dfrac{1}{x\sqrt{(4-x-3x^2)}}$, $(x>0)$
10. $\dfrac{1}{(2x+1)\sqrt{(1-x^2)}}$, $(2x+1>0)$.
11. $x^2\sqrt{(x-2)}$.
12. $\dfrac{x}{\sqrt{(4+x^2)}}$.
13. $\dfrac{x}{\sqrt{(4+x)}}$.
14. $\dfrac{x+1}{\sqrt{(4x^2+16x+25)}}$.
15. $\dfrac{1-2x}{\sqrt{(3x^2+4x-20)}}$.

10.10. Trigonometrical integrals evaluated by substitution

(a) *Functions of $\sin^2 x$ and/or $\cos^2 x$ may sometimes be conveniently integrated by means of the substitution $t=\tan x$.*

Example 22

$$I=\int\frac{dx}{\sin^4 x\cos^4 x}.$$

If $t=\tan x$, $\sin x=t/\sqrt{(1+t^2)}$, $\cos x=1/\sqrt{(1+t^2)}$ and $dx=dt/(1+t^2)$.

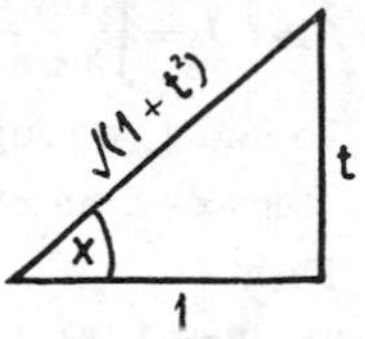

$$\therefore\ I=\int\frac{(1+t^2)^3}{t^4}dt$$

$$=\int(t^2+3+3/t^2+1/t^4)dt$$

$$=\tfrac{1}{3}t^3+3t-3/t-1/3t^3$$

$$=\tfrac{1}{3}(\tan^3 x+9\tan x-9\cot x-\cot^3 x).$$

Fig. 28

(b) *When the integrand is the reciprocal of a linear function of* $\sin x$ *and/or* $\cos x$ *the substitution* $t=\tan\frac{1}{2}x$ *is recommended.*

If $t=\tan\frac{1}{2}x$, $\sin x=2t/(1+t^2)$ and $\cos x=(1-t^2)/(1+t^2)$; hence any rational function of $\sin x$ and $\cos x$ can be expressed as a rational function of t. In particular, $\int\sec x\,dx$, $\int\operatorname{cosec} x\,dx$, and integrals of the form

$$\int\frac{dx}{a+b\sin x},\quad\int\frac{dx}{a+b\cos x},\quad\int\frac{dx}{a+b\cos x+c\sin x}$$

may all be evaluated in this way.

Example 23

$$I=\int\operatorname{cosec} x\,dx.$$

If $t=\tan\frac{1}{2}x$, $dt=\frac{1}{2}\sec^2\frac{1}{2}x\,dx$ so that $dx=2dt/(1+t^2)$,

$$\therefore\ I=\int\frac{dt}{t}$$

$$=\log t$$

$$=\log\tan\tfrac{1}{2}x.$$

Since

$$\sec x=\operatorname{cosec}(\tfrac{1}{2}\pi+x),$$

$$\int\sec x\,dx=\log\tan(\tfrac{1}{4}\pi+\tfrac{1}{2}x).$$

The above results should be compared with those given in Example 8.

(c) *An integral of the form* $\int\frac{a\cos x+b\sin x+c}{A\cos x+B\sin x+C}dx$ *can be evaluated by writing the numerator in the form*

$$p(A\cos x+B\sin x+C)+q\frac{d}{dx}(A\cos x+B\sin x+C)+r,$$

where p, q *and* r *are constants.*

Example 24

Find $I_1=\int\frac{\cos x+8\sin x-2}{3\sin x+2\cos x-1}\,dx$ *and* $I_2=\int\frac{\sin x}{\sin x+\cos x}\,dx.$

To find I_1 we suppose that

$$\cos x+8\sin x-2\equiv p(3\sin x+2\cos x-1)+q(3\cos x-2\sin x)+r.$$

Then

$2p+3q=1$, $3p-2q=8$ and $r-p=-2$; hence $p=2$, $q=-1$ and $r=0$.

$$\therefore\ I_1=\int\left(2-\frac{3\cos x-2\sin x}{3\sin x+2\cos x-1}\right)dx$$
$$=2x-\log(3\sin x+2\cos x-1).$$

To find I_2, we suppose that $\sin x\equiv p(\sin x+\cos x)+q(\cos x-\sin x)$.

Then $p-q=1$ and $p+q=0$ so that $p=-q=\frac{1}{2}$.

$$\therefore\ I_2=\tfrac{1}{2}x-\tfrac{1}{2}\log(\sin x+\cos x).$$

10.11. Products of sines and cosines of multiple angles

The product of two sines, two cosines, or a sine and a cosine may be integrated by expressing the product as a sum by means of the formulae

$$\sin ax\cos bx=\tfrac{1}{2}\{\sin(a+b)x+\sin(a-b)x\},$$
$$\sin ax\sin bx=-\tfrac{1}{2}\{\cos(a+b)x-\cos(a-b)x\},$$
$$\cos ax\cos bx=\tfrac{1}{2}\{\cos(a+b)x+\cos(a-b)x\}.$$

When $a=b$ we obtain, using the preceding formulae, the important results:

$$\int\sin^2 ax\,dx=\tfrac{1}{2}\int(1-\cos 2ax)\,dx=\tfrac{1}{2}\left(x-\frac{1}{2a}\sin 2ax\right),$$
$$\int\cos^2 ax\,dx=\tfrac{1}{2}\int(1+\cos 2ax)\,dx=\tfrac{1}{2}\left(x+\frac{1}{2a}\sin 2ax\right).$$

10.12. Powers of sin *x* and cos *x*

A reduction formula for $\int\sin^m x\cos^n x\,dx$, where m and n are positive integers, is given in § 10.21, but if m and n are small it is possible to evaluate an integral of this type in another way.

(i) When one or other of m and n is odd.

Example 25

$$I=\int\sin^3 x\cos^6 x\,dx$$
$$=-\int(1-\cos^2 x)\cos^6 x\,d(\cos x)$$
$$=\tfrac{1}{9}\cos^9 x-\tfrac{1}{7}\cos^7 x.$$

(ii) When m and n are both even.

Example 26

$$I=\int \sin^2 x \cos^4 x\, dx.$$

We express the integrand in terms of multiple angles by the method of § 7.7 or as follows:

$$\begin{aligned}\sin^2 x \cos^4 x &= \tfrac{1}{8}(1-\cos 2x)(1+\cos 2x)^2\\ &= \tfrac{1}{8}\sin^2 2x(1+\cos 2x)\\ &= \tfrac{1}{16}(1-\cos 4x)(1+\cos 2x)\\ &= \tfrac{1}{16}\{1+\cos 2x-\cos 4x-\tfrac{1}{2}(\cos 6x+\cos 2x)\}. \quad \text{See § 10.11.}\end{aligned}$$

$$\therefore\ I=\tfrac{1}{16}\{x+\tfrac{1}{4}\sin 2x-\tfrac{1}{4}\sin 4x-\tfrac{1}{12}\sin 6x\}.$$

10.13. Powers of tan x and cot x

A reduction formula for $\int \tan^n x\, dx$, where n is a positive integer, is given in § 10.23 but is not required when n is small.

Example 27

$$\begin{aligned}\int \tan^3 x\, dx &= \int \tan x(\sec^2 x-1)\, dx.\\ &= \int \tan x\, d(\tan x)-\int \tan x\, dx\\ &= \tfrac{1}{2}\tan^2 x+\log \cos x.\end{aligned}$$

Powers of cot x may be integrated by similar methods.

10.14. Hyperbolic functions

By using methods similar to those employed for the corresponding circular functions we may integrate hyperbolic functions, but sometimes it is advantageous to express these functions in their exponential form.

Example 28

$$I=\int \operatorname{cosech} x\, dx=\int \frac{dx}{2\sinh \frac{1}{2}x \cosh \frac{1}{2}x}=\frac{1}{2}\int \frac{\operatorname{sech}^2 \frac{1}{2}x}{\tanh \frac{1}{2}x}\, dx.$$

Put $t=\tanh \frac{1}{2}x$. Then $I=\int \frac{dt}{t}=\log t=\log \tanh \frac{1}{2}x.$

Example 29

$$I=\int \operatorname{sech} x\, dx=\int \frac{dx}{\cosh^2 \frac{1}{2}x+\sinh^2 \frac{1}{2}x}=\int \frac{\operatorname{sech}^2 \frac{1}{2}x}{1+\tanh^2 \frac{1}{2}x}\, dx.$$

Put $t=\tanh \frac{1}{2}x$. Then $I=2\int \frac{dt}{1+t^2}=2\tan^{-1} t=2\tan^{-1}(\tanh \frac{1}{2}x).$

But, if we write sech $x=\dfrac{2}{e^x+e^{-x}}$,

$$I=\int\frac{2e^x}{e^{2x}+1}\,dx=\int\frac{2}{u^2+1}\,du,\text{ where }u=e^x$$

and so $\qquad I=2\tan^{-1}e^x.$

This answer differs by a constant ($\frac{1}{2}\pi$) from the answer given above.

Example 30

$$\begin{aligned}
&\int_0^{\log 2}\frac{dx}{5\cosh x-3\sinh x}\\
&=\int_0^{\log 2}\frac{2}{5(e^x+e^{-x})-3(e^x-e^{-x})}dx\\
&=\int_0^{\log 2}\frac{e^x}{e^{2x}+4}\,dx\\
&=\int_1^2\frac{du}{u^2+4}\quad(u=e^x)\\
&=\tfrac{1}{2}\left[\tan^{-1}\tfrac{1}{2}u\right]_1^2\\
&=\tfrac{1}{2}\left(\tan^{-1}1-\tan^{-1}\tfrac{1}{2}\right)\\
&=\tfrac{1}{2}\tan^{-1}\tfrac{1}{3}.
\end{aligned}$$

10.15. Miscellaneous substitutions

We list here suitable substitutions suggested by the presence of certain functions in the integrand.

Type of integrand	*Suggested substitution*
$\sqrt{(a^2-x^2)}$	$x=a\sin\theta$ or $a\tanh\theta$.
$\sqrt{(x^2+a^2)}$	$x=a\tan\theta$ or $a\sinh\theta$.
$\sqrt{(x^2-a^2)}$	$x=a\sec\theta$ or $a\cosh\theta$.
$\dfrac{1}{x(ax^m+b)}$	$x^m=1/t$.
$\dfrac{1}{(x+p)\sqrt{(ax^2+bx+c)}}$	$x+p=1/t$.
Function of x and $\sqrt{(ax+b)}$	$ax+b=u^2$.
Function of x and $\sqrt{\{(x-a)(x-b)\}}$	$x-b=(x-a)u^2$.
Function of x and $\sqrt{\{(x-a)(b-x)\}}$	$b-x=(x-a)u^2$. or $x=a\cos^2\theta+b\sin^2\theta$.
Function of x and $\sqrt{(x^2+bx+c)}$ when x^2+bx+c is not expressible in terms of linear factors with rational coefficients.	$x+\sqrt{(x^2+bx+c)}=u$. (See Example 31.)
Expression containing fractional powers of x.	$x=u^n$, where n is the L.C.M. of the denominators of the fractional indices.

Example 31

$$I=\int\frac{dx}{x^2\sqrt{(x^2+2x-1)}}.$$

Put $x+\sqrt{(x^2+2x-1)}=u$ so that $x=\frac{1}{2}(1+u^2)/(1+u)$.

Then $\left\{1+\dfrac{x+1}{\sqrt{(x^2+2x-1)}}\right\}dx=du$ and so $\dfrac{dx}{\sqrt{(x^2+2x-1)}}=\dfrac{du}{u+1}$.

$$\therefore\ I=\int\frac{4(1+u)}{(1+u^2)^2}\,du$$

$$=\int\frac{4du}{(1+u^2)^2}+2\int\frac{d(1+u^2)}{(1+u^2)^2}$$

$$=4I_1-2/(1+u^2),$$

where $$I_1=\int\frac{du}{(1+u^2)^2}$$

$$=\int\cos^2\theta\,d\theta,\quad (u=\tan\theta)$$

$$=\tfrac{1}{2}(\theta+\tfrac{1}{2}\sin 2\theta)\ \text{by § 10.11}$$

$$=\tfrac{1}{2}\left(\tan^{-1}u+\frac{u}{1+u^2}\right).$$

$$\therefore\ I=2\left(\tan^{-1}u+\frac{u-1}{1+u^2}\right)$$

$$=2\tan^{-1}\{x+\sqrt{(x^2+2x-1)}\}+\frac{x-1+\sqrt{(x^2+2x-1)}}{x\{x+1+\sqrt{(x^2+2x-1)}\}}.$$

10.16. Three important integrals

$$I_1=\int\sqrt{(a^2-x^2)}\,dx$$

$$=a^2\int\cos^2\theta\,d\theta\quad (x=a\sin\theta)$$

$$=\tfrac{1}{2}a^2(\theta+\sin\theta\cos\theta)$$

$$=\tfrac{1}{2}a^2\left\{\sin^{-1}\frac{x}{a}+\frac{x}{a}\sqrt{\left(1-\frac{x^2}{a^2}\right)}\right\}$$

$$=\tfrac{1}{2}\{x\sqrt{(a^2-x^2)}+a^2\sin^{-1}(x/a)\}.$$

Similarly, $$I_2=\int\sqrt{(x^2+a^2)}\,dx$$

$$=\tfrac{1}{2}\{x\sqrt{(x^2+a^2)}+a^2\sinh^{-1}(x/a)\}$$

and $$I_3=\int\sqrt{(x^2-a^2)}dx$$

$$=\tfrac{1}{2}\{x\sqrt{(x^2-a^2)}-a^2\cosh^{-1}(x/a)\}.$$

See also § 10.17, Example 36.

Exercises 10 (*d*)

Integrate the following functions with respect to x:

1. $\sin^3 x$.
2. $\cos^5 x$.
3. $\sinh^2 x$.
4. $\tanh^3 x$.
5. $\operatorname{cosec} 2x$.
6. $\sec 3x$.
7. $\tan^2 \frac{1}{2}x$.
8. $\sin^4 x \cos^3 x$.
9. $\sin^3 x \sec^4 x$.
10. $\sec^4 x$.
11. $\cos^2 x \operatorname{cosec}^4 x$.
12. $\sec^4 x \operatorname{cosec}^2 x$.
13. $\dfrac{1}{1+\cos x}$.
14. $\dfrac{1}{1+\sin x}$.
15. $\sin 2x \cos 4x$.
16. $\sin 4x \sin 6x$.
17. $\cos 3x \cos 5x$.
18. $\dfrac{1}{12+13 \cos x}$.
19. $\dfrac{1}{11+61 \sin x}$.
20. $\dfrac{1}{16 \cos^2 x+9 \sin^2 x}$.
21. $\dfrac{\sin x+\cos x}{\sin x-\cos x}$.
22. $\dfrac{2 \sin x+9 \cos x}{\sin x+2 \cos x}$.
23. $\dfrac{1+\cos x-3 \sin x}{2+2 \cos x-\sin x}$.
24. $\dfrac{2+\sin x}{4+5 \cos x}$.
25. $\sqrt{(4+x^2)}$.
26. $\sqrt{(x^2-9)}$.

10.17. Integration by parts

If u and v are functions of x, the product rule for differentiation gives

$$\frac{d}{dx}(uv)=u\frac{dv}{dx}+v\frac{du}{dx}.$$

Integrating, we have

$$uv=\int u\frac{dv}{dx}\,dx+\int v\frac{du}{dx}\,dx$$

or

$$uv=\int u\,dv+\int v\,du.$$

$$\therefore \int u\,dv=uv-\int v\,du. \tag{10.3}$$

This formula is particularly useful when the integrand contains (i) a product of two factors, (ii) an inverse trigonometrical or hyperbolic function, (iii) a logarithmic function.

Example 32

$$I=\int x^n \log x\,dx.$$

The integrand consists of two factors, one of which, $\log x$, is differentiable but not immediately integrable. We therefore take $u=\log x$, $v=\dfrac{x^{n+1}}{n+1}$.

Then
$$I=\int(\log x)\,d\left(\frac{x^{n+1}}{n+1}\right)$$
$$=\frac{x^{n+1}}{n+1}\log x-\int\frac{x^{n+1}}{n+1}\,d(\log x),$$
$$=\frac{x^{n+1}}{n+1}\log x-\int\frac{x^{n}}{n+1}\,dx$$
$$=\frac{x^{n+1}}{(n+1)^2}\{(n+1)\log x-1\}.$$

Example 33

$I=\int\sin^{-1}x\,dx.$ Let $u=\sin^{-1}x$, $v=x$.

Then
$$I=x\sin^{-1}x-\int x\,d(\sin^{-1}x)$$
$$=x\sin^{-1}x-\int\frac{x}{\sqrt{(1-x^2)}}\,dx$$
$$=x\sin^{-1}x+\sqrt{(1-x^2)}.$$

Example 34

$$I=\int x^2\cos 2x\,dx.$$

Here both factors of the integrand are readily differentiable and integrable. We choose $u=x^2$ so that when we integrate by parts the index of x is reduced by one. A second integration by parts enables us to evaluate the given integral.

Let $u=x^2$, $v=\frac{1}{2}\sin 2x$.

Then
$$I=\int x^2\,d(\tfrac{1}{2}\sin 2x)$$
$$=\tfrac{1}{2}x^2\sin 2x-\int x\sin 2x\,dx.$$

Now
$$\int x\sin 2x\,dx=\int x\,d(-\tfrac{1}{2}\cos 2x)$$
$$=-\tfrac{1}{2}x\cos 2x-\int-\tfrac{1}{2}\cos 2x\,dx$$
$$=-\tfrac{1}{2}x\cos 2x+\tfrac{1}{4}\sin 2x$$
$$\therefore\ I=\tfrac{1}{4}(2x^2\sin 2x+2x\cos 2x-\sin 2x).$$

Example 35

The integrals $C=\int e^{ax}\cos bx\,dx$ and $S=\int e^{ax}\sin bx\,dx$ may be evaluated by integration by parts.

Taking $u=e^{ax}$, $v=\dfrac{1}{b}\sin bx$, we have
$$C=\int e^{ax}\,d\left(\frac{1}{b}\sin bx\right)$$
$$=\frac{1}{b}e^{ax}\sin bx-\int\frac{a}{b}e^{ax}\sin bx\,dx$$
$$\therefore\ bC+aS=e^{ax}\sin bx \qquad \text{(i)}$$

Similarly $$S=-\frac{1}{b}e^{ax}\cos bx+\int\frac{a}{b}e^{ax}\cos bx\,dx$$

$$\therefore\ aC-bS=e^{ax}\cos bx \quad . \quad . \quad . \quad . \quad . \quad . \quad \text{(ii)}$$

Solving (i) and (ii), we may find C and S.

Alternatively, the values of C and S may be found simultaneously by using complex quantities, on the assumption that these may be integrated according to the rules established for real quantities.

$$\begin{aligned} C+iS&=\int e^{ax}(\cos bx+i\sin bx)\,dx \\ &=\int e^{(a+ib)x}\,dx \\ &=\frac{1}{a+ib}\,e^{(a+ib)x} \\ &=\frac{a-ib}{a^2+b^2}\,e^{ax}\,(\cos bx+i\sin bx). \end{aligned}$$

Separating real and imaginary parts on each side of this equation we have

$$C=\frac{e^{ax}}{a^2+b^2}\,(a\cos bx+b\sin bx),$$

$$S=\frac{e^{ax}}{a^2+b^2}\,(a\sin bx-b\cos bx).$$

The method of integration by parts may be applied to the three integrals evaluated in § 10.16.

Example 36

$I=\int\sqrt{(a^2-x^2)}\,dx$. Let $u=\sqrt{(a^2-x^2)}$ and $v=x$.

Then $$\begin{aligned} I&=x\sqrt{(a^2-x^2)}-\int x\,d\{\sqrt{(a^2-x^2)}\} \\ &=x\sqrt{(a^2-x^2)}-\int\frac{-x^2}{\sqrt{(a^2-x^2)}}\,dx \quad . \quad . \quad . \quad \text{(i)} \\ &=x\sqrt{(a^2-x^2)}-\int\frac{(a^2-x^2)-a^2}{\sqrt{(a^2-x^2)}}\,dx \\ &=x\sqrt{(a^2-x^2)}-I+\int\frac{a^2}{\sqrt{(a^2-x^2)}}\,dx \end{aligned}$$

$$\therefore\ I=\tfrac{1}{2}\{x\sqrt{(a^2-x^2)}+a^2\sin^{-1}(x/a)\}.$$

From (i) we deduce that

$$\begin{aligned} \int\frac{x^2}{\sqrt{(a^2-x^2)}}\,dx&=I-x\sqrt{(a^2-x^2)} \\ &=\tfrac{1}{2}\{a^2\sin^{-1}(x/a)-x\sqrt{(a^2-x^2)}\}. \end{aligned}$$

In the same way we may evaluate $\int \sqrt{(a^2+x^2)}\,dx$ and $\int \sqrt{(x^2-a^2)}\,dx$ and deduce the values of $\int \frac{x^2}{\sqrt{(a^2+x^2)}}\,dx$ and $\int \frac{x^2}{\sqrt{(x^2-a^2)}}\,dx$.

Exercises 10 (*e*)

Integrate the following functions with respect to x:

1. $x^3 \log x$.
2. $\frac{\log x}{x^3}$.
3. $x \sin 2x$.
4. xe^{3x}.
5. $x^2 \cos 3x$.
6. $x \tan^{-1} 4x$.
7. $e^{2x} \sin 3x$.
8. $\log (x+2)$.
9. $x \sec^2 x$.
10. $x \cosh x$.
11. $\sinh^{-1} x$.
12. $\cosh^{-1} 2x$.
13. $\sqrt{(9+x^2)}$.
14. $\frac{x^2}{\sqrt{(x^2-9)}}$.
15. $\log (x^2+16)$.

10.18. Infinite or improper integrals

In defining $\int_a^b f(x)\,dx$ as $F(b)-F(a)$ where $F'(x)=f(x)$, we assumed that a and b are finite and that $f(x)$ is continuous in the interval $a \leqslant x \leqslant b$. If, however, a or b is infinite, or if $f(x)$ becomes infinite within or at an extremity of the range of integration, the above integral is said to be *infinite* or *improper* and its meaning must be defined.

Case I. *Infinite limits of integration*

If $f(x)$ is continuous when $x \geqslant a$, and if $\int_a^t f(x)\,dx$ tends to a finite limit L as $t \to \infty$, we write

$$\int_a^\infty f(x)\,dx = \lim_{t\to\infty} \int_a^t f(x)\,dx = L.$$

For example, $\int_0^t e^{-x}\,dx = \Big[-e^{-x}\Big]_0^t = 1-e^{-t} \to 1$ as $t \to \infty$

$$\therefore \int_0^\infty e^{-x}\,dx = 1.$$

This is an example of an infinite integral which exists.

If $\int_a^t f(x)\,dx$ does not tend to a finite limit as $t \to \infty$, the infinite integral $\int_a^\infty f(x)\,dx$ does not exist.

For example, $\int_1^t \frac{dx}{x} = \log t \to \infty$ as $t \to \infty$. Hence $\int_1^\infty \frac{dx}{x}$ does not exist.

In the same way, we define $\int_{-\infty}^{b} f(x)\,dx$ as $\lim\limits_{t_1\to-\infty}\int_{t_1}^{b} f(x)\,dx$. If this limit does not exist, the infinite integral does not exist.

Finally, if $\int_{t_1}^{t} f(x)\,dx$ tends to a limit as $t\to+\infty$ and $t_1\to-\infty$ independently, this limit is denoted by $\int_{-\infty}^{\infty} f(x)\,dx$.

For example, $\int_{t_1}^{t}\frac{dx}{1+x^2}=\tan^{-1} t-\tan^{-1} t_1$.

As $t\to\infty$, $\tan^{-1} t\to\frac{1}{2}\pi$: as $t_1\to-\infty$, $\tan^{-1} t_1=-\frac{1}{2}\pi$.

$$\therefore \int_{-\infty}^{+\infty}\frac{dx}{1+x^2}=\pi.$$

Case II. *Discontinuous integrand*

If $f(x)$ is continuous in the interval $a\leqslant x<b$, and if $|f(x)|\to\infty$ as $x\to b-$, we define $\int_{a}^{b} f(x)\,dx$ as $\lim\limits_{\epsilon\to0+}\int_{a}^{b-\epsilon} f(x)\,dx$ provided that such a limit exists.

If no such limit exists, $\int_{a}^{b} f(x)\,dx$ does not exist.

In the same way, if $f(x)$ is continuous in the interval $a<x\leqslant b$ and if $|f(x)|\to\infty$ as $x\to a+$, we define $\int_{a}^{b} f(x)\,dx$ as $\lim\limits_{\epsilon\to0+}\int_{a+\epsilon}^{b} f(x)\,dx$, provided that such a limit exists.

If $f(x)$ becomes infinite at $x=c$ where $a<c<b$, we define $\int_{a}^{b} f(x)\,dx$ by the relation

$$\int_{a}^{b} f(x)\,dx=\int_{a}^{c} f(x)\,dx+\int_{c}^{b} f(x)\,dx$$

and consider each of the latter integrals separately. If either of these integrals fails to exist, $\int_{a}^{b} f(x)\,dx$ does not exist.

Example 37

$$\int_{0}^{1}\frac{dx}{\sqrt{(1-x)}}=\lim_{\epsilon\to0+}\{-2[\sqrt{(1-x)}]_{0}^{1-\epsilon}\}=2.$$

Example 38

$$\int_{0}^{\frac{1}{2}}\frac{dx}{\sqrt{\{x(1-x)\}}}=\int_{0}^{\frac{1}{2}}\frac{dx}{\sqrt{\{\frac{1}{4}-(x-\frac{1}{2})^2\}}}=\lim_{\epsilon\to0+}[\sin^{-1}(2x-1)]_{\epsilon}^{\frac{1}{2}}=\frac{1}{2}\pi.$$

Exercises 10 (*f*)

Find, when they exist, the values of the integrals in Nos. 1-12:

1. $\displaystyle\int_1^\infty \frac{dx}{x^4}$.
2. $\displaystyle\int_1^\infty \frac{dx}{x^{1/3}}$.
3. $\displaystyle\int_0^\infty \frac{dx}{16+x^2}$.
4. $\displaystyle\int_0^3 \frac{dx}{\sqrt{(9-x^2)}}$.
5. $\displaystyle\int_0^\infty e^{-x}dx$.
6. $\displaystyle\int_0^\infty e^{-x}\cos x\,dx$.
7. $\displaystyle\int_2^4 \frac{dx}{\sqrt{(6x-8-x^2)}}$.
8. $\displaystyle\int_{-2}^2 \frac{dx}{x^2-4}$.
9. $\displaystyle\int_0^{\pi/2} \frac{\sin x}{\sqrt{(\cos x)}}\,dx$.
10. $\displaystyle\int_4^\infty \frac{dx}{x^2+2x+26}$.
11. $\displaystyle\int_0^\infty \frac{x-1}{x^3+1}\,dx$.
12. $\displaystyle\int_1^\infty \frac{x^2}{(1+x^2)^2}\,dx$.

10.19. Miscellaneous examples

Example 39

(i) *Evaluate* $\int \log x\,dx$ *and deduce that*

$$\int \sin\theta \log(1-e\cos\theta)d\theta=(e^{-1}-\cos\theta)\log(e^{-1}-\cos\theta).$$

(ii) *Given that* $x^2+y^2=2ay$, *use the substitution* $y=tx$ *to show that*

$$\int\frac{dx}{y}=-\frac{x}{y}-2\tan^{-1}\frac{y}{x}. \qquad \text{[L.U.]}$$

(i) $$\int \log x\,dx=x\log x-\int x\,d(\log x) \quad \text{by (10.3)}$$

$$=x(\log x-1) \qquad . \quad . \quad . \quad . \quad . \qquad \text{(i)}$$

If $x=1-e\cos\theta$, this result gives

$$e\int \sin\theta\log(1-e\cos\theta)d\theta=(1-e\cos\theta)\{\log(1-e\cos\theta)-\log e\},$$

$$\therefore \int\sin\theta\log(1-e\cos\theta)d\theta=(e^{-1}-\cos\theta)\log(e^{-1}-\cos\theta).$$

(ii) If $x^2+y^2=2ay$, and $y=tx$, $x^2(1+t^2)=2atx$

$$\therefore x=\frac{2at}{1+t^2},\ y=\frac{2at^2}{1+t^2}.$$

Then $I=\displaystyle\int\frac{dx}{y}$, where $dx=2a(1-t^2)/(1+t^2)^2\,dt$

$$\therefore I=\int\frac{1-t^2}{t^2(1+t^2)}dt$$

$$=\int\left(\frac{1}{t^2}-\frac{2}{1+t^2}\right)dt$$

$$=-1/t-2\tan^{-1}$$

$$=-x/y-2\tan^{-1}(y/x).$$

Example 40

(i) *Evaluate*
$$\int \frac{d\theta}{\sin\theta(1+\sin\theta)}.$$

(ii) *Prove that constants* λ, μ, ν *can be found so that*
$$\lambda(ax^2+1)+(\mu x+\nu)\frac{d}{dx}(ax^2+1)\equiv x^2+1, \quad (a\neq 0).$$

Hence show that
$$\int \frac{x^2+1}{(ax^2+1)^2}\,dx$$
can be expressed as a rational function of x *only if* $a=-1$ *or* 0. [L.U.]

(i)
$$I=\int \frac{d\theta}{\sin\theta(1+\sin\theta)}=\int\frac{d\theta}{\sin\theta}-\int\frac{d\theta}{1+\sin\theta}=I_1-I_2 \text{ (say)},$$

where
$$I_1=\log\tan\tfrac{1}{2}\theta$$

and
$$I_2=\int\frac{d\theta}{(\sin\frac{1}{2}\theta+\cos\frac{1}{2}\theta)^2}$$
$$=\int\frac{2d(\tan\frac{1}{2}\theta)}{(1+\tan\frac{1}{2}\theta)^2}$$
$$=\frac{-2}{1+\tan\frac{1}{2}\theta},$$
$$\therefore\ I=\log\tan\tfrac{1}{2}\theta+\frac{2}{1+\tan\frac{1}{2}\theta}.$$

I_2 can also be written in the form $\displaystyle\int\frac{1-\sin\theta}{\cos^2\theta}\,d\theta$
$$\therefore\ I_2=\int\sec^2\theta\,d\theta+\int\frac{d(\cos\theta)}{\cos^2\theta}$$
$$=\tan\theta-\sec\theta.$$

This result differs by a constant (1) from the value of I_2 already found

(ii) Let
$$\lambda(ax^2+1)+(\mu x+\nu)\frac{d}{dx}(ax^2+1)\equiv x^2+1.$$

Then
$$x^2(\lambda a+2\mu a)+2a\nu x+\lambda\equiv x^2+1$$
$$\therefore\ \lambda=1.$$

Also
$$a\nu=0, \text{ so that } \nu=0 \text{ since } a\neq 0,$$

and
$$a(\lambda+2\mu)=1.$$
$$\therefore\ \mu=(1-a)/2a.$$

Hence $$x^2+1 \equiv (ax^2+1)+\frac{1-a}{2a}x\frac{d}{dx}(ax^2+1) \quad . \quad . \quad \text{(i)}$$

Using this result we have

$$I=\int \frac{x^2+1}{(ax^2+1)^2}\,dx$$

$$=\int \frac{dx}{ax^2+1}+\frac{1-a}{2a}I_1 \quad . \quad . \quad . \quad . \quad \text{(ii)}$$

where $$I_1=\int \frac{x\dfrac{d}{dx}(ax^2+1)}{(ax^2+1)^2}\,dx$$

$$=-\int x\,d\left(\frac{1}{ax^2+1}\right)$$

$$=-\frac{x}{ax^2+1}+\int \frac{dx}{ax^2+1}, \text{ integrating by parts.}$$

Thus, by (ii), $$I=\int \frac{dx}{ax^2+1}+\frac{1-a}{2a}\left(\frac{-x}{ax^2+1}+\int \frac{dx}{ax^2+1}\right)$$

$$=\frac{1+a}{2a}\int \frac{dx}{ax^2+1}-\frac{(1-a)}{2a}\frac{x}{ax^2+1} \quad . \quad . \quad . \quad \text{(iii)}$$

When $a=-1$, $I=x/(1-x^2)$, which is a rational function of x.

When $a=0$, (i) is not valid, but $I=\int(x^2+1)\,dx=\frac{1}{3}(x^3+3x)$.

For all other values of a, $I_2 \equiv \int \dfrac{dx}{ax^2+1}$ is not a rational function of x.

When $a>0$, $a=\beta^2$ (say), $I_2=(1/\beta)\tan^{-1}\beta x$. When $a<0$, $a=-\beta^2 \neq -1$, $I_2=\dfrac{1}{2\beta}\log\left(\dfrac{1+\beta x}{1-\beta x}\right)$.

Hence I is expressible as a rational function of x only when $a=0$ or $a=-1$.

Example 41

If $$u_n=\int \frac{\sin(2n-1)x}{\sin x}\,dx, \qquad v_n=\int \frac{\sin^2 nx}{\sin^2 x}\,dx$$

prove that (apart from constants of integration) $nu_{n+1}=nu_n+\sin 2nx$ *and* $v_{n+1}-v_n=u_{n+1}$.

If n is a positive integer, prove that

$$\int_0^{\pi/2} \frac{\sin (2n-1)x}{\sin x}\, dx=\tfrac{1}{2}\pi,$$

and

$$\int_0^{\pi/2} \frac{\sin^2 nx}{\sin^2 x}\, dx=\tfrac{1}{2}n\pi. \qquad \text{[L.U.]}$$

$$n(u_{n+1}-u_n)=n\int \frac{\sin (2n+1)x-\sin (2n-1)x}{\sin x}\, dx$$

$$=2n\int \cos 2nx\, dx \quad (\text{see § 10.11})$$

$$=\sin 2nx \quad . \quad . \quad . \quad . \quad . \quad \text{(i)}$$

$$v_{n+1}-v_n=\tfrac{1}{2}\int \frac{\cos 2nx-\cos 2(n+1)x}{\sin^2 x}\, dx$$

$$=\int \frac{\sin (2n+1)x}{\sin x}\, dx$$

$$=u_{n+1} \quad . \quad . \quad . \quad . \quad . \quad \text{(ii)}$$

If the integrals are taken between the limits 0 and $\frac{1}{2}\pi$, and if n is a positive integer, then from (i)

$$u_{n+1}-u_n=0$$

$$\therefore u_{n+1}=u_n=u_{n-1}=\ldots=u_1=\tfrac{1}{2}\pi \quad . \quad . \quad . \quad \text{(iii)}$$

$$\therefore \int_0^{\pi/2} \frac{\sin (2n-1)x}{\sin x}\, dx=\tfrac{1}{2}\pi.$$

Again, integrating between the limits 0 and $\frac{1}{2}\pi$, we have from (ii)

$$v_{n+1}-v_n=\tfrac{1}{2}\pi, \text{ using (iii)}$$

$$\therefore v_n-v_{n-1}=\tfrac{1}{2}\pi$$

$$. \quad . \quad . \quad . \quad .$$

$$v_2-v_1=\tfrac{1}{2}\pi$$

$$\therefore \text{(adding) } v_n-v_1=\tfrac{1}{2}(n-1)\pi, \text{ and } v_1=\tfrac{1}{2}\pi$$

$$\therefore \int_0^{\pi/2} \frac{\sin^2 nx}{\sin^2 x}\, dx=\tfrac{1}{2}n\pi.$$

Example 42

Evaluate

$$I=\int_0^{\pi} \frac{x}{1+\sin x}\, dx.$$

Put $x=\pi-\theta$; then

$$I=-\int_{\pi}^{0} \frac{\pi-\theta}{1+\sin\theta}\, d\theta=\int_0^{\pi} \frac{\pi}{1+\sin\theta}\, d\theta-I. \quad (\text{See § 10.3, 1 and 2.})$$

$$\therefore 2I=\pi\int_0^{\pi} \frac{d\theta}{1+\sin\theta}$$

$$=\pi\left[\tan\theta-\sec\theta\right]_0^{\pi}, \text{ see Example 40 (i).}$$

$$I=\pi.$$

10.20. Reduction formulae for $\int \sin^n x\,dx$ and $\int \cos^n x\,dx$

It is sometimes possible by an integration by parts to find an expression for one integral in terms of another of similar but simpler form. Such an expression is called a *reduction formula.* For example, if

$$S_n = \int \sin^n x\,dx, \text{ where } n \text{ is a positive integer,}$$

$$\begin{aligned} S_n &= \int \sin^{n-1} x\,d(-\cos x) \\ &= -\sin^{n-1} x \cos x + (n-1)\int \sin^{n-2} x \cos^2 x\,dx \\ &= -\sin^{n-1} x \cos x + (n-1)\{S_{n-2} - S_n\} \end{aligned}$$

$$\therefore nS_n = -\sin^{n-1} x \cos x + (n-1)S_{n-2} \quad . \quad . \quad . \quad . \quad \text{(i)}$$

By successive applications of this reduction formula which expresses S_n in terms of S_{n-2} we can express S_n in terms of S_1 when n is odd and S_0 when n is even.

Similarly, if $\quad C_n = \int \cos^n x\,dx,$

$$nC_n = \cos^{n-1} x \sin x + (n-1)C_{n-2} \quad . \quad . \quad \text{(ii)}$$

10.21. Reduction formula for $\int \sin^m x \cos^n x\,dx$, where m and n are positive integers.

The integral $\int \sin^m x \cos^n x\,dx$ can be readily evaluated by a change of variable when either m or n is odd (see Example 25, page 214). When m and n are both even, the method given in Example 26 may be adopted, but the use of a reduction formula may be preferable when m and n are large.

If $I_{m,n} = \int \sin^m x \cos^n x\,dx$, where m and n are positive integers,

$$\begin{aligned} I_{m,n} &= \int \sin^m x \cos^{n-1} x\,d(\sin x) \\ &= \frac{1}{m+1}\int \cos^{n-1} x\,d(\sin^{m+1} x) \\ &= \frac{1}{m+1}\left\{\sin^{m+1} x \cos^{n-1} x + (n-1)\int \cos^{n-2} x \sin^{m+2} x\,dx\right\} \\ &= \frac{1}{m+1}\left\{\sin^{m+1} x \cos^{n-1} x + (n-1)\int \cos^{n-2} x \sin^m x\,(1-\cos^2 x)dx\right\} \\ &= \frac{1}{m+1}\{\sin^{m+1} x \cos^{n-1} x + (n-1)(I_{m,n-2} - I_{m,n})\} \end{aligned}$$

$$\therefore (m+n)I_{m,n} = \sin^{m+1} x \cos^{n-1} x + (n-1)I_{m,n-2}.$$

This reduction formula which expresses $I_{m,\,n}$ in terms of $I_{m,\,n-2}$ can be used with advantage to find $I_{m,\,n}$ when both m and n are even. In this case by successive applications of the formula $I_{m,\,n}$ may be evaluated in terms of $I_{m,\,0}$ which by the use of formula (i) of § 10.20 may be evaluated in terms of $I_{0,\,0}$ which is $\int 1\,dx$.

Again, writing $I_{m,\,n}=-\dfrac{1}{n+1}\int \sin^{m-1} x\, d(\cos^{n+1} x)$ we obtain

$$(m+n)I_{m,\,n}=-\sin^{m-1} x \cos^{n+1} x+(m-1)I_{m-2,\,n}.$$

This is a reduction formula which when used in conjunction with formula (ii) of § 10.20 enables $I_{m,\,n}$ to be evaluated in terms of $I_{0,\,0}$ when m and n are both even.

10.22. Wallis's formulae

If the integrals considered in § 10.20 and § 10.21 are taken between the limits 0 and $\frac{1}{2}\pi$, m and n being positive integers, we have

$$S_n=\int_0^{\pi/2} \sin^n x\,dx=\frac{n-1}{n} S_{n-2} \text{ if } n>1, \quad . \quad . \quad . \qquad \text{(i)}$$

$$C_n=\int_0^{\pi/2} \cos^n x\,dx=\frac{n-1}{n} C_{n-2} \text{ if } n>1,$$

and
$$I_{m,\,n}=\int_0^{\pi/2} \sin^m x \cos^n x\,dx=\frac{m-1}{m+n} I_{m-2,\,n} \text{ if } m>1, \quad . \qquad \text{(ii)}$$

$$=\frac{n-1}{m+n} I_{m,\,n-2} \text{ if } n>1.$$

By repeated application of formula (i) we have

$$S_n=\frac{n-1}{n} S_{n-2}$$

$$=\frac{n-1}{n}\cdot\frac{n-3}{n-2} S_{n-4}=\dots$$

If n is even, S_n depends ultimately on S_0, i.e. on $\int_0^{\pi/2} 1\,dx$, which is $\frac{1}{2}\pi$.

If n is odd, and greater than 1, S_n depends ultimately on S_1, i.e. on $\int_0^{\pi/2} \sin x\,dx$, which is 1.

In the same way, C_n may be evaluated. The results may be written in the form

$$S_n=C_n=\frac{(n-1)(n-3)\dots 5.3.1}{n(n-2)\dots 6.4.2}\,\frac{\pi}{2} \text{ when } n \text{ is even} \qquad (10.4)$$

$$=\frac{(n-1)(n-3)\dots 6.4.2}{n(n-2)\dots 5.3.1} \text{ when } n \text{ is odd, } n>1 \qquad (10.5)$$

By repeated application of formula (ii) we obtain for values of m and n greater than 1

$$I_{m,n}=\frac{(m-1)(m-3)\ldots(n-1)(n-3)\ldots}{(m+n)(m+n-2)(m+n-4)\ldots}\times p \tag{10.6}$$

where each product is continued until 2 or 1 is reached and $p=\frac{1}{2}\pi$ when m and n are both even, and $p=1$ in all other cases.

Formulae (10.4), (10.5) and (10.6) are known as Wallis's formulae.

Example 43

$$\int_0^{\pi/2} \sin^3 x \cos^5 x\,dx=\frac{2.4.2}{8.6.4.2}=\frac{1}{24}.$$

Example 44

$$\int_0^{\pi/2} \sin^6 x \cos^4 x\,dx=\frac{5.3.1.3.1}{10.8.6.4.2}\cdot\frac{\pi}{2}=\frac{3\pi}{512}.$$

Example 45

$$\int_0^{\pi/2} \sin^8 x\,dx=\frac{7.5.3.1}{8.6.4.2}\cdot\frac{\pi}{2}=\frac{35}{256}\pi.$$

Example 46

$$I=\int_0^{\pi} \sin^9 x \cos^4 x\,dx=\int_0^{\pi/2} \sin^9 x \cos^4 x\,dx+\int_{\pi/2}^{\pi} \sin^9 x \cos^4 x\,dx.$$

In the second integral, put $x=\pi-y$ so that $\sin x=\sin y$, $\cos x=-\cos y$ $dx=-dy$. Then

$$\int_{\pi/2}^{\pi} \sin^9 x \cos^4 x\,dx=\int_0^{\pi/2} \sin^9 y \cos^4 y\,dy$$

$$\therefore\ I=2\int_0^{\pi/2} \sin^9 x \cos^4 x\,dx=2\,.\,\frac{8.6.4.2.3}{13.11.9.7.5.3} \text{ by (10.6)}$$

$$=\frac{256}{15015}.$$

10.23. Reduction formulae for $\int \tan^n x\,dx$ and $\int \sec^n x\,dx$

If
$$I_n=\int \tan^n x\,dx, \text{ where } n\geqslant 2,$$

$$I_n=\int \tan^{n-2} x\,(\sec^2 x-1)\,dx.$$

$$=\frac{\tan^{n-1} x}{n-1}-I_{n-2}.$$

If $$I_n=\int \sec^n x\,dx, \text{ where } n\geqslant 2,$$

$$=\int \sec^{n-2} x\,d(\tan x)$$

$$=\sec^{n-2} x\tan x-\int \tan x\,d(\sec^{n-2} x)$$

$$=\sec^{n-2} x\tan x-(n-2)\int \sec^{n-2} x\tan^2 x\,dx$$

$$=\sec^{n-2} x\tan x-(n-2)(I_n-I_{n-2})$$

$$\therefore\ (n-1)I_n=\sec^{n-2} x\tan x+(n-2)I_{n-2}.$$

Exercises 10 (*g*)

Evaluate the integrals in Nos. 1-8:

1. $\int_0^{\pi/2} \cos^7 x\,dx.$ 2. $\int_0^{\pi/2} \sin^6 x\,dx.$ 3. $\int_0^{\pi/2} \sin^2 x\cos^3 x\,dx.$

4. $\int_0^{\pi/2} \sin^4 x\cos^2 x\,dx.$ 5. $\int_0^{\pi} (1-\cos x)^4\,dx.$ 6. $\int_0^{\pi/2} \sin^4 2x\,dx.$

7. $\int \tan^5 x\,dx.$ 8. $\int \sec^6 x\,dx.$

10.24. Miscellaneous examples

Example 47

If $I_n=\int \dfrac{x^n}{\sqrt{(a^2+x^2)}}\,dx$, *show that* $nI_n=x^{n-1}\sqrt{(a^2+x^2)}-(n-1)a^2I_{n-2}$ *where* $n\geqslant 2$.

Evaluate $$\int_0^2 \frac{x^5}{\sqrt{(5+x^2)}}\,dx.$$ [L.U.]

$$I_n=\int x^{n-1}\,d\{\sqrt{(a^2+x^2)}\}$$

$$=x^{n-1}\sqrt{(a^2+x^2)}-(n-1)\int x^{n-2}\sqrt{(a^2+x^2)}\,dx$$

$$=x^{n-1}\sqrt{(a^2+x^2)}-(n-1)\int \frac{x^{n-2}(a^2+x^2)}{\sqrt{(a^2+x^2)}}\,dx$$

$$=x^{n-1}\sqrt{(a^2+x^2)}-(n-1)(a^2I_{n-2}+I_n)$$

$$\therefore\ nI_n=x^{n-1}\sqrt{(a^2+x^2)}-(n-1)a^2I_{n-2}.$$

When $a^2=5$ and the integration is taken between the limits 0 and 2,

$$I_n=\frac{1}{n}\{3.2^{n-1}-5(n-1)\,I_{n-2}\}$$

and so
$$I_5=\frac{48}{5}-4I_3$$
$$=\frac{48}{5}-4\left\{4-\frac{10}{3}\,I_1\right\}.$$

But
$$I_1=\int_0^2 \frac{x}{\sqrt{(5+x^2)}}\,dx=[\sqrt{(5+x^2)}]_0^2=3-\sqrt{5}$$
$$\therefore\ I_5=\frac{168}{5}-\frac{40\sqrt{5}}{3}.$$

Example 48

If $I_{m,\,n}=\int \cos^m x \sin nx\,dx$, *provc that*

$$(m+n)I_{m,\,n}=-\cos^m x\cos nx+mI_{m-1,\,n-1}.$$

Hence, or otherwise, prove that $\displaystyle\int_0^{\pi/4} \cos^2 x \sin 4x\,dx=\frac{5}{12}.$ [L.U.]

$$I_{m,\,n}=-\frac{1}{n}\int \cos^m x\,d(\cos nx)$$

$$nI_{m,\,n}=-\left\{\cos^m x\cos nx+m\int \cos^{m-1} x\cos nx\sin x\,dx\right\}.$$

But $\sin nx\cos x-\cos nx\sin x=\sin(n-1)x$

$$\therefore\ \cos nx\sin x=\sin nx\cos x-\sin(n-1)x.$$

Hence

$$nI_{m,\,n}=-\cos^m x\cos nx-m\int\{\cos^m x\sin nx-\cos^{m-1} x\sin(n-1)x\}\,dx$$
$$=-\cos^m x\cos nx-m(I_{m,\,n}-I_{m-1,\,n-1})$$

$$\therefore\ (m+n)I_{m,\,n}=-\cos^m x\cos nx+mI_{m-1,\,n-1}.$$

If $\displaystyle\mathcal{I}_{m,\,n}=\int_0^{\pi/4}\cos^m x\sin nx\,dx,$

$$\mathcal{I}_{m,\,n}=\{1-\cos^m(\pi/4)\cos(n\pi/4)+m\mathcal{I}_{m-1,\,n-1}\}/(m+n).$$
$$\therefore\ \mathcal{I}_{2,\,4}=\tfrac{1}{6}\{\tfrac{3}{2}+2\mathcal{I}_{1,\,3}\}=\tfrac{1}{4}+\tfrac{1}{12}\{\tfrac{3}{2}+\mathcal{I}_{0,\,2}\}.$$

But
$$\mathcal{I}_{0,\,2}=\int_0^{\pi/4}\sin 2x\,dx=\tfrac{1}{2}.$$

$$\therefore\ \mathcal{I}_{2,\,4}=\tfrac{5}{12}.$$

Example 49

If $I(p, q)=\int_a^b (x-a)^p(b-x)^q\,dx$ $(b>a)$, *prove that when* $n\geqslant 1$,

(*i*) $I(n, n-1)=I(n-1, n)$,

(*ii*) $2(2n+1)I(n, n)=2n(b-a)I(n, n-1)=n(b-a)^2I(n-1, n-1)$.

Hence obtain the value of $I(n, n)$ *when n is a positive integer.* [L.U.]

(i) $I(n, n-1)=\int_a^b (x-a)^n(b-x)^{n-1}dx$, where $b>a$ and $n\geqslant 1$.

Put $x=a+b-u$; then

$$I(n, n-1)=\int_a^b (b-u)^n(u-a)^{n-1}\,du=I(n-1, n), \text{ see § 10.3, 2.}$$

(ii)
$$I(n, n)=\int_a^b (x-a)^n(b-x)^n\,dx$$
$$=\left[\frac{(x-a)^{n+1}}{n+1}(b-x)^n\right]_a^b+\frac{n}{n+1}\int_a^b (x-a)^{n+1}(b-x)^{n-1}\,dx,$$

on integrating by parts,

$$(n+1)I(n, n)=n\int_a^b \{(x-b)+(b-a)\}(x-a)^n(b-x)^{n-1}dx$$
$$=n\{(b-a)I(n, n-1)-I(n, n)\}$$

$$\therefore\ 2(2n+1)I(n, n)=2n(b-a)I(n, n-1)$$
$$=n(b-a)\{I(n, n-1)+I(n-1, n)\} \quad \text{by (i)}$$
$$=n(b-a)\int_a^b (x-a)^{n-1}(b-x)^{n-1}\{(x-a)+(b-x\ \}\,dx$$
$$=n(b-a)^2I(n-1, n-1)$$

i.e.
$$I(n, n)=\frac{(b-a)^2}{2}\,\frac{n}{2n+1}\,I(n-1, n-1)$$
$$=\frac{(b-a)^4}{2^2}\,\frac{n}{2n+1}\,\frac{n-1}{2n-1}\,I(n-2, n-2)$$
$$\cdot \quad \cdot \quad \cdot \quad \cdot \quad \cdot$$
$$=\frac{(b-a)^{2n}}{2^n}\,\frac{n\,!}{(2n+1)(2n-1)\ldots 3.1}\,I(0, 0).$$

But
$$I(0, 0)=b-a$$
$$\therefore\ I(n, n)=\frac{(b-a)^{2n+1}n\,!\,n\,!}{(2n+1)\,!}.$$

NOTE: $I(n, n)$ may be evaluated directly by substituting
$$x=a\cos^2\theta+b\sin^2\theta.$$

Exercises 10 (*h*)

1. If $f_n(x)=\dfrac{1}{n!}\displaystyle\int_0^x t^n e^{-t}\,dt$, prove that

$$f_n(x)-f_{n-1}(x)=-(x^n e^{-x})/n!.$$

Hence, or otherwise, evaluate the integral $\displaystyle\int_1^2 t^3 e^{-t}\,dt$. [L.U.]

2. (i) If $f(n)=\displaystyle\int_0^\infty x^{n-1}e^{-x}\,dx$, where n is positive, prove that

$$f(n+1)=n\,f(n).$$

Hence evaluate $f(n)$ where n is a positive integer.

(ii) If $\phi(n)=\displaystyle\int_0^{\pi/2}\sin^n x\,dx$, where $n\geqslant 0$, prove that $\phi(n+2)=\dfrac{n+1}{n+2}\,\phi(n)$.

Evaluate $\phi(8)$. [L.U.]

3. If $I_{m,n}=\displaystyle\int \sin^m\theta\cos^n\theta\,d\theta$, where m and n are integers greater than unity, prove that

$$\begin{aligned}(m+n)\,I_{m,n}&=\sin^{m+1}\theta\cos^{n-1}\theta+(n-1)\,I_{m,n-2}\\&=-\sin^{m-1}\theta\cos^{n+1}\theta+(m-1)I_{m-2,n}.\end{aligned}$$

Hence, or otherwise, evaluate the definite integral $\displaystyle\int_0^{\pi/2}\sin^{2p}\theta\cos^{2q}\theta\,d\theta$, where p, q are positive integers. [L.U.]

4. By means of the substitution $1+x=2\cos^2\theta$, or otherwise, evaluate the integral $\displaystyle\int_{-1}^1 (1+x)^m(1-x)^n\,dx$, where m and n are positive integers. [L.U.]

5. If $I_{m,n}=\displaystyle\int_0^{\pi/2}\sin^m\theta\cos^n\theta\,d\theta$, where m and n are positive integers, prove that $(m+n)I_{m,n}=(m-1)I_{m-2,n}$.

Evaluate $\displaystyle\int_0^1 x^2(1-x^2)^{5/2}\,dx$. [L.U.]

6. If $u_n=\displaystyle\int x^n(2ax-x^2)^{1/2}\,dx$, where n is a positive integer, prove that

$$(n+2)u_n-(2n+1)au_{n-1}+x^{n-1}(2ax-x^2)^{3/2}=0.$$

Show that $\displaystyle\int_0^{2a} x^2(2ax-x^2)^{1/2}\,dx=5\pi a^4/8$. [L.U.]

7. Find a reduction formula for $\displaystyle\int (a^2+x^2)^{n/2}\,dx$, where n is an odd positive integer, and explain its usefulness.

Prove that $\displaystyle\int_0^2 (5+x^2)^{3/2}\,dx=(396+75\log 5)/16$. [L.U.]

8. Prove that the integral $I_n=\int(1-x)^{1/2}x^{n+1/2}\,dx$ satisfies the recurrence relation $2(n+2)I_n=(2n+1)I_{n-1}-2(1-x)^{3/2}\,x^{n+1/2}$.

Find $\int_0^1 x^n\sqrt{(x-x^2)}\,dx$, where n is any positive integer. [L.U.]

9. Prove that

$$(n-1)\int\sin n\theta\sec\theta\,d\theta=-2\cos(n-1)\theta-(n-1)\int\sin(n-2)\theta\sec\theta\,d\theta.$$

Hence, or otherwise, evaluate $\int_0^{\pi/2}\cos 5\theta\sin\theta\sec\theta\,d\theta$. [L.U.]

10. If $u_m=\int x^m(a^2-x^2)^{1/2}\,dx$, show that

$$(m+2)u_m=-x^{m-1}(a^2-x^2)^{3/2}+a^2(m-1)u_{m-2}.$$

Evaluate $\int_0^{\pi/2}\sin^{2m}\theta\cos^2\theta\,d\theta$ when m is a positive integer. [L.U.]

11. If $u_n=\int_0^{\pi}e^{-x}\sin^n x\,dx$, show that for $n>1$, $(n^2+1)u_n=n(n-1)u_{n-2}$.

Evaluate $$\int_{-\pi/2}^{\pi/2}e^{-x}\cos^3 x\,dx.$$ [L.U.]

12. (i) If $I_n=\int x^n e^{-x}\,dx$, find a linear relation connecting I_n and I_{n-1}.

(ii) If $I_n=\int_0^{\pi/2}x^n\sin x\,dx$ and $n>1$, prove that

$$I_n+n(n-1)I_{n-2}=n(\tfrac{1}{2}\pi)^{n-1}.$$

Evaluate $$\int_0^{\pi/2}x^5\sin x\,dx.$$ [L.U.]

Miscellaneous Exercises 10

1. Obtain the indefinite integrals of

(i) $\sqrt{(x^2+1)}$ and $x\cosh x$; (ii) $\int_0^{\infty}e^{-x}\sin x\,dx$.

(iii) Show that, if α is an acute angle,

$$\int_0^1\frac{dx}{x^2+2x\cos\alpha+1}=\frac{\alpha}{2\sin\alpha}.$$

What is the value of the integral when $\alpha=0$? [Durham.]

2. Obtain the indefinite integrals of $x(x^2+4x+8)^{-\frac{1}{2}}$, $(x^2+1)(x+1)^{-3}$, $e^{-x}\sin x$, and evaluate $\int_0^{\pi}(\sin x+\cos x)^3\,dx$. [Durham.]

3. (i) Evaluate the integrals

$$\int \frac{dx}{\sqrt{(a^2-b^2x^2)}}; \quad \int \frac{dx}{1+e^x}; \quad \int_0^{\frac{1}{2}} \frac{x\,dx}{(1+x)(1-x)^2}.$$

(ii) By putting $\tan \frac{1}{2}x=t$, or otherwise, evaluate

$$\int_{-\pi/3}^{+\pi/3} \frac{dx}{\cos x}.$$

[Durham.]

4. (i) Evaluate the indefinite integral $\int \frac{dx}{\sin 2x+\sqrt{3}\cos 2x}$.

(ii) Evaluate $\int_0^{\pi} e^{-x} \cos x\,dx$. [Durham.]

5. Find $\int \frac{(x+1)\,dx}{x^2-3x+2}$; $\int x^2 \cosh^{-1} x\,dx$; $\int \frac{(x+1)\,dx}{(x^2+1)^{3/2}}$. [Leeds.]

6. Evaluate (i) $\int_0^{\pi} \frac{dx}{2+\cos x}$; (ii) $\int_0^{\pi} \frac{\cos^2 x\,dx}{2+\cos x}$. [L.U. Anc.]

7. Evaluate (i) $\int_0^1 \frac{\sqrt{x}}{1+x}\,dx$, (ii) $\int_0^{\pi/6} \frac{\cos\theta}{\cos 2\theta}\,d\theta$, by the substitution $\sin\theta=t$, or otherwise. [L.U. Anc.]

8. Evaluate (i) $\int_0^{\tan^{-1} 4/3} \frac{d\theta}{3\cos\theta+4\sin\theta}$; (ii) $\int \frac{\sin(\log x)}{x^2}\,dx$. [L.U.]

9. Evaluate (i) $\int_1^{\sqrt{3}} x\tan^{-1} x\,dx$; (ii) $\int_0^1 \frac{x\,dx}{(x+1)(x^2+1)}$. [Sheffield.]

10. Show that $\int_0^{\pi/4} \frac{3\sin^2\theta+\cos^2\theta}{3\cos^2\theta+\sin^2\theta}\,d\theta=\pi(8\sqrt{3}-9)/36$. [Sheffield.]

11. Evaluate (i) $\int_0^{\pi/4} \sin 2x \cos 3x\,dx$; (ii) $\int_1^2 \frac{(2x-1)\,dx}{\sqrt{(4x^2+4x+2)}}$;

(iii) $\int_0^{\frac{1}{2}} e^{2x} \sin \pi x\,dx$. [L.U.]

12. Find the indefinite integrals of $\frac{1}{x^2(x^4-1)}$, $x^2(\log x)^3$.

Evaluate $\int_0^{\pi} \frac{5-4\cos x}{5+4\cos x}\,dx$. [Sheffield.]

13. Find the indefinite integrals of (i) $\frac{2x+3}{\sqrt{\{(x+1)(2-x)\}}}$; (ii) $x^3(\log x)^2$.

[Sheffield.]

14. (i) By the substitution $x=3\sin^2\theta+\cos^2\theta$, or otherwise, prove that

$$\int_1^3 \sqrt{\left(\frac{3-x}{x-1}\right)}\,dx=\pi.$$

(ii) Prove that $\displaystyle\int_0^a x\sqrt{\left(\frac{a-x}{a+x}\right)}\,dx=(1-\tfrac{1}{4}\pi)a^2.$

(iii) Find $\displaystyle\int \sin x \log(\sin x)\,dx.$ [L.U.]

15. Evaluate the integrals (i) $\displaystyle\int_0^1 x^2\sin^{-1}x\,dx$; (ii) $\displaystyle\int\sqrt{(a^2-x^2)}\,dx$;

(iii) $\displaystyle\int_0^{\pi/2}\frac{dx}{\cos x-2\sin x+3}.$ [Sheffield.]

16. (i) Integrate the following functions with respect to x:

$$\frac{x-1}{\sqrt{(1+x)}};\quad e^{ax}\sin bx\ (a\neq 0);\quad \frac{2}{3+2\cos x}.$$

(ii) By means of the substitution $x=a\cos^2\theta+b\sin^2\theta$, or otherwise,

evaluate $\displaystyle\int_a^b\frac{x\,dx}{\sqrt{\{(b-x)(x-a)\}}},\ (b>a).$ [Sheffield.]

17. (i) Evaluate $\displaystyle\int_0^{\pi/2}\frac{\sin x\,dx}{1+\cos x+\sin x}.$

(ii) By the substitution $t=1/(x+1)$, or otherwise, evaluate

$$\int\frac{dx}{(x+1)^2\sqrt{(2x^2+2x+1)}}.$$ [Sheffield.]

18. Evaluate the integrals

(i) $\displaystyle\int\frac{dx}{(x+1)\sqrt{x}}$; (ii) $\displaystyle\int_0^3 x^2\log(x^2+1)\,dx$; (iii) $\displaystyle\int_0^{\pi/6}\frac{dx}{\cos x\cos 2x}.$

[Sheffield.]

19. Evaluate the integrals

$$\int\frac{dx}{\cosh x};\quad \int_2^{11/4}\frac{x\,dx}{\sqrt{(x^2-4x+5)}};\quad \int_2^3\frac{dx}{(x-1)(x-4)}.$$ [Durham.]

20. Evaluate the integrals

(i) $\displaystyle\int_2^3\frac{2x^3+6x^2+7x+1}{x^4+2x^3-2x-1}\,dx$; (ii) $\displaystyle\int_0^{\pi/4}\frac{dx}{3+\sin 2x}.$

Prove that $\displaystyle\int_0^1\frac{dx}{(1+x)\sqrt{(1+x^2)}}=(\sinh^{-1}1)/\sqrt{2}.$ [Sheffield.]

21. Find (i) $\displaystyle\int\frac{dx}{\sin x(1-\cos x)}$; (ii) $\displaystyle\int_3^8\frac{2x+3}{\sqrt{(16+6x-x^2)}}\,dx.$ [Leeds.]

22. Evaluate to two places of decimals the definite integrals

$$\text{(i)} \int_0^2 \frac{\tan^{-1} x}{1+x^2}\,dx; \quad \text{(ii)} \int_0^5 \frac{x}{\sqrt{(x+4)}-2}\,dx; \quad \text{(iii)} \int_1^{5/3} \frac{2x+1}{\sqrt{(x^2-1)}}\,dx. \quad \text{[L.U.]}$$

23. Evaluate the integrals

$$\text{(i)} \int_0^{\pi} x\cos^2 x\,dx; \quad \text{(ii)} \int_0^2 \frac{x^5\,dx}{(1+x^3)^{\frac{1}{2}}} \text{ by substituting } 1+x^3=z^2;$$

$$\text{(iii)} \int_0^a \sqrt{\left(\frac{a+x}{a-x}\right)}\,dx. \qquad \text{[L.U.]}$$

24. Evaluate

$$\int \frac{3x-1}{(1-x)^2(1+x)}\,dx; \quad \int \frac{dx}{a+b\cos x}\ (a>b); \quad \int_0^{\infty} e^{-ax}\cos bx\,dx\ (a>0).$$

[L.U.]

25. Evaluate $\int \frac{dx}{x+\sqrt{(x^2-1)}}$ and $\int \tan^{-1} x\,dx$. By means of the substitution $x=a\cos^2\theta+b\sin^2\theta$, or otherwise, evaluate

$$\int_a^b \frac{dx}{\sqrt{\{(x-a)(b-x)\}}}\ (a<b). \qquad \text{[L.U.]}$$

26. Find $\int e^{\sqrt{x}}\,dx$, $\int e^{\sin^{-1}\sqrt{x}}\,dx$ and $\int_0^{\pi/2} \frac{d\theta}{\cos\alpha+\cos\theta}$. [L.U.]

27. Evaluate

$$\text{(i)} \int_0^1 \sqrt{\left(\frac{x}{2-x}\right)}\,dx; \quad \text{(ii)} \int_0^{\frac{1}{2}} x^2\sin^{-1} x\,dx;$$

$$\text{(iii)} \int \frac{d\theta}{\sin\theta(1-2\sin\theta)}. \qquad \text{[L.U.]}$$

28. Evaluate

$$\text{(i)} \int x^3\log(1+x^2)\,dx; \quad \text{(ii)} \int_0^{\pi/2} \frac{\sin\theta}{\sin\theta+\cos\theta}\,d\theta;$$

$$\text{(iii)} \int_2^4 \frac{x}{\sqrt{(6x-8-x^2)}}\,dx. \qquad \text{[L.U.]}$$

9. (i) Evaluate the integrals $\int \frac{x+7}{x^2-x-2}\,dx$; $\int_0^1 \frac{dx}{(1+x)\sqrt{(1+2x)}}$.

(ii) If $I_n=\int_0^{\pi/4} \tan^n x\,dx$, prove that for $n\geqslant 2$, $(n-1)(I_n+I_{n-2})=1$, and hence evaluate I_5.

30. Evaluate the integrals

$$\text{(i)} \int \frac{dx}{(1+x)(1-x^2)^{\frac{1}{2}}}; \quad \text{(ii)} \int \frac{x+\sin x}{1+\cos x}\,dx; \quad \text{(iii)} \int_a^b \frac{(b-x)^{\frac{1}{2}}}{(x-a)^{\frac{1}{2}}}\,dx,\ (b>a).$$

[L.U.]

31. Evaluate (i) $\int \sin^4 x \cos^5 x\,dx$; (ii) $\int_2^3 \frac{x^2}{(x-1)(x+2)}\,dx$.

Prove that $\int_0^\infty \frac{dx}{(x+1)^2\sqrt{(x^2+1)}}=(1/\sqrt{2})\log(\sqrt{2}+1)$. [L.U.]

32. Evaluate the integral $\int \frac{(x^2+2)}{x(x^2+4x+6)}\,dx$.

Prove that (i) $\int_8^{15} \frac{dx}{(x-3)\sqrt{(x+1)}}=\tfrac{1}{2}\log(5/3)$;

(ii) $\int_{1/2}^1 \frac{dx}{x\sqrt{(5x^2-4x+1)}}=\log(1+\sqrt{2})$. [L.U.]

33. Evaluate the integrals $\int \frac{dx}{6-x-4x^2-x^3}$; $\int \left(\frac{x}{c-x}\right)^{\frac{1}{2}}dx$.

Prove that

$$\int_0^{\log 2} \frac{dx}{\sinh x+5\cosh x}=\{\tan^{-1}(\sqrt{6})-\tan^{-1}(\tfrac{1}{2}\sqrt{6})\}/\sqrt{6}.$$ [L.U.]

34. Evaluate the integrals

(i) $\int \frac{5-7x}{2x^3-x^2-2x+1}\,dx$; (ii) $\int \frac{\cos\theta-\sin\theta}{\cos\theta+\sin\theta}\,d\theta$.

Prove that $\int_{1/2}^1 \sin^{-1}(\sqrt{x})\,dx=(\pi-1)/4$. [L.U.]

35. (i) Evaluate $\int \frac{5\cos x\,dx}{2\cos x+\sin x+2}$.

(ii) By considering the graph of $\sin x$, show that

$$\sin x \geqslant 2x/\pi \text{ for } 0\leqslant x\leqslant \tfrac{1}{2}\pi.$$

Use this to show that, as R increases without limit through positive values, the value of $R\int_0^{\pi/2} e^{-R\sin x}\,dx$ never exceeds $\pi/2$. [L.U.]

36. (i) Evaluate (a) $\int x\sin x\sec^2 x\,dx$; (b) $\int \sqrt{\left(\frac{x+1}{x-1}\right)}\,dx$.

(ii) By means of the substitution $x^3t+1=0$, or otherwise, show that

$$\int_2^3 \frac{dx}{x(x^3-1)}=\tfrac{1}{3}\log(208/189).$$ [L.U.]

37. Evaluate the integrals

$$\int_0^1 x^2(1-x^2)^{3/2}\,dx;\quad \int \frac{x\sin^{-1}x}{\sqrt{(1-x^2)}}\,dx;\quad \int \frac{d\theta}{1-2\cos\theta}.$$ [L.U.]

38. Evaluate : (i) $\int \sin x \cos x \, e^{\cos^2 x} dx$; (ii) $\int \sqrt{(x^2-1)}dx$.

Prove that
$$\int_0^\pi \frac{\sin^2\theta}{2-\cos\theta}d\theta=(2-\sqrt{3})\pi.$$
[L.U.]

39. (i) Integrate $e^{-x}\sin 2x$ and $\dfrac{x-\sin x}{1-\cos x}$, with respect to x.

(ii) Evaluate $\displaystyle\int_0^a \frac{x(a^2-x^2)^{\frac{1}{2}}}{(a^2+x^2)^{\frac{1}{2}}}dx$ by means of the substitution $x^2=a^2\cos 2\theta$, or by any other method. [L.U.]

40. Find the integrals $\displaystyle\int \frac{dx}{(1+x^2)^{3/2}}$ and $\displaystyle\int x^3\tan^{-1}x\,dx$, and evaluate
$$\int_0^{\pi/2}\frac{a\sin\theta}{(1-a^2\sin^2\theta)^{3/2}}d\theta, \text{ where } 0<a<1.$$
[L.U.]

41. Evaluate the indefinite integrals

(i) $\displaystyle\int\frac{\sqrt{x}}{x+3}dx$; (ii) $\displaystyle\int\frac{d\theta}{\sin 2\theta\sin\theta+\cos^3\theta}$; (iii) $\displaystyle\int\sqrt{\left(\frac{x-1}{x+1}\right)}\,dx$. [L.U.]

42. Find (i) $\int \sin^5 x\cos^3 x\,dx$; (ii) $\displaystyle\int\frac{dx}{1+2\sin x\cos x+\cos^2 x}$.

Prove that
$$\int_0^1\frac{x^3+x}{x^3+1}dx=1-\tfrac{2}{3}\log 2.$$
[L.U.]

43. Evaluate (i) $\int\sqrt{(a^2-x^2)}\,dx$; (ii) $\displaystyle\int\frac{2ax-x^2}{\sqrt{(a^2-x^2)}}dx$.

Prove that
$$\int_0^{a/2}(2ax-x^2)^{3/2}\,dx=(8\pi-9\sqrt{3})a^4/64.$$
[L.U.]

44. Evaluate (i) $\displaystyle\int\frac{x}{\sqrt{(x^2+x+1)}}dx$; (ii) $\displaystyle\int_{\pi/3}^{\pi/2}\operatorname{cosec}x\,dx$.

Show that $\dfrac{d}{dx}\left(\dfrac{b\sin x}{a+b\cos x}\right)=\dfrac{a}{a+b\cos x}-\dfrac{a^2-b^2}{(a+b\cos x)^2}$, and hence, or otherwise, evaluate
$$\int\frac{dx}{(5+4\cos x)^2}$$
[L.U.]

45. Evaluate the integrals

(i) $\int\sec x\,dx$; (ii) $\displaystyle\int\frac{dx}{\sqrt{(4-4x-3x^2)}}$; (iii) $\displaystyle\int_0^{\pi/2}\frac{dx}{5+\cos x}$;

(iv) $\displaystyle\int\frac{2x^2-9x+1}{x^3+1}dx$. [L.U.]

46. Evaluate

$$\text{(i)} \int e^{-5x} \sin 4x \, dx \,; \qquad \text{(ii)} \int \frac{1-2x}{x^3+x^2+x+1} \, dx \,;$$

$$\text{(iii)} \int \frac{dx}{(x+3)\sqrt{(x-1)}} \,; \qquad \text{(iv)} \int \frac{x^2+2}{\sqrt{(x^2+2x-8)}} \, dx.$$ [L.U.]

47. (i) Find the integrals $\int x \tan^{-1}(x+1) \, dx$; $\int \frac{x^5}{(x^3+1)^3} \, dx$.

(ii) Find the value of $\int_0^{\pi/2} \frac{\cos x}{\sqrt{(3+\cos 2x)}} \, dx$. [L.U.]

48. Evaluate the integrals

(i) $\int (x-1)^{-1/2}(x-2)^{-1/2} dx$; (ii) $\int (1+x^2)^{-3/2} dx$; (iii) $\int \sec^3 x \, dx$.

[L.U.]

49. Evaluate (i) $\int \frac{x \, dx}{(x-1)(x^2+1)}$, (ii) $\int \frac{dx}{1+\sin x+\cos x}$.

Use the substitution $x = \tan\theta$ to evaluate $\int \frac{x^2(x+1)}{(x^2+1)^3} \, dx$. [L.U.]

50. Evaluate

$$\int \frac{(x^2-5x+9)}{(x-1)^2(x^2+4)} \, dx \,; \qquad \int \frac{dx}{\sqrt{(x+a)}-\sqrt{(x-a)}} \,;$$

$$\int_0^{\pi} \frac{d\theta}{a^2-2ab\cos\theta+b^2} \; (a>b).$$ [L.U.]

51. Evaluate $\int \frac{2x^3+7}{(2x+1)(x^2+2)} \, dx$; $\int \frac{dx}{x\sqrt{(1+2x-x^2)}}$.

Show that $\int_0^{\pi/2} \frac{d\theta}{3+5\cos\theta} = \frac{1}{4}\log 3$. [L.U.]

52. (i) Evaluate $\int x^3 e^{-x^2} dx$ and $\int_0^{\infty} \frac{dx}{(1+x^2)^3}$.

(ii) Show that $\int_0^{\pi} \frac{d\theta}{(5+3\cos\theta)} = \frac{1}{4}\pi$.

Hence, or otherwise, evaluate $\int_0^{\pi} \frac{(\cos\theta+2\sin\theta)}{(5+3\cos\theta)} \, d\theta$. [L.U.]

53. Integrate with respect to x

(i) $(4-x)/\sqrt{(3+2x-x^2)}$; (ii) $(1/x^2)\log(1+x)$.

Show that $\int_0^a f(x) \, dx = \int_0^a f(a-x) \, dx$ and prove that

$$\int_0^{\pi} \frac{x}{4+\sin^2 x} \, dx = (\pi^2\sqrt{5})/20.$$ [L.U.]

54. Resolve into partial fractions the functions

$$\text{(i)}\ \frac{1-x}{(1+2x)(1+x)} \text{ and (ii) } \frac{x^3}{(x^2+1)(x-2)}.$$

Hence evaluate the integrals of the two functions with respect to x. [L.U.]

55. Evaluate the definite integrals

$$\text{(i)} \int_0^{\pi/2} \cos^3 x\,dx\,; \quad \text{(ii)} \int_0^1 xe^x\,dx\,; \quad \text{(iii)} \int_1^2 \frac{(x+1)}{\sqrt{(x+2)}}\,dx.$$ [L.U.]

56. Evaluate the integrals

$$\text{(i)} \int_0^{\pi/2} x\cos^2 x\,dx\,; \quad \text{(ii)} \int \frac{dx}{(x+1)\sqrt{(x+2)}}\,; \quad \text{(iii)} \int \frac{dx}{\sqrt{\{(x-1)(x-2)\}}}.$$

[L.U.]

57. If $I_{m,n} = \int_0^1 (1-x^m)^n\,dx$, where $m>0$ and $n>0$, prove that

$$(mn+1)I_{m,n} = mnI_{m,n-1}.$$

If, in addition, n is an integer, evaluate $I_{m,n}$. [L.U.]

58. If $u_n = \int_0^{\pi/2} x\cos^n x\,dx$, where $n>1$, show that $n^2u_n = n(n-1)u_{n-2}-1$.

Evaluate u_4 and u_5. [L.U.]

59. Prove that

$$(n-1)\int \cos n\theta \sec\theta\,d\theta = 2\sin(n-1)\theta - (n-1)\int \cos(n-2)\theta\sec\theta\,d\theta.$$

Hence, or otherwise, evaluate $\int_0^{\pi/2} \sin 6\theta \sin\theta \sec\theta\,d\theta$. [L.U.]

60. If $I_m = \int x^m e^{ax}\,dx$, where $a \neq 0$, find a reduction formula giving I_m in terms of I_{m-1}.

Show that if m is a positive integer

$$\int_0^t x^m e^{ax}\,dx = \frac{(-1)^m m!\,e^{at}}{a^{m+1}}\left\{1-\frac{at}{1!}+\frac{(at)^2}{2!}-\frac{(at)^3}{3!}+\ldots+\frac{(-1)^m(at)^m}{m!}\right\} - \frac{(-1)^m m!}{a^{m+1}}.$$ [L.U.]

61. If

$$V_n = \int_0^x \frac{dt}{(t^2+a^2)^n} \text{ and } W_{m,n} = \int_0^x \frac{t^m\,dt}{(t^2+a^2)^n},$$

where m, n are positive integers, obtain the reduction formulae

$$V_{n+1} = \frac{1}{2na^2}\,\frac{x}{(x^2+a^2)^n} + \frac{2n-1}{2na^2}\,V_n,$$

$$W_{m,n} = -\frac{1}{2(n-1)}\,\frac{x^{m-1}}{(x^2+a^2)^{n-1}} + \frac{m-1}{2(n-1)}\,W_{m-2,n-1}.$$

Evaluate $W_{4,4}$. [L.U.]

62. If $I_{n,\,m}=\int_0^{\pi} \sin^n x \sin mx\,dx$ and $J_{n,\,m}=\int_0^{\pi} \sin^n x \cos mx\,dx$, where m and n are positive integers, prove that $(m+n)I_{n,\,m}-nJ_{n-1,\,m-1}=0$, and $(m+n)\,J_{n,\,m}+nI_{n-1,\,m-1}=0$.

Evaluate $I_{4,\,5}$. [L.U.]

63. $I_n=\int_0^1 \frac{x^{2n}}{\sqrt{(1-x^2)}}\,dx$ and $J_n=\int_0^1 \frac{x^{2n}}{(1+x^2)\sqrt{(1-x^2)}}\,dx$, $(n=0, 1, 2, 3, \ldots)$ prove that (i) $J_0=\frac{1}{2}\sqrt{2}I_0$; (ii) $2nI_n=(2n-1)I_{n-1}$.

By considering J_n+J_{n-1}, show how the reduction formula for I_n allows J_n to be evaluated for any particular value of n.

Prove that $J_3=\pi(7-4\sqrt{2})/16$. [L.U.]

64. If $I(p, q)=\int \frac{x^p}{(1+x^2)^q}\,dx$, show that

$$2(q-1)I(p, q)=-x^{p-1}/(1+x^2)^{q-1}+(p-1)I(p-2, q-1).$$

Hence, or otherwise, evaluate $\int_0^1 \frac{x^6}{(1+x^2)^3}\,dx$. [L.U.]

65. If $I_n=\int_0^{\pi/2} x^n \cos x\,dx$, prove that $I_n=(\frac{1}{2}\pi)^n-n(n-1)I_{n-2}$, if $n>1$, and hence evaluate $\int_0^{\pi/2} x^4 \cos x\,dx$. [L.U. Anc.]

66. Prove that, if n is a positive integer,

$$\int_0^1 x^n\sqrt{(1-x)}\,dx=2^{2n+2}\,n!\,(n+1)!/(2n+3)!. \qquad \text{[L.U. Anc.]}$$

67. If $I_n=\int_0^x \frac{t^n}{\sqrt{(1+t+t^2)}}\,dt$, prove that

$$(n+1)I_{n+1}+(n+\tfrac{1}{2})I_n+nI_{n-1}=x^n\sqrt{(1+x+x^2)}.$$

Hence, or otherwise, prove that

$$\int_0^1 \frac{x^3}{\sqrt{(1+x+x^2)}}\,dx=\{21 \log (1+2/\sqrt{3})-2(3\sqrt{3}-1)\}/48. \qquad \text{[L.U.]}$$

68. If $u_n=\int \frac{x^n}{(2ax-x^2)^{\frac{1}{2}}}\,dx$, show that

$$nu_n=-x^{n-1}\,(2ax-x^2)^{\frac{1}{2}}+(2n-1)au_{n-1}.$$

Find the value of u_n when the limits of integration are 0 and $2a$ and n is a positive integer. [L.U.]

69. If $T_{m,\,n}=\int_{-1}^1 (1-x)^m(1+x)^n\,dx$, $(m\geqslant 0, n\geqslant 0)$, show that

(i) $(n+1)T_{m,\,n}=mT_{m-1,\,n+1}$ for $m\geqslant 1$;

(ii) $(m+n+1)T_{m,\,n}=2mT_{m-1,\,n}$ for $m\geqslant 1$.

Evaluate $T_{3,\,2}$. [L.U.]

70. Obtain a reduction formula expressing $\int(x^2+a^2)^{\frac{1}{2}n}\,dx$ in terms of $\int(x^2+a^2)^{\frac{1}{2}n-1}\,dx$.

Prove that $\int_0^a (x^2+a^2)^{3/2}\,dx=\frac{1}{8}a^4\{7\sqrt{2}+3\log(1+\sqrt{2})\}$. [L.U.]

71. If $n!\,y_n=\int (x-a)^n \sin x\,dx$, where n is a positive integer, prove that, when $n>1$, $y_n+y_{n-2}=\dfrac{(x-a)^{n-1}}{(n-1)!}\sin x-\dfrac{(x-a)^n}{n!}\cos x$.

Prove that

$$\int_0^a (x-a)^{2n}\sin x\,dx=(-1)^{n-1}(2n)!\left\{\cos a-1+\frac{a^2}{2!}-\frac{a^4}{4!}+\cdot\quad+(-1)^{n-1}\frac{a^{2n}}{(2n)!}\right\}.$$

[L.U.]

72. If $I_n=\int \sec^n\theta\,d\theta$, show that, when $n\geqslant 1$,

$$(n-1)I_n=\sec^{n-2}\theta\tan\theta+(n-2)I_{n-2}.$$

Show that $8\int_0^{\pi/4}\sec^5\theta\,d\theta=7\sqrt{2}+3\log(1+\sqrt{2})$

and evaluate $\displaystyle\int_0^a \frac{dx}{(2a^2-x^2)^3}$. [L.U.]

73. (i) If $u(n,m)=\int_0^1 x^n(1-x)^m\,dx$, and m, n are positive, prove that

$$(n+1)u(n,m)=mu(n+1,m-1).$$

Using this result, or otherwise, evaluate the integral $\int_0^a x^{5/2}(a-x)^3\,dx$.

(ii) Find a recurrence formula for the integral

$$u_{2n+1}=\int_0^\infty x^{2n+1}e^{-\frac{1}{2}x^2}\,dx$$

and hence show that $u_7=48$. [L.U. Anc.]

74. If $I_{m,n}=\int_0^{\pi/2}\cos^m\theta\sin^n\theta\,d\theta$, where m, n are positive integers with m greater than unity, find $I_{m,n}$ in terms of $I_{m-2,n}$.

Evaluate $\displaystyle\int_0^\infty \frac{t^2}{(1+t^2)^4}\,dt.$ [L.U.]

CHAPTER 11

EXPANSIONS IN SERIES

11.1. Power series

From the identity $1+x+x^2+\ldots+x^{n-1}=(1-x^n)/(1-x)$, $x\neq 1$, we have

$$\frac{1}{1-x}=1+x+x^2+\ldots+x^{n-1}+r_n \quad . \quad . \quad . \quad \text{(i)}$$

i.e. we can represent the function $\frac{1}{1-x}$ by the polynomial

$$1+x+x^2+\ldots+x^{n-1}$$

together with $r_n=x^n/(1-x)$.

If $|x|<1$, r_n may be made as small as we please by making n sufficiently large. Hence in the limit when $n\to\infty$, $r_n\to 0$ and we have

$$\frac{1}{1-x}=1+x+x^2+x^3+\ldots \qquad (-1<x<1).$$

The convergent power series $1+x+x^2+x^3+\ldots$ is known as the expansion of $\frac{1}{1-x}$. It is valid only when $|x|<1$.

Again, by differentiation of (i)

$$\frac{1}{(1-x)^2}=1+2x+3x^2+\ldots+(n-1)x^{n-2}+\rho_{n-1},$$

where $\rho_{n-1}=\frac{d}{dx}(r_n)$ and may be shown to tend to zero when $n\to\infty$ provided that $|x|<1$.

$$\therefore \quad \frac{1}{(1-x)^2}=1+2x+3x^2+\ldots \qquad (-1<x<1).$$

Finally, $$\frac{1}{1+t}=1-t+t^2-t^3+\ldots+(-t)^{n-1}+\frac{(-t)^n}{1+t},$$

and so $$\int_0^x \frac{dt}{1+t}=x-\frac{x^2}{2}+\frac{x^3}{3}-\frac{x^4}{4}+\ldots+(-1)^{n-1}\frac{x^n}{n}+R_n,$$

where $R_n=\int_0^x \frac{(-t)^n}{1+t}dt$ and may be shown to tend to zero when $n\to\infty$ provided that $-1<x\leqslant 1$.

$$\therefore \log(1+x)=x-\frac{x^2}{2}+\frac{x^3}{3}-\frac{x^4}{4}+\dots \qquad (-1<x\leqslant 1).$$

The above examples will suffice to suggest that a function of x may be expanded as an infinite power series in x which in general is valid only for a certain range of values of x. For functions encountered at this stage this range may be assumed to be identical with the interval of convergence of the series.

11.2. Maclaurin's series

Among the properties of power series given in § 4.20 we stated that a power series which converges to $f(x)$ in the interval $-l<x<l$ may be differentiated term by term and the resultant series will converge to $f'(x)$ in the interval $-l<x<l$.

Thus if

$$f(x)=a_0+a_1x+a_2x^2+a_3x^3+\dots+a_nx^n+a_{n+1}x^{n+1}+\dots \qquad \text{(i)}$$

when $-l<x<l$, we obtain by successive differentiation the following results all valid when $-l<x<l$.

$$f'(x)=a_1+2a_2x+3a_3x^2+\dots+na_nx^{n-1}+(n+1)a_{n+1}x^n+\dots$$

$$f''(x)=2a_2+3.2a_3x+\dots+n(n-1)a_nx^{n-2}+(n+1)na_{n+1}x^{n-1}+\dots$$

$$f'''(x)=3.2.1a_3+\dots+n(n-1)(n-2)a_nx^{n-3}+(n+1)n(n-1)a_{n+1}x^{n-2}+\dots$$

$$\cdot \quad \cdot \quad \cdot \quad \cdot \quad \cdot \quad \cdot$$

$$f^{(n)}(x)=n!\,a_n+\{(n+1)(n)(n-1)\dots 3.2\}a_{n+1}x+\dots$$

Substituting $x=0$ in turn in the above results, we get

$$a_0=f(0),\ a_1=f'(0),\ a_2=f''(0)/2!,\ a_3=f'''(0)/3!,\dots,\ a_n=f^{(n)}(0)/n!.$$

Hence if a given function $f(x)$ has a power series expansion of the type (i) in an interval $-l<x<l$, the expansion must be of the form

$$f(x)=f(0)+\frac{f'(0)}{1!}x+\frac{f''(0)}{2!}x^2+\frac{f'''(0)}{3!}x^3+\dots+\frac{f^{(n)}(0)}{n!}x^n+\dots \qquad (11.1)$$

This expansion is known as Maclaurin's series for $f(x)$. It exists only if $f(x)$ and all its derivatives are finite when $x=0$.

11.3. Taylor's series

Taylor's series gives the expansion of $f(x)$ in powers of $x-a$, a being a constant. Suppose that

$$f(x)=b_0+b_1(x-a)+b_2(x-a)^2+b_3(x-a)^3+\dots+b_n(x-a)^n+\dots \quad \text{(ii)}$$

when $a-l<x<a+l$.

Then by successive differentiation as in § 11.2 we obtain series for $f'(x), f''(x)\dots$ all valid when $a-l<x<a+l$. By substituting $x=a$ in each series in turn, we obtain

$$b_0=f(a),\ b_1=f'(a),\ b_2=f''(a)/2!,\ b_3=f'''(a)/3!,\dots,\ b_n=f^{(n)}(a)/n!.$$

Hence, if $f(x)$ has a power series expansion of the type (ii) in an interval $a-l<x<a+l$, the expansion must be of the form

$$f(x)=f(a)+\frac{f'(a)}{1!}(x-a)+\frac{f''(a)}{2!}(x-a)^2+\dots+\frac{f^n(a)}{n!}(x-a)^n+\dots \quad (11.2)$$

(11.2) is known as Taylor's series or the expansion of $f(x)$ in the neighbourhood of $x=a$.

Maclaurin's series, which may be obtained as a special case of Taylor's series by putting $a=0$ in (11.2), is sometimes called the expansion of $f(x)$ in the neighbourhood of $x=0$.

11.4. Another form of Taylor's series

A very convenient form of Taylor's series is obtained by writing $x+h$ for x and x for a in (11.2) The series then becomes

$$f(x+h)=f(x)+hf'(x)+\frac{h^2}{2!}f''(x)+\frac{h^3}{3!}f'''(x)+\dots \quad (11.3)$$

It is beyond the scope of this book to discuss the conditions under which a function $f(x)$ may be validly expanded in an infinite power series. We assume that the functions under discussion may be so expanded and use the foregoing results to find their formal expansions.

11.5. Series for sin x and cos x

If $f(x)=\sin x$, by § 9.16 (iii), $f^{(n)}x=\sin(x+\frac{1}{2}n\pi)$ and $f^{(n)}(0)=\sin\frac{1}{2}n\pi$. Hence, by (11.1),

$$\sin x=\frac{x}{1!}-\frac{x^3}{3!}+\frac{x^5}{5!}-\frac{x^7}{7!}+\dots$$

and, similarly,

$$\cos x=1-\frac{x^2}{2!}+\frac{x^4}{4!}-\frac{x^6}{6!}+\dots$$

The above series are convergent for all values of x, and the expansions are in fact valid for all values of x.

11.6. Series for $(1+x)^n$

If $f(x)=(1+x)^n$, where n is any rational number,

$$f^{(r)}(x)=n(n-1)\dots(n-r+1)(1+x)^{n-r}$$

and so $$f^{(r)}(0)=n(n-1)\dots(n-r+1).$$

Hence, by (11.1),

$$(1+x)^n=1+nx+\frac{n(n-1)}{2!}x^2+\frac{n(n-1)(n-2)}{3!}x^3+\dots$$
$$+\frac{n(n-1)(n-2)\dots(n-r+1)x^r}{r!}+\dots$$

When n is a positive integer, the expansion is finite and contains $(n+1)$ terms. When n is not a positive integer, the series is infinite and converges when $|x|<1$ (see § 4.19, Example 12). For values of x within this range the expansion is valid (cf. § 5.3).

11.7. The series for $\log(1+x)$

It has been shown in § 11.1 that

$$\log(1+x)=x-\frac{x^2}{2}+\frac{x^3}{3}-\frac{x^4}{4}+\dots \text{ when } -1<x\leqslant 1. \qquad (11.4)$$

This result may also be obtained from (11.1). See also § 5.15.

11.8. Other methods of obtaining an expansion of a given function

The main difficulty in using Maclaurin's series to expand a function $f(x)$ lies in the calculation of $f^{(n)}(x)$, which can be found by a simple formula in comparatively few cases (see § 9.16).

Alternative methods of expansion are illustrated below.

Example 1

By integrating the appropriate binomial expansions show that, when x is small,

$$\tan^{-1}x=x-\frac{x^3}{3}+\frac{x^5}{5}-\frac{x^7}{7}+\dots,$$

$$\sin^{-1}x=x+\frac{1}{2}\frac{x^3}{3}+\frac{1.3}{2.4}\frac{x^5}{5}+\frac{1.3.5}{2.4.6}\frac{x^7}{7}+\dots.$$

Evaluate $$\lim_{x\to 0}\frac{2\sin^{-1}x+\tan^{-1}x-3x(1+x^4)^{1/5}}{x^5}.$$ [Sheffield.]

$$\frac{d}{dt}(\tan^{-1}t)=\frac{1}{1+t^2}$$
$$=1-t^2+t^4-t^6+\dots \text{ when } |t|<1$$

by the binomial theorem.

Integrating both sides of this equation between the limits 0 and x, where x is small, we have by § 4.20 (vi)

$$\tan^{-1} x = x - \frac{x^3}{3} + \frac{x^5}{5} - \frac{x^7}{7} + \ldots \quad . \quad . \quad . \quad \text{(i)}$$

This series, known as Gregory's series for $\tan^{-1} x$, can be shown by using the ratio test and tests for the convergence of an alternating series (see § 4.15 and § 4.16) to converge when $-1 \leqslant x \leqslant 1$.

Similarly

$$\frac{d}{dt}(\sin^{-1} t) = \frac{1}{\sqrt{(1-t^2)}} = 1 + \tfrac{1}{2}t^2 + \frac{1.3}{2.4}t^4 + \frac{1.3.5}{2.4.6}t^6 + \ldots, \text{ when } |t| < 1.$$

Integrating between the limits 0 and x, where x is small, we have

$$\sin^{-1} x = x + \frac{1}{2}\frac{x^3}{3} + \frac{1.3}{2.4}\frac{x^5}{5} + \frac{1.3.5}{2.4.6}\frac{x^7}{7} + \ldots \quad . \quad . \quad \text{(ii)}$$

Substitution from (i) and (ii) in $2 \sin^{-1} x + \tan^{-1} x - 3x(1+x^4)^{1/5}$ gives

$$2\left(x + \frac{1}{6}x^3 + \frac{3}{40}x^5 + \ldots\right) + \left(x - \frac{x^3}{3} + \frac{x^5}{5} - \ldots\right) - 3x\left(1 + \frac{x^4}{5} + \ldots\right)$$

$= -\tfrac{1}{4}x^5 +$ terms involving higher powers of x.

Hence by § 4.20 (iv)

$$\lim_{x \to 0} \frac{2 \sin^{-1} x + \tan^{-1} x - 3x(1+x^4)^{1/5}}{x^5} = -\tfrac{1}{4}.$$

Example 2

Find the first four terms of the series for $\log(1+\sin x)$ in powers of x.

By (11.4), $\log(1+\sin x) = \sin x - \tfrac{1}{2}\sin^2 x + \tfrac{1}{3}\sin^3 x - \tfrac{1}{4}\sin^4 x + \ldots$, when $-1 < \sin x \leqslant 1$. Also,

$$\sin x = x - \frac{x^3}{3!} + \frac{x^5}{5!} - \ldots = x\left(1 - \frac{x^2}{6} + \frac{x^4}{120} - \ldots\right) \text{ for all values of } x.$$

Hence, for small x,

$$\log(1+\sin x) = x\left(1 - \frac{x^2}{6} + \frac{x^4}{120} - \ldots\right) - \frac{1}{2}x^2\left(1 - \frac{x^2}{6} + \ldots\right)^2$$

$$+ \frac{1}{3}x^3\left(1 - \frac{x^2}{6} + \ldots\right)^3 - \frac{1}{4}x^4\left(1 - \frac{x^2}{6} + \ldots\right)^4 + \ldots$$

$$= \left(x - \frac{x^3}{6} + \ldots\right) - \left(\frac{1}{2}x^2 - \frac{1}{6}x^4 + \ldots\right) + \left(\frac{1}{3}x^3 - \ldots\right)$$

$$- \left(\frac{1}{4}x^4 + \ldots\right) + \ldots$$

$$= x - \frac{x^2}{2} + \frac{x^3}{6} - \frac{x^4}{12} + \ldots.$$

The result may also be found by using Maclaurin's series.

Example 3

Using the series for sin x and cos x, deduce the first three terms of the series for tan x.

Since tan x is an odd function of x*, we assume that

$$\tan x = \frac{\sin x}{\cos x} = a_1 x + a_3 x^3 + a_5 x^5 + \dots.$$

Then, by § 11.5

$$x - \frac{x^3}{3!} + \frac{x^5}{5!} - \dots = \left(1 - \frac{x^2}{2!} + \frac{x^4}{4!} - \dots\right)(a_1 x + a_3 x^3 + a_5 x^5 + \dots)$$

$$= a_1 x + \left(a_3 - \frac{1}{2}a_1\right)x^3 + \left(a_5 - \frac{1}{2}a_3 + \frac{1}{24}a_1\right)x^5 + \dots.$$

Hence, equating coefficients (cf. § 4.20 (iii)),

$$a_1 = 1, \quad a_3 = \frac{1}{3}, \quad a_5 = \frac{2}{15}.$$

$$\therefore \tan x = x + \frac{1}{3}x^3 + \frac{2}{15}x^5 + \dots.$$

Example 4

Find by Maclaurin's theorem the expansion of sec x in ascending powers of x as far as the term in x^4.

Hence, or otherwise, show that, if x is so small that terms of higher order than x^4 can be neglected,

$$(3 + \sec^2 x)^{1/2} = a + bx^2 + cx^4,$$

where a, b, c are constants and find their values. [L.U. Anc.]

Let $f(x) = \sec x$;

then $f'(x) = \sec x \tan x,$

$f''(x) = 2\sec^3 x - \sec x,$

$f'''(x) = (6\sec^3 x - \sec x)\tan x,$

and $f^{iv}(x) = (6\sec^5 x - \sec^3 x) + \tan^2 x(18\sec^3 x - \sec x).$

Hence $f(0) = 1$, $f''(0) = 1$, $f^{iv}(0) = 5$; $f'(0) = f'''(0) = 0$.

$$\therefore \sec x = 1 + \frac{x^2}{2} + \frac{5x^4}{24} + \dots$$

and so $$3 + \sec^2 x = 4 + x^2 + \tfrac{2}{3}x^4 + \dots \qquad \text{(i)}$$

If $(3 + \sec^2 x)^{1/2} = a + bx^2 + cx^4$, putting $x = 0$ we have $a = 2$.

Also $$3 + \sec^2 x = a^2 + 2abx^2 + x^4(b^2 + 2ac) + \dots \qquad \text{(ii)}$$

Comparing (i) and (ii), we have by § 4.20 (iii) $2ab = 1$ and $b^2 + 2ac = \frac{2}{3}$,

whence $$b = \frac{1}{4} \text{ and } c = \frac{29}{192}.$$

* The function $f(x)$ is odd if $f(-x) = -f(x)$; $f(x)$ is even if $f(-x) = f(x)$. The Maclaurin series for $f(x)$ contains only odd powers of x if $f(x)$ is odd, only even powers if $f(x)$ is even.

Example 5

Find the first six terms of the expansion of $e^x \sin x$ in terms of x.

$$f(x)=e^x \sin x,\ f'(x)=e^x(\sin x+\cos x)=\sqrt{2}e^x \sin (x+\tfrac{1}{4}\pi),$$

and, in general $f^{(n)}x=2^{n/2}e^x \sin (x+\frac{1}{4}n\pi)$, so that $f^n(0)=2^{n/2} \sin \frac{1}{4}n\pi$.

Hence by (11.1)

$$e^x \sin x=x+\frac{2x^2}{2!}+\frac{2x^3}{3!}-\frac{4x^5}{5!}-\frac{8x^6}{6!}-\frac{8x^7}{7!}+\ldots.$$

Exercises 11 (*a*)

1. Write down the power series for e^x and for $\sin x$ and use them to show that $e^{\sin x}=1+x+\frac{1}{2}x^2-\frac{1}{8}x^4+\ldots.$

2. From the series for $\cos x$ deduce that
$$\sec x=1+\frac{1}{2}x^2+\frac{5}{24}x^4+\frac{61}{720}x^6+\ldots.$$

3. Express $1+\cos x$ in terms of $\cos \frac{1}{2}x$ and show that, if x is small, $\log (1+\cos x)=\log 2-\frac{1}{4}x^2-\frac{1}{96}x^4$ approximately.

4. Find the power series expansion of $\{(1-\cos x)/x^2\}^{1/3}$ as far as and including the term in x^4. [Sheffield.]

5. Use the series $\tan x=x+\frac{1}{3}x^3+\frac{2}{15}x^5+\ldots$ to prove that
$$x \cot x=1-\frac{1}{3}x^2-\frac{1}{45}x^4+\ldots$$
and
$$\log \sec x=\frac{1}{2}x^2+\frac{1}{12}x^4+\frac{1}{45}x^6+\ldots.$$

6. Use Maclaurin's theorem to prove that
$$e^x \cos x=1+x-\frac{2x^3}{3!}-\frac{2^2x^4}{4!}-\frac{2^2x^5}{5!}+\frac{2^3x^7}{7!}+\ldots$$
and
$$e^x \sin x=x+x^2+\frac{2x^3}{3!}-\frac{2^2x^5}{5!}-\frac{2^3x^6}{6!}-\frac{2^3x^7}{7!}+\ldots.$$

7. If $f(x)=a_0+a_1x+a_2x^2+\ldots+a_nx^n+\ldots$, show how the Maclaurin formula $a_n=f^{(n)}(0)/n!$ is obtained.

   Starting from an expansion of the function $1/(1+x^2)$, find the expansion in powers of x of the function $\tan^{-1} x$.

   Prove that the tenth derivative of $\tan^{-1} (x^2)$ with respect to x has the value $2(9!)$ when $x=0$. [L.U.]

8. Obtain Gregory's series for $\tan^{-1} x$, stating for what values of x it is convergent.

   Hence, or otherwise, find the expansion of $\cosh (\tan^{-1} x)$ as a series of ascending powers of x, as far as the term in x^4. [Leeds.]

9. By assuming that the binomial expansion of $(1+x)^\alpha$ is valid when $|x|<1$, deduce formally that

$$\log\{x+\sqrt{(1+x^2)}\}=x+\sum_{n=1}^{\infty}(-1)^n\frac{1.3.5\ldots(2n-1)}{2.4.6\ldots2n}\frac{x^{2n+1}}{2n+1}.$$ [Durham.]

10. Prove that $$x/(e^x-1)=1-\frac{1}{2}x+\frac{1}{12}x^2-\frac{1}{720}x^4+\ldots$$

and that $$x/(e^x+1)=\frac{1}{2}x-\frac{1}{4}x^2+\frac{1}{48}x^4-\ldots.$$

11. Prove by Maclaurin's theorem that

$$\sin(x+a)=\sin a+x\cos a-\frac{x^2}{2!}\sin a-\frac{x^3}{3!}\cos a+\ldots$$

and that $$\tan\left(\frac{1}{4}\pi+x\right)=1+2x+2x^2+\frac{8}{3}x^3+\frac{10}{3}x^4+\ldots.$$

12. By integrating the binomial series for $1/\sqrt{(1+x^2)}$ prove that

$$\sinh^{-1}x=x-\frac{1}{2}\frac{x^3}{3}+\frac{1.3}{2.4}\frac{x^5}{5}-\frac{1.3.5}{2.4.6}\frac{x^7}{7}+\ldots.$$

13. If $y=e^x\cos x$, show by induction that

$$\frac{d^ny}{dx^n}=2^{n/2}e^x\cos\left(x+\frac{1}{4}n\pi\right)$$

and prove that $\dfrac{d^4y}{dx^4}+4y=0.$

Expand $e^x\cos x$ formally in a series of powers of x and hence, or otherwise, show that $\cos x\cosh x=\sum_{n=0}^{\infty}\{(-4)^nx^{4n}/(4n)!\}$. [Durham.]

14. Use Maclaurin's expansion to prove that

$$\log(1+e^x)=\log 2+\frac{1}{2}x+\frac{1}{8}x^2-\frac{1}{192}x^4+\ldots.$$

Write down the expansions of $\log(1+e^{ix})$ and $\log(1+e^{-ix})$ and deduce an expansion for $\log(1+\cos x)$ in ascending powers of x.

15. By Maclaurin's theorem, or otherwise, find the expansion of

$$y=\log\{1-\log(1-x)\}$$

as far as the term in x^3. By the substitution $x=t/(1+t)$ deduce the expansion of $\log\{1+\log(1+t)\}$ in powers of t as far as the term in t^3.

[Sheffield.]

16. Find the nth differential coefficient, with respect to x, of the functions $e^{ax}\cos bx$ and $e^{ax}\sin bx$.

Expand the function $e^{x\cos\alpha}\cos(x\sin\alpha)$ in a series of ascending powers of x and find the sum of the series $\sum_{r=1}^{\infty}\{(x^r\sin r\alpha)/r!\}$ [L.U.]

17. If $f(x)$ possesses a Maclaurin series and $f(x)=f(-x)$ prove that the series contains only even powers of x.

Expand x cosec x as far as the fourth power of x inclusive. [L.U.]

18. If $y=\frac{2}{3}\sqrt{3}\tan^{-1}\{x\sqrt{3}/(2+x)\}$, prove that $\frac{dy}{dx}=1/(1+x+x^2)$. If $-1<x<1$, by assuming that y can be expanded in a series of ascending powers of x, and using the equation $(1-x^3)\frac{dy}{dx}=1-x$, or otherwise, prove that

$$y=x-\frac{x^2}{2}+\frac{x^4}{4}-\frac{x^5}{5}+\frac{x^7}{7}-\ldots+\frac{x^{3n+1}}{3n+1}-\frac{x^{3n+2}}{3n+2}+\ldots$$ [L.U.]

19. (i) Obtain the first two terms of the expansion of $\log(1+e^{-x^2})$ as a power series in x.

(ii) Find the coefficient of x^8 in the power series expansion of the function $x^3\sin^3 x$. [L.U.]

20. If n is a real, positive integer, prove that one of the values of

$$(\cos\theta+i\sin\theta)^n \text{ is } (\cos n\theta+i\sin n\theta).$$

If $y=e^x\cos^2 x$, where x is real, expand y in a series of ascending powers of x, and prove that the coefficient of x^n is $\frac{1}{2}(1+5^{n/2}\cos n\theta)/n\,!$, where $\theta=\tan^{-1}2$. [L.U.]

21. Show that the differential coefficient of

$$\tan^{-1}\left(\frac{x\sin\alpha}{1-x\cos\alpha}\right)$$

with respect to x is

$$\frac{1}{2i}\left\{\frac{z}{1-zx}-\frac{1/z}{(1-x/z)}\right\}$$

where $z=\cos\alpha+i\sin\alpha$.

Hence, or otherwise, show that when $|x|<1$, the expansion of

$$\tan^{-1}\left(\frac{x\sin\alpha}{1-x\cos\alpha}\right)$$

as a series in ascending powers of x is $\sum_{n=1}^{\infty}(x^n\sin n\alpha)/n$. [L.U.]

11.9. Use of Leibniz's theorem in the expansion of functions

By means of Leibniz's theorem a relation may be established between successive derivatives of a function and used to determine the expansion of the function.

Example 6

If $y=\cos\log(1+x)$, prove that

$$(1+x)^2y_{n+2}+(2n+1)(1+x)y_{n+1}+(n^2+1)y_n=0,$$

where y_r denotes $\dfrac{d^ry}{dx^r}$.

Show that, if $\cos\log(1+x)$ can be represented by the power series

$$a_0+a_1x+a_2x^2+\ldots+a_nx^n+\ldots,$$

then $(n+1)(n+2)a_{n+2}+(2n+1)(n+1)a_{n+1}+(n^2+1)a_n=0$, and determine the expansion up to and including the term in x^6. [L.U.]

If
$$y=\cos\log(1+x) \quad . \quad . \quad . \quad . \quad \text{(i)}$$

$$(1+x)\frac{dy}{dx}=-\sin\log(1+x) \quad . \quad . \quad . \quad . \quad \text{(ii)}$$

and
$$(1+x)\frac{d^2y}{dx^2}+\frac{dy}{dx}=-\{\cos\log(1+x)\}/(1+x),$$

i.e.
$$(1+x)^2\frac{d^2y}{dx^2}+(1+x)\frac{dy}{dx}+y=0.$$

The result of differentiating this relation n times, using Leibniz's theorem, is

$$(1+x)^2y_{n+2}+(2n+1)(1+x)y_{n+1}+(n^2+1)y_n=0, \text{ where } y_r=\frac{d^ry}{dx^r} \text{ and } n\geqslant 0.$$

When $x=0$, this relation gives

$$b_{n+2}+(2n+1)b_{n+1}+(n^2+1)b_n=0 \quad . \quad . \quad . \quad \text{(iii)}$$

where b_r denotes the value of y_r when $x=0$.

Now if y can be represented by the series $\sum\limits_0^\infty a_rx^r$, $a_r=b_r/r!$ by Maclaurin's theorem, and so from (iii)

$$(n+2)(n+1)a_{n+2}+(2n+1)(n+1)a_{n+1}+(n^2+1)a_n=0 \quad . \quad \text{(iv)}$$

From (i) and (ii), substituting $x=0$, we get $a_0=1$ and $a_1=0$.

From (iv), when $n=0$, $2a_2+a_1+a_0=0$, $\therefore a_2=-\frac{1}{2}$;

and when $n=1$, $6a_3+6a_2+2a_1=0$, $\therefore a_3=\frac{1}{2}$.

Similarly, $a_4=-\dfrac{5}{12}$, $a_5=\dfrac{1}{3}$ and $a_6=-\dfrac{19}{72}$.

Hence $\cos\log(1+x)=1-\dfrac{1}{2}x^2+\dfrac{1}{2}x^3-\dfrac{5}{12}x^4+\dfrac{1}{3}x^5-\dfrac{19}{72}x^6+\ldots$

Example 7

If $x=\cos\theta$, $y=\cos n\theta$, and n is an integer >1, prove that

$$(1-x^2)\frac{d^2y}{dx^2}-x\frac{dy}{dx}+n^2y=0.$$

Assuming that $y=a_0+a_1x+a_2x^2+\ldots+a_nx^n$, where the a_k are constants, prove that $(k+1)(k+2)a_{k+2}+(n^2-k^2)a_k=0$ $(k=0, 1,\ldots, n-2)$.

Show that if n is even and greater than 2, then $a_4=(-1)^{n/2}n^2(n^2-4)/24$.

[L.U.]

Since
$$x=\cos\theta,\quad y=\cos n\theta \quad . \quad . \quad . \quad . \quad \text{(i)}$$

$$\frac{dy}{dx}=\frac{n\sin n\theta}{\sin\theta}$$

and
$$\frac{d^2y}{dx^2}=-\frac{n}{\sin\theta}\left\{\frac{n\cos n\theta\sin\theta-\sin n\theta\cos\theta}{\sin^2\theta}\right\} \text{ see § 9.18 } (a)$$

$$\therefore\ (1-x^2)\frac{d^2y}{dx^2}-x\frac{dy}{dx}+n^2y=0.$$

Differentiating this result k times, we get (using the notation of Example 6)

$$(1-x^2)y_{k+2}-(2k+1)xy_{k+1}+(n^2-k^2)y_k=0,$$

whence by putting $x=0$,

$$b_{k+2}=(k^2-n^2)b_k. \quad . \quad . \quad . \quad . \quad \text{(ii)}$$

But if $y=\sum_0^n a_rx^r$, $a_r=b_r/r\,!$, and so from (ii)

$$(k+2)(k+1)a_{k+2}=(k^2-n^2)a_k \quad . \quad . \quad . \quad \text{(iii)}$$

Since y is a polynomial of degree n, this relation is true for

$$k=0, 1, 2,\ldots(n-2).$$

From (i), when $x=0$, $\theta=\pm\frac{1}{2}\pi$ and $y=\cos\frac{1}{2}n\pi=(-1)^{n/2}$ when n is even,

i.e.
$$a_0=(-1)^{n/2}.$$

From (iii)
$$a_2=-n^2a_0/2$$

and
$$a_4=(4-n^2)a_2/12$$
$$=(-1)^{n/2}n^2(n^2-4)/24.$$

Exercises 11 (*b*)

For brevity, y_r is written for d^ry/dx^r unless otherwise stated.

1. If $y=\sinh^{-1}x$, show that $(1+x^2)y_2+xy_1=0$, and that, for $n\geqslant 2$,

$$(1+x^2)y_n+(2n-3)xy_{n-1}+(n-2)^2y_{n-2}=0.$$

Using this result when $x=0$, or otherwise, obtain the expansion of $\sinh^{-1}x$ in powers of x. [Durham].

2. Show that $e^{\sinh^{-1} x}$ satisfies the differential equation

$$(1+x^2)y_2+xy_1-y=0.$$

By differentiating n times obtain a second order differential equation which is satisfied by $\dfrac{d^n}{dx^n}(e^{\sinh^{-1} x})$.

Expand $e^{\sinh^{-1} x}$ in a series of ascending powers of x, and write down the general term of the series. [Sheffield.]

3. If $y=e^{\tan^{-1} x}$, where $\tan^{-1} x$ lies between $-\frac{1}{2}\pi$, and $+\frac{1}{2}\pi$, show that $(1+x^2)y_1=y$, and, by successive differentiation of this equation, obtain, using Maclaurin's theorem, the expansion of y in ascending powers of x as far as the term in x^5.

Deduce that

$$\lim_{x\to 0} \frac{e^{\tan^{-1} x}-e^x(1-\frac{1}{3}x^3)}{x^5}=\frac{1}{5}. \qquad \text{[L.U. Anc.]}$$

4. If $y=(\sin^{-1} x)^2+a \sin^{-1} x$, where a is a constant, prove that

$$(1-x^2)y_2-xy_1-2=0.$$

Find the expansion of y in positive integral powers of x and give the coefficients of the terms in x^{2p} and x^{2p+1}. [L.U.]

5. If $y=\log \{1+\sqrt{(1-x)}\}$ prove that $4x(1-x)y_2+2(2-3x)y_1+1=0$, and $2x(1-x)y_{n+2}+\{(2n+2)-(4n+3)x\}y_{n+1}-n(2n+1)y_n=0$.

Show that if $|x|<1$,

$$\log\{1+\sqrt{(1-x)}\}=\log 2-\sum_{n=1}^{\infty}\frac{1.3.5\ldots(2n-1)}{2.4.6\ldots 2n}\frac{x^n}{2n}. \qquad \text{[L.U.]}$$

6. If $y=e^{\sin^{-1} x}$, prove that $(1-x^2)y_2-xy_1=y$.

Show that, if y_n denotes the value of $\dfrac{d^ny}{dx^n}$ when $x=0$,

$$y_{n+2}=(n^2+1)y_n,\ n\geqslant 1.$$

Hence, or otherwise, assuming that $e^{\sin^{-1} x}$ can be expanded as a series in ascending powers of x when $|x|<1$, prove that the series is

$$1+x+\frac{x^2}{2!}+\frac{2}{3!}x^3+\frac{5}{4!}x^4+\ldots \qquad \text{[L.U.]}$$

7. If $y=(1-x^2)^{-1/2} \sin^{-1} x$, prove that

(i) $(1-x^2)y_1-xy=1$,

(ii) $(1-x^2)y_{n+1}-(2n+1)xy_n-n^2y_{n-1}=0$ if $n>0$.

Deduce that, as far as the term in x^5, $y=x+\dfrac{2x^3}{3}+\dfrac{8x^5}{15}$.

Hence find $\displaystyle\lim_{x\to 0}\left\{\frac{\sin^{-1} x-x(1-x^2)^{-1/6}}{x^5}\right\}$. [L.U.]

8. If $y=\{x+\sqrt{(1+x^2)}\}^m$, prove that

(i) $$(1+x^2)y_2+xy_1-m^2y=0,$$

(ii) $(1+x^2)y_{n+2}+(2n+1)xy_{n+1}+(n^2-m^2)y_n=0.$

If m is a positive integer (even or odd), prove that the value of y_{m+1} when x is zero is $(2m)!/2^m(m-1)!$. [L.U.]

9. If $y=\cos\{\pi\sqrt{(1+x)}\}$, prove that $4(1+x)y_2+2y_1+\pi^2y=0$, and hence, by differentiating n times, using Leibniz's theorem, obtain a relation between any three successive derivatives of y. Assuming that $y=\sum_0^\infty a_nx^n$, prove that

$$4(n+1)(n+2)a_{n+2}+2(2n+1)(n+1)a_{n+1}+\pi^2a_n=0,$$

and write down the expansion as far as the term in x^4. [L.U.]

10. If $y=\sinh(m\sinh^{-1}x)$, prove that

$$(1+x^2)y_{n+2}+(2n+1)xy_{n+1}+(n^2-m^2)y_n=0.$$

When $m=5$, obtain y as a polynomial in x. [L.U.]

11. If $y=(1+\sinh^{-1}x)^2$, show that $(1+x^2)y_2+xy_1=2$. Prove that for $n>0$,

$$(1+x^2)y_{n+2}+(2n+1)xy_{n+1}+n^2y_n=0.$$

If $y=a_0+a_1x+a_2x^2+\ldots$, find formulae for a_{2n-1} and a_{2n} where $n\geqslant 1$. [L.U.

12. If $y=\sin(m\sin^{-1}x)$, where m is a real constant greater than unity, prove that $(1-x^2)y_2-xy_1+m^2y=0$.

If $1>x>-1$, prove that

$$y=m\left\{x+\frac{(1^2-m^2)}{3!}x^3+\frac{(1^2-m^2)(3^2-m^2)}{5!}x^5+\ldots\right\}$$

and find the term in x^{2n+1}. [L.U.]

13. If $y=[\log\{x+(a^2+x^2)^{\frac{1}{2}}\}]^2$, show that

$$(a^2+x^2)y_2+xy_1=2.$$

Differentiate this equation n times, and deduce, or find by any other means, the expansion of y in terms of positive integral powers of x, giving the general term. [L.U.]

14. If $x=\tanh u$, $y=\operatorname{sech} u$, show that

$$(1-x^2)\frac{d^2y}{dx^2}-x\frac{dy}{dx}+y=0.$$

Hence show that

$$(1-x^2)\frac{d^{n+2}y}{dx^{n+2}}-(2n+1)x\frac{d^{n+1}y}{dx^{n+1}}-(n^2-1)\frac{d^ny}{dx^n}=0,$$

and deduce the expansion of y in ascending powers of x.

Verify as far as the term in x^6 that this expansion is the same as that obtained from the relation $y^2=1-x^2$ using the binomial theorem. [L.U.]

15. If $x=\sinh t$, $y=\sinh pt$, prove that

$$\text{(i)}\quad (x^2+1)\frac{d^2y}{dx^2}+x\frac{dy}{dx}-p^2y=0,$$

$$\text{(ii)}\quad (y^2+1)\frac{d^2x}{dy^2}+y\frac{dx}{dy}-\frac{1}{p^2}x=0.$$

By differentiating (i) n times, show that $y_{n+2}=(p^2-n^2)y_n$, where y_r is the value of $\dfrac{d^ry}{dx^r}$ when $x=0$. Hence obtain the expansion of y in ascending powers of x when $p=7$. [L.U.]

16. If $y=(\sin^{-1}x)^2$, show that $(1-x^2)y_2-xy_1=2$. Apply Leibniz's theorem to this equation and find a relation between y_n, y_{n+1} and y_{n+2}.

Hence show that if $(\sin^{-1}x)^2$ is expanded in a series of ascending powers of x, the coefficient of x^{2n}, $n\geqslant 1$, is $2^{2n-1}\{(n-1)!\}^2/(2n)!$ [L.U.]

17. If $f(x)=\displaystyle\int_0^x (1-t^3)^{-1/2}dt$, prove that $2(1-x^3)f''(x)=3x^2f'(x)$. Deduce that $2f^{n+2}(0)=(2n-1)n(n-1)f^{n-1}(0)$. Obtain the Maclaurin expansion for $f(x)$ as far as the term in x^{10}. [L.U. Anc.]

18. If $y=\cosh(\sin^{-1}x)$ prove

$$\text{(i)}\quad (1-x^2)y_2-xy_1-y=0,$$

$$\text{(ii)}\quad (1-x^2)y_{n+2}-(2n+1)xy_{n+1}=(n^2+1)y_n.$$

If $\cosh(\sin^{-1}x)=a_0+a_1x+a_2x^2+\ldots$, obtain an expression for a_{2n} and show that a_{2n+1} is zero. [L.U.]

11.10. Taylor's theorem applied to the evaluation of limits

Suppose we require to find $\lim\limits_{x\to a}F(x)$ where $F(x)=\dfrac{f(x)}{g(x)}$, and $f(x)$ and $g(x)$ are continuous. Let us assume first that a is finite. Then since the functions are continuous $\lim\limits_{x\to a}f(x)=f(a)$, $\lim\limits_{x\to a}g(x)=g(a)$ and so if $g(a)\neq 0$,

$$\lim_{x\to a}\frac{f(x)}{g(x)}=\frac{f(a)}{g(a)}$$

If $g(a)=0$, two cases arise: either $f(a)\neq 0$ or $f(a)=0$.

In the former case, $\dfrac{f(x)}{g(x)}\to\pm\infty$ as $x\to a$, i.e. $\lim\limits_{x\to a}\dfrac{f(x)}{g(x)}$ does not exist.

If $f(a)=g(a)=0$, $\dfrac{f(x)}{g(x)}$ assumes the indeterminate form 0/0 when $x=a$, and to find the required limit we write

$$\lim_{x\to a}\frac{f(x)}{g(x)}=\lim_{h\to 0}\frac{f(a+h)}{g(a+h)}$$

$$=\lim_{h\to 0}\frac{hf'(a)+\frac{1}{2}h^2f''(a)+\ldots}{hg'(a)+\frac{1}{2}h^2g''(a)+\ldots}\quad\ldots\quad\text{(i)}$$

assuming that, for small values of h, $f(a+h)$ and $g(a+h)$ may be expanded in convergent power series by Taylor's theorem. Then by § 4.20 (iv)

$$\lim_{x\to a}\frac{f(x)}{g(x)}=\frac{f'(a)}{g'(a)}. \qquad (11.5)$$

If $f'(a)=g'(a)=0$, i.e. if $\dfrac{f'(a)}{g'(a)}$ takes the indeterminate form 0/0, then by (i), if $g''(a)\neq 0$, we have

$$\lim_{x\to a}\frac{f(x)}{g(x)}=\frac{f''(a)}{g''(a)},$$

and so on.

Hence, if the fraction $\dfrac{f(x)}{g(x)}$ assumes the indeterminate form 0/0 when $x=a$, then $\lim\limits_{x\to a}\dfrac{f(x)}{g(x)}$ is equal to the first of the expressions

$$\frac{f'(a)}{g'(a)},\quad \frac{f''(a)}{g''(a)},\quad \frac{f'''(a)}{g'''(a)},\ldots$$

which is not indeterminate.

Example 8

Find $\lim\limits_{x\to 0}\dfrac{x-\sin x}{x^3}$.

Here

$f(x)=x-\sin x,\quad f(0)=0;\qquad g(x)=x^3,\quad g(0)=0;$

$f'(x)=1-\cos x,\quad f'(0)=0;\qquad g'(x)=3x^2,\quad g'(0)=0;$

$f''(x)=\sin x,\quad f''(0)=0;\qquad g''(x)=6x,\quad g''(0)=0;$

$f'''(x)=\cos x,\quad f'''(0)=1;\qquad g'''(x)=6,\quad g'''(0)=6.$

Hence the required limit is $\frac{1}{6}$.

Example 9

Find $\lim\limits_{x\to 0}\dfrac{x-\tan x}{x-\sin x}$.

Here

$f(x)=x-\tan x,\quad f(0)=0;\qquad g(x)=x-\sin x,\quad g(0)=0;$

$f'(x)=1-\sec^2 x,\quad f'(0)=0;\qquad g'(x)=1-\cos x,\quad g'(0)=0;$

$f''(x)=-2\sec^2 x\tan x,\quad f''(0)=0;\qquad g''(x)=\sin x,\quad g''(0)=0;$

$f'''(x)=-2(\sec^4 x+2\sec^2 x\tan^2 x),\quad f'''(0)=-2;\qquad g'''(x)=\cos x,\quad g'''(0)=1$

Hence the required limit is -2.

11.11. De l'Hospital's rule

It may also be shown that when $f(a)=g(a)=0$

$$\lim_{x\to a}\frac{f(x)}{g(x)}=\lim_{x\to a}\frac{f'(x)}{g'(x)} \tag{11.6}$$

provided that the second limit exists.

If the value of the right-hand side is not immediately obvious, and if $f'(a)=g'(a)=0$

$$\lim_{x\to a}\frac{f(x)}{g(x)}=\lim_{x\to a}\frac{f''(x)}{g''(x)}$$

if the right-hand limit exists, and so on.

The proof of this result is beyond the scope of this book. It is known as de l'Hospital's rule, and if we use it to evaluate the limits given in Examples 8 and 9 we have

$$\lim_{x\to 0}\frac{x-\sin x}{x^3}=\lim_{x\to 0}\frac{1-\cos x}{3x^2}=\lim_{x\to 0}\frac{1}{6}\left(\frac{\sin x}{x}\right)=\frac{1}{6} \text{ by § 9.3 (2)}$$

$$\lim_{x\to 0}\frac{x-\tan x}{x-\sin x}=\lim_{x\to 0}\frac{1-\sec^2 x}{1-\cos x}=\lim_{x\to 0}\{(-\sec^2 x)(1+\cos x)\}=-2.$$

For the valid application of the above rules the functions $f(x)$ and $g(x)$ are subject to certain restrictions which we shall not discuss here. We shall assume that the rules may be applied to all functions encountered at this stage.

11.12. Other indeterminate forms

It may be shown that if $\dfrac{f(x)}{g(x)}$ assumes the indeterminate form $\dfrac{\infty}{\infty}$ when $x=a$, the above rules may be applied to find $\lim\limits_{x\to a}\dfrac{f(x)}{g(x)}$.

Example 10

By (11.6) $$\lim_{x\to 0}\frac{\log\sin 2x}{\log\sin x}=\lim_{x\to 0}\frac{2\cot 2x}{\cot x}=\lim_{x\to 0}\frac{2\tan x}{\tan 2x} \qquad \left[\text{form } \frac{0}{0}\right]$$

$$=\lim_{x\to 0}\frac{\sec^2 x}{\sec^2 2x}=1.$$

Example 11

By (11.6) $$\lim_{x\to 0}\frac{\log x}{\operatorname{cosec} x}=\lim_{x\to 0}\frac{1/x}{-\operatorname{cosec} x\cot x}=\lim_{x\to 0}\left(\frac{\sin x}{x}\right)(-\tan x)=0$$

by § 9.3(2).

Again, if $\dfrac{f(x)}{g(x)}$ assumes either of the indeterminate forms 0/0 or ∞/∞ when x becomes infinite, rules similar to (11.5) and (11.6) may be used to find $\lim\limits_{x\to\infty} \dfrac{f(x)}{g(x)}$. Alternatively, we may put $x=\dfrac{1}{t}$ and consider $\lim\limits_{t\to 0} \dfrac{f(1/t)}{g(1/t)}$.

Example 12

If $n>0$,
$$\lim_{x\to\infty} \frac{\log x}{x^n} = \lim_{x\to\infty} \frac{1/x}{nx^{n-1}} \quad \text{by (11.6)}$$
$$=0.$$

From this result we deduce that $\log x$ tends to infinity more slowly than any positive power of x. Also, by writing $x=1/t$ we have
$$\lim_{x\to\infty} \frac{\log x}{x^n} = \lim_{t\to 0+} (-t^n \log t) = 0.$$
$$\therefore \quad \lim_{t\to 0+} t^n \log t = 0.$$

Some cases of limits may be reduced to one of the types already considered by rearranging the function. For example:

(i) To find $\lim\limits_{x\to a} uv$ where $u\to 0$ and $v\to\infty$ as $x\to a$, we consider $\lim\limits_{x\to a} \dfrac{u}{1/v}$ or $\lim\limits_{x\to a} \dfrac{v}{1/u}$ and apply (11.5) or (11.6).

Example 13

$$\lim_{x\to\frac{1}{2}} (\tfrac{1}{4}-x^2) \tan \pi x = \lim_{x\to\frac{1}{2}} \frac{(\frac{1}{4}-x^2)}{\cot \pi x} = \lim_{x\to\frac{1}{2}} \frac{-2x}{-\pi \operatorname{cosec}^2 \pi x} = \frac{1}{\pi}.$$

(ii) If $u\to\infty$ and $v\to\infty$ as $x\to a$, $\lim\limits_{x\to a} (u-v)$ may often be rearranged in the form $\dfrac{f(x)}{g(x)}$ where $f(x)$ and $g(x)$ both tend to zero (or to infinity) as $x\to a$.

Example 14

$$\lim_{x\to 0} (2 \operatorname{cosec}^2 x - \tfrac{1}{2} \operatorname{cosec}^2 \tfrac{1}{2}x)$$
$$= \lim_{x\to 0} \left(\frac{1}{2 \sin^2 \frac{1}{2}x \cos^2 \frac{1}{2}x} - \frac{1}{2 \sin^2 \frac{1}{2}x}\right)$$
$$= \lim_{x\to 0} \left(\frac{1-\cos^2 \frac{1}{2}x}{2 \sin^2 \frac{1}{2}x \cos^2 \frac{1}{2}x}\right) \qquad \left[\text{form } \frac{0}{0}\right]$$
$$= \lim_{x\to 0} (\tfrac{1}{2} \sec^2 \tfrac{1}{2}x)$$
$$= \tfrac{1}{2}.$$

Example 15

$$\lim_{y\to 1}\left(\frac{y}{y-1}-\frac{1}{\log y}\right)$$

$$=\lim_{y\to 1}\left(\frac{y\log y-y+1}{(y-1)\log y}\right) \qquad \left[\text{form } \frac{0}{0}\right]$$

$$=\lim_{y\to 1}\frac{\log y}{1-1/y+\log y} \quad \text{by (11.6)} \qquad \left[\text{form } \frac{0}{0}\right]$$

$$=\lim_{y\to 1}\frac{1/y}{1/y^2+1/y} \quad \text{by (11.6)}$$

$$=\tfrac{1}{2}.$$

(iii) If $y=u^v$ where u and v are functions of x, u being positive, then in order to evaluate $\lim_{x\to a} y$ it is convenient to write $y=e^{v\log u}$. If $v\log u\to l$ (a finite limit) as $x\to a$, then $y\to e^l$ since the exponential function is continuous (cf. § 5.8).

Hence $\qquad \lim_{x\to a}(v\log u)=l$ and $\log(\lim_{x\to a} y)=l.$

But $\qquad v\log u=\log y$

$$\therefore \lim_{x\to a}(\log y)=\log(\lim_{x\to a} y)=l.$$

We consider the following cases:

(1) when $u\to 0+$ and $v\to 0$ as $x\to a$ (Example 16),

(2) when $u\to +\infty$ and $v\to 0$ as $x\to a$ (Example 17),

(3) when $u\to 1$ and $v\to \infty$ as $x\to a$ (Example 18).

Example 16

To find $\lim_{x\to 0+} x^x$, let $y=x^x$.

Then $\qquad \log y=x\log x$

and $\qquad \lim_{x\to 0+}(\log y)=0$ (see Example 12) $=\log(\lim_{x\to 0+} y)$.

$$\therefore \lim_{x\to 0+} y=1.$$

Example 17

To find $\lim_{x\to 0+}\left(\frac{1}{x}\right)^{\sin x}$, let $y=\left(\frac{1}{x}\right)^{\sin x}$.

Then
$$\log y=\sin x\,.\log\left(\frac{1}{x}\right)$$
$$=-\sin x\,.\log x$$
$$=-\left(\frac{\sin x}{x}\right)(x\log x).$$

$$\therefore \lim_{x\to 0+}\log y=0=\log(\lim_{x\to 0+} y)$$

and so
$$\lim_{x\to 0+} y=1.$$

Example 18

To find $\lim\limits_{x\to\pi/4}\{(\tan x)^{\tan 2x}\}$, let $y=\tan x^{\tan 2x}$.

Then
$$\log y=\tan 2x\,(\log\tan x)$$
$$=\frac{\log\tan x}{\cot 2x},$$

and this has the form $\frac{0}{0}$ when $x=\pi/4$.

$$\therefore\ \lim_{x\to\pi/4}(\log y)=\lim_{x\to\pi/4}\left(\frac{\sec^2 x/\tan x}{-2\operatorname{cosec}^2 2x}\right)\quad\text{by (11.6)}$$
$$=-1$$
$$\therefore\ \lim_{x\to\pi/4} y=1/e.$$

Example 19

Show that $\lim\limits_{n\to\infty}(1+a/n)^n=e^a$.

If $y=(1+a/n)^n$, $\log y=n\log(1+a/n)$.

Hence
$$\lim_{n\to\infty}\log y=\lim_{n\to\infty} n\log(1+a/n)$$
$$=\lim_{x\to 0}\frac{\log(1+ax)}{x}\qquad\left[\text{form }\frac{0}{0}\right]$$
$$=\lim_{x\to 0}\frac{a}{1+ax}\quad\text{by (11.6)}$$
$$=a$$

i.e.
$$y\to e^a.$$

Hence
$$\lim_{n\to\infty}(1+a/n)^n=e^a.$$

When the functions involved can be expanded in convergent power series it may be easier to evaluate a limit by substituting these series than by using de l'Hospital's rule.

Example 20

Show that for all values of n, $\dfrac{e^x}{x^n}\to\infty$ *as* $x\to\infty$.

If $n<0$, $x^n\to 0$ as $x\to\infty$ and $e^x\neq 0$. Hence $\dfrac{e^x}{x^n}\to\infty$ as $x\to\infty$.

If $n=0$, $\dfrac{e^x}{x^n}=e^x\to\infty$ as $x\to\infty$ (see § 5.8).

If $n>0$, let k be the least integer which is greater than n.

Then when $x>0$,
$$e^x=1+x+\frac{x^2}{2!}+\ldots+\frac{x^{k-1}}{(k-1)!}+\frac{x^k}{k!}+\frac{x^{k+1}}{(k+1)!}+\ldots>\frac{x^k}{k!}$$
$$\therefore\ \frac{e^x}{x^n}>\frac{x^{k-n}}{k!}.$$

As $x \to \infty$, $\dfrac{x^{k-n}}{k!} \to \infty$

$$\therefore \quad \frac{e^x}{x^n} \to \infty \text{ as } x \to \infty \text{ for all values of } n.$$

It follows that e^x tends to infinity more rapidly than any positive power of x; also, by inversion of the above limit, $\lim_{x \to \infty} (e^{-x}x^n) = 0$.

Example 21

Evaluate $$\lim_{x \to 0} \frac{1}{x^4}\left\{\log(1+x) - xe^{-\frac{1}{2}x} - \frac{5}{24}\sin^3 x\right\}. \qquad \text{[L.U. Anc.]}$$

Let $$E = \frac{1}{x^4}\left\{\log(1+x) - xe^{-\frac{1}{2}x} - \frac{5}{24}\sin^3 x\right\}.$$

Then when $|x| < 1$,

$$E = \frac{1}{x^4}\left\{\left(x - \frac{1}{2}x^2 + \frac{1}{3}x^3 - \frac{1}{4}x^4 + \dots\right) - x\left(1 - \frac{1}{2}x + \frac{1}{8}x^2 - \frac{1}{48}x^3 + \frac{x^4}{384} - \dots\right)\right.$$
$$\left. - \frac{5}{24}\left(x - \frac{x^3}{6} + \dots\right)^3\right\}$$

$$= \frac{1}{x^4}\left\{-\frac{11}{48}x^4 + \text{terms involving higher powers of } x\right\}$$

$$= -\frac{11}{48} + \text{terms involving positive powers of } x$$

$$\therefore \quad \lim_{x \to 0} E = -\frac{11}{48}.$$

Exercises 11 (c)

1. Evaluate the following limits :

(i) $\displaystyle\lim_{x \to 0} \frac{\log(1+x)}{x}$, (ii) $\displaystyle\lim_{\phi \to 0} \frac{1 - \cos 2\phi}{1 - \cos 4\phi}$,

(iii) $\displaystyle\lim_{\theta \to 0} \frac{\sin\theta}{\sinh\theta}$, (iv) $\displaystyle\lim_{x \to 0} \frac{\cot 4x}{\cot x}$,

(v) $\displaystyle\lim_{t \to \pi} \frac{1 + \cos t}{\tan^2 t}$, (vi) $\displaystyle\lim_{x \to \infty} x(e^{1/x} - 1)$,

(vii) $\displaystyle\lim_{x \to 0} (1 - x)^{1/x}$, (viii) $\displaystyle\lim_{y \to \infty} y^{1/y}$,

(ix) $\displaystyle\lim_{x \to 1} \frac{1 - x}{\log x}$, (x) $\displaystyle\lim_{x \to \pi/2} (1 + \cot x)^{\tan x}$

(xi) $\displaystyle\lim_{x \to 0} (\operatorname{cosec} x - \cot x)$, (xii) $\displaystyle\lim_{x \to 0} \frac{a^x - 1}{x}$.

2. Expand $e^{\sin x}$ in ascending powers of x as far as x^4.

Find the value of $\lim\limits_{x\to 0} \dfrac{e^x - e^{\sin x}}{x - x\cos x}$. [Durham.]

3. Evaluate $$\lim_{x\to 0} \frac{\sin^2 x - x^2 \cos x}{x^3 \tan x}.$$ [L.U.]

4. Evaluate $$\lim_{x\to 0} \frac{\sin x \sinh x - x^2}{x^2 (\cos x \cosh x - 1)}.$$ [Sheffield.]

5. Expand $\cos^2 x \sin^2 x$ as far as the term involving x^6 and evaluate
$$\lim_{x\to 0} \{\cos^2 x \sin^2 x - x^2(1-x^2)^{4/3}\}/x^6.$$ [Sheffield.]

6. Evaluate $$\lim_{x\to 0} \frac{3x + 2\sin\frac{1}{2}x - 16\sin\frac{1}{4}x}{\sin^5 x}.$$ [Sheffield.]

7. Find the limit of $\dfrac{x^p - 1}{x - 1}$ as x tends to unity, and of $\{(x^2+ax+b)^{\frac{1}{2}} - x\}$ as x tends to infinity. [L.U.]

8. Find the limit as $x\to\infty$ of $\log(1+ax) - 2\log x + \log(a+x)$. [L.U.]

9. (i) Determine
$$(a)\ \lim_{x\to\infty} \frac{\frac{1}{2}\pi - \tan^{-1} x}{\sin(1/x)}; \qquad (b)\ \lim_{x\to 0} \frac{\sinh^{-1}(x) - x}{x^3}.$$

(ii) Prove that $x \log x \to 0$ as $x\to 0$ through positive values.
Hence determine $\lim\limits_{x\to 0} x^x$. [Durham.]

10. Determine

(i) $\lim\limits_{x\to 0} \dfrac{a^x - b^x}{c^x - d^x}$, where a, b, c, d are different positive numbers;

(ii) $\lim\limits_{x\to 0} \dfrac{x \sin^{-1} x}{\log\cos x}$; (iii) $\lim\limits_{x\to\infty} x(\frac{1}{2}\pi - \tan^{-1} x)$.

State the rules you use as precisely as you can. [Durham.]

11. Find $$\lim_{h\to 0} \left\{\frac{\sin(x+h) + \sin(x-h) - 2\sin x}{h^2}\right\}$$

and $$\lim_{h\to 0} \left\{\frac{\log(x+h) + \log(x-h) - 2\log x}{h^2}\right\}.$$ [L.U.]

12. (i) Show that the function $2x - \tan^{-1} x - \log\{x + \sqrt{(x^2+1)}\}$ is positive for positive values of x.

(ii) Find the limit as x tends to zero of the function
$$(x2^x - x)/(1 - \cos x).$$ [L.U.]

13. Write down the series for $\exp x$ (i.e. e^x) and $\sin x$, stating in each case for what range of values of the real variable x the series is convergent. Obtain the expansion of $\exp(\sin x)$ as far as the term in x^4.

Evaluate the limit

$$\lim_{x\to 0}\frac{\exp(\sin x)-1-x}{x^2}.$$ [L.U.]

14. (i) Prove that $$\lim_{x\to 0}\frac{x^2(e^x-e^{-x})}{(1+x^3)^4-(1-x^3)^4}=\tfrac{1}{4}.$$

(ii) Find $$\lim_{x\to 0}\frac{2x-2x^2-\log(1+2x)}{x^2\tan^{-1}x}.$$ [L.U.]

11.13. Newton's approximation to a root of an equation

Suppose that we wish to solve the equation $f(x)=0$ which we know to have a root near x_1. The root will be x_1+h where h is small and

$$f(x_1+h)=0,$$

i.e. $$f(x_1)+hf'(x_1)+\frac{h^2f''(x_1)}{2!}+\ldots=0$$ by Taylor's theorem.

We approximate by neglecting powers of h above the first:

$$f(x_1)+hf'(x_1)\simeq 0$$

$$\therefore\ h\simeq -\frac{f(x_1)}{f'(x_1)}$$

Thus if x_1 is a first approximation and x_2 a second approximation to the root,

$$x_2=x_1-\frac{f(x_1)}{f'(x_1)}. \qquad (11.7)$$

This is Newton's formula for an approximation to the root.

The next approximation x_3 is given by $x_3=x_2-\dfrac{f(x_2)}{f'(x_2)}$, and so on.

Fig 29 (*a*) shows how successive approximations approach the actual value of the root.

The graph of $y=f(x)$ cuts Ox at B so that OB is a root of the equation $f(x)=0$. $OA=x_1$ is the first approximation. If the ordinate at A meets the curve at P and the tangent at P meets Ox at C, $OC=x_2$ as given by (11.7), for

$$OC=x_1-\frac{AP}{\tan\angle PCA}=x_1-\frac{f(x_1)}{f'(x_1)}.$$

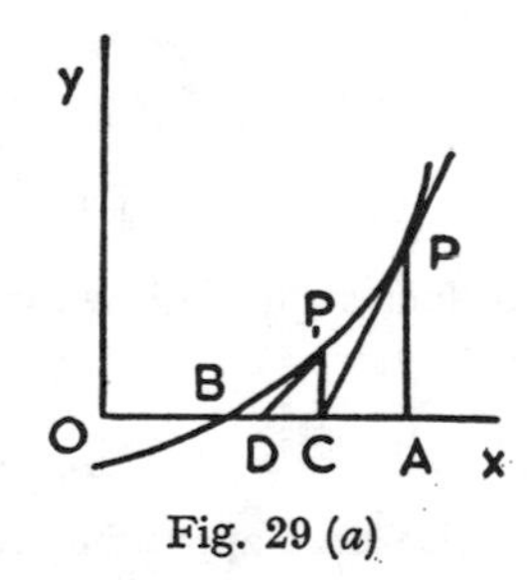

Fig. 29 (*a*)

To obtain a third approximation x_3 we draw the ordinate at C to meet the curve at P_1. Then the tangent at P_1 meets Ox at D where $OD=x_3=x_2-\dfrac{f(x_2)}{f'(x_2)}$.

It will be seen from fig. 29 (*b*) that if we take OA as a first approximation to the root we may obtain a second approximation OC which is on the other side of the root, but this does not matter if subsequent

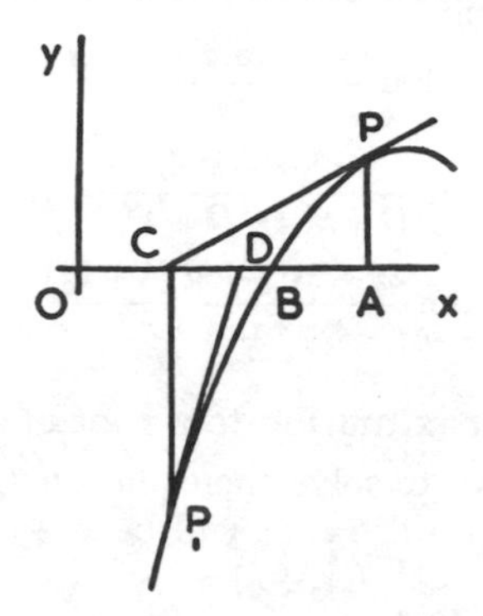

Fig. 29 (*b*)

values given by (11.6) approach closer and closer to the actual root. If a sufficiently close first approximation is taken, a small number of operations will in general give sufficient accuracy.

Example 22

Show that the equation $x^2-2e^{-x}-1{\cdot}4=0$ has a root near $1{\cdot}4$ and find an approximation to this root correct to three significant figures.

Let $$f(x)=x^2-2e^{-x}-1{\cdot}4.$$
Then $$f(1{\cdot}4)=1{\cdot}96-0{\cdot}4932-1{\cdot}4=0{\cdot}0668$$
and $$f(1{\cdot}3)=1{\cdot}69-0{\cdot}545-1{\cdot}4=-0{\cdot}255.$$

Since $f(1{\cdot}4)$ is very small and positive while $f(1{\cdot}3)$ is negative, there is a root of the equation $f(x)=0$ near $1{\cdot}4$

$$f'(x)=2(x+e^{-x})$$
$$f'(1{\cdot}4)=3{\cdot}293.$$

By (11.7), a possibly better approximation to the root is given by

$$x_2=1{\cdot}4-\frac{0{\cdot}0668}{3{\cdot}293}=1{\cdot}4-0{\cdot}0203=1{\cdot}3797=1{\cdot}380$$

to four significant figures.

$$f(1{\cdot}38)=1{\cdot}904-0{\cdot}5032-1{\cdot}4.$$
$$=0{\cdot}0008.$$
$$f'(1{\cdot}38)=2{\cdot}76+0{\cdot}5032=3{\cdot}263.$$

The next approximation will be

$$x_3=1{\cdot}38-\frac{0{\cdot}0008}{3{\cdot}263}=1{\cdot}380$$

to four significant figures. x_3 agrees with x_2 to three significant figures Hence the root is $1{\cdot}38$ to three significant figures.

Example 23

Find to three places of decimals the root of the equation $x = \tan x$ which lies between π and $\frac{3}{2}\pi$.

To facilitate the use of tables, we put $x = \pi + z$ in the equation

$$x - \tan x = 0 \quad . \quad . \quad . \quad . \quad . \qquad \text{(i)}$$

and obtain

$$\pi + z - \tan z = 0.$$

The root of this equation which we require lies between $z = 0$ and $z = \frac{1}{2}\pi$.

From tables or a graph, we see that there is a root near $z = 1{\cdot}35$.

If $f(z) = \pi + z - \tan z$, $f'(z) = -\tan^2 z$, and so if $z_1 = 1{\cdot}35$ is a first approximation to the root, a possibly better approximation is z_2, where by (11.7)

$$z_2 = z_1 - \frac{f(z_1)}{f'(z_1)} = 1{\cdot}35 + \frac{0{\cdot}0364}{19{\cdot}84} = 1{\cdot}3518.$$

The next approximation is z_3, where

$$z_3 = z_2 - \frac{f(z_2)}{f'(z_2)} = 1{\cdot}3518 + \frac{0{\cdot}0003}{20{\cdot}19}.$$

To four places of decimals $z_3 = z_2$, and so to this degree of accuracy $z = 1{\cdot}3518$ Hence the angle between π and $\frac{3}{2}\pi$ which satisfies (i) is $4{\cdot}493$ radians correct to three places of decimals.

11.14. Modification of Newton's formula

In the notation of § 11.13, $f'(x_1)$, $f'(x_2)$, $f'(x_3) \ldots$ are successive approximations to the gradient of the curve $y = f(x)$ at B in figs. 29 (*a*) and (*b*). In the preceding examples the values of these functions do not differ to any great extent and, in general, we may take $f'(x_1)$ or any convenient approximation to it and use it in place of $f'(x_2), f'(x_3) \ldots$.

If in Example 22 we take $f'(1{\cdot}4) = f'(1{\cdot}38) = 3$, we obtain $x_2 = 1{\cdot}3777$ and $x_3 = 1{\cdot}3797$, which give $1{\cdot}380$ to three significant figures.

Exercises 11 (*d*)

1. Show that the equation $x^3 + 3x^2 + 6x - 3 = 0$ has only one real root. Prove that this lies between 0 and 1, and find it correct to one place of decimals. [L.U.]

2. $x = 1{\cdot}2$ is an approximate solution of the equation

$$10 \log x - 3x + 1{\cdot}75 = 0.$$

   Apply Newton's approximation formula to find this solution correct to four significant figures.

3. Prove that the equation $5e^{0\cdot 2x} = 3(1 + x)$ has a root lying between $x = 1$ and $x = 1{\cdot}5$, and find this root correct to three significant figures.

4. Prove that the equation $1{\cdot}4x = \sinh x$ has a root near $x = 1{\cdot}5$. Find this root correct to four significant figures.

5. Show that the equation $\sinh z = 1\cdot25z$ has a root lying between $z=1\cdot1$ and $z=1\cdot2$. Find this root correct to three significant figures.

6. Show that the equation $x+5(\log x)^2=12\cdot8$ is satisfied by a value of x lying between 3·5 and 4, and determine this value correct to three significant figures.

7. Show that the equation $x^2-2e^{-x}=1\cdot5$ has a root near $x=1\cdot4$ and find this root correct to three significant figures.

8. Find, correct to four significant figures, the root of the equation $x^2 \log x=5\cdot812$ given that this root is approximately 2·5.

9. A root of the equation $5x^3+8e^{-3\cdot5x}=2\cdot05$ lies near $x=0\cdot6$. Find this root correct to three significant figures.

10. Show by means of a rough graph that the least positive root of the equation $\tan x=\frac{1}{2}x$ lies between π and $\frac{3}{2}\pi$. Find this least root correct to four significant figures. [L.U.]

11. Prove that the equation $x^3-4x+1=0$ has a root lying between 1 and 2 and find it correct to two decimal places. [L.U.]

12. Show that the equation $x^4-5x+2=0$ has two real roots, both of which are positive.

Find, by Newton's method, or otherwise, the larger root correct to four significant figures. [L.U.]

13. Show that the equation $\log(x+1)=x^2$ has a root between 0·5 and 1, and find its value correct to three decimal places.

14. Show that the function $y=x(x-4)/(x^2+3)^{\frac{1}{2}}$ has one real turning point and find the corresponding value of x correct to two decimal places. [L.U.]

CHAPTER 12

COORDINATE GEOMETRY OF THE STRAIGHT LINE AND CIRCLE

12.1. Useful formulae (revision)

(i) *Distance, gradient and section formulae*

The distance between $A(x_1, y_1)$, $B(x_2, y_2)$ is given by

$$AB = \sqrt{\{(x_2-x_1)^2+(y_2-y_1)^2\}}$$

and the gradient of AB (denoted by m_{AB}) is given by

$$m_{AB} = \frac{y_2-y_1}{x_2-x_1}.$$

The point P which divides AB in the ratio $\lambda : \mu$ is

$$\left(\frac{\lambda x_2+\mu x_1}{\lambda+\mu}, \frac{\lambda y_2+\mu y_1}{\lambda+\mu}\right).$$

P divides AB internally or externally according as the ratio $\lambda : \mu$ is positive or negative.

The coordinates of the mid-point of AB are

$$\tfrac{1}{2}(x_1+x_2),\ \tfrac{1}{2}(y_1+y_2).$$

(ii) *The angle between two given lines*

If θ is the angle between two straight lines of gradients m_1, m_2

$$\tan\theta = \pm\frac{m_1-m_2}{1+m_1m_2}.$$

This formula gives the tangent of the acute or obtuse angle between the lines according as the sign chosen makes the right-hand side positive or negative.

The lines are parallel if $m_1=m_2$; the lines are perpendicular if $m_1m_2=-1$.

(iii) *Area and centroid of a triangle*

The area, Δ, of the triangle whose vertices (arranged in counter-clockwise order) are (x_1, y_1), (x_2, y_2), (x_3, y_3) is given by

$$\Delta = \frac{1}{2}\begin{vmatrix} 1 & 1 & 1 \\ x_1 & x_2 & x_3 \\ y_1 & y_2 & y_3 \end{vmatrix}.$$

The coordinates of the centroid of the same triangle are

$$\tfrac{1}{3}(x_1+x_2+x_3),\ \tfrac{1}{3}(y_1+y_2+y_3).$$

This result may be obtained by regarding the centroid as a point of trisection of a median of the triangle and using the section formula.

(iv) *The equation of the straight line*

In Cartesian coordinates any equation of the first degree in x and y represents a straight line, but the reader should be familiar with the following forms of the equation of a line which correspond to its geometrical properties:

Equation	*Geometrical Property*
$x=\text{constant}$	line parallel to the y-axis
$y=\text{constant}$	line parallel to the x-axis
$y=mx$	line of gradient m through the origin
$y=mx+c$	line of gradient m making an intercept c on Oy
$y-y_1=m(x-x_1)$ . .	line of gradient m passing through the point (x_1, y_1)
$\dfrac{y-y_1}{x-x_1}=\dfrac{y_2-y_1}{x_2-x_1}$ . . .	line joining the points (x_1, y_1), (x_2, y_2)
$\dfrac{x}{a}+\dfrac{y}{b}=1$	line making intercepts a and b on Ox, Oy respectively
$x\cos\alpha+y\sin\alpha=p$ . .	line such that the perpendicular to it from the origin is of length p and makes an angle α with Ox
$ax+by+c+\lambda(Ax+By+C)=0$	line through the point of intersection of the lines $ax+by+c=0$, $Ax+By+C=0$.

(v) *Distance of a point from a straight line and the equations of the bisectors of the angles between two straight lines*

The perpendicular distance of the point $P(x_1, y_1)$ from the line $ax+by+c=0$ is

$$\pm\frac{ax_1+by_1+c}{\sqrt{(a^2+b^2)}}.$$

If the positive sign is chosen when c is positive and the negative sign when c is negative, this formula gives a positive result when P

lies on the same side of the line as the origin, a negative result when P lies on the opposite side from the origin.

From the above formula we deduce that the equations of the bisectors of the angles between the lines $ax+by+c=0$, $Ax+By+C=0$ are

$$\frac{ax+by+c}{\sqrt{(a^2+b^2)}}=\frac{Ax+By+C}{\pm\sqrt{(A^2+B^2)}}.$$

12.2. Parametric equations of a straight line

Let a straight line AB (fig. 30) drawn through the fixed point $A(x_1, y_1)$ make an angle θ with Ox; let $P(x, y)$ be any point on AB, or on AB produced, and let $AP=r$.

Then
$$x=x_1+r\cos\theta,$$
$$y=y_1+r\sin\theta.$$

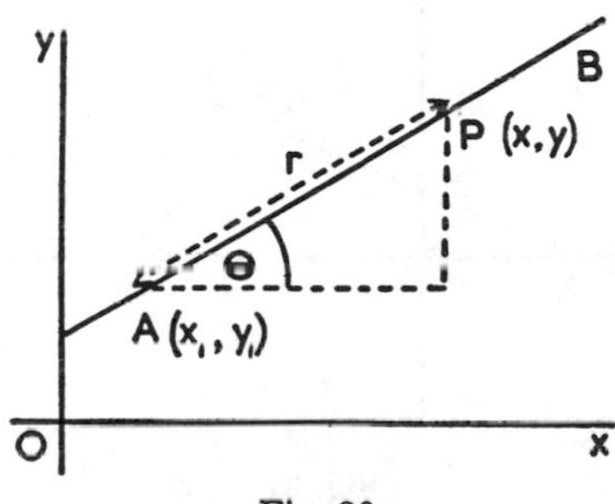

Fig. 30

These equations, which give the coordinates of any point on the line in terms of the single variable (parameter) r, are the parametric equations of the line. They may also be written in the form

$$\frac{x-x_1}{\cos\theta}=\frac{y-y_1}{\sin\theta}=r.$$

Note that points on BA produced correspond to negative values of r

Example 1

Find the equation of the straight line drawn through the point $P(h, k)$ such that, if it meets the axes of coordinates in the points A and B, P is the middle point of AB.

If any straight line drawn through P meets the axis of x at the point X, the axis of y at the point Y and the parallel through the origin to the straight line AB at the point Q, show that $2/PQ=1/PX+1/PY$, the lengths being measured algebraically. [L.U.]

If, in fig. 31, $P(h, k)$ is the mid-point of AB, $A \equiv (2h, 0)$, $B \equiv (0, 2k)$, the equation of AB is

$$\frac{x}{2h}+\frac{y}{2k}=1,$$

and the equation of the line drawn parallel to AB through the origin is

$$\frac{x}{h}+\frac{y}{k}=0 \quad . \quad . \quad . \quad . \quad . \quad \text{(i)}$$

The equation of any line through P may be taken as

$$\frac{x-h}{\cos\theta}=\frac{y-k}{\sin\theta}=r \quad . \quad . \quad . \quad . \quad \text{(ii)}$$

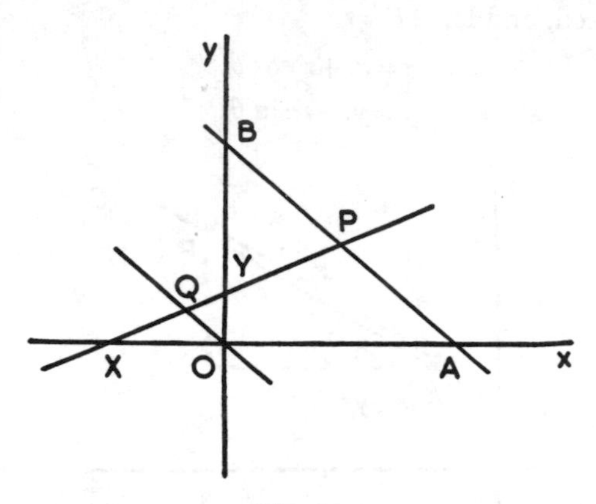

Fig. 31

This line meets Ox at X where $y=0$ and $r=PX$

$$\therefore \; PX=-k/\sin\theta \quad . \quad . \quad . \quad . \quad \text{(iii)}$$

Similarly,
$$PY=-h/\cos\theta \quad . \quad . \quad . \quad . \quad \text{(iv)}$$

Lines (i) and (ii) meet at Q where

$$\frac{h+r\cos\theta}{h}+\frac{k+r\sin\theta}{k}=0 \text{ and } r=PQ.$$

Hence
$$\frac{\cos\theta}{h}+\frac{\sin\theta}{k}=-\frac{2}{PQ}$$

and so by (iii) and (iv)
$$\frac{1}{PX}+\frac{1}{PY}=\frac{2}{PQ}.$$

12.3. Change of axes

It is often possible to simplify the equation of a given locus by referring it to new axes. Any desired change of axes may be effected by using either (or both) of the following methods. The axes are assumed in all cases to be rectangular.

(a) *Change of origin* (translation of axes)

In this case the new axes O_1X, O_1Y (fig. 32) are drawn through O_1, parallel to the original axes Ox, Oy and in the same sense. Then if

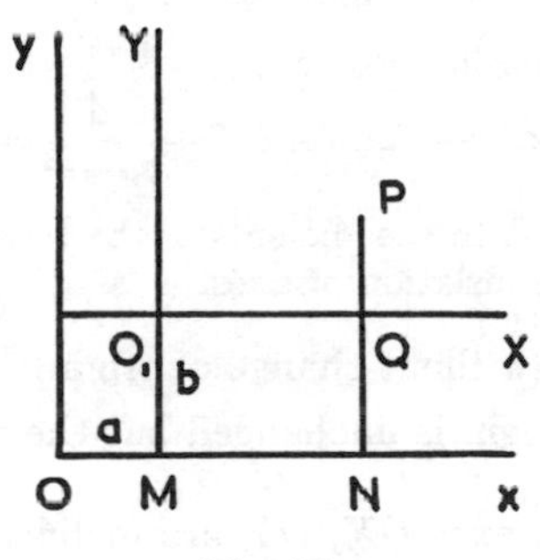

Fig. 32

$P \equiv (x, y)$ and $O_1 \equiv (a, b)$ referred to the original axes Ox, Oy, and if $P \equiv (X, Y)$ referred to the new axes, we have

$$x = X + a, \quad y = Y + b.$$

Thus if the equation of a locus referred to Ox, Oy is $f(x, y) = 0$, the equation referred to parallel axes through (a, b) is

$$f(X+a, Y+b) = 0. \tag{12.1}$$

Example 2

Show that by a suitable translation of axes the equation

$$ax^2 + 2hxy + by^2 + 2gx + 2fy + c = 0$$

may be expressed in the form $aX^2 + 2hXY + bY^2 + \Delta/(ab - h^2) = 0$, *where* $\Delta = abc + 2fgh - af^2 - bg^2 - ch^2$ *and* $ab - h^2 \neq 0$.

Transfer the origin to the point $(\bar{x}, \bar{y})$. Then by **(12.1)** the given equation becomes

$$a(X+\bar{x})^2 + 2h(X+\bar{x})(Y+\bar{y}) + b(Y+\bar{y})^2 + 2g(X+\bar{x}) + 2f(Y+\bar{y}) + c = 0,$$

i.e. $$aX^2 + 2hXY + bY^2 + 2X(a\bar{x} + h\bar{y} + g) + 2Y(h\bar{x} + b\bar{y} + f) + k = 0 \quad . \tag{i}$$

where $$\begin{aligned} k &= a\bar{x}^2 + 2h\bar{x}\bar{y} + b\bar{y}^2 + 2g\bar{x} + 2f\bar{y} + c \\ &= \bar{x}(a\bar{x} + h\bar{y} + g) + \bar{y}(h\bar{x} + b\bar{y} + f) + (g\bar{x} + f\bar{y} + c). \end{aligned}$$

(i) will be reduced to the required form if we can choose $(\bar{x}, \bar{y})$ so that

and $$\left.\begin{aligned} a\bar{x} + h\bar{y} + g &= 0 \\ h\bar{x} + b\bar{y} + f &= 0 \end{aligned}\right\} \quad . \quad . \quad . \quad . \quad . \tag{ii}$$

If $ab-h^2\neq 0$ these equations are satisfied by

$$\bar{x}=\frac{hf-bg}{ab-h^2},\quad \bar{y}=\frac{gh-af}{ab-h^2}\quad . \quad . \quad . \qquad \text{(iii)}$$

With these values $\qquad k=g\bar{x}+f\bar{y}+c$ by (ii)

$$=\frac{\Delta}{ab-h^2},\text{ substituting from (iii).}$$

Hence the given equation becomes

$$aX^2+2hXY+bY^2+\frac{\Delta}{ab-h^2}=0.$$

It should be noted that the coefficients of the terms of the highest degree are unchanged by a translation of axes.

(b) *Rotation of axes* (without change of origin)

In this case the origin is unchanged but the direction of the axes is altered.

In fig. 33 the new axes OX, OY are inclined at an angle θ to the

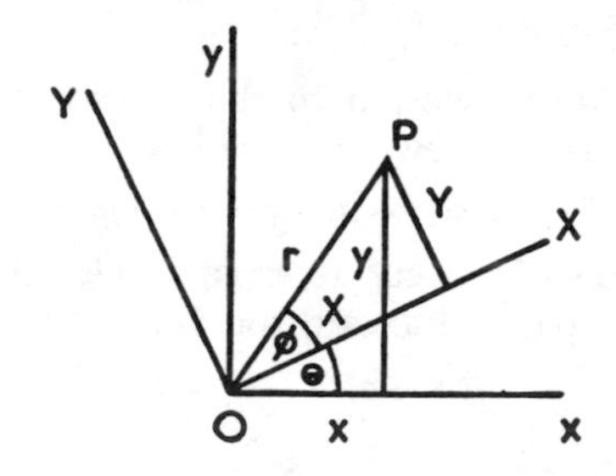

Fig. 33

original axes Ox, Oy, $P\equiv(x, y)$ referred to Ox, Oy and $P\equiv(X, Y)$ referred to OX, OY. Then if $OP=r$ and $\angle XOP=\phi$,

$$\begin{aligned} x&=r\cos(\theta+\phi)\\ &=(r\cos\phi)\cos\theta-(r\sin\phi)\sin\theta\\ &=X\cos\theta-Y\sin\theta. \end{aligned}$$

Similarly, $\qquad y=X\sin\theta+Y\cos\theta.$

Thus, if the equation of a locus referred to Ox, Oy is $f(x, y)=0$, referred to OX, OY it is

$$f(X\cos\theta-Y\sin\theta,\ X\sin\theta+Y\cos\theta)=0. \qquad (12.2)$$

Example 3

Show that, by a suitable rotation of axes, the equation $ax^2+2hxy+by^2=1$, $h\neq 0$, *may be transformed into an equation of the form* $a'X^2+b'Y^2=1$.

On rotating the axes as above, the given equation becomes

$$a(X\cos\theta - Y\sin\theta)^2 + 2h(X\cos\theta - Y\sin\theta)(X\sin\theta + Y\cos\theta) + b(X\sin\theta + Y\cos\theta)^2 = 1$$

i.e.

$$a'X^2 + 2h'XY + b'Y^2 = 1$$

where

$$a' = a\cos^2\theta + 2h\cos\theta\sin\theta + b\sin^2\theta,$$
$$h' = h(\cos^2\theta - \sin^2\theta) - (a-b)\sin\theta\cos\theta,$$
$$b' = a\sin^2\theta - 2h\sin\theta\cos\theta + b\cos^2\theta.$$

If h' is to be zero

$$h(\cos^2\theta - \sin^2\theta) = (a-b)\sin\theta\cos\theta,$$
$$h\cos 2\theta = \tfrac{1}{2}(a-b)\sin 2\theta$$
$$\therefore\ \tan 2\theta = 2h/(a-b).$$

It is possible to find a value of θ satisfying this equation whatever be the values of a, b and h. (When $b=a$ we take $2\theta = \frac{1}{2}\pi$.) Hence it is always possible by rotation of axes to reduce the equation

$$ax^2 + 2hxy + by^2 = 1 \text{ to the form } a'X^2 + b'Y^2 = 1.$$

PAIRS OF STRAIGHT LINES

12.4. The homogeneous equation of the second degree in x and y

The standard form of this equation is

$$ax^2 + 2hxy + by^2 = 0 \quad . \quad . \quad . \quad . \quad \text{(i)}$$

which may be written

$$b(y/x)^2 + 2h(y/x) + a = 0 \quad . \quad . \quad . \quad \text{(ii)}$$

When $b \neq 0$, solving this quadratic for y/x we obtain

$$y/x = \{-h \pm \sqrt{(h^2 - ab)}\}/b \quad . \quad . \quad . \quad \text{(iii)}$$

i.e. the equations of two straight lines passing through the origin.

If $b=0$ and $a \neq 0$, (i) may be written $x(ax+2hy)=0$ and represents the lines $x=0$ and $ax+2hy=0$.

If $a=b=0$ and $h \neq 0$, (i) may be written $2hxy=0$, which gives the lines $x=0$ and $y=0$.

Thus (i) always represents a pair of straight lines through the origin, but throughout this chapter we shall assume that $b \neq 0$.

The nature of the pair of lines depends on the nature of the roots of (ii). The lines are real and distinct if $h^2 > ab$, real and coincident if $h^2 = ab$, and imaginary if $h^2 < ab$.

(i) is called the *combined equation* of the pair of lines whose separate equations are given by (iii). These separate equations may be written

$$y = m_1 x,\ y = m_2 x$$

where

$$m_1 + m_2 = -2h/b \text{ and } m_1 m_2 = a/b. \qquad (12.3)$$

Hence if the equation $ax^2 + 2hxy + by^2 = 0$ represents the lines $y = m_1 x$, $y = m_2 x$, then m_1 and m_2 are connected by the relations (12.3).

12.5. The angle between the lines $ax^2+2hxy+by^2=0$

Let the given equation represent the lines $y=m_1x$, $y=m_2x$ and let θ be an angle between these lines. Then

$$\tan\theta=\pm\frac{m_1-m_2}{1+m_1m_2}=\pm\frac{\sqrt{\{(m_1+m_2)^2-4m_1m_2\}}}{1+m_1m_2}$$

$$=\pm\frac{2\sqrt{(h^2-ab)}}{a+b}, \text{ by (12.3).}$$

The lines are coincident if $h^2=ab$ and perpendicular if

$$a+b=0. \tag{12.4}$$

Example 4

Prove that the equation of a pair of straight lines drawn through the point (α, β) parallel to the lines $ax^2+2hxy+by^2=0$ is

$$a(x-\alpha)^2+2h(x-\alpha)(y-\beta)+b(y-\beta)^2.$$

In the notation of § 12.5 the equations of the parallels through (α, β) are

$$y-\beta=m_1(x-\alpha),\ y-\beta=m_2(x-\alpha).$$

Their combined equation is

$$\{(y-\beta)-m_1(x-\alpha)\}\{(y-\beta)-m_2(x-\alpha)\}=0$$

i.e. $$(y-\beta)^2-(m_1+m_2)(x-\alpha)(y-\beta)+m_1m_2(x-\alpha)^2=0$$

$$(y-\beta)^2+(2h/b)(x-\alpha)(y-\beta)+(a/b)(x-\alpha)^2=0, \text{ by (12.3)}$$

$$a(x-\alpha)^2+2h(x-\alpha)(y-\beta)+b(y-\beta)^2=0.$$

Example 5

(a) *Find the equation of the lines through the origin perpendicular to the lines $ax^2+2hxy+by^2=0$.*

(b) *Show that the numerical value of the product of the lengths of the perpendiculars from the point (α, β) to the lines $ax^2+2hxy+by^2=0$* is

$$(a\alpha^2+2h\alpha\beta+b\beta^2)/\{(a-b)^2+4h^2\}^{\frac{1}{2}}.$$

The circle on the line joining the origin O to the point $P(2, 3)$ as diameter cuts the lines $5x^2-12xy+3y^2=0$ at Q and R. Find the combined equation of the lines PQ, PR. [L.U.]

(*a*) If the given equation represents the lines $y=m_1x$, $y=m_2x$, then the equations of the perpendiculars through the origin are $m_1y+x=0$, $m_2y+x=0$, and their combined equation is

$$(m_1y+x)(m_2y+x)=0$$

i.e. $$bx^2-2hxy+ay^2=0 \quad \text{by (12.3).}$$

(*b*) The perpendicular distances of (α, β) from the two given lines are

$$\frac{\beta-m_1\alpha}{\pm\sqrt{(1+m_1^2)}} \text{ and } \frac{\beta-m_2\alpha}{\pm\sqrt{(1+m_2^2)}}.$$

Numerically their product is

$$\frac{\beta^2-(m_1+m_2)\alpha\beta+m_1m_2\alpha^2}{\sqrt{\{(m_1m_2-1)^2+(m_1+m_2)^2\}}}$$

$$=\frac{a\alpha^2+2h\alpha\beta+b\beta^2}{\sqrt{\{(a-b)^2+4h^2\}}} \quad \text{by (12.3).}$$

In fig. 34, P is the point (2, 3) and O is the origin. The circle on OP as diameter cuts at Q and R the lines whose combined equation is $5x^2-12xy+3y^2=0$, hence PQ, PR are perpendicular to OQ, OR respectively.

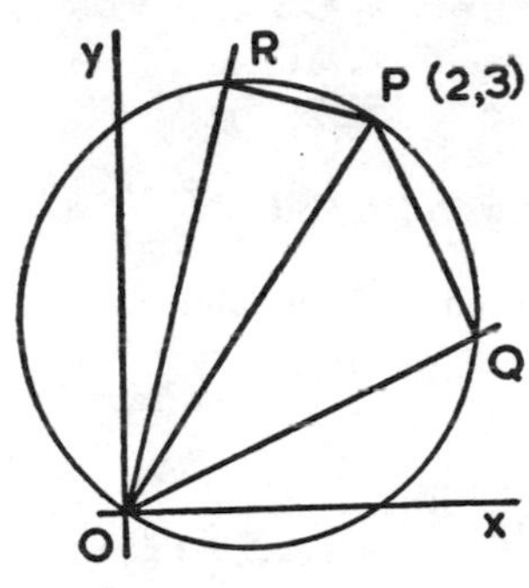

Fig. 34

By (*a*), the combined equation of the perpendiculars through O to OQ, OR is $3x^2+12xy+5y^2=0$ and, from Example 4, the equation of the parallels to these lines through P is

$$3(x-2)^2+12(x-2)(y-3)+5(y-3)^2=0$$

i.e. $$3x^2+12xy+5y^2-48x-54y+129=0.$$

12.6. The bisectors of the angles between the lines $ax^2+2hxy+by^2=0$

If the given equation represents the lines $y=m_1x$, $y=m_2x$, the equations of the required bisectors are, by § 12.1(v),

$$(y-m_1x)/\sqrt{(1+m_1^2)}=\pm(y-m_2x)/\sqrt{(1+m_2^2)}.$$

Squaring, we have

$$(1+m_2^2)(y-m_1x)^2=(1+m_1^2)(y-m_2x)^2$$

$$(m_1^2-m_2^2)(x^2-y^2)=2xy\{(m_1-m_2)-m_1m_2(m_1-m_2)\}$$

$$(m_1+m_2)(x^2-y^2)=2xy(1-m_1m_2)$$

i.e. $$h(x^2-y^2)=(a-b)xy, \text{ by (12.3).}$$

If neither h nor $a-b$ is zero, this result may be written

$$\frac{x^2-y^2}{a-b}=\frac{xy}{h}. \tag{12.5}$$

Example 6

Find the condition for the lines $a'x^2+2h'xy+b'y^2=0$ *to have the same bisectors as the lines* $ax^2+2hxy+by^2=0$, *and show that every such pair of lines can be represented by an equation of the form*

$$ax^2+2hxy+by^2+\lambda(x^2+y^2)=0.$$

Find the equation of the pair of lines, one of which passes through the point (α, β), *and whose bisectors are* $x^2-y^2=0$. [L.U.]

The equation of the bisectors of the angles between the lines

$$a'x^2+2h'xy+b'y^2=0 \quad . \quad . \quad . \quad . \quad \text{(i)}$$

is

$$\frac{x^2-y^2}{a'-b'}=\frac{xy}{h'}.$$

This equation is identical with (12.5) if $\dfrac{a'-b'}{a-b}=\dfrac{h'}{h}$.

This is the required condition. If it is fulfilled,

i.e. if

$$\frac{a'-b'}{a-b}=\frac{h'}{h}=k \text{ (say)},$$

(i) may be written in the form

$$(a'-b')x^2+2h'xy+b'(x^2+y^2)=0$$

$$k(a-b)x^2+2khxy+b'(x^2+y^2)=0$$

$$k(ax^2+2hxy+by^2)+(b'-kb)(x^2+y^2)=0$$

i.e.

$$ax^2+2hxy+by^2+\lambda(x^2+y^2)=0 \quad . \quad . \quad . \quad \text{(ii)}$$

where

$$\lambda=(b'-kb)/k.$$

Hence every line pair which has (12.5) as the equation of its bisectors can be represented by an equation of the form (ii).

The lines $x^2-y^2=0$ are the bisectors of the angles between the axes whose combined equation is $xy=0$. Hence any line pair whose bisectors are $x^2-y^2=0$ has an equation of the form

$$xy+\lambda(x^2+y^2)=0.$$

(α, β) lies on one of these lines, hence $\lambda=-\alpha\beta/(\alpha^2+\beta^2)$, and so the equation of the required line pair is

$$(\alpha^2+\beta^2)xy-\alpha\beta(x^2+y^2)=0.$$

12.7. The condition that the general equation of the second degree in x and y should represent two straight lines

The standard form of the general equation of the second degree is

$$ax^2+2hxy+by^2+2gx+2fy+c=0 \quad . \quad . \quad \text{(i)}$$

If this equation represents two straight lines, the left-hand side must break up into two linear factors.

Writing (i) as a quadratic equation in y, and solving, we obtain

$$by^2+2y(hx+f)+(ax^2+2gx+c)=0$$

and so if $b\neq 0$, $y=[-(hx+f)\pm\sqrt{\{(hx+f)^2-b(ax^2+2gx+c)\}}]/b$.

Now y is expressible in the form $Ax+B$ if, and only if,

$$(hx+f)^2-b(ax^2+2gx+c)$$

is a perfect square,

i.e. if
$$(hf-bg)^2=(h^2-ab)(f^2-bc)$$

or
$$b(abc+2fgh-af^2-bg^2-ch^2)=0.$$

Since $b\neq 0$, the required condition is

$$abc+2fgh-af^2-bg^2-ch^2=0 \quad . \quad . \quad . \quad \text{(ii)}$$

or, in determinant form,

$$\begin{vmatrix} a & h & g \\ h & b & f \\ g & f & c \end{vmatrix}=0. \qquad (12.6)$$

When $b=0$ and $a\neq 0$, the above condition is obtained by solving (i) for x. If $a=b=0$ and $h\neq 0$, (i) becomes

$$2hxy+2gx+2fy+c=0$$

or
$$hx(2y+2g/h)+f(2y+c/f)=0$$

The left side has linear factors only if

$$2g/h=c/f$$

i.e. if
$$c=2fg/h.$$

This is condition (ii) when $a=b=0$, so that in all cases (12.6) is the condition for (i) to represent two straight lines.

Example 7

Find by change of origin the condition that the equation

$$ax^2+2hxy+by^2+2gx+2fy+c=0$$

should represent two straight lines and show that, if $h^2\neq ab$, the lines meet at a point whose coordinates satisfy the equations $ax+hy+g=0$, $hx+by+f=0$.

Suppose that the given equation represents a pair of straight lines which intersect at $P(\bar{x}, \bar{y})$, and transfer the origin to P. The equation of the line pair referred to the new axes is given in § 12.3 Example 2, and since it represents a pair of lines through the new origin it must reduce to $aX^2+2hXY+bY^2=0$.

$$\therefore\ a\bar{x}+h\bar{y}+g=0 \quad . \quad . \quad . \quad . \quad . \quad \text{(iii)}$$

$$h\bar{x}+b\bar{y}+f=0 \quad . \quad . \quad . \quad . \quad . \quad \text{(iv)}$$

and $k=0$, i.e.
$$g\bar{x}+f\bar{y}+c=0 \quad . \quad . \quad . \quad . \quad . \quad \text{(v)}$$

by virtue of (iii) and (iv).

By (3.11) (page 42) we obtain (12.6) by eliminating $\bar{x}$ and $\bar{y}$ between (iii), (iv) and (v).

From (iii) and (iv), the coordinates of P are

$$\left(\frac{hf-bg}{ab-h^2}, \frac{gh-af}{ab-h^2}\right).$$

If $ab=h^2$, the lines (i) are coincident or parallel.

From Example 4 it follows that when (i) represents a pair of straight lines they are parallel to the lines $ax^2+2hxy+by^2=0$.

NOTE: If (i) is written in the form $S=0$, P satisfies the equations

$$\frac{\partial S}{\partial x}=0 \text{ and } \frac{\partial S}{\partial y}=0,$$

see Chapter 19, § 3.

Example 8

Show that

$$x^2+4xy-2y^2+6x-12y-15=0$$

represents a pair of straight lines, and that these lines together with the pair of lines $x^2+4xy-2y^2=0$ form a rhombus. [L.U.]

The equation $$x^2+4xy-2y^2+6x-12y-15=0 \quad . \quad . \quad . \quad \text{(i)}$$

represents a pair of lines since

$$\begin{vmatrix} 1 & 2 & 3 \\ 2 & -2 & -6 \\ 3 & -6 & -15 \end{vmatrix}=0.$$

These lines together with the lines

$$x^2+4xy-2y^2=0 \quad . \quad . \quad . \quad . \quad \text{(ii)}$$

form a parallelogram. They will form a rhombus if the diagonals of the parallelogram are perpendicular.

The point of intersection of line-pair (i) satisfies the equations

$$x+2y+3=0 \text{ and } x-y-3=0.$$

It is $(1, -2)$; lines (ii) intersect at the origin. Hence the equation of the diagonal through the origin is

$$y+2x=0 \quad . \quad . \quad . \quad . \quad . \quad \text{(iii)}$$

The second diagonal joins the two points which simultaneously satisfy (i) and (ii). Hence its equation is

$$6x-12y-15=0 \quad . \quad . \quad . \quad . \quad \text{(iv)}$$

(iii) and (iv) are perpendicular lines, hence line-pairs (i) and (ii) form a rhombus.

12.8. The equation of the pair of straight lines joining the origin to the two points at which a given straight line meets the curve (or line-pair) represented by the general equation of the second degree in x and y

Suppose that the line whose equation is

$$lx+my+n=0 \quad . \quad . \quad . \quad . \quad \text{(i)}$$

meets the curve

$$ax^2+2hxy+by^2+2gx+2fy+c=0 \quad . \quad . \quad \text{(ii)}$$

at A and B, and consider the equation obtained by making (ii) homogeneous by means of (i) :

$$ax^2+2hxy+by^2+2(gx+fy)\left(\frac{lx+my}{-n}\right)+c\left(\frac{lx+my}{-n}\right)^2=0,$$

i.e. $$n^2(ax^2+2hxy+by^2)-2n(gx+fy)(lx+my)+c(lx+my)^2=0 \quad . \quad \text{(iii)}$$

Since equation (iii) is homogeneous and of the second degree in x and y it represents a pair of straight lines through the origin ; it is also satisfied by points which simultaneously satisfy (i) and (ii), i.e. by A and B.

Hence (iii) represents the pair of lines which join the origin to the points of intersection of (i) and (ii) when these are real.

Example 9

Find the equation of the straight lines joining the origin to the points of intersection of $x^2+2hxy-y^2+2gx+2fy+c=0$ and $lx+my+1=0$, and find the condition that these lines should be perpendicular.

If this condition is satisfied, show that the locus of the foot of the perpendicular from the origin to the line $lx+my+1=0$, as l and m vary, is $2gx+2fy+c=0$.

[L.U.]

The equation of the required line-pair is

$$(x^2+2hxy-y^2)+2(gx+fy)(-lx-my)+c(-lx-my)^2=0$$

i.e. $$x^2(cl^2-2gl+1)+y^2(cm^2-2fm-1)+2xy(clm-gm-fl+h)=0.$$

By (12.4), these lines are perpendicular if

$$(cl^2-2gl+1)+(cm^2-2fm-1)=0$$

i.e. if $$c(l^2+m^2)=2(gl+fm) \quad . \quad . \quad . \quad . \quad \text{(i)}$$

The equation of the perpendicular from the origin to the line

$$lx+my+1=0 \quad . \quad . \quad . \quad . \quad \text{(ii)}$$

is $$mx-ly=0 \quad . \quad . \quad . \quad . \quad \text{(iii)}$$

and, solving (ii) and (iii), we obtain the coordinates of the foot of this perpendicular :

$$x=-l/(l^2+m^2),\ y=-m/(l^2+m^2) \quad . \quad . \quad . \quad \text{(iv)}$$

Eliminating l and m between (i) and (iv), we obtain $2gx+2fy+c=0$, which is the equation of the required locus.

Example 10

Find the equation of the pair of straight lines joining the origin of coordinates to the intersection of the circle $x^2+y^2+2gx+2fy+c=0$ *and the straight line*

$$lx+my=1.$$

Hence, or otherwise, find the coordinates of the circumcentre of the triangle formed by the lines

$$ax^2+2hxy+by^2=0, \; lx+my=1.$$

If the lines $ax^2+2hxy+by^2=0$ *vary in such a way that they are equally inclined to the axes but are not at right angles, show that the circumcentre moves on a line through the origin.* [L.U.]

The equation of the pair of straight lines joining the origin to the points of intersection of the line

$$lx+my=1 \quad . \quad . \quad . \quad . \quad \text{(i)}$$

and the circle

$$x^2+y^2+2gx+2fy+c=0 \quad . \quad . \quad . \quad . \quad \text{(ii)}$$

is

$$x^2+y^2+2(gx+fy)(lx+my)+c(lx+my)^2=0,$$

i.e.

$$x^2(1+2gl+cl^2)+2xy(fl+gm+clm)+y^2(1+2fm+cm^2)=0. \quad . \quad \text{(iii)}$$

The circumcircle of the triangle formed by the line (i) and the line pair

$$ax^2+2hxy+by^2=0 \quad . \quad . \quad . \quad . \quad \text{(iv)}$$

passes through the origin. Hence if (ii) is its equation, $c=0$.

Also (iii) must be identical with (iv)

i.e.

$$\frac{1+2gl}{a}=\frac{fl+gm}{h}=\frac{1+2fm}{b}.$$

Solving these equations for $-g$ and $-f$, we have

$$-g=\tfrac{1}{2}\{l(b-a)-2hm\}/(am^2-2hlm+bl^2)$$

and

$$-f=\tfrac{1}{2}\{m(a-b)-2hl\}/(am^2-2hlm+bl^2).$$

But by § 12.9 (ii) these are the coordinates of C the centre of (ii) and the circumcentre of the triangle formed by (i) and (iv).

If the lines (iv) are equally inclined to the axes, and are not at right angles, $a=b$.

The coordinates of C are then

$$x=-hm/\{a(l^2+m^2)-2hlm\}, \; y=-hl/\{a(l^2+m^2)-2hlm\}$$

and the locus of C is $y/x=l/m$, which is a straight line through the origin.

Exercises 12 (*a*)

1. The fixed line $x/a+y/b=1$ meets the axis of x at X and the axis of y at Y. Any straight line perpendicular to this straight line meets the axis of x at X' and the axis of y at Y'. Prove that the locus of the intersection of the straight lines XY' and $X'Y$ is the circle $x^2+y^2=ax+by$. [L.U.]

2. Show that, the axes being rectangular, the area of the triangle whose vertices are the points $(0, 0)$, (x_1, y_1), (x_2, y_2) is $\pm\frac{1}{2}(x_1y_2-x_2y_1)$.

If O is the origin, and if the line $lx+my=1$ meets the pair of lines whose joint equation is

$$ax^2+2hxy+by^2=0$$

in $P(x_1, y_1)$ and $Q(x_2, y_2)$, prove that the area of the triangle OPQ is

$$\pm(h^2-ab)^{\frac{1}{2}}/(am^2-2hlm+bl^2). \qquad \text{[L.U.]}$$

3. From a point $P(p, q)$ perpendiculars PM, PN are drawn to the straight lines given by the equation $ax^2+2hxy+by^2=0$.

Show that if O is the origin of coordinates, the area of the triangle OMN is

$$\{(aq^2-2hpq+bp^2)(h^2-ab)^{1/2}\}/\{(a-b)^2+4h^2\}. \qquad \text{[L.U.]}$$

4. Prove that

$$y^2-4xy+x^2-10y+8x+13=0$$

represents a pair of straight lines; find their point of intersection and the angle between them. [L.U.]

5. Show that the pair of straight lines joining the origin O to the intersections A and B of the line $lx+my=1$ with the conic $ax^2+by^2=1$ has the equation $(a-l^2)x^2-2lmxy+(b-m^2)y^2=0$.

Deduce that if AOB is a right angle, then the line AB touches the circle $(a+b)(x^2+y^2)=1$. [L.U.]

6. Form the equation of the straight lines joining the origin to the points given by the equations $ax^2+2hxy+by^2+2gx+2fy+c=0$ and $px+qy+r=0$, and write down the condition that these lines should be at right angles.

If this condition is satisfied, show that the locus of the foot of the perpendicular from the origin to the line $px+qy+r=0$ is $(a+b)(x^2+y^2)+2gx+2fy+c=0$. [L.U.]

7. One of the medians of the triangle formed by the straight lines $ax^2+2hxy-ay^2=0$ and the line $px+qy=r$ lies along the y-axis. If a and r are both different from zero, prove that $ap+hq=0$. [L.U.]

8. Prove that the equation

$$ax^2+2hxy+by^2=0$$

represents a pair of straight lines through the origin, and that the sine of the angle between them is

$$2(h^2-ab)^{\frac{1}{2}}/\{(a-b)^2+4h^2\}^{\frac{1}{2}}.$$

Prove that the length intercepted by these lines on the line $lx+my+n=0$ is

$$2n\{(l^2+m^2)(h^2-ab)\}^{\frac{1}{2}}/\{am^2-2hlm+bl^2\}. \qquad \text{[L.U.]}$$

9. If the lines $ax^2+2hxy+by^2=0$ meet the line $qx+py=pq$ in points which are equidistant from the origin, prove that

$$h(p^2-q^2)=pq(b-a).$$ [L.U.]

10. Lines through a point A perpendicular to the lines $ax^2+2hxy+by^2=0$ cut the x-axis at L and M, and the y-axis at P and Q. The mid-point of LM is N; the mid-point of PQ is R. The mid-point of NR is B, and O is the origin. If O, A, B are collinear, show that A lies on the lines $bx^2=ay^2$. [L.U.]

11. Prove that if the points

$$(x_1, y_1), (x_2, y_2), (x_3, y_3), (x_4, y_4)$$

are the vertices of a parallelogram taken in order,

$$x_1+x_3=x_2+x_4 \text{ and } y_1+y_3=y_2+y_4.$$

Two sides of a parallelogram lie along the straight lines

$$ax^2+2hxy+by^2=0,$$

and the diagonal which does not pass through the origin lies along the straight line $lx+my+n=0$.

Find the coordinates of the vertex opposite to the origin and prove that the figure is a rhombus if $h(l^2-m^2)=(a-b)lm$. [L.U.]

12. Show that the equation $3x^2-4xy-4y^2+14x+12y-5=0$ represents two straight lines, and find the combined equation of the bisectors of the angles between them. [L.U.]

13. Prove that

$$x^2-y^2+2xy \sinh \theta+2ax \cosh \theta+a^2=0$$

represents a pair of straight lines for all values of θ, and show that the locus of their point of intersection is the circle $x^2+y^2=a^2$. [L.U.]

14. Find the condition that the equation

$$ax^2+2hxy+by^2+2gx+2fy+c=0$$

represents two straight lines.

Assuming this condition is satisfied, A is the point of intersection of the straight lines. Parallel straight lines are drawn through the origin O, and they intersect the other lines in B and C. Find the equations of the diagonals OA and BC of the parallelogram formed, and show that the parallelogram is a square if

$$a+b=0 \text{ and } h(g^2-f^2)=fg(a-b).$$ [L.U.]

15. Find the equation of the pair of lines joining the origin to the points in which the pair of lines

$$4x^2-15xy-4y^2+39x+65y-169=0$$

are met by the line $x+2y-5=0$.

Show that the quadrilateral having the first pair, and also the second pair, as adjacent sides is cyclic, and find the equation of its circumcircle. [L.U.]

16. Prove that, if $\mu \leqslant 169/56$, there are two finite real values of λ for which the equation
$$9x^2+\lambda xy+\mu y^2-45x+13y+14=0$$
represents a pair of straight lines. If $\mu=-1$, find the separate equations of the lines constituting *one* of these pairs. [L.U.]

17. Prove that the equation $my^2-m^3x^2-(m^2+1)y-m(m^2-1)x+m=0$ represents two straight lines.
Find the point of intersection of these straight lines, and show that, for different values of m, the locus of the point of intersection is $(2x+1)y^2=(x+1)^2$. [L.U.]

18. Find the equation of the pair of lines obtained by rotating the lines represented by $ax^2+2hxy+by^2=0$ about the origin through a positive angle of 60°. Write down the equation which corresponds to the case $a=0, b=1, 2h=\sqrt{3}$, and sketch the two pairs of lines in this case. [L.U.]

19. Find the area enclosed by the pentagon $ABCDE$ whose vertices are respectively (1, 3), (4, 1), (5, 3), (3, 2) and (2, 4). [L.U.]

20. Show that the equation
$$(x^2-y^2)\cos\alpha+2xy\sin\alpha=\lambda(x^2+y^2),$$
where $0<\lambda<1$, represents a pair of straight lines and find the acute angle between them.
Show that, for every value of λ, the lines given by this equation are equally inclined to a fixed line. [L.U.]

21. The locus Γ represented by the equation
$$3xy-4y^2-6x-2y+c=0$$
meets the line represented by the equation
$$3x+y-5=0$$
in the two points A and B. Find a value of c for which the lines joining the origin O to A and B are at right angles, and show that in this case Γ is a pair of lines which meet in a point P, and that the diagonals of the quadrilateral $OAPB$ are perpendicular. [L.U.]

THE CIRCLE

12.9. Useful formulae (revision)

(i) The equation of the circle with centre (α, β) and radius a is $(x-\alpha)^2+(y-\beta)^2=a^2$.

(ii) The equation $x^2+y^2+2gx+2fy+c=0$ represents a circle with centre $(-g, -f)$ and radius $\sqrt{(g^2+f^2-c)}$. This circle is real if $g^2+f^2 \geqslant c$.

(iii) The equation of the circle whose diameter is the line joining the points (x_1, y_1) and (x_2, y_2) is $(x-x_1)(x-x_2)+(y-y_1)(y-y_2)=0$.

(iv) The circles $x^2+y^2+2gx+2fy+c=0$, $x^2+y^2+2g'x+2f'y+c'=0$ cut orthogonally if $2gg'+2ff'=c+c'$.

12.10. The equation of the tangent to a circle at a given point

Let (x_1, y_1) be a point on the circle

$$x^2+y^2+2gx+2fy+c=0 \quad . \quad . \quad . \quad \text{(i)}$$

Differentiating (i) with respect to x, we have

$$2x+2y\frac{dy}{dx}+2g+2f\frac{dy}{dx}=0$$

so that the gradient of the circle at the point (x, y) is given by

$$\frac{dy}{dx}=-\frac{x+g}{y+f}.$$

Hence the equation of the tangent at (x_1, y_1) is

$$y-y_1=-\frac{x_1+g}{y_1+f}(x-x_1)$$

$$xx_1+yy_1+gx+fy=x_1^2+y_1^2+gx_1+fy_1$$

$$\therefore\ xx_1+yy_1+g(x+x_1)+f(y+y_1)+c=x_1^2+y_1^2+2gx_1+2fy_1+c$$

i.e.
$$xx_1+yy_1+g(x+x_1)+f(y+y_1)+c=0 \qquad (12.7)$$

since (x_1, y_1) lies on (i).

12.11. Tangents to the circle $x^2+y^2=a^2$ which are parallel to $y=mx$

The line
$$y=mx+c \quad . \quad . \quad . \quad . \quad \text{(i)}$$

meets the circle $x^2+y^2=a^2$ in points whose abscissae are the roots of the equation

$$x^2+(mx+c)^2=a^2$$

$$x^2(1+m^2)+2cmx+(c^2-a^2)=0. \quad . \quad . \quad . \quad \text{(ii)}$$

Since (i) is a tangent to the circle, the roots of (ii) must be equal

$$\therefore\ 4c^2m^2=4(1+m^2)(c^2-a^2),$$

which leads to
$$c^2=a^2(1+m^2)$$

or
$$c=\pm a\sqrt{(1+m^2)}.$$

Thus the two tangents to the circle $x^2+y^2=a^2$ with gradient m are

$$y=mx\pm a\sqrt{(1+m^2)}.$$

12.12. The chord of contact of tangents drawn to a circle from the point (x_1, y_1)

In fig. 35 tangents from $T(x_1, y_1)$ touch the circle

$$x^2+y^2+2gx+2fy+c=0$$

at $P(x_2, y_2)$ and $Q(x_3, y_3)$.

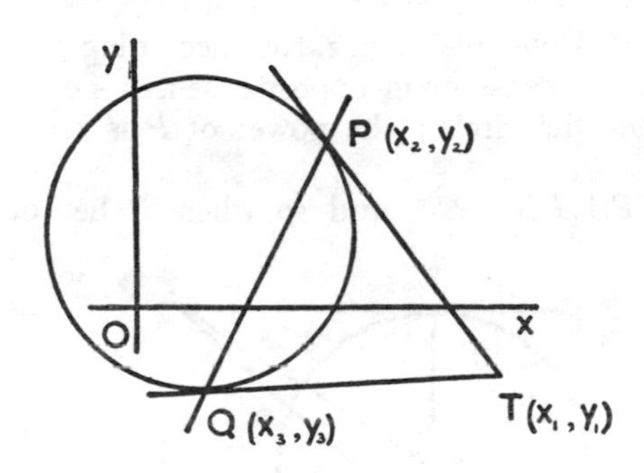

Fig. 35

Then by (12.7) the equation of PT is

$$xx_2+yy_2+g(x+x_2)+f(y+y_2)+c=0$$

and T lies on this line

$$\therefore\ x_1x_2+y_1y_2+g(x_1+x_2)+f(y_1+y_2)+c=0.$$

Similarly, by considering the tangent QT, we have

$$x_1x_3+y_1y_3+g(x_1+x_3)+f(y_1+y_3)+c=0.$$

Thus the points $P(x_2, y_2)$, $Q(x_3, y_3)$ both satisfy the equation

$$xx_1+yy_1+g(x+x_1)+f(y+y_1)+c=0 \qquad (12.8)$$

and this is the equation of a straight line ; hence it is the equation of the line joining P and Q.

PQ is known as the *polar* of T with respect to the circle, and T is called the *pole* of PQ.

12.13. The power of a point

If any line drawn through a fixed point P meets a circle at A and B, the product $PA.PB$ is constant and is called the *power* of P with respect to the circle.

Let the equation of the circle be $x^2+y^2+2gx+2fy+c=0$, let $P\equiv(x_1,y_1)$ and take the equation of PAB in the form

$$\frac{x-x_1}{\cos\theta}=\frac{y-y_1}{\sin\theta}=r \quad . \quad . \quad . \quad . \qquad (i)$$

Then by substituting for x and y from (i) in the equation of the circle we obtain the quadratic equation in r

$$(x_1+r\cos\theta)^2+(y_1+r\sin\theta)^2+2g(x_1+r\cos\theta)+2f(y_1+r\sin\theta)+c=0$$

i.e. $r^2+2r\{(x_1+g)\cos\theta+(y_1+f)\sin\theta\}+x_1^2+y_1^2+2gx_1+2fy_1+c=0.$

The roots r_1, r_2 of this equation are the measures of PA, PB.

Hence $$PA.PB=r_1r_2=x_1^2+y_1^2+2gx_1+2fy_1+c. \qquad (12.9)$$

The power of P is positive or negative according as PA and PB are drawn in the same sense or in opposite senses, i.e. according as P is outside or inside the circle; the power of P is zero when P lies on the circle.

In fig. 36, $PA.PB=PT^2$, and so when P lies outside the circle

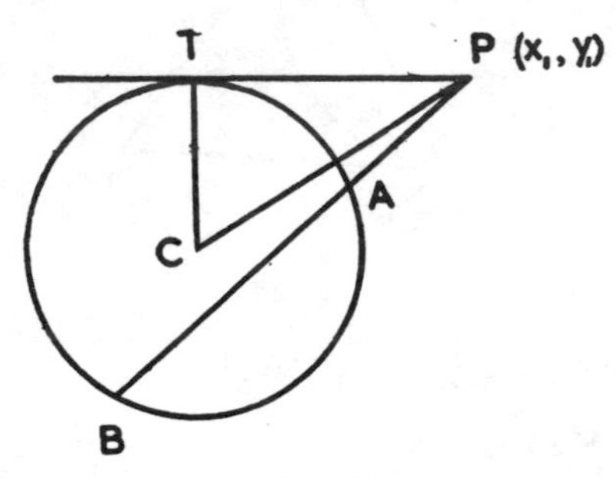

Fig. 36

(12.9) gives the square of the length of the tangent drawn from (x_1, y_1)—a result which can also be obtained from the relation $PT^2=CP^2-CT^2$, where C is the centre of the circle.

12.14. The radical axis of two circles

The locus of points whose powers with respect to two circles are equal is a straight line called the *radical axis* of the circles.

If $P(x_1, y_1)$ is any point on the radical axis of the circles

$$x^2+y^2+2gx+2fy+c=0$$

and $$x^2+y^2+2g'x+2f'y+c'=0,$$

then, by definition, using (12.9), we have

$$x_1^2+y_1^2+2gx_1+2fy_1+c=x_1^2+y_1^2+2g'x_1+2f'y_1+c',$$

i.e. P lies on the line

$$2(g-g')x+2(f-f')y+c-c'=0. \qquad (12.10)$$

If we write $S\equiv x^2+y^2+2gx+2fy+c$ and $S'\equiv x^2+y^2+2g'x+2f'y+c'$, this result may be written

$$S-S'=0.$$

The radical axis is perpendicular to the line of centres of the circles.

If the circles intersect in real points A and B, these points are on the radical axis since A and B have zero powers with respect to both circles. In this case the radical axis is AB, the common chord of the circles. If the circles touch, the radical axis is the common tangent at their point of contact.

Example 11

Prove that, in general, the radical axes of three circles taken in pairs are concurrent and that the point of concurrence is the centre of a circle which cuts all three circles orthogonally.

Find the equation of the circle which is orthogonal to all three circles

$$x^2+y^2+2x+4y+3=0,\ x^2+y^2+3x+5y=0 \text{ and } x^2+y^2+4x+5y-1=0.$$

[L.U.]

Let the equations of the circles be $S_1=0$, $S_2=0$, $S_3=0$, where

$$S_r \equiv x^2+y^2+2g_rx+2f_ry+c_r.$$

The radical axis of $S_1=0$ and $S_2=0$ is $S_1-S_2=0$ . . . (i)

The radical axis of $S_2=0$ and $S_3=0$ is $S_2-S_3=0$ . . . (ii)

Where (i) and (ii) meet, $S_1-S_3=0$, and this is the equation of the radical axis of $S_1=0$ and $S_3=0$. Thus, in general, the three radical axes meet at a point C (known as the radical centre of the three circles).

The radical centre C' of the three given circles is the meet of the lines $x+y-3=0$ and $2x+y-4=0$; i.e. C' is the point $(1, 2)$. The power of C' with respect to each circle is 18 and so C' lies outside the circles, and from C' a tangent of length $3\sqrt{2}$ can be drawn to each circle. The circle with centre C' and radius $3\sqrt{2}$ cuts all three circles orthogonally. Its equation is

$$(x-1)^2+(y-2)^2=18$$

or

$$x^2+y^2-2x-4y-13=0.$$

12.15. Coaxal circles

If a system of circles is such that the radical axis of one pair is the same as that of any other pair, the circles are said to form a *coaxal system*. For example, all circles which pass through two fixed points A and B form a coaxal system with AB as radical axis.

Since the radical axis of two circles is perpendicular to their line of centres, it follows that the centres of circles of a coaxal system are collinear.

Take as x-axis the line of centres and as y-axis the common radical axis. Then the equations of any two circles of the system may be taken as

$$x^2+y^2+2g_1x+c_1=0,\ x^2+y^2+2g_2x+c_2=0.$$

Their radical axis is $2(g_1-g_2)x+c_1-c_2=0$ and this will be the y-axis if $c_1=c_2$.

Thus the equation $x^2+y^2+2\lambda x+c=0$ represents for varying λ and constant c a coaxal system of circles with centres on Ox, and with Oy as radical axis.

12.16. Intersecting and non-intersecting systems of coaxal circles

The radical axis $x=0$ meets every circle of the system

$$x^2+y^2+2\lambda x+c=0 \quad . \quad . \quad . \quad \text{(i)}$$

in the points $\{0, \pm\sqrt{(-c)}\}$.

These points are real if $c\leqslant 0$ and imaginary if $c>0$. Hence equation (i) represents a coaxal system intersecting in real points if $c<0$, and a non-intersecting system if $c>0$.

If $c=0$, every circle of the system touches the y-axis at the origin and the system is said to be tangential.

12.17. Limiting points of a non-intersecting system of coaxal circles

The radius of the circle $x^2+y^2+2\lambda x+c=0$ is $\sqrt{(\lambda^2-c)}$.

If $c>0$ (i.e. if the system is non-intersecting),

$$\lambda^2-c=0 \text{ when } \lambda=\pm\sqrt{c}.$$

Thus there are two circles of the system which have zero radius These point circles of the system occur at $(\pm\sqrt{c}, 0)$ and are called the *limiting points* of the system. They are equidistant from the radical axis.

Example 12

Show that any circle which passes through the limiting points of a coaxal system cuts every circle of the system orthogonally.

The limiting points of the non-intersecting coaxal system

$$x^2+y^2+2\lambda x+a^2=0 \quad . \quad . \quad . \quad . \quad \text{(i)}$$

are L $(a, 0)$ and $L'(-a, 0)$.

Let the circle $x^2+y^2+2gx+2fy+c=0$ pass through L and L'. Then $a^2+2ga+c=0$ and $a^2-2ga+c=0$ so that $g=0$ and $c=-a^2$. Hence the equation of any circle through L and L' is

$$x^2+y^2+2\mu y-a^2=0 \quad . \quad . \quad . \quad . \quad \text{(ii)}$$

where μ varies.

By § **12.9** (iv), circles (i) and (ii) are orthogonal for all values of λ and μ. Thus each circle of the intersecting coaxal system formed by the circles which pass through L and L' cuts every circle of the given system orthogonally.

12.18. Equations of the form $S+\lambda S'=0$, $S+\lambda l=0$

Let $S=0$, $S'=0$ be the equations of two circles as in § 12.14, and let $l\equiv px+qy+r=0$ be the equation of a straight line.

For all values of λ (except $\lambda=-1$) the equation

$$S+\lambda S'=0 \quad . \quad . \quad . \quad . \quad \text{(i)}$$

represents a circle; for the coefficients of x^2 and y^2 are equal and the coefficient of xy is zero. Hence for varying λ, (i) represents a system of circles. We may show that the system is coaxal by finding the equation of the radical axis of any two circles of the system. Let the equations of the circles be

$$S+\lambda_1 S'=0 \qquad . \quad . \quad . \quad . \qquad \text{(ii)}$$

and

$$S+\lambda_2 S'=0 \qquad . \quad . \quad . \quad . \qquad \text{(iii)}$$

To find their radical axis we eliminate the terms of the second degree between these equations by multiplying (ii) by $(1+\lambda_2)$, (iii) by $(1+\lambda_1)$ and subtracting. This gives

$$S-S'=0.$$

Hence the radical axis of any two circles of system (i) is the radical axis of the circles $S=0$, $S'=0$. It follows that when $\lambda \neq -1$, (i) represents a circle of the coaxal system defined by the circles $S=0$, $S'=0$. When these circles intersect in real points, (i) represents a circle which passes through these points.

In the same way we may show that the equation

$$S+\lambda l=0 \qquad \text{(12.11)}$$

represents for all values of λ a circle of a coaxal system of which the line $l=0$ is the radical axis and the circle $S=0$ is a member. If the line $l=0$ intersects the circle $S=0$ in real points, (12.11) represents a circle through these points.

It is useful to note that the equation of any circle which passes through the points of intersection of the circles $S=0$ and $S'=0$ is of the form (i) for we can choose λ to make (i) pass through any given point which does not lie on either $S=0$ or $S'=0$.

In the same way it can be shown that the equation of any circle which passes through the points of intersection of the circle $S=0$ and the line $l=0$ is of the form (12.11).

Example 13

The circle $x^2+y^2+2x-4y-11=0$ and the line $x-y+1=0$ intersect at A and B. Find the equation of the circle on AB as diameter and the equation of the circle through A, B orthogonal to the given circle. [L.U.]

By (12.11) the equation

$$x^2+y^2+2x-4y-11+\lambda(x-y+1)=0 \qquad . \quad . \qquad \text{(i)}$$

represents a circle through A and B. If its centre $(-1-\frac{1}{2}\lambda,\ 2+\frac{1}{2}\lambda)$ lies on the line $x-y+1=0$, $\lambda=-2$. Hence the equation of the circle on AB as diameter is

$$x^2+y^2-2y-13=0 \quad . \quad . \quad . \quad . \qquad \text{(ii)}$$

Circle (i) is orthogonal to the circle $x^2+y^2+2x-4y-11=0$ if, by § 12.9 (iv)

$$(2+\lambda)+2(4+\lambda)=\lambda-22$$

or

$$\lambda=-16.$$

Hence the equation of the orthogonal circle is $x^2+y^2-14x+12y-27=0$.

Example 14

Find the limiting points of the system of coaxal circles

$$(x-1)^2+(y-2)^2+\lambda(x^2+y^2+2x+5)=0. \qquad \text{[L.U.]}$$

The equation of the system can be written

$$(x^2+y^2)(\lambda+1)+2x(\lambda-1)-4y+5(\lambda+1)=0.$$

The values of λ corresponding to the limiting points of the system (i.e. to the circles of the system which have zero radius) are given by

$$\frac{(\lambda-1)^2}{(\lambda+1)^2}+\frac{4}{(\lambda+1)^2}-5=0,$$

which reduces to $\lambda^2+3\lambda=0$, so that $\lambda=0, -3$.

Hence the equations of the circles with zero radius are $(x-1)^2+(y-2)^2=0$ and $(x+2)^2+(y+1)^2=0$ i.e. the limiting points are (1, 2) and $(-2, -1)$.

Example 15

Show that when k is any constant, the equation

$$x^2+y^2+2gx+2fy+c+k\{xx_1+yy_1+g(x+x_1)+f(y+y_1)+c\}=0$$

represents a circle which touches the circle $x^2+y^2+2gx+2fy+c=0$ at the point (x_1, y_1).

Find the equation of the circle which touches the circle $x^2+y^2+8x+14=0$ at the point $(-5, 1)$ and passes through the point (1, 3).

Find also the equation of the circle which touches these two circles, one internally and the other externally, and has its centre collinear with their centres.

The equation of the tangent to the circle $x^2+y^2+2gx+2fy+c=0$ (i) at the point (x_1, y_1) is, by (12.7),

$$xx_1+yy_1+g(x+x_1)+f(y+y_1)+c=0 \quad . \quad . \quad \text{(ii)}$$

Also, by (12.11), the equation

$$x^2+y^2+2gx+2fy+c+k\{xx_1+yy_1+g(x+x_1)+f(y+y_1)+c\}=0 \quad . \quad \text{(iii)}$$

represents a circle passing through the points of intersection of (i) and (ii). These points coincide at the point (x_1, y_1). Hence (iii) represents for all values of k a circle touching (i) at (x_1, y_1).

The tangent at $P(-5, 1)$ to the circle

$$x^2+y^2+8x+14=0 \quad . \quad . \quad . \quad . \quad \text{(iv)}$$

is, by (12.7), $x-y+6=0$, and so the equation of the required circle is of the form $x^2+y^2+8x+14+k(x-y+6)=0$.

But this circle goes through (1, 3) so that $k=-8$ and the equation of the circle is

$$x^2+y^2+8y-34=0 \quad . \quad . \quad . \quad . \quad \text{(v)}$$

C, the centre of circle (iv), is $(-4, 0)$. Hence A (fig. 37) the extremity of the diameter through $P(-5, 1)$, is $(-3, -1)$ since C is the mid-point of

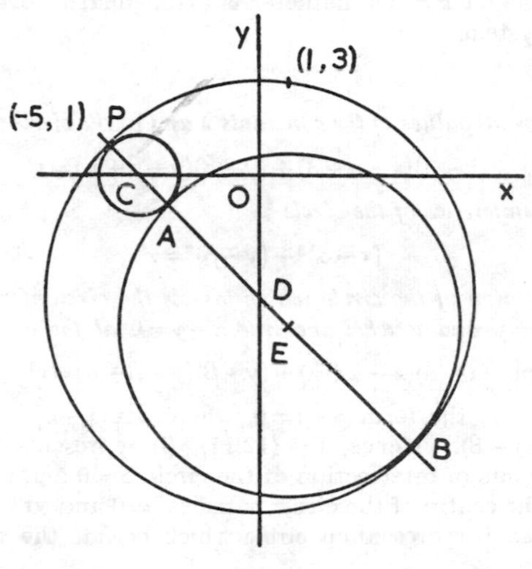

Fig. 37

AP. Similarly, D the centre of circle (v) is $(0, -4)$ and B, the extremity of the diameter of this circle drawn through P is $(5, -9)$.

The required circle which touches (iv) externally and (v) internally is the circle on AB as diameter. Its equation is

$$(x+3)(x-5)+(y+1)(y+9)=0, \quad \text{by § 12.9 (iii)}$$

i.e. $$x^2+y^2-2x+10y-6=0.$$

12.19. Miscellaneous examples

Example 16

Write down the equation of the polar of the point $P(h, k)$ with respect to the circle $x^2+y^2+2\lambda x+c=0$, and show that for every value of λ the polar passes through a fixed point P'. Find the coordinates of P'.

Show that the circle on PP' as diameter is orthogonal to every circle of the given coaxal system. [L.U.]

The polar of $P(h, k)$ with respect to the circle

$$x^2+y^2+2\lambda x+c=0 \quad . \quad . \quad . \quad . \quad \text{(i)}$$

is, by (12.8), $$hx+ky+c+\lambda(x+h)=0.$$

But for all values of λ this equation is satisfied by the point which simultaneously satisfies the equations $hx+ky+c=0$ and $x+h=0$, i.e. by the fixed point $P'\{-h, (h^2-c)/k\}$.

The equation of the circle on PP' as diameter is

$$(x-h)(x+h)+(y-k)\{y-(h^2-c)/k\}=0,$$

i.e.

$$x^2+y^2-(h^2+k^2-c)y/k-c=0 \quad . \quad . \quad \text{(ii)}$$

By § 12.9 (iv), circles (i) and (ii) cut orthogonally for all values of λ Hence the circle on PP' as diameter is orthogonal to every circle of the given coaxal system.

Example 17

Prove that, for all values of the constants λ and μ, the circle whose equation is

$$(x-\alpha)(x-\alpha+\lambda)+(y-\beta)(y-\beta+\mu)=r^2$$

bisects the circumference of the circle

$$(x-\alpha)^2+(y-\beta)^2=r^2.$$

Find the equation of the circle which bisects the circumference of the circle $x^2+y^2+2y-3=0$ and touches the line $x-y=0$ at the origin. [L.U.]

The equation $(x-\alpha)(x-\alpha+\lambda)+(y-\beta)(y-\beta+\mu)=r^2 \quad . \quad . \quad \text{(i)}$ may be written in the form $S+l=0$, where $S\equiv(x-\alpha)^2+(y-\beta)^2-r^2$ and $l\equiv\lambda(x-\alpha)+\mu(y-\beta)$. Hence, by (12.11), (i) represents a circle passing through the points of intersection of the circle $S=0$ and the line $l=0$.

But (α, β), the centre of the circle, satisfies $l=0$ and so $l=0$ is a diameter of $S=0$. Hence (i) represents a circle which bisects the circumference of $S=0$.

If we apply this result, the equation of a circle which bisects the circumference of the circle $x^2+(y+1)^2=4$ is of the form

$$x(x+\lambda)+(y+1)(y+1+\mu)=4 \quad . \quad . \quad . \quad \text{(ii)}$$

It passes through the origin, hence $\mu=3$ and (ii) becomes

$$x^2+y^2+\lambda x+5y=0.$$

The tangent to this circle at the origin is, by (12.7),

$$\lambda x+5y=0,$$

which reduces to $x-y=0$ if $\lambda=-5$.

Hence the equation of the required circle is

$$x^2+y^2-5x+5y=0.$$

Exercises 12 (b)

1. Find the equations of the two circles of radius $\sqrt{2}$ with their centres on the x-axis which touch the line $x+y+1=0$. [Durham.]

2. A circle passes through the points of intersection of the circles

$$3x^2+3y^2-6x-1=0 \text{ and } x^2+y^2+2x-4y+1=0$$

and also passes through the centre of the first circle. Find its equation and verify that it cuts the second circle orthogonally. [Durham.]

3. Find the equation of the circle on the join of (1, 3), (2, 4) as diameter, and obtain the equation of another circle with centre $(-1, 2)$ which meets the first circle orthogonally. [Durham.]

4. The positive x-axis cuts the circle $x^2+y^2=4$ at A, and B is a point on the circle such that the angle AOB is 60°, where O is the origin. A point P on OB is such that the circle centre P which touches the given circle internally also touches externally the circle on OA as diameter. Show that $OP=8/5$ and find the equation of this circle centre P. [Durham.]

5. Find the value of k if the equation
$$2x^2-5xy+2y^2-7x+11y+k=0$$
represents two straight lines.

For this value of k show that the two lines intersect at a point on the circle
$$x^2+y^2-12x+6y+20=0,$$
and find the equation of the tangent to the circle at this point. [L.U.]

6. If the length of the tangent from a point P to the circle $x^2+y^2=r^2$ is n times the length of the tangent from P to the circle $(x-2r)^2+y^2=4r^2$, prove that the locus of P is a circle. If the radius of this circle is $2r$, show that $n=\sqrt{(\frac{3}{7})}$. [L.U.]

7. Find the radical centre of the three circles
$$x^2+y^2+2x+4y-1=0,$$
$$x^2+y^2+10x+8y-13=0,$$
$$x^2+y^2-6x+6y+5=0,$$
and show that there is a circle with this point as centre which cuts all three circles orthogonally. [Durham.]

8. Show that the equation
$$x^2+y^2+2(1-k)x+2(1-2k)y-1=0,$$
where k is variable, represents a system of coaxal circles, and find the equation of the circle of the system which cuts orthogonally the circle
$$x^2+y^2+10y-9=0.$$
Find the radical centre of the circles
$$x^2+y^2=16,$$
$$x^2+y^2+4x+8y+16=0,$$
$$x^2+y^2+4x+14y+52=0.$$

Find also the equation of the circle which cuts all three circles orthogonally.

9. Three circles of a coaxal system have centres A, B, C and the lengths of the tangents to these circles from any given point are a, b, c respectively. Prove that
$$a^2BC+b^2CA+c^2AB=0,$$
the sign of the line segments being taken into account. [Sheffield.]

10. Prove that the origin is outside the circle

$$x^2+y^2+2gx+2fy+c=0$$

if $c>0$, and interpret the value of c geometrically when this happens.

The equation of a circle s_1 is $x^2+y^2-4x+3=0$; find the equation of the circle s_2 which cuts s_1 orthogonally and has its centre at the origin.

If s_3 is any circle such that l, the chord common to s_3 and s_1 goes through the origin, prove that s_3 cuts s_2 orthogonally. [L.U.]

11. By using the fact that a limiting point of a system of coaxal circles may be regarded as a point-circle belonging to the system, or otherwise, prove that, if (h, k) is a limiting point of a coaxal system of which the circle $x^2+y^2+2gx+2fy+c=0$ is a member, then the equation of the radical axis of the system is

$$2(g+h)x+2(f+k)y+(c-h^2-k^2)=0.$$

The circle $x^2+y^2+x-5y+9=0$ is a member of a coaxal system of which $(1, 2)$ is a limiting point. Find the coordinates of the other limiting point. [Leeds.]

12. A coaxal system of circles contains the circle

$$x^2+y^2+2gx+2fy+c=0,$$

and one of its limiting points is $(-g, 0)$. Find the equation of the radical axis and the coordinates of the other limiting point. Show that the system of circles orthogonal to this system is represented by the equation

$$f(x^2+y^2)+2fgx-(g^2-c)y+fg^2+\mu(x+g)=0,$$

where μ is a parameter. [L.U.]

13. The limiting points of a family of coaxal circles are the points $(\pm 2, 0)$ and l is the line whose equation is $2x+3y=2$. Find the equation of that circle of the family which cuts from l a chord subtending a right angle at the origin.

Find the equation of that circle of the orthogonal coaxal system which also cuts from l a chord subtending a right angle at the origin. [L.U.]

14. The circles

$$x^2+y^2+y-4=0, \quad x^2+y^2-3x-5y+2=0$$

are two members of a coaxal system; the circle

$$x^2+y^2-y+2=0$$

is a member of another coaxal system, of which the radical axis is the line $x+y+1=0$. Show that the two systems have a common circle, and find its equation. [Leeds.]

15. Find the limiting points of the system given by

$$x^2+y^2-2\lambda(x+y-4)-6=0.$$

Determine the equations of the circles which pass through these points and have radius 2 units. [L.U.]

16. Find the polar of a point P with respect to any member of the family of circles $x^2+y^2+2\mu x-c^2=0$, and show that for all values of μ, these polars are concurrent at a point P_1. Show that PP_1 is a diameter of a circle of the system $x^2+y^2+2\lambda x+c^2=0$. [L.U.]

17. Given two circles; a tangent to one of the given circles at any point P on it meets the polar of P with respect to the other in P_1. Prove that the circles on PP_1 as diameter form a coaxal system, and find the limiting points of this system. [L.U.]

18. Ox and Oy are two perpendicular lines, P is a variable point on Ox, and the polar of P with respect to a given fixed circle in the plane Oxy intersects Oy at Q. Prove that for different positions of P on Ox, the circles on PQ as diameter form a coaxal system whose radical axis passes through the centre of the given fixed circle. [L.U.]

19. The polar of a point P with respect to a fixed circle whose equation is $x^2+y^2+2gx+2fy+f^2=0$ cuts the axes at the points A and B. If P is on the y-axis, find the equation of the circle on AB as diameter, and show that as P varies along the y-axis, these circles form a coaxal system. Find the common points of the system. [L.U.]

20. Show that, in general, two circles of the coaxal system

$$x^2+y^2+2\lambda x-a^2=0,$$

where λ is an arbitrary parameter, touch an arbitrary straight line and that, if these circles cut orthogonally the rectangle contained by the perpendiculars drawn to the straight line from the common points of the system is equal to a^2. [L.U.]

21. Show that the circles

$$x^2+y^2+2g_1x+2f_1y+c_1=0 \text{ and } x^2+y^2+2g_2x+2f_2y+c_2=0$$

are orthogonal if $2g_1g_2+2f_1f_2=c_1+c_2$.

Assuming that these circles are orthogonal, intersecting at C and D, and that their centres are A and B respectively, show that the equation of the circle through A, B, C, D is

$$2(x^2+y^2)+2(g_1+g_2)x+2(f_1+f_2)y+c_1+c_2=0.$$

Assuming that the equation of the circle on CD as diameter is written the form

$$x^2+y^2+2g_1x+2f_1y+c_1+\lambda\{2(g_1-g_2)x+2(f_1-f_2)y+c_1-c_2\}=0,$$

show that $\lambda=-r_1^2/AB^2$ where r_1 is the radius of the first circle. [L.U.]

22. If three circles with centres A, B, and C cut each other orthogonally in pairs, prove that the polar of A with respect to the circle centre B passes through C. [L.U.]

23. Show that the equation

$$(x^2+2hxy+y^2)+(\lambda x+\mu y)(x+y+1)=0$$

represents a conic passing through the vertices of the triangle formed by the lines $x^2+2hxy+y^2=0$ and $x+y+1=0$.

Deduce the equation of the circumcircle of this triangle, and show that this circle is orthogonal to the circumcircle of the triangle formed by the lines $ax^2+2kxy+ay^2=0$ and $x-y+1=0$. [L.U.]

24. Show that the coordinates of any point on the circle

$$x^2+y^2+2gx+2fy+c=0$$

can be expressed in the form

$$(-g+r\cos\theta,\ -f+r\sin\theta),$$

where $r=\sqrt{(g^2+f^2-c)}$ and θ is a varying parameter. Find the equation of the tangent to the circle at this point.

Tangents are drawn from the fixed point $(a, 0)$ to the system of circles given by the equation

$$x^2+y^2-2p(x+y)=0,$$

where p is a varying parameter. Prove that the equation of the locus of the points of contact is

$$(x^2+y^2)(x+y+a)-2ax(x+y)=0.$$ [L.U.]

25. Find the condition that the circle whose equation is

$$x^2+y^2+2g_1x+2f_1y+c_1=0$$

should cut the circle

$$x^2+y^2+2gx+2fy+c=0$$

at the ends of a diameter of the latter circle.

Find the locus of the centre of a circle which cuts the circles $x^2+y^2=25$ and $x^2+y^2-2x-4y-11=0$ at the ends of diameters of the latter circles. [L.U.]

26. The tangents drawn from a point P on the x-axis to the circle

$$x^2+y^2-2ay+2a-1=0 \quad (a>1)$$

touch the circle at Q and R. Prove that the point P can be found such that QR subtends a right angle at the origin only if $a \geqslant 2+\sqrt{2}$, and show that when $a=4$ this point is at a distance $\sqrt{2}$ from the origin. [L.U.]

27. Prove that the equation of the circle having the points (x_1, y_1) and (x_2, y_2) as extremities of a diameter is

$$(x-x_1)(x-x_2)+(y-y_1)(y-y_2)=0.$$

The line $x\sin\alpha+y\cos\alpha=1$ intersects the conic $ax^2+2hxy+by^2=1$ at points P, Q which lie at finite distances from the origin. Find the equation of the circle on PQ as diameter and prove that, if this circle passes through the origin and has its centre on the x-axis, then

$$a+b=1 \text{ and } h=a\cot\alpha.$$ [L.U.]

28. Find the equations of the two circles which pass through the points of intersection of the circle $4(x^2+y^2)=5p^2$ and the line $x \cos \alpha+y \sin \alpha=p$ and touch the line $x \cos (\alpha+60°)+y \sin (\alpha+60°)=p$. Show that the larger of the two circles also passes through the points of intersection of the circle $4(x^2+y^2)=p^2$ and the line $x \cos \alpha+y \sin \alpha+\frac{1}{2}p=0$. [L.U.]

29. (i) A variable circle passes through a fixed point and cuts a fixed circle at the ends of a variable diameter. Show that the locus of the centre of the variable circle is a straight line.

(ii) The polar of the origin with respect to the circle

$$x^2+y^2+2gx+2fy+c=0$$

intersects the circle at P and Q. Show that the equation of the circle on PQ as diameter is

$$(f^2+g^2)(x^2+y^2)+2cgx+2cfy+2c^2-c(f^2+g^2)=0. \qquad \text{[L.U.]}$$

30. Find the values of the constants λ and μ for the equation

$$\lambda y(3x-y)+\mu y(3x+y-6)+(3x-y)(3x+y-6)=0$$

to represent a circle. Hence find the centre and radius of the circumcircle OAB whose sides OA, OB, AB have the equations

$$y=0,\ y=3x,\ 3x+y=6$$

respectively.

Obtain also the coordinates of the centre of the circle which cuts this circle orthogonally at O and A. [L.U.]

CHAPTER 13

THE PARABOLA

13.1. Conic sections

A *conic section*, or, more briefly, a *conic*, may be defined as follows: If S is a fixed point and l is a fixed line which does not pass through S, the locus of a point which moves in the plane of S and l so that its distance from S is in a constant ratio to its distance from l, is called a conic. The fixed point is called a *focus* of the conic, the fixed line a *directrix*, and the constant ratio (denoted by e) the *eccentricity* of the conic. The conic is called a *parabola*, *ellipse* or *hyperbola* according as e is equal to, less than or greater than unity.

13.2. The parabola (revision)

The standard equation of the parabola is $y^2=4ax$. When the equation of the curve is in this form,

(i) the focus S is the point $(a, 0)$,

(ii) the directrix ZM is the line $x=-a$,

(iii) the vertex O is the origin of co-ordinates,

(iv) the x-axis is the axis of symmetry of the curve and is called the axis of the parabola,

(v) the y-axis is the tangent at the vertex,

(vi) the latus rectum LL_1 (the double ordinate through the focus) is of length $4a$.

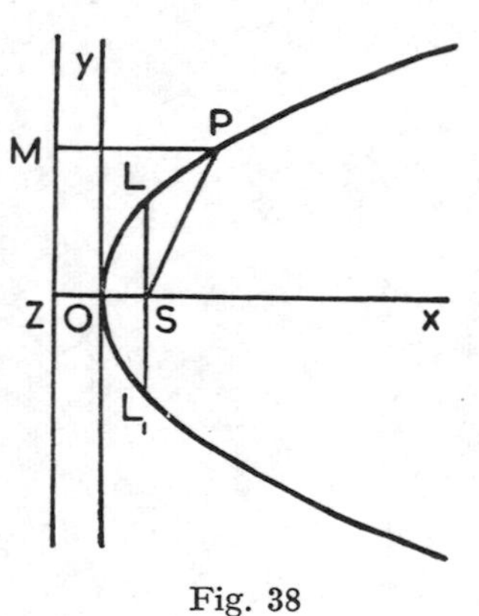

Fig. 38

The parabola $y^2=4ax$ is shown in fig. 38.

The point whose coordinates are given in terms of the single variable t by the equations

$$x=at^2, \quad y=2at \quad . \quad . \quad . \quad . \quad \text{(i)}$$

lies on the parabola $y^2=4ax$ for all values of t, and equations (i) may be taken as the parametric equations of the parabola $y^2=4ax$. As t varies from $-\infty$ to $+\infty$, the point given by (i) describes the parabola completely. The point $(at^2, 2at)$ is referred to as the point of parameter

t, or the point $[t]$; the chord joining the points $[t_1]$ and $[t_2]$ on the parabola is called the chord $[t_1, t_2]$.

The equation of the tangent at (x_1, y_1) on the parabola may be found by the method of § 12.10 to be

$$yy_1 = 2a(x + x_1).$$

The equation of the normal at the same point is

$$y - y_1 = -\frac{y_1}{2a}(x - x_1).$$

13.3. Chord of contact

By the method of § 12.12 we may show that the equation of PQ, the chord of contact of tangents drawn to the parabola $y^2 = 4ax$ from the point $T(x_1, y_1)$, is

$$yy_1 = 2a(x + x_1).$$

PQ is called the polar of T with respect to the parabola, and T is called pole of PQ.

13.4. The chord $[t_1, t_2]$; focal chords

The gradient m of the line joining $(at_1^2, 2at_1)$ and $(at_2^2, 2at_2)$ is given by

$$m = \frac{2a(t_2 - t_1)}{a(t_2^2 - t_1^2)} = \frac{2}{t_1 + t_2} \qquad (t_1 \neq t_2).$$

Hence by joining pairs of points $[t_1]$ and $[t_2]$ on the parabola such that $t_1 + t_2 = 2/m =$ constant, we obtain a set of parallel chords of gradient m.

The equation of the chord $[t_1, t_2]$ is

$$\frac{y - 2at_1}{x - at_1^2} = \frac{2}{t_1 + t_2},$$

i.e.
$$2x - (t_1 + t_2)y + 2at_1t_2 = 0. \tag{13.1}$$

This chord passes through the focus $(a, 0)$ of the parabola if

$$t_1t_2 = -1. \tag{13.2}$$

Thus if the point $(at_1^2, 2at_1)$ is one extremity of a focal chord, $(a/t_1^2, -2a/t_1)$ is the other.

13.5. The tangent and normal at $[t]$

If, in (13.1), we write $t_1 = t_2 = t$, we obtain the equation of the tangent to the parabola at $[t]$:

$$x - ty + at^2 = 0.$$

This equation may be found directly by substituting $x_1=at^2$, $y_1=2at$ in $yy_1=2a(x+x_1)$, the equation of the tangent at (x_1, y_1) ; alternatively, from the parametric equations $x=at^2$, $y=2at$ we have

$$\frac{dy}{dx}=\left(\frac{dy}{dt}\right)\bigg/\left(\frac{dx}{dt}\right)=\frac{1}{t}.$$

Hence the equation of the tangent at $[t]$ is

$$y-2at=\frac{1}{t}(x-at^2),$$

$$y=\frac{x}{t}+at. \qquad (13.3)$$

The equation of the normal at $[t]$ is

$$y-2at=-t(x-at^2),$$

$$y+tx=2at+at^3. \qquad (13.4)$$

13.6. The point of intersection of the tangents at $[t_1]$ and $[t_2]$

From (13.3) the equations of the tangents at $[t_1]$ and $[t_2]$ are

$$t_1y=x+at_1^2,$$

$$t_2y=x+at_2^2.$$

These tangents meet at the point

$$\{at_1t_2,\ a(t_1+t_2)\} \qquad (13.5)$$

The tangents are perpendicular if $t_1t_2=-1$, which is the same condition as (13.2). It follows that tangents at the extremities of a focal chord are perpendicular and intersect on the line $x=-a$, i.e. on the directrix of the parabola.

13.7. The tangent of gradient m

Suppose that the line $y=mx+c$ touches the parabola at $[t]$. Then this line is the same as (13.3) and so corresponding coefficients in the two equations are proportional, i.e. $m=1/t$ and

$$c=at=a/m.$$

Hence the line $y=mx+a/m$ touches the parabola for all values of m. The point of contact is $(a/m^2, 2a/m)$.

13.8. Locus of mid-points of parallel chords of the parabola

The coordinates of M, the mid-point of the chord $[t_1, t_2]$, are

$$x=\tfrac{1}{2}a(t_1^2+t_2^2),\ y=a(t_1+t_2).$$

If $t_1+t_2=$constant, $y_M=$constant and M lies on a straight line parallel to the axis of the parabola. But (see § 13.4) by joining points

$[t_1]$ and $[t_2]$ on the curve such that $t_1+t_2=$constant we obtain a set of parallel chords. Hence the mid-points of parallel chords of a parabola lie on a line parallel to its axis. Such a line is known as a *diameter* of the curve. The diameter bisecting chords of gradient m is $y=2a/m$.

By regarding a tangent as the limiting case of a chord, it follows that the tangent at the extremity of a diameter of a parabola is parallel to the chords which the diameter bisects.

13.9. Conormal points on the parabola $y^2=4ax$

By (13.4) the normal at $[t]$ will pass through (X, Y), if

$$Y+tX=2at+at^3,$$

i.e. if

$$at^3+t(2a-X)-Y=0 \quad . \quad . \quad . \quad \text{(i)}$$

The roots t_1, t_2, t_3 of this cubic equation are the values of the parameters of the points on the curve the normals at which pass through (X, Y).

Three points on a parabola the normals at which meet in a point are called conormal points.

There is no term in t^2 in (i) and so (see § 2.2)

$$t_1+t_2+t_3=0.$$

Thus the algebraic sum of the ordinates of three conormal points on a parabola is zero.

From (i) we also have

$$t_1t_2t_3=Y/a \quad . \quad . \quad . \quad . \quad \text{(ii)}$$

Now suppose that the normals at $[t_1]$, $[t_2]$ meet at a point on the parabola. In such a case (X, Y) must coincide with $[t_3]$ and so $Y=2at_3$. Substitution of this value in (ii) gives

$$t_1t_2=2.$$

This result shows that for normals at $[t_1]$, $[t_2]$ to intersect on the parabola a necessary condition is that $t_1t_2=2$. That this condition is also sufficient can be deduced using the method of Example 5 of § 13.11.

13.10. Concyclic points on the parabola $y^2=4ax$

The circle $x^2+y^2+2gx+2fy+c=0$ meets the parabola $y^2=4ax$ at the point $[t]$ if

$$a^2t^4+4a^2t^2+2gat^2+4fat+c=0,$$

i.e. if

$$a^2t^4+2t^2(2a^2+ga)+4fat+c=0 \quad . \quad . \quad . \quad \text{(i)}$$

The roots t_1, t_2, t_3, t_4 of this quartic equation in t are the values of the parameters of the four points, not necessarily all real, in which the circle meets the parabola.

Since in (i) the coefficient of t^3 is zero, $t_1+t_2+t_3+t_4=0$ (see § 2.2).

Hence the algebraic sum of the ordinates of the four points of intersection of a circle and a parabola is zero.

If, in addition, $[t_1]$, $[t_2]$, $[t_3]$ are conormal points, $t_4=0$. Hence the circle through three conormal points passes through the vertex of the parabola.

13.11. Miscellaneous examples

Example 1

Show that the equation

$$y^2+8x-2y-23=0$$

represents a parabola with latus rectum of length 8 *and focus* (1, 1). *Find the coordinates of its vertex and the equation of its directrix.*

The given equation may be written

$$(y-1)^2=-8(x-3)$$

and simplified by the substitutions

$$X=x-3, \quad Y=y-1,$$

i.e. by transferring the origin to (3, 1) as in § 12.3 (*a*).

The equation then reduces to $Y^2=-8X$. This is the standard equation of the parabola with $a=-2$ (see § 13.2). The length of the latus rectum ($4a$ numerically) is 8; the coordinates of the focus are $X=a$, $Y=0$, i.e. $x=1$, $y=1$; the vertex, $X=0$, $Y=0$ is the point (3, 1), and the equation of the directrix $X=-a$ is $x=5$.

Example 2

Find the equations of the tangent and normal to the parabola $y^2=4ax$ at the point $(at^2, 2at)$.

PQ, a variable chord of the parabola $y^2=4ax$, subtends a right angle at the vertex; TP, TQ are tangents; NP, NQ are normals. Show that the locus of the mid-point of TN is a parabola. [L.U.]

For the equations of the tangent and normal at $[t]$, see § 13.5.

Let $P\equiv(at_1^2, 2at_1)$ and $Q\equiv(at_2^2, 2at_2)$.

Then, by (13.5), the tangents at P and Q meet at T $\{at_1t_2, a(t_1+t_2)\}$.

The equations of the normals at P and Q are respectively

$$y+t_1x=2at_1+at_1^3$$

and

$$y+t_2x=2at_2+at_2^3.$$

The normals intersect at N where

$$x(t_1-t_2)=2a(t_1-t_2)+a(t_1^3-t_2^3)$$

i.e.

$$x=a(2+t_1^2+t_1t_2+t_2^2) \text{ since } t_1\neq t_2$$

$$\therefore y=-at_1t_2(t_1+t_2).$$

The coordinates of M, the mid-point of TN, are

$$x=\tfrac{1}{2}(x_T+x_N)=\tfrac{1}{2}a\{2+(t_1+t_2)^2\}$$

$$y=\tfrac{1}{2}(y_T+y_N)=\tfrac{1}{2}a(t_1+t_2)(1-t_1t_2) \quad . \quad . \quad \text{(i)}$$

Now the gradients of OP and OQ are $2/t_1$, $2/t_2$, and PQ subtends a right angle at the vertex O,

$$\therefore\ t_1t_2=-4.$$

Substituting this value for t_1t_2 in (i), we have

$$y=\tfrac{5}{2}a(t_1+t_2)$$

and

$$x-a=\tfrac{1}{2}a(t_1+t_2)^2.$$

$$\therefore\ 2y^2=25a(x-a).$$

Thus the locus of M is a parabola with vertex at $(a, 0)$ and latus rectum of length $25a/2$.

Example 3

The chord PQ joining the points $P(at_1^2, 2at_1)$ and $Q(at_2^2, 2at_2)$ on the parabola $y^2=4ax$ passes through the focus. Find the relation between t_1 and t_2.

Prove that the circle on PQ as diameter touches the directrix of the parabola, and that its point of contact is the point of intersection of the tangents to the parabola at P and Q. [L.U.]

By (13.2), $t_1t_2=-1$, and so (see § 13.6) the tangents at P and Q meet at right angles on the directrix at $T\{-a, a(t_1+t_2)\}$, i.e. the circle on PQ as diameter passes through T, and C its centre is the mid-point of PQ. It follows that the ordinate of C is $a(t_1+t_2)$ and this is also the ordinate of T. Hence the directrix is perpendicular to TC and so touches the circle at T.

Example 4

If the normal at $P(at^2, 2at)$ on the parabola $y^2=4ax$ meets the curve again at $Q(at_1^2, 2at_1)$, show that $t^2+tt_1+2=0$.

If the tangents at P and Q intersect at R, show that the line through R parallel to the axis of the parabola meets the parabola in P', where PP' is a focal chord. [L.U.]

The equation of the normal at P is $y+tx=2at+at^3$. It meets the curve at $Q[t_1]$

$$\therefore\ 2at_1+att_1^2=2at+at^3,$$

$$(t-t_1)(t^2+tt_1+2)=0,$$

$$\therefore\ t^2+tt_1+2=0 \qquad \text{(i)}$$

since $t\neq t_1$.

By (13.5) the tangents at P and Q meet at $R\{att_1, a(t+t_1)\}$ and so the equation of the line through R parallel to the axis of the curve is $y=a(t+t_1)$.

It meets the curve at $P'(aT^2, 2aT)$ where

$$2aT=a(t+t_1),$$

or

$$2tT=t^2+tt_1.$$

$$\therefore\ tT=-1 \text{ by (i).}$$

Thus, by (13.2), PP' is a focal chord.

Example 5

The normals at P $(at_1^2, 2at_1)$ *and* $Q(at_2^2, 2at_2)$ *intersect on the parabola at* $R(aT^2, 2aT)$. *Show that* t_1 *and* t_2 *are the roots of the equation* $t^2+tT+2=0$.
Show that for all values of T the locus of the mid-point of the chord PQ is a parabola. [L.U.]

The normal at $P[t_1]$ meets the parabola again at $R[T]$,

$$\therefore t_1^2+t_1T+2=0 \quad \text{(see Example 4).}$$

Similarly, $$t_2^2+t_2T+2=0.$$

Hence t_1 and t_2 are the roots of the equation

$$t^2+tT+2=0$$

and so, cf. § 13.9, $$t_1t_2=2 \quad . \quad . \quad . \quad . \quad \text{(i}$$

This is the condition that the normals at $[t_1]$ and $[t_2]$ should intersect on the parabola.

The coordinates of M, the mid-point of PQ, are

$$x=\tfrac{1}{2}a(t_1^2+t_2^2), \quad y=a(t_1+t_2).$$

To find the locus of M we eliminate t_1 and t_2 between these equations using (i).

$$\begin{aligned} y^2 &= a^2(t_1^2+t_2+2t_1t_2) \\ &= a^2(t_1^2+t_2^2+4) \\ &= 2a(x+2a). \end{aligned}$$

This is the equation of a parabola with vertex at $(-2a, 0)$ and latus rectum of length $2a$.

Example 6

Show that the equation of the chord of the parabola $y^2=4ax$ *which is bisected at* (α, β) *is*

$$2ax-\beta y=2a\alpha-\beta^2.$$

Find the locus of the middle points of chords of the parabola $y^2=4ax$ *which touch the parabola* $y^2+4ax=0$. [L.U.]

Any chord through the point (α, β) is of the form

$$y-\beta=m(x-\alpha) \quad . \quad . \quad . \quad . \quad \text{(i)}$$

This chord meets the parabola at the point $(at^2, 2at)$,

where $$2at-\beta=m(at^2-\alpha),$$

i.e. $$mat^2-2at+(\beta-m\alpha)=0.$$

If t_1 and t_2 are the roots of this equation,

$$t_1+t_2=2/m \quad . \quad . \quad . \quad . \quad . \quad \text{(ii)}$$

But t_1 and t_2 are the parameters of the points where the chord meets the curve, and so if (α, β) is the mid-point of the chord,

$$\beta=a(t_1+t_2)=2a/m \text{ by (ii)}$$
$$\therefore m=2a/\beta.$$

Substituting this value in (i), we obtain the equation of the chord bisected at (α, β):

$$2ax - \beta y = 2a\alpha - \beta^2 \quad . \quad . \quad . \quad . \quad \text{(iii)}$$

Identifying (iii) with the line $y = nx - a/n$ which touches the parabola $y^2 = -4ax$ for all values of n, we have

$$2a/\beta = n \quad \text{and} \quad (\beta^2 - 2a\alpha)/\beta = -a/n.$$

Elimination of n between these equations gives

$$3\beta^2 = 4a\alpha.$$

Hence the locus of (α, β) is $3y^2 = 4ax$.

Exercises 13

1. Find the slope of the tangent to the curve $y^2 = 4ax$ at the point $P(at^2, 2at)$.
   The line PA, where A is $(a, 2a)$, meets Ox at B. The line through A parallel to Ox meets OP at C. Show that BC is bisected by Oy and is parallel to the tangent to the curve at P. [Liverpool.]

2. The tangent at a point P on the parabola $y^2 = 4ax$ meets the directrix in Z; a parallel to the axis of the parabola through Z meets the normal at P in R. Prove that the locus of R is another parabola with the same axis as before and vertex at the point $(3a, 0)$. [L.U.]

3. A point P moves on the parabola $y^2 = 4ax$, and B is the point $(2a, 0)$. Show that the locus of the middle point of BP is a parabola whose focus is the point $(3a/2, 0)$. [Sheffield.]

4. Find the equation of the normal to the parabola $x^2 = 4ay$ at the point $P(2a, a)$ on it. If the normal cuts the parabola again in Q find the coordinates of Q and the angle subtended at the origin by PQ. [Sheffield.]

5. The normal at the point $P(at^2, 2at)$ on the parabola $y^2 = 4ax$ meets the parabola again in Q and the line through Q parallel to the axis of the parabola meets the tangent at P in T. Prove that PT is divided by the directrix in the ratio $1:3$. [Durham.]

6. Show that the line $y = mx + c$ touches the parabola $y^2 = 4ax$ if $c = a/m$.
   Hence find the gradients of the two common tangents of the above parabola and the circle $x^2 + y^2 = a^2$. [Leeds.]

7. Prove that the line $y = mx + a/m$ is a tangent to the parabola $y^2 = 4ax$, and that the point of contact, P, is the point $(a/m^2, 2a/m)$. Another tangent to the parabola is drawn parallel to OP, where O is the origin. Prove that the two tangents meet in the point Q whose coordinates are $(a/2m^2, 3a/2m)$. Deduce the equation of the locus of Q as m varies. [Sheffield.]

8. A line through Q $(-2a, 0)$ cuts the parabola $y^2=4ax$ at R_1, R_2 and the tangents at these points meet at P. If A is the vertex, prove that AP is inclined to the y-axis at the same angle that QR_1R_2 is inclined to the x-axis.
   Prove also that the normals at R_1, R_2 meet on the curve. [L.U.]

9. The tangents to the parabola $y^2=4ax$ at the points $R(at^2, 2at)$, $S(au^2, 2au)$ intersect at P. Find the coordinates of P and prove that, if the angle between the tangents to the parabola at R and S is α, P lies on the curve $y^2-4ax=(x+a)^2 \tan^2 \alpha$. [Durham.]

10. Prove that two perpendicular tangents of a parabola intersect upon the directrix, and touch the curve at the ends of a focal chord.
    Prove that the mid-point of a variable focal chord describes a parabola having a latus rectum equal to half that of the given parabola. [Leeds.]

11. Prove that the chord joining the points of contact of perpendicular tangents to a parabola passes through the focus.
    Prove also that the orthocentre of the triangle formed by any three tangents to the parabola lies on the directrix. [L.U.]

12. Find the coordinates of the point of intersection of the tangents at the points $(at_1^2, 2at_1)$, $(at_2^2, 2at_2)$ to the parabola $y^2=4ax$.
    Tangents are drawn from a point on the parabola $y^2=4bx$ to the parabola $y^2=4ax$. Prove that the locus of the intersection of corresponding normals to $y^2=4ax$ is the curve
    $$y^2(4b-a)^3=4ab(x-2a)^3. \qquad \text{[L.U.]}$$

13. Prove that the line $x-ty+at^2=0$ touches the parabola $y^2=4ax$ for any value of t.
    Find the equation of the circumcircle of the triangle formed by the y-axis and two other tangents to the parabola, and show that this circle passes through the focus. [L.U.]

14. A, B are the extremities of one of a family of parallel chords of a parabola, and the normals to the curve at A, B meet in P. Show that the locus of P is a straight line which is normal to the curve. [Leeds.]

15. Prove that the locus of mid-points of normal chords of the parabola $y^2=4ax$ is the curve $y^4-2a(x-2a)y^2+8a^4=0$. [Sheffield.]

16. Find the condition that the line $lx+my+n=0$ should touch the parabola $y^2=4ax$.
    A tangent to the parabola $y^2=4ax$ meets the parabola $y^2=4bx$ at the points P, Q, and the tangents at P, Q meet at R. Show that the locus of R is the parabola $ay^2=4b^2x$, and find the locus of the point of intersection of the normals at P, Q. [L.U.]

17. A point P moves on the line $x=h$ and the three normals from P meet the parabola at Q_1, Q_2, Q_3. Show that the centroid of the triangle $Q_1Q_2Q_3$ remains at a fixed point on the x-axis. [Durham.]

18. If two variable points P, Q, on the parabola $y^2=4ax$ subtend a right angle at the vertex, show that PQ meets the x-axis at a fixed point.

Show also that the locus of the mid-point of PQ is a parabola whose vertex is at the point $(4a, 0)$. [L.U.]

19. Two of the normals to the parabola $y^2=4ax$ from the point $P(\lambda, \mu)$ are perpendicular. Show that P lies on the parabola $y^2=a(x-3a)$ and that the length of the chord joining the feet of the perpendicular normals is $\lambda+a$. [L.U.]

20. If P is the point $(at^2, 2at)$ and the normal at P meets the parabola $y^2=4ax$ again at the point Q whose coordinates are $(at_1^2, 2at_1)$, show that $t^2+tt_1+2=0$.

If M is the mid-point of PQ and N is the mid-point of PM, show that, as P varies on the parabola, the locus of N is the parabola $y^2=a(x-3a)$. [L.U.]

21. The normals to the parabola $y^2=4ax$ at $P(at_1^2, 2at_1)$ and $Q(at_2^2, 2at_2)$ intersect on the parabola at $R(aT^2, 2aT)$. Show that t_1 and t_2 are the roots of the equation $t^2+tT+2=0$.

Show that for all values of T the locus of the mid-point of the chord PQ is a parabola. [L.U.]

22. Q is the variable point $(aT^2, 2aT)$ on the parabola $y^2=4ax$. The normals at points P and P' on the curve pass through Q. The tangents at P and Q intersect at R, and the tangents at P' and Q intersect at R'. Show that the locus of the mid-point of RR' is a parabola. [L.U.]

23. P is the point $(aT^2, 2aT)$ on the parabola $y^2=4ax$. Show that if $T^2>8$, there are two real normals to the curve which pass through P in addition to the normal at P.

Q and R are the points at which the normals will pass through P. M is the mid-point of QR and N is the mid-point of PM. Show that, as P moves along the parabola, N describes a parabola and QR passes through a fixed point. [L.U.]

24. If the normal at the point $P(at^2, 2at)$ on the parabola $y^2=4ax$, meets the parabola again in Q, show that the length of PQ is $4a(1+t^2)^{3/2}/t^2$.

Prove also that, if a normal chord PQ subtends a right angle at the focus S, then $PQ=5a\sqrt{5}$. [L.U.]

25. P is any point of a parabola whose vertex is A, and the normals at points Q, R of the curve meet at P. Show that, as P varies, AP and QR intersect on a fixed straight line perpendicular to the x-axis. [L.U.]

26. P and Q are the points of contact of the tangents drawn from a point T to a parabola whose focus is S. If R is the middle point of PQ, prove that $TP.TQ=2TS.TR$. [L.U.]

27. A is the point $(4a, 4a)$ and P is a variable point on the parabola $y^2=4ax$; the chord AQ is drawn parallel to the normal at P. If the tangents at P and Q intersect at R, prove that the locus of R is the hyperbola
$$x^2+2xy+8ax+4ay+4a^2=0.$$ [L.U.]

28. A variable straight line, whose direction is fixed, cuts a parabola in points P and Q. Prove that the locus of the point of intersection of the normals at P and Q to the parabola is a straight line which is itself a normal to the parabola. [L.U.]

29. P, Q, R are the vertices of a triangle inscribed in a parabola. The sides QR, RP, PQ meet the axis of the parabola at the points L, M, N respectively. Show that the perpendiculars drawn from L, M, N to the tangents at P, Q, R respectively meet at a point on the tangent at the vertex of the parabola. [L.U.]

30. The normal at P to the parabola $y^2=4ax$ cuts the x-axis at G, and Q is the point on the tangent at P, on the same side of the ordinate at P as G, such that $PQ=PG$. Prove that the locus of Q is a parabola. [L.U.]

31. If the normals at two points of the parabola $y^2=4ax$ make complementary angles with the axis, show that their point of intersection lies on the parabola $y^2=a(x-a)$. [L.U.]

32. If the tangents at two points on the parabola $y^2=4ax$ intersect at Q and the normals at the same points intersect at P, express the coordinates of P in terms of X and Y, the coordinates of Q. Hence show that (i) if the locus of P is a line parallel to the x-axis, the locus of Q is a hyperbola ; (ii) if the locus of P is a line parallel to the y-axis, the locus of Q is a parabola. [L.U.]

33. Tangents at the points $(at_1^2, 2at_1)$, $(at_2^2, 2at_2)$ of the parabola $y^2=4ax$ meet the x-axis at A and B, and the y-axis at C and D. Find the equation of the circles on AB as diameter and on CD as diameter.

If the two tangents are perpendicular, prove that, as they vary in position, the circles on AB as diameter form a coaxal system with real limiting points and that the circles on CD as diameter form the orthogonal coaxal system. [L.U.]

34. Normals to the parabola $y^2=4ax$ at the points P and Q meet on the fixed straight line

$$lx+my+n=0.$$

Prove that the tangents to the parabola at the points P and Q meet on the hyperbola

$$mxy-ly^2+alx-2a^2l-an=0. \qquad \text{[L.U.]}$$

35. The curves p and q are arcs of parabolas. The equation of p is $y=+2\sqrt{(ax)}$ and that of q is $y=+2\sqrt{(-ax)}$, a being a positive constant. The tangent to p at the point $(at^2, 2at)$ intersects q at Q, and the tangent to q at Q intersects p at the point $(at_1^2, 2at_1)$. Prove that $t_1=(\sqrt{2}-1)^2t$. [L.U.

36. A tangent to the parabola $y^2+n^2ax=0$, where n is positive, intersects the parabola $y^2=4ax$ at P and Q. Prove that the locus of the mid-point of PQ is another parabola and that, if the latus rectum of this parabola is $16a/17$, then $n=3$. [L.U.]

37. Two triangles are formed respectively by the tangents and normals to the parabola $y^2=4ax$ at any three of its points. Prove that the line joining their orthocentres is parallel to the axis of the parabola. [L.U.]

38. Find the equation of the tangent to the parabola $y^2=4ax$ at the point $(at^2, 2at)$.

Find the coordinates of the centre of the circle which touches the parabola at this point and passes through the focus, and show that, as t varies, the locus of the centres of these circles is the curve

$$27ay^2=(2x-a)(5a-x)^2.$$ [L.U.]

39. If the tangents drawn to a parabola from a point T are equidistant from a point K on its axis, show that T lies on the circle which passes through K and has its centre at the focus of the parabola. [L.U.]

40. The tangent at any point $P(at^2, 2at)$ of the parabola $y^2=4ax$ is denoted by l. Find the coordinates of the point Q on the parabola $x^2=4by$ such that the tangent m to this parabola at Q is perpendicular to l.

Prove also that the locus of the points of intersection of l and m is the curve $(x^2+y^2)(ax+by)+(bx-ay)^2=0.$ [Sheffield.]

41. A circle is drawn with its centre at the focus S of the parabola $y^2=4ax$ to touch the parabola at the vertex. A radius SQ of the circle meets the parabola at a point $P(a\cot^2\theta, 2a\cot\theta)$. Show that the coordinates of Q are $(2a\cos^2\theta, 2a\sin\theta\cos\theta)$.

If the tangent at the point P of the parabola intersects the tangent at the vertex in the point R, prove that the line RQ is perpendicular to the line SP. [L.U.]

42. The points $P(ap^2, 2ap)$ and $R(ar^2, 2ar)$ lie on the parabola γ with the equation $y^2=4ax$. Find the equation of the line PR in its simplest form, and obtain the condition which must be satisfied by p and r if PR is to pass through the focus of γ.

PR and QS are focal chords of γ. Show that both pairs of opposite sides of the quadrilateral $PQRS$ intersect on the directrix of γ. [L.U.]

43. Prove that three normals to the parabola pass through a general point (X, Y), and that the parameters of the feet of the normals satisfy the relation $t_1+t_2+t_3=0$.

Prove also that the equation of the circle through the feet of these normals is $x^2+y^2-(X+2a)x-\frac{1}{2}Yy=0.$ [L.U.]

44. The circle through the origin O with centre (h, k) meets the parabola again at points with the parameter values t_1, t_2, t_3. Show that the normals at these three points are concurrent and find the coordinates of their common point in terms of h and k. [L.U.]

45. Obtain the equation of the chord QR of the parabola $y^2=4ax$ whose mid-point is $P(h, k)$.

Find the locus of P when the tangents to the parabola at Q and R are perpendicular; and find also, in this case, the locus of the point of intersection of the normals to the parabola at Q and R. [L.U.]

46. Find the angle between the two lines whose combined equation is $ax^2+2hxy+by^2=0$, and show that the lines

$$(m+1)y^2-(m^2+2m-1)xy+(m^2-m)x^2=0$$

intersect each other at an angle of $\frac{1}{4}\pi$ radians for all values of m.

Find the equations of the chords of the parabola $y^2=4ax$, which pass through the point $(-6a, 0)$ and which subtend an angle of $\frac{1}{4}\pi$ radians at the vertex. [L.U.]

47. The circle $x^2+y^2=a^2$ intersects the x-axis at the points $A(a, 0)$ and $B(-a, 0)$ and P is the point $(a \cos \theta, a \sin \theta)$. Prove that the equations of the lines AP and BP are

$$(x-a) \cos (\theta/2)+y \sin (\theta/2)=0$$

and

$$(x+a) \sin (\theta/2)-y \cos (\theta/2)=0.$$

The lines AB and CD are two perpendicular diameters of a circle, and P is a variable point on its circumference. The straight line AP meets CD in Q and the straight line through Q parallel to AB meets BP in R. Prove that the locus of the point R is a parabola passing through the points B, C and D. What position of the point P corresponds to the point at infinity on the parabola ? [L.U.]

48. Prove that the equation

$$2yt=x+(t^2-1)a, \quad . \quad . \quad . \quad . \quad (1)$$

where a and t are parameters, represents a family of parallel straight lines when t is fixed and a is variable, and write down the equation of that straight line which passes through the origin.

When a is fixed and t is variable, prove that equation (1) represents a family of tangents to the parabola $y^2=a(x-a)$.

When a and t both vary, find the relation which must hold between a and t in order that equation (1) shall represent the family of tangents to the circle $x^2+y^2=1$. [L.U.]

49. Prove that for all values of t and c

$$(1+t^2)(y^2-4ax)+(x-ty+at^2)(x+ty+c)=0$$

is the equation of a circle which touches $y^2-4ax=0$.

PFQ is a focal chord of a parabola. Circles are drawn through the focus F to touch the parabola at P and Q respectively. Prove that these circles cut orthogonally. [L.U.]

50. P and Q are points on the parabola $y^2=4ax$. The perpendicular bisector of PQ meets the axis of the parabola at R, and the perpendicular to the axis drawn through the mid-point V of PQ meets the axis at M. Prove that MR is of constant length for all positions of P and Q.

If a parabola, its focus and its axis are given, show how to construct the chord of the parabola which is bisected at a given point V. [L.U.]

CHAPTER 14

THE ELLIPSE AND HYPERBOLA

14.1. The ellipse (revision)

An ellipse is a conic whose eccentricity e is less than unity. The standard equation of the ellipse is

$$\frac{x^2}{a^2}+\frac{y^2}{b^2}=1, \tag{14.1}$$

where $b^2=a^2(1-e^2)$.

When the equation of the curve is in this form,

(i) the foci S, S_1 are the points $(\pm ae, 0)$,

(ii) the directrices ZM, Z_1M_1 are the lines $x=\pm a/e$,

(iii) the eccentricity e, less than unity, is given by $e^2=1-b^2/a^2$,

(iv) the centre O is the origin of coordinates,

(v) the major axis AA_1 is of length $2a$ and lies along the x-axis,

(vi) the minor axis BB_1 is of length $2b$ and lies along the y-axis,

(vii) each latus rectum is of length $2b^2/a$.

The ellipse $\frac{x^2}{a^2}+\frac{y^2}{b^2}=1$ is shown in fig. 39. Since the major and minor axes (i.e. the principal axes) of the curve lie along the axes of co-ordinates (14.1) is sometimes called the equation of the ellipse referred to its principal axes.

The sum of the focal distances SP, S_1P of any point P on the ellipse is constant and equal to the length of the major axis. Hence the ellipse may also be defined as the locus of a point which moves in a plane so that the sum of its distances from two fixed points is constant.

The equation of the tangent at (x_1, y_1) to the ellipse $\frac{x^2}{a^2}+\frac{y^2}{b^2}=1$ may be found by the method of § 12.10 to be

$$\frac{xx_1}{a^2}+\frac{yy_1}{b^2}=1.$$

The equation of the normal at the same point is

$$\frac{a^2x}{x_1}-\frac{b^2y}{y_1}=a^2-b^2.$$

By the method of § 12.12 we may show that the equation of PQ, the chord of contact of tangents drawn to the ellipse $\frac{x^2}{a^2}+\frac{y^2}{b^2}=1$ from the point $T(x_1, y_1)$, is

$$\frac{xx_1}{a^2}+\frac{yy_1}{b^2}=1 \tag{14.2}$$

PQ is called the polar of T with respect to the ellipse and T is called the pole of PQ.

14.2. The auxiliary circle and eccentric angle

The circle described on the major axis of an ellipse as diameter is called the auxiliary circle of the ellipse.

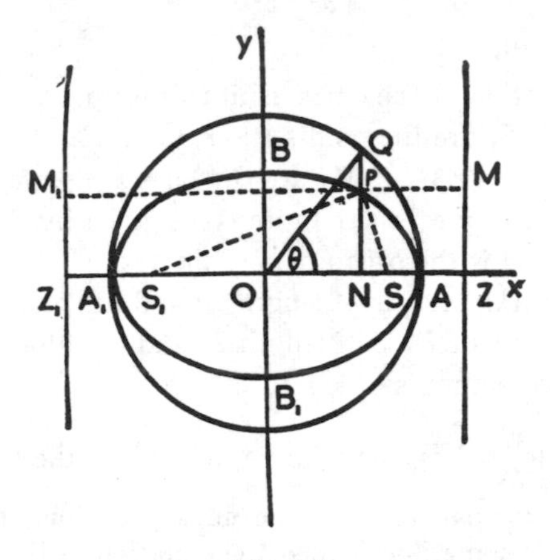

Fig. 39

If NP (fig. 39), any ordinate of the ellipse, is pro duced to meet the auxiliary circle at Q, Q is said to be the point on the auxiliary circle corresponding to P on the ellipse. If $\angle xOQ=\theta$, Q is the point $(a \cos \theta, a \sin \theta)$ and from (14.1) P is $(a \cos \theta, b \sin \theta)$. θ is called the eccentric angle of P, and P is referred to as the point $[\theta]$. The chord joining $[\theta]$ and $[\phi]$ on the ellipse is called the chord $[\theta, \phi]$.

The equations

$$x=a \cos \theta, \; y=b \sin \theta$$

may be taken as parametric equations of the ellipse.

Since $NP : NQ=b/a$, NP can be obtained as the orthogonal projection of NQ on a plane inclined at an angle $\cos^{-1}(b/a)$ to NQ. Thus an ellipse is the orthogonal projection of its auxiliary circle on a plane drawn through A_1A (or through any line parallel to A_1A) which makes an angle $\cos^{-1}(b/a)$ with the plane of the auxiliary circle.

14.3. The chord $[\theta, \phi]$

The gradient m of the line joining $(a \cos \theta, b \sin \theta)$ to $(a \cos \phi, b \sin \phi)$ is given by

$$m = \frac{b(\sin \theta - \sin \phi)}{a(\cos \theta - \cos \phi)}$$
$$= -(b/a) \cot \tfrac{1}{2}(\theta + \phi).$$

It follows that, by joining pairs of points $[\theta]$ and $[\phi]$ on the ellipse such that $\theta + \phi = 2\alpha = \text{constant}$, a system of parallel chords of gradient $-(b/a) \cot \alpha$ is obtained.

The equation of the chord $[\theta, \phi]$ is

$$\frac{y - b \sin \phi}{x - a \cos \phi} = -\frac{b \cos \frac{1}{2}(\theta + \phi)}{a \sin \frac{1}{2}(\theta + \phi)},$$

i.e. $$bx \cos \tfrac{1}{2}(\theta + \phi) + ay \sin \tfrac{1}{2}(\theta + \phi)$$
$$= ab\{\cos \tfrac{1}{2}(\theta + \phi) \cos \phi + \sin \tfrac{1}{2}(\theta + \phi) \sin \phi\}.$$

This reduces to

$$\frac{x}{a} \cos \tfrac{1}{2}(\theta + \phi) + \frac{y}{b} \sin \tfrac{1}{2}(\theta + \phi) = \cos \tfrac{1}{2}(\theta - \phi) \qquad (14.3)$$

By making $\phi \rightarrow \theta$ in (14.3) we obtain the equation of the tangent at $[\theta]$:

$$\frac{x}{a} \cos \theta + \frac{y}{b} \sin \theta = 1.$$

14.4. Tangent and normal at $[\theta]$

From the parametric equations $x = a \cos \theta$, $y = b \sin \theta$ of the ellipse we have

$$\frac{dy}{dx} = \left(\frac{dy}{d\theta}\right) \Big/ \left(\frac{dx}{d\theta}\right) = -(b/a) \cot \theta.$$

Hence the equation of the tangent at $[\theta]$ is

$$(y - b \sin \theta) = -\frac{b \cos \theta}{a \sin \theta}(x - a \cos \theta),$$

$$\frac{x}{a} \cos \theta + \frac{y}{b} \sin \theta = 1 \qquad (14.4)$$

(as in § 14.3).

The equation of the normal at $[\theta]$ is

$$(y - b \sin \theta) = \frac{a \sin \theta}{b \cos \theta}(x - a \cos \theta),$$

i.e. $$\frac{ax}{\cos \theta} - \frac{by}{\sin \theta} = a^2 - b^2.$$

The equations of the tangent and normal may also be found from those given in § 14.1 by substituting $x_1 = a \cos \theta$, $y_1 = b \sin \theta$.

14.5. The point of intersection of the tangents at $[\theta]$ and $[\phi]$

The equations of the tangents at $[\theta]$ and $[\phi]$ are respectively:

$$bx\cos\theta+ay\sin\theta=ab$$
$$bx\cos\phi+ay\sin\phi=ab.$$

By (3.5), page 40, the coordinates of their point of intersection are given by the equations

$$\frac{x}{\begin{vmatrix} ab & a\sin\theta \\ ab & a\sin\phi \end{vmatrix}}=\frac{y}{\begin{vmatrix} b\cos\theta & ab \\ b\cos\phi & ab \end{vmatrix}}=\frac{1}{\begin{vmatrix} b\cos\theta & a\sin\theta \\ b\cos\phi & a\sin\phi \end{vmatrix}},$$

i.e. $$\frac{x}{2a\cos\frac{1}{2}(\phi+\theta)\sin\frac{1}{2}(\phi-\theta)}=\frac{y}{-2b\sin\frac{1}{2}(\theta+\phi)\sin\frac{1}{2}(\theta-\phi)}$$
$$=\frac{1}{2\sin\frac{1}{2}(\phi-\theta)\cos\frac{1}{2}(\phi-\theta)}.$$

Hence the point of intersection is

$$\left(a\frac{\cos\frac{1}{2}(\theta+\phi)}{\cos\frac{1}{2}(\theta-\phi)},\ b\frac{\sin\frac{1}{2}(\theta+\phi)}{\cos\frac{1}{2}(\theta-\phi)}\right).$$

14.6. The tangents to the ellipse which are parallel to the diameter $y=mx$

Suppose that the line $y=mx+c$ touches the ellipse at $[\theta]$. Then this line is the same as (14.4), and so corresponding coefficients in the two equations are proportional.

Hence $$\frac{\cos\theta}{-am}=\frac{\sin\theta}{b}=\frac{1}{c}.$$

These equations for θ have a solution if

$$\cos\theta=-am/c,\quad \sin\theta=b/c,$$

i.e. if $\quad c^2=a^2m^2+b^2, \quad (\text{since } \sin^2\theta+\cos^2\theta=1).$

It follows that for all values of m, the lines

$$y=mx\pm\sqrt{(a^2m^2+b^2)}$$

touch the ellipse.

14.7. The director circle

A tangent perpendicular to

$$y-mx=\sqrt{(a^2m^2+b^2)}$$

is $$my+x=\sqrt{(b^2m^2+a^2)},$$

and the point of intersection of these tangents satisfies the equation

$$(y-mx)^2+(my+x)^2=(1+m^2)(a^2+b^2)$$

i.e.

$$x^2+y^2=a^2+b^2.$$

Hence perpendicular tangents to an ellipse intersect on a circle concentric with the ellipse and with radius $\sqrt{(a^2+b^2)}$. This circle is known as the director circle of the ellipse.

14.8. The locus of the mid-points of parallel chords of an ellipse

The coordinates of M, the mid-point of the chord $[\theta, \phi]$ are

$$x=\tfrac{1}{2}a(\cos\theta+\cos\phi)=a\cos\tfrac{1}{2}(\theta+\phi)\cos\tfrac{1}{2}(\theta-\phi),$$
$$y=\tfrac{1}{2}b(\sin\theta+\sin\phi)=b\sin\tfrac{1}{2}(\theta+\phi)\cos\tfrac{1}{2}(\theta-\phi).$$

If $\theta+\phi=2\alpha=$constant, the coordinates of M become

$$x=a\cos\alpha\cos\tfrac{1}{2}(\theta-\phi),\quad y=b\sin\alpha\cos\tfrac{1}{2}(\theta-\phi).$$

Eliminating the variable $\cos\frac{1}{2}(\theta-\phi)$ between these equations, we see that the locus of M is the straight line

$$y/x=(b/a)\tan\alpha \quad . \quad . \quad . \quad . \quad \text{(i)}$$

which, since it passes through the centre of the curve, is called a *diameter* of the ellipse.

But (see § 14.3) by joining points $[\theta]$ and $[\phi]$ on the ellipse such that $\theta+\phi=2\alpha$ we obtain a set of chords parallel to the diameter $y/x=-(b/a)\cot\alpha$. Hence the mid-points of these parallel chords lie on (i).

It follows that the mid-points of chords parallel to the diameter $y=mx$ lie on the diameter $y=m'x$ where

$$mm'=-b^2/a^2. \tag{14.5}$$

14.9. Conjugate diameters

From the symmetry of (14.5), if a diameter of gradient m' bisects all chords of the ellipse of gradient m, the diameter of gradient m will bisect all chords of gradient m'.

The diameters $y=mx$, $y=m'x$ which are such that each bisects all chords parallel to the other are called *conjugate diameters* and their gradients are connected by the relation (14.5).

If $A[\theta]$ and $B[\phi]$ are the extremities of conjugate diameters of an ellipse, centre O, the gradients of OA and OB are $(b/a)\tan\theta$, $(b/a)\tan\phi$, where by (14.5)

$$(b^2/a^2)\tan\theta\tan\phi=-b^2/a^2$$
$$\cos\theta\cos\phi+\sin\theta\sin\phi=0,$$
$$\cos(\theta-\phi)=0,$$
$$\theta-\phi=\pm\tfrac{1}{2}\pi.$$

Hence the eccentric angles of extremities of conjugate diameters differ by $\frac{1}{2}\pi$.

Thus if AOA', BOB' are conjugate diameters, we may take θ, $\theta+\pi$, $\theta+\frac{1}{2}\pi$, $\theta-\frac{1}{2}\pi$ as the eccentric angles of A, A', B, B' respectively.

If, in addition, $OA=OB$, the diameters are said to be equiconjugate and

$$a^2\cos^2\theta+b^2\sin^2\theta=a^2\sin^2\theta+b^2\cos^2\theta,$$

$$\therefore\ (a^2-b^2)(\cos^2\theta-\sin^2\theta)=0,$$

$$\tan\theta=\pm 1.$$

Hence the equations of the equiconjugate diameters are $y=\pm(b/a)x$.

The axes of the ellipse are perpendicular conjugate diameters.

14.10 The chord of the ellipse which is bisected at (α, β)

The line

$$\frac{x-\alpha}{\cos\theta}=\frac{y-\beta}{\sin\theta}=r \quad . \quad . \quad . \quad . \quad \text{(i)}$$

meets the ellipse $b^2x^2+a^2y^2=a^2b^2$ in points A and B whose distances r_1 and r_2 from P (α, β) are the roots of the equation

$$b^2(\alpha+r\cos\theta)^2+a^2(\beta+r\sin\theta)^2=a^2b^2,$$

$$r^2(b^2\cos^2\theta+a^2\sin^2\theta)+2r(b^2\alpha\cos\theta+a^2\beta\sin\theta)+(b^2\alpha^2+a^2\beta^2-a^2b^2)=0.$$

If P is mid-point of AB, these roots will be equal in magnitude and opposite in sign

$$\therefore\ b^2\alpha\cos\theta+a^2\beta\sin\theta=0 \quad . \quad . \quad . \quad \text{(ii)}$$

Eliminating θ between (i) and (ii), we obtain the equation of the chord bisected at P:

$$\frac{\alpha x}{a^2}+\frac{\beta y}{b^2}=\frac{\alpha^2}{a^2}+\frac{\beta^2}{b^2}.$$

14.11. Miscellaneous examples

Example 1

Show that the equation

$$20x^2+36y^2+40x-108y-79=0$$

represents an ellipse, and find its eccentricity and the coordinates of its centre and foci.

The given equation may be written

$$20(x+1)^2+36(y-\tfrac{3}{2})^2=180,$$

and reduced to

$$X^2/9+Y^2/5=1 \quad . \quad . \quad . \quad \text{(i)}$$

by the substitutions

$$X=x+1,\quad Y=y-\tfrac{3}{2},$$

i.e. by transferring the origin to $(-1, \frac{3}{2})$ as in § 12.3 (a).

(i) is the equation of an ellipse whose eccentricity e is given by

$$5=9(1-e^2),$$
$$e=\tfrac{2}{3}$$

The centre is $X=0,\ Y=0,$

i.e. $x=-1,\ y=\tfrac{3}{2}.$

The coordinates of the foci are

$$X=\pm ae,\ Y=0$$
$$x+1=\pm 2,\ y-\tfrac{3}{2}=0.$$

Hence the centre is $(-1, \tfrac{3}{2})$ and the foci are $(1, \tfrac{3}{2})$, $(-3, \tfrac{3}{2})$.

Example 2

Common tangents are drawn to the ellipse $x^2/a^2+y^2/b^2=1$ and the circle $x^2+y^2=c^2$, where $0<b<c<a$. Prove that the eccentric angles of the points where the tangents touch the ellipse are given by the equation

$$\tan^2\theta=b^2(c^2-a^2)/a^2(b^2-c^2).$$ [Sheffield.]

The equation of the tangent at $[\theta]$ to the ellipse $\dfrac{x^2}{a^2}+\dfrac{y^2}{b^2}=1$ is

$$\frac{x}{a}\cos\theta+\frac{y}{b}\sin\theta=1 \quad . \quad . \quad . \quad \text{(i)}$$

The equation of the tangent at $[\phi]$ to the circle $\dfrac{x^2}{c^2}+\dfrac{y^2}{c^2}=1$ is

$$\frac{x}{c}\cos\phi+\frac{y}{c}\sin\phi=1 \quad . \quad . \quad . \quad . \quad \text{(ii)}$$

(i) and (ii) represent the same line if

$$\frac{c\cos\theta}{a\cos\phi}=\frac{c\sin\theta}{b\sin\phi}=1,$$

i.e. if $\cos\phi=(c/a)\cos\theta,\ \sin\phi=(c/b)\sin\theta.$

Hence

$$(\cos^2\theta)/a^2+(\sin^2\theta)/b^2=1/c^2,$$
$$(1/b^2-1/c^2)\tan^2\theta=1/c^2-1/a^2,$$
$$\tan^2\theta=b^2(c^2-a^2)/a^2(b^2-c^2). \quad . \quad . \quad \text{(iii)}$$

The eccentric angles of the points where the common tangents touch the ellipse are given by (iii).

Example 3

P and Q are variable points of the ellipse $x^2/a^2+y^2/b^2=1$ such that the lines joining them to the centre are perpendicular, and the tangents at P and Q meet at T. Find the locus of T. [L.U.]

Let T be the point (x_1, y_1); then by (14.2) the equation of PQ is

$$\frac{xx_1}{a^2}+\frac{yy_1}{b^2}=1$$

and by § 12,8 the equation of the lines joining the centre of the ellipse to P and Q is

$$\frac{x^2}{a^2}+\frac{y^2}{b^2}-\left(\frac{xx_1}{a^2}+\frac{yy_1}{b^2}\right)^2=0.$$

By (12.4), page 276, these lines are perpendicular if

$$\left(\frac{1}{a^2}-\frac{x_1^2}{a^4}\right)+\left(\frac{1}{b^2}-\frac{y_1^2}{b^4}\right)=0,$$

i.e. if T lies on the locus

$$\frac{x^2}{a^4}+\frac{y^2}{b^4}=\frac{1}{a^2}+\frac{1}{b^2}.$$

Example 4

An ellipse with principal axes 2a, 2b slides between two fixed straight lines which are at right angles to one another. Show that the locus of its centre is a circle. [L.U.]

The two perpendicular lines touch the ellipse and so P, their point of intersection, lies on the director circle whose centre C is the centre of the ellipse, and whose radius is $\sqrt{(a^2+b^2)}$.

$$\therefore CP=\sqrt{(a^2+b^2)}=\text{constant}.$$

But P is fixed, hence C lies on a circle, centre P and radius $\sqrt{(a^2+b^2)}$.

Exercises 14 (*a*)

1. The foci of an ellipse of eccentricity 12/13 are the points $(0, \pm 12)$. Show that the equation of the ellipse is $x^2/5^2+y^2/13^2=1$.

   Find the equations of the tangent and normal to the ellipse at the point $(5 \cos \phi, 13 \sin \phi)$ on it. If the tangent and normal meet the x-axis in the points T, N, show that $ON.OT$ is constant, O being the origin. [Sheffield.]

2. The tangent and normal to an ellipse at a point T cut the minor axis at points P, Q respectively. Show that the circle described on PQ as diameter passes through the foci of the ellipse. [Durham.]

3. If the point P on the ellipse $x^2/a^2+y^2/b^2=1$ has the eccentric angle θ, find the equations of the lines PA, PA', joining P to the ends A, A' of the major axis.

   The lines through P perpendicular to PA, PA' meet the major axis in K, K'; show that the length KK' is constant. [L.U.]

4. The tangent at the point $P(a \cos \theta, b \sin \theta)$ on the ellipse $x^2/a^2+y^2/b^2=1$ meets the tangents at the ends of the major axis at M and M'. Show that the two foci of the ellipse lie on the circle with MM' as diameter, and that its area is $\pi (a^2+b^2 \cot^2 \theta)$. [Durham.]

5. A straight rod PQ has length $a+b$, and R is the point on it such that $QR=a$, $RP=b$. The rod moves so that its ends P, Q slide on two perpendicular straight lines OX, OY respectively. Find with respect to axes OX, OY, the equation of the curve traced by R.

   Show that the rod is tangential to the curve only when it is inclined to OX at an angle $\tan^{-1}\sqrt{(b/a)}$. [Sheffield.]

6. Show that tangents at corresponding points on an ellipse and its auxiliary circle meet on the major axis of the ellipse, while normals at the same points intersect on a circle concentric with the ellipse.

7. If two variable points $P[\theta]$, $Q[\phi]$ move on the ellipse $x^2/a^2+y^2/b^2=1$ in such a way that $(\theta-\phi)$ is constant and equal to 2α, show that the locus of the mid-point of PQ is the ellipse $x^2/a^2+y^2/b^2=\cos^2\alpha$. Show also that the tangents to the given ellipse at P and Q intersect on the ellipse $x^2/a^2+y^2/b^2=\sec^2\alpha$.

8. The distance between the foci of an ellipse is 8 units and that between the directrices is 18 units. Find the equation of the ellipse referred to its principal axes.

   A tangent to this ellipse which is equally inclined to the axes meets them in A and B. Prove that $AB=4\sqrt{7}$. [L.U. Anc.]

9. The normal at the point $P(a\cos\theta, b\sin\theta)$ on the ellipse $x^2/a^2+y^2/b^2=1$ meets the major axis at Q, and R is the point of the straight line through P and Q such that $PR=kPQ$, where k may be negative. Prove that the locus of R is another ellipse, except for $k=1$ or a^2/b^2, when the locus is a straight line. Show also that the locus is a circle for $k=\pm a/b$. [Sheffield.]

10. PFQ and $PF'Q'$ are two focal chords of an ellipse, the foci being F, F', and the eccentric angles of Q and Q' are ϕ and ϕ'. Show that the ratio $\tan\frac{1}{2}\phi : \tan\frac{1}{2}\phi'$ is constant for all positions of P. [L.U.]

11. The normal to an ellipse meets the major axis at G and the minor axis at g. Show that the locus of the mid-point of Gg is the ellipse
$$4(a^2x^2+b^2y^2)=(a^2-b^2)^2.$$

12. Show that the straight line $x\cos\alpha+y\sin\alpha=p$ touches the ellipse $x^2/a^2+y^2/b^2=1$ if $p^2=a^2\cos^2\alpha+b^2\sin^2\alpha$. Hence show that the foot of the perpendicular drawn from the centre of the ellipse to the tangent at any point lies on the curve whose polar equation is
$$r^2=a^2\cos^2\theta+b^2\sin^2\theta.$$

13. Find the condition that the line $y=mx+c$ touches the ellipse
$$x^2/a^2+y^2/b^2=1.$$

   If H, H' be points on the minor axis of an ellipse such that $HC=CH'=CS$, where C is the centre and S is a focus, show that the sum of the squares of the perpendiculars from H and H' on any tangent to the ellipse is constant. [Leeds.]

14. Prove that the feet of the perpendiculars drawn from the foci of an ellipse to any tangent lie on the auxiliary circle of the ellipse.

15. The line $lx+my=1$ meets the ellipse $x^2/a^2+y^2/b^2=1$ in points A, B. Show that the equation $x^2(l^2-1/a^2)+2lmxy+y^2(m^2-1/b^2)=0$ represents the pair of straight lines joining A, B to the origin.

   If the chord AB subtends a right angle at the origin, and if the tangents at A, B to the ellipse meet in P, show that the locus of P is an ellipse. [Leeds.]

16. TP and TQ are tangents drawn to the ellipse $x^2/a^2+y^2/b^2=1$. If M is the mid-point of PQ, show that the centre of the ellipse lies on TM. Show also that, if M moves on the ellipse $x^2/a^2+y^2/b^2=1/c^2$, the locus of T is the ellipse $x^2/a^2+y^2/b^2=c^2$.

17. Find the equation of the chord joining two points on the ellipse $x^2/a^2+y^2/b^2=1$ with eccentric angles θ_1 and θ_2, respectively, and deduce that, for a system of parallel chords, the sum of the eccentric angles at the extremities of any chord is constant.

Two points P, P' on the ellipse $x^2/a^2+y^2/b^2=1$ have eccentric angles θ and $2\alpha-\theta$, respectively. The tangents at P and P' meet at Q. Show that the coordinates of Q are

$$\{a \cos \alpha/\cos (\alpha-\theta),\ b \sin \alpha/\cos (\alpha-\theta)\}.$$

Deduce that, for a system of parallel chords of an ellipse, the tangents at the extremities of the chords intersect on a straight line through the centre of the ellipse. [Sheffield.]

18. The normal at a point P on an ellipse meets the major axis at G. Prove that $SG=e\,SP$ and $S'G=e\,S'P$, where S and S' are the foci. Deduce that SP, $S'P$ are equally inclined to the tangent (or normal) at P.

19. The point $P(x, y)$ lies on the ellipse

$$x^2/a^2+y^2/(a^2-c^2)=1 \quad (0<c<a)$$

and S is the point $(c, 0)$. Show that $SP=(a^2-cx)/a$ and hence prove that the sum of the focal distances of a point on the ellipse is equal to $2a$.

Show that the product of the focal distances of P is

$$2a^2-c^2-x^2-y^2.$$ [Durham.]

20. Find the equation of the normal to the ellipse $x^2/a^2+y^2/b^2=1$ at the point $P(a \cos \theta, b \sin \theta)$. This normal meets the ellipse again at Q and the tangents at P and Q meet $R(\xi, \eta)$. By comparing the two forms of the equation of PQ, in terms of θ and in terms of ξ, η, respectively, or otherwise, prove that the locus of R, when P varies on the ellipse, is the curve

$$x^2y^2(a^2-b^2)^2=a^6y^2+b^6x^2.$$ [Sheffield.]

21. The foci of an ellipse are S, S', and P is any point on the curve. If the normal at P intersects the line SS' at G, prove that

$$PG^2=SP.S'P(1-e^2)$$

where e is the eccentricity of the ellipse. [L.U.]

22. The perpendicular from the centre of the ellipse to the tangent at a variable point P meets the line joining P to the focus $S(ae, 0)$ in G. Show that the locus of G is a circle of radius a whose centre is at S. [L.U. Anc.]

23. A point P moves so that the chord of contact of the tangents from P to the ellipse $b^2x^2+a^2y^2=a^2b^2$ touches the ellipse $4(b^2x^2+a^2y^2)=a^2b^2$. Find the locus of P. [L.U.]

24. If the normal at a variable point P on the central conic $ax^2+by^2=1$ meets Ox, Oy in G, g respectively, and if a point Q is taken on the normal so that $GQ : Qg=k : l$, prove that the locus of Q is the conic

$$k^2bx^2+l^2ay^2=k^2l^2(a-b)^2/ab(k+l)^2.$$ [L.U.]

25. The normal at P to the ellipse $x^2/a^2+y^2/b^2=1$ meets the x-axis at G. If Q is the point of intersection of the line through G parallel to the y-axis and the line joining P to the centre of the ellipse, show that the equation of the locus of Q is

$$x^2/a^2+y^2/b^2=(a^2-b^2)^2/a^4.$$ [L.U.]

26. The normal at the point $P(a\cos\theta, b\sin\theta)$ on the ellipse $x^2/a^2+y^2/b^2=1$ meets the x-axis at G and the tangent meets the y-axis at T. Find the coordinates of G and T and show that the locus of the circumcentre of the triangle OGT is

$$16a^2x^2y^2-4(a^2-b^2)^2y^2+b^2(a^2-b^2)^2=0.$$ [L.U.]

27. Find the equations of the tangents to the ellipse

$$x^2/a^2+y^2/b^2=1$$

which are parallel to the line $y=mx$.

Prove that any point from which the tangents drawn to the above ellipse are equally inclined to the line $y=x\tan\alpha$ lies on the hyperbola

$$x^2-2xy\cot 2\alpha-y^2=a^2-b^2.$$ [L.U.]

28. Show that the line $y=mx+c$ touches the ellipse

$$x^2/a^2+y^2/b^2=1$$

if $c^2=a^2m^2+b^2$. Hence, or otherwise, show that the equation of the pair of tangents from the point $P(\alpha, \beta)$ may be expressed in the form

$$(\beta x-\alpha y)^2=a^2(y-\beta)^2+b^2(x-\alpha)^2.$$

These tangents cut the x-axis at the points A and B.

(i) If PA and PB are perpendicular, find the locus of P.

(ii) If the mid-point of AB is the fixed point $(k, 0)$ show that the locus of P is the parabola $ky^2=b^2(k-x)$. [L.U.]

29. If CP, CQ are conjugate semi-diameters of the ellipse $b^2x^2+a^2y^2=a^2b^2$, show that

(i) $CP^2+CQ^2=a^2+b^2$;

(ii) the smallest possible value of the *acute* angle between CP and CQ is $\tan^{-1}\{2ab/(a^2-b^2)\}$. [Durham.]

30. Prove that the line $lx+my+n=0$ touches the ellipse $x^2/a^2+y^2/b^2=1$ if $a^2l^2+b^2m^2=n^2$.

Lines are drawn through the origin perpendicular to the tangents from a point P to the above ellipse. If the lines are conjugate diameters of the ellipse, prove that P lies on the curve

$$a^2x^2+b^2y^2=a^4+b^4.$$ [L.U.]

31. The centre of the ellipse $x^2/a^2+y^2/b^2=1$ is C and CP, CQ are a pair of conjugate semi-diameters. If the eccentric angle at P is θ and the chord PQ passes through the point $(\sqrt{2}a, 0)$, prove that
$$\theta=\pm\pi/12, \quad \pm 7\pi/12.$$ [L.U.]

32. Define conjugate diameters of an ellipse and prove that if PCP', DCD' are conjugate diameters the eccentric angles of P and D differ by a right angle.

Prove that, as P moves round the ellipse $x^2/a^2+y^2/b^2=1$, the locus of the intersection of the normals at P and D is the curve
$$2(a^2x^2+b^2y^2)^3=(a^2-b^2)^2(a^2x^2-b^2y^2)^2.$$ [L.U.]

33. D_1 and D_2 are two points on the ellipse $x^2/a^2+y^2/b^2=1$ and the tangents at these points meet at P. If P is the point (h, k), obtain the combined equation of the diameters CD_1 and CD_2. Find the locus of P (i) when the diameters are perpendicular; (ii) when they are conjugate. Show that the two loci meet in four points, which are the vertices of a rectangle. [L.U.]

34. If P is a point on the director circle of the ellipse $x^2/a^2+y^2/b^2=1$ and if the polar of P with respect to the ellipse meets the ellipse in M and N, show that the mid-point of MN lies on the curve
$$(x^2/a^2+y^2/b^2)^2=(x^2+y^2)/(a^2+b^2).$$ [L.U.]

35. Show that the straight line
$$x\cos\alpha+y\sin\alpha=\sqrt{(a^2\cos^2\alpha+b^2\sin^2\alpha)}$$
is a tangent to the ellipse
$$b^2x^2+a^2y^2=a^2b^2.$$

Find the coordinates of the point of contact and the equation of the corresponding normal.

Show that the rectangle contained by the perpendiculars drawn from the foci of an ellipse to the normal at the point P of the ellipse is proportional to the square of the perpendicular drawn from P to the major axis. [L.U.]

36. Show that the equation of the chord of the ellipse $b^2x^2+a^2y^2=a^2b^2$ whose mid-point P is (α, β) is $x\alpha/a^2+y\beta/b^2=\alpha^2/a^2+\beta^2/b^2$.

Show that, if the pole of this chord lies on the circle $x^2+y^2=a^2$, then P is on the curve $a^2b^4(x^2+y^2)=(b^2x^2+a^2y^2)^2$ and find the point P corresponding to the point $(a/\sqrt{2}, a/\sqrt{2})$ on the circle. [L.U.]

37. Find the equation of the chord of the ellipse $x^2/a^2+y^2/b^2=1$ whose mid-point is (ξ, η).

Show that if the mid-point of the chord lies on the fixed line $lx+my+n=0$, the locus of the pole of the chord is the ellipse $n(x^2/a^2+y^2/b^2)+lx+my=0$. [L.U.]

38. Show that the locus of the mid-point of a chord of the ellipse which subtends a right angle at the centre is
$$(x^2/a^2+y^2/b^2)^2(1/a^2+1/b^2)=x^2/a^4+y^2/b^4.$$ [L.U.]

39. Find the equation of the normal to the ellipse $x^2/a^2+y^2/b^2=1$ at the point $(a\cos\theta, b\sin\theta)$, and find the pole of this normal.

Denoting this pole by P, and the pole of the same normal with respect to the ellipse $x^2/(a^2+n^2b^2)+y^2/(b^2+n^2a^2)=1$ by Q, show that for all values of n, the straight line PQ touches the ellipse whose equation is

$$a^2x^2+b^2y^2=(a^2+b^2)^2.$$ [L.U.]

40. Tangents are drawn to an ellipse of eccentricity e from a point on a concentric circle. Prove that the chord of contact of the tangents touches a concentric coaxal ellipse of eccentricity $e\sqrt{(2-e^2)}$. [L.U.]

41. If the polar of a point P with respect to the ellipse $x^2/a^2+y^2/b^2=1$ is a tangent to the circle $x^2+y^2=b^2$, show that the locus of P is a concentric ellipse with the same minor axis, and find the points P, for which the tangent makes a positive angle $\frac{1}{4}\pi$ with the x-axis. [L.U.]

42. If the ellipse $x^2/a^2+y^2/b^2=1$ be denoted by S, and the ellipse $x^2/a^4+y^2/b^4=1/c^2$ be denoted by S', show that the polars of points on S' with respect to S touch the circle $x^2+y^2=c^2$, and also that the polars of points on this circle with respect to S touch the ellipse S'. [L.U.]

43. A straight line moves so that its intercept between the axes is always of length c. Show that the locus of its pole with respect to the ellipse $x^2/a^2+y^2/b^2=1$ has the equation $a^4/x^2+b^4/y^2=c^2$. [Leeds.]

44. Find the equation of the chord of the ellipse $(x/a)^2+(y/b)^2=1$ which is bisected at the point (h, k), and find also the coordinates of its pole with respect to the ellipse.

Find the locus of this pole if (h, k) lies on the ellipse $x^2/a^2+y^2/b^2=c^2$. [L.U.]

45. If a chord AB of the ellipse $x^2/a^2+y^2/b^2=1$ cuts the x-axis at P and the y-axis at Q, and if $a^2/OP+b^2/OQ=$constant, where O is the origin, find the locus of the mid-point of AB. [L.U.]

46. Any point P is taken on an ellipse whose foci are A, B. Q is a point on the bisector of the angle APB and L is the foot of the perpendicular from P on to AB, produced if necessary. From Q the perpendicular QM is drawn to PA. If $PL=QM$ prove that the length of PM has the constant value b^2/ae where a, b are the principal semi-axes and e is the eccentricity. [L.U.]

47. Show that λ can be found so that

$$x^2/a^2+y^2/b^2-1+\lambda(lx+my-1)(lx-my-c)=0$$

represents, for any value of c, a circle through the intersections of the ellipse $x^2/a^2+y^2/b^2=1$ and the straight line $lx+my=1$.

Show that, if P and Q are the points of contact of the tangents to an ellipse drawn from a point L on an equiconjugate diameter produced, then the circle through P, Q and the centre of the ellipse also passes through L. [L.U.]

48. Express $\cos\theta$ and $\sin\theta$ in terms of $t=\tan\frac{1}{2}\theta$ and deduce that, in general, four normals can be drawn to an ellipse from a given point.
If the given point is $(0, k)$ and four real normals can be drawn, prove that the two normals, which do not lie along the minor axis, cut the ellipse at points on the line

$$(a^2-b^2)y+b^2k=0.$$ [L.U.]

49. If PP' is a chord of the ellipse drawn parallel to the y-axis, and the normal at P meets the diameter through P' at Q, prove that the locus of the mid-point of PQ is the ellipse

$$x^2/a^6+y^2/b^6=1/(a^2+b^2)^2.$$ [L.U.]

50. Find the equation of the two lines joining the origin O to the points of intersection P, Q of the line $lx+my=1$ and the ellipse $x^2-xy+y^2=1$.
Prove that, if the angle POQ is a right angle, the line PQ touches the circle $x^2+y^2=\frac{1}{2}$. [L.U.]

14.12. The hyperbola (revision)

A hyperbola is a conic whose eccentricity e is greater than unity. The standard equation of the hyperbola is

$$\frac{x^2}{a^2}-\frac{y^2}{b^2}=1 \tag{14.6}$$

where $b^2=a^2(e^2-1)$.

When the equation of the curve is in this form,

(i) the foci S, S_1 are the points $(\pm ae, 0)$,

(ii) the directrices ZM, Z_1M_1 are the lines $x=\pm a/e$,

(iii) the eccentricity e, greater than unity, is given by $e^2=1+b^2/a^2$,

(iv) the centre O is the origin of coordinates,

(v) the transverse axis AA_1 is of length $2a$ and lies along the x-axis,

(vi) the conjugate axis BB_1 (which is not a chord of the hyperbola) is of length $2b$ and lies along the y-axis, with its centre at the origin,

(vii) the asymptotes of the curve are the lines $\dfrac{x^2}{a^2}-\dfrac{y^2}{b^2}=0$,

(viii) each latus rectum is of length $2b^2/a$.

The hyperbola $\dfrac{x^2}{a^2}-\dfrac{y^2}{b^2}=1$ is shown in fig. 40.

The difference of the focal distances of a point on a hyperbola is constant and equal to the length of the transverse axis. Hence the

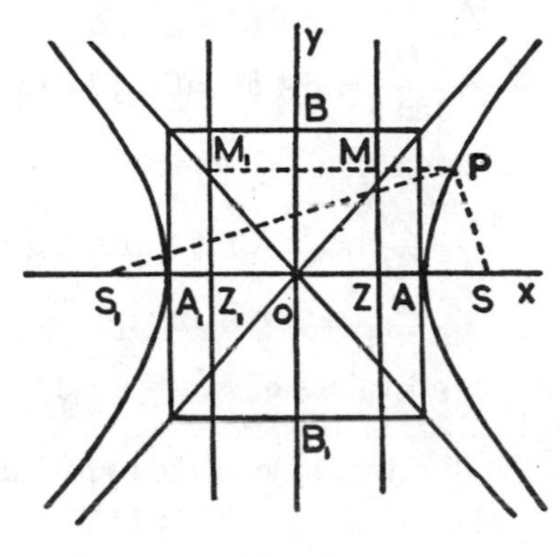

Fig. 40

hyperbola may also be defined as the locus of a point which moves in a plane so that the difference of its distances from two fixed points is constant.

14.13. Parametric representation

The point whose coordinates are

$$x=a \sec \theta, \quad y=b \tan \theta \quad . \quad . \quad . \quad . \quad \text{(i)}$$

lies on the hyperbola $\dfrac{x^2}{a^2}-\dfrac{y^2}{b^2}=1$ for all values of θ, and so equations (i) may be taken as the parametric equations of the hyperbola. We shall use $[\theta]$ to denote the point $(a \sec \theta, b \tan \theta)$ and the chord $[\theta, \phi]$ to denote the chord joining $[\theta]$ and $[\phi]$.

Other parametric equations of the hyperbola are

$$x= \pm a \cosh u, \quad y=b \sinh u$$

and

$$x=\tfrac{1}{2}a(t+1/t), \quad y=\tfrac{1}{2}b(t-1/t).$$

14.14. Standard results

The reader should establish the following results, using the methods indicated.

The equation of the chord $[\theta, \phi]$ is

$$\frac{x}{a} \cos \tfrac{1}{2}(\theta-\phi)-\frac{y}{b} \sin \tfrac{1}{2}(\theta+\phi)=\cos \tfrac{1}{2}(\theta+\phi). \quad \text{(Cf. § 14.3.)}$$

The equations of the tangent and normal at $[\theta]$ are respectively

$$\frac{x}{a}\sec\theta-\frac{y}{b}\tan\theta=1$$

and
$$\frac{ax}{\sec\theta}+\frac{by}{\tan\theta}=a^2+b^2. \quad \text{(Cf. § 14.4.)}$$

The equation of the tangent at (x_1, y_1) is

$$\frac{xx_1}{a^2}-\frac{yy_1}{b^2}=1. \quad \text{(Cf. § 12.10.)}$$

The lines
$$y=mx\pm\sqrt{(a^2m^2-b^2)}$$
touch the hyperbola for all values of m. (Cf. § 14.6.)

The lines are real only if $m^2\geqslant b^2/a^2$.

The equation of the director circle of the hyperbola is

$$x^2+y^2=a^2-b^2. \quad \text{(Cf. § 14.7.)}$$

The circle is real only if $b<a$.

The equation of PQ, the chord of contact of tangents drawn to the hyperbola from $T(x_1, y_1)$, is

$$\frac{xx_1}{a^2}-\frac{yy_1}{b^2}=1. \quad \text{(Cf. § 12.12.)}$$

PQ is called the polar of T with respect to the hyperbola, and T is called the pole of PQ.

14.15. Conjugate diameters of a hyperbola

The equation of the chord of the hyperbola which is bisected at (α, β) is

$$\frac{\alpha x}{a^2}-\frac{\beta y}{b^2}=\frac{\alpha^2}{a^2}-\frac{\beta^2}{b^2}. \quad \text{(Cf. § 14.10.)}$$

The gradient m of this chord is given by

$$m=b^2\alpha/a^2\beta.$$

$\therefore$ (α, β), the mid-point of the chord, lies on the line

$$y=\frac{b^2}{a^2m}x,$$

which, since it passes through the centre of the curve, is called a *diameter* of the hyperbola.

Hence the mid-points of all chords of gradient m lie on the diameter $y=m'x$ where

$$mm'=b^2/a^2. \qquad (14.7)$$

From the symmetry of this relation it follows that the mid-points of chords of gradient m' lie on the diameter $y=mx$.

The diameters $y=mx$, $y=m'x$ are said to be *conjugate* and their gradients are connected by the relation (14.7).

14.16. The conjugate hyperbola

Two hyperbolas are said to be *conjugate* if the transverse and conjugate axes of the one coincide respectively with the conjugate and transverse axes of the other.

The hyperbola conjugate to

$$\frac{x^2}{a^2}-\frac{y^2}{b^2}=1$$

is

$$\frac{x^2}{a^2}-\frac{y^2}{b^2}=-1.$$

Conjugate hyperbolas have the same centre and asymptotes, and diameters conjugate for one are conjugate for the other.

Two conjugate hyperbolas are shown in fig. 41.

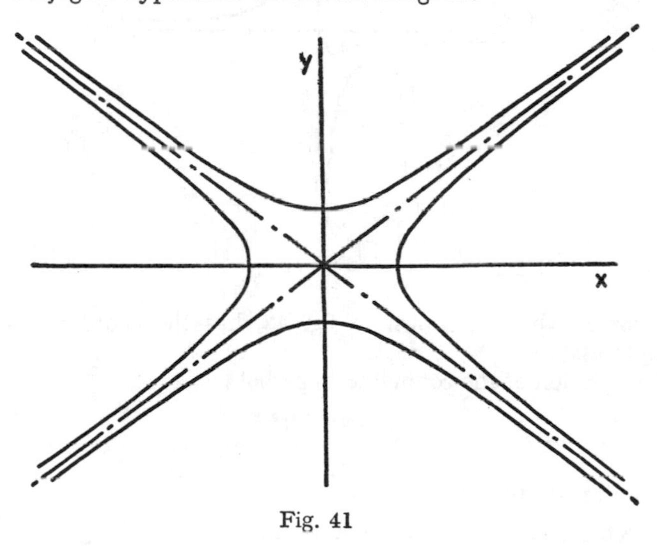

Fig. 41

14.17. The rectangular hyperbola

A hyperbola is said to be rectangular if its asymptotes are perpendicular.

The asymptotes $y=\pm(b/a)x$ of the hyperbola $\dfrac{x^2}{a^2}-\dfrac{y^2}{b^2}=1$ are perpendicular if $b^2=a^2$. In this case, the eccentricity of the hyperbola is $\sqrt{2}$, its equation is $x^2-y^2=a^2$ and its asymptotes $y=\pm x$ bisect the angles between the axes.

14.18. The rectangular hyperbola referred to its asymptotes as axes

If we rotate the coordinate axes through $-45°$ keeping the origin unchanged, the equation $x^2-y^2=a^2$ becomes by (12.2), page 274,

$$\{x \cos(-45°)-y \sin(-45°)\}^2-\{x \sin(-45°)+y \cos(-45°)\}^2=a^2,$$

$$\text{i.e. } 2xy=a^2$$

or

$$xy=c^2 \quad (\text{writing } c^2=\tfrac{1}{2}a^2).$$

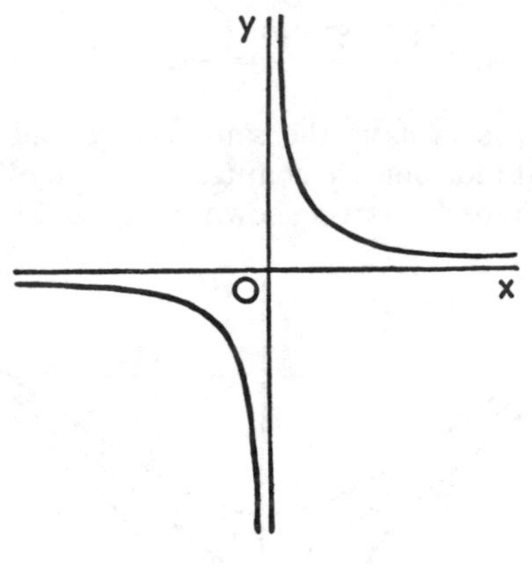

Fig. 42

This curve, which is shown in fig. 42, has the coordinate axes as asymptotes.

The equation of the conjugate hyperbola is

$$xy=-c^2.$$

14.19. Parametric representation

Since the point whose coordinates are

$$x=ct, \quad y=c/t \quad . \quad . \quad . \quad . \quad \text{(i)}$$

lies on the rectangular hyperbola $xy=c^2$ for all values of t except $t=0$, and since the point given by (i) describes the complete hyperbola as t varies, equations (i) are suitable parametric equations of this curve.

It is left as an exercise to the reader to show that the equation of the chord $[t_1, t_2]$ is

$$x+t_1t_2y=c(t_1+t_2).$$

14.20. The tangent and normal at (x_1, y_1) on the hyperbola $xy=c^2$

Using the product rule to differentiate the equation

$$xy=c^2 \quad . \quad . \quad . \quad . \quad . \quad . \qquad \text{(i)}$$

with respect to x, we have

$$x\frac{dy}{dx}+y=0,$$

so that the gradient of the curve at the point (x, y) is given by

$$\frac{dy}{dx}=-\frac{y}{x}.$$

Hence the equation of the tangent at $P(x_1, y_1)$ is

$$y-y_1=-\frac{y_1}{x_1}(x-x_1),$$

i.e. $\qquad xy_1+yx_1=2c^2$, since P lies on (i).

The equation of the normal at P is

$$yy_1-y_1^2=xx_1-x_1^2.$$

Substituting $x_1=ct$, $y_1=c/t$, we obtain the equations of the tangent and normal respectively at $[t]$:

$$x+t^2y=2ct,$$

$$tx-y/t=c(t^2-1/t^2).$$

These may also be obtained by the method shown in § 14.4.

14.21. Conormal points on the rectangular hyperbola $xy=c^2$

The normal at $[t]$ passes through (X, Y) if

$$tX-Y/t=c(t^2-1/t^2).$$

i.e. if $ct^4-Xt^3+Yt-c=0.$

The roots t_1, t_2, t_3, t_4 of this quartic equation are connected by the relation $t_1t_2t_3t_4=-1$ (see § 2.2); hence the product of the abscissae or of the ordinates of four conormal points on a rectangular hyperbola is equal to $-c^4$. Also, the point of concurrence of the normals is

$$X=c(t_1+t_2+t_3+t_4),$$

$$Y=-ct_1t_2t_3t_4(1/t_1+1/t_2+1/t_3+1/t_4)=c(1/t_1+1/t_2+1/t_3+1/t_4).$$

14.22. Concyclic points on the rectangular hyperbola $xy=c^2$

The circle $x^2+y^2+2gx+2fy+k=0$ cuts the hyperbola at $[t]$ if

$$c^2t^2+c^2/t^2+2gct+2fc/t+k=0,$$

i.e. if $\quad c^2t^4+2gct^3+kt^2+2fct+c^2=0.$

The roots t_1, t_2, t_3, t_4 of this equation are connected by the relation $t_1t_2t_3t_4=1$ (see § 2.2), and so the product of the abscissae or of the ordinates of four concyclic points on the hyperbola is c^4.

14.23. Miscellaneous examples

Example 5

A variable chord of the hyperbola $x^2/a^2-y^2/b^2=1$ is a tangent to the circle $x^2+y^2=c^2$. Prove that the locus of the mid-point of the chord is the curve

$$(x^2/a^2-y^2/b^2)^2=c^2(x^2/a^4+y^2/b^4). \qquad \text{[L.U.]}$$

The chord

$$\frac{\alpha x}{a^2}-\frac{\beta y}{b^2}=\frac{\alpha^2}{a^2}-\frac{\beta^2}{b^2} \quad . \quad . \quad . \quad . \quad \text{(i)}$$

of the hyperbola is bisected at (α, β). It is a tangent to the circle $x^2+y^2=c^2$ if the perpendicular distance of the origin from (i) is equal to c

i.e. if
$$\frac{\alpha^2}{a^2}-\frac{\beta^2}{b^2}=\pm c\sqrt{\left(\frac{\alpha^2}{a^4}+\frac{\beta^2}{b^4}\right)}, \text{ by (v) of § 12.1.}$$

Hence (α, β) lies on the curve

$$\left(\frac{x^2}{a^2}-\frac{y^2}{b^2}\right)^2=c^2\left(\frac{x^2}{a^4}+\frac{y^2}{b^4}\right).$$

Example 6

Prove that the line $lx+my+n=0$ touches the hyperbola $b^2x^2-a^2y^2=a^2b^2$ if $a^2l^2-b^2m^2=n^2$.

The circle $x^2+y^2-2\alpha y+\beta=0$ cuts one asymptote of the hyperbola in the distinct points P, P′, and the other in Q, Q′, where P and Q are on the same side of the y axis. If PQ is a tangent to the hyperbola, prove that the circle must pass through the foci of the hyperbola. [L.U.]

The tangent at $[\theta]$ to the hyperbola is $bx \sec\theta-ay\tan\theta=ab$. This equation represents the same line as $lx+my+n=0$ if

$$\frac{b\sec\theta}{l}=-\frac{a\tan\theta}{m}=-\frac{ab}{n},$$

i.e. if $\sec\theta=-al/n$, $\tan\theta=bm/n$.

There is a value of θ satisfying these equations if, and only if,

$$a^2l^2-b^2m^2=n^2 \quad . \quad . \quad . \quad . \quad . \quad \text{(i)}$$

This is the required condition.

Let the equation of PQ (fig. 43) be

$$lx+my+n=0 \quad . \quad . \quad . \quad . \quad \text{(ii)}$$

Then the equation of the line pair OP, OQ joining the origin to the points of intersection of (ii) with the circle

$$x^2+y^2-2\alpha y+\beta=0 \quad . \quad . \quad . \quad . \quad \text{(iii)}$$

is, by § 12.8, $x^2+y^2+2\alpha y(lx+my)/n+\beta(lx+my)^2/n^2=0$,

i.e. $$x^2(n^2+\beta l^2)+y^2(n^2+2\alpha mn+\beta m^2)+2lxy(\alpha n+\beta m)=0 \quad . \quad \text{(iv)}$$

These lines will be asymptotes of the hyperbola if (iv) is identical with

$$b^2x^2-a^2y^2=0,$$

i.e. if $(n^2+\beta l^2)/b^2=-(n^2+2\alpha mn+\beta m^2)/a^2$ and if $\alpha n+\beta m=0$, since $l\neq 0$.

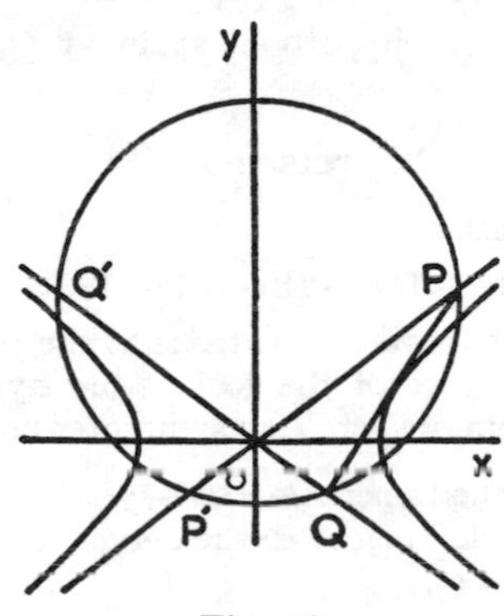

Fig. 43

These relations give

$$a^2(n^2+\beta l^2)+b^2(n^2-\beta m^2)=0,$$
$$n^2(a^2+b^2)+\beta(a^2l^2-b^2m^2)=0$$

i.e. $$n^2a^2e^2+\beta n^2=0$$

by (i), since PQ touches the hyperbola. But $n\neq 0$, since PQ does not pass through the origin.

$$\therefore\ a^2e^2+\beta=0.$$

This is the condition that circle (iii) should pass through the foci $(\pm ae, 0)$ of the given hyperbola, and so if PQ is a tangent to the hyperbola, circle (iii) passes through its foci.

Example 7

$P_i(i=1, 2, 3)$ are the points $(kt_i, k/t_i)$ on the rectangular hyperbola $xy=k^2$. Show that the circumcircle of the triangle $P_1P_2P_3$ cuts the hyperbola at a fourth point P_4 with parameter

$$t_4=1/t_1t_2t_3.$$

$Q_i(i=1, 2, 3, 4)$ are the other points at which the circles of curvature at the points P_i cut the hyperbola. Show that $Q_1Q_2Q_3Q_4$ is a cyclic quadrilateral.

[L.U.]

If the points $P_i[t_i]$, $i=1, 2, 3, 4$ are concyclic, by § 14.22 $t_1t_2t_3t_4=1$. Hence the circle $P_1P_2P_3$ cuts the hyperbola $xy=k^2$ in a fourth point $P_4[t_4]$ where

$$t_4=1/t_1t_2t_3 \quad . \quad . \quad . \quad . \quad . \qquad \text{(i)}$$

The circle of curvature at $P_1[t_1]$ on the hyperbola meets the curve in three coincident points at P_1 (see § 22.10) and at another point Q_1 whose parameter, obtained by substituting $t_2=t_3=t_1$ in (i), is $1/t_1^3$.

Similarly, the parameters of Q_2, Q_3, Q_4 are $1/t_2^3$, $1/t_3^3$, $1/t_4^3$, respectively.

The circle $Q_1Q_2Q_3$ meets the hyperbola at a fourth point $[T]$ where by § 14.22

$$T/t_1^3t_2^3t_3^3=1$$

$$\therefore\ T=t_1^3t_2^3t_3^3=1/t_4^3 \text{ by (i)},$$

i.e. circle $Q_1Q_2Q_3$ cuts the hyperbola again at Q_4. Hence quadrilateral $Q_1Q_2Q_3Q_4$ is cyclic.

Exercises 14 (*b*)

1. Show that the equation

$$9x^2-16y^2-18x+64y-199=0$$

represents a hyperbola whose centre is at the point (1, 2).

Find the co-ordinates of the foci of the hyperbola and the y-co-ordinates of the points where its asymptotes cut the y-axis. [Leeds.]

2. From a focus S of the hyperbola $x^2/a^2-y^2/b^2=1$ the perpendicular SN is drawn to a variable tangent to the curve. Show that the locus of N is the circle $x^2+y^2=a^2$. [Leeds.]

3. (i) The tangent at any point P of the hyperbola

$$x^2/a^2-y^2/b^2=1$$

meets the asymptotes in X and Y. Show that, if O is the origin, the area of the triangle OXY is ab.

(ii) A straight line meets a rectangular hyperbola in the points M, N and its asymptotes in A, B. Prove that $AM=BN$. [Sheffield.]

4. The tangent at any point on the hyperbola $x^2/a^2-y^2/b^2=1$ meets the asymptotes in P, Q. Prove that $OP.OQ=a^2+b^2$, where O is the centre of the hyperbola. Hence, or otherwise, show that P, Q and the foci are concyclic.

5. Prove that, as u varies, the locus of the point $(a \cosh u, b \sinh u)$ is the hyperbola $x^2/a^2-y^2/b^2=1$ and find from first principles the equation of the tangent at that point.

If the tangent meets the lines $x/a \pm y/b=0$ at L and M, find the area of the triangle CLM where C is the centre of the hyperbola.

What is the locus of the mid-point of LM? [L.U.]

6. If the tangent and normal at any point of the hyperbola $x^2/a^2-y^2/b^2=1$ meet the axis of y at P and Q, show that the circle described on PQ as diameter passes through the foci of the hyperbola. [Durham.]

7. Find the coordinates of the points of contact of the tangents from the point $(1, -1)$ to the hyperbola $x^2-15y^2=21$. Also find the coordinates of the pole, with respect to this hyperbola, of the line joining the point $(1,-1)$ to the point $(-4, 11)$. [Leeds.]

8. The normal at the point $(ct_1, c/t_1)$ on the rectangular hyperbola $xy=c^2$ meets the hyperbola again at the point $(ct_2, c/t_2)$. Prove that $t_1^3 t_2=-1$; hence or otherwise show that there is only one line which is normal to the hyperbola at both points of intersection. [L.U.]

9. The tangent to the curve $xy=c^2$ at a point P meets OX, OY at points A, B respectively. Through A and B perpendiculars are drawn to AB and meet OY, OX respectively at C, D. Prove that the line CD is a tangent to the hyperbola. [Durham.]

10. Given a positive constant m, prove that there are two points A, A' on the rectangular hyperbola $xy=c^2$ at which the normals have slope m. Show also that all chords of $xy=c^2$ with slope m subtend a right angle at both A and A'. [Sheffield.]

11. The chord LM of the hyperbola $xy=c^2$ subtends a right angle at a third point N on the hyperbola. Show that the tangent at N is perpendicular to LM. [L.U. Anc.]

12. Prove that the equation of the line joining the points $(ct_1, c/t_1)$, $(ct_2, c/t_2)$ on the rectangular hyperbola $xy=c^2$ is $x+t_1t_2y-c(t_1+t_2)=0$.

A, B, C, D are four points on a rectangular hyperbola. If AB is perpendicular to CD prove that AC is perpendicular to BD and AD is perpendicular to BC. Prove further that three of the points are on one branch of the hyperbola and one on the other branch. [L.U. Anc.]

13. A and B are two points on the hyperbola $xy=c^2$ whose centre is O. The mid-point of AB is M, and the mid-point of OM is N. If N is on the hyperbola, show that the chord AB touches the hyperbola $xy=4c^2$. [L.U.]

14. The tangent to the hyperbola $xy=c^2$ at the point $(ct, c/t)$ intersects the hyperbola $xy=-c^2$ at the points M and N. Show that the tangents at M and N to the hyperbola $xy=-c^2$ intersect on the hyperbola $xy=c^2$. [L.U.]

15. The tangent at any point P of the hyperbola $x^2/a^2-y^2/a^2=1$ meets one of the asymptotes in Q, and M, N are the feet of the perpendiculars from Q on the axes. Prove that MN passes through P. [Sheffield.]

16. Prove that the equation of the tangent to the hyperbola $x^2/a^2-y^2/b^2=1$ at the point $\{\frac{1}{2}a(t+t^{-1}), \frac{1}{2}b(t-t^{-1})\}$ is $x(t^2+1)/a-y(t^2-1)/b=2t$.

A variable tangent to the above hyperbola cuts the asymptotes at L and M. Prove that the locus of the centre of the circle OLM, where O is the origin, is given by the equation

$$4(a^2x^2-b^2y^2)=(a^2+b^2)^2.$$

[L.U.]

17. If A is a fixed point (h, k), not lying on the hyperbola, prove that, in general, four normals can be drawn from A to the hyperbola $x^2/a^2-y^2/b^2=1$, and that their feet all lie on the curve

$$(a^2+b^2)xy-b^2kx-a^2hy=0.$$

Prove that this curve is a rectangular hyperbola, and find its asymptotes. [L.U.]

18. Prove that the point P whose coordinates are

$$x=a \sec \theta, \quad y=a \tan \theta,$$

lies on the rectangular hyperbola $x^2-y^2=a^2$.

If Q is the point whose parameter is $(\theta+\pi/2)$ and $R(x_1, y_1)$ is the mid-point of PQ, prove that

$$y_1/x_1= \sin \theta+\cos \theta$$

and find the locus of R. L.U.]

19. Find the equation of a circle which has the points $P(x_1, y_1)$ and $Q(x_2, y_2)$ as extremities of one diameter.

A and B are points on opposite branches of the rectangular hyperbola $xy=c^2$; the circle on AB as diameter cuts the hyperbola again at C and D.

Prove that CD is a diameter of the hyperbola. [L.U.]

20. Prove that the equation of the normal at the point $(ct, c/t)$ on the rectangular hyperbola $xy=c^2$ is

$$xt^3-yt+c(1-t^4)=0.$$

Prove also that this normal meets the hyperbola again at the point with parameter $-1/t^3$.

Hence, or otherwise, show that the locus of the mid-points of normal chords of the hyperbola is given by the equation $c^2(x^2-y^2)^2+4x^3y^3=0$. [L.U.]

21. A rectangular hyperbola is cut by any circle in four points. Prove that the sum of the squares of the distances of these four points from the centre of the hyperbola is equal to the square on the diameter of the circle [L.U.]

22. AB is a chord of a rectangular hyperbola which subtends a right angle at a fixed point P of the hyperbola. Show that AB must be parallel to a fixed direction, and that the circle on AB as diameter is one of a system of coaxal circles. [L.U.]

23. The position of a point on a rectangular hyperbola is determined by the parameter θ, where the coordinates of the point are $(c \tan \theta, c \cot \theta)$. Find the point of intersection of the tangents at the points whose parameters are θ and $(\theta+\alpha)$, and if $\alpha=\frac{1}{2}\pi$, show that its locus is $x+y=0$. [L.U.]

24. A chord PQ of the rectangular hyperbola $xy=c^2$ meets the asymptotes at R and S. Prove that $PR=SQ$.

A variable tangent to the parabola $y^2=4ax$ meets the hyperbola $xy=c^2$ in two points P, Q. Prove that the locus of the mid-point of PQ is a parabola. [L.U.]

CHAPTER 15

THE STRAIGHT LINE, CIRCLE AND CONIC IN POLAR COORDINATES

15.1. Polar coordinates

Let O be a fixed point and Ox a fixed straight line in a plane. Then the position of any other point P in the plane is uniquely determined by the coordinates $r \equiv OP$ and $\theta \equiv \angle xOP$ which are called the *polar coordinates of P*. Ox is termed the ***initial line*** and O the *pole*; OP is called the ***radius vector*** and θ the ***vectorial angle*** of P.

The angle θ is positive or negative according as it is measured counter-clockwise or clockwise from Ox. If OM makes an angle θ with Ox, the point $P(r, \theta)$ lies on OM at a distance r from O when $r>0$; P lies on MO produced at a distance $|r|$ from O when $r<0$. The convention is sometimes made, notably in connection with the study of complex numbers in the Argand diagram, that only positive values of r shall be used, but this is not convenient in geometry.

In fig. 44, the points $P(1, 60°)$, $Q(-2, 60°)$ and $R(-2, -60°)$ are shown. Note that each of these points may be expressed in terms of other polar coordinates. For example, R is the point $(2, 120°)$ and also $(-2, 300°)$.

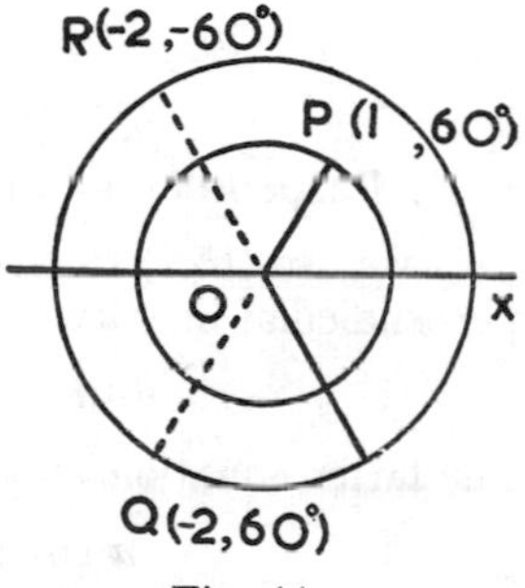

Fig. 44

15.2. Relations between cartesian and polar coordinates

If the cartesian origin and x-axis (fig. 45) are chosen as the pole and initial line respectively of polar coordinates, the cartesian and polar coordinates of a point are connected by the relations

$$x=r \cos \theta, \quad y=r \sin \theta \qquad (15.1)$$

$$r^2=x^2+y^2, \quad \tan \theta=y/x \qquad (15.2)$$

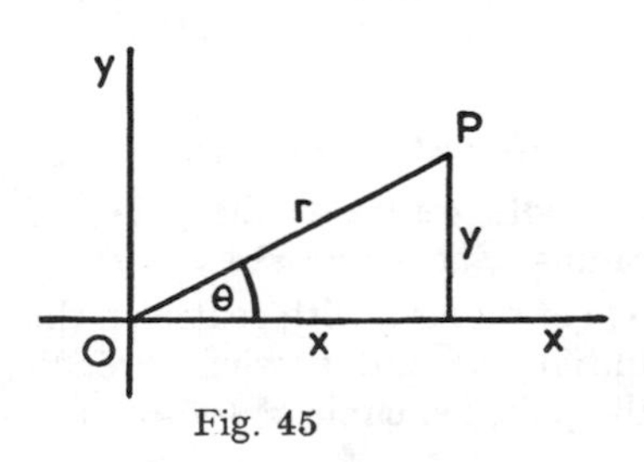

Fig. 45

Equations (15.2) do not determine r, θ uniquely. To obtain unique values of r and θ for a given x and y we may take $r=+\sqrt{(x^2+y^2)}$ and θ as the angle which satisfies the relations $\cos \theta=x/r$, $\sin \theta=y/r$ such that $-\pi<\theta \leqslant \pi$.

15.3. The straight line in polar coordinates

Equation	*Geometric properties*
(i) $\theta=\alpha$	line through the pole inclined at an angle α to the initial line
(ii) $r=a \operatorname{cosec} \theta$	line drawn parallel to the initial line at a distance a from the pole, i.e. $y=a$
(iii) $r=b \sec \theta$	line drawn perpendicular to the initial line at a distance b from the pole, i.e. $x=b$
(iv) $r \cos (\theta-\alpha)=p$	line shown in fig. 46 where $P \equiv (r, \theta)$, the perpendicular $OM=p$ and $\angle xOM=\alpha$ so that $OP \cos (\theta-\alpha)=OM$. The corresponding cartesian equation is $x \cos \alpha+y \sin \alpha=p$.

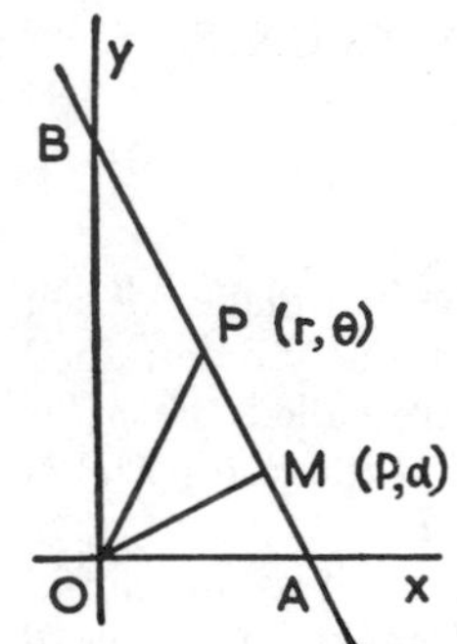

Fig. 46

15.4. Perpendicular lines

If we use (15.1) to change to polar coordinates the equations of the perpendicular lines $ax+by=c$ and $bx-ay=d$, we obtain

$$a \cos \theta+b \sin \theta=c/r \text{ and } b \cos \theta-a \sin \theta=d/r.$$

The latter equation may be written in the form

$$a \cos (\theta+\tfrac{1}{2}\pi)+b \sin (\theta+\tfrac{1}{2}\pi)=d/r.$$

It follows that the equations of a pair of perpendicular lines may be taken as

$$\left.\begin{aligned} A \cos \theta+B \sin \theta&=l/r \\ \text{and} \qquad A \cos (\theta+\tfrac{1}{2}\pi)+B \sin (\theta+\tfrac{1}{2}\pi)&=k/r \end{aligned}\right\} \qquad (15.3)$$

where l and k are any constants.

15.5. The circle

Equation	*Geometric properties*
(i) $r=a=\text{constant}$	Circle with centre at the pole and radius a, i.e. circle $x^2+y^2=a^2$
(ii) $r=2a \cos \theta$	Circle of radius a with centre on the initial line and passing through the pole, i.e. circle $x^2+y^2=2ax$

Equation	*Geometric properties*
(iii) $r=2a \sin \theta$ . . .	Circle of radius a touching the initial line at the pole, i.e. circle $x^2+y^2=2ay$
(iv) $r^2-2\rho r \cos (\theta-\alpha)+\rho^2=a^2$	Circle shown in fig. 47 with centre $C(\rho, \alpha)$ and radius a. If $P(r, \theta)$ is any point on the circle, then $PC^2 = OP^2+OC^2-2OP.OC \cos (\theta-\alpha)$.

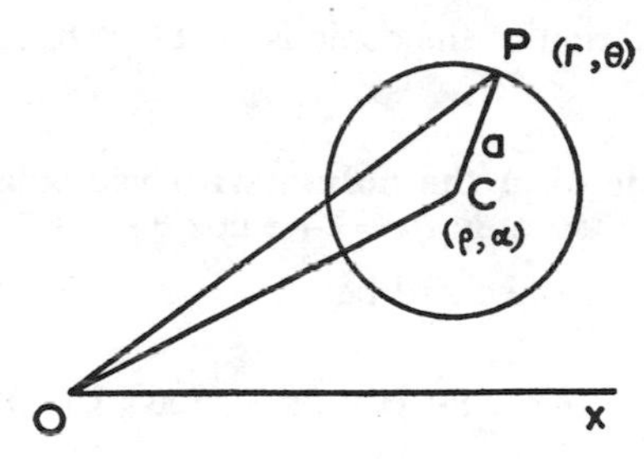

Fig. 47

15.6. The conic

The simplest equation of a conic in polar coordinates is obtained by taking the pole at a focus S and taking as initial line the perpendicular SZ from S to the corresponding directrix DD' of the conic. In fig. 48, $P(r, \theta)$ is any point on a conic which has eccentricity e and semi-latus rectum $LS=l$. LK, PM are drawn perpendicular to DD' and PN is perpendicular to SZ.

Then, by the focus-directrix property of the conic

$$LS=l=e.LK \quad . \quad . \quad \text{(i)}$$

and

$$PS=e.PM \quad . \quad . \quad \text{(ii)}$$

i.e.

$$r=e(LK-SN),$$

$$r=l-er \cos \theta \quad \text{by (i)}$$

$$\therefore \frac{l}{r}=1+e \cos \theta.$$

Fig. 48

When $e=1$, the conic is a parabola and its equation is

$$r \cos^2 \tfrac{1}{2}\theta=a, \text{ where } 2a=l.$$

When $e>1$, the conic is a hyperbola one branch of which is given by the values of θ for which $\cos \theta>-1/e$, the other by values of θ for which $\cos \theta<-1/e$.

Since $SZ=LK=l/e$, it follows from § 15.3 (iii) that the equation of the directrix DD' is $r=(l/e)\sec\theta$,

i.e. $$\frac{l}{r}=e\cos\theta. \qquad (15.4)$$

If, in fig. 48, ZS produced is taken as initial line,

$$SN=r\cos(\pi-\theta)=-r\cos\theta$$

and by (ii) the equation of the conic is $\frac{l}{r}=1-e\cos\theta$.

15.7. The chord joining the points with vectorial angles $(\alpha+\beta)$, $(\alpha-\beta)$ on the conic $l/r=1+e\cos\theta$

Let the equation of the chord be

$$\frac{l}{r}=A\cos\theta+B\sin\theta. \quad \text{(See § 15.4.)}$$

Then if the point $(r_1, \alpha+\beta)$ lies on this chord and on the conic,

$$\frac{l}{r_1}=A\cos(\alpha+\beta)+B\sin(\alpha+\beta)=1+e\cos(\alpha+\beta)$$

$$\therefore\ (A-e)\cos(\alpha+\beta)+B\sin(\alpha+\beta)=1 \qquad \text{(i)}$$

Similarly, $$(A-e)\cos(\alpha-\beta)+B\sin(\alpha-\beta)=1 \qquad \text{(ii)}$$

Solving (i) and (ii) for $(A-e)$ and B, by (3.5) page 40, we have

$$\frac{A-e}{\begin{vmatrix}1 & \sin(\alpha-\beta)\\ 1 & \sin(\alpha+\beta)\end{vmatrix}}=\frac{B}{\begin{vmatrix}\cos(\alpha-\beta) & 1\\ \cos(\alpha+\beta) & 1\end{vmatrix}}=\frac{1}{\begin{vmatrix}\cos(\alpha-\beta) & \sin(\alpha-\beta)\\ \cos(\alpha+\beta) & \sin(\alpha+\beta)\end{vmatrix}},$$

$$\frac{A-e}{2\cos\alpha\sin\beta}=\frac{B}{2\sin\alpha\sin\beta}=\frac{1}{\sin 2\beta}$$

$$\therefore\ A-e=\cos\alpha\sec\beta, \qquad B=\sin\alpha\sec\beta.$$

Thus the equation of the required chord is

$$\frac{l}{r}=(e+\cos\alpha\sec\beta)\cos\theta+\sin\alpha\sec\beta\sin\theta$$

i.e. $$\frac{l}{r}=e\cos\theta+\sec\beta\cos(\theta-\alpha). \qquad (15.5)$$

15.8. The tangent to the conic $l/r=1+e\cos\theta$ at the point of vectorial angle α

If in (15.5) we make $\beta\to 0$, we obtain the equation of the tangent to the given conic at the point of vectorial angle α:

$$\frac{l}{r}=e\cos\theta+\cos(\theta-\alpha). \tag{15.6}$$

15.9. Miscellaneous examples

Example 1

An ellipse of semi-latus rectum l, eccentricity e and foci S and S′ is intersected at P and Q by a parabola of semi-latus rectum el whose axis lies along SS′ and whose focus is at S. The vertex of the parabola lies on the same side of S as does S′. Prove that the area of triangle PSQ is

$$l^2(1-e)\{e(2-e+e^2)\}^{1/2}/(1+e)^{3/2}. \qquad \text{[L.U.]}$$

If S (fig. 49) is chosen as pole, the equation of the ellipse may be taken as

$$l/r=1+e\cos\theta \quad . \quad . \quad . \quad . \quad . \quad . \quad \text{(i)}$$

and the equation of the parabola is then

$$(el)/r=1-\cos\theta \quad . \quad . \quad . \quad . \quad . \quad . \quad \text{(ii)}$$

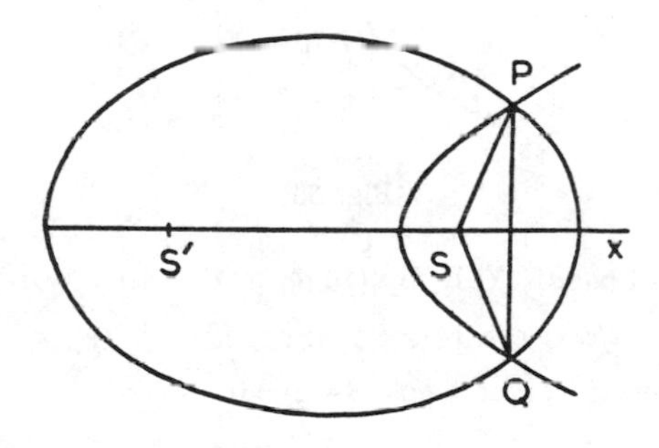

Fig. 49

The curves intersect at points P and Q, whose vectorial angles are given by

$$e(1+e\cos\theta)=1-\cos\theta$$

$$\cos\theta=(1-e)/(1+e^2) \quad . \quad . \quad . \quad . \quad . \quad \text{(iii)}$$

Substituting this value in (i), we obtain

$$SP=SQ=l(1+e^2)/(1+e).$$

The area of $\Delta PSQ=\frac{1}{2}SP.SQ\sin 2\theta$

$$=\{l^2(1+e^2)^2\sin\theta\cos\theta\}/(1+e)^2.$$

But, from (iii), $\sin\theta=\{e(e+1)(2-e+e^2)\}^{1/2}/(1+e^2)$,

hence $\Delta PSQ=l^2(1-e)\{e(2-e+e^2)\}^{1/2}/(1+e)^{3/2}$.

Example 2

P and Q are two points with vectorial angles α, $(\alpha-\frac{1}{3}\pi)$ *where* $\frac{1}{2}\pi<\alpha<\pi$ *on the conic* $l=r(1+e\cos\theta)$, *with focus S. The tangent at P and the chord PQ meet the corresponding directrix at T and D respectively. SQ meets the tangent at P in R. Show that*

$$1/SD+1/SR-\sqrt{3}/ST=1/2l \qquad \text{[L.U.]}$$

The equation of the directrix (fig. 50) corresponding to S is, by (15.4),

$$l/r=e\cos\theta \quad . \quad . \quad . \quad . \quad . \quad \text{(i)}$$

The equation of the tangent at P is, by (15.6),

$$l/r=e\cos\theta+\cos(\theta-\alpha) \quad . \quad . \quad . \quad . \quad \text{(ii)}$$

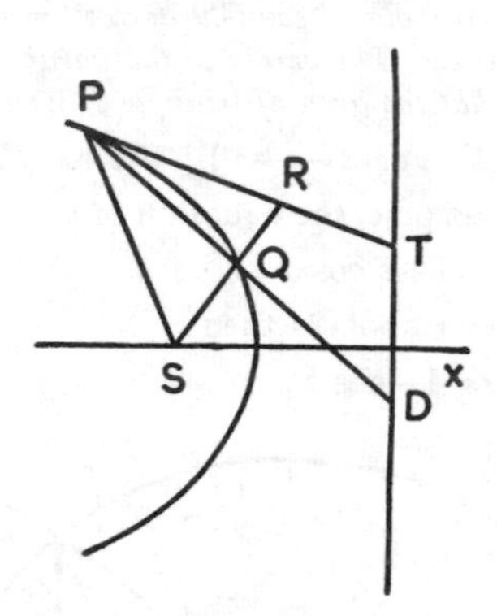

Fig. 50

The equation of chord PQ is, by (15.5),

$$l/r=e\cos\theta+\sec(\tfrac{1}{6}\pi)\cos(\theta-\alpha+\tfrac{1}{6}\pi) \quad . \quad . \quad \text{(iii)}$$

(i) and (ii) meet at T, where $\cos(\theta-\alpha)=0$

$$\theta=\alpha\pm\tfrac{1}{2}\pi.$$

From the figure, $\theta=\alpha-\frac{1}{2}\pi$ at T and

$$l/ST=e\sin\alpha \quad . \quad . \quad . \quad \text{(iv)}$$

(i) and (iii) meet at D, where $\cos(\theta-\alpha+\frac{1}{6}\pi)=0$

$$\theta=\alpha-\tfrac{1}{6}\pi\pm\tfrac{1}{2}\pi.$$

Again from the figure, $\theta=\alpha-\frac{2}{3}\pi$ and

$$l/SD=e\cos(\alpha-\tfrac{2}{3}\pi)=\tfrac{1}{2}e(\sqrt{3}\sin\alpha-\cos\alpha) \quad . \quad . \quad \text{(v)}$$

SQ, the line whose equation is $\theta=\alpha-\frac{1}{3}\pi$, meets (ii) at R, where

$$l/SR=e\cos(\alpha-\tfrac{1}{3}\pi)+\cos\tfrac{1}{3}\pi=\tfrac{1}{2}e(\cos\alpha+\sqrt{3}\sin\alpha)+\tfrac{1}{2} \quad . \quad \text{(vi)}$$

From (iv), (v) and (vi)

$$1/SD+1/SR-\sqrt{3}/ST=1/2l.$$

Example 3

Show that the polar equation of the chord joining the points of the conic $l/r=1+e\cos\theta$ *whose vectorial angles are* $(\alpha\pm\beta)$ *is*

$$l/r=e\cos\theta+\sec\beta\cos(\theta-\alpha).$$

A chord of this conic subtends a right angle at the pole. Show that the locus of the foot of the perpendicular from the pole on the chord is, in general, a circle, but reduces to a straight line when the conic is a rectangular hyperbola.

[L.U.]

The first part of the question is proved in § 15.7.

Let P and Q (fig. 51) be points on the conic with vectorial angles $\alpha-\frac{1}{4}\pi$, $\alpha+\frac{1}{4}\pi$ respectively, and let N be the foot of the perpendicular from the pole S to PQ.

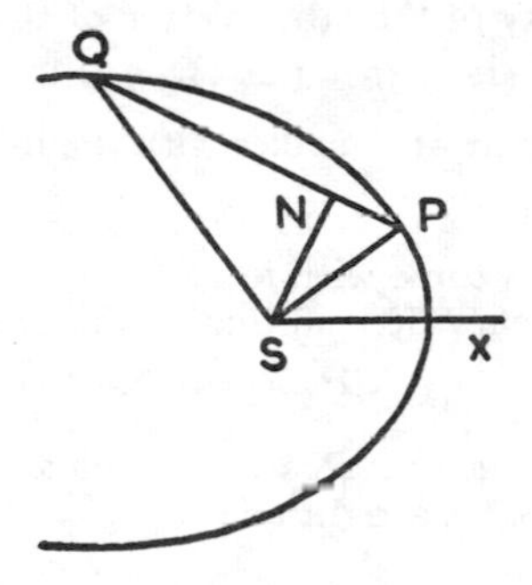

Fig. 51

Then the equation of PQ is by (15.5)

$$l/r=e\cos\theta+\sqrt{2}\cos(\theta-\alpha) \quad . \quad . \quad . \quad \text{(i)}$$

By (15.3) the equation of SN is of the form

$$k/r=e\cos(\theta+\tfrac{1}{2}\pi)+\sqrt{2}\cos(\theta-\alpha+\tfrac{1}{2}\pi)$$

i.e.
$$k/r=-e\sin\theta-\sqrt{2}\sin(\theta-\alpha),$$

where k is a constant.

This equation must reduce to the form $\theta=$constant (see § 15.3 (i)) and this is only possible if $k=0$. Hence

$$0=e\sin\theta+\sqrt{2}\sin(\theta-\alpha) \quad . \quad . \quad . \quad \text{(ii)}$$

Eliminating α from (i) and (ii) we have

$$(l/r-e\cos\theta)^2+e^2\sin^2\theta=2,$$

$$r^2(2-e^2)+2elr\cos\theta=l^2.$$

This is the equation of a circle with centre on the initial line except when $e=\sqrt{2}$, in which case the conic is a rectangular hyperbola and the locus is the straight line $r=(\frac{1}{2}l\sqrt{2})\sec\theta$ which is parallel to the directrices of the hyperbola.

Exercises 15

1. F is a focus and PFP' is a focal chord of the ellipse $r(1+e\cos\theta)=l$. Prove that $1/r+1/r'=2/l$, where r and r' are the distances of P and P' from F respectively. What is the harmonic mean of r and r'? [Durham.]

2. If PSP' and QSQ' are two mutually perpendicular focal chords of a conic, prove that $1/(PS.SP')+1/(QS.SQ')$ is constant.

3. Prove that mutually perpendicular focal chords of a rectangular hyperbola are equal in length.

4. If P is an extremity of the latus rectum of the conic
$$l/r=1-e\cos\theta,$$
show that the tangent at P makes with the initial line an angle whose tangent is e.

5. P, Q are points on a conic with focus S. TP, TQ are tangents. Show that TS bisects angle PSQ and that, if the conic is a parabola,
$$SP.SQ=ST^2.$$

6. Show that, if the tangent at P, any point on a conic, meets the directrix at K, the angle KSP is a right angle.

7. Show that the line
$$l/r=a\cos\theta+b\sin\theta$$
touches the conic $l/r=1-e\cos\theta$, if $(a+e)^2+b^2=1$.

8. If the tangents at $P(r_1, \alpha)$ and $Q(r_2, \beta)$ on the parabola $l/r=1+\cos\theta$ intersect at $T(r_3, \gamma)$, prove that $\alpha+\beta=2\gamma$. If the point T is on the latus rectum $\theta=\frac{1}{2}\pi$, prove that $1/r_1+1/r_2=2/l$. [L.U.]

9. Show that the equation of the tangent to the conic $l/r=1+e\cos\theta$ at the point $\theta=\alpha$ is given by $l/r=\cos(\theta-\alpha)+e\cos\theta$. Show that the two conics
$$l\sqrt{3}=r(\sqrt{3}+\cos\theta)$$
and
$$l\sqrt{3}=2r\{\sqrt{3}+\cos(\theta+\tfrac{1}{3}\pi)\}$$
touch where $\theta=\frac{1}{2}\pi$. [L.U.]

10. Show that the equation $r/l=\sin\theta-\cos\theta$ is that of a circle which passes through the origin and which touches the conic $l/r=1+\cos\theta$ at $\theta=\frac{1}{2}\pi$. [L.U.]

11. If a chord PQ of the conic $l/r=1+e\cos\theta$ subtends a constant angle 2γ at the focus, show that the locus of the point of intersection of the tangents at P and Q is a conic with the same focus and directrix as the given conic but with eccenticity $e\sec\gamma$. [L.U.]

12. A chord PQ of a rectangular hyperbola subtends a right angle at the focus S. The tangents at P and Q intersect at T, and ST intersects PQ at R. Prove that the locus of R is a parabola. [L.U.]

13. Variable points P and Q on one branch of a rectangular hyperbola are such that PQ subtends a right angle at the focus S. Show that the locus of the foot of the perpendicular from S on PQ is a straight line parallel to the directrix of the hyperbola. [L.U.]

14. Find the points of intersection of the straight line

$$2 \cos \theta + \tan \alpha \sin \theta = l/r$$

and the conic $1 + \cos \theta = l/r$ and show that the tangents to the conic at these points cut each other at an angle α. [L.U.]

15. An ellipse of eccentricity e, and a parabola intersect at the ends L, and L' of a common latus rectum, and S is their common focus. If a common tangent touches them at P and Q respectively, show that SP and SQ are both inclined to LL' at an angle $\sin^{-1}\{\frac{1}{2}(1 \pm e)\}$. [L.U.]

CHAPTER 16

COORDINATE GEOMETRY OF THREE DIMENSIONS: THE PLANE AND STRAIGHT LINE

16.1. Coordinates of a point in space

In the rectangular cartesian system of coordinates, the position of a point is fixed by its perpendicular distances from three mutually perpendicular planes. Three such planes intersecting in three mutually perpendicular lines $x'Ox$, $y'Oy$, $z'Oz$ are shown in fig. 52. Their point of intersection O is the *origin*, the lines $x'Ox$, $y'Oy$, $z'Oz$ are the *co-*

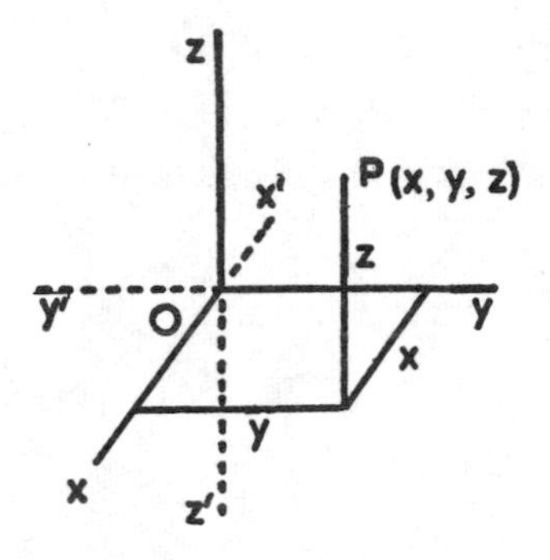

Fig. 52

ordinate axes and the planes yOz, zOx, xOy (known respectively as the yz-, zx-, xy-planes) are the *coordinate planes*. The point $P(x, y, z)$ lies at perpendicular distances x, y, z from the yz-, zx-, xy-planes respectively, x being positive when measured in the direction Ox, negative if measured in the direction Ox'. Similar sign conventions hold for y and z. The coordinate planes divide space into eight regions known as *octants*. The octant $Oxyz$, in which x, y, z are all positive, is called the *positive octant*. In the octant $Oxyz'$, x and y are positive and z is negative, and so on.

16.2. Section formula

Let $P(x, y, z)$ divide the join of $A(x_1, y_1, z_1)$ and $B(x_2, y_2, z_2)$ in the ratio $\lambda : \mu$ (fig. 53).

Let A', P', B' be the projections of A, P and B respectively on the xy-plane and draw AC, PD parallel to $A'B'$.

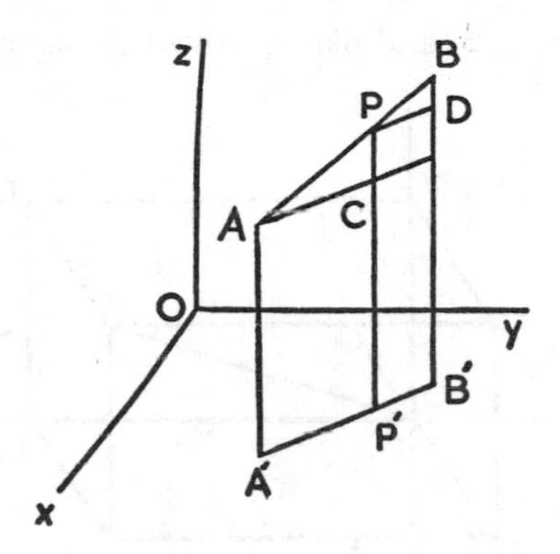

Fig. 53

Then $CP=z-z_1$, $DB=z_2-z$ and, by similar triangles,

$$\frac{CP}{DB}=\frac{AP}{PB}=\frac{\lambda}{\mu}$$

$$\therefore \quad \frac{z-z_1}{z_2-z}=\frac{\lambda}{\mu}$$

i.e.

$$z=\frac{\lambda z_2+\mu z_1}{\lambda+\mu}.$$

Similarly, $$x=\frac{\lambda x_2+\mu x_1}{\lambda+\mu} \text{ and } y=\frac{\lambda y_2+\mu y_1}{\lambda+\mu}.$$

Hence P is the point

$$\left(\frac{\lambda x_2+\mu x_1}{\lambda+\mu}, \frac{\lambda y_2+\mu y_1}{\lambda+\mu}, \frac{\lambda z_2+\mu z_1}{\lambda+\mu}\right). \tag{16.1}$$

P divides AB internally or externally according as the ratio $\lambda : \mu$ is positive or negative. In particular, the coordinates of the mid-point of AB are

$$\tfrac{1}{2}(x_1+x_2), \quad \tfrac{1}{2}(y_1+y_2), \quad \tfrac{1}{2}(z_1+z_2). \tag{16.2}$$

16.3. Direction cosines of a straight line

If lengths measured along a line are reckoned positive in one direction and negative in the opposite direction, the line is called a *directed line* and the direction in which the lengths are measured positive is called the direction of the directed line.

The direction of a directed straight line in space is fixed by means of the angles which it makes with the positive directions of the coordinate axes.

In fig. 54, $P \equiv (x, y, z)$ and planes drawn through P parallel to the

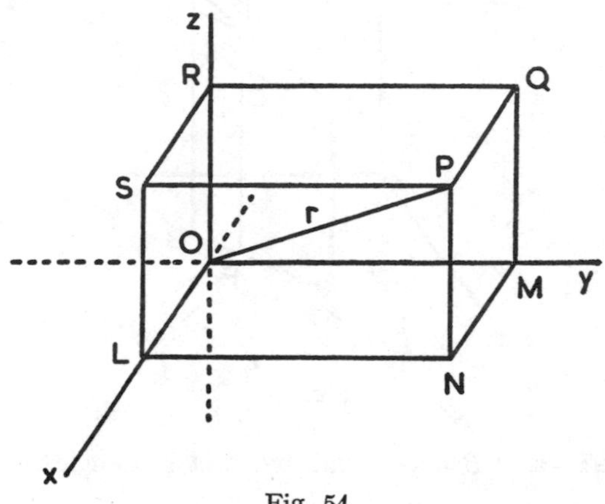

Fig. 54

coordinate planes form the rectangular prism $PQRS$, $NMOL$.

If OP is of length r and makes angles α, β, γ with Ox, Oy, Oz respectively,

$$\left.\begin{aligned} OL&=x=r\cos\alpha,\\ OM&=y=r\cos\beta,\\ OR&=z=r\cos\gamma \end{aligned}\right\} \quad . \quad . \quad . \quad . \qquad \text{(i)}$$

$\cos\alpha$, $\cos\beta$, $\cos\gamma$ are called the *direction cosines* (D.C.s) of the line OP and are usually denoted for brevity by $[l, m, n]$, square brackets being used to distinguish D.C.s from coordinates.

With this notation, $OL=lr$, $OM=mr$ and $OR=nr$.

But from fig. 54, $\qquad OP^2=ON^2+NP^2=OL^2+OM^2+OR^2$

$$r^2=x^2+y^2+z^2,$$

i.e.

$$r^2=r^2(l^2+m^2+n^2)$$

$$\therefore\ l^2+m^2+n^2=1. \tag{16.3}$$

Again, from (i) we have

$$x=lr,\quad y=mr,\quad z=nr \tag{16.4}$$

and so

$$x/l=y/m=z/n=r. \tag{16.5}$$

Equations (16.4) give the coordinates of the point P distant r from O in the direction $[l, m, n]$, and since as r varies from $-\infty$ to $+\infty$ they give the coordinates of every point on the line, these equations may be regarded as the parametric equations of the straight line drawn through the origin with D.C.s $[l, m, n]$, r being the parameter. The equations of the line are given in *symmetrical* (or standard) form by (16.5).

Any directed line in space drawn in the direction of OP (i.e. parallel to OP and in the same sense as OP) has the same D.C.s as OP. The D.C.s of PO are $[-l, -m, -n]$.

The angle between two directed straight lines AB, CD is defined as the angle between lines OP, OP' drawn in the directions of AB and CD respectively.

16.4. Length, direction and equations of the line joining the points $(x_1, y_1\ z_1)$ and (x_2, y_2, z_2)

Through $P(x_1, y_1, z_1)$ and $Q(x_2, y_2, z_2)$ planes are drawn parallel to the coordinate planes to form the rectangular prism $PADB$, $CRQS$ (fig. 55). Let PQ be of length d and make angles α, β, γ with the directions of Ox, Oy, Oz respectively.

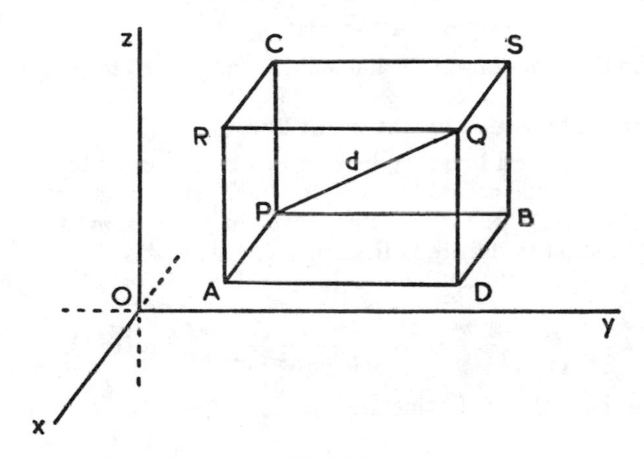

Fig. 55

Then $$PA = x_2 - x_1,\ PB = y_2 - y_1,\ PC = z_2 - z_1$$
and $$PQ^2 = PA^2 + PB^2 + PC^2.$$
Hence $$d = \sqrt{\{(x_2 - x_1)^2 + (y_2 - y_1)^2 + (z_2 - z_1)^2\}} \tag{16.6}$$
Also, $\angle APQ = \alpha$, $\angle BPQ = \beta$, $\angle CPQ = \gamma$

$$\therefore \frac{PA}{PQ} = \cos\alpha, \quad \frac{PB}{PQ} = \cos\beta, \quad \frac{PC}{PQ} = \cos\gamma.$$

Hence, if the D.C.s of PQ are $[l, m, n]$,

$$l = \frac{x_2 - x_1}{d}, \quad m = \frac{y_2 - y_1}{d}, \quad n = \frac{z_2 - z_1}{d}. \tag{16.7}$$

These equations which give the D.C.s of PQ in terms of the coordinates of P and Q show that the coordinates of Q, the point distant d from $P(x_1, y_1, z_1)$ in the direction $[l, m, n]$ are (x_1+ld, y_1+md, z_1+nd).

The coordinates of the point distant r from P along PQ are given by the equations

$$x=x_1+lr, \quad y=y_1+mr, \quad z=z_1+nr \tag{16.8}$$

which may be regarded as the parametric equations of the line drawn through the point (x_1, y_1, z_1) in the direction $[l, m, n]$, r being the parameter. Alternatively, we may write equations (16.8) in the symmetrical (or standard) form

$$\frac{x-x_1}{l}=\frac{y-y_1}{m}=\frac{z-z_1}{n}=r. \tag{16.9}$$

Substituting from (16.7) for l, m and n we obtain

$$\frac{x-x_1}{x_2-x_1}=\frac{y-y_1}{y_2-y_1}=\frac{z-z_1}{z_2-z_1}=\rho \text{ (say)}. \tag{16.10}$$

These are the equations of the line joining $P(x_1, y_1, z_1)$ to $Q(x_2, y_2, z_2)$.

16.5. Direction ratios of a straight line

The direction of a line may be specified by means of three numbers proportional to the actual D.C.s of the line. Such numbers are called *direction-ratios* (D.R.s) of the line. We shall use $[l, m, n]$ to denote D.C.s, $[\lambda : \mu : \nu]$ to denote D.R.s.

If $[\lambda : \mu : \nu]$ are the D.R.s of a line, its D.C.s $[l, m, n]$ are given by

$$\frac{l}{\lambda}=\frac{m}{\mu}=\frac{n}{\nu}=\frac{1}{\sqrt{\{\lambda^2+\mu^2+\nu^2\}}}, \text{ from (16.3)}.$$

Hence the D.C.s of the line are

$$\left[\frac{\lambda}{\sqrt{\{\lambda^2+\mu^2+\nu^2\}}}, \frac{\mu}{\sqrt{\{\lambda^2+\mu^2+\nu^2\}}}, \frac{\nu}{\sqrt{\{\lambda^2+\mu^2+\nu^2\}}}\right]$$

and we denote them briefly by $\left[\dfrac{\lambda, \mu, \nu}{\sqrt{\{\lambda^2+\mu^2+\nu^2\}}}\right]$. (16.11)

It follows from (16.7) that the D.R.s of the line joining (x_1, y_1, z_1) and (x_2, y_2, z_2) are

$$[x_2-x_1 : y_2-y_1 : z_2-z_1]. \tag{16.12}$$

Also, from (16.9) the equations of the line through the point (x_1, y_1, z_1) with D.R.s $[\lambda : \mu : \nu]$ are

$$\frac{x-x_1}{\lambda}=\frac{y-y_1}{\mu}=\frac{z-z_1}{\nu}=\rho \text{ (say)}. \tag{16.13}$$

Example 1

Find the equations of the line joining $A(1, 2, 4)$ and $B(2, 4, 2)$. Show that AB meets the xy-plane in a point of trisection of the line joining $C(6, 7, 1)$, $D(-3, 4, -2)$.

By (16.10) the equations of AB are

$$\frac{x-1}{1}=\frac{y-2}{2}=\frac{z-4}{-2}.$$

AB meets the xy-plane at P, where $z=0$ and

$$\frac{x-1}{1}=\frac{y-2}{2}=2;$$

hence P is the point $(3, 6, 0)$.

By (16.1) the point which divides CD in the ratio $1:2$ is

$$\{\tfrac{1}{3}(12-3),\ \tfrac{1}{3}(14+4),\ \tfrac{1}{3}(2-2)\},$$

i.e. the point $(3, 6, 0)$.

Hence P is a point of trisection of CD.

Example 2

The line joining $A(1, 8, -1)$ and $B(4, -4, 2)$ meets the xz- and yz-planes at P and Q respectively. Find the coordinates of P and Q and the ratios in which they divide AB.

The coordinates of the point which divides AB in the ratio $k:1$ are

$$\frac{4k+1}{k+1},\ \frac{8-4k}{k+1},\ \frac{2k-1}{k+1} \quad . \quad . \quad . \quad . \qquad \text{(i)}$$

If this point lies on the plane $y=0$, its y-coordinate is zero and so $k=2$. Hence $P\equiv(3, 0, 1)$, and since k is positive P divides AB internally in the ratio $2:1$.

If the point given by (i) lies on the plane $x=0$, $k=-\frac{1}{4}$. Hence

$$Q\equiv(0, 12, -2),$$

and since k is negative Q divides AB externally in the ratio $1:4$.

16.6. Note on projection

Let a segment PQ of a directed line be considered positive or negative according as PQ points in the direction of the line or in the opposite direction.

The projection of a segment AB of a directed line k upon a directed line k' is the segment $A'B'$, where A' and B' are the feet of the perpendiculars drawn to k' from A and B respectively. Then

$$A'B'=AB\cos\theta,$$

where θ is the angle between the directions of k and k'.

The projection of a segment AB on a directed line k' is the algebraic sum of the projections on k' of any series of segments which form a continuous path from A to B. Two particular cases should be noted :

(a) If P is (x, y, z) and O is the origin (fig. 54), the projection of OP on any line k is equal to the sum of the projections of the segments OL, LN, NP on k. But OL is of length x and lies along Ox so that if k has D.C.s $[l, m, n]$, the projection of OL on k is lx. Similarly, the projections of LN, NP on k are my and nz respectively, and so the projection of OP on k is $lx+my+nz$.

(b) In fig. 55 the projection of PQ on any line k is the sum of the projections of PA, AD and DQ on k. These segments are of lengths (x_2-x_1), (y_2-y_1) and (z_2-z_1) respectively, and are parallel to Ox, Oy, Oz respectively, so that, as in (a), if k has D.C.s $[l, m, n]$ the projection of PQ on k is

$$l(x_2-x_1)+m(y_2-y_1)+n(z_2-z_1).$$

16.7. The angle between two straight lines

Let OP, OP' (fig. 56) be two lines drawn through the origin in the direction of two given lines whose D.C.s are $[l, m, n]$, $[l', m', n']$ respectively. Then the angle POP' is equal to the angle θ between the given lines. Let $OP=r$, draw PN perpendicular to the xy-plane and NL perpendicular to Ox. Then the projection of OP on OP' is equal to the sum of the projections of OL, LN and NP on OP'. Now $OL=lr$, $LN=mr$, $NP=nr$ and OP' has D.C.s $[l', m', n']$. Hence, as in § 16.6 (a),

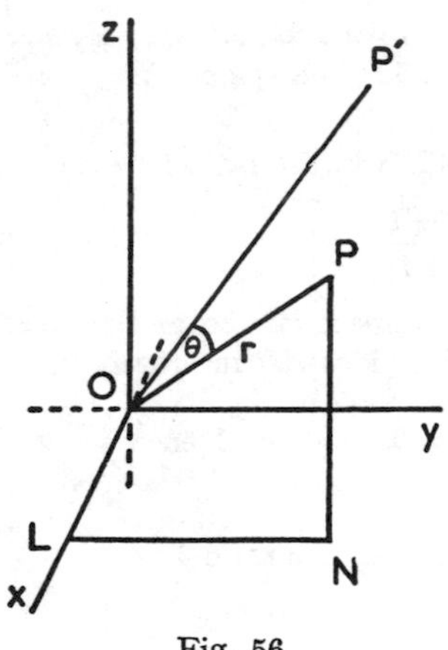

Fig. 56

$$OP \cos\theta=(rl)l'+(rm)m'+(rn)n'$$

$$\therefore \cos\theta=ll'+mm'+nn'. \quad (16.14)$$

The condition for perpendicular lines is

$$\cos\theta=0$$

or

$$ll'+mm'+nn'=0. \quad (16.15)$$

If the given lines have D.R.s $[\lambda:\mu:\nu]$, $[\lambda':\mu':\nu']$, we deduce from (16.14), using (16.11), that the angle θ between them is given by

$$\cos\theta=(\lambda\lambda'+\mu\mu'+\nu\nu')/\sqrt{\{(\lambda^2+\mu^2+\nu^2)(\lambda'^2+\mu'^2+\nu'^2)\}}. \quad (16.16)$$

The condition for perpendicular lines is

$$\lambda\lambda'+\mu\mu'+\nu\nu'=0. \quad (16.17)$$

Example 3

Show that the points $A(2, 4, 3)$, $B(4, 1, 9)$, $C(10, -1, 6)$ are the vertices of an isosceles right-angled triangle.

By (16.12), the D.R.s of AB, BC, AC are $[2 \quad -3:6]$, $[6:-2:-3]$ and $[8:-5:3]$ respectively.

By (16.17), the lines AB and BC are perpendicular; by (16.16) $\cos \angle CAB = 1/\sqrt{2}$.

Hence $\angle CAB = 45°$ and so ABC is a right-angled isosceles triangle.

Example 4

If A, B, C, D are four points in space such that AB is perpendicular to CD and AC is perpendicular to BD, prove that AD is perpendicular to BC.

[L.U.]

Let A, B, C, D be the points (x_1, y_1, z_1), (x_2, y_2, z_2), (x_3, y_3, z_3) and (x_4, y_4, z_4) respectively.

Then by (16.12) the D.R.s of AB and CD are $[x_2-x_1 : y_2-y_1 : z_2-z_1]$ and $[x_4-x_3 : y_4-y_3 : z_4-z_3]$ respectively, and since AB is perpendicular to CD, by (16.17),

$$(x_2-x_1)(x_4-x_3)+(y_2-y_1)(y_4-y_3)+(z_2-z_1)(z_4-z_3)=0. \qquad \text{(i)}$$

Similarly, since AC is perpendicular to BD,

$$(x_3-x_1)(x_4-x_2)+(y_3-y_1)(y_4-y_2)+(z_3-z_1)(z_4-z_2)=0. \qquad \text{(ii)}$$

Subtracting corresponding sides of (i) and (ii), we have

$$(x_4-x_1)(x_3-x_2)+(y_4-y_1)(y_3-y_2)+(z_4-z_1)(z_3-z_2)=0.$$

This is the condition that the lines whose D.R.s are $[x_4-x_1 : y_4-y_1 : z_4-z_1]$ and $[x_3-x_2 : y_3-y_2 : z_3-z_2]$ should be perpendicular, i.e. that AD should be perpendicular to BC.

Exercises 16 (*a*)

1. Find the distance between the points $P(-2, 4, 3)$ and $Q(0, 1, -3)$. If PQ is produced to R so that $PQ = QR$ find the coordinates of R. Find also the equations of the line PQ.

2. The point $P(x, y, z)$ moves so that its distance from $A(1, 2, 3)$ is equal to its distance from $B(-2, 3, 4)$. Find the equation of the locus of P. What is represented by this equation?

   If P moves so that its distance from A is twice its distance from B, what is the equation of the locus of P?

3. Find the coordinates of the centroid of the triangle whose vertices are $(1, 3, -4)$, $(-4, 2, -6)$, $(-3, 1, 1)$.

4. Show that the points $(4, 2, 3)$, $(1, 4, 9)$, $(-1, 10, 6)$ are the vertices of a right-angled isosceles triangle and find the equations of its longest side.

5. Find the coordinates of the point P in which the line joining the points $A(1, -2, 6)$ and $B(2, -4, 3)$ meets the xy-plane. In what ratio does P divide AB?

6. The projections of a straight line on the coordinate axes are 3, 6 and 2 respectively. Find the length of the line.

7. A straight line drawn through the point $P(-2, 1, 4)$ has D.R.s $[6 : -2 : 3]$. Find the D.C.s of the line and also the coordinates of the points which lie on the line at a distance of 7 units from P.

8. If a straight line makes an angle of 60° with each of the x- and y-axes, what angle does it make with the z-axis?

9. Calculate the lengths of the sides and the sizes of the angles of the triangle whose vertices are (1, 2, 3), (3, 3, 5), (3, 0, 5).

10. If A is the point (3, 7, 5) and B is the point (−3, 2, 6), find the length of the projection of AB on the straight line which joins the points (7, 9, 4) and (4, 5, −8).

11. The straight line $(x-6)/2=(y-4)/4=(z+6)/3$ meets the yz-, zx-, xy-planes at P, Q and R respectively. Find the coordinates of the centroid of triangle PQR.

16.8. The equation of a plane

We establish two standard forms of the equation of a plane:

(a) *The perpendicular form*

In fig. 57, $P(x, y, z)$ is any point on the plane ABC, and OQ, the perpendicular drawn from the origin to the plane is of length p and has D.C.s $[l, m, n]$.

By (16.4), Q is the point (pl, pm, pn) and by (16.12), PQ has D.R.s $[x-pl : y-pm : z-pn]$. But OQ is perpendicular to QP so that by (16.17) $$l(x-pl)+m(y-pm)+n(z-pn)=0.$$
Hence by (16.3), $$lx+my+nz=p. \tag{16.18}$$
This is the equation of the plane which lies at a perpendicular distance p, $p>0$, from the origin and whose normals drawn in the direction of OQ (i.e. *from* the origin *to* the plane) have D.C.s $[l, m, n]$.

(b) *The intercept form*

The volume of the tetrahedron $OABC$ (fig. 57) is equal to the sum of the volumes of the tetrahedra $OPBC$, $OPCA$, $OPAB$. Thus

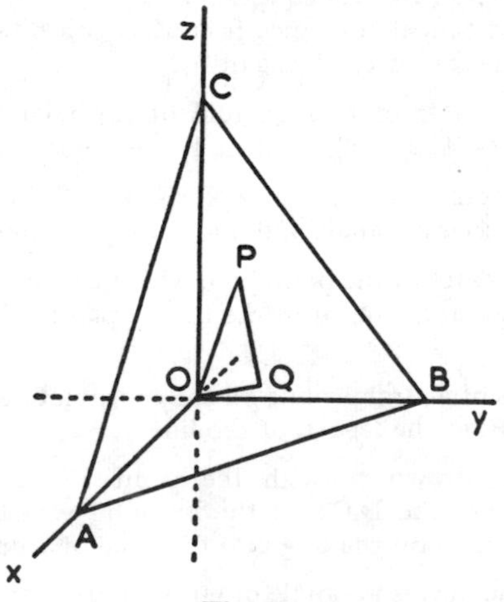

Fig. 57

if A, B, C are the points $(a, 0, 0)$, $(0, b, 0)$ and $(0, 0, c)$ respectively,

$$\tfrac{1}{6}abc=\tfrac{1}{6}(xbc+yca+zab)$$

$$\therefore \frac{x}{a}+\frac{y}{b}+\frac{z}{c}=1. \tag{16.19}$$

This is the equation of the plane which makes intercepts of lengths a, b and c on Ox, Oy, Oz respectively.

16.9. The general equation of a plane

From (16.18) it can be seen that the general equation of a plane is of the form

$$Ax+By+Cz+D=0. \tag{16.20}$$

Comparing coefficients in this equation with those in (16.18), we have

$$l/A=m/B=n/C=-p/D$$

i.e.

$$A:B:C=l:m:n.$$

Hence the D.R.s of the normal to the plane $Ax+By+Cz+D=0$ are

$$[A:B:C]. \tag{16.21}$$

(16.20) may be written in the perpendicular form by dividing throughout by $\pm\sqrt{\{A^2+B^2+C^2\}}$ to give

$$\frac{Ax}{\pm\sqrt{\{A^2+B^2+C^2\}}}+\frac{By}{\pm\sqrt{\{A^2+B^2+C^2\}}}+\frac{Cz}{\pm\sqrt{\{A^2+B^2+C^2\}}}$$
$$=\frac{-D}{\pm\sqrt{\{A^2+B^2+C^2\}}} \tag{16.22}$$

Then $p=\dfrac{-D}{\pm\sqrt{\{A^2+B^2+C^2\}}}$, the sign of the radical being conventionally chosen so that p is positive. With this choice of sign, the coefficients of x, y and z on the left-hand side of (16.22) are the D.C.s of the normal drawn from the origin to the plane. For example, the perpendicular form of the plane

$$2x-y+2z+6=0$$

is

$$-\tfrac{2}{3}x+\tfrac{1}{3}y-\tfrac{2}{3}z=2.$$

The D.C.s of the perpendicular drawn from the origin to the plane are $[-\frac{2}{3}, \frac{1}{3}, -\frac{2}{3}]$; the length of the perpendicular is 2.

16.10. The line of intersection of two planes

Suppose that the planes $Ax+By+Cz+D=0$ . . . (i)

and $A'x+B'y+C'z+D'=0$ . . . (ii)

intersect in a line L whose D.R.s are $[\lambda:\mu:\nu]$. Then the normal to plane (i) is perpendicular to L, and by (16.17)

$$A\lambda+B\mu+C\nu=0.$$

Similarly, $$A'\lambda+B'\mu+C'\nu=0.$$

Solving for λ/ν and μ/ν, we have by (3.5), page 40,

$$\lambda:\mu:\nu=(BC'-B'C):(CA'-C'A):(AB'-A'B). \quad (16.23)$$

Example 5

Find in symmetrical form the equations of the line of intersection of the planes $x+y+z=5$, $4x+y+2z=15$.

1st Method. The D.R.s of the required line satisfy the equations

$$\lambda+\mu+\nu=0,\quad 4\lambda+\mu+2\nu=0 \quad \therefore \lambda:\mu:\nu=1:2:-3.$$

Before we can use (16.13) we must find a point on the line of intersection of the given planes. The x- and y-coordinates of the point in which this line meets the plane $z=0$ are given by the equations $x+y=5$, $4x+y=15$. Hence the point is $(\frac{10}{3}, \frac{5}{3}, 0)$ and by (16.13) the equations of the required line are $(x-\frac{10}{3})/1=(y-\frac{5}{3})/2=z/(-3)$.

2nd Method. Eliminating y and z in turn between the equations of the given planes we obtain

$$3x+z=10, \quad x=\tfrac{1}{3}(10-z);$$

$$2x-y=5, \quad x=\tfrac{1}{2}(5+y).$$

Hence the equations of the line of intersection of the planes may be taken as

$$x=(y+5)/2=(z-10)/(-3).$$

Example 6

Through the point $A(-1, 1, 2)$ a line is drawn parallel to the line of intersection of the planes $x-2y+z=3$ and $x+6y-5z=0$. This line cuts the plane $x-3y+2z=2$ in B. Find the equations of the line AB and the coordinates of B.

The planes $x-2y+z=3$ and $x+6y-5z=0$ intersect in a line L whose D.R.s $[\lambda:\mu:\nu]$ are found from the equations $\lambda-2\mu+\nu=0$, $\lambda+6\mu-5\nu=0$ to be $[2:3:4]$.

The equations of the line AB drawn through $A(-1, 1, 2)$ parallel to L are by (16.13)

$$(x+1)/2=(y-1)/3=(z-2)/4=\rho \quad . \quad . \quad . \quad \text{(i)}$$

since parallel lines have the same D.R.s.

The coordinates of B, the point in which this line meets the plane $x-3y+2z=2$ are $(2\rho-1, 3\rho+1, 4\rho+2)$ where

$$(2\rho-1)-3(3\rho+1)+2(4\rho+2)=2.$$

Hence $\rho=2$ and $B\equiv(3, 7, 10)$.

It is worthy of note that if a line is drawn in the plane $z=c$ parallel to the line $y=mx$, $z=0$ in the plane xOy, its equations are

$$y=mx, \quad z=c \qquad \text{(ii)}$$

But this line passes through the point $(0, 0, c)$ and is perpendicular to Oz.

Hence by (16.13) its equations are

$$\frac{x}{1}=\frac{y}{m}=\frac{z-c}{0} \qquad \text{(iii)}$$

Equations (iii) are taken as the symmetrical form of equations (ii).

In the same way, the symmetrical form of the equations of the line drawn through (a, b, c) parallel to Oz is

$$\frac{x-a}{0}=\frac{y-b}{0}=\frac{z-c}{1}$$

i.e. $x=a$, $y=b$.

16.11. The equation of the plane which is parallel to a given plane and passes through a given point

Let the given point be (x_1, y_1, z_1) and the equation of the given plane be

$$Ax+By+Cz+D=0 \qquad \text{(i)}$$

Since the required plane is parallel to (i) the D.R.s of its normal are the same as the D.R.s $[A : B : C]$ of the normal to (i).

Hence the equation of the plane is of the form

$$Ax+By+Cz+k=0 \qquad \text{(ii)}$$

where k is a constant to be found. Plane (ii) passes through (x_1, y_1, z_1)

$$\therefore Ax_1+By_1+Cz_1+k=0 \qquad \text{(iii)}$$

Eliminating k between (ii) and (iii), we obtain the equation of the required plane

$$A(x-x_1)+B(y-y_1)+C(z-z_1)=0. \qquad (16.24)$$

Comparing (i) and (ii), we note that the equations of parallel planes differ only by a constant.

(16.24) is also the equation of the plane through (x_1, y_1, z_1) whose normal has D.R.s $[A : B : C]$.

Example 7

The equation of the plane which passes through the point (1, 2, 3) and is parallel to the plane $2x-3y+4z=12$ is by (16.24)

$$2x-3y+4z=8.$$

16.12. The equation of a plane through three given points

The plane $$Ax+By+Cz+D=0 \quad . \quad . \quad . \qquad \text{(i)}$$
will pass through the three points (x_r, y_r, z_r), $r=1, 2, 3,$
if $$Ax_r+By_r+Cz_r+D=0, \quad r=1, 2, 3.$$

Solving for A, B, C in terms of D we obtain the equation of the required plane.

Example 8

Find the equation of the plane which passes through the points (3, 4, 1), (1, 1, −7) *and* (2, 2, −4).

The plane $$Ax+By+Cz+D=0$$
passes through the given points if

$$3A+4B+C+D=0 \quad . \quad . \quad . \quad . \qquad \text{(i)}$$
$$A+B-7C+D=0 \quad . \quad . \quad . \quad . \qquad \text{(ii)}$$
$$2A+2B-4C+D=0 \quad . \quad . \quad . \quad . \qquad \text{(iii)}$$

From (ii) and (iii) $$C=\tfrac{1}{10}D,$$
and from (i) and (ii) $$B=-\tfrac{2}{10}D, \quad A=-\tfrac{1}{10}D.$$

Hence the equation of the required plane is
$$x+2y-z=10.$$

16.13. The equation of a plane through the line of intersection of two given planes

Let the planes
$$\left.\begin{aligned} Ax+By+Cz+D&=0 \\ A'x+B'y+C'z+D'&=0 \end{aligned}\right\} \quad . \quad . \quad . \qquad \text{(i)}$$
intersect in a line l. Then the equation
$$Ax+By+Cz+D+k(A'x+B'y+C'z+D')=0 \qquad (16.25)$$
where k is any constant, is of the first degree in x, y, z and so represents a plane. But (16.25) is satisfied by points which simultaneously satisfy equations (i), i.e. by points on l.

Hence for all values of k, (16.25) represents a plane which passes through the line of intersection of planes (i). Also, the equation of any plane through l (other than $A'x+B'y+C'z+D'=0$) is of the form (16.25) for we can choose k to make plane (16.25) pass through any given point not on l.

Example 9

Find the equation of the plane which contains the line
$$(x-4)/2=(y-3)/5=(z+1)/(-2)$$
and passes through the point (2, −4, 2).

The line $$(x-4)/2=(y-3)/5=(z+1)/(-2) \quad . \quad . \quad \text{(i)}$$
may be regarded as the line of intersection of the planes
$$(x-4)/2=(y-3)/5, \quad \text{i.e. } 5x-2y=14$$
and
$$(y-3)/5=(z+1)/(-2), \quad 2y+5z=1.$$

But by (16.25) the plane
$$5x-2y-14+k(2y+5z-1)=0 \quad . \quad . \quad . \quad \text{(ii)}$$
passes through line (i) and will contain the point $(2, -4, 2)$ if $k=-4$.

Hence the equation of the required plane is
$$x-2y-4z=2.$$

Example 10

The two planes $2x-y-z=3$ and $2x+y-2z=1$ meet in a line l. Find (a) *the equation of the plane through the line $\frac{1}{2}x=y+1=1-z$ parallel to l; and* (b) *the equation of the plane through l which passes through the origin.*

(*a*) As in § 16.10, the D.R.s $[\lambda : \mu : \nu]$ of l are given by
$$2\lambda-\mu-\nu=0, \quad 2\lambda+\mu-2\nu=0.$$

Hence
$$\lambda : \mu : \nu=3 : 2 : 4.$$

The line $\frac{1}{2}x=y+1=1-z$ is the line of intersection of the planes
$$x-2y-2=0 \text{ and } y+z=0,$$
and by (16.25) the equation of any plane through this line is of the form
$$x-2y-2+k(y+z)=0,$$
i.e.
$$x+(k-2)y+kz=2 \quad . \quad . \quad . \quad . \quad \text{(i)}$$
This plane is parallel to l if its normal is perpendicular to l, i.e. if the line with D.R.s $[1 : k-2 : k]$ is perpendicular to l. This condition is fulfilled if, by (16.17),
$$3+2(k-2)+4k=0.$$

Hence $k=\frac{1}{6}$ and the equation of the required plane is
$$6x-11y+z=12.$$

(*b*) By (16.25), the plane
$$2x-y-z-3+k(2x+y-2z-1)=0$$
passes through l. It will pass through the origin if $k=-3$.

Hence the equation of the required plane is
$$4x+4y-5z=0.$$

16.14. Other forms of the equation of a plane

It should be noted that an equation of the form
$$By+Cz+D=0$$
represents a plane parallel to Ox since normals to this plane have D.R.s $[0 : B : C]$ and hence are perpendicular to Ox.

An equation of the form $Cz+D=0$, i.e. $z=$constant, represents a plane parallel to the xy-plane.

16.15. The angle between two planes

The angle θ between two planes is the angle between their respective normals.

By (16.16) and (16.21) the angle θ between the planes

$$Ax+By+Cz+D=0 \text{ and } A'x+B'y+C'z+D'=0$$

is given by

$$\cos\theta=(AA'+BB'+CC')/\sqrt{\{(A^2+B^2+C^2)(A'^2+B'^2+C'^2)\}} \quad (16.26)$$

16.16. The perpendicular distance of a point from a plane

Let the equation of the plane be $lx+my+nz=p$, $p>0$, and let P be the point (x_1, y_1, z_1). To find the distance of P from the plane, change the origin to (x_1, y_1, z_1). Then by an extension of (12.1), page 273, the equation of the plane becomes

$$l(x+x_1)+m(y+y_1)+n(z+z_1)=p$$

or

$$lx+my+nz=p',$$

where $p'\equiv p-(lx_1+my_1+nz_1)$ is the distance of the plane from P, the new origin.

If P is on the same side of the plane as the original origin O, $[l, m, n]$ are still the D.C.s of the normal from the new origin P to the plane; hence $p'>0$. If P and O are on opposite sides of the plane, $[l, m, n]$ are the D.C.s of the normal from the plane to P; hence $p'<0$.

It follows from § 16.9 that the distance of (x_1, y_1, z_1) from the plane $Ax+By+Cz+D=0$ is

$$\frac{Ax_1+By_1+Cz_1+D}{\pm\sqrt{(A^2+B^2+C^2)}}. \quad (16.27)$$

If the positive sign is chosen when D is positive, and the negative sign when D is negative, this formula gives a positive result when (x_1, y_1, z_1) and the origin lie on the same side of the plane, a negative result if they are on opposite sides.

Exercises 16 (*b*)

1. Find the point of intersection of the straight line

$$(x-1)/2=(y+3)/(-1)=(z-1)/3$$

and the plane $3x+4y-z+11=0$.

2. Find the distance of the point (1, 3, 5) from the plane

$$2x+y-3z+30=0$$

measured parallel to the straight line $x/3=y/2=z/6$.

3. Find the equation of the plane which passes through the points (1, 2, 0), (3, 4, 2) and (5, −3, 1).

4. Find the equations of the three planes which pass through the points (2, 3, −4) and (4, −1, 8) and are parallel to Ox, Oy and Oz respectively.

5. Find the equation of the plane which makes intercepts 2, 3 and 4 on the x-, y- and z-axes respectively. What is the perpendicular distance of the origin from this plane ?

6. Find the coordinates of the point at which the line joining the points $(3, 1, 4)$, $(-2, 6, 1)$ meets the plane $2x+y-3z=3$.

7. Find the equation of the plane which passes through the point $(1, 0, 6)$ and is parallel to the plane $2x+3y+6z=7$. What is the acute angle between this plane and the y-axis ?

8. Find the equation of the plane which contains the x-axis and passes through the point $(2, 1, 4)$.

9. Find the equation of the plane which bisects at right angles the straight line joining the points $(3, 4, 8)$ and $(5, -2, 4)$.

10. Find the D.C.s of the line of intersection of the planes $x+y+z=1$, $4x+y+2z=3$, and prove that it is perpendicular to the line $x=y=z$.

11. Find the equations of the line which passes through the point $(2, 3, 4)$ and which is parallel to the line of intersection of the planes $x+y-2z=1$ and $2x-3y+z=-3$.

12. Find the equation of the plane through the point $A(2, -3, 5)$ normal to the line joining $P(3, 2, -1)$ and $Q(2, -1, 1)$. Find also the D.R.s of the perpendicular drawn from A to the line PQ.

13. Find the equation of the plane through the points $(2, 3, 1)$, $(1, 1, 3)$ and $(2, 2, 3)$. Find also the perpendicular distance from the point $(5, 6, 7)$ to this plane.

14. Find the equation of the plane which passes through the origin and contains the line of intersection of the planes

$$2x-y+3z=4 \text{ and } 5x-7y-9z+2=0.$$

15. Find the equation of the plane through the line

$$(x-1)/2=(y-2)/1=(z-3)/2$$

which is parallel to the line $x/3=y/1=z/(-2)$.

16. Show that the two lines

$$(x-1)/2=(y+2)/1=z/3 \text{ and } (x-2)/3=(y-2)/(-2)=(z-4)/2$$

have a common point.

Find the perpendicular distance between the first line and the parallel line $(x-1)/2=(y-7)/1=(z-1)/3$. [L.U.]

17. Show that the plane π through the point $A(1, 2, -1)$ which contains the line of intersection of the planes $x-y+2z=0$ and $3x+y+2z=4$ has the equation $2x+y+z=3$.

Find the equations of the line through A perpendicular to π, and if this line meets the given planes at B and C, show that the length of BC is $(8\sqrt{6})/9$. [L.U.]

18. Show that the planes $2x+y-3z+5=0$, $5x-7y+2z+3=0$ and $x+10y-11z+12=0$ have a common line p which is equally inclined to the axes.

Find the equations of the line q through the origin, parallel to the first of the given planes and perpendicular to the line p. Find also the shortest distance between the lines p and q. [L.U.]

19. Show that, for all values of λ, the point $(3+\lambda, 5+2\lambda, 2+3\lambda)$ is on the line through $A(3, 5, 2)$ perpendicular to the plane $x+2y+3z-5=0$. Perpendiculars AP and BQ are drawn through the points $A(3, 5, 2)$ and $B(-7, -1, 0)$ to the given plane. Find the coordinates of P and Q. If M is the mid-point of PQ, find the equations of the line in the plane, passing through M and perpendicular to PQ.

Find also the equations of the reflection of AB in the plane. [L.U.]

20. A plane is drawn through the line of intersection of the planes $x+2y+2z=1$, $x+y-z=-1$, and is at a distance 1 from the point $(4, -2, 1)$. Prove that there are two such planes and find their equations.

Prove also that the planes intersect at an angle $\cos^{-1}(17/35)$. [L.U.]

16.17. The condition for coplanar lines

In general, two lines in space do not intersect; if they do, the lines are coplanar.

Suppose that the lines are

$$\frac{x-a}{\lambda}=\frac{y-b}{\mu}=\frac{z-c}{\nu} \quad . \quad . \quad . \quad \text{(i)}$$

and

$$\frac{x-a'}{\lambda'}=\frac{y-b'}{\mu'}=\frac{z-c'}{\nu'} \quad . \quad . \quad . \quad \text{(ii)}$$

and that they are not parallel. To find the condition that these lines are coplanar we first find the equation of the plane through the point (α, β, γ) parallel to each of the lines (i) and (ii). The equation of this plane is of the form

$$A(x-\alpha)+B(y-\beta)+C(z-\gamma)=0 \quad . \quad . \quad \text{(iii)}$$

where

$$A\lambda+B\mu+C\nu=0, \quad . \quad . \quad . \quad \text{(iv)}$$

and

$$A\lambda'+B\mu'+C\nu'=0. \quad . \quad . \quad . \quad \text{(v)}$$

(iv) and (v) determine $A:B:C$ (since $\lambda:\mu:\nu\neq\lambda':\mu':\nu'$). Substitution in (iii) then gives the required equation. The result of this elimination of A, B, C from (iii), (iv) and (v) is, by § 3.5,

$$\begin{vmatrix} x-\alpha & y-\beta & z-\gamma \\ \lambda & \mu & \nu \\ \lambda' & \mu' & \nu' \end{vmatrix}=0,$$

and this is the equation of the plane.

In particular the plane through line (ii) parallel to (i) is

$$\begin{vmatrix} x-a' & y-b' & z-c' \\ \lambda & \mu & \nu \\ \lambda' & \mu' & \nu' \end{vmatrix} = 0. \qquad (16.28)$$

The condition for lines (i) and (ii) to be coplanar is that (a, b, c) should lie in plane (16.28), namely

$$\begin{vmatrix} a-a' & b-b' & c-c' \\ \lambda & \mu & \nu \\ \lambda' & \mu' & \nu' \end{vmatrix} = 0. \qquad (16.29)$$

When this condition is satisfied, (16.28) is the equation of the plane containing lines (i) and (ii).

When condition (16.29) is satisfied we find the common point of lines (i) and (ii) as follows: The points of parameters ρ, ρ' respectively on lines (i) and (ii) are $(a+\lambda\rho,\ b+\mu\rho,\ c+\nu\rho)$, $(a'+\lambda'\rho',\ b'+\mu'\rho'$ $c'+\nu'\rho')$. At the common point $a+\lambda\rho=a'+\lambda'\rho'$

i.e. $\lambda\rho-\lambda'\rho'+a-a'=0$

and, similarly, $\mu\rho-\mu'\rho'+b-b'=0$

and $\nu\rho-\nu'\rho'+c-c'=0$

By (16.29) these equations for ρ and ρ' are consistent (see § 3.6), and so ρ, ρ' (and hence the coordinates of the common point) may be found by solving two of these equations.

Example 11

Find the equations of the line through the point (1, 2, 3) *which intersects at right angles the line*

$$(x-2)/1=(y-1)/2=z/3.$$

Determine the coordinates of their point of intersection and the equation of the plane containing them. [L.U.]

The coordinates of P, any point on the line

$$(x-2)/1=(y-1)/2=z/3=\rho \qquad . \quad . \quad . \quad \text{(i)}$$

are

$$(\rho+2,\ 2\rho+1,\ 3\rho). \qquad . \quad . \quad . \quad . \quad \text{(ii)}$$

If Q is the point (1, 2, 3), the D.R.s of PQ are by (16.12)

$$[\rho+1 : 2\rho-1 : 3\rho-3]. \qquad . \quad . \quad . \quad . \quad \text{(iii)}$$

PQ is perpendicular to (i) if, by (16.17),

$$(\rho+1)+2(2\rho-1)+3(3\rho-3)=0,$$

$$\rho=5/7.$$

Substituting this value in (ii), we obtain the coordinates (19/7, 17/7, 15/7) of the foot of the perpendicular drawn from Q to line (i). From (iii) the

D.R.s of this perpendicular are [4 : 1 : −2] and so by (16.13) its equations are

$$(x-1)/4=(y-2)/1=(z-3)/(-2). \quad . \quad . \quad . \quad \text{(iv)}$$

By (16.28) the equation of the plane containing (i) and (iv) is

$$\begin{vmatrix} x-2 & y-1 & z \\ 1 & 2 & 3 \\ 4 & 1 & -2 \end{vmatrix}=0,$$

which reduces to $x-2y+z=0$.

Example 12

Prove that the straight line

$$(x-4)/3=(y-1)/2=(z-3)/1$$

intersects the line of intersection of the planes $x+y+2z=4$ *and* $3x-2y-z=3$, *and find the equation of the plane which contains these two lines.* [L.U.]

The straight line

$$(x-4)/3=(y-1)/2=(z-3)/1=\rho \quad . \quad . \quad . \quad \text{(i)}$$

meets the plane $\qquad x+y+2z=4 \quad . \quad . \quad . \quad . \quad \text{(ii)}$

where $\qquad (3\rho+4)+(2\rho+1)+2(\rho+3)=4,$

$$\rho=-1,$$

i.e. at the point (1, −1, 2).

This point satisfies the equation

$$3x-2y-z=3 \quad . \quad . \quad . \quad . \quad \text{(iii)}$$

Hence (i) meets (ii) and (iii) in a point common to these planes, i.e. at a point on their line of intersection.

Any plane through the line of intersection of (ii) and (iii) is of the form

$$x+y+2z-4+k(3x-2y-z-3)=0. \quad . \quad . \quad . \quad \text{(iv)}$$

It will contain (i) if the point (4, 1, 3) satisfies (iv), i.e. if

$$k=-\tfrac{7}{4}.$$

Hence the equation of the required plane is

$$17x-18y-15z=5.$$

16.18. Length of the perpendicular from a given point to a given line

In fig. 58, PM is the perpendicular from the point $P(x_1, y_1, z_1)$ to the line AB drawn through $A(\alpha, \beta, \gamma)$ with D.C.s $[l, m, n]$.

The equations of AB are $\dfrac{x-\alpha}{l}=\dfrac{y-\beta}{m}=\dfrac{z-\gamma}{n}$.

If $AM=r$, by (16.8) M is the point $(\alpha+lr, \beta+mr, \gamma+nr)$ and by (16.12) the D.R.s of PM are $[\alpha+lr-x_1 : \beta+mr-y_1 : \gamma+nr-z_1]$.

But PM is perpendicular to AB so that by (16.17)

$$l(\alpha+lr-x_1)+m(\beta+mr-y_1)+n(\gamma+nr-z_1)=0$$

giving $r=l(x_1-\alpha)+m(y_1-\beta)+n(z_1-\gamma)$, since $l^2+m^2+n^2=1$.

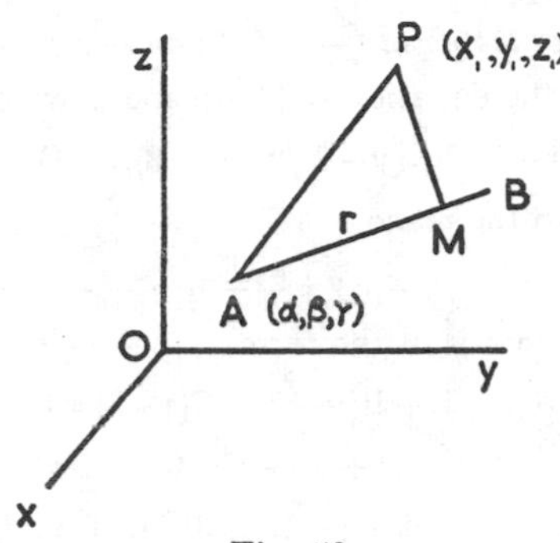

Fig. 58

Now

$$\begin{aligned}PM^2&=AP^2-AM^2=AP^2-r^2\\&=(x_1-\alpha)^2+(y_1-\beta)^2+(z_1-\gamma)^2-\{l(x_1-\alpha)+m(y_1-\beta)+n(z_1-\gamma)\}^2\\&=(m^2+n^2)(x_1-\alpha)^2+(n^2+l^2)(y_1-\beta)^2+(l^2+m^2)(z_1-\gamma)^2\\&\qquad-2lm(x_1-\alpha)(y_1-\beta)-2mn(y_1-\beta)(z_1-\gamma)-2nl(z_1-\gamma)(x_1-\alpha)\\&=\Sigma\{(y_1-\beta)n-(z_1-\gamma)m\}^2\end{aligned} \tag{16.30}$$

16.19. Miscellaneous examples

Example 13

Find the equation of the right circular cylinder of radius r which has the line $y/m=z/n$, $x=0$ as its axis. Find also the equation of the curve in which this cylinder cuts the plane $z=0$.

Any point $P(x, y, z)$ on the cylinder lies at a perpendicular distance r from its axis which passes through the origin and has D.C.s

$$\left[\frac{0,\ m,\ n}{\sqrt{(m^2+n^2)}}\right].$$

Hence by (16.30) the equation of the cylinder is

$$(ny-mz)^2+x^2(m^2+n^2)=r^2(m^2+n^2).$$

The cylinder meets the plane $z=0$ in the curve

$$\frac{x^2}{r^2}+\frac{n^2y^2}{r^2(m^2+n^2)}=1,\ z=0$$

which is an ellipse.

Any plane perpendicular to the axis of the cylinder cuts it in a circle. The fact that the plane $z=0$ cuts the cylinder in an ellipse shows that an ellipse can be orthogonally projected into a circle.

Example 14

Find the equations in standard form of the projection of the straight line

$$(x+1)/3=(y-2)/2=(z-3)/(-1)$$

on the plane $x+y+2z=4$.

Find the projection of the point $(-1, 2, 3)$ *on this plane.* [L.U.]

We must first find the equation of the plane π which contains the line

$$(x+1)/3=(y-2)/2=(z-3)/(-1) \quad . \quad . \quad . \quad \text{(i)}$$

and is perpendicular to the plane

$$x+y+2z=4. \quad . \quad . \quad . \quad . \quad \text{(ii)}$$

The equation of plane π is of the form

$$A(x+1)+B(y-2)+C(z-3)=0 \quad . \quad . \quad . \quad \text{(iii)}$$

where $$3A+2B-C=0 \quad . \quad . \quad . \quad . \quad \text{(iv)}$$

since π passes through the point $(-1, 2, 3)$ on (i) and the line $[A:B:C]$ is perpendicular to (i).

Also, the angle between planes (ii) and (iii) is 90° and so by (16.26)

$$A+B+2C=0. \quad . \quad . \quad . \quad . \quad \text{(v)}$$

Eliminating A, B and C between (iii), (iv) and (v), we obtain the equation of π:

$$\begin{vmatrix} x+1 & y-2 & z-3 \\ 3 & 2 & -1 \\ 1 & 1 & 2 \end{vmatrix}=0,$$

giving $$5x-7y+z+16=0. \quad . \quad . \quad . \quad . \quad \text{(vi)}$$

Planes (ii) and (vi) define l, the projection of (i) on (ii).

The D.R.s of l, $[\lambda:\mu:\nu]$, are found as in § 16.10 from the equations

$$\lambda+\mu+2\nu=0 \text{ and } 5\lambda-7\mu+\nu=0.$$

Hence $\lambda:\mu:\nu=5:3:-4$.

Eliminating z between (ii) and (vi), we have $9x-15y+36=0$, and this equation is satisfied when $x=-4$ and $y=0$. Substituting these values in (ii) or (vi), we see that $(-4, 0, 4)$ lies on l, and so the equations of l, the projection of (i) on (ii), are

$$(x+4)/5=y/3=(z-4)/(-4).$$

The projection of $P(-1, 2, 3)$ on plane (ii) is the point P' in which the normal to (ii) through $(-1, 2, 3)$ meets (ii).

The equations of this normal are

$$(x+1)/1=(y-2)/1=(z-3)/2=\rho$$

and it meets plane (ii) at the point $(\rho-1, \rho+2, 2\rho+3)$ where

$$(\rho-1)+(\rho+2)+2(2\rho+3)=4.$$

Hence $\rho=-\frac{1}{2}$ and $P'\equiv(-\frac{3}{2}, \frac{3}{2}, 2)$.

Miscellaneous Exercises 16

1. Show that the points (1, 1, 1), (0, 2, 3), (3, 0, 2) are the vertices of an isosceles triangle and find its area.

   Prove that the straight line whose equation is $x/3=y/5=z/(-1)$ is perpendicular to the plane of the triangle.

2. Find the equation of the plane through the point $(2, -3, 1)$ normal to the line joining $P(3, 4, -1)$ and $Q(2, -1, 5)$.

3. Find the equation of the plane through the points (2, 3, 6), (3, 6, 2), (6, 2, 3) and determine the perpendicular distance from (5, 6, 7) to this plane.

4. A plane meets the coordinate axes at A, B, C, and the foot of the perpendicular from the origin O to the plane is P. If $OA=a$, $OB=b$, $OC=c$, find the coordinates of P.

   (i) Prove that, if P is the centroid of the triangle ABC, then $|a|=|b|=|c|$.

   (ii) If the plane varies so that $1/a^2+1/b^2+1/c^2=1/R^2$ where R is constant, show that P describes the sphere $x^2+y^2+z^2=R^2$. [Sheffield.]

5. Find the equation of the plane which passes through the point (5, 1, 2) and is perpendicular to the line $2x-4=y-4=z-5$. Find also the coordinates of the point in which the line cuts the plane.

6. Find the D.C.s of the line of intersection of the planes $x+y+z=4$, $4x+y+6z=3$, and prove that it is perpendicular to the line $x=y=z$.

7. The plane $x/a+y/b+z/c=1$ meets the axes of x, y, z in A, B, C respectively. Find the D.C.s of the side AB of triangle ABC and prove that if angle A is equal to 60° then $3a^4=b^2c^2+c^2a^2+a^2b^2$.

8. Find the ratio in which N, the foot of the perpendicular from the origin, divides the line joining the points $A(4, 6, 0)$ and $B(1, 2, -1)$ and prove that N does not lie between A and B. [L.U.]

9. A plane is drawn through the line $x+y=1$, $z=0$ to make an angle $\sin^{-1}(\frac{1}{3})$ with the plane $x+y+z=0$.

   Prove that two such planes can be drawn and find their equations. Prove also that the angle between these planes is $\cos^{-1}\frac{7}{9}$.

10. The foot of the perpendicular from the point $P(4, 7, -9)$ to the line $(x-2)/2=(y+1)/(-2)=(z+3)/1$ is Q.

    Find the length of PQ and the coordinates of Q. [Leeds.]

11. Show that the lines $(x-6)/3=(y-3)/2=(z-2)$

    and $$5x+4y+7z-26=2x+3y+2z-11=0$$

    are coplanar. Find the coordinates of the common point and the equation of the plane which contains these two lines. [L.U.]

12. The feet of the perpendiculars from $P(x_1, y_1, z_1)$ to the line $x=a, y=z$ and to the plane $x+y+z=3a$ are Q and R respectively; find the coordinates of Q and R.
If QR is parallel to the plane $3x-y+z=0$, prove that P must lie on the plane $x-y=0$.
Also find the locus of P when $PQ=PR$. [L.U.]

13. Obtain the equation of the plane $\pi(A, B)$ which bisects at right angles the straight segment AB, where A and B are the points (a, b, c), (d, e, f). Show that, for any three points A, B, C, the planes $\pi(A, B)$, $\pi(B, C)$, $\pi(C, A)$ have, in general, a straight line, λ, in common. Show that, if A, B, C are the points $(1, 0, 1)$, $(0, 1, 1)$, $(2, 2, 0)$, the distance of the origin from λ is $3\sqrt{22}/11$. [L.U.]

14. Find the equations of the planes bisecting the angles between the planes whose equations are $ax+by+cz+d=0$ and $a'x+b'y+c'z+d'=0$.
Show that the locus of the centres of spheres which touch the three planes whose equations are $x+2y+2z-3=0$, $x+2y+2z+3=0$ and $6x+3y+2z-7=0$ consists of two parallel lines, and that the equation of one of them is $x/2=y/(-10)=z/9$. [L.U.

15. A straight line is drawn through $P(1, 3, 4)$ to meet the line $x=-1$, $y=1$ in Q and the line $x=3$, $z=3$ in R. Prove that the equations of the line PQR are $(x+1)/2=(y-1)/2=(z-5)/(-1)$.
Show that P is the mid-point of QR.

16. A line PQR is drawn parallel to the line $x=\frac{1}{2}y=\frac{1}{3}z$ and cuts the coordinate planes $x=0$, $y=0$, $z=0$ in P, Q, R respectively, and is such that Q is the mid-point of PR. Prove that the locus of Q is the line $3x+z=0$, $y=0$ and that the locus of the line PQR is the plane $3x-3y+z=0$.

17. Find the equation of the plane which passes through the origin and contains the line $(x-4)/2=(y-4)/6=(z-17)/(-9)$.
Find the point at which this plane meets the line

$$(x-6)/2=(y-42)/(-2)=(z-18)/1,$$

and prove that the line which is drawn through the origin to meet each of the two given lines is their common perpendicular. [Sheffield.]

18. The plane $x/a+y/b+z/c=1$ meets the axes Ox, Oy, Oz in A, B, C respectively, and L, M, N are mid-points of BC, CA, AB respectively.
Prove that the three planes OAL, OBM, OCN, meet in the line $x/a=y/b=z/c$. [L.U.]

19. If A, B, C, D are four points such that $AB^2+CD^2=AC^2+BD^2$, prove that BC is perpendicular to AD. [L.U.]

20. B and C are the points $(2, 1, 0)$ and $(1, 0, 2)$ respectively. If A is a point in the plane $x=0$ such that the triangle ABC is equilateral, prove that there are two possible positions of A and that the line joining them passes through the origin O. If the two positions of A are A_1 and A_2, show that the acute angle between the planes A_1A_2B and A_1A_2C is $\cos^{-1}\frac{2}{7}$. [L.U.]

21. Find the equations of the line drawn from any point of the line $x=1$, $z=0$ to intersect the two lines $x=0$, $y+z=0$; $x+y=0$, $z=1$, and show that all such lines lie on the surface whose equation is

$$x^2+z^2+yz+zx+xy=x+y+z.$$ [L.U.]

22. Find the equations of the line which passes through the origin and intersects each of the lines $(x-1)/(-1)=(y+2)/2=(z-3)/1$ and $(x+1)/3=(y-1)/2=(z+1)/(-1)$.

Find also the coordinates of its point of intersection with the first of the given lines. [L.U.]

23. The straight line joining the point $P(a, b, c)$ to a point Q is bisected perpendicularly by the line $2x=y=z$.

Prove that the coordinates of Q are

$$\tfrac{1}{9}(-7a+4b+4c), \quad \tfrac{1}{9}(4a-b+8c), \quad \tfrac{1}{9}(4a+8b-c).$$

If P moves along the line $x=-y=z$, find the equations of the locus of Q. [L.U.]

24. Show that the lines (a) $15x=5y=3z$
and (b) $(x+2)/3=(y+1)/4=z/5$
intersect. If line (b) is the orthogonal projection of (a) on the plane π, find the equation of π. [L.U.]

25. Prove that the lines

$$(x-3)/1=(y-1)/2=(z+1)/3$$

and

$$(x-2)/1=(y-5)/(-1)=z/1$$

intersect and find the coordinates of their common point and the equation of their common plane. [L.U.]

26. An arbitrary straight line is drawn through the point $(0, 0, a)$ at right angles to the axis of z. A second straight line at right angles to the first and also at right angles to the axis of z is drawn through the point $(0, 0, -a)$. If the straight line $(x-\alpha)/l=(y-\beta)/m=(z-\gamma)/n$ intersects both of these straight lines, show that

$$(n\alpha-l\gamma)^2+(n\beta-m\gamma)^2=a^2(l^2+m^2).$$ [L.U.]

27. A straight line is drawn through the point $(1, 3, 2)$ and meeting each of the lines $x-1=y+1=z/2$, $x=(y+2)/3=(z-1)/2$. Determine its direction cosines. [Leeds.]

28. Show that the lines

(a) $(x-2)/2=(y-3)/(-1)=(z+4)/3$

and (b) $(x-3)/1=(y+1)/3=(z-1)/(-2)$

intersect. Find their common point and the equation of the plane containing them. Find also the equation of the plane perpendicular to this plane and passing through line (a). [L.U.]

29. Find the equation of the plane which passes through the origin and contains the straight line $(x-1)/2=(y-2)/1=(z-3)/(-2)$.

    Find the equations of the straight line meeting the axis of x at right angles, whose orthogonal projection on this plane coincides with the above straight line. [L.U.]

30. Find the point of intersection of the plane π whose equation is $2x-y-z+3=0$ with the line L whose equations are $2x+y-4=0$ and $y+2z-8=0$. Find the equations defining L', the projection of the line L in π, and the angle between L and L'. If L' is the line of greatest slope in the plane π, and the angle between L' and the vertical is 60° find the direction cosines of either possible vertical. [L.U.]

31. Show that the line $(x-1)/2=(y-2)/3=(z-3)/4$ lies in the plane $x+2y-2z+1=0$ and that the line $(x-3)/3=(y-2)/2=(z-1)/4$ lies in the plane $2x+y-2z-6=0$.

    If, in addition, these lines are lines of greatest slope to the horizontal for the planes in which they lie, find the direction cosines of the vertical. [L.U.]

32. Prove that any plane through the line L common to the two non-parallel planes $ax+by+cz+d=0$, $a'x+b'y+c'z+d'=0$ has the equation $\lambda(ax+by+cz+d)+\mu(a'x+b'y+c'z+d')=0$, where λ, μ are finite constants and not both zero.

    Find the equation of the plane through the line L which is parallel to the line $x/l=y/m=z/n$ and, hence or otherwise, show that the two lines are coplanar if, and only if, $d(la'+mb'+nc')=d'(la+mb+nc)$. [L.U.

16.20. The shortest distance between two skew lines

Two straight lines are said to be skew if they are not coplanar, i.e. if they neither intersect nor are parallel.

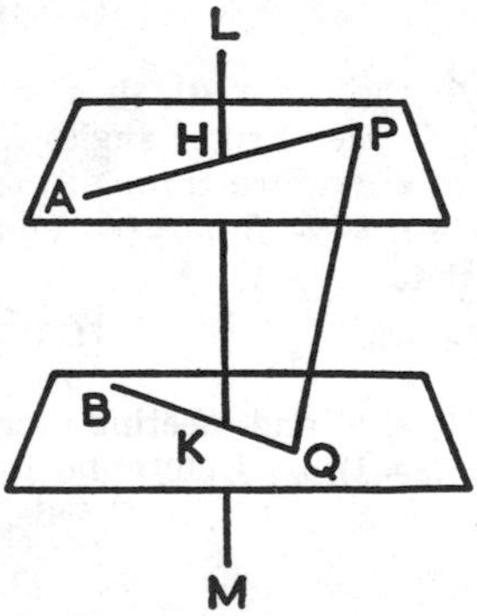

Fig. 59 shows two skew lines AP, BQ which pass through the points $P(x_1, y_1, z_1)$, $Q(x_2, y_2, z_2)$. The shortest distance between them is the intercept HK which they make on their common perpendicular LM.

Fig. 59

Since LM is perpendicular to AP and BQ, HK is the projection of PQ on LM; hence if $[l, m, n]$ are the D.C.s of LM, by § 16.6 (b),

$$HK=|l(x_2-x_1)+m(y_2-y_1)+n(z_2-z_1)| \quad \text{(i)}$$

But if the D.R.s of AP and BQ are $[\lambda_1:\mu_1:\nu_1]$ and $[\lambda_2:\mu_2:\nu_2]$ respectively, the D.R.s $[\lambda:\mu:\nu]$ of their common perpendicular (found from the relations

$$\lambda\lambda_1+\mu\mu_1+\nu\nu_1=0,\ \lambda\lambda_2+\mu\mu_2+\nu\nu_2=0)$$

are given by $\lambda:\mu:\nu=\mu_1\nu_2-\mu_2\nu_1:\nu_1\lambda_2-\nu_2\lambda_1:\lambda_1\mu_2-\lambda_2\mu_1$

and so the D.C.s of HK are

$$\left[\frac{(\mu_1\nu_2-\mu_2\nu_1),\ (\nu_1\lambda_2-\nu_2\lambda_1),\ (\lambda_1\mu_2-\lambda_2\mu_1)}{\sqrt{\{(\mu_1\nu_2-\mu_2\nu_1)^2+(\nu_1\lambda_2-\nu_2\lambda_1)^2+(\lambda_1\mu_2-\lambda_2\mu_1)^2\}}}\right].$$

Substituting these values for l, m, n in (i) we have

$$HK=\left|\frac{(x_2-x_1)(\mu_1\nu_2-\mu_2\nu_1)+(y_2-y_1)(\nu_1\lambda_2-\nu_2\lambda_1)+(z_2-z_1)(\lambda_1\mu_2-\lambda_2\mu_1)}{\sqrt{\{(\mu_1\nu_2-\mu_2\nu_1)^2+(\nu_1\lambda_2-\nu_2\lambda_1)^2+(\lambda_1\mu_2-\lambda_2\mu_1)^2\}}}\right| \tag{16.31}$$

The above method may be used when the length of the shortest distance is required. A different method is adopted when the equations as well as the length of the line of shortest distance must be found (see Example 16 below).

Example 15

Show that the length of the common perpendicular to the lines whose equations referred to rectangular axes are

$$(x-5)/1=y/2=(z+1)/(-1),$$

and
$$(x-2)/1=(y-4)/(-1)=z/1,$$

is $\sqrt{14}$. [Sheffield.]

The line $(x-5)/1=y/2=(z+1)/(-1)$ passes through $P(5,\ 0,\ -1)$; the line $(x-2)/1=(y-4)/(-1)=z/1$ passes through $Q(2,\ 4,\ 0)$. If the common perpendicular to the given lines has D.R.s $[\lambda : \mu : \nu]$

$$\lambda+2\mu-\nu=0 \text{ and } \lambda-\mu+\nu=0 \quad \therefore\ \lambda : \mu : \nu=1 : -2 : -3.$$

Hence the D.C.s of the common perpendicular are

$$[1/\sqrt{14},\ -2/\sqrt{14},\ -3\sqrt{14}].$$

The projection of PQ on this line is, by § 16.6 (*b*),

$$3(1/\sqrt{14})+(-4)(-2/\sqrt{14})+(-1)(-3/\sqrt{14})=\sqrt{14},$$

and this is the shortest distance between the given lines.

Example 16

Find the length and the equations of the shortest distance between the lines $x=y-1=4-z$ *and* $x-2y+9=0,\ x+z-10=0$. [L.U.]

Writing the lines in standard form we have

$$x/1=(y-1)/1=(z-4)/(-1)=\rho \quad . \quad . \quad . \quad \text{(i)}$$

$$x/2=(y-\tfrac{9}{2})/1=(z-10)/(-2)=\tau \quad . \quad . \quad \text{(ii)}$$

The coordinates of any point P on (i) may be taken as $(\rho,\ 1+\rho,\ 4-\rho)$ and similarly Q, any point on (ii) is $(2\tau,\ \tau+\tfrac{9}{2},\ 10-2\tau)$.

By (16.12) the D.R.s of PQ are $[\rho-2\tau : \rho-\tau-\frac{7}{2} : 2\tau-\rho-6]$, and so, by (16.17), PQ will be the common perpendicular to (i) and (ii) if

$$(\rho-2\tau)+(\rho-\tau-\tfrac{7}{2})-(2\tau-\rho-6)=0$$

and
$$2(\rho-2\tau)+(\rho-\tau-\tfrac{7}{2})-2(2\tau-\rho-6)=0$$

i.e. if
$$3\rho-5\tau+\tfrac{5}{2}=0 \text{ and } 5\rho-9\tau+\tfrac{17}{2}=0$$

giving
$$\rho=10, \quad \tau=\tfrac{13}{2}.$$

With these values P is the point $(10, 11, -6)$ and Q is $(13, 11, -3)$.

By (16.6) $PQ^2=18$, $PQ=3\sqrt{2}$.

(This result may be verified by the method of Example 15.)

By (16.10) the equations of PQ are $(x-10)/3=(y-11)/0=(z+6)/3$, i.e. $x-z=16$, $y=11$.

Exercises 16 (*c*)

1. Find the length and equations of the line of shortest distance between the lines

$$(x-5)/1=(y-4)/(-2)=(z-4)/1, \quad (x-1)/7=(y+2)/(-6)=(z+4)/1$$

[Sheffield.]

2. Find the shortest distance between the straight lines

$$(x+1)/3=(y-4)/(-2)=(z-3)/(-1)$$

and
$$x/1=(y-4)/(-3)=(z+1)/2.$$

Find also the coordinates of the points in which it meets them. [L.U.]

3. Show that the shortest distance between the lines

$$x/2=y/(-3)=z/1$$

and
$$(x-2)/3=(y-1)/(-5)=(z+1)/2 \text{ is } \tfrac{2}{3}\sqrt{3}.$$

Show also that the shortest distance lies along the line

$$3x-23=3y+92=3z.$$

[L.U.]

4. Find the length and equations of the shortest distance between the lines $(x-6)/2=(y+4)/5=(z-2)/1$, $(x+1)/(-4)=(y-9)/5=(z-5)/7$.

[L.U.]

5. Find the equations and the magnitude of the shortest distance between the two lines

$$(x-1)/4=(y-1)/3=(z-2)/(-2)$$

and
$$x/4=(y-5)/0=(z-15)/(-1).$$

[L.U.]

6. Find the magnitude and direction of the shortest distance between the lines

$$(x+7)/(-8)=(y-5)/3=(z-4)/1$$

and
$$(x+4)/4=y/3=(z-19)/(-2).$$

[L.U.]

7. Find the length and the equations of the line of shortest distance between the lines

$$(x-2)/0=(y-5)/2=(z-1)/1, \quad (x-8)/2=(y-4)/2=(z+1)/(-1).$$

[Sheffield.]

8. Show that the shortest distance between the lines $x/2=y/(-3)=z/1$ and $(x-2)/3=(y-1)/(-5)=(z+2)/2$ is $\frac{1}{3}\sqrt{3}$.

Show also that the shortest distance lies along the intersection of the planes $4x+y-5z=0$ and $7x+y-8z=31$. [L.U.]

9. Two lines are given by the equations

$$x=2, \quad y-z=1 \text{ and } 6x=3(y-1)=2(z-2).$$

Find the length of their common perpendicular and the coordinates of its feet. [L.U.]

10. Find the length and the equations of the shortest distance between the lines $x+y=0$, $z=4$; and $(x-1)/4=(y-2)/3=(z-36)/(-6)$. [L.U.]

11. Find the coordinates of the point P on the intersection of the two planes $2x+2y-z=0$, $x+y-3z=0$, which has minimum distance from the straight line L passing through the points $(-2, 3, -2)$, $(-5, 5, -3)$.

Find the coordinates of the point Q on L which is nearest to P. [L.U.]

12. Show that the shortest distance between the axis of z and the straight line joining the point $P(r_1, r_1, c)$ on the line $x-y=0$, $z=c$ to the point $Q(r_2, -r_2, -c)$ on the line $x+y=0$, $z=-c$ divides PQ in the ratio $r_1^2 : r_2^2$.

Show also that the straight line along which this distance lies makes an angle $\tan^{-1}(r_1/r_2)$ with the first line and an angle $\tan^{-1}(r_2/r_1)$ with the second line. [L.U.]

13. P and Q are the points $(r \cos \alpha, r \sin \alpha, 0)$ and $(r \cos \beta, r \sin \beta, 0)$ respectively, α and β being acute angles. Find the equations of the line PK, drawn from P perpendicular to OP at an acute angle γ with the z axis and at an acute angle with the x-axis. If also the line QL is perpendicular to OQ, makes an acute angle γ with the z-axis and an acute angle with the x-axis, show that the direction ratios of the common normal of PK and QL are

$$[-\cos \gamma \sin \tfrac{1}{2}(\alpha+\beta) : \cos \gamma \cos \tfrac{1}{2}(\alpha+\beta) : \sin \gamma \cos \tfrac{1}{2}(\alpha-\beta)].$$

Hence find the shortest distance between PK and QL. [L.U.]

14. Two skew lines have the equations $(x-x_1)/l_1=(y-y_1)/m_1=(z-z_1)/n_1$ and $(x-x_2)/l_2=(y-y_2)/m_2=(z-z_2)/n_2$ and $[\lambda : \mu : \nu]$ are direction ratios of their line of shortest distance.

Prove that

$$\begin{vmatrix} x-x_1 & y-y_1 & z-z_1 \\ l_1 & m_1 & n_1 \\ \lambda & \mu & \nu \end{vmatrix} = 0$$

is the equation of the plane containing the first line and the line of shortest distance between the two given lines.

Find the coordinates of the point in which the line of shortest distance between

$$(x+1)/1=(y-1)/2=(z-2)/(-1) \text{ and } (x-1)/2=y/1=(z+1)/(-3)$$

cuts the plane $y=0$. [L.U.]

15. If (a, b, c) are the rectangular cartesian coordinates of a point P and if X, Y, Z are the feet of the perpendiculars from P to the axes, find the equation of the plane XYZ.

If the line l through P perpendicular to the plane XYZ meets this plane at D and the coordinate planes at A, B, C prove that

$$1/PA+1/PB+1/PC=2/PD.$$

Prove that the shortest distance between l and the x-axis is

$$|b^2-c^2|/\sqrt{(b^2+c^2)}. \qquad \text{[L.U.]}$$

16.21. Simplest form of the equations of two skew lines

If the common perpendicular PQ to two skew lines AB and CD (fig. 60) is taken as z-axis and O the mid-point of PQ is taken as origin, the lines AB and CD will lie in the planes $z=c$, $z=-c$ where $2c=PQ$, the shortest distance between the lines. The lines $A'OB'$, $C'OD'$ drawn through the origin parallel to AB, CD respectively define a plane perpendicular to PQ whose equation is $z=0$. Take as Ox and Oy respectively the internal and external bisectors of the angle $C'OB'$. Then, if the angle between AB and CD is 2α, $\angle C'OB'=2\alpha$, $\angle xOB'=\alpha$ and the equations of $A'B'$ and $C'D'$ are

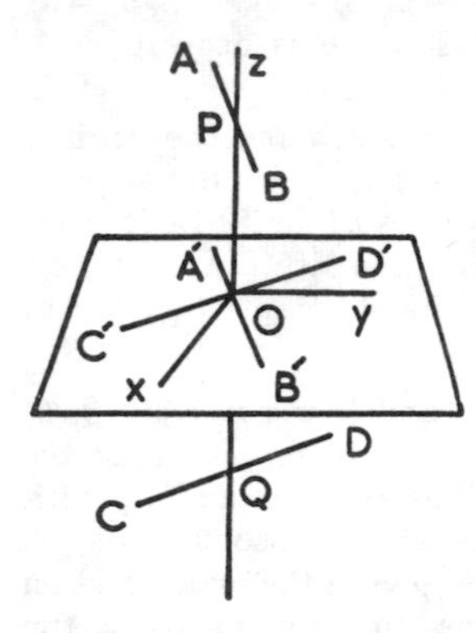

Fig. 60

$$y=x\tan\alpha,\ z=0 \text{ and } y=-x\tan\alpha,\ z=0$$

respectively. It follows that the equations of AB and CD are respectively

$$y=x\tan\alpha,\ z=c$$

and

$$y=-x\tan\alpha,\ z=-c,$$

or, putting $\tan\alpha=m$, $y=mx$, $z=c$ and $y=-mx$, $z=-c$,

Example 17

P and P' are the feet of the common perpendicular to two skew straight lines, Q and Q' are two other points on the respective lines and R is the mid-point of QQ'. If $PQ^2+P'Q'^2$ is constant, show that the locus of R is an ellipse.

[L.U.]

Choose the axes as in § 16.21. Then the equations of PQ, $P'Q'$ are

$$y=mx,\ z=c \quad\text{and}\quad y=-mx,\ z=-c \quad \text{respectively.}$$

Then P is the point $(0, 0, c)$, P' is $(0, 0, -c)$ and we may take Q as the point $(\lambda, m\lambda, c)$ and Q' as $(\mu, -m\mu, -c)$.

Then by (16.6), $PQ^2=\lambda^2(1+m^2)$ and $P'Q'^2=\mu^2(1+m^2)$.

If $\qquad PQ^2+P'Q'^2=k^2$, where k is constant,

$$(\lambda^2+\mu^2)(1+m^2)=k^2$$

$$\therefore\ \lambda^2+\mu^2=k^2/(1+m^2)=2a^2 \text{ (say)} \quad . \quad . \quad . \quad \text{(i)}$$

By (16.2), the coordinates of R, the mid-point of QQ', are

$$x=\tfrac{1}{2}(\lambda+\mu), \quad y=\tfrac{1}{2}m(\lambda-\mu), \quad z=0.$$

Thus R moves on a curve in the xy-plane given by

$$2x=\lambda+\mu,\ 2y/m=\lambda-\mu, \text{ where, by (i), } \lambda^2+\mu^2=2a^2.$$

Eliminating λ and μ from these equations we have

$$4x^2+4y^2/m^2=2(\lambda^2+\mu^2)=4a^2.$$

Hence R traces out in the xy-plane the ellipse whose equation is

$$x^2/a^2+y^2/a^2m^2=1.$$

Example 18

Prove that the equations of any two straight lines may be expressed in the form $y=kx,\ z=c;\ y=-kx,\ z=-c$.

If a variable line intersects both these lines and is equally inclined to them, prove that it lies on one or other of the surfaces $yz=kcx,\ kzx=cy$. [L.U.]

For the first part of this question see § 16.21.

Consider the line PQ joining $P(\lambda, k\lambda, c)$ on the line

$$y=kx, \quad z=c \qquad . \quad . \quad . \quad . \qquad \text{(i)}$$

to $Q\ (\mu, -k\mu, -c)$ on the line

$$y=-kx, \quad z=-c \qquad . \quad . \quad . \quad . \qquad \text{(ii)}$$

The D.C.s of PQ are $[(\lambda-\mu)/p,\ k(\lambda+\mu)/p,\ 2c/p]$, where

$$p=\sqrt{\{(\lambda-\mu)^2+k^2(\lambda+\mu)^2+4c^2\}},$$

and the D.C.s of lines (i) and (ii) are respectively

$$[1/q,\ k/q,\ 0] \text{ and } [1/q,\ -k/q,\ 0],$$

where $q=\sqrt{(1+k^2)}$.

Hence, by (16.14), PQ makes equal angles with (i) and (ii) if

$$(\lambda-\mu)+k^2(\lambda+\mu)=\pm\{(\lambda-\mu)-k^2(\lambda+\mu)\}$$

i.e. if $\lambda-\mu=0$ or if $\lambda+\mu=0$.

Now the equations of PQ are

$$\frac{x-\mu}{\lambda-\mu}=\frac{y+k\mu}{k(\lambda+\mu)}=\frac{z+c}{2c}.$$

If $\lambda-\mu=0$, the equations of PQ are

$$\frac{y+k\mu}{k\mu}=\frac{z+c}{c}, \quad x=\mu,$$

and, eliminating μ, we obtain the equation of the surface on which PQ lies:

$$kx(z+c)=c(y+kx),$$
$$kxz=cy.$$

If $\lambda+\mu=0$, the equations of PQ are

$$\frac{x-\mu}{-\mu}=\frac{z+c}{c}, \quad y=-k\mu$$

and, eliminating μ, we have $yz=kcx$.

Hence if PQ is equally inclined to the two given lines, PQ lies on one or other of the surfaces $yz=kcx,\ kzx=cy$.

Exercises 16 (*d*)

1. Show that coordinate axes can be chosen in relation to a pair of skew lines so that these lines have equations of the form

$$x/1=y/\pm k=(z\mp c)/0$$

for some constants k and c satisfying $0<k<1$, $c>0$.

Points P, Q are taken one on each of these lines so that PQ has constant length $c\sqrt{8}$. Prove that the locus of the mid-point of PQ is an ellipse with eccentricity $\sqrt{(1-k^4)}$. [Sheffield.]

2. The shortest distance between two skew lines is AB. P is a variable point on the line on which A lies, and Q is a variable point on the line on which B lies, so that AQ and BP are at right angles. Prove that the locus of the mid-point of PQ is a hyperbola whose asymptotes are parallel to the given skew lines, provided that the latter are not perpendicular. [L.U.]

3. Show that by a suitable choice of axes the equations of any two non-intersecting straight lines can be written in the form

$$y=x\tan\alpha,\ z=c;\ y=-x\tan\alpha,\ z=-c.$$

Variable points P and Q are taken one on each of the above lines such that the line PQ makes with the z-axis an acute angle θ which is constant. Show that PQ meets the plane $z=0$ in points which lie on an ellipse whose semi-axes are $c\tan\theta\tan\alpha$ and $c\tan\theta\cot\alpha$. [L.U.]

4. Show that the equations of two non-intersecting lines p, p' may simultaneously be expressed in the forms

$$y=mx,\ z=c \quad\text{and}\quad y=-mx,\ z=-c.$$

The common perpendicular to p, p' meets them at A, A' respectively. Points P, P' are taken on p, p' respectively so that $AP.A'P'$ is constant. Prove that the locus of the mid-point of the segment PP' is a pair of conjugate hyperbolas whose common asymptotes are parallel to p, p'. [L.U.]

5. Two lines, L and L', whose equations are respectively $y=x\tan\alpha$, $z=c$ and $y=-x\tan\alpha$, $z=-c$, are met by their shortest distance in A on L and B on L'. Points P and Q are taken on L and L' respectively such that $PQ=AP+BQ$. Show that PQ makes an angle $(\frac{1}{2}\pi-\alpha)$ with the y-axis, and that the locus of the point of intersection of PQ and the plane $y=0$ is a circle on AB as diameter. [L.U.]

6. Two straight lines are met by their line of shortest distance in A and A' and points P and P' are taken on the lines respectively so that $AP.A'P'=c^2$, where c is a constant. If Q divides PP' in the fixed ratio $m:n$, show that Q lies in a plane perpendicular to AA' and find its locus. [L.U.]

7. Two fixed non-intersecting lines meet their common perpendicular at A and A'. P is any point on the line through A, P' is any point on the line through A', and Q is a point on PP' such that

$$AP/A'P'=PQ/QP'.$$

If $AP+A'P'=PP'$, show that $AP.A'P'$ is constant, and that the locus of Q is a circle on AA' as diameter. [L.U.]

8. Show that, by a suitable choice of coordinate axes, the equations of any two skew lines in space can be written in the form

$$y=mx,\ z=c;\quad y=-mx,\ z=-c.$$

If points P and P' are taken on these lines such that $OP=k.OP'$ where O is the origin and k is a positive constant different from 1, prove that the locus of intersection of the line PP' with the plane $z=0$ is a conic.

Show that for $m=1$, $c=1$, $k=3$ the equation of this conic is

$$2x^2-5xy+2y^2+1=0.$$ [L.U.]

9. If r is any transversal of the three lines

$$y=mx,\ z=c;\quad y=0,\ z=0;\quad y=-mx,\ z=-c$$

and s is any transversal of the three lines

$$y=mz,\ x=c;\quad y=0,\ x=0;\quad y=-mz,\ x=-c,$$

prove that r meets s. [L.U.]

10. Show that a rectangular cartesian coordinate system may be chosen so that the equations of two given skew lines take the forms $y=mx$, $z=c$ and $y=-mx$, $z=-c$.

Using this coordinate system, find the equation of the surface traced by a line which meets the two given lines and the x-axis. [L.U.]

CHAPTER 17

THE SPHERE

17.1. The equation of a sphere

If $P(x, y, z)$ is any point on a sphere, centre $C(\alpha, \beta, \gamma)$ and radius a,

$$CP^2 = a^2$$

$$\therefore \quad (x-\alpha)^2 + (y-\beta)^2 + (z-\gamma)^2 = a^2. \tag{17.1}$$

The equation of a sphere with centre the origin and radius a is

$$x^2 + y^2 + z^2 = a^2.$$

The general equation of a sphere may be taken as

$$x^2 + y^2 + z^2 + 2ux + 2vy + 2wz + d = 0, \tag{17.2}$$

i.e. $$(x+u)^2 + (y+v)^2 + (z+w)^2 = u^2 + v^2 + w^2 - d.$$

Comparing this equation with (17.1), we see that (17.2) represents a sphere with centre $(-u, -v, -w)$ and radius $\sqrt{\{u^2+v^2+w^2-d\}}$.

17.2. The diametral form of the equation of a sphere

If $P(x, y, z)$ is any point on a sphere and $A(x_1, y_1, z_1)$, $B(x_2, y_2, z_2)$ are the extremities of a diameter, PA is perpendicular to PB and so by (16.12), page 350, and (16.17), page 352,

$$(x-x_1)(x-x_2) + (y-y_1)(y-y_2) + (z-z_1)(z-z_2) = 0.$$

This is the equation of the sphere on AB as diameter.

17.3. Tangent plane to a sphere

If P is any point on a sphere with centre C, all lines drawn through P perpendicular to CP touch the sphere and all such tangent lines lie in the plane through P perpendicular to CP. This plane is called the *tangent plane* at P.

If (17.2) is the equation of the sphere and P is the point (x_1, y_1, z_1), the D.R.s of CP are $[x_1+u : y_1+v : z_1+w]$ and the equation of the tangent plane at P is by (16.24), page 357,

$$(x_1+u)(x-x_1) + (y_1+v)(y-y_1) + (z_1+w)(z-z_1) = 0,$$

$$xx_1 + yy_1 + zz_1 + ux + vy + wz = x_1^2 + y_1^2 + z_1^2 + ux_1 + vy_1 + wz_1$$

i.e. $$xx_1 + yy_1 + zz_1 + u(x+x_1) + v(y+y_1) + w(z+z_1) + d = 0 \tag{17.3}$$

since P lies on sphere (17.2).

17.4. Condition that a plane should touch a sphere

The perpendicular distance of the centre of a sphere from a tangent plane is equal to the radius of the sphere.

Hence the plane $lx+my+nz=p$ will touch sphere (17.2) if, by (16.27), page 360,

$$(lu+mv+nw+p)/\sqrt{(l^2+m^2+n^2)}=\pm\sqrt{(u^2+v^2+w^2-d)},$$

or

$$(lu+mv+nw+p)^2=(l^2+m^2+n^2)(u^2+v^2+w^2-d). \qquad (17.4)$$

17.5. The intersection of a plane and a sphere

In subsequent articles we shall use the notation

$$S\equiv x^2+y^2+z^2+2ux+2vy+2wz+d,$$

$$S'\equiv x^2+y^2+z^2+2u'x+2v'y+2w'z+d', \quad P\equiv lx+my+nz-p.$$

The section of a sphere by a plane is a circle whose centre is the foot of the perpendicular from C, the centre of the sphere, to the plane, and whose radius r is given by

$$r^2=R^2-h^2, \qquad (17.5)$$

where R is the radius of the sphere and h the distance of the plane from C.

The equation

$$S+kP=0, \qquad (17.6)$$

where k is any constant is seen to represent a sphere. It is satisfied by the simultaneous equations $S=0$ and $P=0$ and so represents for all values of k a sphere passing through the circle of intersection of $S=0$ and $P=0$ when this is real.

17.6. The power of a point

If any straight line drawn through a fixed point P cuts a sphere in A and B, the product $PA.PB$ is constant and is called the *power* of P with respect to the sphere.

Let $P\equiv(x_1, y_1, z_1)$, let the equation of the sphere be $S=0$, and suppose that the D.C.s of PAB are $[l, m, n]$. Then the equations of PAB are

$$\frac{x-x_1}{l}=\frac{y-y_1}{m}=\frac{z-z_1}{n}=r \quad . \quad . \quad . \qquad (i)$$

By substituting for x, y, z from (i) in $S=0$ we obtain the quadratic equation

$$r^2+2r\{(x_1+u)l+(y_1+v)m+(z_1+w)n\}$$
$$+x_1^2+y_1^2+z_1^2+2ux_1+2vy_1+2wz_1+d=0$$

whose roots r_1, r_2 are the measures of PA, PB.

Hence $$PA\,.\,PB = x_1^2+y_1^2+z_1^2+2ux_1+2vy_1+2wz_1+d. \qquad (17.7)$$

Since the right-hand side is independent of l, m, n, $PA\,.\,PB$ is constant.

When P lies outside the sphere, (17.7) gives the square of the length of the tangent drawn from P to the sphere.

17.7. The radical plane

The *radical plane* of two spheres is defined as the locus of points whose powers with respect to the spheres are equal.

If $S=0$ and $S'=0$, are the equations of the spheres, from (17.7) the equation of the radical plane is

$$2x(u-u')+2y(v-v')+2z(w-w')+d-d'=0 \qquad (17.8)$$

i.e. $$S-S'=0$$

The radical plane is perpendicular to the line of centres of the spheres and if the spheres intersect in real points it contains their circle of intersection.

17.8. Orthogonal spheres

Two spheres are *orthogonal* if the tangent planes at any point of their circle of intersection are perpendicular. This condition implies that the square of the distance between the centres of the spheres is equal to the sum of the squares of their radii; hence the spheres $S=0$, $S'=0$ are orthogonal if

$$(u-u')^2+(v-v')^2+(w-w')^2=(u^2+v^2+w^2-d)+(u'^2+v'^2+w'^2-d'),$$
$$2uu'+2vv'+2ww'=d+d'.$$

17.9. Pencils of spheres

If $S=0$, $S'=0$ are the equations of two spheres, for all values of the constant k (except $k=-1$), the equation

$$S+kS'=0 \quad . \quad . \quad . \quad . \quad . \qquad \text{(i)}$$

represents a sphere, and for varying k (i) represents a pencil (or system) of *coaxal* spheres, that is a system of spheres any two of which have the same radical plane (cf. § 12.15 and § 12.18). When $k=-1$, (i) gives $S-S'=0$, the equation of the radical plane of the pencil, which is also the radical plane of the spheres $S=0$, $S'=0$.

If the spheres $S=0$, $S'=0$ intersect, (i) is the equation of the pencil of spheres through their circle of intersection.

It is useful to note that the equation of any sphere which passes through the circle of intersection of the spheres $S=0$, $S'=0$ is of the form (i).

17.10. Polar plane

The *polar plane* of an external point P with respect to a sphere is the plane which contains the points of contact of all the tangent lines drawn from P to the sphere.

Suppose that PA, any tangent line drawn from $P(x_1, y_1, z_1)$, touches the sphere $S=0$ at $A(x_2, y_2, z_2)$. Then by (17.3) the tangent plane at A is

$$xx_2+yy_2+zz_2+u(x+x_2)+v(y+y_2)+w(z+z_2)+d=0,$$

and P lies on this plane

$$\therefore\ x_1x_2+y_1y_2+z_1z_2+u(x_1+x_2)+v(y_1+y_2)+w(z_1+z_2)+d=0.$$

This shows that A lies on the fixed plane

$$xx_1+yy_1+zz_1+u(x+x_1)+v(y+y_1)+w(z+z_1)+d=0. \qquad (17.9)$$

But A is the point of contact of *any* tangent line drawn from P to the sphere and so (17.9) contains the points of contact of all such tangent lines, i.e. (17.9) is the polar plane of P with respect to the sphere.

17.11. Polar lines

Suppose that a line L (which does not pass through the origin) has equations

$$\frac{x-x_1}{l}=\frac{y-y_1}{m}=\frac{z-z_1}{n}=\lambda.$$

Then the polar plane of $P(x_1+\lambda l,\ y_1+\lambda m,\ z_1+\lambda n)$, any point on L with respect to the sphere $x^2+y^2+z^2=a^2$ is by (17.9)

$$(x_1+\lambda l)x+(y_1+\lambda m)y+(z_1+\lambda n)z=a^2,$$

i.e.
$$(xx_1+yy_1+zz_1-a^2)+\lambda(lx+my+nz)=0.$$

By (16.25), page 358, this equation represents for all values of λ a plane passing through L', the line of intersection of the planes

$$xx_1+yy_1+zz_1-a^2=0$$

and
$$lx+my+nz=0.$$

L and L' are called *polar lines* with respect to the sphere; they possess the property that the polar plane of any point on one of them passes through the other.

If L passes through the origin, the polar planes of all points on L are parallel.

17.12. Miscellaneous examples

Example 1

Find the centre and the radius of the sphere whose equation is

$$x^2+y^2+z^2-2x-4y-6z-2=0.$$

Show that the intersection of this sphere and the plane $x+2y+2z-20=0$ *is a circle whose centre is the point* (2, 4, 5), *and find the radius of this circle.* [L.U.]

The sphere has centre $C(1, 2, 3)$ and radius 4.

The normal through C to the given plane has equations

$$(x-1)/1=(y-2)/2=(z-3)/2=\rho$$

and meets the plane at A where

$$(\rho+1)+2(2\rho+2)+2(2\rho+3)-20=0,$$
$$\rho=1.$$

Hence A, the centre of the circle of section, is the point (2, 4, 5).

By (16.6), page 349, $AC=3$, and by (17.5) the radius of the circle is $\sqrt{7}$.

Example 2

Show that the line $l: (x-7)/2=(y-4)/7=(z-13)/10$ *touches the sphere* S : $x^2+y^2+z^2-6x+2y-4z+5=0$.

Find the coordinates of P, *its point of contact with* S. *Find also the equation of the sphere* T *which touches* S *at* P *and passes through the centre of* S. [L.U.]

The line $\qquad (x-7)/2=(y-4)/7=(z-13)/10=\rho$

meets the sphere S whose equation is

$$x^2+y^2+z^2-6x+2y-4z+5=0$$

where

$$(2\rho+7)^2+(7\rho+4)^2+(10\rho+13)^2-6(2\rho+7)+2(7\rho+4)-4(10\rho+13)+5=0.$$

This equation reduces to $(\rho+1)^2=0$, and so the line l meets the sphere S in two coincident points at $P(5, -3, 3)$, i.e. l touches S at P.

Sphere T, which touches S at P and passes through $C(3, -1, 2)$, the centre of S, is the sphere on PC as diameter. Its equation (see § 17.2) is

$$(x-5)(x-3)+(y+3)(y+1)+(z-3)(z-2)=0$$

i.e.
$$x^2+y^2+z^2-8x+4y-5z+24=0.$$

Example 3

Find the equation of the family of spheres through the points $(a, 0, 0)$, $(0, a, 0)$, $(0, 0, a)$.

Prove that one of these spheres passes through the point $(2a, 0, 0)$ *and find the tangent plane at this point. Prove that another sphere of the family touches the same plane at a different point.* [Sheffield.]

The three given points lie on the plane $x+y+z=a$ and also on the sphere $x^2+y^2+z^2=a^2$.

Hence by (17.6) the equation of the required family of spheres is of the form

$$x^2+y^2+z^2-a^2+k(x+y+z-a)=0 \quad . \quad . \quad \text{(i)}$$

The sphere of family (i) which passes through $(2a, 0, 0)$ is given by $k=-3a$. Its equation is

$$x^2+y^2+z^2-3a(x+y+z)+2a^2=0. \quad . \quad . \quad \text{(ii)}$$

By (17.3), the tangent plane to this sphere at $(2a, 0, 0)$ is

$$x-3y-3z=2a. \qquad \text{(iii)}$$

By (17.4), (iii) will touch (i) if

$$(\tfrac{5}{2}k-2a)^2=19(\tfrac{3}{4}k^2+ak+a^2),$$
$$8k^2+29ak+15a^2=0,$$
$$k=-3a, \quad -5a/8.$$

$k=-3a$ gives sphere (ii) ; $k=-5a/8$ gives

$$8x^2+8y^2+8z^2-5ax-5ay-5az-3a^2=0. \qquad \text{(iv)}$$

The point $(2a, 0, 0)$ does not lie on this sphere and so (iv) touches (iii) at another point.

Exercises 17

1. Determine the radius and the centre of the circle of intersection of the two spheres
$$x^2+y^2+z^2-2x+4y-164=0,$$
$$x^2+y^2+z^2-10x-4y+14z-82=0. \qquad \text{[L.U.]}$$

2. Show that the plane (a) $3y+4z-37=0$ touches the sphere (b) $x^2+y^2+z^2-6x-8y=0$, and find the point of contact.
   Find also the equations of the planes which are parallel to (a) and cut (b) in a circle of radius 4 units. [L.U.]

3. Prove that the sphere $x^2+y^2+z^2-2x-2y-2z+1=0$ touches the coordinate axes and find the coordinates of the points of contact. Find also the centre and radius of the circle formed by the intersection of the sphere and the plane through these points of contact. [L.U.]

4. A plane equally inclined to the coordinate axes cuts a sphere which passes through the origin in a circle of radius 2 and centre $(1, 2, -1)$.
   Find the distance between the centres of the circle and the sphere. Find also the equation of the tangent plane through the origin.

5. Find the centre and radius of the circle of intersection of the sphere $x^2+y^2+z^2+12x-12y-16z+111=0$, and the plane $2x+2y+z-17=0$.
   Show that there exist two planes through the origin which meet the above plane at right angles and touch the sphere. [L.U.]

6. Find the condition that $lx+my+nz+p=0$ should be a tangent plane to the sphere $x^2+y^2+z^2+2ux+2vy+2wz+d=0$.
   Find the equations of the tangent planes to the sphere
$$x^2+y^2+z^2-2x-4y+2z-219=0$$
   which intersect in the line $3(x-10)=-4(y-14)=-6(z-2)$. [L.U.]

7. Prove that the plane $x+2y+2z=8$ touches the sphere
$$x^2+y^2+z^2-2x-4y+6z+5=0.$$
   Find the equation of the other tangent plane through the line of intersection of the plane $x+2y+2z=8$ and the plane $x=0$. Find also the coordinates of its point of contact. [L.U.]

8. Find the equations of the two spheres which pass through the circle

$$x^2+y^2+z^2+6x-2y-4z=0, \quad 4x+2y+2z-5=0$$

and touch the plane $z=0$.

If P and Q are the points of contact of the spheres and the plane $z=0$, show that the plane of the circle bisects PQ. [L.U.]

9. Find the equations of the two spheres which touch the plane $z=4a$ and which intersect the plane $x+y+z=3a$ in a circle of radius a and centre (a, a, a). [L.U.]

10. Find the condition that the plane $lx+my+nz+p=0$ should touch the sphere $(x-a)^2+(y-b)^2+(z-c)^2=r^2$.

Three spheres S_1, S_2, S_3 have centres (0, 0, 0), (3, 0, 0), (0, 30, 0) and radii 1, 1, 19 respectively. Find the equations of all common tangent planes π of the three spheres such that S_1 and S_3 lie on opposite sides, S_2 and S_3 on the same side of π. Show that there are two such planes and that the acute angle between them is ϕ, where $\cos\phi=7/9$. [L.U.]

11. A variable sphere touches the xy-plane at P and passes through the points (0, 0, 2), (2, 0, 1). Prove that the locus of P is a circle, and find its centre and radius. [Sheffield.]

12. Find the equations of the tangent planes to the sphere

$$x^2+y^2+z^2-4x+6y-2z=0$$

which pass through the line given by

$$2x-y+z=0 \text{ and } 4x+y+5z+18=0.$$

Find also the coordinates of the point of contact of one of the tangent planes. [L.U.]

13. Three spheres have centres (0, 0, 0), $(3a, 0, 0)$, $(0, 4a, 0)$ and radii a, $2a$, $3a$ respectively, and two planes, making an acute angle ϕ with each other, are such that every one of the spheres touches the two planes. Show that $\cos\phi=5/18$. [L.U.]

14. Find the equation of the sphere which has its centre at the point $(2, 3, -1)$ and touches the line $(x-13)/10=(y-8)/3=(z+7)/(-8)$.

Find also the equation of the tangent plane to the sphere which contains the above tangent line. [L.U.]

15. Find the equation of the sphere with centre $(a, 2a\ 4a)$ which touches the line $2x=y=z$.

Find also the radius and centre of the circle in which the polar plane of the origin with respect to the above sphere cuts the sphere. [L.U.]

16. A sphere is drawn through the points (2, 0, 0) and (4, 0, 0) to touch the straight line $y=z$, $x=0$.

Prove that its centre must lie on one or other of two parallel straight lines, and find the equations of those lines. [Leeds.]

17. A straight line L passes through the origin and is a tangent to each of the spheres $x^2+y^2+z^2+2ax+p=0$, $x^2+y^2+z^2+2by+q=0$.
Show that the angle ϕ between L and the z-axis is given by
$$\sin^2\phi=(p/a^2)+(q/b^2).$$
Also find the distance between the points of contact of L with the spheres. [L.U.]

18. Find the condition that the line joining the points $(\lambda, 0, c)$, $(0, \mu, -c)$ should touch the sphere $x^2+y^2+z^2=c^2$.
Variable points P, Q are taken one on each of two given perpendicular skew lines; the common perpendicular to these lines meets them at A and B, and the line PQ touches the sphere on AB as diameter. Prove that the mid-point of PQ lies on one of two fixed hyperbolas. [Sheffield.]

19. Find the condition that the line $x/l=y/m=z/n$ should touch the sphere $x^2+y^2+z^2+2ax+2by+2cz+g=0$, and find also the coordinates of the point of contact.
Hence show that the equation of the tangent cone from the origin is $g(x^2+y^2+z^2)=(ax+by+cz)^2$. [L.U.]

20. A line with direction ratios $[l:m:n]$ is drawn through the fixed point $(0, 0, a)$ to touch the sphere $x^2+y^2+z^2=2ax$. Prove that $m^2+2nl=0$.
Find the coordinates of the point P in which this line meets the plane $z=0$ and prove that, as the line varies, P traces out the parabola $y^2=2ax$, $z=0$. [L.U.]

21. Show that the equations of any two straight lines can be put in the form $y=x\tan\alpha$, $z=a$ and $y=-x\tan\alpha$, $z=-a$.
The above two lines are intersected by their shortest distance at A and A_1, and the mid-point of AA_1 is O. P is a point on the line through A, and P_1 is a point on the line through A_1. If $AP.A_1P_1$ has either of the values $a^2\sec^2\alpha$ or $-a^2\operatorname{cosec}^2\alpha$, show that PP_1 touches the sphere whose centre is O and whose radius is a. [L.U.]

22. (i) Show that the sphere
$$(x-\alpha-p)(x-\alpha)+(y-\beta-q)(y-\beta)+(z-\gamma-r)(z-\gamma)=a^2$$
intersects the sphere $(x-\alpha)^2+(y-\beta)^2+(z-\gamma)^2=a^2$ along a great circle of the second sphere.
(ii) Find the equation of the sphere of minimum radius which belongs to the coaxal system defined by the spheres
$$x^2+y^2+z^2+2x-2y+4z+2=0,$$
$$x^2+y^2+z^2+4x+2y-4z=0.$$ [L.U.]

23. Find the coordinates of the centre of the circle circumscribing the triangle whose vertices are (2, 0, 4) (2, 4, 2) (0, 2, 4).
Find also the equation of the sphere which passes through these three points and also through the origin. [L.U.]

24. Show that the square of the length of the perpendicular from the origin to the straight line $(x-a)/l=(y-b)/m=(z-c)/n$ is

$$\{(mc-nb)^2+(na-lc)^2+(lb-ma)^2\}/(l^2+m^2+n^2).$$

If the straight line

$$y-mx+\lambda(z-c)=0,$$
$$y+mx+\mu(z+c)=0$$

touches the sphere $x^2+y^2+z^2=c^2$, show that

$$\lambda\mu=-1 \text{ or } \lambda\mu=m^2.$$ [L.U.]

25. Show that the line $(x-6)/3=(y-7)/4=(z-3)/5$ touches the sphere $x^2+y^2+z^2=2x+4y+4$ and find the coordinates of the point of contact.

Find the equations of the two tangent planes to this sphere which contain the line $(x-7)/5=(y-11)/6=(z-3)/(-2)$. [L.U.]

26. The plane of intersection of a sphere S and the sphere

$$x^2+y^2+z^2-4x+6y+4z-8=0$$

is $3x-12y-4z=25$. If the spheres intersect orthogonally, find the equation of S. [Leeds.]

27. Find the equation of the sphere which passes through the points $(0, 0, 0)$, $(2a, 0, 0)$, $(0, 2b, 0)$, $(0, 0, 2c)$. Determine its centre and radius.

Find the equations of the two spheres each of which passes through the points $(0, 0, 0)$, $(2, 0, 0)$, $(0, 2, 0)$ and touches the line $x=y=z-1$.

Show that the two spheres intersect orthogonally. [L.U.]

28. Prove that there exist eight spheres, of radius 5 units, which are orthogonal to the sphere $x^2+y^2+z^2=16$, touch the x-axis, and cut off a segment of length 2 units from the y-axis. [L.U.]

29. Prove that the polar planes of all points on the straight line

$$(x-3)/2=(y-2)/1=(z+1)/3$$

with respect to the sphere $x^2+y^2+z^2+2x-4y-6z+10=0$ pass through a fixed line which is perpendicular to the given line. [L.U.]

30. Find the polar plane of $P(1, 2, 3)$ with respect to the sphere whose centre is $C(2, -2, 1)$ and whose radius is 3. Find also the foot, Q, of the perpendicular drawn from C to this plane, and verify that $CP.CQ=9$. [L.U.]

31. Find the equation of the surface on which a point must lie if its polar plane with respect to the sphere $x^2+y^2+z^2-2x-4y-8z+3=0$ is a tangent plane to the sphere $x^2+y^2+z^2-x-y-z=0$. [L.U.]

32. Find the equation of the polar plane of the point (x', y', z') with respect to the sphere $x^2+y^2+z^2=a^2$.

Prove that the polar planes of points on a line l all pass through a line l' perpendicular to l.

Show that if l is the variable line $(x-at)/1=(y-at)/(-1)=z/2t$, where t is a parameter, then the line l' lies on the surface $x^2-y^2+2az=0$.

[Leeds.]

33. Show that, by a suitable choice of axes, the equations of any two non-intersecting straight lines can be written in the form

$$y=mx,\ z=c\ ;\ y=-mx,\ z=-c.$$

The shortest distance between two skew lines is AB. P is any point on the line through A, and M is the mid-point of AP. Q is any point on the line through B, and N is the mid-point of BQ. Prove that if OM, ON are perpendicular, where O is the mid-point of AB, then AQ and BP are perpendicular. Show also that the radical plane of the spheres with diameters MN and PQ is the polar plane of O with respect to the second sphere. [L.U.]

34. Show that rectangular axes of coordinates may be chosen so that the equations of two given skew lines take the forms

$$y-mx=0=z-c,\quad y+mx=0=z+c.$$

Show that the centre of a sphere, to which each of the lines is a tangent, lies on the surface $mxy+(1+m^2)cz=0$. [L.U.]

35. Two spheres, of equal radius a, lie in the positive octant of the co-ordinate frame. One sphere touches the three coordinate axes and the other touches the three coordinate planes. Obtain the equations of the two spheres.

Prove that the circle in which the two spheres intersect lies in the plane $x+y+z=3a(2+\sqrt{2})/4$, and find the coordinates of the centre and the radius of this circle. [L.U.]

36. Find the coordinates of the centre and the radius of the sphere

$$x^2+y^2+z^2-2x-4y+6z=2.$$

Show that the intersection of the above sphere and the sphere

$$x^2+y^2+z^2-4x-6y+4z+4=0$$

is a circle lying in the plane $x+y+z=3$.

Find the coordinates of the centre and the radius of this circle. [L.U.]

37. Show that if the plane $px+qy+z=5$ touches the sphere $x^2+y^2+z^2=16$, then $16(p^2+q^2)=9$.

Find the condition for this plane to touch the sphere

$$(x+5)^2+y^2+z^2=25.$$

Hence find the equations of the real common tangent planes of these two spheres which pass through the point (0, 0, 5). [L.U.]

38. If the plane $ax+by+cz+d=0$ touches the sphere

$$(x-1)^2+y^2+(z-1)^2=1,$$

prove that $a^2+b^2+c^2=(a+c+d)^2$.

If this plane also passes through the point (0, 0, 2), show that it cuts the plane $z=0$ in a line which touches the parabola $y^2=4x$, $z=0$. [L.U.]

CHAPTER 18

THE QUADRIC

18.1. The equation of a surface

An equation of the first degree in x, y and z represents a plane. We now interpret other equations in x, y and z.

Consider first the equation

$$f(x, y)=0 \quad . \quad . \quad . \quad . \quad \text{(i)}$$

in which z is absent. Points in the xy-plane which satisfy (i) lie on a curve C. If through any point P $(x_0, y_0, 0)$ on this curve a line PP' is drawn parallel to the z-axis, the coordinates of Q any point on the line PP' may be taken as (x_0, y_0, z_0). Then since P lies on (i), $f(x_0, y_0)=0$ and it follows, since (i) is independent of z, that the coordinates of Q also satisfy (i) no matter what value z_0 may have. Hence the coordinates of any point on PP' satisfy (i). But P is any point on C and so (i) is the equation of the surface of the cylinder generated by straight lines drawn parallel to Oz through points on C. For example, the equation $x^2+y^2=a^2$ represents a right circular cylinder of radius a with axis along Oz. In the same way the equation $\dfrac{x^2}{a^2}+\dfrac{z^2}{c^2}=1$ represents an elliptic cylinder with axis along Oy.

We now consider the more general equation

$$F(x, y, z)=0 \quad . \quad . \quad . \quad . \quad \text{(ii)}$$

Points in the plane $z=k$ whose coordinates satisfy (ii) have x and y coordinates such that $F(x, y, k)=0$. This means that these points lie on the curve C_k (say) in which the plane $z=k$ cuts the cylinder $F(x, y, k)=0$. As k varies the curve C_k also varies and generates a surface of which (ii) is the equation.

If $F(x, y, z)=0$ is a homogeneous equation of degree $n(n>0)$ in x, y and z, O the origin lies on the surface and

$$F(kx, ky, kz)=k^nF(x, y, z)$$

for all values of k. (See § 19.12.)

It follows that if the point $P(x, y, z)$ lies on the surface, $Q(kx, ky, kz)$, any point on the straight line OP also lies on the surface. The surface is therefore a cone with vertex at the origin.

18.2. Quadric surfaces

All surfaces represented by equations of the second degree in x, y, and z are known as *quadric surfaces* or *quadrics*. It can be shown that any plane section of a quadric is a conic and so quadrics are also given the name *conicoids*.

18.3. Central quadrics

If $P(\alpha, \beta, \gamma)$ lies on the surface

$$ax^2+by^2+cz^2=1 \quad . \quad . \quad . \quad . \quad \text{(i)}$$

$Q(-\alpha, -\beta, -\gamma)$ also lies on the surface, and O the origin is the mid-point of PQ. Hence all chords of (i) which pass through O are bisected at O. For this reason (i) is called a *central quadric*, O is called its *centre* and a chord through O is called a *diameter*.

Surface (i) is symmetrical about each of the coordinate planes, for if the point (α, β, γ) lies on (i) so do $(-\alpha, \beta, \gamma)$, $(\alpha, -\beta, \gamma)$ and $(\alpha, \beta, -\gamma)$. These three planes of symmetry are called the *principal planes* and the coordinate axes the *principal axes* of the quadric.

18.4. Equations of central quadrics

If in (i) we write $a=1/A^2$, $b=1/B^2$, $c=1/C^2$ we obtain the equation

$$\frac{x^2}{A^2}+\frac{y^2}{B^2}+\frac{z^2}{C^2}=1,$$

which is the standard form of the equation of an ellipsoid (fig. 61 (*a*)).

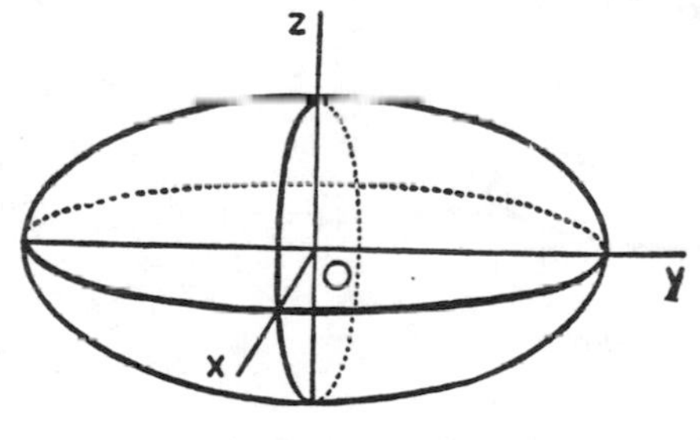

Fig. 61 (*a*)

Sections of this surface parallel to the coordinate planes are ellipses. For example, the equations of the section of this surface by the plane $z=k$ are

$$\frac{x^2}{A^2}+\frac{y^2}{B^2}=1-\frac{k^2}{C^2}, \quad z=k$$

and these are the equations of an ellipse when $k^2 \leqslant C^2$. When $k^2 > C^2$ the plane does not cut the surface in real points. Similarly sections parallel to the other coordinate planes are ellipses. The surface meets the coordinate axes in the points $(\pm A, 0, 0)$, $(0, \pm B, 0)$ $(0, 0, \pm C)$.

If $A=B=C$, the surface is a sphere.

If in (i) we write $a=1/A^2$, $b=1/B^2$, $c=-1/C^2$, we obtain the equation

$$\frac{x^2}{A^2}+\frac{y^2}{B^2}-\frac{z^2}{C^2}=1,$$

which is the standard form of the equation of a hyperboloid of one sheet (fig. 61 (*b*)).

Sections of the surface parallel to the yz- and zx-planes are hyperbolas; **those parallel to** the xy-plane are ellipses. For example, the section by the plane $z=k$ is given by

$$\frac{x^2}{A^2}+\frac{y^2}{B^2}=1+\frac{k^2}{C^2}, \; z=k$$

and for all values of k, this is an ellipse.

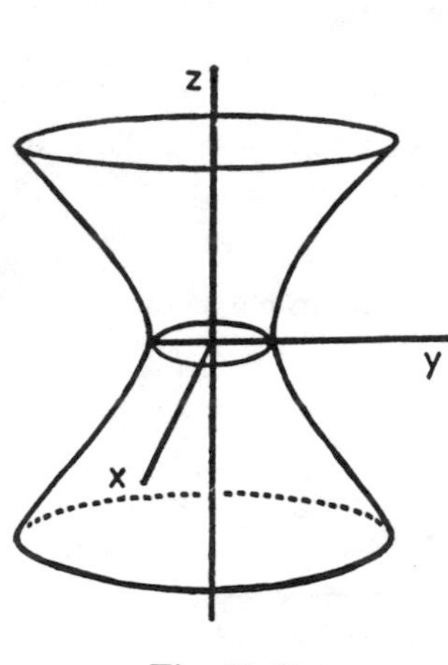

Fig. 61 (*b*)

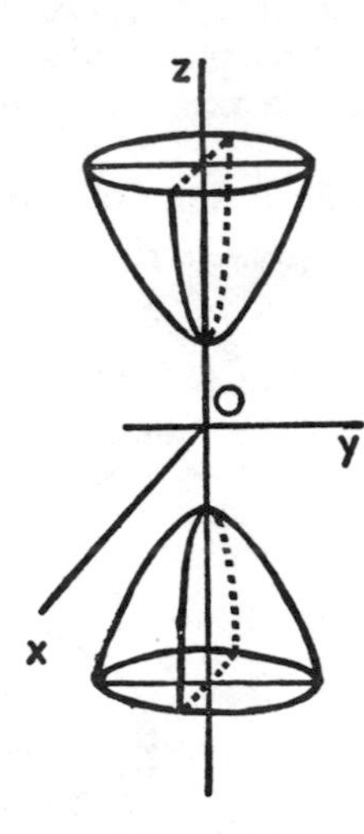

Fig. 61 (*c*)

The section by the plane $y=k$ is given by

$$\frac{x^2}{A^2}-\frac{z^2}{C^2}=1-\frac{k^2}{B^2}, \; y=k,$$

which is a hyperbola except in the case where $k=\pm B$, when it is a pair of straight lines.

Similarly, sections parallel to the yz-plane are hyperbolas.

When $a=-1/A^2$, $b=-1/B^2$, $c=1/C^2$, we obtain the equation

$$-\frac{x^2}{A^2}-\frac{y^2}{B^2}+\frac{z^2}{C^2}=1,$$

which is the standard form of the equation of a hyperboloid of two sheets (fig. 61 (*c*)).

The equations of the section of this surface by the plane $z=k$ are

$$\frac{x^2}{A^2}+\frac{y^2}{B^2}=\frac{k^2}{C^2}-1,\ z=k.$$

These are the equations of an ellipse when $k^2 \geqslant C^2$. When $k^2 < C^2$ the plane does not cut the surface in real points. Sections parallel to the yz- and zx-planes are hyperbolas.

18.5. The intersections of a line and a quadric

Suppose that the straight line

$$\frac{x-x_1}{\lambda}=\frac{y-y_1}{\mu}=\frac{z-z_1}{\nu}=r \tag{18.1}$$

drawn through (x_1, y_1, z_1) in the direction $[\lambda : \mu : \nu]$ meets the quadric

$$ax^2+by^2+cz^2=1 \tag{18.2}$$

at the point $(x_1+\lambda r, y_1+\mu r, z_1+\nu r)$. Then the parameter r is given by the quadratic equation

$$a(x_1+\lambda r)^2+b(y_1+\mu r)^2+c(z_1+\nu r)^2=1,$$

$$r^2(a\lambda^2+b\mu^2+c\nu^2)+2r(a\lambda x_1+b\mu y_1+c\nu z_1)+ax_1^2+by_1^2+cz_1^2-1=0. \tag{18.3}$$

Hence, in general, a given line meets the quadric in two points, and each plane section of the quadric is therefore a conic.

18.6. Tangent plane and normal to a quadric

Suppose that the line (18.1) is drawn through $P(x_1, y_1, z_1)$ to meet the quadric (18.2) in A and B. Then if

$$ax_1^2+by_1^2+cz_1^2=1, \quad . \quad . \quad . \quad . \tag{i}$$

P lies on the quadric (18.2), one of the roots of (18.3) is zero and P coincides with either A or B. If, in addition,

$$a\lambda x_1+b\mu y_1+c\nu z_1=0, \quad . \quad . \quad . \tag{ii}$$

(18.3) has a second zero root, A and B coincide at $P(x_1, y_1, z_1)$ and (18.1) touches the quadric at P.

Now (ii) is the condition that (18.1) should be perpendicular to the *fixed* direction which has D.R.s $[ax_1 : by_1 : cz_1]$; hence all tangent lines at P lie in a plane through P perpendicular to this direction.

This plane is known as the tangent plane at P and by (16.24), page 357, its equation is

$$ax_1(x-x_1)+by_1(y-y_1)+cz_1(z-z_1)=0,$$

i.e. by (i),

$$axx_1+byy_1+czz_1=1. \tag{18.4}$$

The equations of the line through P perpendicular to this plane are

$$\frac{x-x_1}{ax_1}=\frac{y-y_1}{by_1}=\frac{z-z_1}{cz_1}. \tag{18.5}$$

This line is the normal at P

The D.C.s of the normal at P are often taken as $[pax_1, pby_1, pcz_1]$. Then $p^2(a^2x_1^2+b^2y_1^2+c^2z_1^2)=1$, and by (16.27), page 360, p is the perpendicular distance of the origin from the tangent plane at P.

18.7. The condition for a plane to touch a quadric

If the plane
$$lx+my+nz=p \quad . \quad . \quad . \quad \text{(i)}$$
touches the quadric (18.2) at $P(x_1, y_1, z_1)$, equations (i) and (18.4) represent the same plane

$$\therefore \; ax_1/l=by_1/m=cz_1/n=1/p.$$

i.e. $$x_1=l/ap, \quad y_1=m/bp, \quad z_1=n/cp.$$

But $ax_1^2+by_1^2+cz_1^2=1$, since P lies on the quadric.

$$\therefore \; p^2=l^2/a+m^2/b+n^2/c, \qquad (18.6)$$
$$p=\pm\sqrt{(l^2/a+m^2/b+n^2/c)}.$$

Hence there are two tangent planes to the quadric which are parallel to the plane $lx+my+nz=0$.

Example 1

The normal at the point P of the ellipsoid

$$x^2+4y^2+\tfrac{1}{4}z^2=1$$

meets the ellipsoid again at Q. The foot of the perpendicular from O to the tangent plane at P is M. If $PQ.OM^3=-2$, show that P lies on the ellipsoid

$$x^2+64y^2+\tfrac{1}{64}z^2=1. \qquad \text{[L.U.]}$$

If the equations of the normal at $P(x_1, y_1, z_1)$ are taken in the form

$$\frac{x-x_1}{px_1}=\frac{y-y_1}{4py_1}=\frac{4(z-z_1)}{pz_1}=r$$

where $$p^2(x_1^2+16y_1^2+\tfrac{1}{16}z_1^2)=1 \quad . \quad . \quad . \quad . \quad \text{(i)}$$

by (16.8), page 350, r is the distance between the point (x, y, z) and P.

The normal meets the ellipsoid

$$x^2+4y^2+\tfrac{1}{4}z^2=1 \quad . \quad . \quad . \quad . \quad \text{(ii)}$$

where $$x_1^2(1+pr)^2+4y_1^2(1+4pr)^2+\tfrac{1}{4}z_1^2(1+\tfrac{1}{4}pr)^2=1,$$

i.e. $$p^2r^2(x_1^2+64y_1^2+\tfrac{1}{64}z_1^2)+2pr(x_1^2+16y_1^2+\tfrac{1}{16}z_1^2)=0,$$

since P lies on (ii).

The non-zero root of this equation measures the distance between P and the point Q at which the normal meets (ii) again

$$\therefore \; PQ=-2/\{p^3(x_1^2+64y_1^2+\tfrac{1}{64}z_1^2)\} \text{by (i); and } p=OM. \text{ (See § 18.6.}$$

Hence, if $$PQ.OM^3=-2, \quad x_1^2+64y_1^2+\tfrac{1}{64}z_1^2=1$$

i.e. P lies on the ellipsoid $x^2+64y^2+\tfrac{1}{64}z^2=1$.

Example 2

P is a point on the ellipsoid $\frac{1}{5}x^2+\frac{1}{2}y^2+z^2=a^2$. The normal at P to the ellipsoid intersects the plane $z=0$ at the point Q, and the tangent plane at P intersects this plane in the line (l). Show that if Q lies on the hyperbola $xy=a^2$, $z=0$, then the line (l) touches the same hyperbola. [L.U.]

The equations of the normal at $P(x_1, y_1, z_1)$ to the given ellipsoid are

$$\frac{5(x-x_1)}{x_1}=\frac{2(y-y_1)}{y_1}=\frac{z-z_1}{z_1}.$$

This normal meets the plane $z=0$ at the point $Q(\frac{4}{5}x_1, \frac{1}{2}y_1, 0)$ which lies on the hyperbola

$$xy=a^2, \quad z=0 \quad \quad . \quad . \quad . \quad . \quad \text{(i)}$$

if

$$2x_1y_1=5a^2. \quad \quad . \quad . \quad . \quad . \quad \text{(ii)}$$

The tangent plane at P meets the plane $z=0$ in the line (l) given by

$$\tfrac{1}{5}xx_1+\tfrac{1}{2}yy_1=a^2, \quad z=0.$$

This line meets the hyperbola (i) in points whose abscissae are given by

$$2x(5a^2-xx_1)/5y_1=a^2,$$

i.e.

$$2x_1x^2-10a^2x+5a^2y_1=0. \quad \quad . \quad . \quad . \quad \text{(iii)}$$

(l) touches the hyperbola if (iii) has two equal roots, i.e. if $2x_1y_1=5a^2$, which is condition (ii).

Hence if Q lies on the hyperbola $xy=a^2$, $z=0$, the line (l) touches the same hyperbola.

Example 3

Find the equations of the tangent planes to the ellipsoid $3x^2+4y^2+5z^2=48$ which pass through the line $(x-2)/2=(y-3)/(-3)=(5z-6)/6$. Find the point of contact in each case.

The given line l is the intersection of the planes

$$3x+2y-12=0 \text{ and } 2y+5z-12=0,$$

and by (16.25), page 358, any plane through l is of the form

$$3x+2y-12+k(2y+5z-12)=0$$

i.e.

$$3x+2(1+k)y+5kz=12(1+k).$$

By (18.6) this plane will touch the given ellipsoid if

$$144+48(1+k)^2+240k^2=144(1+k)^2$$

from which

$$k=\tfrac{1}{3} \text{ or } 1.$$

Hence the equations of the required planes are

$$9x+8y+5z=48$$

and

$$3x+4y+5z=24.$$

Comparing these results with the plane $3xx_1+4yy_1+5zz_1=48$ which touches the ellipsoid at (x_1, y_1, z_1) we see that the points of contact are (3, 2, 1) and (2, 2, 2) respectively.

Exercises 18 (*a*)

1. Find the equations of the tangent plane and normal to the ellipsoid $ax^2+by^2+cz^2=1$ at the point (x_0, y_0, z_0).

   If the normal to this surface at P meets the coordinate planes in E, F and G, show that $PE:PF:PG=1/a:1/b:1/c$.

   If further, $PE^2+PF^2+PG^2=l^2$, where l is constant, show that P also lies on the ellipsoid

$$a^2x^2+b^2y^2+c^2z^2=a^2b^2c^2l^2/(b^2c^2+c^2a^2+a^2b^2).$$

[L.U.]

2. Determine the equations of the tangent planes to the ellipsoid $x^2+2y^2+4z^2=25$ at the points $A(1, 2, 2)$, $B(3, 0, 2)$ and show that the angle θ between the line AB and the line of intersection of these tangent planes is given by $\cos\theta=4/\sqrt{178}$. [L.U. Anc.]

3. Prove that the normals to the ellipsoid $x^2/3+y^2/2+z^2=1$ at its points of intersection with the cylinder $16x^2+9y^2=54$, meet the plane $z=0$ in points lying on a circle. [L.U.]

4. Normals are drawn to the ellipsoid $x^2/3+y^2/2+z^2=1$ at its intersections with the ellipsoid $x^2+y^2/2+z^2/3=1$.

   Show that the locus of the points at which they meet the plane $z=0$ is the ellipse $3x^2+2y^2=1$, $z=0$. [L.U.]

5. At the points where the plane $x+y+z=0$ intersects the ellipsoid $x^2+2y^2+3z^2=6$, normals to the ellipsoid are drawn. Show that these normals meet the plane $z=0$ on the ellipse $3x^2+9xy+15y^2=2$. [L.U.]

6. The point P on the ellipsoid $3x^2+2y^2+4z^2=a^2$ is such that the normal at P intersects the plane $z=0$ at a point lying on the parabola $y^2=4hx$. Show that the tangent plane at P intersects the plane $z=0$ in a line which touches the parabola $4hy^2+3a^2x=0$. [L.U.]

7. The normal at P to the ellipsoid $3x^2+2y^2+z^2=1$ intersects the plane $z=0$ at the point N. If the tangent plane at P touches the sphere whose centre is $(0, 1, 0)$ and whose radius is $2\sqrt{2}$, show that as P varies, the locus of N is the circle $12x^2+12y^2-4y+7=0$, $z=0$. [L.U.]

8. Prove that the normals to the ellipsoid $2x^2+3y^2+6z^2=6$ at its points of intersection with the plane $x+y+z=0$ meet the plane $z=0$ in points lying on an ellipse. [L.U.]

9. Prove that if the normal at a point on the conicoid $ax^2+by^2+cz^2=1$ touches the sphere $x^2+y^2+z^2=r^2$ the point must also lie on the surface $(x^2+y^2+z^2-r^2)(a^2x^2+b^2y^2+c^2z^2)=1$. [L.U.]

10. A family of ellipsoids is determined by varying the parameter λ in the equation $x^2/a^2+y^2/b^2+z^2/c^2=\lambda$ and the line through the fixed point $(\alpha, \beta, 0)$ parallel to the z-axis meets one of these surfaces at the point P. Prove that the normal at P passes through a fixed point on the plane $z=0$ as P moves on this line. [L.U.]

11. Write down the equation of the tangent plane at the point $P(x_1, y_1, z_1)$ to the ellipsoid $x^2/a^2+y^2/b^2+z^2/c^2=1$ and show that, if the normal at P meets the plane $x=0$ at G, then PG is of length a^2/p, where p is the distance of the tangent plane from the centre of the ellipsoid.

If through G a line GH is drawn parallel to the x-axis so that $PG=GH$ show that the locus of H is the conicoid

$$x^2/a^2-y^2/(a^2-b^2)-z^2/(a^2-c^2)=1. \qquad \text{[L.U.]}$$

12. Find the equation of the tangent plane to the ellipsoid

$$x^2/a^2+y^2/b^2+z^2/c^2=1$$

at the point (x', y', z') and deduce the condition that the plane

$$lx+my+nz+1=0$$

may be a tangent plane to this ellipsoid.

Show that the two tangent planes to the ellipsoid that can be drawn through a tangent to the ellipse

$$x^2/(a^2+c^2)+y^2/(b^2+c^2)=1, \quad z=0$$

are at right angles. [L.U.]

13. Find the equations of the two tangent planes to the ellipsoid

$$x^2+y^2+\tfrac{1}{2}z^2=1$$

which pass through the line $z=0$, $x+y=10$. [L.U.]

14. Find the equations of the tangent planes to the ellipsoid

$$x^2/36+y^2/9+z^2/4=1$$

which pass through the two points $(9, 0, 0)$ and $(7, 4, -2)$. [Leeds.]

15. Find the equations of the tangent planes to $x^2+y^2+4z^2=1$ which intersect in the line whose equations are $12x-3y-5=0$, $z=1$. [L.U.]

16. Prove that the plane $lx+my+nz=1$ touches the ellipsoid

$$x^2/a^2+y^2/b^2+z^2/c^2=1$$

at the point $P(a^2l, b^2m, c^2n)$ where

$$a^2l^2+b^2m^2+c^2n^2=1.$$

The normal to the ellipsoid at P meets the plane through the origin parallel to the tangent plane at P at Q. Show that the coordinates of Q are $\{(a^2-\theta)l, (b^2-\theta)m, (c^2-\theta)n\}$, where $\theta(l^2+m^2+n^2)=1$.

If the tangent plane to the ellipsoid at P also touches the sphere $x^2+y^2+z^2=r^2$, show that Q lies on the ellipsoid

$$x^2/(a^2-r^2)^2+y^2/(b^2-r^2)^2+z^2/(c^2-r^2)^2=1/r^2. \qquad \text{[L.U.]}$$

17. Find the equations of the two tangent planes to the ellipsoid

$$x^2+y^2+2z^2=a^2$$

which intersect in the line

$$y=\alpha x+\beta, \quad z=0.$$

Prove that if these planes are perpendicular, this line touches a circle of radius $a\sqrt{(3/2)}$ whose centre is the centre of the ellipsoid. [L.U.]

18. Find the equations of the two tangent planes to the ellipsoid $x^2/a^2+y^2/b^2+z^2/c^2=1$ which pass through the line $ay=bx$, $z=3c$ and show that their points of contact lie on the line $ay+bx=0$, $3z=c$. [L.U.]

19. Find the equations of the tangent planes to the ellipsoid $x^2+2y^2+3z^2=6$ which intersect in the line $x=3-y=3z$. Find also the coordinates of their points of contact. [L.U.]

20. The point $P(p, q, r)$ lies on the ellipsoid E whose equation is $ax^2+by^2+cz^2=1$. Find the condition that the straight line through P with direction ratios $[l : m : n]$ should touch E.

If P also lies on two other ellipsoids with the equations $a^2x^2+b^2y^2+c^2z^2=1$ and $a^3x^2+b^3y^2+c^3z^2=1$, and the normal to E at P meets E again at Q, prove that PQ has length 2. If PQ is also normal to E at Q, prove that E is a sphere, provided that P does not lie in any coordinate plane. [L.U.]

21. Write down the equation of a plane which makes intercepts a, a and c on the x, y and z axes respectively. If this plane touches the ellipsoid $x^2+4y^2+4z^2=16$, show that $c=\pm 2a(a^2-20)^{-\frac{1}{2}}$ and prove that the coordinates (x_1, y_1, z_1) of the point of contact P, are given by $ax_1=4ay_1=4cz_1=16$.

Show that as a varies the locus of the foot of the perpendicular from P upon the plane Oyz is the ellipse $5y^2+z^2=4$. [L.U. Anc.]

22. The normal at $P(1, 1, 1)$ to the ellipsoid $x^2+2y^2+3z^2=6$ meets the plane $z=0$ at A. Find the pole of the plane which bisects AP at right angles and show that it is on the line joining P to the origin. [L.U.]

23. If P is any point on the ellipse in which the ellipsoid

$$x^2/a^2+y^2/b^2+z^2/c^2=1 \quad (a>b)$$

is cut by the plane $z=k$, and Q is the point in which the normal to the ellipsoid at P cuts the plane $x=0$, prove that as P describes the ellipse, Q oscillates in a straight line parallel to the y-axis, the amplitude of the oscillation being $\{(a^2-b^2)\sqrt{(c^2-k^2)}\}/bc$. [L.U.]

24. A tangent plane to the ellipsoid $x^2+2y^2+3z^2=4/9$ at a point in the positive octant where this surface is cut by the plane $y=z$ touches the sphere $x^2+y^2+z^2=4/17$. Find the equation of the plane and the equations of the normals at the points of contact on the ellipsoid and the sphere. [L.U.]

25. Find the equations of the two planes containing the line $x+y=0=3-z$ which touch the ellipsoid $3x^2+4y^2+6z^2=12$.

Show that the equations of the line joining the points at which these planes touch the ellipsoid may be expressed in the form $3x-4y=0$, $z=2/3$. [L.U.]

26. Show that the plane $3x+2y+z=p$ touches the ellipsoid $3x^2+4y^2+z^2=20$ if $p=\pm 10$, and find the length of the chord of contact between the two tangent planes. What is the angle between this chord and the common normal to the planes ? [L.U.]

27. Show that the distance between the points of contact of the two tangent planes to the conicoid $x^2+y^2-2z^2=1$ passing through the line $(x+1)/4=(y+1)/12=(z+1)/9$ is $5\sqrt{2}$ units. [L.U.]

18.8. The plane containing all chords bisected at a given point

If, in the notation of § 18.6, P is the mid-point of AB, the roots of (18.3) are equal in magnitude but opposite in sign.

$$\therefore\ a\lambda x_1+b\mu y_1+c\nu z_1=0. \quad . \quad . \quad . \qquad \text{(i)}$$

Eliminating λ, μ, ν between (i) and (18.1), we have

$$ax_1(x-x_1)+by_1(y-y_1)+cz_1(z-z_1)=0, \qquad (18.7)$$

which is the equation of the plane containing all chords of the quadric which are bisected at (x_1, y_1, z_1).

From (i) it follows that the mid-points of all chords of the quadric drawn parallel to the diameter $[\lambda : \mu : \nu]$ lie in the plane

$$a\lambda x+b\mu y+c\nu z=0. \qquad (18.8)$$

Tangent lines parallel to the diameter $[\lambda : \mu : \nu]$ may be regarded as the limiting case of these chords and so the plane (18.8) passes through points of contact of all such tangent lines. This plane is known as the diametral plane conjugate to the diameter $[\lambda : \mu : \nu]$.

18.9. Polar planes and polar lines

The locus of points of contact of tangents drawn from the point P (x_1, y_1, z_1) to the quadric $ax^2+by^2+cz^2=1$ can be shown by the method of § 17.10 to be the plane

$$axx_1+byy_1+czz_1=1.$$

This is known as the polar plane of P with respect to the quadric. The equations of polar lines with respect to the quadric may be found as in § 17.11.

18.10. Note on ruled surfaces

The equation of the hyperboloid of one sheet

$$\frac{x^2}{a^2}+\frac{y^2}{b^2}-\frac{z^2}{c^2}=1 \quad . \quad . \quad . \quad . \qquad \text{(i)}$$

may be written in the form

$$\frac{x^2}{a^2}-\frac{z^2}{c^2}=1-\frac{y^2}{b^2},$$

i.e.

$$\left(\frac{x}{a}+\frac{z}{c}\right)\left(\frac{x}{a}-\frac{z}{c}\right)=\left(1+\frac{y}{b}\right)\left(1-\frac{y}{b}\right).$$

From this we see that any point which satisfies simultaneously the equations

$$\frac{x}{a}+\frac{z}{c}=\lambda\left(1+\frac{y}{b}\right) \text{ and } \frac{x}{a}-\frac{z}{c}=\frac{1}{\lambda}\left(1-\frac{y}{b}\right) \quad . \quad . \quad \text{(ii)}$$

where λ is any constant, lies on the hyperboloid. But equations (ii) define a straight line all points on which lie on the quadric. Such a line is called a *generator* of the quadric.

Similarly the line given by the equations

$$\frac{x}{a}-\frac{z}{c}=\mu\left(1+\frac{y}{b}\right), \quad \frac{x}{a}+\frac{z}{c}=\frac{1}{\mu}\left(1-\frac{y}{b}\right), \quad . \quad . \quad \text{(iii)}$$

where μ is any constant is a generator of the quadric.

For varying λ and μ, (ii) and (iii) each define a system of generators of the hyperboloid. It can be shown that no two generators of the same system intersect and that every number of the λ-system meets each member of the μ-system.

A surface which is generated by a family of straight lines is called a *ruled surface*.

18.11. Miscellaneous examples

Example 4

Prove that the plane $lx+my+nz=p$ cuts the conicoid $ax^2+by^2+cz^2=1$ in a conic whose centre (α, β, γ) is given by the equations

$$a\alpha/l=b\beta/m=c\gamma/n=(a\alpha^2+b\beta^2+c\gamma^2)/p.$$

Deduce that the plane touches the conicoid if

$$l^2/a+m^2/b+n^2/c=p^2.$$

Find the locus of the centres of the sections of the conicoid $ax^2+by^2+cz^2=1$ made by tangent planes to the conicoid $Ax^2+By^2+Cz^2=1$. [L.U.]

The section of the quadric $ax^2+by^2+cz^2=1$ of which the point $C(\alpha, \beta, \gamma)$ is centre is by (18.7)

$$a\alpha(x-\alpha)+b\beta(y-\beta)+c\gamma(z-\gamma)=0.$$

This plane is identical with

$$lx+my+nz=p \quad . \quad . \quad . \quad . \quad \text{(i)}$$

if

$$a\alpha/l=b\beta/m=c\gamma/n=(a\alpha^2+b\beta^2+c\gamma^2)/p. \quad . \quad . \quad \text{(ii)}$$

If C lies on the conicoid,

$$a\alpha^2+b\beta^2+c\gamma^2=1 \quad . \quad . \quad . \quad . \quad \text{(iii)}$$

and (i) is a tangent plane. Eliminating α, β and γ between (ii) and (iii) we obtain the condition that (i) should touch the quadric:

$$l^2/a+m^2/b+n^2/c=p^2.$$

Applying this result, (i) will touch the conicoid $Ax^2+By^2+Cz^2=1$

$$l^2/A+m^2/B+n^2/C=p^2.$$

With this value of p^2 we have from (ii)

$$a^2\alpha^2/A+b^2\beta^2/B+c^2\gamma^2/C=(a\alpha^2+b\beta^2+c\gamma^2)^2(l^2/A+m^2/B+n^2/C)/p^2.$$
$$=(a\alpha^2+b\beta^2+c\gamma^2)^2,$$

$\therefore$ (α, β, γ), the centre of the section made by (i) on the conicoid

$$ax^2+by^2+cz^2=1$$

lies on the surface $a^2x^2/A+b^2y^2/B+c^2z^2/C=(ax^2+by^2+cz^2)^2$.

Example 5

Show that the locus of all points on the conicoid $ax^2+by^2+cz^2=1$ through which a tangent line to the conicoid can be drawn with given direction ratios $[l:m:n]$ is the conic obtained as intersection of the conicoid and the plane $alx+bmy+cnz=0$.

Tangents are drawn to the sphere $x^2+y^2+z^2=1$ which have direction ratios [1 : 1 : 2]. *Show that the locus of the mid-points of the segments intercepted on these tangents by the ellipsoid $2x^2+y^2+z^2=2$ is an ellipse whose semi-axes are of lengths 1, $\frac{3}{7}\sqrt{6}$ respectively.* [L.U.]

The first part of the question is answered in § 18.8. Applying this result, we see that the points of contact of tangents drawn in the direction [1 : 1 : 2] lie on the circle C in which the plane $x+y+2z=0$ cuts the sphere $x^2+y^2+z^2=1$.

The ellipsoid $2x^2+y^2+z^2=2$ intercepts on these tangents segments whose mid-points by (18.8) lie on some curve E in the plane $2x+y+2z=0$.

But C is the orthogonal projection of E on the plane of C, since all tangent lines to the sphere with points of contact on C are perpendicular to the plane of C. Hence E is an ellipse (see § 16.19 Example 13).

If θ is the angle between the planes of C and E, by (16.26), page 360,

$$\cos\theta=\frac{7}{3\sqrt{6}}.$$

The semi-minor axis of E=radius of C=1.

The semi-major axis of E=(radius of C)/cos θ=$(3\sqrt{6})/7$.

Example 6

Find the equations of the generators of the hyperboloid $x^2/a^2+y^2/b^2-z^2/c^2=1$ which pass through the point $(a\cos\theta, b\sin\theta, 0)$.

Suppose that the equations of a generator through the given point are

$$\frac{x-a\cos\theta}{al}=\frac{y-b\sin\theta}{bm}=\frac{z}{cn}=\rho \quad . \quad . \quad . \quad \text{(i)}$$

Then since every point on (i) must lie on the hyperboloid

$$x^2/a^2+y^2/b^2-z^2/c^2=1,$$
$$\rho^2(l^2+m^2-n^2)+2\rho(l\cos\theta+m\sin\theta)=0$$

for all values of ρ.

$$\therefore\ l^2+m^2-n^2=0 \text{ and } l\cos\theta+m\sin\theta=0.$$

These relations are satisfied if $l=\sin\theta$, $m=-\cos\theta$, $n=\pm 1$, and so the equations of the required generators are

$$\frac{x-a\cos\theta}{a\sin\theta}=\frac{y-b\sin\theta}{-b\cos\theta}=\pm\frac{z}{c}.$$

Exercises 18 (*b*)

1. Prove that all chords of the ellipsoid
$$x^2/a^2+y^2/b^2+z^2/c^2=1$$
which are bisected at the point (α, β, γ) lie in the plane
$$\alpha(x-\alpha)/a^2+\beta(y-\beta)/b^2+\gamma(z-\gamma)/c^2=0.$$
Prove also that the centre of the conic in which the above ellipsoid is cut by any plane through the point (a, b, c) lies on the surface
$$x^2/a^2+y^2/b^2+z^2/c^2=x/a+y/b+z/c.$$ [L.U.]

2. Prove that all chords of the surface $ax^2+by^2+cz^2=1$ which are bisected by the point (α, β, γ) lie in the plane $a\alpha(x-\alpha)+b\beta(y-\beta)+c\gamma(z-\gamma)=0$.
Prove also that the centre of the conic in which the above surface is cut by any tangent plane to the sphere $x^2+y^2+z^2=r^2$ lies on the surface $(ax^2+by^2+cz^2)^2=r^2(a^2x^2+b^2y^2+c^2z^2)$. [L.U.]

3. Show that the plane sections of the ellipsoid $x^2+2y^2+3z^2=1$, whose centres lie on the line $(x-1)/1=(y-1)/(-1)=z/1$, are parallel to the line $-2x=4y=3z$. [L.U.]

4. Find the locus of the centres of the ellipses in which the ellipsoid $x^2+2y^2+3z^2=16$ is cut by the tangent planes to the sphere
$$x^2+y^2+z^2=1.$$ [L.U.]

5. Planes passing through the line $x-2=y=z$ cut the conicoid $4x^2+3y^2+z^2=1$; prove that the locus of the centres of the plane sections so formed is a conic.
By transferring the origin of coordinates to the point (1, 0, 0), or otherwise, find the position of the centre of this conic. [L.U.]

6. Show that all chords of the ellipsoid
$$x^2/a^2+y^2/b^2+z^2/c^2=1$$
which are bisected at $A(\alpha, \beta, \gamma)$ lie in the plane
$$\alpha x/a^2+\beta y/b^2+\gamma z/c^2=\alpha^2/a^2+\beta^2/b^2+\gamma^2/c^2.$$
Find the pole B of this plane with respect to the ellipsoid.
If O is the origin show that O, A, B are collinear and that, if the ratio $OA : OB$ is constant, B must lie on a concentric ellipsoid. [L.U.]

7. Prove that the middle points of a set of parallel chords of an ellipsoid are coplanar.
If the equation of the ellipsoid is $7x^2+3y^2+4z^2=1$ and the direction ratios of the chords are [2 : 5 : 4], find the inclination of the chords to the plane that bisects them. [L.U.]

8. Show that the polar planes of points on the line
$$3(x-1)=6(y+1)=2(z-2)$$
with respect to the ellipsoid $x^2+3y^2+2z^2=1$ all pass through the line whose equations can be put in the form
$$3(x+3)=-45y=-10(z-1).$$ [L.U.]

9. If the polar plane of a point P with respect to the ellipsoid $ax^2+by^2+cz^2=1$ touches the sphere with unit radius and centre the origin, prove that P lies on the ellipsoid $a^2x^2+b^2y^2+c^2z^2=1$. [L.U.]

10. Prove that the points of contact of the tangent planes to the quadric $ax^2+by^2+cz^2=1$ which contain the point P (x_1, y_1, z_1) lie in the plane $axx_1+byy_1+czz_1=1$.

Find the locus of P if this plane contains the fixed line

$$(x-f)/l=(y-g)/m=(z-h)/n.$$ [L.U.]

11. Prove that the locus of the pole with respect to the ellipsoid $x^2/a+y^2/b+z^2/c=1$, of a tangent plane to the ellipsoid

$$x^2/a^2+y^2/b^2+z^2/c^2=1,$$

is a sphere. [L.U.]

12. Show that the polar planes of all points on the line

$$(x-2)/1=(y-1)/2=(z+1)/3$$

with respect to the ellipsoid $x^2+2y^2+3z^2=6$ pass through a fixed line and find the equations of the fixed line in standard form. [Leeds.]

13. Prove that there are two and only two points on the ellipsoid

$$x^2/a^2+y^2/b^2+z^2/c^2=1$$

at which the normals are equally inclined to the positive directions of the coordinate axes and find their coordinates.

The normal, equally inclined to the positive directions of the coordinate axes, is drawn from a point in the positive octant on the ellipsoid $3x^2+4y^2+12z^2=24$. Prove that the polar planes of all points on this normal pass through the line $(x-16)/8=(y+12)/(-9)=z$. [L.U.]

14. Show that the line $x/a+z/c=\lambda(1+y/b)$, $\lambda(x/a-z/c)=1-y/b$ is a generator of the quadric $x^2/a^2+y^2/b^2-z^2/c^2=1$ for any value of λ, and write down equations for the second family of generators.

Prove that the generators of the quadric $5x^2-5y^2+3z^2=27$ through the point $(2, 1, 2)$ are perpendicular. [Leeds.]

15. Find the equations of the two generators of the hyperboloid

$$(x^2+y^2)/a^2-z^2/c^2=1$$

that pass through the point $(a \cos\theta, a \sin\theta, 0)$.

Prove that they cut at an angle independent of θ, and find the condition that this angle should be a right angle. [Leeds.]

16. Show that for all values of the constants λ and μ the lines

$$(x+a)/(z+y)=(z-y)/(x-a)=\lambda, \quad (x+a)/(z-y)=(z+y)/(x-a)=\mu,$$

lie on the surface $x^2+y^2-z^2=a^2$.

Prove that, if these lines intersect at right angles, $\lambda=-\mu$ and the locus of their point of intersection, for different values of λ, is a circle in the plane $z=0$. [L.U.]

17. If P is a variable point on the line r whose equations are

$$(x-a)/l=(y-b)/m=(z-c)/n$$

prove that the polar plane of P with respect to the ellipsoid

$$x^2+2y^2+3z^2=1$$

turns about a fixed line r' when r does not pass through the origin.

If r touches the ellipsoid and (a, b, c) is the point of contact, prove that r' also touches the ellipsoid at the same point. [L.U.]

CHAPTER 19

PARTIAL DIFFERENTIATION

19.1. Continuous functions of several variables

In this chapter we are concerned with the rates of change of functions of several variables. For simplicity we shall deal in general with functions of two variables, but the arguments used are quite general and the results obtained may be extended to functions of more than two variables. We shall assume that the functions under consideration are continuous. A function $f(x, y)$ of two independent variables is said to be continuous for $x=a$, $y=b$ if it is defined for these values and if

$$\lim \{f(x, y)-f(a, b)\}=0$$

when $x \to a$ and $y \to b$ in any manner whatsoever; or, more precisely, if for every positive number ϵ we can find a number η such that

$$|f(x, y)-f(a, b)|<\epsilon \text{ when } |x-a| \leqslant \eta \text{ and } |y-b| \leqslant \eta.$$

19.2. Partial derivatives

Let z be a function of two independent variables given by the equation $z=f(x, y)$, and let δz be the increase in z due to an increment δx in x, while y remains constant. Then $\delta z=f(x+\delta x, y)-f(x, y)$ and $\dfrac{\delta z}{\delta x}=\dfrac{f(x+\delta x, y)-f(x, y)}{\delta x}$. If $\lim\limits_{\delta x \to 0} \dfrac{f(x+\delta x, y)-f(x, y)}{\delta x}$ exists, it is called the *first partial derivative* of z with respect to x and denoted by $\dfrac{\partial z}{\partial x}$ or f_x; $\dfrac{\partial z}{\partial y}$ or f_y is defined as $\lim\limits_{\delta y \to 0} \dfrac{f(x, y+\delta y)-f(x, y)}{\delta y}$.

19.3. Calculation of partial derivatives

To find $\dfrac{\partial z}{\partial x}$ when $z=f(x,y)$ we differentiate z with respect to x treating y as a constant. No new principle is involved and, in general, the rules of differentiation given in § 9.2 remain valid. An exception is Rule VI, for, in general, $\dfrac{\partial z}{\partial x}$ and $\dfrac{\partial x}{\partial z}$ are not reciprocals of each other.

A slight modification is required in Rule V: if z is a function of u, where u is a function of x and y, $\dfrac{\partial z}{\partial x}=\dfrac{dz}{du}\dfrac{\partial u}{\partial x}$ and $\dfrac{\partial z}{\partial y}=\dfrac{dz}{du}\dfrac{\partial u}{\partial y}$ where $\dfrac{dz}{du}$ is the ordinary derivative of z with respect to u.

Example 1

If $z=y/x+x/y$, $\dfrac{\partial z}{\partial x}=-y/x^2+1/y$, $\dfrac{\partial z}{\partial y}=1/x-x/y^2$.

Example 2

If $z=e^{x^2+y^2}$, $\dfrac{\partial z}{\partial x}=2xe^{x^2+y^2}$, $\dfrac{\partial z}{\partial y}=2ye^{x^2+y^2}$.

Example 3

If $z=y\sin(2x+3y)$, $\dfrac{\partial z}{\partial x}=2y\cos(2x+3y)$,

$$\frac{\partial z}{\partial y}=3y\cos(2x+3y)+\sin(2x+3y).$$

Example 4

If $z=y+x\log(x/y)$, $\dfrac{\partial z}{\partial x}=\log(x/y)+1$, $\dfrac{\partial z}{\partial y}=1-x/y$.

Example 5

If $z=\tan^{-1}(y/x)$, we write $z=\tan^{-1}u$, where $u=y/x$. Then

$$\frac{\partial z}{\partial x}=\frac{dz}{du}\frac{\partial u}{\partial x}=-y/(x^2+y^2) \text{ and } \frac{\partial z}{\partial y}=x/(x^2+y^2).$$

Example 6

If $x^2+2y^2+3z^2=1$, $2x+6z\dfrac{\partial z}{\partial x}=0$. Hence $\dfrac{\partial z}{\partial x}=-\tfrac{1}{3}x/z$ and similarly $\dfrac{\partial z}{\partial y}=-\tfrac{2}{3}y/z$.

Example 7

If $z=x^nf(y/x)$ where f denotes an arbitrary function, prove that

$$x\frac{\partial z}{\partial x}+y\frac{\partial z}{\partial y}=nz.$$

If $$z=x^nf(y/x),$$

$$\log z=n\log x+\log f(y/x).$$

$$\therefore\ \frac{1}{z}\frac{\partial z}{\partial x}=\frac{n}{x}+\frac{1}{f}\frac{\partial f}{\partial x} \text{ and } \frac{1}{z}\frac{\partial z}{\partial y}=\frac{1}{f}\frac{\partial f}{\partial y}.$$

Put $$y/x=u;$$

then $$\frac{\partial f}{\partial x}=\frac{df}{du}\cdot\frac{\partial u}{\partial x}=f'(u)(-y/x^2), \text{ and } \frac{\partial f}{\partial y}=\frac{df}{du}\cdot\frac{\partial u}{\partial y}=f'(u)(1/x).$$

Hence $$\frac{1}{z}\frac{\partial z}{\partial x}=\frac{n}{x}-\frac{f'(u)}{f(u)}\frac{y}{x^2} \text{ and } \frac{1}{z}\frac{\partial z}{\partial y}=\frac{f'(u)}{f(u)}\frac{1}{x}.$$

$$\therefore\ x\frac{\partial z}{\partial x}+y\frac{\partial z}{\partial y}=nz.$$

This result is known as Euler's first theorem for a homogeneous function of the nth degree in two independent variables (see § 19.12).

19.4. Geometric interpretation

If $z=f(x, y)$ is a function of two independent variables x and y, the partial derivatives may be interpreted in terms of the geometry of the surface $z=f(x, y)$ in the neighbourhood of the point $P(x, y, z)$.

The plane through P parallel to the plane zOx cuts the surface in a curve along which y is constant. Hence the gradient of this curve at P is $\dfrac{\partial z}{\partial x}$.

Similarly, the plane through P parallel to the plane zOy cuts the surface in a curve along which x is constant. The gradient of this curve at P is $\dfrac{\partial z}{\partial y}$.

19.5. Partial derivatives of higher orders

In general, if $z=f(x, y)$, $\dfrac{\partial z}{\partial x}$ and $\dfrac{\partial z}{\partial y}$ are themselves functions of x and y and the partial derivatives of the second order are defined as follows:

$$\frac{\partial^2 z}{\partial x^2}=\frac{\partial}{\partial x}\left(\frac{\partial z}{\partial x}\right) \quad . \quad . \quad \text{(i)} \qquad \frac{\partial^2 z}{\partial y^2}=\frac{\partial}{\partial y}\left(\frac{\partial z}{\partial y}\right) \quad . \quad . \quad \text{(ii)}$$

$$\frac{\partial^2 z}{\partial x \partial y}=\frac{\partial}{\partial x}\left(\frac{\partial z}{\partial y}\right) \quad . \quad . \quad \text{(iii)} \qquad \frac{\partial^2 z}{\partial y \partial x}=\frac{\partial}{\partial y}\left(\frac{\partial z}{\partial x}\right) \quad . \quad . \quad \text{(iv)}$$

provided that the limits implied in these definitions exist. The second order derivatives (i), (ii), (iii) and (iv) are also denoted by f_{xx}, f_{yy}, f_{xy} and f_{yx} respectively.

Subject to certain conditions involving the continuity of z and its derivatives it may be shown that

$$\frac{\partial^2 z}{\partial x\, \partial y}=\frac{\partial^2 z}{\partial y\, \partial x}. \qquad (19.1)$$

This result, known as the commutative property of partial derivatives, may be assumed to be true for all functions encountered at this stage.

Example 8

If $\quad z=(x^2+y^2)\tan^{-1}(y/x)$

$$\frac{\partial z}{\partial x}=2x\tan^{-1}(y/x)-y \text{ and } \frac{\partial z}{\partial y}=2y\tan^{-1}(y/x)+x \text{ (see Example 5).}$$

$$\therefore \frac{\partial^2 z}{\partial x^2}=2\tan^{-1}(y/x)-2xy/(x^2+y^2) \text{ and } \frac{\partial^2 z}{\partial y^2}=2\tan^{-1}(y/x)+2xy/(x^2+y^2).$$

Also $\quad \dfrac{\partial}{\partial y}\left(\dfrac{\partial z}{\partial x}\right)=2x^2/(x^2+y^2)-1$ and $\dfrac{\partial}{\partial x}\left(\dfrac{\partial z}{\partial y}\right)=-2y^2/(x^2+y^2)+1$

$$\therefore \frac{\partial^2 z}{\partial y \partial x}=\frac{\partial^2 z}{\partial x \partial y}=(x^2-y^2)/(x^2+y^2).$$

Example 9

If $z=f(x+cy)+\phi(x-cy)$, where f and ϕ are arbitrary functions, show that $\dfrac{\partial^2 z}{\partial y^2}=c^2\dfrac{\partial^2 z}{\partial x^2}$.

Put $x+cy=u$ and $x-cy=v$.

Then
$$z=f(u)+\phi(v),$$

$$\frac{\partial z}{\partial x}=\frac{df}{du}\cdot\frac{\partial u}{\partial x}+\frac{d\phi}{dv}\cdot\frac{\partial v}{\partial x}=f'(u)+\phi'(v),$$

and
$$\frac{\partial z}{\partial y}=\frac{df}{du}\cdot\frac{\partial u}{\partial y}+\frac{d\phi}{dv}\cdot\frac{\partial v}{\partial y}=c\{f'(u)-\phi'(v)\}.$$

Similarly,
$$\frac{\partial^2 z}{\partial x^2}=f''(u)\frac{\partial u}{\partial x}+\phi''(v)\frac{\partial v}{\partial x}=f''(u)+\phi''(v),$$

and
$$\frac{\partial^2 z}{\partial y^2}=c\left\{f''(u)\frac{\partial u}{\partial y}-\phi''(v)\frac{\partial v}{\partial y}\right\}=c^2\{f''(u)+\phi''(v)\}.$$

$$\therefore\ \frac{\partial^2 z}{\partial y^2}=c^2\frac{\partial^2 z}{\partial x^2}.$$

Example 10

If $U=y\log(y+r)-r$, where $r^2=x^2+y^2$, prove that

$$\frac{\partial^2 U}{\partial x^2}+\frac{\partial^2 U}{\partial y^2}=\frac{1}{y+r}. \qquad \text{[L.U.]}$$

Since
$$r^2=x^2+y^2 \quad . \quad . \quad . \quad . \quad . \quad . \quad \text{(i)}$$

$$r\frac{\partial r}{\partial x}=x \text{ and } r\frac{\partial r}{\partial y}=y \quad . \quad . \quad . \quad . \quad . \quad \text{(ii)}$$

Also
$$U=y\log(y+r)-r$$

$$\therefore\ \frac{\partial U}{\partial x}=y\frac{\partial}{\partial x}\{\log(y+r)\}-\frac{\partial r}{\partial x}$$

$$=y\left\{\frac{1}{y+r}\frac{\partial r}{\partial x}\right\}-\frac{\partial r}{\partial x}$$

$$=\left(\frac{y}{y+r}-1\right)\frac{\partial r}{\partial x}$$

$$=\frac{-x}{y+r}, \text{ by (ii)},$$

and so
$$\frac{\partial^2 U}{\partial x^2} = -\left\{\frac{y+r-x\dfrac{\partial r}{\partial x}}{(y+r)^2}\right\}$$
$$= -\left\{\frac{r(y+r)-x^2}{r(y+r)^2}\right\}, \text{ by (ii)},$$
$$= \frac{-y}{r(y+r)}, \text{ by (i)}.$$

Also,
$$\frac{\partial U}{\partial y} = \log(y+r) + y\frac{\partial}{\partial y}\{\log(y+r)\} - \frac{\partial r}{\partial y}$$
$$= \log(y+r) + \frac{y}{y+r}\left(1+\frac{\partial r}{\partial y}\right) - \frac{\partial r}{\partial y}$$
$$= \log(y+r), \text{ by (ii)},$$

and so
$$\frac{\partial^2 U}{\partial y^2} = \frac{1+\dfrac{\partial r}{\partial y}}{y+r} = \frac{1}{r}, \text{ by (ii)},$$
$$\therefore \frac{\partial^2 U}{\partial x^2} + \frac{\partial^2 U}{\partial y^2} = \frac{1}{y+r}.$$

Exercises 19 (a)

Find f_x, f_y for the the functions $f(x, y)$ in Nos. 1-6:

1. $x^2/y - y^2/x$.
2. $(x-y)/(x+y)$.
3. xe^{2x+3y}.
4. $y\sqrt{(x^2-y^2)}$.
5. $\tan^{-1}(x/y)$.
6. $e^{x-y}\log(y-x)$.

Find f_{xx}, f_{yy} for the functions $f(x, y)$ in Nos. 7-9 and verify in each case that $f_{xy} = f_{yx}$:

7. $x^2 \sin y + y^2 \cos x$.
8. $(y/x)\log x$.
9. $\sin^{-1}(y/x)$.

10. If $z = f(x^2+y^2)$ prove that $x\dfrac{\partial z}{\partial y} = y\dfrac{\partial z}{\partial x}$.

11. If $z = f(y/x)$, prove that $x\dfrac{\partial z}{\partial x} + y\dfrac{\partial z}{\partial y} = 0$.

12. If $z = y^3 f(x/y)$, prove that $x^2\dfrac{\partial^2 z}{\partial x^2} + 2xy\dfrac{\partial^2 z}{\partial x\,\partial y} + y^2\dfrac{\partial^2 z}{\partial y^2} = 6z$.

19.6. Total variation

Suppose that $z = f(x, y)$ is a continuous function whose partial derivatives of the first order are also continuous functions of x and y.

Then the change δz in z corresponding to simultaneous increments δx in x and δy in y is given by

$$\begin{aligned}\delta z &= f(x+\delta x, y+\delta y) - f(x, y) \\ &= \{f(x+\delta x, y+\delta y) - f(x, y+\delta y)\} + \{f(x, y+\delta y) - f(x, y)\} \\ &= \frac{f(x+\delta x, y+\delta y) - f(x, y+\delta y)}{\delta x}\,\delta x + \frac{f(x, y+\delta y) - f(x, y)}{\delta y}\,\delta y \quad . \quad \text{(i)}\end{aligned}$$

If we regard $f(x, y+\delta y)$ as a function of x, y and δy being fixed, then by the mean value theorem (see § 9.21) there is a number ξ between x and $x+\delta x$ such that

$$\frac{f(x+\delta x, y+\delta y) - f(x, y+\delta y)}{\delta x} = f_x(\xi, y+\delta y),$$

where $f_x(x, y)$ denotes $\frac{\partial}{\partial x} f(x, y)$.

Also, $f_x(\xi, y+\delta y) \to f_x(x, y)$ as δx and $\delta y \to 0$, since f_x is a continuous function ; and so we may write

$$f_x(\xi, y+\delta y) = f_x(x, y) + \eta_1,$$

where $\eta_1 \to 0$ as δx and $\delta y \to 0$. Hence

$$\frac{f(x+\delta x, y+\delta y) - f(x, y+\delta y)}{\delta x} = f_x(x, y) + \eta_1,$$

where $\eta_1 \to 0$ as δx and $\delta y \to 0$.

Similarly, $$\frac{f(x, y+\delta y) - f(x, y)}{\delta y} = f_y(x, y) + \eta_2,$$

where $\eta_2 \to 0$ as $\delta y \to 0$ and $f_y(x, y)$ denotes $\frac{\partial}{\partial y} f(x, y)$.

Hence by (i) we have

$$\delta z = \left(\frac{\partial f}{\partial x} + \eta_1\right)\delta x + \left(\frac{\partial f}{\partial y} + \eta_2\right)\delta y, \qquad (19.2)$$

where η_1 and $\eta_2 \to 0$ when δx and $\delta y \to 0$.

δz as defined by (19.2) is called the *total variation* (*or increment*) of z.

When δx and δy are small and $\frac{\partial f}{\partial x} \neq 0$, $\frac{\partial f}{\partial y} \neq 0$,

$$\delta z \simeq \frac{\partial z}{\partial x}\,\delta x + \frac{\partial z}{\partial y}\,\delta y$$

since each term omitted is the product of two small quantities. This result is useful in calculating approximately the error δz in z resulting from small errors δx and δy in the measurements of x and y respectively. The fraction $\delta z/z$ is called the proportional (or relative) error in z.

Example 11

The diameter of a circle from which a segment has been cut is determined from the length of the chord a and the maximum height b of the segment. If the measurements of a and b are slightly inaccurate each to an extent of p per cent., find the approximate percentage error in the calculated value of the diameter. Prove that this error is also p per cent., provided (i) that a and b are measured both in excess or both in defect of their actual values, or (ii) that, if a and b are measured one in excess and the other in defect of their actual values, the segment is a semicircle. [L.U.]

Let C (fig. 62) be the mid-point of the arc AB of a circle and let CD, the diameter through C, meet AB at O. Then OC is the maximum height of the segment ABC and

$$OC.OD=OA.OB.$$

If $AB=a$, $OC=b$ and $CD=x$ we have

$$b(x-b)=\tfrac{1}{4}a^2,$$

$$x=b+\tfrac{1}{4}a^2/b \quad . \quad . \quad . \quad (1)$$

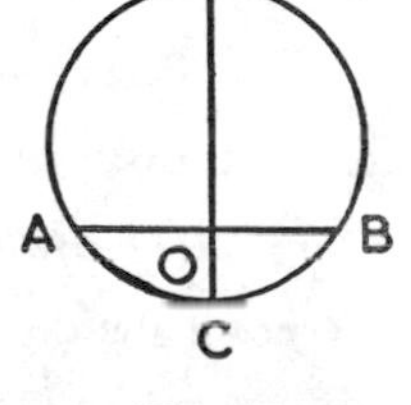

Fig. 62

If δx is the error in x caused by small errors δa and δb in a and b respectively, we have

$$\delta x \simeq \frac{\partial x}{\partial a}\delta a+\frac{\partial x}{\partial b}\delta b$$

i.e. $$\delta x \simeq \tfrac{1}{2}(a/b)\delta a+(1-\tfrac{1}{4}a^2/b^2)\delta b.$$

If the percentage error in a and b is p per cent., then $100(\delta a/a)=\pm p$, $100(\delta b/b)=\pm p$, and ϵ, the percentage error in x, is given by

$$\epsilon=100(\delta x/x)=\pm\tfrac{1}{4}p\{2a^2\pm(4b^2-a^2)\}/bx \text{ approximately.}$$

If a and b are measured both in excess or both in defect of their actual values, δa and δb have the same signs

$$\therefore\ \epsilon\simeq\pm\tfrac{1}{4}p\{2a^2+(4b^2-a^2)\}/bx=\pm p\{b+\tfrac{1}{4}a^2/b\}/x$$

i.e. $$\epsilon\simeq\pm p, \text{ by (i).}$$

If a and b are measured one in excess and the other in defect of their actual values, δa and δb have opposite signs and

$$\epsilon\simeq\pm\tfrac{1}{4}p\{3a^2-4b^2\}/bx.$$

If, in addition, the segment ACB is a semicircle $x=a=2b$ and so

$$\epsilon\simeq\pm p.$$

Example 12

The points A and B, at a distance a apart on a horizontal plane, are in line with the base C of a vertical tower and on the same side of C. The elevations of the top of the tower from A and B are observed to be α and β(α<β). Show that the distance BC is a sin α cos β cosec (β−α).

If the observations of the angles of elevation are uncertain by 4 minutes, show that the maximum possible percentage error in the calculated value of BC is approximately

$$\pi \sin(\alpha+\beta)/\{27 \sin\alpha \cos\beta \tan(\beta-\alpha)\}. \quad \text{[L.U.]}$$

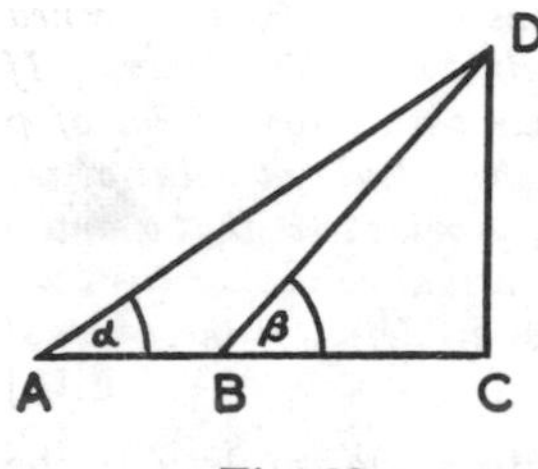

Fig. 63

In fig. 63, $\frac{BC}{AB}=\frac{BC}{BD}\cdot\frac{BD}{AB}=\cos\beta\frac{\sin\alpha}{\sin(\beta-\alpha)}$

$$\therefore BC=a\cos\beta\sin\alpha\operatorname{cosec}(\beta-\alpha),$$

and denoting BC by x we have

$$\log x=\log a+\log\cos\beta+\log\sin\alpha+\log\operatorname{cosec}(\beta-\alpha) \quad . \quad \text{(i)}$$

If δx is the error in x caused by small errors $\delta\alpha$ and $\delta\beta$ in α and β respectively, the proportional error in x is $(\delta x/x)$ where, by (i),

$$\delta x/x \simeq -\tan\beta\,\delta\beta+\cot\alpha\,\delta\alpha-(\delta\beta-\delta\alpha)\cot(\beta-\alpha)$$

i.e. $$\delta x/x \simeq \{\cot\alpha+\cot(\beta-\alpha)\}\delta\alpha-\{\tan\beta+\cot(\beta-\alpha)\}\delta\beta.$$

Now $\delta\alpha=\pm\pi/2700$ radians and $\delta\beta=\pm\pi/2700$ radians.

Hence if ϵ is the percentage error in x,

$$\epsilon=100(\delta x/x)\simeq\pm\frac{\pi}{27}[\{\cot\alpha+\cot(\beta-\alpha)\}\mp\{\tan\beta+\cot(\beta-\alpha)\}].$$

Since α and β are acute and $\beta>\alpha$, the maximum numerical value of ϵ is obtained by taking the positive sign inside the square brackets.

$$\therefore \epsilon_{\max}=\frac{\pi}{27}\left\{\frac{\cos\alpha}{\sin\alpha}+\frac{\sin\beta}{\cos\beta}+\frac{2}{\tan(\beta-\alpha)}\right\},$$

$$=\pi\{\sin(\beta-\alpha)+2\sin\alpha\cos\beta\}/27\sin\alpha\cos\beta\tan(\beta-\alpha),$$

$$=\pi\sin(\alpha+\beta)/\{27\sin\alpha\cos\beta\tan(\beta-\alpha)\} \text{ approximately.}$$

Exercises 19 (*b*)

1. In a triangle ABC, the angle A is accurately known, but the measurement of the side b is in error to the extent δb, and that of the side c to an extent δc. Find the error in calculating the value of a from b, c and A.

   What is the best shape for triangle ABC in order to minimise as much as possible the effect of the error δb? [L.U.]

2. The side BC of a triangle ABC is to be determined from measurements of the sides AB and AC and of the angle BAC. The measured values of the sides are liable to a small proportional error θ and the angle BAC to a small absolute error δA. Show that the calculated value of BC is liable to a proportional error $\theta+\{(bc/a^2) \sin A\}\delta A$.

The measured values of b, c and A are 4, 5 and 120° respectively and are liable to errors of $\frac{1}{2}\%$, $\frac{1}{2}\%$ and 1° respectively. Show that the calculated value of a is liable to an error of approximately 1%. [L.U.]

3. Two triangles have equal bases, each of length a, and their base angles are B, C and $B+\delta B$, $C+\delta C$ respectively, where δB and δC are small. Prove that their areas differ approximately by

$$\tfrac{1}{2}a^2(\sin^2 C\delta B+\sin^2 B\delta C) \operatorname{cosec}^2 A.$$

The angles of a triangle whose sides are proportional to 3:4:5 are 36° 52′, 53° 8′ and 90°. Given that a radian is 57° 18′, show that the area of a triangle, whose base is 60 yards and whose base angles are 54° and 89° 8′, is approximately 2,427 square yards. [L.U.]

4. Find the diameter D of the circumcircle of the triangle with sides a, a, $2b$. Calculate, to the first order of small quantities, the change in D due to small changes δa and δb in the values of a and b respectively. Deduce that, if $a=\sqrt{3}b$, there is no change in the value of D when a slightly increases and b slightly decreases in the same ratio. [L.U.]

5. Show that the volume of a segment of a sphere is $\frac{1}{6}\pi h(h^2+3R^2)$, where h is the height of the segment and R is the radius of its base.

If the measurement of h is too large by a small amount α, and that of R is too small by an equal amount, show that the calculated volume is too large by an amount $\frac{1}{2}\pi\alpha(h-R)^2$ approximately.

If the segment is a hemisphere, show that the error in the calculated volume is $\frac{2}{3}\pi\alpha^3$ exactly. [L.U.]

6. (i) If $u(x, y)=x^2-y^2$, find a function $v(x, y)$ such that $\dfrac{\partial v}{\partial x}=-\dfrac{\partial u}{\partial y}$ and $\dfrac{\partial v}{\partial y}=\dfrac{\partial u}{\partial x}$ for all x and y.

(ii) If $f(x, y)=xe^{xy}$, and the values of x and y are slightly changed from 1 and 0 to $1+\delta x$ and δy respectively so that δf, the change in f, is very nearly $3\delta x$, show that δy must be very nearly $2\delta x$. [L.U.]

7. (a) If $z=(x+y)/\sqrt{(x^2+y^2)}$, find $x\dfrac{\partial z}{\partial x}+y\dfrac{\partial z}{\partial y}$.

(b) If $z=\sin\theta \sin\phi/\sin\psi$ and z is calculated for the values $\theta=30°$, $\psi=60°$, $\phi=45°$, find approximately the change in the value of z if each of the angles θ and ψ is increased by the same small angle $\alpha°$ and ϕ is decreased by $\frac{1}{2}\alpha°$. [L.U.]

8. (i) If $f(x, y)=\log(x^2+y^2)$, prove that

$$\frac{\partial^2 f}{\partial x^2}+\frac{\partial^2 f}{\partial y^2}=0.$$

(ii) The area Δ of a triangle ABC is calculated from measurements of the sides b, c, with a possible error of $\frac{1}{2}\%$ in each, and of the angle A, correct to the nearest half degree. Find an approximate expression for the proportional error in Δ in terms of the errors δb, δc, δA in the measured values.

If the measured value of A is $60°$, determine approximately the maximum proportional error in Δ. [L.U.]

19.7. Differentials

If $y=f(x)$, and $f'(x)$ exists, then in the usual notation we have

$$\frac{\delta y}{\delta x}=\frac{f(x+\delta x)-f(x)}{\delta x}\to f'(x), \text{ as } \delta x\to 0,$$

so that

$$\frac{\delta y}{\delta x}=f'(x)+\eta,$$

$$\delta y=f'(x)\,\delta x+\eta\,\delta x \quad . \quad . \quad . \quad . \quad . \quad \text{(i)}$$

where $\eta\to 0$ as $\delta x\to 0$.

The first term on the right of (i), $f'(x)\,\delta x$, is called the *differential* of y and denoted by dy. Hence

$$dy=f'(x)\,\delta x \quad . \quad . \quad . \quad . \quad \text{(ii)}$$

This equation holds for any differentiable function $f(x)$, and so in particular for the case when $f(x)=x$. For this particular function (ii) gives

$$dy=\delta x.$$

But in this case

$$y=x \text{ and so}$$

$$dx=\delta x \quad . \quad . \quad . \quad . \quad . \quad \text{(iii)}$$

From (ii) and (iii) we have, in general,

$$dy=f'(x)\,dx. \qquad (19.3)$$

The geometrical significance of differentials is shown in fig. 64, where $P(x, y)$ and $Q(x+\delta x, y+\delta y)$ are points on the curve $y=f(x)$ and R, T are the points at which the ordinate at Q meets the parallel to Ox through P and the tangent at P respectively.

Then if

$$\angle RPT=\psi, \quad (-\pi/2<\psi<\pi/2),$$

$$f'(x)=\tan\psi.$$

Also

$$PR=\delta x=dx, \text{ by (iii)}, \quad RQ=\delta y$$

and

$$RT=PR\tan\psi=f'(x)\,dx=dy, \text{ by (19.3)}.$$

Hence T is the point $(x+dx, y+dy)$.

It follows that δy is the increment in ordinate as we move along the curve from P and dy is the increment as we move along the tangent at P.

In Chapter 9 we defined $\frac{dy}{dx}$ as $\lim_{\delta x \to 0} \frac{\delta y}{\delta x}$, and its value when $y=f(x)$ was denoted by $f'(x)$. (19.3) gives us a meaning for dy standing alone,

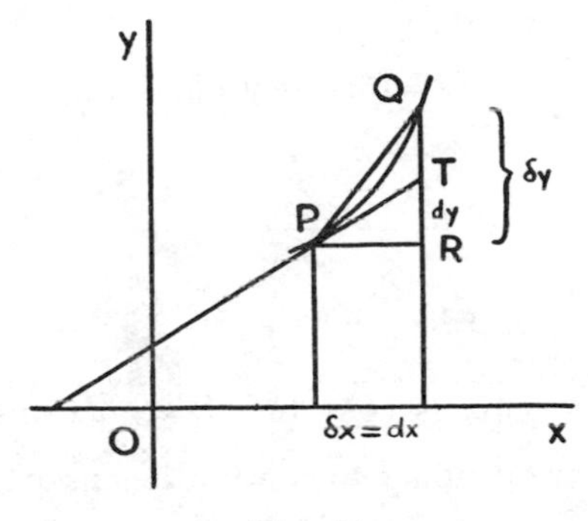

Fig. 64

and from it we see that the ratio of the differentials dy and dx is $f'(x)$.

Thus the relation $\frac{dy}{dx}=f'(x)$ is true whether we regard $\frac{dy}{dx}$ as the limiting value of $\frac{\delta y}{\delta x}$ or as the quotient of the differentials dy and dx. The advantage of the latter interpretation lies in the fact that $\frac{dy}{dx}$ is an ordinary algebraic fraction with a numerator and denominator which can be treated as separate entities. On the basis of (19.3) any formula for the derivative of a function of x becomes a formula for its differential by multiplying throughout by dx. For example:

$$d(\sin x) = \cos x\, dx$$

$$d(\log x) = \frac{dx}{x}$$

$$d(uv) = \left(u\frac{dv}{dx} + v\frac{du}{dx}\right)dx$$

$$= u\,dv + v\,du.$$

19.8. Total differential of a function of two variables

If $z=f(x, y)$, where x and y are independent variables, we define the differential dz by the equation

$$dz=\frac{\partial z}{\partial x}\delta x+\frac{\partial z}{\partial y}\delta y \quad . \quad . \quad . \quad . \quad \text{(i)}$$

Here z can be any function with continuous partial derivatives. Taking z to be the function x, we have

$$dx=1\delta x+0\delta y=\delta x.$$

Similarly $\qquad dy=\delta y.$

Hence we may write (i) in the form

$$dz=\frac{\partial z}{\partial x}dx+\frac{\partial z}{\partial y}dy. \tag{19.4}$$

If x varies while y remains constant, $dz=\frac{\partial z}{\partial x}dx$ and so $\frac{\partial z}{\partial x}dx$ is the differential of z corresponding to a variation in x alone. Similarly $\frac{\partial z}{\partial y}dy$ is the differential of z corresponding to a variation in y alone. These terms are called the *partial differentials* of z and their sum dz is called the *total differential* of z.

19.9. Total derivative

If, in § 19.6, x and y are both differentiable functions of an independent variable t and δx, δy are the respective increments in x and y corresponding to an increment δt in t, then z is also a function of t, and by (19.2) its derivative $\frac{dz}{dt}$ (known as the *total derivative* of z with respect to t) is easily seen to be given by

$$\frac{dz}{dt}=\lim_{\delta t\to 0}\frac{\delta z}{\delta t}=\frac{\partial z}{\partial x}\frac{dx}{dt}+\frac{\partial z}{\partial y}\frac{dy}{dt}. \tag{19.5}$$

Written in differential form, this result gives

$$dz=\frac{\partial z}{\partial x}dx+\frac{\partial z}{\partial y}dy$$

and comparison of this equation with (19.4) shows that the expression for the differential dz of a function of two variables x and y is the same when x and y are functions of a single independent variable t as when x and y are independent.

When x and y are both functions of two independent variables u and v, z may be expressed in terms of u and v and $\frac{\partial z}{\partial u}, \frac{\partial z}{\partial v}$ found by the ordinary rules of partial differentiation. Alternatively, in the same way as we proved (19.5) we may show that

$$\left.\begin{aligned}\frac{\partial z}{\partial u}=\frac{\partial z}{\partial x}\frac{\partial x}{\partial u}+\frac{\partial z}{\partial y}\frac{\partial y}{\partial u}\\ \frac{\partial z}{\partial v}=\frac{\partial z}{\partial x}\frac{\partial x}{\partial v}+\frac{\partial z}{\partial y}\frac{\partial y}{\partial v}\end{aligned}\right\}. \qquad (19.6)$$

We now prove that in this case also, $dz=\frac{\partial z}{\partial x}dx+\frac{\partial z}{\partial y}dy$.

By definition, since u and v are independent variables,

$$\begin{aligned}dz&=\frac{\partial z}{\partial u}du+\frac{\partial z}{\partial v}dv\\ &=\left(\frac{\partial z}{\partial x}\frac{\partial x}{\partial u}+\frac{\partial z}{\partial y}\frac{\partial y}{\partial u}\right)du+\left(\frac{\partial z}{\partial x}\frac{\partial x}{\partial v}+\frac{\partial z}{\partial y}\frac{\partial y}{\partial v}\right)dv, \text{ by (19.6)}\\ &=\frac{\partial z}{\partial x}\left(\frac{\partial x}{\partial u}du+\frac{\partial x}{\partial v}dv\right)+\frac{\partial z}{\partial y}\left(\frac{\partial y}{\partial u}du+\frac{\partial y}{\partial v}dv\right)\\ &=\frac{\partial z}{\partial x}dx+\frac{\partial z}{\partial y}dy.\end{aligned}$$

By an obvious extension of (19.5) to a function V of three variables x, y, z, all functions of the independent variable t, we have

$$\frac{dV}{dt}=\frac{\partial V}{\partial x}\frac{dx}{dt}+\frac{\partial V}{\partial y}\frac{dy}{dt}+\frac{dV}{\partial z}\frac{dz}{dt}. \qquad (19.7)$$

Example 13

If $z=f(x, y)$ and $x=\frac{1}{2}(u^2-v^2)$, $y=uv$, show that

(i) $u\frac{\partial z}{\partial v}-v\frac{\partial z}{\partial u}=2\left(x\frac{\partial z}{\partial y}-y\frac{\partial z}{\partial x}\right)$;

(ii) $\frac{\partial^2 z}{\partial u^2}+\frac{\partial^2 z}{\partial v^2}=(u^2+v^2)\left(\frac{\partial^2 z}{\partial x^2}+\frac{\partial^2 z}{\partial y^2}\right)$. [L.U.]

If $\qquad x=\frac{1}{2}(u^2-v^2)$ and $y=uv$ . . . . . (i)

$$\frac{\partial x}{\partial u}=u,\ \frac{\partial x}{\partial v}=-v,\ \frac{\partial y}{\partial u}=v \text{ and } \frac{\partial y}{\partial v}=u.$$

By (19.6), $\dfrac{\partial z}{\partial u}=u\dfrac{\partial z}{\partial x}+v\dfrac{\partial z}{\partial y}$ and $\dfrac{\partial z}{\partial v}=-v\dfrac{\partial z}{\partial x}+u\dfrac{\partial z}{\partial y}$

$$\therefore\ u\frac{\partial z}{\partial v}-v\frac{\partial z}{\partial u}=(u^2-v^2)\frac{\partial z}{\partial y}-2uv\frac{\partial z}{\partial x}=2\left(x\frac{\partial z}{\partial y}-y\frac{\partial z}{\partial x}\right),\quad \text{by (i)}$$

$$\frac{\partial^2 z}{\partial u^2}=\frac{\partial}{\partial u}\left(u\frac{\partial z}{\partial x}+v\frac{\partial z}{\partial y}\right)$$

$$=\frac{\partial z}{\partial x}+u\frac{\partial}{\partial u}\left(\frac{\partial z}{\partial x}\right)+v\frac{\partial}{\partial u}\left(\frac{\partial z}{\partial y}\right) \quad . \quad . \quad . \quad \text{(ii)}$$

Now if V is any function of x and y, by (19.6)

$$\frac{\partial V}{\partial u}=\frac{\partial V}{\partial x}\frac{\partial x}{\partial u}+\frac{\partial V}{\partial y}\frac{\partial y}{\partial u}=u\frac{\partial V}{\partial x}+v\frac{\partial V}{\partial y}.$$

Put $V=\dfrac{\partial z}{\partial x}$; then $\dfrac{\partial}{\partial u}\left(\dfrac{\partial z}{\partial x}\right)=u\dfrac{\partial^2 z}{\partial x^2}+v\dfrac{\partial^2 z}{\partial y\partial x}.$

Put $V=\dfrac{\partial z}{\partial y}$; then $\dfrac{\partial}{\partial u}\left(\dfrac{\partial z}{\partial y}\right)=u\dfrac{\partial^2 z}{\partial x\,\partial y}+v\dfrac{\partial^2 z}{\partial y^2}.$

Substituting these values in (ii), we have

$$\frac{\partial^2 z}{\partial u^2}=\frac{\partial z}{\partial x}+u^2\frac{\partial^2 z}{\partial x^2}+2uv\frac{\partial^2 z}{\partial x\,\partial y}+v^2\frac{\partial^2 z}{\partial y^2}.$$

Similarly,

$$\frac{\partial^2 z}{\partial v^2}=-\frac{\partial z}{\partial x}+v^2\frac{\partial^2 z}{\partial x^2}-2uv\frac{\partial^2 z}{\partial x\,\partial y}+u^2\frac{\partial^2 z}{\partial y^2}.$$

$$\therefore\ \frac{\partial^2 z}{\partial u^2}+\frac{\partial^2 z}{\partial v^2}=(u^2+v^2)\left(\frac{\partial^2 z}{\partial x^2}+\frac{\partial^2 z}{\partial y^2}\right).$$

19.10. Applications of differentials

(a) *If u and v are given in terms of x and y, to find*

$$\frac{\partial x}{\partial u},\ \frac{\partial x}{\partial v},\ \frac{\partial y}{\partial u}\ \textit{and}\ \frac{\partial y}{\partial v}.$$

If $u=\phi(x, y)$, $v=\psi(x, y)$, it may be difficult to find expressions for x and y in terms of u and v from which the required derivatives would be directly obtainable. However by the use of differentials these derivatives may be found in terms of x and y.

We have $$du=\frac{\partial u}{\partial x}dx+\frac{\partial u}{\partial y}dy$$

and $$dv=\frac{\partial v}{\partial x}dx+\frac{\partial v}{\partial y}dy.$$

Solving these equations for dx and dy we obtain expressions for dx and dy in terms of du and dv and from these we may find

$$\frac{\partial x}{\partial u}, \frac{\partial x}{\partial v}, \frac{\partial y}{\partial u} \text{ and } \frac{\partial y}{\partial v}.$$

Example 14

If $u=x+y$ and $v=xy$, find $\partial x/\partial u$, $\partial x/\partial v$, $\partial y/\partial u$ and $\partial y/\partial v$.

$$du=dx+dy$$
$$dv=y\,dx+x\,dy.$$
$$\therefore (x-y)dx=xdu-dv \quad . \quad . \quad . \quad . \quad \text{(i)}$$

and

$$(x-y)dy=dv-y\,du \quad . \quad . \quad . \quad \text{(ii)}$$

Keeping v constant, so that $dv=0$, we obtain

$$\frac{\partial x}{\partial u}=x/(x-y), \ \frac{\partial y}{\partial u}=-y/(x-y).$$

Keeping u constant, so that $du=0$, we have

$$\frac{\partial x}{\partial v}=-1/(x-y), \ \frac{\partial y}{\partial v}=1/(x-y).$$

(b) *To find $\frac{dy}{dx}$ when the relation between x and y is given in the form $f(x, y)=0$.*

Since

$$f(x, y)=0 \quad . \quad . \quad . \quad \text{(i)}$$
$$df=0$$
$$\therefore f_x\,dx+f_y\,dy=0 \quad . \quad . \quad . \quad . \quad \text{(ii)}$$

Thus, if we interpret (i) as giving y implicitly in terms of x we have) from (ii), $dy/dx=-f_x/f_y$, it being assumed that f_x and f_y are not both zero.

Example 15

Find dy/dx at any point of the curve $x^3+y^3-3xy=a^3$.

Here

$$f(x, y)\equiv x^3+y^3-3xy-a^3$$
$$\therefore f_x=3(x^2-y), \ f_y=3(y^2-x),$$

and so

$$\frac{dy}{dx}=-(x^2-y)/(y^2-x).$$

(c) *To find $\frac{\partial z}{\partial x}, \frac{\partial z}{\partial y}$ when the relation between x, y and z is given in the, form $F(x, y, z)=0$.*

$$dF=0$$
$$\therefore F_x\,dx+F_y\,dy+F_z\,dz=0 \quad . \quad . \quad . \quad \text{(i)}$$

If we interpret $F(x, y, z)=0$ as giving z implicitly in terms of the independent variables x and y, by keeping y constant, so that $dy=0$, we obtain, from (i),

$$\frac{\partial z}{\partial x}=-F_x/F_z$$

and, by keeping x constant,

$$\frac{\partial z}{\partial y}=-F_y/F_z.$$

Example 16

If $x^3+y^3+z^3-3xyz=a^3$, find $\partial y/\partial x$. [L.U. Anc.]

Here $$F(x, y, z)\equiv x^3+y^3+z^3-3xyz-a^3.$$

$$\therefore\ F_x=3(x^2-yz),\ F_y=3(y^2-xz),\ F_z=3(z^2-xy).$$

But $dF=0$;

$$\therefore\ (x^2-yz)\,dx+(y^2-xz)\,dy+(z^2-xy)\,dz=0.$$

Keeping z constant, so that $dz=0$, we have

$$\frac{\partial y}{\partial x}=-(x^2-yz)/(y^2-xz).$$

(d) *To find $\dfrac{dy}{dx}$ from the relations $f(x, y, z)=0$ and $F(x, y, z)=0$.*

We may find $\dfrac{dy}{dx}$ by eliminating z between the two given equations and establishing a relation between x and y. Alternatively, we may proceed as follows:

$$f_x\,dx+f_y\,dy+f_z\,dz=0$$

and $$F_x\,dx+F_y\,dy+F_z\,dz=0.$$

Hence, as in § 16.10, we obtain

$$\frac{dx}{\begin{vmatrix} f_y & f_z \\ F_y & F_z \end{vmatrix}}=\frac{-dy}{\begin{vmatrix} f_x & f_z \\ F_x & F_z \end{vmatrix}}=\frac{dz}{\begin{vmatrix} f_x & f_y \\ F_x & F_y \end{vmatrix}}.$$

$$\frac{dy}{dx}=-\begin{vmatrix} f_x & f_z \\ F_x & F_z \end{vmatrix}\div\begin{vmatrix} f_y & f_z \\ F_y & F_z \end{vmatrix}.$$

Example 17

If the variables x, y and z are connected by the equations $f(x, y, z)=0$ $1/x+1/y+1/z=$constant, find dy/dx.

We have $$f_x\,dx+f_y\,dy+f_z\,dz=0$$

and $$(1/x^2)\,dx+(1/y^2)\,dy+(1/z^2)\,dz=0.$$

$$\therefore\ \frac{dy}{dx}=-\left(\frac{1}{z^2}f_x-\frac{1}{x^2}f_z\right)\bigg/\left(\frac{1}{z^2}f_y-\frac{1}{y^2}f_z\right).$$

19.11. Miscellaneous examples

Example 18

If z is a function of x and y and y=ux, prove that

$$\left(\frac{\partial z}{\partial x}\right)_{y\text{ constant}}=\left(\frac{\partial z}{\partial x}\right)_{u\text{ constant}}-\frac{u}{x}\left(\frac{\partial z}{\partial u}\right)_{x\text{ constant}}. \qquad \text{[L.U.]}$$

Since z is a function of x and y,

$$dz=\left(\frac{\partial z}{\partial x}\right)dx+\left(\frac{\partial z}{\partial y}\right)dy \quad . \quad . \quad . \quad \text{(i)}$$

where

$\left(\frac{\partial z}{\partial x}\right)$ $\left(\frac{\partial z}{\partial y}\right)$ denote $\left(\frac{\partial z}{\partial x}\right)_{y\text{ constant}}$ and $\left(\frac{\partial z}{\partial y}\right)_{x\text{ constant}}$ respectively.

But
$$y=ux$$
$$\therefore dy=u\,dx+x\,du$$

and so from (i)
$$dz=\left(\frac{\partial z}{\partial x}\right)dx+\left(\frac{\partial z}{\partial y}\right)(u\,dx+x\,du)$$
$$=\left(\frac{\partial z}{\partial x}+u\frac{\partial z}{\partial y}\right)dx+x\frac{\partial z}{\partial y}du.$$

When u is constant, $du=0$

$$\therefore \left(\frac{\partial z}{\partial x}\right)_{u\text{ constant}}=\left(\frac{\partial z}{\partial x}\right)+u\left(\frac{\partial z}{\partial y}\right).$$

When x is constant, $dx=0$

$$\therefore \left(\frac{\partial z}{\partial u}\right)_{x\text{ constant}}=x\left(\frac{\partial z}{\partial y}\right).$$

Hence
$$\left(\frac{\partial z}{\partial x}\right)_{u\text{ constant}}-\frac{u}{x}\left(\frac{\partial z}{\partial u}\right)_{x\text{ constant}}=\left(\frac{\partial z}{\partial x}\right)_{y\text{ constant}}.$$

Example 19

If $\phi(z)=y/x$, prove that
$$\frac{\partial z}{\partial x}+\phi(z)\cdot\frac{\partial z}{\partial y}=0,$$

and that
$$\left(\frac{\partial z}{\partial y}\right)^2\cdot\frac{\partial^2 z}{\partial x^2}+\left(\frac{\partial z}{\partial x}\right)^2\cdot\frac{\partial^2 z}{\partial y^2}=2\frac{\partial z}{\partial x}\cdot\frac{\partial z}{\partial y}\cdot\frac{\partial^2 z}{\partial x\,\partial y}, \qquad \text{[L.U.]}$$

Since
$$\phi(z)=y/x,\ \phi'(z)\frac{\partial z}{\partial x}=-y/x^2 \text{ and } \phi'(z)\frac{\partial z}{\partial y}=1/x.$$

Hence
$$\phi'(z)\left\{\frac{\partial z}{\partial x}+\frac{y}{x}\frac{\partial z}{\partial y}\right\}=0$$

i.e.
$$\frac{\partial z}{\partial x}+\phi(z)\frac{\partial z}{\partial y}=0 \quad . \quad . \quad . \quad \text{(i)}$$

Partially differentiating (i) first with respect to x and then with respect to y, we have

$$\frac{\partial^2 z}{\partial x^2}+\phi(z)\frac{\partial^2 z}{\partial x\,\partial y}+\phi'(z)\frac{\partial z}{\partial x}\cdot\frac{\partial z}{\partial y}=0 \quad . \quad . \quad . \quad \text{(ii)}$$

and

$$\frac{\partial^2 z}{\partial y\,\partial x}+\phi(z)\frac{\partial^2 z}{\partial y^2}+\phi'(z)\left(\frac{\partial z}{\partial y}\right)^2=0 \quad . \quad . \quad . \quad \text{(iii)}$$

Eliminating $\phi'(z)$ between (ii) and (iii), we obtain

$$\left(\frac{\partial z}{\partial y}\right)\frac{\partial^2 z}{\partial x^2}+\frac{\partial^2 z}{\partial x\,\partial y}\left\{\phi(z)\frac{\partial z}{\partial y}-\frac{\partial z}{\partial x}\right\}-\phi(z)\frac{\partial z}{\partial x}\frac{\partial^2 z}{\partial y^2}=0.$$

But by (i),

$$\phi(z)=-\frac{\partial z}{\partial x}\Big/\frac{\partial z}{\partial y},$$

whence

$$\left(\frac{\partial z}{\partial y}\right)^2\frac{\partial^2 z}{\partial x^2}+\left(\frac{\partial z}{\partial x}\right)^2\frac{\partial^2 z}{\partial y^2}=2\frac{\partial z}{\partial x}\cdot\frac{\partial z}{\partial y}\cdot\frac{\partial^2 z}{\partial x\,\partial y}.$$

Example 20

If $x=f(u, v)$, $y=g(u, v)$, show that the curves given by taking u=constant and v=constant respectively intersect orthogonally if

$$\frac{\partial x}{\partial u}\frac{\partial x}{\partial v}+\frac{\partial y}{\partial u}\frac{\partial y}{\partial v}=0.$$

If $x+iy=e^{u+iv}$, show that the curves u=constant v=constant are orthogonal. [L.U.]

We may regard the equations $x=f(u, v)$, $y=g(u, v)$ as determining the curve C traced out by the point $P(x, y)$ in the (xy) plane corresponding to a point $P'(u, v)$ which traces out a curve C' in the (u, v) plane. The gradient $\frac{dy}{dx}$ at any point of C may be calculated in terms of the gradient $\frac{dv}{du}$ at any point of C' for

$$dx=\frac{\partial x}{\partial u}du+\frac{\partial x}{\partial v}dv \text{ and } dy=\frac{\partial y}{\partial u}du+\frac{\partial y}{\partial v}dv.$$

$$\therefore \frac{dy}{dx}=\left(\frac{\partial y}{\partial u}du+\frac{\partial y}{\partial v}dv\right)\Big/\left(\frac{\partial x}{\partial u}du+\frac{\partial x}{\partial v}dv\right).$$

If u=constant, $du=0$; hence $\frac{dy}{dx}=\frac{\partial y}{\partial v}\Big/\frac{\partial x}{\partial v}$.

If v=constant, $dv=0$; hence $\frac{dy}{dx}=\frac{\partial y}{\partial u}\Big/\frac{\partial x}{\partial u}$.

The curves in the xy plane corresponding to u=constant, v=constant will intersect orthogonally if

$$\left(\frac{\partial y}{\partial v}\Big/\frac{\partial x}{\partial v}\right)\left(\frac{\partial y}{\partial u}\Big/\frac{\partial x}{\partial u}\right)=-1,$$

i.e. if

$$\frac{\partial x}{\partial u}\frac{\partial x}{\partial v}+\frac{\partial y}{\partial u}\frac{\partial y}{\partial v}=0 \quad . \quad . \quad . \quad \text{(i)}$$

If $x+iy=e^{u+iv}=e^u(\cos v+i\sin v)$,

$$x=e^u\cos v,\ y=e^u\sin v$$

$$\therefore\ \frac{\partial x}{\partial u}=e^u\cos v,\ \frac{\partial x}{\partial v}=-e^u\sin v,$$

$$\frac{\partial y}{\partial u}=e^u\sin v,\ \frac{\partial y}{\partial v}=e^u\cos v;$$

and so condition (i) is fulfilled and the curves $u=$constant, $v=$constant are orthogonal.

Note that

$$x^2+y^2=e^{2u}$$

$$y/x=\tan v$$

so that the curves $u=$constant are concentric circles with centres at the origin, and the curves $v=$constant are straight lines through the origin which, being diameters of the concentric circles, cut them orthogonally.

Example 21

If $u=y^3-3x^2y$, prove that

$$\frac{\partial^2 u}{\partial x^2}+\frac{\partial^2 u}{\partial y^2}=0.$$

Prove also that, if $z=r^nu$ where $r^2=x^2+y^2$, then

$$r^2\left\{\frac{\partial^2 z}{\partial x^2}+\frac{\partial^2 z}{\partial y^2}\right\}=(n^2+6n)z. \qquad \text{[L.U.]}$$

If $u=y^3-3x^2y$,

$$\frac{\partial u}{\partial x}=-6xy,\ \frac{\partial u}{\partial y}=3(y^2-x^2). \quad \text{(i)}$$

$$\frac{\partial^2 u}{\partial x^2}=-6y,\ \frac{\partial^2 u}{\partial y^2}=6y$$

$$\therefore\ \frac{\partial^2 u}{\partial x^2}+\frac{\partial^2 u}{\partial y^2}=0. \quad \text{(ii)}$$

If $z=r^nu$, where $r^2=x^2+y^2$,

$$\frac{\partial z}{\partial x}=r^n\frac{\partial u}{\partial x}+u\left(nr^{n-1}\frac{\partial r}{\partial x}\right),\ \text{and}\ \frac{\partial r}{\partial x}=\frac{x}{r}$$

$$\therefore\ \frac{\partial z}{\partial x}=r^n\frac{\partial u}{\partial x}+nr^{n-2}xu. \quad \text{(iii)}$$

From (iii),

$$\frac{\partial^2 z}{\partial x^2}=\frac{\partial}{\partial x}\left(r^n\frac{\partial u}{\partial x}\right)+n\frac{\partial}{\partial x}(r^{n-2}xu). \quad \text{(iv)}$$

Now

$$\frac{\partial}{\partial x}\left(r^n\frac{\partial u}{\partial x}\right)=r^n\frac{\partial^2 u}{\partial x^2}+nr^{n-2}x\frac{\partial u}{\partial x}$$

and

$$n\frac{\partial}{\partial x}(r^{n-2}xu)=n\left\{(n-2)r^{n-3}xu\frac{\partial r}{\partial x}+r^{n-2}u+r^{n-2}x\frac{\partial u}{\partial x}\right\}$$

$$=n\left\{(n-2)r^{n-4}x^2u+r^{n-2}u+r^{n-2}x\frac{\partial u}{\partial x}\right\}.$$

Hence, from (iv),

$$\frac{\partial^2 z}{\partial x^2}=r^n\frac{\partial^2 u}{\partial x^2}+2nr^{n-2}x\frac{\partial u}{\partial x}+nr^{n-4}u\{(n-2)x^2+r^2\}.$$

Similarly, $$\frac{\partial^2 z}{\partial y^2}=r^n\frac{\partial^2 u}{\partial y^2}+2nr^{n-2}y\frac{\partial u}{\partial y}+nr^{n-4}u\{(n-2)y^2+r^2\}$$

$$\therefore\ \frac{\partial^2 z}{\partial x^2}+\frac{\partial^2 z}{\partial y^2}=2nr^{n-2}\left(x\frac{\partial u}{\partial x}+y\frac{\partial u}{\partial y}\right)+n^2r^{n-2}u,\quad \text{using (ii).}$$

But $$x\frac{\partial u}{\partial x}+y\frac{\partial u}{\partial y}=3(y^3-3x^2y)=3u$$

$$\therefore\ r^2\left\{\frac{\partial^2 z}{\partial x^2}+\frac{\partial^2 z}{\partial y^2}\right\}=(6n+n^2)r^n u=(n^2+6n)z.$$

In examples of this type lengthy working may sometimes be avoided by using operators, as in the following examples.

Example 22

If $V=f(x, y)$ and $x=e^u\cos v$, $y=e^u\sin v$, prove that

$$\frac{\partial^2 V}{\partial u^2}+\frac{\partial^2 V}{\partial v^2}=e^{2u}\left(\frac{\partial^2 V}{\partial x^2}+\frac{\partial^2 V}{\partial y^2}\right).$$ [Leeds.]

$$\frac{\partial V}{\partial u}=\frac{\partial V}{\partial x}\frac{\partial x}{\partial u}+\frac{\partial V}{\partial y}\frac{\partial y}{\partial u}$$

$$=e^u\cos v\frac{\partial V}{\partial x}+e^u\sin v\frac{\partial V}{\partial y}$$

$$\therefore\ \frac{\partial V}{\partial u}=x\frac{\partial V}{\partial x}+y\frac{\partial V}{\partial y}\quad \text{(i)}$$

Similarly $$\frac{\partial V}{\partial v}=-y\frac{\partial V}{\partial x}+x\frac{\partial V}{\partial y}.$$

The symbol $\frac{\partial}{\partial u}$ may be regarded as an operator which obtains from V its derivative $\frac{\partial V}{\partial u}$. If we write (i) in the form

$$\frac{\partial}{\partial u}(V)=\left(x\frac{\partial}{\partial x}+y\frac{\partial}{\partial y}\right)V,$$

we see that the operator $\frac{\partial}{\partial u}$ is equivalent to the operator $\left(x\frac{\partial}{\partial x}+y\frac{\partial}{\partial y}\right)$ and, in the same way, the operator $\frac{\partial}{\partial v}$ is equivalent to $\left(-y\frac{\partial}{\partial x}+x\frac{\partial}{\partial y}\right)$.

Again $$\frac{\partial^2 V}{\partial u^2}=\frac{\partial}{\partial u}\left(\frac{\partial V}{\partial u}\right)$$

$$=\left(x\frac{\partial}{\partial x}+y\frac{\partial}{\partial y}\right)\left(x\frac{\partial V}{\partial x}+y\frac{\partial V}{\partial y}\right)$$

$$=x\left(x\frac{\partial^2 V}{\partial x^2}+\frac{\partial V}{\partial x}+y\frac{\partial^2 V}{\partial x\,\partial y}\right)+y\left(x\frac{\partial^2 V}{\partial y\partial x}+y\frac{\partial^2 V}{\partial y^2}+\frac{\partial V}{\partial y}\right)$$

$$=x^2\frac{\partial^2 V}{\partial x^2}+2xy\frac{\partial^2 V}{\partial x\,\partial y}+y^2\frac{\partial^2 V}{\partial y^2}+x\frac{\partial V}{\partial x}+y\frac{\partial V}{\partial y}.$$

Similarly, $$\frac{\partial^2 V}{\partial v^2}=y^2\frac{\partial^2 V}{\partial x^2}-2xy\frac{\partial^2 V}{\partial x\,\partial y}+x^2\frac{\partial^2 V}{\partial y^2}-x\frac{\partial V}{\partial x}-y\frac{\partial V}{\partial y},$$

$$\therefore\ \frac{\partial^2 V}{\partial u^2}+\frac{\partial^2 V}{\partial v^2}=(x^2+y^2)\left(\frac{\partial^2 V}{\partial x^2}+\frac{\partial^2 V}{\partial y^2}\right)=e^{2u}\left(\frac{\partial^2 V}{\partial x^2}+\frac{\partial^2 V}{\partial^2 y}\right).$$

Example 23

If V is a function of x and y and $x=r\cos\theta$, $y=r\sin\theta$, prove that

$$\frac{\partial^2 V}{\partial x^2}+\frac{\partial^2 V}{\partial y^2}=\frac{\partial^2 V}{\partial r^2}+\frac{1}{r}\frac{\partial V}{\partial r}+\frac{1}{r^2}\frac{\partial^2 V}{\partial \theta^2}.$$

We have $$\frac{\partial^2 V}{\partial x^2}+\frac{\partial^2 V}{\partial y^2}=\left(\frac{\partial}{\partial x}-i\frac{\partial}{\partial y}\right)\left(\frac{\partial V}{\partial x}+i\frac{\partial V}{\partial y}\right) \quad . \quad . \quad \text{(i)}$$

Now $$\frac{\partial V}{\partial x}=\frac{\partial V}{\partial r}\frac{\partial r}{\partial x}+\frac{\partial V}{\partial \theta}\frac{\partial \theta}{\partial x},$$

and since $r^2=x^2+y^2$, $\theta=\tan^{-1} y/x$

$$\therefore\ \frac{\partial r}{\partial x}=\frac{x}{r}=\cos\theta,\quad \frac{\partial \theta}{\partial \mathrm{x}}=\frac{-y}{r^2}=\frac{-\sin\theta}{\mathrm{r}}.$$

Hence $$\frac{\partial V}{\partial x}=\cos\theta\frac{\partial V}{\partial r}-\frac{\sin\theta}{r}\frac{\partial V}{\partial \theta}.$$

Similarly, $$\frac{\partial V}{\partial y}=\sin\theta\frac{\partial V}{\partial r}+\frac{\cos\theta}{r}\frac{\partial V}{\partial \theta}$$

$$\therefore\ \frac{\partial V}{\partial x}+i\frac{\partial V}{\partial y}=e^{i\theta}\left(\frac{\partial V}{\partial r}+\frac{i}{r}\frac{\partial V}{\partial \theta}\right) \quad . \quad . \quad . \quad . \quad \text{(ii)}$$

and $$\frac{\partial V}{\partial x}-i\frac{\partial V}{\partial y}=e^{-i\theta}\left(\frac{\partial V}{\partial r}-\frac{i}{r}\frac{\partial V}{\partial \theta}\right) \quad . \quad . \quad . \quad \text{(iii)}$$

since $\cos\theta+i\sin\theta=e^{i\theta}$.

From (i), (ii) and (iii)

$$\frac{\partial^2 V}{\partial x^2}+\frac{\partial^2 V}{\partial y^2}=e^{-i\theta}\left(\frac{\partial}{\partial r}-\frac{i}{r}\frac{\partial}{\partial \theta}\right)\left\{e^{i\theta}\left(\frac{\partial V}{\partial r}+\frac{i}{r}\frac{\partial V}{\partial \theta}\right)\right\}$$

$$=\left(\frac{\partial^2 V}{\partial r^2}-\frac{i}{r^2}\frac{\partial V}{\partial \theta}+\frac{i}{r}\frac{\partial^2 V}{\partial r\,\partial \theta}\right)+\frac{1}{r}\left(\frac{\partial V}{\partial r}+\frac{i}{r}\frac{\partial V}{\partial \theta}\right)-\frac{i}{r}\left(\frac{\partial^2 V}{\partial \theta\,\partial r}+\frac{i}{r}\frac{\partial^2 V}{\partial \theta^2}\right)$$

$$=\frac{\partial^2 V}{\partial r^2}+\frac{1}{r}\frac{\partial V}{\partial r}+\frac{1}{r^2}\frac{\partial^2 V}{\partial \theta^2}.$$

19.12. Euler's theorems on homogeneous functions

A function $F(x, y, z, \ldots)$ is said to be a *homogeneous function* of degree n if, for all positive values of t,

$$F(tx, ty, tz, \ldots)=t^n F(x, y, z, \ldots).$$

By putting $t=1/x$, we get

$$F(1, y/x, z/x, \ldots)=\frac{1}{x^n} F(x, y, z, \ldots),$$

i.e. $$F(x, y, z, \ldots)=x^n f(y/x, z/x \ldots)$$

where we write $F(1, y/x, z/x, \ldots)=f(y/x, z/x, \ldots)$.

Hence a homogeneous function of the nth degree is of the form $x^n f(y/x, z/x, \ldots)$ if $x>0$.

Euler's first theorem on homogeneous functions states that if $F(x, y)$ is a homogeneous function of the nth degree in x and y

$$x\frac{\partial F}{\partial x}+y\frac{\partial F}{\partial y}=nF(x, y). \tag{19.8}$$

This result has been proved in § 19.3 Example 7.

From (19.8) we see that the operator $\left(x\frac{\partial}{\partial x}+y\frac{\partial}{\partial y}\right)$ applied to $F(x, y)$, a homogeneous function of the nth degree in x and y has the effect of multiplying $F(x, y)$ by n. Also, from (19.8) it follows that $x\frac{\partial F}{\partial x}+y\frac{\partial F}{\partial x}$ is a homogeneous function of the nth degree in x and y since n is a constant.

$$\therefore \left(x\frac{\partial}{\partial x}+y\frac{\partial}{\partial y}\right)\left(x\frac{\partial F}{\partial x}+y\frac{\partial F}{\partial y}\right)=n\{nF(x, y)\}.$$

On performing the operations indicated on the left-hand side, we obtain

$$x^2\frac{\partial^2 F}{\partial x^2}+2xy\frac{\partial^2 F}{\partial x\,\partial y}+y^2\frac{\partial^2 F}{\partial y^2}+\left(x\frac{\partial F}{\partial x}+y\frac{\partial F}{\partial y}\right)=n^2F(x, y)$$

and so by (19.8)

$$x^2\frac{\partial^2 F}{\partial x^2}+2xy\frac{\partial^2 F}{\partial x\,\partial y}+y^2\frac{\partial^2 F}{\partial y^2}=n(n-1)F(x, y). \qquad (19.9)$$

This is Euler's second theorem.

Example 24

If $f(x, y)=x^3 \tan^{-1}(y/x)-y^3\tan^{-1}(x/y)$, *evaluate* $x^2 f_{xx}+2xy\,f_{xy}+y^2 f_{yy}$.

In this case, $f(x, y)$ is a homogeneous function of the third degree. Hence, by (19.9)

$$x^2 f_{xx}+2xy\,f_{xy}+y^2 f_{yy}=6f(x, y).$$

19.13. Note on envelopes

Suppose that a family of curves C is defined by the equation $f(x, y, \theta)=0$ where θ is a parameter fixed for each member of the family but varying from curve to curve, for example, the family $y=\theta x+\dfrac{a}{\theta}$, which is a family of straight lines touching the parabola $y^2=4ax$ (see § 13.7). If a curve E exists which touches every member of the family C, E is called the *envelope* of family C. Let us assume that E exists, that its equation is $F(x, y)=0$ and that the curve $f(x, y, \alpha)=0$, i.e. the curve of family C with parameter α touches E at (ξ, η). Then $F(\xi, \eta)=0$ and $f(\xi, \eta, \alpha)=0$ so that ξ and η are functions of α.

If we denote the values of $\dfrac{\partial f}{\partial x}$ and $\dfrac{\partial f}{\partial y}$ at (ξ, η) by $\dfrac{\partial f}{\partial \xi}$, $\dfrac{\partial f}{\partial \eta}$ and the value of $\dfrac{\partial f}{\partial \theta}$ when $\theta=\alpha$ by $\dfrac{\partial f}{\partial \alpha}$ we have by (19.6) and (19.7), page 415,

$$\frac{\partial F}{\partial \xi}\frac{d\xi}{d\alpha}+\frac{\partial F}{\partial \eta}\frac{d\eta}{d\alpha}=0 \quad . \quad . \quad . \quad \text{(i)}$$

and

$$\frac{\partial f}{\partial \xi}\frac{d\xi}{d\alpha}+\frac{\partial f}{\partial \eta}\frac{d\eta}{d\alpha}+\frac{\partial f}{\partial \alpha}=0. \quad . \quad . \quad . \quad \text{(ii)}$$

Now the gradients of the curves $f(x, y, \alpha)=0$ and $F(x, y)=0$ at (ξ, η) are equal since the curves touch at that point; hence by § 19.10 (*b*)

$$-\frac{\partial f}{\partial \xi}\Big/\frac{\partial f}{\partial \eta}=-\frac{\partial F}{\partial \xi}\Big/\frac{\partial F}{\partial \eta}=\frac{d\eta}{d\alpha}\Big/\frac{d\xi}{d\alpha}, \text{ by (i).}$$

$$\therefore \frac{\partial f}{\partial \xi}\frac{d\xi}{d\alpha}+\frac{\partial f}{\partial \eta}\frac{d\eta}{d\alpha}=0. \quad . \quad . \quad . \quad . \quad \text{(iii)}$$

Comparing (ii) and (iii), we see that

$$\frac{\partial f}{\partial \alpha}=0$$

and this equation is satisfied by (ξ, η).

It follows that (ξ, η) satisfies the equations $f(x, y, \alpha)=0$ and $\dfrac{\partial f}{\partial \alpha}=0$. Hence we find the envelope of the family $f(x, y, \theta)$ by eliminating θ between the equations

$$\left.\begin{aligned} f(x, y, \theta)=0 \\ \frac{\partial}{\partial\theta}f(x, y, \theta)=0 \end{aligned}\right\}. \qquad (19.10)$$

Example 25

The envelope of the lines

$$x \cos \theta+y \sin \theta=p, \quad . \quad . \quad . \quad . \quad \text{(i)}$$

where θ is a parameter, is obtained by eliminating θ between (i) and

$$-x \sin \theta+y \cos \theta=0.$$

Fig. 65

The envelope is the circle $x^2+y^2=p^2$ and the given lines which lie at a perpendicular distance p from the origin all touch this circle (see fig. 65).

Example 26

The envelope of the circles

$$(x-a)^2+y^2=r^2 \quad . \quad . \quad . \quad . \quad \text{(i)}$$

where a is a parameter is given by (i) and the equation

$$x-a=0.$$

This gives the two straight lines $y=\pm r$ which are parallel to the line of centres of the family of circles (fig. 66).

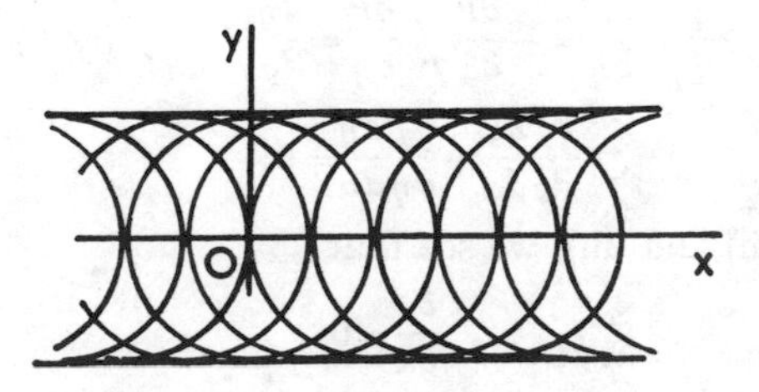

Fig. 66

Example 27

The envelope of the normals to the parabola $y^2=4ax$ is given by the equations

$$y+tx=2at+at^3$$

and $$x=2a+3at^2 \quad . \quad . \quad . \quad . \quad \text{(i)}$$

whence $$y=-2at^3. \quad . \quad . \quad . \quad . \quad . \quad \text{(ii)}$$

Eliminating t between (i) and (ii) we have

$$27ay^2=4(x-2a)^3.$$

The envelope of the normals to a curve is called the *evolute* of the curve.

Exercises 19 (c)

1. Show that $u=\log(x^2+y^2)$ satisfies the differential equation

$$\frac{\partial^2 u}{\partial x^2}+\frac{\partial^2 u}{\partial y^2}=0.$$

Hence, or otherwise, prove that $xy/(x^2+y^2)^2$ also satisfies the equation. [Sheffield.]

2. If $u=x^n\{f(y+x)+g(y-x)\}$, where f and g are arbitrary functions, prove that

$$\frac{\partial^2 u}{\partial x^2}-\frac{\partial^2 u}{\partial y^2}-\frac{2n}{x}\frac{\partial u}{\partial x}+\frac{n(n+1)}{x^2}u=0. \qquad \text{[L.U.]}$$

3. If $u=f(z)$ and $v=\phi(z)$ where $z=px^2+2qxy+ry^2$ and p, q, r are constants, show that

$$\frac{\partial u}{\partial x}\frac{\partial v}{\partial y}=\frac{\partial u}{\partial y}\frac{\partial v}{\partial x}. \qquad \text{[L.U.]}$$

4. (i) If $x(1/y+1/z)=\text{constant}$, prove that

$$z^2\left(\frac{\partial y}{\partial x}\right)_z=y^2\left(\frac{\partial z}{\partial x}\right)_y,$$

the suffix indicating the quantity which is kept constant.

(ii) If $V=(Ar+B/r)f(\theta)$, where A and B are constants, satisfies the equation

$$r^2\frac{\partial^2 V}{\partial r^2}+r\frac{\partial V}{\partial r}+\frac{\partial^2 V}{\partial \theta^2}=0,$$

find the form of the function $f(\theta)$. [L.U. Anc.]

5. (i) If $z=f(x, y)+g(u)$, where $u=xy$, and f and g are arbitrary functions, show that $w\equiv x\dfrac{\partial z}{\partial x}-y\dfrac{\partial z}{\partial y}$ is independent of the choice of g. Find w when $f(x, y)=xye^{x-y}$.

(ii) Show that at the point (a, a) on the curve $x^3+y^3-axy=a^3$, the value of $\dfrac{d^2y}{dx^2}=-7/a$. [L.U.]

6. If $\phi=f(\rho)$, where f is an arbitrary function and $\rho=(x^2+y^2)^{n/2}$, prove that

$$\text{(i)} \qquad x\frac{\partial\phi}{\partial x}+y\frac{\partial\phi}{\partial y}=n\rho f'(\rho);$$

$$\text{(ii)} \quad (x^2+y^2)\left(\frac{\partial^2\phi}{\partial x^2}+\frac{\partial^2\phi}{\partial y^2}\right)=n^2\rho\frac{d}{d\rho}\left(\rho\frac{d\phi}{d\rho}\right). \qquad \text{[L.U.]}$$

7. If $z=(x+y)\phi(y/x)$, where ϕ is an arbitrary function, prove that

$$x\frac{\partial z}{\partial x}+y\frac{\partial z}{\partial y}=z$$

and that

$$x^2\frac{\partial^2 z}{\partial x^2}+2xy\frac{\partial^2 z}{\partial x\,\partial y}+y^2\frac{\partial^2 z}{\partial y^2}=0. \qquad \text{[L.U.]}$$

8. If $V=f(x^2+y^2)$, where f is any function, show that

$$y\frac{\partial V}{\partial x}-x\frac{\partial V}{\partial y}=0$$

and $$y^2\frac{\partial^2 V}{\partial x^2}-2xy\frac{\partial^2 V}{\partial x\,\partial y}+x^2\frac{\partial^2 V}{\partial y^2}=x\frac{\partial V}{\partial x}+y\frac{\partial V}{\partial y}. \qquad \text{[L.U.]}$$

9. If $u=(x^2-y^2)f(t)$, where $t=xy$, prove that

$$\frac{\partial^2 u}{\partial x\,\partial y}=(x^2-y^2)\{tf''(t)+3f'(t)\}.$$

Find $f(t)$ if $$\frac{\partial^2 u}{\partial x\,\partial y}=0. \qquad \text{[L.U.]}$$

10. If $U=f(x^2+y^2+z^2)$, prove that

$$\frac{\partial^2 U}{\partial x^2}+\frac{\partial^2 U}{\partial y^2}+\frac{\partial^2 U}{\partial z^2}=4(x^2+y^2+z^2)f''(x^2+y^2+z^2)+6f'(x^2+y^2+z^2). \qquad \text{[L.U.]}$$

11. If $u=\frac{1}{r}f(ct-r)+\frac{1}{r}F(ct+r)$, where f and F denote arbitrary functions, show that

$$\frac{\partial^2 u}{\partial t^2}=c^2\left(\frac{\partial^2 u}{\partial r^2}+\frac{2}{r}\frac{\partial u}{\partial r}\right).$$

If u is also of the form $\{(\cos r)/r\}\phi(t)$, find the form of $\phi(t)$. [L.U.]

12. (i) If $z=f(x+y)g(x-y)$, where f and g are arbitrary functions, prove that

$$z\frac{\partial^2 z}{\partial x^2}-z\frac{\partial^2 z}{\partial y^2}=\left(\frac{\partial z}{\partial x}\right)^2-\left(\frac{\partial z}{\partial y}\right)^2.$$

(ii) If the variables x, y, z are connected by the equations

$$f(x, y, z)=0, \quad x^2+y^2+z^2=\text{constant},$$

prove that $$dy/dx=-(zf_x-xf_z)/(zf_y-yf_z). \qquad \text{[L.U.]}$$

13. Prove that the partial differential equation

$$x^2\frac{\partial^2 z}{\partial x^2}+2xy\frac{\partial^2 z}{\partial x\,\partial y}+y^2\frac{\partial^2 z}{\partial y^2}+x\frac{\partial z}{\partial x}+y\frac{\partial z}{\partial y}=z$$

is satisfied by $z=x\phi(y/x)+y^{-1}\psi(y/x)$, where ϕ and ψ denote arbitrary functions. [L.U.]

14. If $u=x^nF(x/y)$, where F denotes an arbitrary function, show that

$$x\frac{\partial u}{\partial x}+y\frac{\partial u}{\partial y}=nu,$$

and hence that

$$x^2\frac{\partial^2 u}{\partial x^2}+2xy\frac{\partial^2 u}{\partial x\,\partial y}+y^2\frac{\partial^2 u}{\partial y^2}=n(n-1)u.$$ [L.U.]

15. (i) If $r=\sqrt{(x^2+y^2)}$ and $u=f(r)$, show that

$$\frac{\partial^2 u}{\partial x^2}+\frac{\partial^2 u}{\partial y^2}=f''(r)+(1/r)f'(r)$$

and find u in terms of r if $\dfrac{\partial^2 u}{\partial x^2}+\dfrac{\partial^2 u}{\partial y^2}=0.$

(ii) If $x=r\cos\theta,\ y=r\sin\theta,\ u=r^\alpha\cos\theta,$ and $\dfrac{\partial^2 u}{\partial x^2}+\dfrac{\partial^2 u}{\partial y^2}=0,$ find the possible values of the constant α. [L.U.]

16. (i) If z is a function of x and y and if $x=e^{\theta+i\phi},\ y=e^{\theta-i\phi}$, prove that

$$4xy\frac{\partial^2 z}{\partial x\,\partial y}=\frac{\partial^2 z}{\partial \theta^2}+\frac{\partial^2 z}{\partial \phi^2}.$$

(ii) If u, v are functions of X and Y, and X, Y are functions of x and y, prove that

$$\frac{\partial u}{\partial x}\frac{\partial v}{\partial y}-\frac{\partial u}{\partial y}\frac{\partial v}{\partial x}=\left(\frac{\partial u}{\partial X}\frac{\partial v}{\partial Y}-\frac{\partial u}{\partial Y}\frac{\partial v}{\partial X}\right)\left(\frac{\partial X}{\partial x}\frac{\partial Y}{\partial y}-\frac{\partial X}{\partial y}\frac{\partial Y}{\partial x}\right).$$

[Leeds.]

17. If $z=f(x, y)$ and $u=x+y,\ v=y(x+y)$, prove that

$$u\frac{\partial z}{\partial u}=\left(u+\frac{v}{u}\right)\frac{\partial z}{\partial x}-\frac{v}{u}\frac{\partial z}{\partial y},$$

$$u\frac{\partial z}{\partial v}=\frac{\partial z}{\partial y}-\frac{\partial z}{\partial x}.$$ [Leeds.]

18. If $z(x, y)=f(u, v)$ where $u=x\phi(y/x),\ v=y\phi(x/y)$, prove that

(i) $$x\frac{\partial z}{\partial x}+y\frac{\partial z}{\partial y}=u\frac{\partial f}{\partial u}+v\frac{\partial f}{\partial v},$$

(ii) $$x^2\frac{\partial^2 z}{\partial x^2}+2xy\frac{\partial^2 z}{\partial x\,\partial y}+y^2\frac{\partial^2 z}{\partial y^2}=u^2\frac{\partial^2 f}{\partial u^2}+2uv\frac{\partial^2 f}{\partial u\,\partial v}+v^2\frac{\partial^2 f}{\partial v^2}.$$

[Durham.]

19. If $x=f(u, v)$ and $y=g(u, v)$, express $\dfrac{\partial u}{\partial x}$ in terms of the partial derivatives of f and g.

If $f=u^2+v^2$ and $g=u^3+v^3$, prove that

$$\frac{u_x+v_x}{u_y+v_y}=\frac{u_x}{u_y}+\frac{v_x}{v_y},$$

where $$u_x=\frac{\partial u}{\partial x}, \text{ etc.}$$ [Leeds.]

20. If $u(x, y)=\log r$ and $v(x, y)=\Theta$, where r and Θ are polar coordinates and $r\neq 0$, show that

$$\frac{\partial u}{\partial x}=\frac{\partial v}{\partial y} \quad\text{and}\quad \frac{\partial u}{\partial y}=-\frac{\partial v}{\partial x}.$$

Deduce that both $\Delta u=0$ and $\Delta v=0$, where Δ is the operator

$$\frac{\partial^2}{\partial x^2}+\frac{\partial^2}{\partial y^2}.$$

$\left[\text{You may assume that } \dfrac{\partial^2 u}{\partial x\,\partial y}=\dfrac{\partial^2 u}{\partial y\,\partial x} \text{ and, similarly, for } v.\right]$ [Durham.]

1. The pairs of variables x, y and u, v are connected by the relations

$$x=\frac{au+bv}{u^2+v^2}, \quad y=\frac{bu-av}{u^2+v^2}.$$

Prove that

$$v\frac{\partial x}{\partial u}-u\frac{\partial x}{\partial v}=-y,$$

and, by expressing u, v in terms of x, y, or otherwise, show that

$$\frac{\partial^2 u}{\partial x^2}+\frac{\partial^2 u}{\partial y^2}=0.$$ [Durham.]

22. (i) If $z=x\log(x^2+y^2)-2y\tan^{-1}(y/x)$, show that

$$\frac{\partial^2 z}{\partial x^2}+\frac{\partial^2 z}{\partial y^2}=0.$$

(ii) If x, y, u, v are variables connected by the equations

$x^2=au^{\frac{1}{2}}+bv^{\frac{1}{2}}$, $y^2=au^{\frac{1}{2}}-bv^{\frac{1}{2}}$, where a, b are constants, show that

$$\left(\frac{\partial u}{\partial x}\right)_y\left(\frac{\partial x}{\partial u}\right)_v=\frac{1}{2}=\left(\frac{\partial v}{\partial y}\right)_x\left(\frac{\partial y}{\partial v}\right)_u,$$

where the suffix indicates the variable which remains constant in each partial differentiation. [L.U. Anc.]

23. (i) If $f(x, y) \equiv F(u, v)$, where $u=x^2-y^2$ and $v=2xy$, show that

$$\left(\frac{\partial f}{\partial x}\right)^2+\left(\frac{\partial f}{\partial y}\right)^2=4(u^2+v^2)^{\frac{1}{2}}\left[\left(\frac{\partial F}{\partial u}\right)^2+\left(\frac{\partial F}{\partial v}\right)^2\right],$$

and
$$\frac{\partial^2 f}{\partial x^2}+\frac{\partial^2 f}{\partial y^2}=4(u^2+v^2)^{\frac{1}{2}}\left[\frac{\partial^2 F}{\partial u^2}+\frac{\partial^2 F}{\partial v^2}\right].$$

(ii) If x, y and z satisfy the relations $f(x, y, z)=\text{constant}$, and $xyz=\text{constant}$, show that

$$\frac{dy}{dx}=-\frac{y}{x}\left(x\frac{\partial f}{\partial x}-z\frac{\partial f}{\partial z}\right)\bigg/\left(y\frac{\partial f}{\partial y}-z\frac{\partial f}{\partial z}\right).$$ [L.U.]

24. (i) If $\xi=x+y$, $\eta=\sqrt{(xy)}$, and z is a function of x and y, show that

$$x\frac{\partial z}{\partial x}+y\frac{\partial z}{\partial y}=\xi\frac{\partial z}{\partial \xi}+\eta\frac{\partial z}{\partial \eta}.$$

(ii) If $z=f(x+y)+g(xy)$, prove that

$$(y-x)\left[x\frac{\partial^2 z}{\partial x^2}-(x+y)\frac{\partial^2 z}{\partial x\,\partial y}+y\frac{\partial^2 z}{\partial y^2}\right]=(x+y)\left(\frac{\partial z}{\partial y}-\frac{\partial z}{\partial x}\right).$$ [L.U.]

25. If (r, θ) are the polar coordinates of a point in a plane, show that $V=r\cos\theta$ and $V=(\cos\theta)/r$ both satisfy the equation

$$\frac{\partial^2 V}{\partial x^2}+\frac{\partial^2 V}{\partial y^2}=0.$$

If $V=V_0=br\cos\theta+c\,(\cos\theta)/r$ when $r\geqslant a$, and $V=V_1=dr\cos\theta$ when $r\leqslant a$, where a and b are known constants, find the values of c and d, being given that when $r=a$

(i) $V_0=V_1$; (ii) $4\dfrac{\partial V_0}{\partial r}=3\dfrac{\partial V_1}{\partial r}$ for all values of θ. [L.U.]

26. If $V=e^{(r-x)/l}$, where $r=x^2+y^2$ and l is constant, prove

(i)
$$\left(\frac{\partial V}{\partial x}\right)^2+\left(\frac{\partial V}{\partial y}\right)^2+\frac{2V}{l}\frac{\partial V}{\partial x}=0,$$

(ii)
$$\frac{\partial^2 V}{\partial x^2}+\frac{\partial^2 V}{\partial y^2}+\frac{2}{l}\frac{\partial V}{\partial x}=\frac{V}{lr}.$$ [L.U.]

27. Express $xy/(x^2+y^2)^2$ in polar coordinates, and show that in the cartesian form the function satisfies the equation

$$\frac{\partial^2 V}{\partial x^2}+\frac{\partial^2 V}{\partial y^2}=0$$

and in the polar form it satisfies the equation

$$\frac{\partial^2 V}{\partial r^2}+\frac{1}{r}\frac{\partial V}{\partial r}+\frac{1}{r^2}\frac{\partial^2 V}{\partial \theta^2}=0.$$ [L.U.]

28. Transform the partial differential equation $\frac{\partial^2 V}{\partial x^2}+\frac{\partial^2 V}{\partial y^2}=0$ to the form

$$\frac{\partial^2 V}{\partial r^2}+\frac{1}{r}\frac{\partial V}{\partial r}+\frac{1}{r^2}\frac{\partial^2 V}{\partial \theta^2}=0,$$

where $x=r\cos\theta$, $y=r\sin\theta$.

(i) If V is a function of r only, find the most general form of V satisfying the equation.

(ii) If $V=r^n f(\theta)$, where n is a constant, find the most general form of V satisfying the equation. [L.U.]

29. If $V=x^n f(y/x)$, where f is any function, prove that

$$x\frac{\partial V}{\partial x}+y\frac{\partial V}{\partial y}=nV.$$

If $\frac{\partial^2 V}{\partial x\,\partial y}=0$, and u denotes y/x, show that $(n-1)\frac{df}{du}=u\frac{d^2 f}{du^2}$, and hence verify that V is of the form ax^n+by^n, where a and b are constants. [L.U.]

30. If $u=f(y/x)+2xy$, where f denotes an arbitrary function, prove that

$$x^2\frac{\partial^2 u}{\partial x^2}+2xy\frac{\partial^2 u}{\partial x\,\partial y}+y^2\frac{\partial^2 u}{\partial y^2}=4xy.$$ [L.U.]

31. If $V=x^n f(Y, Z)$, where $Y=y/x$ and $Z=z/x$, prove that

$$x\frac{\partial V}{\partial x}+y\frac{\partial V}{\partial y}+z\frac{\partial V}{\partial z}=nV.$$ [L.U.]

32. If $v=ze^{ax+by}$, where z is a homogeneous function of degree n in x and y, prove that

$$x\frac{\partial v}{\partial x}+y\frac{\partial v}{\partial y}=(ax+by+n)v.$$ [L.U.]

33. If V is a homogeneous function of x, y and z of degree n, show that

$$x\frac{\partial V}{\partial x}+y\frac{\partial V}{\partial y}+z\frac{\partial V}{\partial z}=nV.$$

If V also satisfies

$$\frac{\partial^2 V}{\partial x^2}+\frac{\partial^2 V}{\partial y^2}+\frac{\partial^2 V}{\partial z^2}=0,$$

and $r^2=x^2+y^2+z^2$, show that $\phi=r^{-(2n+1)}V$ satisfies the latter equation. [L.U. Anc.]

34. If $f(x, y)=\phi(x, y)+\psi(x, y)$, where $\phi(x, y)$ is a homogeneous polynomial in x and y of degree p and $\psi(x, y)$ is one of degree q, show that

$$\phi(x, y)=\frac{1}{p(p-q)}\left\{x^2\frac{\partial^2 f}{\partial x^2}+2xy\frac{\partial^2 f}{\partial x\,\partial y}+y^2\frac{\partial^2 f}{\partial y^2}-(q-1)\left(x\frac{\partial f}{\partial x}+y\frac{\partial f}{\partial y}\right)\right\}.$$

[L.U.]

35. If V is a homogeneous polynomial of degree n in the variables x and y, prove that $x\dfrac{\partial V}{\partial x}+y\dfrac{\partial V}{\partial y}=nV$.

If $r^2=x^2+y^2$, prove that

$$\frac{\partial^2}{\partial x^2}\{r^m V\}=r^m\frac{\partial^2 V}{\partial x^2}+2mr^{m-2}x\frac{\partial V}{\partial x}+mr^{m-2}V+m(m-2)r^{m-4}x^2V$$

and deduce that

$$\frac{\partial^2}{\partial x^2}\left(\frac{V}{r^{2n}}\right)+\frac{\partial^2}{\partial y^2}\left(\frac{V}{r^{2n}}\right)=0 \text{ if } \frac{\partial^2 V}{\partial x^2}+\frac{\partial^2 V}{\partial y^2}=0.$$ [L.U.]

36. (i) If $w=f(y-z,\ z-x,\ x-y)$, prove that

$$\frac{\partial w}{\partial x}+\frac{\partial w}{\partial y}+\frac{\partial w}{\partial z}=0.$$

(ii) If V is a function of x and y, and if x and y are functions of u and v such that

$$\frac{\partial x}{\partial u}=\frac{\partial y}{\partial v},\quad \frac{\partial x}{\partial v}=-\frac{\partial y}{\partial u},$$

prove that

$$\frac{\partial^2 V}{\partial u^2}+\frac{\partial^2 V}{\partial v^2}=\left\{\left(\frac{\partial x}{\partial u}\right)^2+\left(\frac{\partial x}{\partial v}\right)^2\right\}\left\{\frac{\partial^2 V}{\partial x^2}+\frac{\partial^2 V}{\partial y^2}\right\}.$$ [L.U.]

37. (i) If $z^2=xy\,F(x^2-y^2)$, prove that

$$2xy\left(x\frac{\partial z}{\partial y}+y\frac{\partial z}{\partial x}\right)=z(x^2+y^2).$$

(ii) If $V=f(xz,\ y/z)$, prove that

$$z\frac{\partial V}{\partial z}=x\frac{\partial V}{\partial x}-y\frac{\partial V}{\partial y}.$$ [L.U.]

38. Find the envelopes of the following for different values of θ:

(i) the straight lines $y=\theta^2x-2\theta^3$,

(ii) the parabolas $y^2=\theta(x-\theta)$,

(iii) the circles $(x-\theta)^2+y^2=4\theta$,

(iv) the conics $x^2\cos\theta+y^2\sin\theta=a^2$.

19.14. The tangent to a space curve

Suppose that a curve in space is given parametrically by the equations

$$x=f(t),\ y=g(t),\ z=h(t).$$

Let $A(x, y, z)$ and $B(x+\delta x, y+\delta y, z+\delta z)$ be the points on this curve corresponding respectively to the values t and $t+\delta t$ of the parameter.

Then by (16.12) the D.R.s of AB may be taken as

$$[\delta x : \delta y : \delta z]$$

or as

$$\left[\frac{\delta x}{\delta t} : \frac{\delta y}{\delta t} : \frac{\delta z}{\delta t}\right].$$

As $\delta t \rightarrow 0$, δx, δy and δz all tend to zero, B approaches A along the curve and the chord AB approaches the tangent at A as a limiting position. Hence the D.R.s of the tangent at A are the limiting values of $\frac{\delta x}{\delta t}, \frac{\delta y}{\delta t}, \frac{\delta z}{\delta t}$ as $\delta t \rightarrow 0$, i.e.

$$\left[\frac{dx}{dt} : \frac{dy}{dt} : \frac{dz}{dt}\right]. \tag{19.11}$$

19.15. Tangent lines to a surface

If a chord AB joining two neighbouring points on a surface tends to a limiting position AT as B is made to approach A along a curve on the surface, AT is said to be a *tangent line to the surface at A*. We shall show that all tangent lines to the surface at A lie in a plane known as the *tangent plane* to the surface at A; the line drawn through A perpendicular to this tangent plane is called the *normal* to the surface at A.

19.16. The equations of the tangent plane and normal to a surface

Let the curve

$$x=f(t),\ y=g(t),\ z=h(t) \quad . \quad . \quad . \quad \text{(i)}$$

lie on the surface whose equation is

$$F(x, y, z)=0 \quad . \quad . \quad . \quad . \quad \text{(ii)}$$

Then (ii) is satisfied for all values of t by equations (i); also $\frac{dF}{dt}=0$ so that, by (19.7),

$$\frac{\partial F}{\partial x}\frac{dx}{dt}+\frac{\partial F}{\partial y}\frac{dy}{dt}+\frac{\partial F}{\partial z}\frac{dz}{dt}=0 \quad . \quad . \quad . \quad \text{(iii)}$$

Now by (19.11) $\left[\frac{dx}{dt} : \frac{dy}{dt} : \frac{dz}{dt}\right]$ are the D.R.s of the tangent line to curve (i) at $A(x, y, z)$; hence by virtue of (16.17) page 352, we conclude from (iii) that the line whose D.R.s are

$$\left[\frac{\partial F}{\partial x} : \frac{\partial F}{\partial y} : \frac{\partial F}{\partial z}\right] \quad . \quad . \quad . \quad . \quad \text{(iv)}$$

is perpendicular to this tangent line at A and, since its D.R.s are independent of f, g, h, to all tangent lines to surface (ii) at A.

The line which passes through A and has D.R.s given by (iv) is defined to be the *normal* at A. Since all the tangent lines to the surface at A are perpendicular to the normal at A they lie in a plane which is the *tangent plane* at A.

The equations of the normal at P (α, β, γ) to the surface $F(x, y, z)=0$ are

$$\frac{x-\alpha}{\dfrac{\partial F}{\partial \alpha}}=\frac{y-\beta}{\dfrac{\partial F}{\partial \beta}}=\frac{z-\gamma}{\dfrac{\partial F}{\partial \gamma}} \tag{19.12}$$

and the equation of the tangent plane at P is

$$(x-\alpha)\frac{\partial F}{\partial \alpha}+(y-\beta)\frac{\partial F}{\partial \beta}+(z-\gamma)\frac{\partial F}{\partial \gamma}=0 \tag{19.13}$$

where $\dfrac{\partial F}{\partial \alpha}, \dfrac{\partial F}{\partial \beta}, \dfrac{\partial F}{\partial \gamma}$ denote respectively the values of $\dfrac{\partial F}{\partial x}, \dfrac{\partial F}{\partial y}, \dfrac{\partial F}{\partial z}$ at P (α, β, γ).

Example 28

Find the equation of the tangent plane at the point (α, β, γ) *on the surface*

$$xyz=a(x^2-y^2).$$

Show that all planes drawn through either of the lines

$$y+x=0,\ z=0;\quad y-x=0,\ z=0,$$

touch the surface at some point on these lines. [L.U.]

If $\qquad F(x, y, z)=xyz-a(x^2-y^2)$,

$$F_x=yz-2ax,\ F_y=xz+2ay,\ F_z=xy,$$

and so, by (19.13), the equation of the tangent plane at $P(\alpha, \beta, \gamma)$ on the surface

$$xyz=a(x^2-y^2) \quad . \quad . \quad . \quad . \quad \text{(i)}$$

is $\qquad (x-\alpha)(\beta\gamma-2a\alpha)+(y-\beta)(\alpha\gamma+2a\beta)+(z-\gamma)(\alpha\beta)=0,$

i.e. $\qquad (\beta\gamma-2a\alpha)x+(\alpha\gamma+2a\beta)y+\alpha\beta z=\alpha\beta\gamma \quad . \quad . \quad . \quad \text{(ii)}$

since $\alpha\beta\gamma=a(\alpha^2-\beta^2)$ by (i).

The line

$$x+y=0,\ z=0 \quad . \quad . \quad . \quad . \quad \text{(iii)}$$

lies on the given surface; the coordinates of any point Q on this line may be taken as $(\lambda, -\lambda, 0)$ and by (ii) the equation of the tangent plane to (i) at Q is

$$x+y+\lambda z/2a=0 \quad . \quad . \quad . \quad . \quad \text{(iv)}$$

But by (16.25) page 358, this equation represents, as λ varies, all planes which pass through the line (iii); and for any particular value of λ it represents a plane which touches the given surface at $(\lambda, -\lambda, 0)$, a point on line (iii).

Similarly, all planes drawn through the line $y-x=0$, $z=0$ touch the surface at some point on the line.

Example 29

Find the equations of the tangent plane and normal to the surface

$$1/x+1/y+1/z=1/a$$

at the point whose coordinates are (x_1, y_1, z_1).

If P is a point on the line of intersection of this surface and the plane $z=a$, *show that the locus of the point of intersection of the normal at P and the plane* $z=0$ *is* $(x+y)(x-y)^2+8a^3=0$, $z=0$. [L.U.]

By (19.13) the equation of the tangent plane at $(x_1, y_1\ z_1)$ on the surface

$$\frac{1}{x}+\frac{1}{y}+\frac{1}{z}=\frac{1}{a} \qquad \text{(i)}$$

is

$$\frac{1}{x_1^2}(x-x_1)+\frac{1}{y_1^2}(y-y_1)+\frac{1}{z_1^2}(z-z_1)=0,$$

i.e.

$$\frac{x}{x_1^2}+\frac{y}{y_1^2}+\frac{z}{z_1^2}=\frac{1}{a} \qquad \text{(ii)}$$

The equations of the normal at the same point are, by (19.12),

$$x_1^2(x-x_1)=y_1^2(y-y_1)=z_1^2(z-z_1) \qquad \text{(iii)}$$

The plane $z=a$ meets (i) in the straight line $x+y=0$, $z=a$ any point P on which may be taken as $(\lambda, -\lambda, a)$. By (iii) the equations of the normal at P are

$$\lambda^2(x-\lambda)=\lambda^2(y+\lambda)=a^2(z-a).$$

This normal meets the plane $z=0$ at a point Q whose coordinates are

$$x=\lambda-a^3/\lambda^2,\ y=-\lambda-a^3/\lambda^2,\ z=0.$$

The locus of Q is found by eliminating λ between these equations, and since $x+y=-2a^3/\lambda^2$ and $x-y=2\lambda$ the locus is

$$(x+y)(x-y)^2+8a^3=0,\ z=0.$$

Example 30

Prove that the point $P(\theta^3+3\lambda\theta^2,\ \theta^2+2\lambda\theta,\ \theta+\lambda)$ *lies on the surface* $(yz-x)^2=4(y-z^2)(xz-y^2)$, *and show that the equation of the tangent plane at P is*

$$x-3y\theta+3z\theta^2-\theta^3=0.$$

Deduce that, if θ *remains constant but* λ *varies, the locus of P is a line at every point of which the surface has the same tangent plane.* [L.U.]

The coordinates of P identically satisfy the equation

$$(yz-x)^2-4(y-z^2)(xz-y^2)=0 \qquad \text{(i)}$$

and so P lies on the surface (i).

We may use (i) to find the equation of the tangent plane at P.

Alternatively, denote surface (i) by $f(x, y, z)=0$. Then since x, y, and z are functions of λ and θ given by the equations $x=\theta^3+3\lambda\theta^2$, $y=\theta^2+2\lambda\theta$, $z=\theta+\lambda$, then

$$\frac{\partial f}{\partial x}\frac{\partial x}{\partial \lambda}+\frac{\partial f}{\partial y}\frac{\partial y}{\partial \lambda}+\frac{\partial f}{\partial z}\frac{\partial z}{\partial \lambda}=0$$

i.e.

$$3\theta^2\frac{\partial f}{\partial x}+2\theta\frac{\partial f}{\partial y}+\frac{\partial f}{\partial z}=0 \quad . \quad . \quad . \quad \text{(ii)}$$

Also

$$\frac{\partial f}{\partial x}\frac{\partial x}{\partial \theta}+\frac{\partial f}{\partial y}\frac{\partial y}{\partial \theta}+\frac{\partial f}{\partial z}\frac{\partial z}{\partial \theta}=0$$

$$\therefore\ 3\theta(\theta+2\lambda)\frac{\partial f}{\partial x}+2(\theta+\lambda)\frac{\partial f}{\partial y}+\frac{\partial f}{\partial z}=0 \quad . \quad . \quad . \quad \text{(iii)}$$

Solving (ii) and (iii) by determinants, we have

$$\frac{\partial f}{\partial x}:\frac{\partial f}{\partial y}:\frac{\partial f}{\partial z}=1:-3\theta:3\theta^2$$

and the equation of the tangent plane to (i) at P is, by (19.13),

$$\{x-(\theta^3+3\lambda\theta^2)\}-3\theta\{y-(\theta^2+2\lambda\theta)\}+3\theta^2\{z-(\theta+\lambda)\}=0,$$

$$x-3\theta y+3\theta^2 z-\theta^3=0 \quad . \quad . \quad \text{(iv)}$$

If θ remains constant, $\theta=k$, say, while λ varies, all three coordinates of P are linear functions of λ given by

$$x=k^2(k+3\lambda), \quad y=k(k+2\lambda), \quad z=k+\lambda$$

and so as λ varies P moves on the straight line

$$\frac{x-k^3}{3k^2}=\frac{y-k^2}{2k}=\frac{z-k}{1}=\lambda\,. \quad . \quad . \quad . \quad \text{(v)}$$

But (iv) is independent of λ and so the tangent plane to surface (i) at every point of line (v) is the plane

$$x-3ky+3k^2z=k^3.$$

Exercises 19 (*d*)

1. Show that

   (*a*) any tangent plane to the surface $xyz=a^3$ and the coordinate planes bound a tetrahedron of constant volume;

   (*b*) if the normal at any point P of the ellipsoid

   $$x^2/a^2+y^2/b^2+z^2/c^2=1$$

   meets the coordinate planes in G_1, G_2, G_3, then the ratios $PG_1:PG_2:PG_3$ are constant. [L.U. Anc.]

2. Find the equation of the tangent plane at any point of the surface $x^4+y^4+z^4=3a^4$.

   Show that the points on this surface at which the normals pass through O lie on the sphere $x^2+y^2+z^2=3a^2$. [L.U.]

3. Find the equations of the tangent plane and the normal at the point P (α, β, γ) on the surface $xyz = a^3$.

Show that if the tangent plane at P cuts the coordinate axes at A, B, C, then P is at the intersection of the medians of triangle ABC.

Show also that if the normal at P cuts the coordinate planes $x=0$, $y=0$, $z=0$ at L, M, N respectively, $MN : NL : LM = \beta^2-\gamma^2 : \gamma^2-\alpha^2 : \alpha^2-\beta^2$. [L.U.]

4. Find the equation of the tangent plane at any poin (x_1, y_1, z_1) on the surface $x^3+y^3+z^3=3a^3$.

P is the point (a, a, a). Show that the normal at P passes through the origin, and that P is the centroid of the triangle whose vertices are the intersections of the tangent plane at P with the coordinate axes. [L.U.]

5. Find the equation of the tangent plane and the equations of the normal at the point (x_1, y_1, z_1) on the surface $xy+yz+zx=0$.

If the normal at any point P on the surface meets the surface again at Q and the normal at Q meets the surface again at R, prove that RP passes through the origin O and that $OR=9OP$. [L.U.]

6. Find the equations of the tangent plane and normal at the point (x_1, y_1, z_1) of the surface $z(x^2-y^2)=2kxy$ $(k>0)$.

Show that the normal at the point P $(2k\sqrt{3}, 0, 0)$ makes an angle of 30° with the z-axis and that it meets the surface again in points Q, R such that $PQ=PR=8k$. [L.U.]

7. Find the equations of the tangent plane and normal to the surface $5z^2+4x^2y-6xz^2=3$ at the point P $(1, 1, 1)$.

If the normal at the point P meets the surface again at A and B, and if C is the mid-point of AB, show that the length of PC is $\frac{1}{4}(31\sqrt{6})$. [L.U.]

8. Obtain the equations of the tangent plane and of the normal to the surface $z^2-x^2y+zy^2=1$ at the point (α, β, γ).

Show that the normal at the point $A(1, -1, 0)$ meets the surface again at two points distant $\sqrt{6}$ and $(6\sqrt{6})/5$ from A. [L.U.]

9. Find the equation of the tangent plane at the point

$$\{-2a/(1+\lambda),\ -2a/(1-\lambda),\ a/2\}$$

of the surface whose equation is $1/x+1/y+1/z=1/a$.

Show that the intersection of the tangent plane with the plane $z=0$ touches the hyperbola given by $xy=a(x+y)$, $z=0$. [L.U.]

10. Show that the equation of the tangent plane at (α, β, γ) on the surface $xy=cz$ is $\beta x+\alpha y-cz=c\gamma$, and find the equations of the normal at this point.

Show that this tangent plane touches the sphere of radius r with centre at the origin if $r^2(\alpha^2+\beta^2+c^2)=c^2\gamma^2$, and that the normal at (α, β, γ) to the given surface is a tangent line to the sphere of radius R with centre at the origin if

$$R^2(\alpha^2+\beta^2+c^2)=(\alpha^2+\beta^2)(\alpha^2+\beta^2+\gamma^2+c^2).$$ [L.U.]

11. Find the equations of the two tangent planes to the surface

$$x^2+y^3+z^4=108$$

which pass through the z-axis, and determine the point of contact.

Find the y and z coordinates of all points P on the surface, not in any coordinate plane, such that the normal to the surface at P passes through the origin. [L.U.]

12. Find the equation of the tangent plane at any point (α, β, γ) on the surface $x^2y-a^2z=0$, and prove that the tangent planes at all points on the surface which lie in the plane $x=\alpha$ intersect the plane $x=0$ in parallel lines. [L.U.]

13. Show that for the surface $xyz=a^3$, one and only one, real normal can be drawn parallel to the line $8x=y=z$; find its equations and the points where it meets the surface again. [L.U.]

14. Prove that all points on the line $x=y$, $z=0$ lie on the surface

$$(y-1)(z-c)^2-(x-1)(z+c)^2=0,$$

and that the normals to the surface at these points lie on the surface

$$(x-y)(x+y-2)-cz=0.$$ [L.U.]

15. Find the equation of the tangent plane to the surface

$$z^2(x^2+z^2-1)=y^2(x^2+z^2) \text{ at } (a, b, c).$$

If P is any point common to the surface $x^2+z^2=1$ and the plane $y=0$, show that P also lies on the surface $z^2(x^2+z^2-1)=y^2(x^2+z^2)$ and that the surfaces have a common tangent plane at P. [L.U.]

16. Show that the tangent planes to the surface $xy+yz+zx=r^2$ at the points on the intersection of the surface with the plane $x+y+z=r\sqrt{3}$ are also tangent planes to the sphere $x^2+y^2+z^2=r^2$. [L.U.]

17. Write down the equations of the tangent plane and normal at any point $P(x_1, y_1, z_1)$ of the surface $z(x^2-y^2)-axy=0$.

Prove that

(i) If the tangent plane at P meets the z-axis in Q, then PQ is perpendicular to the z-axis.

(ii) The normals at points on the x-axis lie on the surface

$$xy+az=0.$$ [L.U.]

18. Show that the tangent plane at the point P $(\sin\theta, \cos\theta, c)$ to the surface $z^2(x^2+y^2)=c^2$ meets the plane $z=0$ in a line which touches the circle $x^2+y^2=4$, $z=0$.

Show that one of the points where the normal at P meets the surface again lies on the sphere $c^2(x^2+y^2+z^2)=1+c^6$. [L.U.]

19. Show that the normal at the point $P(a, 2c, a)$ to the surface $y^2z=4c^2x$ lies in the surface $z^2-x^2=4c(y-2c)$.

If O is the origin, show that the acute angle θ between this normal and OP satisfies the inequality $\frac{1}{3}\pi \leqslant \theta \leqslant \frac{1}{2}\pi$.

The tangent plane at P to the first surface meets the axes of co-ordinates A, B, C and r is the radius of the sphere $OABC$. Prove that $OP^2=r^2+3c^2$. [L.U.]

20. Write down the equations of the tangent plane and of the normal at the point (x_1, y_1, z_1) to the surface $x^2+2yz=2$.

Find the equations of the tangent planes to this surface which are parallel to the plane $4x+y-7z=0$.

Find also the coordinates of the point in which the normal at $(2, 1, -1)$ meets the surface again. [L.U.]

21. Obtain the equations of the tangent plane and normal at each of the points $(-a, -a, a)$ and $(-a, a, -a)$ on the surface $xyz+x^2(y+z)=a^3$.

Prove that the normals intersect and find the equation of the plane in which they both lie. Show that the tangent planes intersect in the line $x+3a=2y=2z$ [L.U.]

CHAPTER 20

APPLICATIONS OF INTEGRATION —CARTESIAN COORDINATES

20.1. The definite integral as the limit of a sum

Fig. 67 shows part of the graph of the function $y=\phi(x)$ which, for simplicity, we shall assume to be not only continuous, but also positive and steadily increasing in the interval $a \leqslant x \leqslant b$.

We investigate the area $AUVB$ enclosed between the curve $y=\phi(x)$, the x-axis, and the ordinates $x=a$, $x=b$ by dividing it into n strips

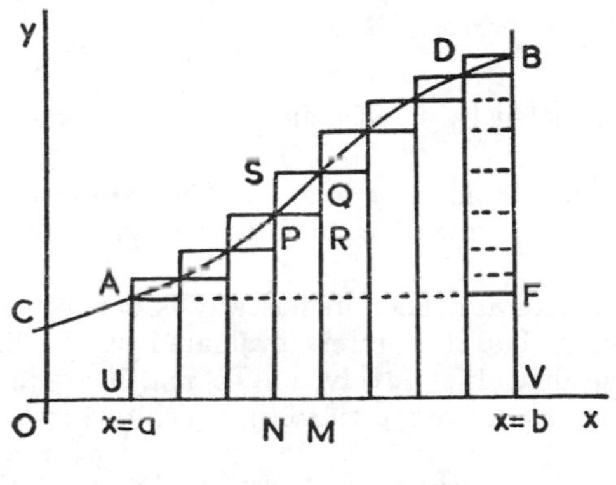

Fig. 67

by means of equally spaced ordinates δx apart and constructing a set of inscribed and circumscribed rectangles as shown in fig. 67.

We consider a typical strip $PNMQ$, P and Q being the points (x, y) and $(x+\delta x, y+\delta y)$ respectively. We assume that the area $CONP$, bounded by the axes, the ordinate PN, and the curve $y=\phi(x)$, is some function $A(x)$ of x, and that, when x increases to $x+\delta x$, the area $CONP$ increases to area $COMQ$ which we take to be $A(x)+\delta A(x)$ where $\delta A(x)=$ area $PNMQ$. Then from the figure we see that

$$\text{rect. } PM < \delta A(x) < \text{rect. } QN$$

i.e.

$$y\,\delta x < \text{area } PNMQ < (y+\delta y)\,\delta x \qquad . \quad . \qquad \text{(i)}$$

Summing over the n strips, we have

$$\sum_{x=a}^{x=b} y\,\delta x < \text{area } AUVB < \sum_{x=a}^{x=b} (y+\delta y)\,\delta x \quad . \quad . \quad \text{(ii)}$$

Now $\sum_{x=a}^{x=b} y\,\delta x$ is the sum of the inner rectangles, $\sum_{x=a}^{x=b} (y+\delta y)\,\delta x$ is the sum of the outer rectangles, and the difference between these sums is the sum of the n rectangles such as SR. If we slide these n rectangles parallel to the x-axis until they lie between BV and the preceding ordinate, we see that they make up the rectangle DF whose height BF is $\phi(b)-\phi(a)$ and whose width is $\delta x=(b-a)/n$. Now rectangle DF may be made as small as we please by making n sufficiently large, that is, by making δx sufficiently small. Hence, as $\delta x \to 0$, the sums of both sets of rectangles (inner and outer) tend to the same limit; by virtue of (ii) we define the area $AUVB$ as this common limit and write

$$\text{area } AUVB = \lim_{\delta x \to 0} \sum_{x=a}^{x=b} y\,\delta x \quad . \quad . \quad . \quad \text{(iii)}$$

This limit is denoted by $\int_a^b y\,dx$, and is called the definite integral of y with respect to x from $x=a$ to $x=b$. The letters a and b show the range of values of x from UA to VB over which the summation is made.

The definite integral defined in this way is independent of the idea of differentiation, but it is rarely evaluated as the limit of a sum because of the difficulties involved. To find a practical method of evaluating the integral we return to (i), which gives

$$y\,\delta x < \delta A(x) < (y+\delta y)\,\delta x,$$

$$y < \frac{\delta A(x)}{\delta x} < y+\delta y,$$

and so

$$\frac{dA(x)}{dx} = y$$

since, by continuity of $\phi(x)$, $\delta y \to 0$ with δx.

Thus $A(x)$ is the indefinite integral of y with respect to x.

Now $$A(x) = \text{area } CONP.$$

Hence $$\text{area } COVB = A(b), \text{ area } COUA = A(a)$$

and so $$\text{area } AUVB = A(b) - A(a) \quad . \quad . \quad . \quad \text{(iv)}$$

Combining (iii) and (iv), we have

$$\text{area } AUVB = \lim_{\delta x \to 0} \sum_{x=a}^{x=b} y\,\delta x = \int_a^b y\,dx = A(b) - A(a) \qquad (20.1)$$

where $A(x)$ is the indefinite integral of y with respect to x. This definition of a definite integral is consistent with the one given in Chapter 10 and gives us the geometrical significance of the process of definite integration.

NOTE.—In subsequent sections we shall assume that if a number can be shown to lie between two sums of the form $\sum_{x=a}^{x=b} f(x)\,\delta x$ and $\sum_{x=a}^{x=b} f(x+\delta x)\,\delta x$ its value is given by $\lim_{\delta x \to 0} \sum_{x=a}^{x=b} f(x)\,\delta x$ and that this limit may be evaluated as $\int_a^b f(x)\,dx$.

If in the interval $a \leqslant x \leqslant b$, $y = \phi(x)$ is a decreasing function, the inequalities (i) and (ii) are reversed, but this does not alter the result.

If in the interval $a \leqslant x \leqslant b$, $y = \phi(x)$ is negative, $\sum_{x=a}^{x=b} y\,\delta x$ is negative, and hence $\int_a^b \phi(x)\,dx$ gives a negative result which is numerically equal to the area enclosed between the curve, the x-axis and the ordinates $x=a$ and $x=b$.

If in the interval $a < x < b$ the curve $y = \phi(x)$ crosses the x-axis, the area enclosed between the curve, the x-axis and the ordinates $x=a$ and $x=b$ lies partly above and partly below the x-axis and $\int_a^b y\,dx$ gives the *algebraic* sum of these areas. To obtain the numerical value of such an area the negative portions must be found separately and their numerical values added to the positive areas.

By a similar argument we may show that the area enclosed by a curve, the y-axis and the lines $y=c$, $y=d$ where $c<d$ is given by $\int_c^d x\,dy$.

If the equation of the curve is given parametrically by the equations $x = x(t)$, $y = y(t)$, we write $\int y\,dx = \int y(t)\frac{dx}{dt}\,dt$, and integrate with respect to t between appropriate limits. In the same way we may use $\int x\,dy$ in the form $\int x\frac{dy}{dt}\,dt$. (See also § 21.9.)

20.2. Mean value

Let $y_1, y_2, \ldots, y_n$ be the values of the function $y=\phi(x)$ corresponding to the values $x=a$, $x=a+\delta x, \ldots, x=a+(n-1)\delta x$, where $n\delta x=b-a$ as shown in fig. 68.

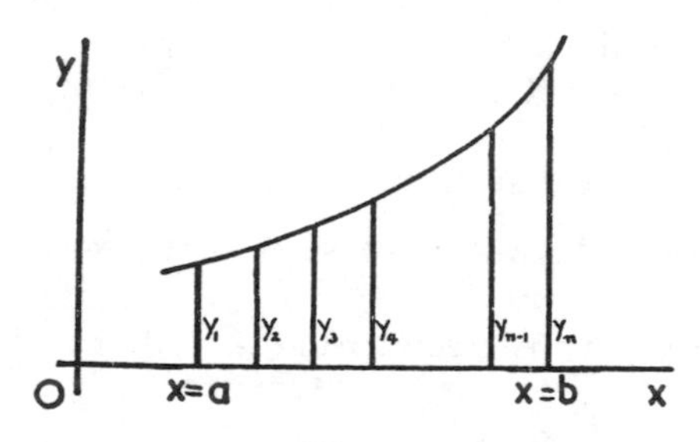

Fig. 68

Then the average (or mean) of these values of the function is given by

$$\frac{(y_1+y_2+\ldots+y_n)}{n}$$

$$=\frac{y_1\delta x+y_2\delta x+\ldots y_n\delta x}{n\,\delta x}$$

$$=\frac{1}{b-a}\sum_{x=a}^{x=b} y\,\delta x, \text{ since } n\,\delta x=b-a.$$

If this expression tends to a limit as $n\to\infty$ (i.e. as $\delta x\to 0$), the limit is

$$\frac{1}{b-a}\int_a^b y\,dx \tag{20.2}$$

and it is called the *mean value* of y with respect to x in the interval from $x=a$ to $x=b$.

20.3. Volumes

By arguments similar to those used in § 20.1 it can be shown that if the area of the section of a solid by a plane perpendicular to Ox at a distance x from O is a function $S(x)$ of x, the volume V of the solid enclosed between planes perpendicular to Ox at $x=a$, $x=b$ is given by

$$V=\int_a^b S(x)\,dx. \tag{20.3}$$

When the solid is generated by the revolution about the x-axis of the part of the curve $y=\phi(x)$ which lies between the ordinates $x=a$ and

$x=b$, the cross-section at a distance x from O is a circle of radius y where $y=\phi(x)$

$$\therefore\ V=\lim_{\delta x\to 0}\sum_{x=a}^{x=b}\pi y^2\,\delta x=\int_a^b \pi y^2\,dx=\pi\int_a^b \{\phi(x)\}^2\,dx. \tag{20.4}$$

The corresponding formula for a solid generated by rotating a curve about Oy is $\pi\int x^2\,dy$ taken between appropriate limits.

If a curve is given parametrically by the equations $x=x(t)$, $y=y(t)$, we may use the above results in the forms

$$\left.\begin{aligned} &\pi\int \{y(t)\}^2\,\frac{dx}{dt}\,dt \\ \text{and}\qquad &\pi\int \{x(t)\}^2\,\frac{dy}{dt}\,dt \end{aligned}\right\} \tag{20.5}$$

taken between appropriate limits.

20.4. Curve sketching from cartesian equations

In problems which deal with areas and volumes it is often necessary to make a rough sketch of the curves involved. One variable can usually be found explicitly in terms of the other from the cartesian equation of the curve and certain features of the curve readily deduced.

I. Symmetry

A curve is symmetrical about Ox if its equation contains only even powers of y; it is symmetrical about Oy if its equation contains only even powers of x; it is symmetrical about the origin if its equation is unaltered when both x and y are changed in sign.

II. Points on the coordinate axes

The curve cuts Ox where $y=0$ and Oy where $x=0$.

III. Restrictions on the ranges of x and y

These may be revealed by expressing y in terms of x, or x in terms of y. For example, if $y^2=4x$, y is real only when $x\geqslant 0$.

If $y^2=x(x^2-1)$, y is real only when $-1\leqslant x\leqslant 0$ when $x\geqslant 1$.

If $y=2x/(x^2+1)$, $yx^2-2x+y=0$ and so $x=1\pm\sqrt{(1-y^2)}$.

Hence x is real only when $|\,y\,|\leqslant 1$.

IV. Slope at any point

The gradient of the curve at any point is given by dy/dx.

At stationary points $dy/dx=0$.

V. Form at the origin

When the equation of the curve has been rationalised and cleared of fractions, for small values of x and y the terms of higher degree are negligible compared with those of low degree. Hence if a curve passes through the origin, a first approximation to its form may be obtained by retaining only the terms of lowest degree in its equation. For example, a first approximation to the curve $y=2x/(x^2+1)$ at the origin is the line $y=2x$.

To study the form of a curve at any point we may change the origin to that point and apply the above considerations.

VI. Asymptotes

An asymptote to a curve is a straight line to which the shape of the curve approximates at a great distance from the origin. Accordingly if, as a point moves along a curve, its abscissa x approaches a value a and at the same time its ordinate y becomes either positively or negatively infinite, the vertical line $x=a$ is called an asymptote of the curve. If, as a point moves along a curve, its ordinate y approaches a value b, and at the same time its abscissa x becomes either positively or negatively infinite, the horizontal line $y=b$ is called an asymptote of the curve.

If $y=f(x)/F(x)$, where $f(x)$ and $F(x)$ are polynomials and their quotient has been reduced to its lowest terms, the vertical asymptotes of the curve are given by $F(x)=0$. The limiting values of y as $x \to \pm\infty$ depend on the degree of $f(x)$ and $F(x)$. There are three cases:

(a) If $f(x)$ is of lower degree than $F(x)$, i.e. if $f(x)/F(x)$ is a proper fraction, $y \to 0$ as $x \to \pm\infty$ and so the x-axis is an asymptote of the curve.

(b) If $f(x)$ and $F(x)$ are of the same degree, there is an symptote parallel to Ox. If we express y by division as the sum of a constant k and a proper fraction $P(x)$, then, as in (a), $P(x) \to 0$ as $x \to \pm\infty$ and so $y=k$ is the equation of the asymptote. For example if $y=x(x-2)/(x^2-1)=1-(2x-1)/(x^2-1)$, y becomes infinite as $x \to \pm 1$ and $y \to 1$ as $x \to \pm\infty$. Hence the lines $x=\pm 1$ are vertical asymptotes and $y=1$ is the horizontal asymptote.

(c) If $f(x)$ is of higher degree than $F(x)$, y becomes infinite as $x \to \pm\infty$. If we express y by division as the sum of a polynomial $\phi(x)$ and a proper fraction $Q(x)$, then, as in (a), $Q(x) \to 0$ as $x \to \pm\infty$ and $y-\phi(x) \to 0$. If $\phi(x)$ is of the first degree, $y=\phi(x)$ is the equation of the oblique asymptote of the curve. For example, if $y=(x-1)(x-3)/(x-2)=x-2-\{1/(x-2)\}$, y becomes infinite as $x \to 2$ and $y-(x-2) \to 0$ as $x \to \pm\infty$. Hence $x=2$ is a vertical asymptote and $y=x-2$ is an oblique asymptote. We note that $y<x-2$ when $x>2$ and $y>x-2$ when $x<2$. Hence the curve lies below the asymptote as $x \to +\infty$ and lies above it as $x \to -\infty$.

Example 1

Sketch the form of the curve whose equation is $8ay^2=x(2a-x)^2$, $a>0$. *Prove that the area of the loop is* $16a^2/15$. *Show that if the tangent at any point P of the curve meets the axis of y at the point T, then* $OP=2OT$, *O being the origin.* [L.U.]

For the sketch of the curve we note the following :

(1) The curve is symmetrical about Ox.

(2) It cuts Ox at $x=0$ and at $x=2a$; it cuts Oy at the origin only.

(3) Since $a>0$, x can take positive values only.

(4) Since $y=\pm\frac{1}{2}(2ax^{1/2}-x^{3/2})/\sqrt{(2a)}$, $\dfrac{dy}{dx}=\pm\frac{1}{4}(2a-3x)/\sqrt{(2ax)}$.

The curve has tangents parallel to Ox at the points $(\frac{2}{3}a, \pm\frac{2}{9}a\sqrt{3})$; when $x=2a$, $\dfrac{dy}{dx}=\pm\frac{1}{2}$.

(5) The tangent at the origin is $x=0$, and the curve approximates at the origin to the parabola $y^2=\frac{1}{2}ax$.

(6) As $x\to\infty$, $y\to\pm\infty$.

The form of the curve is shown in fig. 69.

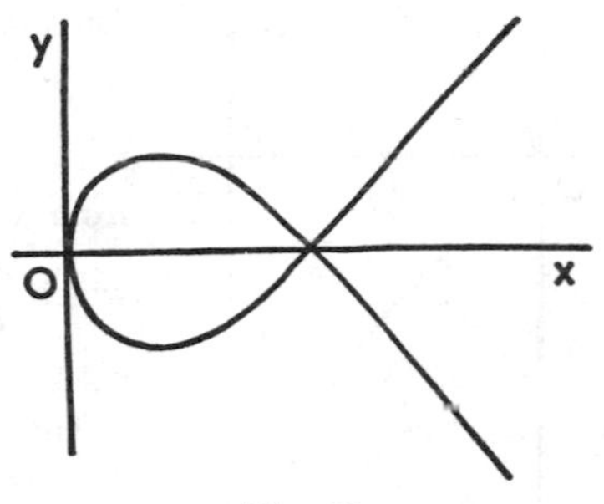

Fig. 69

The area A of the loop is twice the area enclosed by the upper part of the curve and the x-axis.

By (20.1),

$$A=\frac{1}{\sqrt{(2a)}}\int_0^{2a}(2a-x)\sqrt{x}\,dx=\frac{1}{\sqrt{(2a)}}\left[\frac{4}{3}ax^{3/2}-\frac{2}{5}x^{5/2}\right]_0^{2a}=16a^2/15.$$

The equation of the tangent at $P(x_1, y_1)$ on the curve is by (4)

$$(y-y_1)=\frac{\pm\frac{1}{4}(2a-3x_1)}{\sqrt{(2ax_1)}}(x-x_1)\,.$$

If P lies on the branch of the curve corresponding to the positive sign, i.e. if

$$y_1=\frac{\frac{1}{2}(2a-x_1)\sqrt{x_1}}{\sqrt{(2a)}} \quad . \quad . \quad . \quad . \quad \text{(i)}$$

the tangent at P meets Oy at T, where $x=0$ and

$$y=y_1-\frac{\frac{1}{4}(2a-3x_1)\sqrt{x_1}}{\sqrt{(2a)}}$$

i.e.
$$OT=\frac{\frac{1}{4}(2a+x_1)\sqrt{x_1}}{\sqrt{(2a)}} \text{ by (i).}$$

But
$$OP^2=x_1^2+y_1^2=\frac{x_1(2a+x_1)^2}{8a} \text{ by (i).}$$

Hence
$$OP=2OT.$$

By symmetry the same result is obtained when P lies on the other branch of the curve.

Example 2

Sketch the curve given by the equation

$$y^2=a^2x/(2a-x).$$

Prove that the area enclosed by the above curve and the line $x=a$ is $(\pi-2)a^2$, and that the volume traced out by rotating this area about the axis $y=0$ through two right angles is

$$\pi(\log 4-1)a^3. \qquad \text{[L.U.]}$$

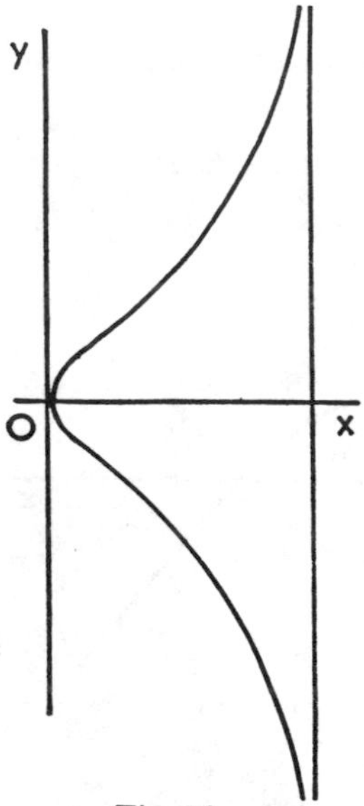

This curve which is symmetrical about Ox touches Oy at the origin, and has $x=2a$ as vertical asymptote. There are no real values of y when $x<0$ or when $x>2a$. Logarithmic differentiation gives

$$\frac{dy}{dx}=\frac{ay}{x(2a-x)}=\pm\frac{a^2}{\sqrt{\{x(2a-x)^3\}}},$$

from which we see that there is no point at which $\frac{dy}{dx}=0$. The form of the curve when $a>0$ is shown in fig. 70.

Fig. 70

The area enclosed by the curve and the line $x=a$ is given by

$$I_1=2\int_0^a \frac{a\sqrt{x}}{\sqrt{(2a-x)}}\,dx.$$

Put $x=2a\sin^2\theta$, $dx=4a\sin\theta\cos\theta d\theta$.

Then $$I_1=8a^2\int_0^{\pi/4}\sin^2\theta d\theta=4a^2[\theta-\tfrac{1}{2}\sin 2\theta]_0^{\pi/4}=(\pi-2)a^2.$$

The volume traced out by rotating this area about Ox is given by

$$I_2=\pi\int_0^a \frac{a^2x}{2a-x}\,dx$$
$$=\pi a^2\int_a^{2a}\left(\frac{2a}{t}-1\right)dt, \quad t=2a-x$$
$$=\pi a^2[2a\log t-t]_a^{2a}$$
$$=\pi a^3(\log 4-1).$$

Example 3

A parabola is drawn having for its vertex the centre of an ellipse of major axis $2a$ and eccentricity $\frac{1}{2}$ and having the minor axis as the tangent at its vertex ; also, it passes through the ends of a latus rectum of the ellipse. Show *that it divides the elliptic area into two parts, the ratio of whose areas is* $(4\pi+\sqrt{3}) : (8\pi-\sqrt{3})$.

If the smaller of these areas revolves through four right angles about the minor axis of the ellipse show that the volume generated is $21\pi a^3/20$. [L.U.]

The equation of an ellipse of major axis $2a$ and eccentricity $e=\frac{1}{2}$ is $x^2/a^2+y^2/b^2=1$, where $b^2=a^2(1-e^2)=\frac{3}{4}a^2$.

Thus the equation of the given ellipse is

$$3x^2+4y^2=3a^2 \quad . \quad . \quad . \quad . \quad \text{(i)}$$

The equation of a parabola with vertex at the centre of (i) and the minor axis of (i) as tangent at its vertex is of the form

$$y^2=4px.$$

Since it passes through the ends of the latus rectum of the ellipse, i.e. the points $(\frac{1}{2}a, \pm\frac{3}{4}a)$, $9a^2/16=2ap$, i.e. $p=9a/32$.

Hence the equation of the parabola is $8y^2=9ax$.

The form of the curves is shown in fig. 71. The area enclosed by the

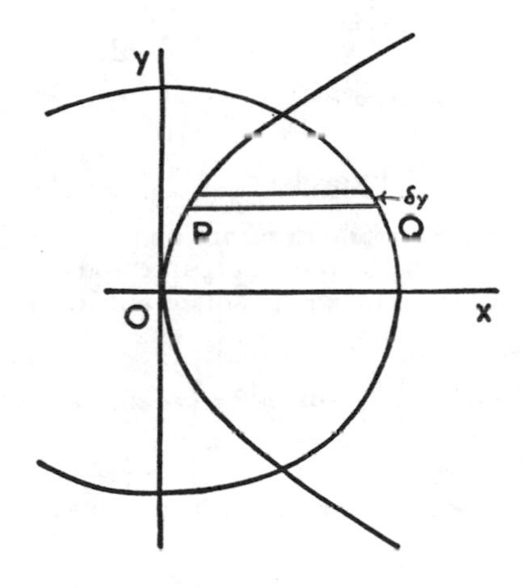

Fig. 71

ellipse and the parabola may be divided into elementary strips parallel to Ox, the area of each strip being $(x_2-x_1)\delta y$, where x_1 and x_2 are the abscissae of the points P and Q respectively.

Hence the required area is

$$\lim_{\delta y\to 0}\left\{2\sum_{y=0}^{y=\frac{3}{4}a}(x_2-x_1)\delta y\right\}=2\int_0^{\frac{3}{4}a}\left\{\sqrt{\left(a^2-\frac{4}{3}y^2\right)}-\frac{8y^2}{9a}\right\}dy \quad . \quad \text{(i)}$$

Now $\int_0^{\frac{3}{2}a} \sqrt{(a^2-\frac{4}{3}y^2)}\,dy=(\frac{1}{2}\sqrt{3})a^2\int_0^{\pi/3}\cos^2\theta d\theta,\ y=(\frac{1}{2}\sqrt{3})a\sin\theta,$

$$=(\tfrac{1}{4}\sqrt{3})a^2\Big[\theta+\tfrac{1}{2}\sin 2\theta\Big]_0^{\pi/3}=(\tfrac{1}{4}\sqrt{3})a^2\{\tfrac{1}{3}\pi+\tfrac{1}{4}\sqrt{3}\},$$

and
$$\int_0^{\frac{3}{2}a}\frac{8}{9a}y^2\,dy=\tfrac{1}{8}a^2.$$

Hence from (i) the required area is $\dfrac{\sqrt{3}a^2}{24}(4\pi+\sqrt{3})$.

The total area of the ellipse is $(\frac{1}{2}\sqrt{3})a^2\pi$ and so the ratio of the two parts into which the parabola divides the ellipse is $(4\pi+\sqrt{3}):(8\pi-\sqrt{3})$.
If the smaller area rotates about Oy, the volume generated is

$$V=\pi\int_{-\frac{3}{2}a}^{\frac{3}{2}a}(x_2^2-x_1^2)dy$$

$$=2\pi\int_0^{\frac{3}{2}a}\left(a^2-\frac{4}{3}y^2-\frac{64y^4}{81a^2}\right)dy$$

$$=2\pi\left[a^2y-\frac{4}{9}y^3-\frac{64}{405}\frac{y^5}{a^2}\right]_0^{\frac{3}{2}a}$$

$$=21\pi a^3/20.$$

Exercises 20 (a)

1. Sketch the curve $ay^2=x^2(a-x)$, where $a>0$.
   Find the area of the loop of the curve and the volume generated when the loop rotates through π radians about the axis of x.

2. Sketch the curves
$$ay^2=(x-a)(x-2a)^2 \text{ and } ay^2=(x-a)(x-2a)(x-3a)$$
for a positive value of the constant a.
   Find the area of the loop of the first curve, and the volume of the solid formed by the revolution of the loop of the latter curve round the axis of x. [L.U.]

3. Prove that the parabola $y=a(x-a)^2$ touches the curve $27y=4x^3$, and find the equation of the tangent at the point of contact. Show that the area bounded by this tangent and the parabola $y=4a(x-4a)^2$ is $18a^4$. [L.U.]

4. Prove that, if $a>0$, the curves $y^2=4ax$ and $27ay^2=4(x-2a)^3$ intersect where $x=8a$, and that the area bounded by the curves is $352\sqrt{2}a^2/15$. If this area revolves through two right angles about the axis of x prove that the volume generated is $80\pi a^3$. [L.U.]

5. Sketch the curve $y^2(a+x)=x^2(3a-x)$, and show that the area of the loop is $9a^2/\sqrt{3}$. [Sheffield.]

6. Sketch the curve $y^2=x^2(a-x)/(a+x)$, $a>0$, and show that the tangents to the curve at the origin are perpendicular. Show also that the area enclosed by the loop of the curve is $(4-\pi)a^2/2$.

   Show that the volume generated when the area enclosed by the loop is rotated through π about the x-axis is $\pi a^3(6 \log 2-4)/3$. [L.U.]

7. Draw a sketch of the curve $y^2=x^3-2x$, showing that it consists of a closed oval S and an open branch S_1.

   If the area bounded by S and the area bounded by S_1 and the line $x=+(2+\frac{4}{3}\sqrt{3})^{\frac{1}{2}}$ are rotated about the axis of x through two right angles, prove that the separate volumes swept out are in the ratio 3 : 4. [L.U.]

8. (i) Find the area between the curve $a(a-x)y=x^3$, the axis of x and the line $2x=a$.

   (ii) Prove that the volume obtained by revolving the cycloid
$$x=a(\theta-\sin\theta),\ y=a(1-\cos\theta)$$
   about the axis of x, between the points where $x=0$ and $x=2\pi$ is $5\pi^2a^3$. [L.U.]

9. The curve $y=ae^{-x/a}$ passes through the point (b, c). The area bounded by this curve, the axes, and the line $x=b$ is rotated through four right angles about the line $x=b$. Show that the volume swept out is $2\pi a^2(c+b-a)$. [L.U. Anc.]

10. Show that if the tangent at the point (x_0, y_0) on the curve
$$x=a\cos^2 t\sin t,\ y=a\cos t\sin^2 t$$
    meets the x-axis at the point $(x_1, 0)$, then x_1 does not lie between 0 and $\frac{1}{2}x_0$.

    Show also that the loop of the curve corresponding to the range $0\leqslant t\leqslant\frac{1}{2}\pi$ has an area $\pi a^2/32$. [L.U.]

11. Sketch roughly the curve whose equation is $y^2=x^3/(a-x)$ where $a>0$, and find the area included between the curve and its asymptote. Find also the volume generated on revolving the curve about its asymptote. [L.U.]

12. Sketch the graph of the curve given by the equation $y^2(1-x)(x-2)=x^2$, and prove that the area between the curve and its asymptotes is 3π. [L.U.]

13. Sketch in the same diagram the curves $xy^2=a^2(a-x)$ and $(a-x)y^2=a^2x$. Prove that they enclose an area $(\pi-2)a^2$, and find the volume of the solid obtained by rotating this area through two right angles about the line $y=0$. [L.U.]

14. Prove that the area bounded by the line $y=1$ and the curve $y=\tanh x$ in the region defined by $x\geqslant 0$ is log 2. [L.U.]

15. Find the volume generated when the area contained between the ellipse $x^2+2y^2=2$ and the two branches of the hyperbola $2x^2-2y^2=1$ revolves through two right angles about the y-axis. [L.U. Anc.]

16. Show that the curve whose equation is $y^2=a^2(x-a)/x$, where $a>0$, has the lines $y=\pm a$, $x=0$ as asymptotes. Sketch the curve.

Find the equations of the tangents which pass through the origin, and deduce, or prove otherwise, that the equation in x, $\lambda x^3=a^2(x-a)$ has three real roots if $0<\lambda<4/27$, but only one real root if λ does not lie between those limits. [L.U.]

17. The smaller segment of the ellipse $3x^2+4y^2=1$ cut off by one latus rectum is rotated through four right angles about the other latus rectum. Find the volume of the annular solid so formed. [L.U.]

18. If $f(x)$ is a positive and continuous function which decreases as x increases, prove geometrically that

$$f(1)+f(2)+\ldots+f(m)>\int_1^{m+1} f(x)\,dx.$$

Deduce or otherwise prove that $\sum\limits_{n=1}^{\infty} \dfrac{1}{n \log 2n}$ is divergent.

19. Prove that, if $a>0$, the curve $y=e^{-x} \sin ax$ has an inflexion where $ax=(2 \tan^{-1} a)+k\pi$ for any integral value of k.

Find the area bounded by the curve and the segment of the x-axis between $x=k\pi/a$ and $x=(k+1)\pi/a$.

Hence, or otherwise, evaluate

$$\int_0^{\infty} e^{-x} \sin ax\,dx.$$ [L.U.]

20.5. Length of arc and surface area ; definitions

Let us consider the n-sided open polygon $HP_1P_2\ldots P_{n-1}K$ inscribed in the arc HK of a continuous curve (fig. 72). Let the perimeter (the sum of the n sides $HP_1, P_1P_2, \ldots, P_{n-1}K$) of the polygon be denotedby s_n.

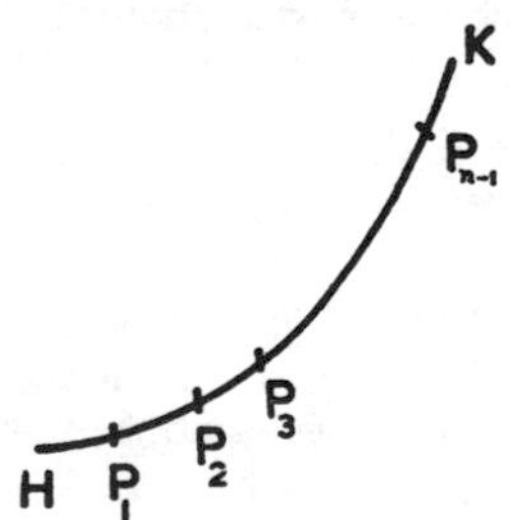

Fig. 72

If, as $n\to\infty$ (H and K remaining fixed), each of the chords $HP_1, P_1P_2, \ldots, P_{n-1}K$ tends to zero and s_n tends to a definite limit s, we say that the arc HK is of length s.

Hence arc HK

$$=\lim_{n\to\infty} \Sigma\,(HP_1+P_1P_2+\ldots+P_{n-1}K)=s.$$

In the same way, the surface area of the solid obtained by rotating the arc HK about any axis is defined as the limit of the sum of the surface areas obtained by rotating each of the chords $HP_1, P_1P_2, \ldots, P_{n-1}K$ about this axis.

20.6. Length of arc

To obtain formulae for the length of an arc and for the surface area of a solid of revolution, let us consider an arc of the curve $y=\phi(x)$ whose end points A and B have abscissae $x=a$, $x=b$ respectively (fig. 73). We shall assume that the angle ψ between the positive tangent at (x, y) and the x-axis is not only a continuous function of x but is also acute and increases steadily as (x, y) moves along the curve from A to B. Suppose that arc AB is divided into n parts by

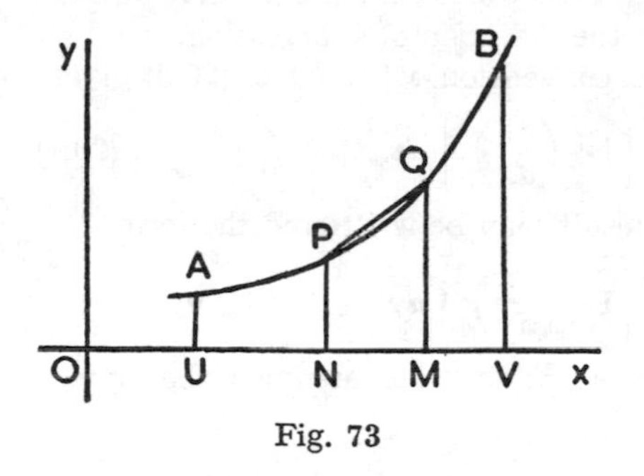

Fig. 73

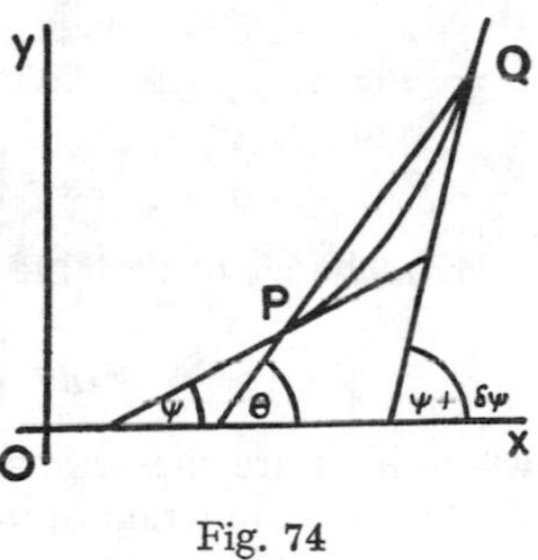

Fig. 74

equally spaced ordinates δx apart and consider a typical chord joining $P(x, y)$ to $Q(x+\delta x, y+\delta y)$ on the arc and making an angle θ with OX.

Then $$PQ=\sec\theta \, . \, \delta x.$$

Now if the tangents at P and Q make angles ψ and $(\psi+\delta\psi)$ with Ox (fig. 74),

$$\psi<\theta<\psi+\delta\psi,$$

$$\therefore \sec\psi\,\delta x<\sec\theta\,\delta x<\sec(\psi+\delta\psi)\,\delta x,$$

i.e. $$\sec\psi\,\delta x<PQ<\sec(\psi+\delta\psi)\,\delta x \quad . \quad . \quad . \quad \text{(i)}$$

$$\therefore \sum_{x=a}^{x=b}\sec\psi\,\delta x<\sum_{x=a}^{x=b}PQ<\sum_{x=a}^{x=b}\sec(\psi+\delta\psi)\,\delta x$$

$$\therefore \text{arc } AB=\lim_{n\to\infty}\sum_{x=a}^{x=b}PQ=\lim_{\delta x\to 0}\sum_{x=a}^{x=b}\sec\psi\,\delta x=\int_a^b\sec\psi\,dx.$$ (Cf. note on p. 443.)

But $$\sec^2\psi=1+\tan^2\psi=1+\left(\frac{dy}{dx}\right)^2.$$

Hence, denoting arc AB by s_{AB} we have

$$s_{AB}=\pm\int_a^b\sqrt{\left\{1+\left(\frac{dy}{dx}\right)^2\right\}}\,dx. \tag{20.6}$$

If ψ steadily decreases along the arc AB, the above inequalities are reversed but the formula obtained for s is still valid.

20.7. Sign convention for s

If an arc of a curve is measured from a point A (fig. 75), it is customary to attach a plus or a minus sign to the arc AP according as P lies on the portion AC or the portion AB of the curve.

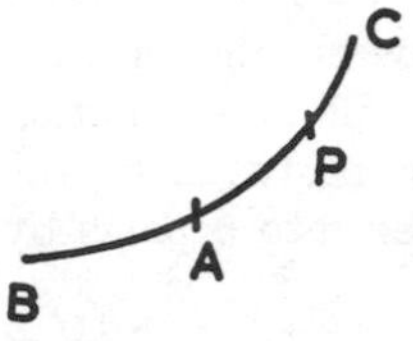

Fig. 75

When the cartesian equation of the curve is given in the form $y=\phi(x)$ it is usual to measure s so that it increases with x. It follows from the definition given in § 9.19 that the positive tangent is drawn in the direction of s increasing.

With this convention, when $b>a$, (20.6) gives

$$s_{AB}=\int_a^b \sqrt{\left\{1+\left(\frac{dy}{dx}\right)^2\right\}}\, dx. \tag{20.7}$$

By a change of variable this result may be written in the form

$$s_{AB}=\int_\alpha^\beta \sqrt{\left\{1+\left(\frac{dx}{dy}\right)^2\right\}}\, dy$$

where α, β are the ordinates of A, B, it being assumed that x is a single-valued function of y.

20.8. Surface area

As in § 20.6, by considering the conical frustum formed by rotating the chord PQ, it can be shown that the surface area S_{AB} of the solid generated when the arc AB revolves about Ox is given by

$$S_{AB}=2\pi\int_a^b y\sqrt{\left\{1+\left(\frac{dy}{dx}\right)^2\right\}}\, dx. \tag{20.8}$$

By changing the variable we may write this result in the form

$$S_{AB}=2\pi\int_\alpha^\beta y\sqrt{\left\{1+\left(\frac{dx}{dy}\right)^2\right\}}\, dy, \tag{20.9}$$

it being assumed that x is a single-valued function of y.

20.9. Differential relations

If s (fig. 73) denotes the length of the arc of the curve $y=\phi(x)$ from A to P (x, y)

$$s=\int_a^x \sec\psi\, dx$$

and (although the validity of the differentiation is beyond the scope of this book) it follows that

$$\left.\begin{aligned} &\frac{ds}{dx}=\sec\psi=\sqrt{\left\{1+\left(\frac{dy}{dx}\right)^2\right\}} \\ \text{and}\quad &\frac{ds}{dy}=\frac{ds}{dx}\bigg/\frac{dy}{dx}=\operatorname{cosec}\psi=\sqrt{\left\{1+\left(\frac{dx}{dy}\right)^2\right\}} \end{aligned}\right\} \tag{20.10}$$

These results enable us to write (20.8) and (20.9) in the forms

$$S_{AB}=2\pi\int_a^b y\frac{ds}{dx}\,dx=2\pi\int_\alpha^\beta y\frac{ds}{dy}\,dy=2\pi\int_{S_A}^{S_B} y\,ds. \tag{20.11}$$

Again, from (20.10)

$$\frac{dx}{ds}=\cos\psi,\quad \frac{dy}{ds}=\sin\psi. \tag{20.12}$$

$$\therefore\ \left(\frac{dx}{ds}\right)^2+\left(\frac{dy}{ds}\right)^2=1$$

giving the differential relation

$$(dx)^2+(dy)^2=(ds)^2. \tag{20.13}$$

The above results may be memorised with the help of fig. 64, p. 413, in which $\angle RPT=\psi$, $dx=PR$, $dy=RT$ and, by (20.13), $ds=PT$.

20.10. Curves given in parametric form

The positive tangent at any point on the curve $x=x(t)$, $y=y(t)$ is drawn in the direction of s increasing, and s is conventionally measured so as to increase with the parameter t.

If dots denote differentiation with respect to t we have from (20.13)

$$\dot{s}^2=\dot{x}^2+\dot{y}^2. \tag{20.14}$$

Hence if t increases steadily from t_1 at A to t_2 at B

$$s_{AB}=\int_{t_1}^{t_2}\sqrt{\{\dot{x}^2+\dot{y}^2\}}\,dt. \tag{20.15}$$

The formula corresponding to (20.11) is

$$S_{AB}=2\pi\int_{t_1}^{t_2} y\frac{ds}{dt}\,dt=2\pi\int_{t_1}^{t_2} y\sqrt{\{\dot{x}^2+\dot{y}^2\}}\,dt \tag{20.16}$$

Example 4

Prove that the length of the arc of the plane curve

$$y=(x+1)(x+2)-\tfrac{1}{8}\log(2x+3)$$

between the points for which $x=1$ and $x=2$ respectively is $6+\frac{1}{8}\log 7/5$.

[L.U.]

If
$$y=(x+1)(x+2)-\tfrac{1}{8}\log(2x+3)\quad .\quad .\quad .\quad \text{(i)}$$

$$\frac{dy}{dx}=2x+3-\frac{1}{4(2x+3)}$$

and
$$1+\left(\frac{dy}{dx}\right)^2=\left\{(2x+3)+\frac{1}{4(2x+3)}\right\}^2.$$

Thus by (20.7) the length of arc of curve (i) between $x=1$ and $x=2$ is

$$\int_1^2\left\{(2x+3)+\frac{1}{4(2x+3)}\right\}dx=\left[x^2+3x+\frac{1}{8}\log(2x+3)\right]_1^2=6+\frac{1}{8}\log 7/5.$$

Example 5

If s, the arc OP of a curve, measured from a fixed point O on it, is $f(\psi)$, where ψ is the angle which the tangent at P makes with the positive direction of the x-axis, show how to find the cartesian coordinates (x, y) of P in terms of ψ.

The intrinsic equation of a curve is

$$5s=4a(5+\tan^2\tfrac{1}{2}\psi)\sqrt{(\tan\tfrac{1}{2}\psi)}.$$

If the axes are chosen so that x, y, s, and ψ vanish simultaneously find the cartesian coordinates of any point on the curve in terms of ψ, and verify that

$$5x^2+9y^2=5s^2. \qquad \text{[L.U.]}$$

The equation which connects s and ψ in any curve is known as the *intrinsic* equation of the curve.

Let $P(x, y)$ be any point on the curve whose intrinsic equation is $s=f(\psi)$.

Then
$$\frac{dx}{d\psi}=\frac{dx}{ds}\cdot\frac{ds}{d\psi}=\cos\psi\, f'(\psi) \quad \text{by (20.12)}$$

$$\therefore\ x=\int\cos\psi\, f'(\psi)\,d\psi \quad . \quad . \quad . \quad . \qquad \text{(i)}$$

and in the same way,
$$y=\int\sin\psi\, f'(\psi)\,d\psi \quad . \quad . \quad . \quad . \qquad \text{(ii)}$$

the integration in each case being taken between appropriate limits.

Now if

$$s=f(\psi)=\tfrac{4}{5}a(5+\tan^2\tfrac{1}{2}\psi)\sqrt{(\tan\tfrac{1}{2}\psi)} \quad . \quad . \quad . \quad . \quad . \qquad \text{(iii)}$$

$$f'(\psi)=\tfrac{4}{5}a\{\tan^2\tfrac{1}{2}\psi\sec^2\tfrac{1}{2}\psi+\tfrac{1}{4}\sec^2\tfrac{1}{2}\psi(5+\tan^2\tfrac{1}{2}\psi)\}/\sqrt{(\tan\tfrac{1}{2}\psi)}$$
$$=a\sec^4\tfrac{1}{2}\psi/\sqrt{(\tan\tfrac{1}{2}\psi)}.$$

Hence

$$\cos\psi\, f'(\psi)=\frac{1-\tan^2\frac{1}{2}\psi}{1+\tan^2\frac{1}{2}\psi}\cdot\frac{a\sec^4\frac{1}{2}\psi}{\sqrt{(\tan\frac{1}{2}\psi)}}=\frac{a(1-\tan^2\frac{1}{2}\psi)\sec^2\frac{1}{2}\psi}{\sqrt{(\tan\frac{1}{2}\psi)}},$$

and so, from (i)

$$x=2a\int\frac{1-t^2}{\sqrt{t}}\,dt, \quad t=\tan\tfrac{1}{2}\psi$$
$$=2a\left\{2t^{1/2}-\frac{2}{5}t^{5/2}+c\right\}.$$

But when $\psi=0$ (so that $t=0$), $x=0$ and so $c=0$,

$$\therefore\ x=\tfrac{4}{5}at^{1/2}(5-t^2) \quad . \quad . \quad . \quad . \qquad \text{(iv)}$$

Also

$$\sin\psi\, f'(\psi)=\frac{2\tan\frac{1}{2}\psi}{1+\tan^2\frac{1}{2}\psi}\cdot\frac{a\sec^4\frac{1}{2}\psi}{\sqrt{(\tan\frac{1}{2}\psi)}}=2a(\sec^2\tfrac{1}{2}\psi)\sqrt{(\tan\tfrac{1}{2}\psi)}$$

and so, from (ii)

$$y=4a\int\sqrt{t}\,dt, \quad t=\tan\tfrac{1}{2}\psi$$
$$=\tfrac{8}{3}at^{3/2}+c'.$$

But when $\psi=0$, $y=0$ and so $c'=0$

$$\therefore\ y=\tfrac{8}{3}at^{3/2} \quad . \quad . \quad . \quad . \quad . \quad . \quad (v)$$

From (iv) and (v)

$$5x^2+9y^2=\tfrac{16}{5}a^2(5+\tan^2\tfrac{1}{2}\psi)^2\tan\tfrac{1}{2}\psi$$
$$=5s^2 \quad \text{by (iii).}$$

Example 6

The region bounded by a quadrant of a circle of radius a, and the tangents at its extremities, revolves through 360° about one of these tangents. Prove that the volume of the solid thus generated is $(\tfrac{5}{3}-\tfrac{1}{2}\pi)\pi a^3$, *and the area of its curved surface is* $\pi(\pi-2)a^2$. [L.U.]

Let AC (fig. 76) be the tangent about which the quadrant of the circle rotates. From the points $P(x, y)$ and $Q(x+\delta x,\ y+\delta y)$ on the circle draw perpendiculars PM, QN to AC. Then V, the volume required, is the limit of the sum of the volumes generated by rotating figures such as $QPMN$ about AC.

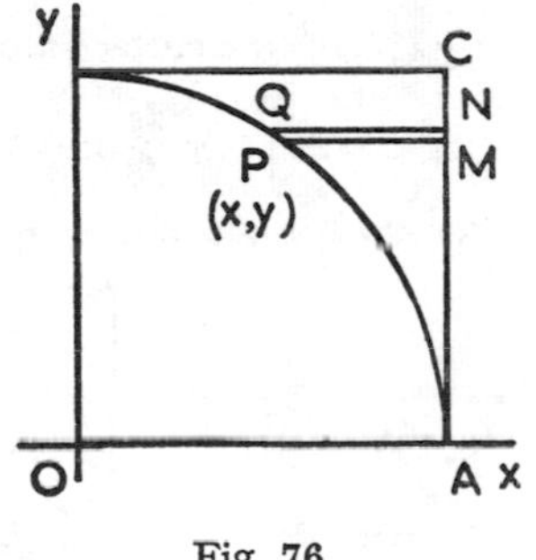

Fig. 76

$$\text{Hence } V=\lim_{\delta y\to 0}\sum_{y=0}^{y=a}\pi PM^2.MN$$
$$=\lim_{\delta y\to 0}\sum_{y=0}^{y=a}\pi(a-x)^2\delta y$$
$$=\pi\int_0^a (a-x)^2\,dy.$$

The parametric equations of the circle are

$$x=a\cos\theta, \quad y=a\sin\theta;$$

and so
$$V=\pi a^3\int_0^{\pi/2}(1-\cos\theta)^2\cos\theta\,d\theta$$
$$=\pi a^3\int_0^{\pi/2}(\cos\theta-2\cos^2\theta+\cos^3\theta)\,d\theta$$
$$=(\tfrac{5}{3}-\tfrac{1}{2}\pi)\pi a^3.$$

The surface area $S=2\pi\displaystyle\int_0^{\pi/2}(a-x)\frac{ds}{d\theta}\,d\theta,$

and $\dfrac{ds}{d\theta}=a$, s being measured to increase with θ

$$\therefore\ S=2\pi a^2\int_0^{\pi/2}(1-\cos\theta)\,d\theta$$
$$=\pi(\pi-2)a^2.$$

Example 7

The area in the first quadrant bounded by the axes and the curve $x=a\cos^3\theta$, $y=a\sin^3\theta$ is rotated through four right angles about Ox. Show that the area of the surface generated is $\frac{6}{5}\pi a^2$, and find the volume contained within this surface. [L.U.]

The curve is shown in fig. 101 (*b*), page 494.

$$x=a\cos^3\theta, \qquad y=a\sin^3\theta,$$

$$\frac{dx}{d\theta}=-3a\cos^2\theta\sin\theta, \qquad \frac{dy}{d\theta}=3a\sin^2\theta\cos\theta.$$

By (20.14),

$$\left(\frac{ds}{d\theta}\right)^2=\left(\frac{dx}{d\theta}\right)^2+\left(\frac{dy}{d\theta}\right)^2=9a^2\cos^2\theta\sin^2\theta(\cos^2\theta+\sin^2\theta)$$

and so, for θ between 0 and $\frac{1}{2}\pi$,

$$\frac{ds}{d\theta}=3a\cos\theta\sin\theta, \text{ } s \text{ being measured to increase with } \theta.$$

By (20.16) the surface area required is

$$S=2\pi\int_0^{\pi/2} y\frac{ds}{d\theta}\,d\theta=6\pi a^2\int_0^{\pi/2}\sin^4\theta\cos\theta\,d\theta=\tfrac{6}{5}\pi a^2.$$

By (20.4) the volume required is

$$\begin{aligned}V&=\pi\int_0^a y^2\,dx\\ &=\pi\int_{\pi/2}^0 y^2\frac{dx}{d\theta}\,d\theta\\ &=3\pi a^3\int_0^{\pi/2}\sin^7\theta\cos^2\theta\,d\theta\\ &=3\pi a^3\,\tfrac{6}{9}\,\tfrac{4}{7}\cdot\tfrac{2}{5}\cdot\tfrac{1}{3} \quad \text{by (10.6), page 229,}\\ &=16\pi a^3/105.\end{aligned}$$

Exercises 20 (*b*)

1. A function $y=f(x)$ is defined in the interval $a\leqslant x\leqslant b$. Write down formulae for the length, s, of its graph and for the area, A, of the surface obtained by rotating this graph about the x-axis.

   In the case where $f(x)=\frac{1}{2}x^2$, $a=0$ and $b=\sinh c\,(c>0)$, prove that

   $$s=(\sinh 2c+2c)/4 \text{ and } A=\pi(\sinh 4c-4c)/32. \qquad \text{[Durham.]}$$

2. Trace roughly the curve $8a^2y^2=x^2(a^2-2x^2)$, and show that its whole length of arc is πa.

   Show that the area enclosed by the curve is two-thirds of that of the circumscribing rectangle whose sides are parallel to the axes of coordinates [L.U.]

3. The area bounded by the two parabolas $y^2=4ax$ and $x^2=4ay$ is rotated through 360° about the x-axis. Find the superficial area and the volume of the solid so formed. [L.U.]

4. A is the vertex and $LL'=4a$, the latus rectum of a parabola. Find the area generated when the arc AL is rotated through four right angles (i) about the axis of the parabola, and (ii) about LL'. [L.U.]

5. Sketch the curve whose equation is $y=e^{-x}\sin x$, and show that the areas included between the axis of x and semi-undulations of the curve form a decreasing geometric progression.

   The curve $y=\sin x$ is rotated about the x-axis, find the superficial area generated by the portion of the curve lying between the lines $x=0$ and $x=\pi$. [L.U.]

6. The curve C is represented by the equation

$$9ay^2=(a-x)(x+2a)^2 \quad (a>0).$$

   Find the coordinates of the points on C at which the tangent is parallel to the x-axis, and also of the point at which C has two distinct tangents. Give a sketch of the curve.

   Find the area of the surface of revolution obtained by rotating the closed portion of this curve through two right angles about the x-axis. [L.U.]

7. The curve $y=c\cosh(x/c)$ cuts the axis Oy in the point C and the straight line $y=2c$ in the points A and B. Prove that the volume of the solid formed by rotating the area ABC about the line AB through four right angles is

$$\pi\{9\log(2+\sqrt{3})-6\sqrt{3}\}c^3.$$

   Prove also that the area of the surface of the solid is

$$\pi\{4\sqrt{3}-2\log(2+\sqrt{3})\}c^2. \qquad \text{[L.U.]}$$

8. Sketch the curve $y=-\log(1-x^2)$, and find its length from the origin to the point where $x=x_1$ $(0<x_1<1)$.

   Find also the area bounded by the curve, the x-axis, and the ordinate $x=x_1$. Show that, as $x_1 \to 1$, this area tends to the limit $2-2\log 2$. [L.U.]

9. A and B are the points on the curve $y=c\cosh(x/c)$ at which x has the values a and b respectively $(a<b)$. The region bounded by the arc AB, the ordinates at A and B, and the x-axis is rotated through 360° about the x-axis. If V is the volume of the solid generated and S is its curved surface area, show that

$$V=\tfrac{1}{2}cS=\tfrac{1}{2}\pi c^2\{b-a+\tfrac{1}{2}c\sinh(2b/c)-\tfrac{1}{2}c\sinh(2a/c)\}. \qquad \text{[L.U.]}$$

10. Sketch the curve $16y^2=x^2(2-x^2)$ and find the area of one loop. Show that the total length of the curve is 2π. [L.U.]

20.11. Centres of mass

In text-books on mechanics it is shown that if $m_1, m_2, \ldots, m_n$ are the masses of a system of n particles situated at

$$P_1(x_1, y_1, z_1), P_2(x_2, y_2, z_2), \ldots, P_n(x_n, y_n, z_n)$$

the resultant weight of the system acts at a fixed point G $(\bar{x}, \bar{y}, \bar{z})$ where

$$\bar{x}=\frac{m_1x_1+m_2x_2+\ldots+m_nx_n}{m_1+m_2+\ldots+m_n}, \quad \bar{y}=\frac{m_1y_1+m_2y_2+\ldots+m_ny_n}{m_1+m_2+\ldots+m_n},$$

$$\bar{z}=\frac{m_1z_1+m_2z_2+\ldots+m_nz_n}{m_1+m_2+\ldots+m_n}.$$

G is known as the *centre of mass* or *centre of gravity* of the system, and the coordinates of G may be written in the form

$$\bar{x}=N_{yz}/M, \ \bar{y}=N_{zx}/M, \ \bar{z}=N_{xy}/M$$

where $$M=m_1+m_2+\ldots+m_n=\Sigma m,$$

$$N_{yz}=m_1x_1+m_2x_2+\ldots+m_nx_n=\Sigma mx,$$

$$N_{zx}=\Sigma my \text{ and } N_{xy}=\Sigma mz,$$

the sign Σ denoting summation over all the masses of the system. N_{yz}, N_{zx}, N_{xy} are called the first moments of the system with respect to the $yz-$, $zx-$ and $xy-$ planes respectively.

If the particles all lie in the plane $z=0$, the coordinates of G are

$$\bar{x}=N_y/M, \quad \bar{y}=N_x/M \tag{20.17}$$

where $N_y \equiv \Sigma mx$ is the first moment of the system about the axis Oy, and $N_x \equiv \Sigma my$ is the first moment of the system about the axis Ox.

In the case of a continuous body, we replace the above system of particles by the elements of the body and use limiting sums (i.e. integrals) in place of summations.

20.12. Centroid of a uniform plane lamina, and solid of revolution

The centre of mass G of a plane lamina lies in the plane of the lamina. If the lamina is uniform, G is called the *centroid* of the area enclosed by the lamina. If this area is symmetrical about any straight line, the centroid lies on that line, e.g. a circle is symmetrical about any diameter, hence its centroid is the point common to all diameters, i.e. the centre of the circle. The centroid of a uniform rectangular lamina is the point of intersection of its diagonals.

In the case of the uniform lamina $AUVB$ bounded by the curve $y=\phi(x)$, the x-axis and the ordinates $x=a$, $x=b$ (fig. 77), we divide the area into equal strips and consider a typical strip $PNMQ$, P and Q

being the points (x, y) and $(x+\delta x, y+\delta y)$ respectively. We complete rectangles $PNMR$ and $SNMQ$; then if σ is the surface-density of the lamina, the masses of these rectangles are $\sigma y\,\delta x$ and $\sigma(y+\delta y)\,\delta x$

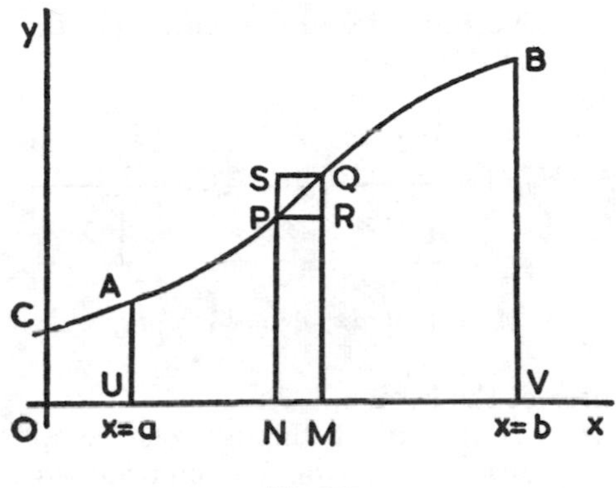

Fig. 77

respectively. We suppose that $N_y(x)$, $\delta N_y(x)$ are the moments of the area $CONP$ and the strip $PNMQ$ respectively about the axis Oy.

Then $$x\sigma y\,\delta x < \delta N_y(x) < (x+\delta x)\sigma(y+\delta y)\,\delta x.$$

This inequality leads in the usual way to

$$\frac{dN_y(x)}{dx} = x\sigma y, \text{ and so to } N_y(x) = \int_a^x x\sigma y\,dx.$$

Hence the first moment N_y of the whole lamina about Oy is

$$N_y = \int_a^b x\sigma y\,dx = \sigma\int_a^b xy\,dx,$$

σ being constant.

The centres of mass of the rectangles $PNMR$, $SNMQ$ are at distances $\frac{1}{2}y$ and $\frac{1}{2}(y+\delta y)$ respectively from Ox. Hence if $\delta N_x(x)$ is the moment of the strip $PNMQ$ about Ox,

$$\tfrac{1}{2}y\,.\,\sigma y\,\delta x < \delta N_x(x) < \tfrac{1}{2}(y+\delta y)\sigma(y+\delta y)\,\delta x.$$

This relation leads to

$$\frac{dN_x(x)}{dx} = \tfrac{1}{2}\sigma y^2, \text{ and so to } N_x(x) = \tfrac{1}{2}\sigma\int_a^x y^2\,dx.$$

Hence the moment N_x of the whole lamina about Ox is given by

$$N_x = \tfrac{1}{2}\sigma\int_a^b y^2\,dx.$$

The total mass M of the lamina (σ times its area) is given by

$$M=\sigma\int_a^b y\,dx.$$

Hence if $(\bar{x}, \bar{y})$ are the coordinates of the centroid of the lamina, equations equivalent to (20.17) give

$$\bar{x}=N_y/M=\frac{\int_a^b xy\,dx}{\int_a^b y\,dx},\ \bar{y}=N_x/M=\frac{\frac{1}{2}\int_a^b y^2\,dx}{\int_a^b y\,dx} \tag{20.18}$$

or

$$A\bar{x}=\int_a^b xy\,dx,\ A\bar{y}=\tfrac{1}{2}\int_a^b y^2\,dx,$$

where A is the area of the lamina.

If the area $AUVB$ is rotated through one complete revolution about Ox, G' the centre of mass of the solid formed lies on Ox by symmetry and if the solid is of uniform volume density, G' is called the centroid of the volume generated.

The cross-sections of the solid by planes perpendicular to Ox drawn through P and Q are circles of radii y and $y+\delta y$ respectively. If $\delta N(x)$ is the moment about Oy of the element of volume bounded by these planes and if ρ is the volume density of the solid

$$x\,.\,\rho\pi y^2\,\delta x<\delta N(x)<(x+\delta x)\,.\,\rho\pi(y+\delta y)^2\delta x.$$

This leads to

$$\frac{dN(x)}{dx}=x\rho\pi y^2$$

and so to

$$N(x)=\pi\rho\int_a^x xy^2\,dx.$$

It follows that the moment N of the whole solid of revolution about Oy is given by

$$N=\pi\rho\int_a^b xy^2\,dx,$$

and the total mass M of the solid (ρ times its volume) is given by

$$M=\pi\rho\int_a^b y^2\,dx.$$

Hence the abscissa of G', the centroid of the volume, is

$$\bar{x}=\frac{N}{M}=\frac{\int_a^b xy^2\,dx}{\int_a^b y^2\,dx}.$$

20.13. Theorems of Pappus

I. If a plane arc is revolved through an angle θ radians about a coplanar axis which does not cross the arc, the area of the surface generated is equal to the product of the length of the arc and the length of the circular path described by the centroid of the arc.

In fig. 78, Ox is the axis of rotation and G $(\bar{x}, \bar{y})$ is the centroid of

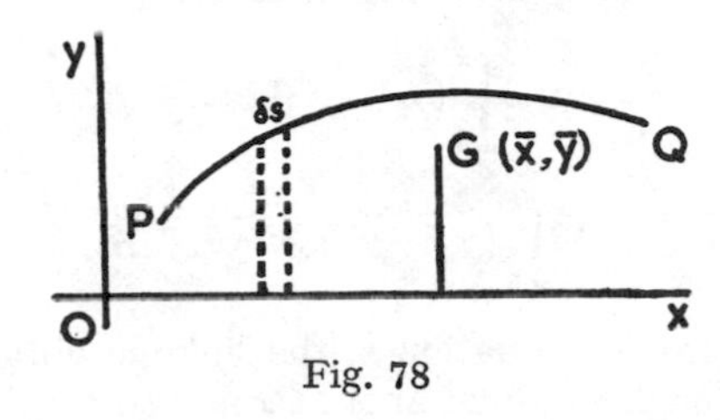

Fig. 78

the arc PQ of length l. The element δs may be taken as lying at a distance y from Ox.

Then
$$\Sigma y\,\delta s = \bar{y}\Sigma\,\delta s$$

i.e.
$$\int_{s_P}^{s_Q} y\,ds = \bar{y}l$$

$$\therefore\ \theta\int_{s_P}^{s_Q} y\,ds = \theta\bar{y}l.$$

But $\theta\int_{s_P}^{s_Q} y\,ds$ is the area of the surface swept out by PQ in rotating through an angle θ, see (20.11), page 455; and so the theorem is proved.

II. If a plane area is revolved through an angle θ radians about a coplanar axis which does not divide the area into two parts, the volume of the solid generated is equal to the product of the area and the length of the circular path described by its centroid.

In fig. 79, Ox is the axis of rotation, $G(\bar{x}, \bar{y})$ is the centroid of the area A bounded by the curve $HPKQ$, and $x=h$, $x=k$ are tangents to the

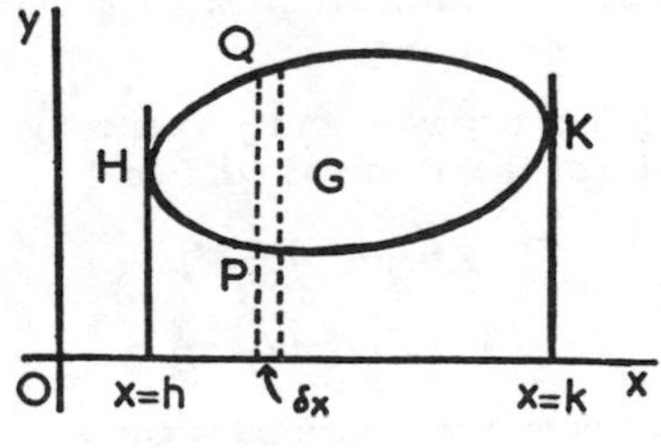

Fig. 79

curve parallel to Oy. Then, if y_1 and y_2 are the ordinates of P and Q respectively, and δx is the width of the elementary strip shown, we have

$$\Sigma\tfrac{1}{2}(y_2+y_1)\delta A=\bar{y}\Sigma\delta A$$

i.e.

$$\Sigma\tfrac{1}{2}(y_2+y_1)(y_2-y_1)\delta x=\bar{y}\Sigma\delta A,$$

the summation extending over the whole area,

$$\therefore \tfrac{1}{2}\int_h^k (y_2^2-y_1^2)dx=\bar{y}.A$$

and so

$$\tfrac{1}{2}\theta\int_h^k (y_2^2-y_1^2)dx=\theta.\bar{y}.A.$$

But the integral in this equation is the volume obtained by rotating the given area through θ radians about Ox and so the theorem is proved.

Example 8

OB is the middle radius which divides into two quadrants a semicircle ABC with diameter AC of length 2a. Find the position of G, the centroid of the semicircular area ABC, and also of G_1, the centroid of the quadrant OBC.

By symmetry G lies on OB (fig. 80) and if the semicircle rotates about AC its bounding diameter, we have by Pappus' second theorem

$$2\pi OG(\tfrac{1}{2}\pi a^2)=\text{volume of sphere}=\tfrac{4}{3}\pi a^3,$$

$$\therefore OG=4a/3\pi.$$

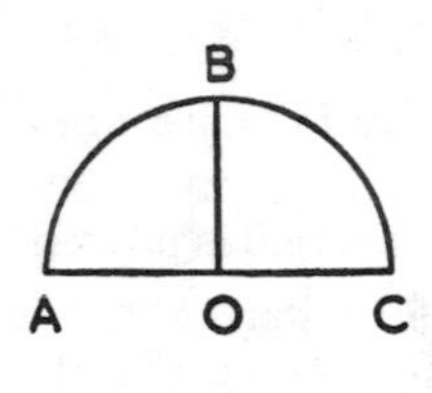

Fig. 80

The position of G_1 may be found in the same way, but the centroids G_1 and G_2 of the two quadrants must be such that G_1G_2 passes through G and is parallel to AC. Hence the distance of G_1, from OC is $4a/3\pi$, and by symmetry its distance from OB is the same.

The above results may also be found by integration using (20.18).

Example 9

If OP and OQ are bounding radii of a quadrant of a circle, and the semicircle with OP as diameter is cut away from the quadrant, find the position of the centroid of the area remaining. [L.U.]

If we take OP, OQ (fig. 81) as x- and y-axes respectively and if $OP=a$ the equation of the arc of the quadrant QKP is

$$y=+\sqrt{(a^2-x^2)} \quad . \quad . \quad . \quad . \quad \text{(i)}$$

and that of arc OHP is

$$y=+\sqrt{\{x(a-x)\}} \quad . \quad . \quad . \quad . \quad \text{(ii)}$$

The area remaining when the semicircle is cut away from the quadrant may be divided into strips such as HK, of width δx and parallel to Oy

the area of each strip being $(y_2-y_1)\delta x$, where y_2, y_1 are the ordinates of K and H respectively.

If x is the abscissa of H and K, the centroid of the strip HK may be taken as the point $(x, \frac{1}{2}\{y_2+y_1\})$ and the moments of the strip HK about Oy,

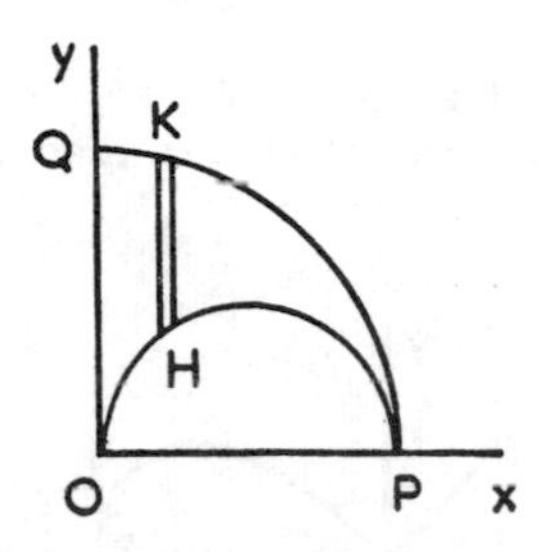

Fig. 81

Ox may be taken as $x(y_2-y_1)\delta x$ and $\frac{1}{2}(y_2+y_1)(y_2-y_1)\delta x$ respectively. Hence if $(\bar{x}, \bar{y})$ is the centroid of the area $OHPKQ$ and A is its area

$$A\bar{x}=\lim_{\delta x\to 0}\sum_{x=0}^{x=a} x(y_2-y_1)\delta x=\int_0^a x\{\sqrt{(a^2-x^2)}-\sqrt{[x(a-x)]}\}\,dx$$

$$\text{and } A\bar{y}=\lim_{\delta x\to 0}\sum_{x=0}^{x=a} \tfrac{1}{2}({y_2}^2-{y_1}^2)\delta x=\tfrac{1}{2}\int_0^a a(a-x)\,dx \text{ by (i) and (ii).}$$

Now
$$\int_0^a \sqrt{[x^3(a-x)]}\,dx=2a^3\int_0^{\pi/2}\sin^4\theta\cos^4\theta\,d\theta,\quad x=a\sin^2\theta$$
$$=\pi a^3/16 \quad \text{by (10.6), page 229,}$$

$A=\pi a^2/8$ and the other integrals are easily evaluated. Hence we find

$$\bar{x}=a\left(\frac{8}{3\pi}-\frac{1}{2}\right),\ \bar{y}=\frac{2a}{\pi}.$$

The method given above is of general application, but in this particular example we know from Example 8 the positions of the centroids of the quadrant and the semicircle, and the principle of moments applied to each of the above areas states that the moment of the quadrant about any axis is equal to the sum of the moments of the semicircle and the area $OHPKQ$ about that axis. Hence if $(\bar{x}, \bar{y})$ is the centroid of the area $OHPKQ$ and Oy is taken as axis,

$$\frac{\pi a^2}{4}\left(\frac{4a}{3\pi}\right)=\frac{\pi a^2}{8}\left(\frac{a}{2}\right)+\frac{\pi a^2}{8}\bar{x};\quad \bar{x}=a\left(\frac{8}{3\pi}-\frac{1}{2}\right).$$

If Ox is taken as axis,

$$\frac{\pi a^2}{4}\left(\frac{4a}{3\pi}\right)=\frac{\pi a^2}{8}\left(\frac{2a}{3\pi}\right)+\frac{\pi a^2}{8}\bar{y};\quad \bar{y}=\frac{2a}{\pi}.$$

Example 10

Sketch the curve $ay^2=x^2(a-x)$ and find the area of its loop. Find also the position of the centroid of this area and deduce the volume of the solid formed when this area is rotated through 2π radians about a tangent to the curve at the origin. [L.U.]

The curve is sketched for $a>0$ in fig. 82. It cuts Ox at $x=0$ and at $x=a$; also x cannot take values greater than a. The lines $y=\pm x$ are tangents at the origin.

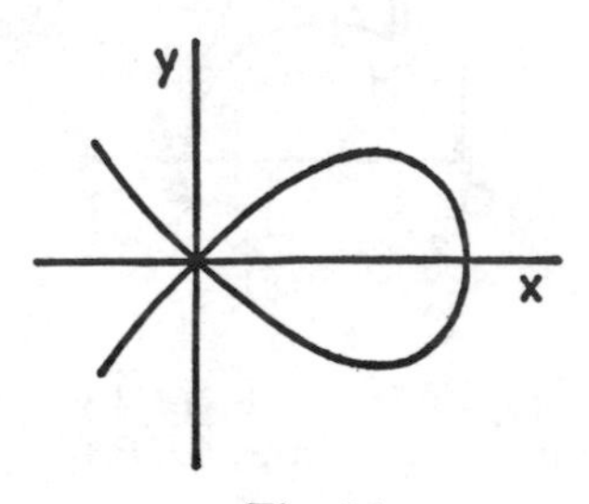

Fig. 82

The area of the loop is given by

$$\frac{2}{\sqrt{a}}\int_0^a x\sqrt{(a-x)}\,dx=\frac{2}{\sqrt{a}}\int_0^a (a-u)\sqrt{u}\,du, \quad \text{where } u=a-x,$$

$$=\frac{2}{\sqrt{a}}\left[\frac{2}{3}au^{3/2}-\frac{2}{5}u^{5/2}\right]_0^a$$

$$=8a^2/15.$$

By symmetry, the centroid G lies on Ox at a distance $\bar{x}$ from O, where by (20.18)

$$\bar{x}=\frac{\int_0^a xy\,dx}{\int_0^a y\,dx}=\frac{15}{4a^{5/2}}\int_0^a x^2\sqrt{(a-x)}\,dx$$

$$=\frac{15}{4a^{5/2}}\int_0^a (a^2-2au+u^2)\sqrt{u}\,du, \quad u=a-x,$$

$$=\frac{15}{4a^{5/2}}\left[\frac{2}{3}a^2u^{3/2}-\frac{4}{5}au^{5/2}+\frac{2}{7}u^{7/2}\right]_0^a$$

$$=\frac{4}{7}a.$$

We now apply Pappus' second theorem to obtain the required volume.

The distance of the centroid $G\left(\frac{4a}{7}, 0\right)$ from the tangent $y=x$ is $4a/7\sqrt{2}$. Thus the path of G is of length $8\pi a/7\sqrt{2}$ and the required volume is $\frac{8}{15}a^2\left(\frac{8\pi a}{7\sqrt{2}}\right)$ i.e. $64\pi a^3/105\sqrt{2}$.

Exercises 20 (*c*)

Find the centroids of the areas in Nos. 1-5 :

1. The area bounded by the curve $y=e^x$, the coordinate axes and the line $x=1$.

2. The area under the curve $y=\sin x$ from $x=0$ to $x=\frac{1}{2}\pi$.

3. The area bounded by the parabola $4y=x^2$, the x-axis and the line $x=2$.

4. The area bounded by the parabolas $y^2=4x$, $x^2=4y$.

5. The area bounded by the hyperbola $xy=4$ and the line $x+y=5$.

6. Find the area and the centroid of the portion of a plane bounded by the parabola $y^2=ax$, the line $x=b$ and the axis $y=0$. The area is revolved about the axis of y so as to form a solid ring. Find the volume of the ring. [L.U.]

7. Find the area of the loop of the curve whose equation is
$$ay^2=(x-a)(x-5a)^2.$$
Find also the distance of the centroid of this area from the y-axis.

Each of the areas described in Nos. 8-10 is revolved about Ox. Find the centroid of the solid of revolution generated.

8. The area bounded by Ox, the curve $y=1/x$ and the ordinates $x=1$, $x=4$.

9. The area of the ellipse $x^2/a^2+y^2/b^2=1$ which lies in the first quadrant.

10. The area in the first quadrant bounded by the parabola $y^2=4x$, the x-axis and the line $x=4$.

11. A regular hexagon is inscribed in the circle $x^2+(y-2)^2=1$ and is rotated about the x-axis. Find the volume and the surface area of the solid so formed. [Durham.]

12. The altitude from a vertex A of an equilateral triangle of side a makes an angle α with a line l through A in the plane of the triangle, the triangle lying on one side of l. Find the volume V and the surface area S of the body obtained by rotating the triangle about l. For what values of α are V and S largest ? [Durham.]

13. Find the centroid of the semicircular *arc* $(x-r)^2+y=r^2$, $y \geqslant 0$.
If this arc is rotated about the line $mx+y=0$, where $m>0$, determine the generated surface area A and show that A is a maximum when $m=\frac{1}{2}\pi$. [Durham.]

14. Show that the mean centre of a semicircular area is distant $4a/3\pi$ from the bounding diameter.
A kite-shaped area consists of an isosceles triangle OAB (where $OA=OB$) and a semicircle described on AB as diameter. If $AB=2a$, and the angle AOB is 2α, prove that the volume generated by one revolution of the area about OA is equal to

$$\pi^2a^3 \cos \alpha+\tfrac{4}{3}\pi a^3 \operatorname{cosec} \alpha. \qquad \text{[Sheffield.]}$$

15. The area bounded by $y=0$, $x=0$, $y=\cos x \left(0 \leqslant x \leqslant \frac{\pi}{2}\right)$ is rotated about the line $x=\frac{1}{2}\pi$. Prove that the volume swept out in one revolution is 2π. [L.U.]

16. Sketch the curve $x=a(t-\sin t)$, $y=a(1-\cos t)$ in the range $0 \leqslant t \leqslant 2\pi$. Show that
   (i) the gradient at any point is $\cot \frac{1}{2}t$;
   (ii) the area enclosed by the arc and the x-axis is $3\pi a^2$;
   (iii) the centroid of the area is at the point $(\pi a, \frac{5}{6}a)$. [L.U. Anc.]

17. The smaller segment of the ellipse $3x^2+4y^2=1$ cut off by one latus rectum is rotated through four right angles about the other latus rectum. Find the volume of the annular solid so formed. [L.U.]

18. The radii of the upper and lower faces of the frustum of a right circular cone are 3 in. and 6 in. respectively and the altitude is 8 in. Find the position of the centroid.

20.14. Moments of inertia

The product of the mass of a particle and the square of its distance from a line or from a plane is called the *second moment* or *moment of inertia* (M.I.) of the particle with respect to the line or plane.

If $m_1, m_2, \ldots, m_n$ are the masses of a system of n particles situated at distances $r_1, r_2, \ldots, r_n$ respectively from a given straight line a, then the sum

$$I_a=m_1r_1^2+m_2r_2^2+\ldots+m_nr_n^2=\Sigma mr^2$$

is defined to be the moment of inertia of the system about the axis a. The sum Σmr^2 is also called the second moment of the system about the axis. The total mass of the system is M where

$$M=m_1+m_2+\ldots+m_n=\Sigma m,$$

and if

$$I_a=Mk^2$$

then k is called the *radius of gyration* of the system about the axis a.

20.15. Moments of inertia about perpendicular axes

Let Ox, Oy, Oz be three mutually perpendicular axes and let $m_1, m_2, \ldots, m_n$ be the masses of a system of n particles situated at $P_1\ (x_1, y_1, z_1)$, $P_2\ (x_2, y_2, z_2), \ldots, P_n\ (x_n, y_n, z_n)$ distant $r_1, r_2, \ldots, r_n$ respectively from O. Then if I_{Ox}, I_{Oy}, I_{Oz} denote the moments of inertia of the system about Ox, Oy, Oz respectively

$$I_{Ox}=\Sigma m(y^2+z^2),\ I_{Oy}=\Sigma m(z^2+x^2),\ I_{Oz}=\Sigma m(x^2+y^2)$$

$$\therefore\ I_{Ox}+I_{Oy}+I_{Oz}=2\Sigma m(x^2+y^2+z^2)=2\Sigma mr^2. \qquad (20.19)$$

When the system of particles lies, say, in the plane xOy, then

$$I_{Ox}=\Sigma my^2,\ I_{Oy}=\Sigma mx^2,\ I_{Oz}=\Sigma m(x^2+y^2)$$

$$\therefore\ I_{Oz}=I_{Ox}+I_{Oy}. \qquad (20.20)$$

Example 11

Find the M.I. of a thin uniform spherical shell of mass M and radius a about a diameter.

Let O be the centre of the shell and Ox, Oy, Oz be three mutually perpendicular radii. Then by symmetry

$$I_{Ox}=I_{Oy}=I_{Oz}$$

and, in the notation of (20.19), $\Sigma mr^2=\Sigma ma^2=Ma^2$, since every particle of the shell is distant a from O. Hence the required M.I. is $\frac{2}{3}Ma^2$.

Example 12

Find the M.I. about a diameter of a thin uniform wire of mass M in the form of a circle of radius a.

Let O be the centre of the circle, Ox and Oy any two radii at right angles, and Oz perpendicular to the plane of the wire.

Then $I_{Oz}=\Sigma ma^2=Ma^2$, since every particle of the ring is distant a from Oz.

But by symmetry $I_{Ox}=I_{Oy}$ and the ring lies in the plane xOy.

Hence by (20.20) $I_{Oz}=2I_{Ox}=2I_{Oy}$

$$\therefore\ I_{Ox}=\tfrac{1}{2}Ma^2.$$

20.16. The principle of parallel axes

An important relation exists between the moment of inertia of a system of particles about any axis and the moment of inertia of the system about a parallel axis through the centre of mass of the system.

Let G (fig. 83) be the centre of mass of the system and Oz be any given axis. Draw GZ parallel to Oz and let O be the foot of the perpendicular from G to Oz. Produce OG to y and draw GX and Ox perpendicular to plane $zOGZ$. Then Ox, Oy, Oz form a set of rectangular axes through O and GX, Gy, GZ are a parallel set through G.

If a typical particle of the system with mass m is situated at the point P whose coordinates are x, y, z referred to the axes through O and X, Y, Z referred to the axes through G, then

$$x = X, \quad y = Y + h, \quad z = Z,$$

where $h = OG$, the perpendicular distance between the parallels Oz and GZ. Hence if I_{Oz} is the M.I. of the system about Oz

$$\begin{aligned} I_{Oz} &= \Sigma m(x^2 + y^2) \\ &= \Sigma m(X^2 + Y^2 + 2hY + h^2) \\ &= \Sigma m(X^2 + Y^2) + 2h\Sigma mY + h^2\Sigma m. \end{aligned}$$

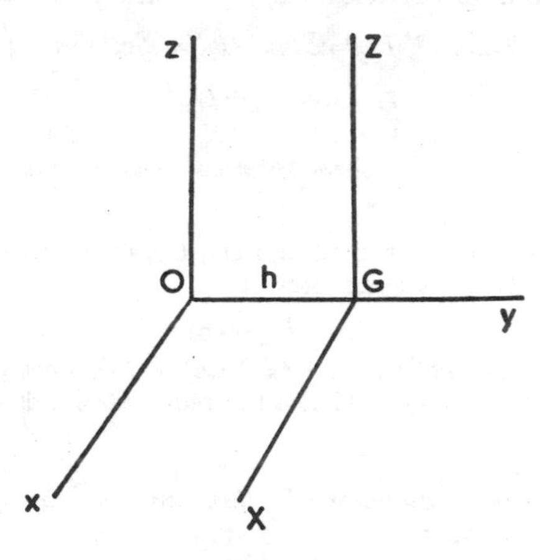

Fig. 83

But $\Sigma m(X^2 + Y^2) = I_{GZ}$ and $\Sigma mh^2 = h^2\Sigma m = Mh^2$, where M is the total mass of the system. Also (see § 20.11) $\bar{Y}$, the distance of G from the plane XGZ is given by $\dfrac{\Sigma mY}{\Sigma m}$, and $\bar{Y} = 0$; hence $\Sigma mY = 0$.

$$\therefore I_{Oz} = I_{GZ} + Mh^2,$$

i.e. the M.I. of a system of particles about any axis is equal to the M.I. of the system about a parallel axis through its centre of mass, together with the product of the total mass of the system and the square of the distance between the two axes.

20.17. Calculation of moments of inertia

In practice we frequently have to deal with solid bodies rather than with a system of separate particles. We replace the above system of particles by the elements of the body and carry out the summations involved by integration as in the following examples.

20.18. A thin uniform rod

To find the M.I. of a uniform rod of mass M and length l about an axis through one end perpendicular to the rod, we suppose the rod lies along Ox with one end at O and we take Oy as the axis. We assume that the rod is so thin that it may be regarded as made up of particles uniformly distributed along Ox and that the line-density of the rod is λ. Then the mass of an element whose ends are x and $(x+\delta x)$ from O is $\lambda\delta x$ and the moment of inertia $\delta I(x)$ of this element about Oy satisfies the relation

$$x^2\lambda\,\delta x < \delta I(x) < (x+\delta x)^2\lambda\,\delta x,$$

which leads to $\dfrac{dI(x)}{dx} = x^2\lambda$ and so to $I(x) = \int_0^x \lambda x^2\,dx.$

Hence the total moment of inertia I is given by

$$I = \int_0^l \lambda x^2\,dx = \tfrac{1}{3}\lambda l^3.$$

But $M=\lambda l$ $\qquad \therefore\ I=\frac{1}{3}Ml^2$ (20.21)

From the relation $Mk^2=I$ we see that the radius of gyration k is given by $k^2=\frac{1}{3}l^2$.

Treating the problem in the same way or using the principle of parallel axes, we may show that the M.I. of the same rod about an axis which bisects it at right angles, i.e. a parallel axis through G, the centre of mass, is $\frac{1}{12}Ml^2$.

20.19. A rectangular lamina

$ABCD$ (fig. 84) is a uniform rectangular lamina of mass M with $AB=l$ and $BC=b$. By drawing straight lines parallel to AB and sufficiently close together we may divide the lamina into strips so narrow that each may be regarded as a thin rod. If m is the mass of any one of these strips, its M.I. about the side AD is $\frac{1}{3}ml^2$ and, by addition, it follows for the whole rectangular lamina that $\quad I_{AD}=\frac{1}{3}Ml^2$

and, similarly, $\quad I_{AB}=\frac{1}{3}Mb^2$.

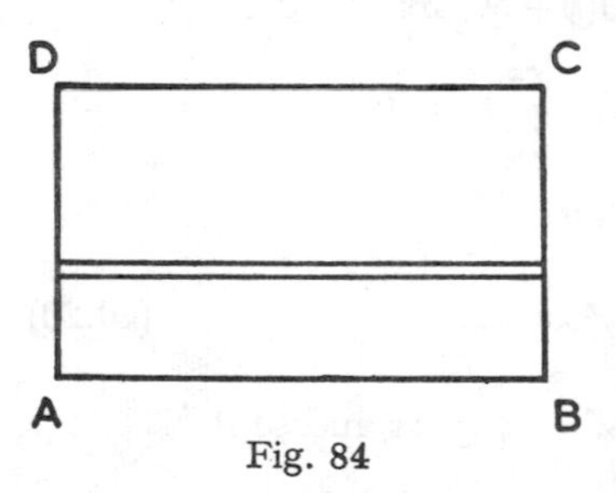

Fig. 84

By similar considerations, or by using the principle of parallel axes we may show that the M.I. of the rectangle about an axis bisecting AB at right angles (i.e. an axis parallel to AD through the centroid G) is $\frac{1}{12}Ml^2$.

By (20.20) it follows that the M.I. of the rectangle about an axis through any vertex perpendicular to its plane is $\frac{1}{3}M(l^2+b^2)$.

20.20. A right rectangular prism

If the prism has dimensions l, b and h, we may divide the solid into thin rectangular plates of mass m parallel to the faces of area lb. The M.I. of each plate about an edge of the prism perpendicular to the plate is $\frac{1}{3}m(l^2+b^2)$ and by addition we obtain for the whole solid $\frac{1}{3}M(l^2+b^2)$.

Similarly the M.I. of the prism about an edge perpendicular to the faces of area bh is $\frac{1}{3}M(b^2+h^2)$ and about an edge perpendicular to the faces of area lh the M.I. is $\frac{1}{3}M(l^2+h^2)$.

20.21. Plane area

Let us consider the uniform lamina $AUVB$ of fig. 77. The M.I. $\delta I_y(x)$ about Oy of the typical element $PNMQ$ satisfies the relation

$$x^2(\sigma y\,\delta x) < \delta I_y(x) < (x+\delta x)^2\sigma(y+\delta y)\,\delta x$$

where σ is the surface density of the lamina.

Hence $$\frac{dI_y(x)}{dx}=x^2\sigma y, \quad I_y(x)=\int_a^x x^2\sigma y\,dx$$

and so the total M.I. about Oy is given by

$$I_y=\sigma\int_a^b x^2y\,dx. \tag{20.22}$$

By (20.21) the M.I. about Ox of rectangle $PNMR$ is $\frac{1}{3}\{\sigma y\delta x\}y^2$ and the M.I. about Ox of the rectangle $SNMQ$ is $\frac{1}{3}\{\sigma(y+\delta y)\delta x\}(y+\delta y)^2$. Hence if $\delta I_x(x)$ is the M.I. about Ox of the element $PNMQ$

$$\tfrac{1}{3}\sigma y^3\delta x < \delta I_x(x) < \tfrac{1}{3}\sigma(y+\delta y)^3\delta x$$

$$\therefore\ \frac{dI_x(x)}{dx}=\tfrac{1}{3}\sigma y^3, \quad I_x(x)=\int_a^x \tfrac{1}{3}\sigma y^3\,dx,$$

and so the total M.I. about Ox is given by

$$I_x=\tfrac{1}{3}\sigma\int_a^b y^3\,dx. \tag{20.23}$$

The mass M of the lamina is given by $M=\sigma\int_a^b y\,dx$ and so if k_x, k_y are the radii of gyration of the lamina about Ox, Oy respectively, we have from (20.22) and (20.23)

$$Ak_x^2=\tfrac{1}{3}\int_a^b y^3\,dx\,; \quad Ak_y^2=\int_a^b x^2y\,dx \tag{20.24}$$

where A is the area of the lamina.

20.22. Circular lamina

To find the M.I. of a uniform circular disc of surface-density σ and radius a about an axis Oz through the centre O perpendicular to the plane of the disc, we consider the ring element bounded by circles of radii r and $r+\delta r$ shaded in fig. 85. The mass of the ring is approximately $2\pi\sigma r\,\delta r$ and all points within the ring lie at distances from the axis Oz of between r and $r+\delta r$. Hence if δI_r is the M.I. of the elementary ring

$$r^2(2\pi\sigma r\,\delta r) < \delta I_r < (r+\delta r)^2(2\pi\sigma r\,\delta r).$$

Hence $$\frac{dI_r}{dr} = 2\pi\sigma r^3, \quad I_r = \int_0^r 2\pi\sigma r^3\,dr,$$

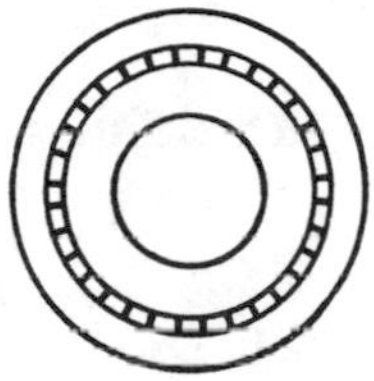

and so if a is the radius of the lamina and I is its M.I. about Oz

$$I = 2\pi\sigma\int_0^a r^3\,dr = \tfrac{1}{2}\pi\sigma a^4.$$

Fig. 85

If M is the total mass of the lamina, $M = \pi\sigma a^2$ and so

$$I = \tfrac{1}{2}Ma^2. \qquad (20.25)$$

If Ox and Oy are perpendicular radii of the lamina, by symmetry

$$I_{Ox} = I_{Oy} \text{ and } I_{Oz} = I_{Ox} + I_{Oy} \text{ by } (20.20)$$

hence the M.I. of the lamina about any diameter is given by

$$I = \tfrac{1}{4}Ma^2. \qquad (20.26)$$

The M.I. of a solid cylinder of mass M and radius a about its axis is found by dividing the cylinder into thin circular laminae by planes perpendicular to the axis. Then using (20.25) we find by addition that the M.I. is $\tfrac{1}{2}Ma^2$. If the cylinder is hollow, by using the result of Example 12, we find that the M.I. of the cylinder about its axis is Ma^2.

20.23. Solid of revolution

Let us consider the solid obtained by rotating through one complete revolution about Ox the area $AUVB$ of fig. 77. The masses of the discs generated by rotating the rectangles $PNMR$ and $SNMQ$ about Ox are $\rho\pi y^2\,\delta x$ and $\rho\pi(y+\delta y)^2\,\delta x$, ρ being the volume-density of the solid. By (20.25) $\delta I_x(x)$, the M.I. about Ox of the typical solid generated by $PNMQ$ satisfies the relation

$$\tfrac{1}{2}y^2(\rho\pi y^2\,\delta x) < \delta I_x(x) < \tfrac{1}{2}(y+\delta y)^2\{\rho\pi(y+\delta y)^2\,\delta x\}.$$

Hence $$\frac{dI_x(x)}{dx} = \tfrac{1}{2}\rho\pi y^4, \quad I_x(x) = \tfrac{1}{2}\rho\pi\int_a^x y^4\,dx$$

and so I_x, the M.I. of the solid about Ox, is given by

$$I_x = \tfrac{1}{2}\rho\pi\int_a^b y^4\,dx. \qquad (20.27)$$

In the same way, using (20.26), we may show that δI, the M.I. of the typical solid about a diameter parallel to Oy satisfies the relation

$$\tfrac{1}{4}y^2 . \rho\pi y^2 \delta x < \delta I < \tfrac{1}{4}(y+\delta y)^2 . \rho\pi(y+\delta y)^2 \delta x$$

and hence, by the principle of parallel axes $\delta I_y(x)$, the M.I. of the same element about Oy satisfies the relation

$$(\tfrac{1}{4}y^2+x^2) . \rho\pi y^2 \delta x < \delta I_y(x) < \{\tfrac{1}{4}(y+\delta y)^2+(x+\delta x)^2\} . \rho\pi(y+\delta y)^2 \delta x$$

and this leads to $\dfrac{dI_y(x)}{dx} = \rho\pi y^2(\tfrac{1}{4}y^2+x^2)$, $I_y(x) = \rho\pi \displaystyle\int_a^x y^2(\tfrac{1}{4}y^2+x^2)\,dx.$

Hence I_y, the M.I. of the solid about Oy, is given by

$$I_y = \rho\pi \int_a^b y^2(\tfrac{1}{4}y^2+x^2)\,dx.$$

20.24. Solid sphere

By (20.27) the M.I. about Ox of the solid sphere obtained by rotating through one complete revolution about Ox the circle $x^2+y^2=a^2$ is given by

$$\begin{aligned} I &= \tfrac{1}{2}\rho\pi \int_{-a}^{a} (a^2-x^2)^2\,dx \\ &= \rho\pi \int_0^a (a^4-2a^2x^2+x^4)\,dx. \\ &= \tfrac{8}{15}\rho\pi a^5 \\ &= \tfrac{2}{5}Ma^2 \end{aligned}$$

where $M = \tfrac{4}{3}\rho\pi a^3$ is the mass of the sphere.

Example 13

Show that the centroid of the area bounded by the x-axis and the arc of the curve $y=a \sin x$ between the points (0, 0), (π, 0) is the point $\left(\dfrac{\pi}{2}, \dfrac{\pi a}{8}\right)$. Show also that the radius of gyration of this area about the x-axis is $(a\sqrt{2})/3$.

[L.U.]

The area A bounded by the curve $y=a \sin x$ between the given points is

$$A = \int_0^\pi a \sin x\,dx = -[a \cos x]_0^\pi = 2a$$

and if $(\bar{x}, \bar{y})$ is the centroid of this area, by symmetry $\bar{x} = \tfrac{1}{2}\pi$ and

$$A\bar{y} = \tfrac{1}{2}\int_0^\pi y^2\,dx = \tfrac{1}{2}a^2 \int_0^\pi \sin^2 x\,dx = \tfrac{1}{4}\pi a^2.$$

$$\therefore \bar{y} = \tfrac{1}{8}\pi a.$$

By (20.24) k_x, the radius of gyration of the area about Ox, is given by

$$Ak_x^2=\tfrac{1}{3}\int_0^{\pi} y^3\,dx=\tfrac{1}{3}a^3\int_0^{\pi}\sin^3 x\,dx$$

$$=\tfrac{1}{3}a^3\int_0^{\pi}(1-\cos^2 x)\sin x\,dx$$

$$=4a^3/9.$$

$$\therefore\ k_x^2=2a^2/9,\ k_x=(a\sqrt{2})/3.$$

Example 14

Find the moments of inertia of an elliptic lamina about its principal axes and also about the axis through the centre perpendicular to the plane of the lamina. [L.U.]

Suppose that the equation of the ellipse bounding the lamina is

$$\frac{x^2}{a^2}+\frac{y^2}{b^2}=1.$$

Then the principal axes are the axes of coordinates and by (20.23), taking the symmetry of the figure into account, we have

$$I_x=\tfrac{2}{3}\sigma\int_{-a}^{a} y^3\,dx=\tfrac{4}{3}\sigma\int_0^{a} y^3\,dx.$$

The substitutions $x=a\cos\theta$, $y=b\sin\theta$ give

$$I_x=\tfrac{4}{3}ab^3\sigma\int_0^{\pi/2}\sin^4\theta\,d\theta=\tfrac{1}{4}\pi ab^3\sigma \text{ by (10.4), page 228.}$$

The mass M of the lamina is given by $4\sigma\int_0^a y\,dx$ and the same substitutions give

$$M=4ab\sigma\int_0^{\pi/2}\sin^2\theta\,d\theta=\pi ab\sigma$$

$$\therefore\ I_x=\tfrac{1}{4}Mb^2.$$

By symmetry $I_y=\tfrac{1}{4}Ma^2$ and by (20.20) the moment of inertia of the lamina about the axis through the centre perpendicular to the plane of the lamina is $\tfrac{1}{4}M(a^2+b^2)$.

Exercises 20 (*d*)

For each of the areas described in Nos. 1-5 find the radius of gyration (a) about Ox; (b) about Oy:

1. The area under the curve $y=\sin x$ from $x=0$ to $\tfrac{1}{2}\pi$.

2. The area under the curve $y=x^2$ from $x=0$ to $x=1$.

3. The area under the curve $y=e^x$ from $x=0$ to $x=1$.

4. The area under the curve $xy=4$ from $x=2$ to $x=4$.

5. The area in the first quadrant bounded by the parabola $y^2=4x$, the x-axis and the line $x=4$.

For each of the volumes described in Nos. 6-8 find the radii of gyration (a) about Ox; and (b) about Oy:

6. The volume generated by revolving about Ox the area under the curve $y=e^x$ from $x=0$ to $x=1$.

7. The volume generated by revolving about Ox the area under the curve $xy=4$ from $x=1$ to $x=2$.

8. The volume generated by revolving about Ox the area under the curve $y=\sin x$ from $x=0$ to $x=\pi$.

CHAPTER 21

APPLICATIONS OF INTEGRATION—POLAR COORDINATES

21.1. Polar coordinates

In Chapter 15 we defined the polar coordinates of a point in a plane and showed that if the cartesian origin and x-axis are taken respectively as the pole and initial line of polar coordinates, the two systems of coordinates are connected by the relations

$$x=r\cos\theta,\ y=r\sin\theta \tag{21.1}$$

and $$r^2=x^2+y^2,\ \tan\theta=y/x.\quad \text{(See § 15.2.)}$$

21.2. Length of arc of a polar curve

From (21.1) $$dx=-r\sin\theta\, d\theta+\cos\theta\, dr,$$
$$dy=r\cos\theta\, d\theta+\sin\theta\, dr.$$

Hence $(dx)^2+(dy)^2=r^2(d\theta)^2+(dr)^2$

i.e. $$(ds)^2=r^2(d\theta)^2+(dr)^2 \quad \text{by (20.13), page 455,}$$

or $$\left(\frac{ds}{d\theta}\right)^2=r^2+\left(\frac{dr}{d\theta}\right)^2. \tag{21.2}$$

Hence the length of the arc AB of a polar curve is given by

$$s=\int_{\theta_A}^{\theta_B}\sqrt{\left\{r^2+\left(\frac{dr}{d\theta}\right)^2\right\}}d\theta,\quad \theta_B>\theta_A$$

where s is measured to increase with θ.

21.3. Tangents in polar coordinates

The positive tangent at any point $P(r, \theta)$ on the curve $r=f(\theta)$ is drawn in the direction of s increasing, and since s is conventionally measured so as to increase with θ, the positive tangent is drawn in the direction of θ increasing. The positive normal at P makes an angle $+\frac{1}{2}\pi$ with the positive tangent at P. In fig. 86 $\overline{TP}$ and $\overline{PN}$ are the positive tangent and normal respectively.

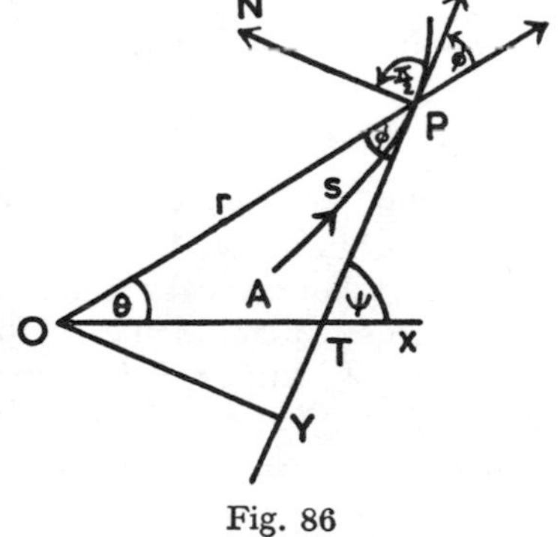

Fig. 86

The angle ψ between the positive tangent at P and the initial line is given by

$$\psi=\theta+\phi \tag{21.3}$$

where ϕ is the angle measured counter-clockwise from the radius vector OP to the positive tangent TP at P (fig. 86).

Here ψ is not restricted to lie between $\pm\frac{1}{2}\pi$ as on page 189, but it is not difficult to see that (20.12), page 455, remains valid.

From (21.3) we may deduce formulae for $\sin\phi$, $\cos\phi$ and $\tan\phi$.

We have
$$\cos\phi=\cos(\psi-\theta)$$
$$=\cos\psi\cos\theta+\sin\psi\sin\theta$$
$$=\frac{dx}{ds}\cdot\frac{x}{r}+\frac{dy}{ds}\cdot\frac{y}{r} \quad \text{by (20.12) and (21.1).}$$

Hence
$$\cos\phi=\frac{x\,dx+y\,dy}{r\,ds}$$

and, similarly,
$$\sin\phi=\frac{x\,dy-y\,dx}{r\,ds}.$$

But $x^2+y^2=r^2$ so that $x\,dx+y\,dy=r\,dr$

$$\therefore \cos\phi=\frac{dr}{ds} \tag{21.4}$$

Also $\frac{y}{x}=\tan\theta$, so that $\frac{x\,dy-y\,dx}{x^2}=\sec^2\theta\,d\theta$

$$\text{i.e. } x\,dy-y\,dx=r^2\,d\theta. \tag{21.5}$$

$$\therefore \sin\phi=r\frac{d\theta}{ds} \tag{21.6}$$

From (21.4) and (21.6)
$$\tan\phi=r\frac{d\theta}{dr}. \tag{21.7}$$

This result is proved independently in § 21.7, Example 4.

21.4. The perpendicular from the pole to a tangent

If p is the length of the perpendicular OY drawn from O to the tangent at P (fig. 86),

$$p=r\sin\phi \tag{21.8}$$

$$\therefore \frac{1}{p^2}=\frac{1}{r^2}\operatorname{cosec}^2\phi=\frac{1}{r^2}(1+\cot^2\phi) \tag{21.9}$$

and so, by (21.7)
$$\frac{1}{p^2}=\frac{1}{r^2}+\frac{1}{r^4}\left(\frac{dr}{d\theta}\right)^2 \tag{21.10}$$

If we eliminate θ between (21.10) and $r=f(\theta)$, the equation of the given curve, we obtain a relation between p and r known as the p, r or *pedal* equation of the curve. In practice, when a simple relation exists between θ and ϕ, the p, r equation is established by using (21.8). In other cases we use (21.9), since $\cot\phi$ is immediately obtained in terms of θ by differentiating logarithmically with respect to θ the equation $r=f(\theta)$.

(21.8) may be written in the form $p=r^2\,d\theta/ds$ using (21.6), and so with the usual convention as to the sign of s, p is always positive.

By writing $u=1/r$ we obtain (21.10) in the form

$$\frac{1}{p^2}=u^2+\left(\frac{du}{d\theta}\right)^2.$$

21.5. Curve sketching from polar equations

In general, it is useful to tabulate some values of r and θ when a curve is given by an equation of the form $r=f(\theta)$, but much labour may be saved by applying the following considerations:

I. Symmetry

If the equation of the curve is unaltered when θ is replaced by $(-\theta)$, the curve is symmetrical about the line $\theta=0$. In particular, if r is a function of $\cos\theta$ alone, the curve is symmetrical about the initial line.

If the value of r is altered in sign but not in magnitude when θ is replaced by $(-\theta)$ in the equation of the curve, the curve is symmetrical about the line $\theta=\frac{1}{2}\pi$. For example, the curve $r=a\tanh\theta$ is symmetrical about $\theta=\frac{1}{2}\pi$. Again, if the equation of the curve is unaltered when $(\pi-\theta)$ is substituted for θ, the curve is symmetrical about $\theta=\frac{1}{2}\pi$. In particular, if r is a function of $\sin\theta$ only, the curve is symmetrical about $\theta=\frac{1}{2}\pi$.

If only even powers of r occur in its equation, the curve is symmetrical about the pole.

II. Form of the curve at the pole

In general, if the curve $r=f(\theta)$ passes through the pole, the directions of tangents to the curve at the pole are found by solving the equation $f(\theta)=0$, since as $r\to 0$ the curve approaches the pole.

III. Limitations on the value of r and θ

These are readily seen in an equation of the form $r=f(\theta)$. For example, the curve $r=2+\sin\theta$ lies entirely within the concentric circles $r=1$ and $r=3$, while for the curve $r^2=a^2\cos 2\theta$, $r^2\leqslant a^2$ so that the curve lies wholly within the circle $r=a$; also when $\cos 2\theta<0$, i.e. when $\frac{1}{4}\pi<\theta<\frac{3}{4}\pi$ and when $\frac{5}{4}\pi<\theta<\frac{7}{4}\pi$ there is no real value of r.

IV Direction of the tangent

The relation $\tan\phi=r\dfrac{d\theta}{dr}$ determines the angle between the radius vector and the tangent to the curve at any point.

V. Asymptotic circles

In some curves, r approaches a limit as θ tends to infinity; for example, the curve $r=a\tanh\theta$ tends to the circle $r=a$ as θ tends to infinity and is said to have an asymptotic circle $r=a$.

Linear asymptotes to a polar curve are in general difficult to deal with, and the problem of finding them is not considered here.

21.6. Some well-known polar curves

The curves sketched on page 481 are standard polar curves.

I. The limaçon and cardioid

By drawing the circle $r=a\cos\theta$ and extending the radius vector corresponding to each value of θ by an amount b we construct the curve (known as the limaçon)

$$r=b+a\cos\theta \quad . \quad . \quad . \quad . \quad \text{(i)}$$

In this formula a and b can be either positive or negative. Fig. 87 (*a*) shows the limaçon when $a>b>0$; fig. 87 (*b*) shows the limaçon when $0<a<b$. Curves of this type are also obtained from an equation of the form $r=b+a\sin\theta$.

If in (i) we write $b=a$, we obtain the equation $r=a(1+\cos\theta)$. This curve, known as the cardioid, is shown in fig. 88.

The equation of a cardioid may appear in the forms

$$r=a(1-\cos\theta)\ ;\ r=a(1+\sin\theta) \text{ and } r=a(1-\sin\theta)$$

and the corresponding graphs are obtained by rotating fig. 88 in its own plane about the pole through π, $+\frac{1}{2}\pi$ and $-\frac{1}{2}\pi$ radians respectively.

II. The lemniscate $r^2=a^2\cos 2\theta$

Since $\cos 2\theta$ may be expressed as a function of $\cos\theta$ only, or as a function of $\sin\theta$ only, this curve is symmetrical about the lines $\theta=0$, $\theta=\frac{1}{2}\pi$ and we need consider only values of θ between 0 and $\frac{1}{2}\pi$. As θ increases from 0 to $\frac{1}{4}\pi$, r decreases from a to 0, and the line $\theta=\frac{1}{4}\pi$ is a tangent to the curve at the pole. Between $\theta=\frac{1}{4}\pi$ and $\theta=\frac{1}{2}\pi$ where $\cos 2\theta$ (and hence r^2) is negative, the curve does not exist. At the point $(a, 0)$, $\phi=\frac{1}{2}\pi$. The curve is shown in fig. 89. The lemniscate $r^2=a^2\sin 2\theta$ is obtained by rotating fig. 89 in its own plane through $\frac{1}{2}\pi$ radians about O.

III. The n-leaved rose

The curves $r=a\sin n\theta$, $r=a\cos n\theta$ consist of n leaves if n is odd and $2n$ leaves if n is even. The leaves are equal in size and are spaced at equal intervals round the pole.

Fig. 90 (*a*) shows the curve $r=a\sin 3\theta$; fig. 90 (*b*) shows the curve $r=a\cos 2\theta$.

IV. Spirals

These are curves in which, as θ increases without limit, r either steadily increases or steadily decreases so that the curves wind round and round the pole.

The Archimedean spiral $r=a\theta$ $(a>0)$ is shown in fig. 91 (*a*).

The logarithmic or equiangular spiral $r=e^{a\theta}$ $(a>0)$ is shown in fig. 91 (*b*).

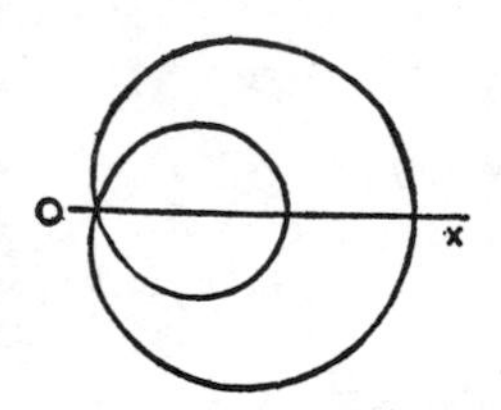

Fig. 87 (*a*)

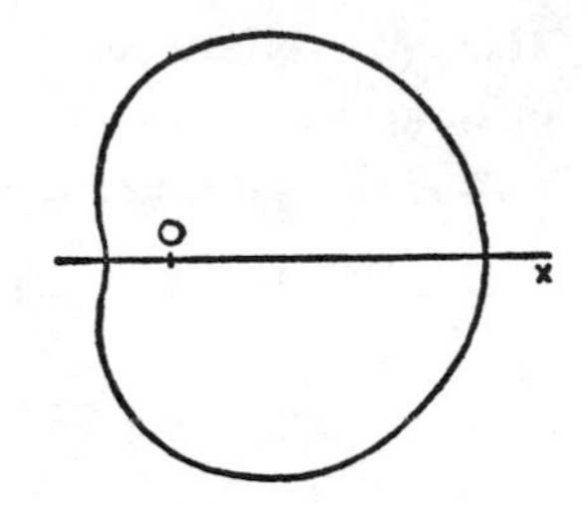

Fig. 87 (*b*)

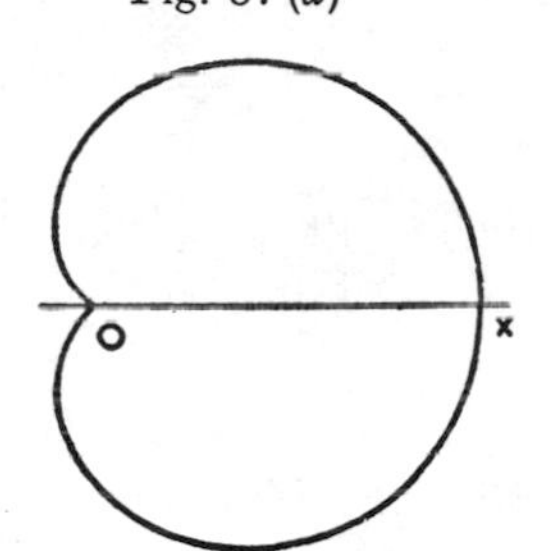

Fig. 88

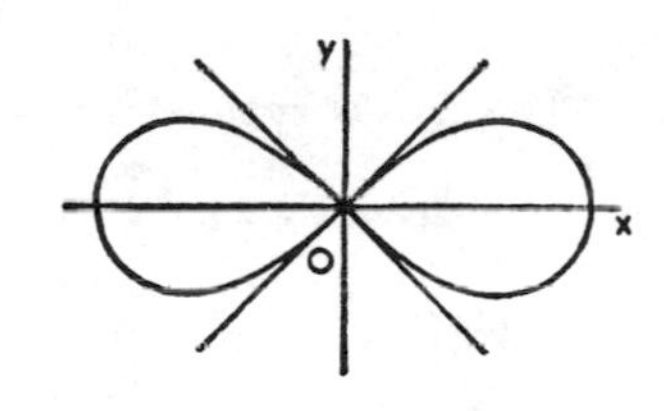

Fig. 89

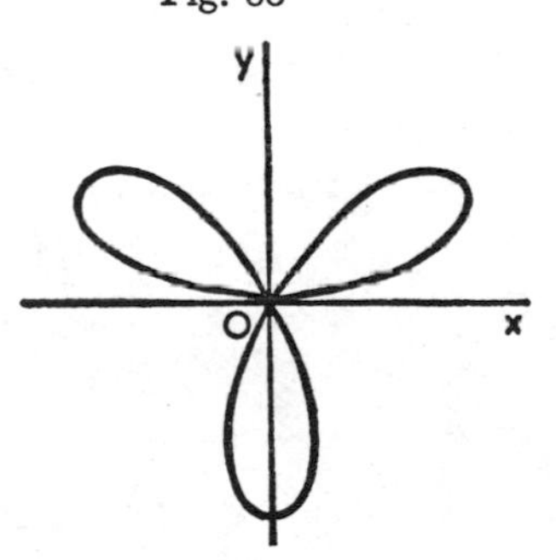

Fig. 90 (*a*)

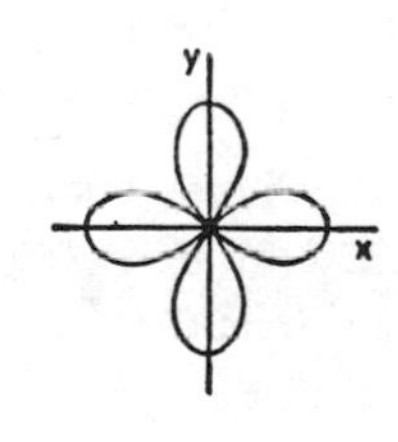

Fig. 90 (*b*)

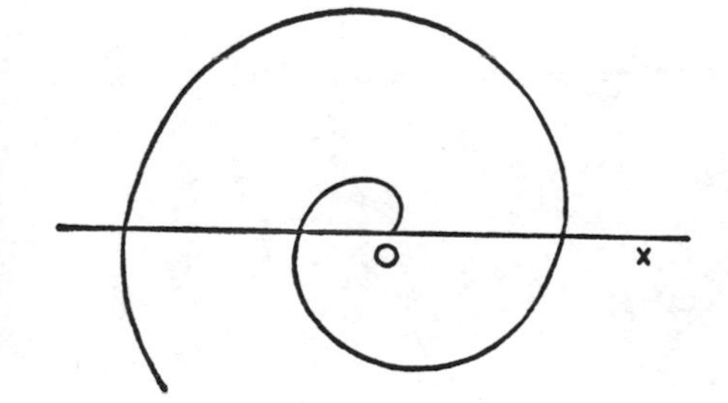

Fig. 91 (*a*)

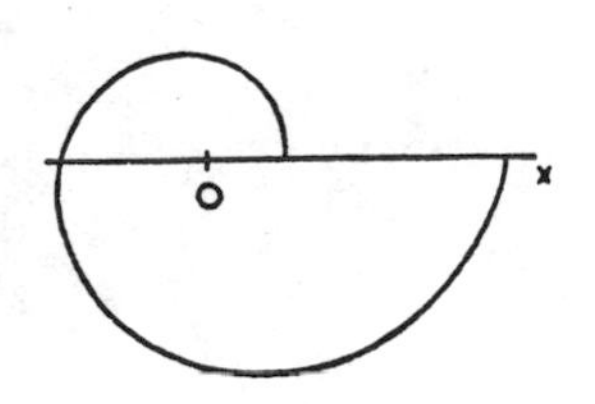

Fig. 91 (*b*)

21.7. Miscellaneous examples

Example 1

Find the total length of arc of the cardioid $r=a(1+\cos\theta)$.

$$r=a(1+\cos\theta),\quad \frac{dr}{d\theta}=-a\sin\theta,$$

$$\therefore\quad \left(\frac{ds}{d\theta}\right)^2=r^2+\left(\frac{dr}{d\theta}\right)^2=a^2(2+2\cos\theta)=4a^2\cos^2\tfrac{1}{2}\theta.$$

Hence, with the usual convention as to the sign of s,

$$\frac{ds}{d\theta}=2a\cos\tfrac{1}{2}\theta.\quad (0\leqslant\theta\leqslant\pi).$$

The curve is symmetrical about the initial line and so its total length of arc is given by $2\int_0^\pi 2a\cos\frac{1}{2}\theta\,d\theta=8a$.

Example 2

Show that, for the cardioid $r=a(1+\cos\theta)$, $\phi=\frac{1}{2}(\theta\pm\pi)$ and deduce the p, r equation of the curve.

If
$$r=a(1+\cos\theta)\quad\ldots\ldots\quad\text{(i)}$$
$$\log r=\log a+\log(1+\cos\theta)$$
and
$$\frac{1}{r}\frac{dr}{d\theta}=\frac{-\sin\theta}{1+\cos\theta}$$
i.e.
$$\cot\phi=-\tan\tfrac{1}{2}\theta\quad\text{by (21.7)}$$
and we may take
$$\phi=\tfrac{1}{2}(\theta\pm\pi).$$
By (21.8)
$$p=r\sin\phi=\pm r\cos\tfrac{1}{2}\theta$$
and, from (i),
$$r=2a\cos^2\tfrac{1}{2}\theta.$$
Hence
$$r^3=2ap^2.$$

Note: Since $p>0$ we must take $\phi=\frac{1}{2}(\theta+\pi)$ when $0\leqslant\theta\leqslant\pi$, and $\phi=\frac{1}{2}(\theta-\pi)$ when $\pi\leqslant\theta\leqslant 2\pi$.

Example 3

Find the p, r equation of the spiral $r=a\theta$ where a is constant.

If
$$r=a\theta\quad\ldots\ldots\quad\text{(i)}$$
$$\log r=\log a+\log\theta,$$
and by (21.7)
$$\cot\phi=1/\theta=a/r\text{ by (i).}$$
But
$$\frac{1}{p^2}=\frac{1}{r^2}(1+\cot^2\phi)=\frac{1}{r^2}\left(1+\frac{a^2}{r^2}\right)$$
i.e.
$$p^2=\frac{r^4}{a^2+r^2}.$$

Example 4

$P(r, \theta)$ and $Q(r+\delta r, \theta+\delta\theta)$ are two points on a curve given in polar coordinates by the equation $r=f(\theta)$. If PN is perpendicular to OQ, where O is the pole (origin) of coordinates, prove that $\cot PQN=\tan \frac{1}{2}\delta\theta+\delta r/(r \sin \delta\theta)$, and deduce that, if ϕ is the angle between the radius vector to P and the tangent at P, $\tan \phi=r d\theta/dr$.

If a line through O meets again at R and S the curve given by $r=a(1+\cos \theta)$, prove that RS is constant and that the tangents at R and S meet at right angles. [L.U.]

In fig. 92, $OP=r$, $OQ=r+\delta r$ and $\angle POQ=\delta\theta$. PT is the tangent at P and $\angle OPT=\phi$

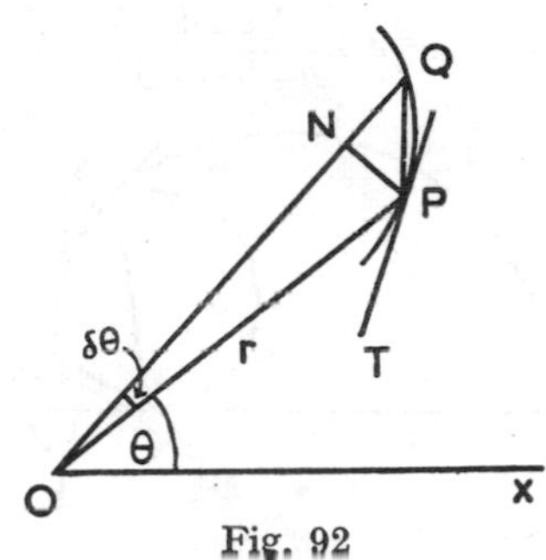

Fig. 92

$$\cot PQN=\frac{NQ}{PN}$$

$$=\frac{(r+\delta r)-r \cos \delta\theta}{r \sin \delta\theta}$$

$$=\frac{(1-\cos \delta\theta)}{\sin \delta\theta}+\frac{\delta r}{r \sin \delta\theta}$$

$$=\tan \tfrac{1}{2}\delta\theta+\frac{1}{r}\frac{\delta r}{\delta\theta}\left(\frac{\delta\theta}{\sin \delta\theta}\right).$$

As $\delta\theta \to 0$, Q moves along the curve towards P, the chord PQ approaches its limiting position PT and $\angle PQN \to \phi$,

Hence $$\tan \phi=r\frac{d\theta}{dr} \text{ since } \lim_{\delta\theta\to 0}\frac{\delta\theta}{\sin \delta\theta}=1.$$

In Example 2 we showed that for the cardioid $r=a(1+\cos \theta)$, $\phi=\frac{1}{2}(\theta \pm \pi)$; and so by (21.3),

$$\psi=\tfrac{3}{2}\theta \pm \tfrac{1}{2}\pi \quad . \quad . \quad . \quad . \quad . \qquad \text{(i)}$$

If $\theta=\alpha$ $(0 \leqslant \alpha \leqslant \pi)$ at R and $\theta=\alpha+\pi$ at S, $OR=a(1+\cos \alpha)$ and $OS=a(1-\cos \alpha)$. Hence $RS=2a=$constant.

Also, from (i) the tangents at R and S make angles $\frac{3}{2}\alpha+\frac{1}{2}\pi$, $\frac{3}{2}\alpha+\pi$ respectively with Ox and so these tangents meet at right angles.

Example 5

The limaçon $r=a-b\cos\theta\,(a>b)$ is cut at the points A and B by a circle touching the line $\theta=0$ at the origin. Prove that the tangents to the limaçon at A and B intersect at a point on the circle. [L.U.]

Let the equation of the circle be $r=k\sin\theta$. It meets the limaçon at A and B where

$$k\sin\theta=a-b\cos\theta \quad . \quad . \quad . \quad . \quad \text{(i)}$$

On the limaçon $\quad r=a-b\cos\theta, \quad (a>b)$

$$\log r=\log(a-b\cos\theta)$$

and so, differentiating with respect to θ, we obtain

$$\cot\phi=\frac{b\sin\theta}{a-b\cos\theta}.$$

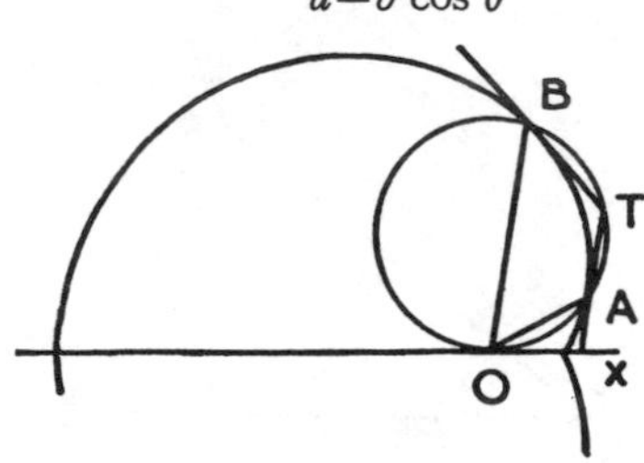

Fig. 93

Hence, by (i), at A and B

$$\cot\phi=b/k=\text{constant}$$

$$\therefore\ \phi_A=\phi_B=\alpha \text{ (say).}$$

If the tangents to the limaçon at A and B meet at T (fig. 93),

$$\angle OAT=\pi-\phi_A=\pi-\alpha \text{ and } \angle OBT=\phi_B=\alpha.$$

Hence quadrilateral $OATB$ is cyclic and the tangents TA, TB meet on circle OAB.

Exercises 21 (*a*)

1. Find the total length of arc of the cardioid $r=a(1-\cos\theta)$.

2. Find the length of arc of the curve $r=a\cos^3\frac{1}{3}\theta$ from $\theta=0$ to $\theta=3\pi$.

3. Find the length of arc of the equiangular spiral $r=ae^{\theta\cot\alpha}$ between the radii vectores whose lengths are r_1 and $r_2 (r_1<r_2)$, and show that in this spiral the inclination of the tangent to the radius vector is constant.

4. In the following curves find ϕ, the angle between the tangent and the radius vector, in terms of θ and deduce the p, r equation of each curve:

(i) $r=\frac{1}{2}a\sec^2\frac{1}{2}\theta$, (ii) $r^2\cos 2\theta=a^2$,
(iii) $r^2=a^2\sin 2\theta$, (iv) $r=a\sin^n(\theta/n)$,
(v) $r^n=a^n\sin n\theta$, (vi) $r^3=\cos 3\theta$.

5. Discuss the shape of the lemniscate $r^2=\cos 2\theta$, and determine the ranges of θ corresponding to the two loops of this curve.

If ϕ is the angle between the tangent and the radius vector, show that $\phi=2\theta+\frac{1}{2}\pi$, and hence, or otherwise, determine the greatest width of a loop, measured at right angles to the line $\theta=0$. [Durham.]

6. P is any point on the cardioid whose equation is $r=a(1-\cos\theta)$. Find two other points Q and R on the cardioid such that the tangents at P, Q and R are all parallel. Show that the sum of the ordinates of the points P, Q and R is zero. [L.U.]

7. P is the point (r, θ) on the cardioid whose equation is $r=a(1+\cos\theta)$ and O is the origin. Find (i) the length of the arc OP in terms of a and θ and (ii) the angle between OP and the tangent at P.

A circle is drawn touching the cardioid at P and passing through O. Show that, for varying positions of P on the cardioid, the locus of the centre is a circle. [L.U.]

8. Prove that any two straight lines drawn through the origin intercept arcs of the same length on the parabola $ay=x^2$ and the curve $r=a\sec^2\theta$.

Find this common length when the straight lines are the axis of x and the line $y=x$. [L.U.]

9. The polar equation of a plane curve is $r=f(\theta)$. OP is the radius vector to the point P on the curve, and PT, PN are respectively the tangent and normal to the curve at P. If a perpendicular to OP through O intersects PT at A and PN at B, prove that $\dfrac{OB}{OA}=\dfrac{1}{r^2}\left(\dfrac{dr}{d\theta}\right)^2$.

A curve is such that (i) $OB/OA=\theta^2$, (ii) r is infinite when θ is infinite, and (iii) $r=a$ when $\theta=0$. Find the equation of the curve and prove that the area of triangle APB is $\frac{1}{2}a^2e^{\theta^2}(\theta+1/\theta)$. [L.U.]

10. Show that the equation of the tangent to the cardioid $r=a(1+\cos\theta)$ at the point of vectorial angle α is $r\cos(\theta-\frac{3}{2}\alpha)=2a\cos^3\frac{1}{2}\alpha$.

If PQ is a chord of the cardioid which passes through O, the origin, prove that the tangents at P and Q are at right angles, and that if T is their point of intersection, then $OT^2=OP^2+OQ^2-OP.OQ$. [L.U.]

11. P and P' are points on the curve $r^2=a^2\cos 2\theta$ whose vectorial angles θ, θ' $(\theta'>\theta)$ are each positive and less than $\frac{1}{4}\pi$. If the tangents at P and P' intersect at right angles at T, show that $\theta'=\theta+\frac{1}{6}\pi$ and that $OT^2=\frac{3}{4}\sqrt{3}a^2\cos(2\theta+\frac{1}{6}\pi)$, where O is the origin. [L.U.]

12. The equation of a plane curve is $x=f(\theta)$, where x is a cartesian coordinate and (r, θ) are polar coordinates and $x=r\cos\theta$, $y=r\sin\theta$. Prove that the length of the arc of the curve included between the radii vectores $\theta=\theta_1$ and $\theta=\theta_2$ is

$$\int_{\theta_1}^{\theta_2}\sec\theta\sqrt{[\{f'(\theta)+f(\theta)\tan\theta\}^2+\{f(\theta)\}^2]}\,d\theta.$$

Verify the correctness of the formula by finding, in the following simple cases: (i) the length of the curve $x=a\cot\theta$ between $\theta=\theta_1$ and $\theta=\theta_2$; and (ii) the length of the closed curve $x=a\cos\theta$. [L.U.]

21.8. Areas in polar coordinates

In fig. 94, HK is an arc of the curve $r=f(\theta)$ and we shall suppose that r is a continuous function which increases steadily as θ increases from α at H to β at K.

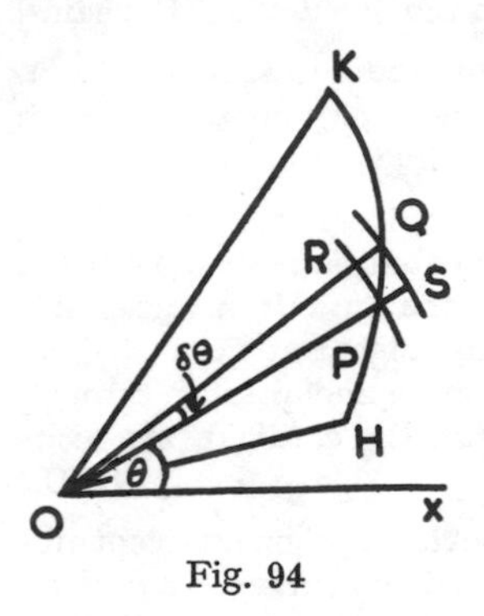

Fig. 94

$P(r, \theta)$ and $Q(r+\delta r, \theta+\delta\theta)$ are two points on the arc HK. With centre O and radius OP draw an arc of a circle to cut OQ in R; with centre O and radius OQ draw an arc of a circle to cut OP in S. Then δA, the area of the elementary sector OPQ, lies between the areas of the two circular sectors OPR and OQS, i.e.

$$\tfrac{1}{2}r^2\delta\theta < \delta A < \tfrac{1}{2}(r+\delta r)^2\delta\theta.$$

By dividing the sector OHK into sectors like OPQ and summing over them all we have

$$\sum_{\theta=\alpha}^{\theta=\beta} \tfrac{1}{2}r^2\,\delta\theta < \text{area of sector } OHK < \sum_{\theta=\alpha}^{\theta=\beta} \tfrac{1}{2}(r+\delta r)^2\delta\theta.$$

Hence (see note in § 20.1)

$$\text{area of sector } OHK = \lim_{\delta\theta\to 0}\sum_{\theta=\alpha}^{\theta=\beta} \tfrac{1}{2}r^2\,\delta\theta = \int_\alpha^\beta \tfrac{1}{2}r^2\,d\theta. \qquad (21.11)$$

By a similar argument we may establish this formula for an arc along which r decreases steadily as θ increases, and so (21.11) is valid for any arc which is divisible into a finite number of arcs for each of which r increases or decreases steadily as θ increases.

21.9. Areas of closed curves in polar coordinates and in parametric form

From (21.5) $\qquad r^2\,d\theta = x\,dy - y\,dx.$

Thus, if a curve is given parametrically by the equations

$$x=x(t),\ y=y(t) \quad . \quad . \quad . \quad . \quad \text{(i)}$$

and if t_1 and t_2 are the values of the parameter at H and K (fig. 94)

$$\text{area of sector } HOK = \int_\alpha^\beta \tfrac{1}{2}r^2\,d\theta$$

$$= \int_{t_1}^{t_2} \frac{1}{2}\left(x\frac{dy}{dt} - y\frac{dx}{dt}\right)dt. \qquad (21.12)$$

Formulae (21.11) and (21.12) may be used to find the area within a closed curve. There are three cases:

Case I

If the pole O lies within a closed polar curve as in fig. 95, the area enclosed is, by (21.11),

$$\int_0^{2\pi} \tfrac{1}{2}r^2\,d\theta.$$

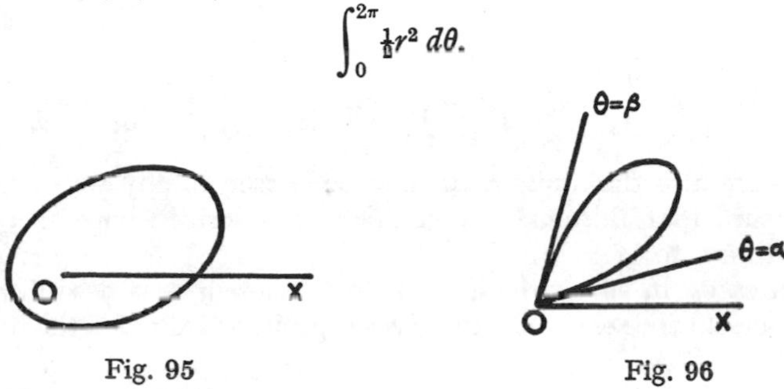

Fig. 95 Fig. 96

Case II

If the pole O lies on the curve as in fig. 96, and if α and β are the angles made with the initial line by the tangents to the curve at the pole, the area enclosed is

$$\int_\alpha^\beta \tfrac{1}{2}r^2\,d\theta.$$

Case III

If the pole O lies outside the curve as in fig. 97, let the tangents OA and OB make angles α and β respectively with the initial line and

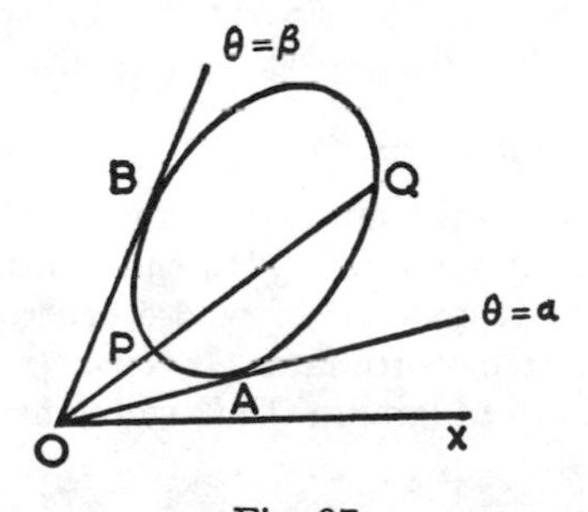

Fig. 97

let the radius vector OPQ be inclined at an angle θ to the initial line. Then if $OQ=r_1$ and $OP=r_2$ the area enclosed by the curve is by (21.11)

$$\int_\alpha^\beta \tfrac{1}{2}r_1^2\,d\theta-\int_\alpha^\beta \tfrac{1}{2}r_2^2\,d\theta=\int_\alpha^\beta \tfrac{1}{2}r_1^2\,d\theta+\int_\beta^\alpha \tfrac{1}{2}r_2^2\,d\theta.$$

In Cases I and II, if the curve is given parametrically by equations (i), formula (21.12) may be used to find the area enclosed by the curve if, as t takes values from t_1 to t_2, $t_1 \neq t_2$, the curve is described completely once.

In Case III we suppose that as the arc AQB is described t varies from t_1 to τ and as arc BPA is described t varies from τ to t_2. Then the enclosed area is

$$\int_{t_1}^{\tau} \tfrac{1}{2}r^2(t)\,\frac{d\theta}{dt}\,dt+\int_{\tau}^{t_2} \tfrac{1}{2}r^2(t)\,\frac{d\theta}{dt}\,dt=\int_{t_1}^{t_2} \tfrac{1}{2}r^2(t)\,\frac{d\theta}{dt}\,dt=\int_{t_1}^{t_2} \tfrac{1}{2}\left(x\frac{dy}{dt}-y\frac{dx}{dt}\right)dt.$$

In this case also the limits t_1, t_2 may be chosen in any way provided they ensure that the curve is completely described once as t takes values from t_1 to t_2.

In all cases, if, as t varies from t_1 to t_2, the curve is described in a counter-clockwise sense, (21.12) gives a positive value for the area.

21.10. Surface area in polar coordinates

By (20.11) the formula $S=2\pi\int y\,ds$ taken between suitable limits gives the surface area of the solid generated when the arc AB of the curve $y=f(x)$ revolves about Ox. By subctituting $y=r\sin\theta$, $ds=\sqrt{\left\{r^2+\left(\frac{dr}{d\theta}\right)^2\right\}}\,d\theta$ we obtain a formula applicable to the solid generated when the arc AB of the curve $r=f(\theta)$ revolves about the initial line :

$$S=2\pi\int_{\theta_A}^{\theta_B} r\sin\theta\sqrt{\left\{r^2+\left(\frac{dr}{d\theta}\right)^2\right\}}\,d\theta. \qquad (21.13)$$

21.11. Centroid of a plane area

To find the centroid of the uniform lamina OHK shown in fig. **94** we consider the elementary sector OPQ which is to the first order of small quantities a triangle of area $\frac{1}{2}r^2\,\delta\theta$ whose centroid (two-thirds along its median) has cartesian coordinates $(\frac{2}{3}r\cos\theta, \frac{2}{3}r\sin\theta)$ approximately. If $(\bar{x}, \bar{y})$ is the centroid of lamina OHK whose area is A,

$$A\bar{x}=\lim_{\delta\theta\to 0}\sum_{\theta=\alpha}^{\theta=\beta}(\tfrac{1}{2}r^2\,\delta\theta)(\tfrac{2}{3}r\cos\theta)=\tfrac{1}{3}\int_{\alpha}^{\beta} r^3\cos\theta\,d\theta$$

$$A\bar{y}=\lim_{\delta\theta\to 0}\sum_{\theta=\alpha}^{\theta=\beta}(\tfrac{1}{2}r^2\,\delta\theta)(\tfrac{2}{3}r\sin\theta)=\tfrac{1}{3}\int_{\alpha}^{\beta} r^3\sin\theta\,d\theta.$$

where by (21.11) $\qquad A=\tfrac{1}{2}\int_{\alpha}^{\beta} r^2\,d\theta.$

Example 6

Find the centroid of the area bounded by the cardioid $r=a(1-\cos\theta)$ and the lines $\theta=0$ and $\theta=\frac{1}{2}\pi$.

If A is the area enclosed by the curve and the given lines

$$A=\tfrac{1}{2}\int_0^{\pi/2} r^2\,d\theta=\tfrac{1}{2}\int_0^{\pi/2} a^2(1-\cos\theta)^2\,d\theta=\tfrac{1}{8}a^2(3\pi-8).$$

If the centroid is $(\bar{x}, \bar{y})$

$$A\bar{x}=\tfrac{1}{3}a^3\int_0^{\pi/2}(1-\cos\theta)^3\cos\theta\,d\theta$$

$$=\tfrac{1}{16}a^3(16-5\pi) \quad \text{by (10.4), page 228.}$$

$$\therefore\ \bar{x}=(16-5\pi)a/(6\pi-16).$$

$$A\bar{y}=\tfrac{1}{3}a^3\int_0^{\pi/2}(1-\cos\theta)^3\sin\theta\,d\theta$$

$$=\tfrac{1}{3}a^3\int_0^{1}u^3\,du$$

$$=\tfrac{1}{12}a^3.$$

$$\bar{y}=\tfrac{2}{3}a/(3\pi-8).$$

Example 7

Sketch the curve $r=a\sin^2\theta$, showing that it consists of two loops. Find the area enclosed by one loop of the curve. Show that the volume of the solid formed by rotating the upper loop through 2π about the tangent at the point $r=a$, $\theta=\frac{1}{2}\pi$ is

$$2\pi a^3\left(\frac{3\pi}{16}-\frac{32}{105}\right).$$

Find also the length of one loop of the curve. [L.U.]

The curve is symmetrical about the lines $\theta=0$ and $\theta=\frac{1}{2}\pi$ and so we need consider only values of θ between 0 and $\frac{1}{2}\pi$.

The line $\theta=0$ is a tangent at the origin, and as θ varies from 0 to $\frac{1}{2}\pi$, r increases from 0 to its maximum value a. The form of the curve is shown in fig. 98.

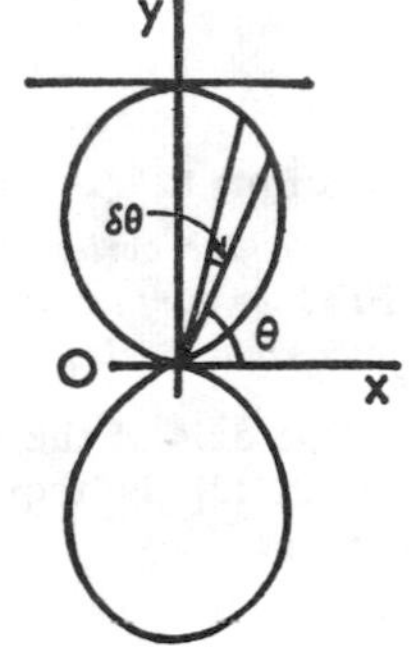

Fig. 98

The area of the upper loop

$$=\tfrac{1}{2}\int_0^{\pi} r^2\,d\theta$$

$$=\int_0^{\pi/2} a^2\sin^4\theta\,d\theta \quad \text{by symmetry,}$$

$$=3\pi a^2/16 \quad \text{by (10.4).}$$

To find the volume of the solid formed by rotating the upper loop through 2π radians about the tangent at $r=a$, $\theta=\frac{1}{2}\pi$, we consider the volume generated by rotating the elementary area bounded by the radii vectores joining O, the pole, to the points (r, θ),

$(r+\delta r,\ \theta+\delta\theta)$ on the curve. The centroid of this elementary area may be taken to lie at a distance $(a-\frac{2}{3}r\sin\theta)$ from the axis of rotation and its area may be taken as $\frac{1}{2}r^2\,\delta\theta$.

By Pappus' second theorem, the volume generated by rotating this elementary area through 2π radians about the tangent at $(a, \frac{1}{2}\pi)$ is

$$2\pi(a-\tfrac{2}{3}r\sin\theta)(\tfrac{1}{2}r^2\delta\theta)$$

and the volume obtained by rotating one loop of the curve is the limit of the sum of all such volumes

i.e.
$$V=\lim_{\delta\theta\to 0}\sum_{\theta=0}^{\theta=\pi}\pi r^2(a-\tfrac{2}{3}r\sin\theta)\delta\theta=\pi\int_0^{\pi}r^2(a-\tfrac{2}{3}r\sin\theta)d\theta$$

$$=2\pi a^3\left\{\frac{3\pi}{16}-\frac{32}{105}\right\}\quad\text{by (10.4) and (10.5).}$$

The length s of the upper loop $=2\int_0^{\pi/2}\sqrt{\left\{r^2+\left(\frac{dr}{d\theta}\right)^2\right\}}d\theta$ by symmetry

and since
$$r^2+\left(\frac{dr}{d\theta}\right)^2=a^2\sin^2\theta(1+3\cos^2\theta)$$

$$s=2a\int_0^{\pi/2}\sin\theta\sqrt{(1+3\cos^2\theta)}\,d\theta.$$

By the substitution $\sqrt{3}\cos\theta=\sinh\phi$, we have

$$s=\frac{2a\sqrt{3}}{3}\int_0^{\sinh^{-1}\sqrt{3}}\cosh^2\phi\,d\phi$$

$$=\frac{a\sqrt{3}}{3}\int_0^{\sinh^{-1}\sqrt{3}}(1+\cosh 2\phi)\,d\phi$$

$$=\frac{a\sqrt{3}}{3}\Big[\phi+\sinh\phi\cosh\phi\Big]_0^{\sinh^{-1}\sqrt{3}}$$

$$=\frac{a\sqrt{3}}{3}\{(\sinh^{-1}\sqrt{3})+2\sqrt{3}\}$$

$$=\frac{a\sqrt{3}}{3}\{\log(\sqrt{3}+2)+2\sqrt{3}\}.$$

Example 8

A is the vertex and LL′ is the latus rectum of the parabola $r=a\sec^2\frac{1}{2}\theta$. Find the surface area of the solid generated when the arc AL is rotated through four right angles about the axis of the parabola.

The axis of the parabola is the initial line. The vertex is at $r=a$ and by (21.13) the required surface area S is given by

$$S=2\pi\int_0^{\pi/2}r\sin\theta\sqrt{\left\{r^2+\left(\frac{dr}{d\theta}\right)^2\right\}}d\theta$$

where $r=a\sec^2\frac{1}{2}\theta$.

Now $$r^2+\left(\frac{dr}{d\theta}\right)^2=a^2(\sec^4\tfrac{1}{2}\theta+\sec^4\tfrac{1}{2}\theta\tan^2\tfrac{1}{2}\theta)=a^2\sec^6\tfrac{1}{2}\theta$$

$$\therefore\ S=4\pi a^2\int_0^{\pi/2}\sec^3\tfrac{1}{2}\theta\tan\tfrac{1}{2}\theta\,d\theta$$
$$=8\pi a^2\int_0^{\pi/2}\sec^2\tfrac{1}{2}\theta\,d(\sec\tfrac{1}{2}\theta)$$
$$=\tfrac{8}{3}\pi a^2\Big[\sec^3\tfrac{1}{2}\theta\Big]_0^{\pi/2}$$
$$=\tfrac{8}{3}\pi a^2(\sqrt{8}-1)$$

21.12. Roulettes

A roulette is the curve traced out by a point carried by a plane curve which rolls without slipping on a fixed curve in its own plane. We consider here some important examples of curves of this type.

I. The cycloid

The cycloid is the curve traced out by a point on the circumference of a circle which rolls without slipping along a fixed straight line. The parametric equations of this curve can be found from fig. 99, which

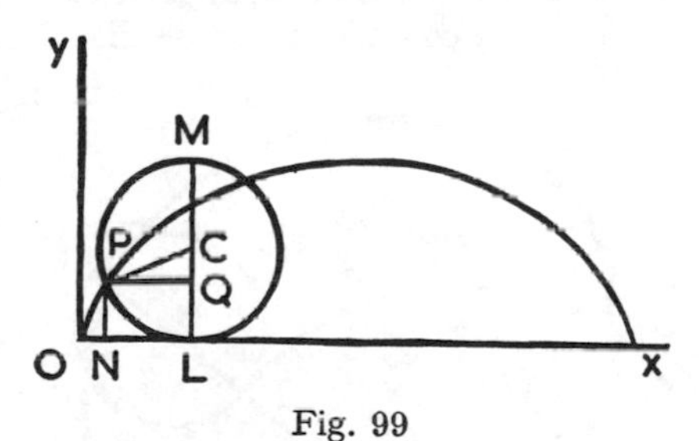

Fig. 99

shows a circle with centre C and radius a moving along a horizontal line Ox which we take as x-axis. P, the tracing point, originally coincided with the origin O, but the circle has rolled through an angle θ so that $\angle PCL=\theta$, where L is the point of the circle now in contact with Ox. Because there is no slipping, $OL=\text{arc } PL=a\theta$ and if P is the point (x, y)

$$x=OL-PQ=a(\theta-\sin\theta),\ y=CL-CQ=a(1-\cos\theta).$$

These are the parametric equations of the cycloid. From them we obtain

$$\frac{dy}{dx}=\frac{\sin\theta}{1-\cos\theta}=\cot\tfrac{1}{2}\theta$$

and so the tangent to the cycloid at P makes with Ox an angle $\frac{1}{2}(\pi-\theta)$. But if P is joined to M, the other extremity of the diameter LC,

$\angle PML = \frac{1}{2}\angle PCL = \frac{1}{2}\theta$ and so PM makes with Ox an angle $\frac{1}{2}(\pi - \theta)$. It follows that PM is the tangent to the cycloid at P and PL is the normal at P. One arch of the cycloid is shown in fig. 99. This arch is repeated for each revolution of the generating circle.

The parametric equations of the curve traced out by a point Q on CP or CP produced at a distance k from C are

$$x = a\theta - k \sin \theta, \quad y = a - k \cos \theta.$$

This curve is known as a trochoid.

II. The epicycloid and the hypocycloid

When one of two coplanar circles rolls without slipping on the other circle which is fixed, any point on the circumference of the rolling circle traces an epicycloid or a hypocycloid according as the rolling circle is outside or inside the fixed circle. An epicycloid in which the rolling circle surrounds the fixed circle is sometimes called a pericycloid.

Let P (fig. 100 (a)) be the tracing point on a circle, centre B and radius b, which rolls on the outside of a fixed circle, centre O and radius a, I being the point of contact. Suppose that when rolling start

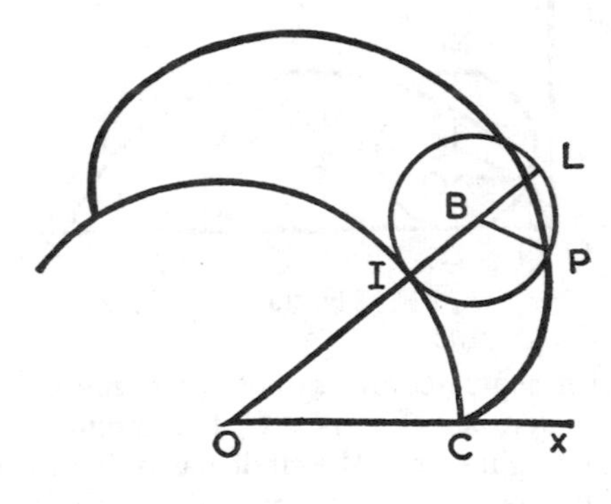

Fig. 100 (a)

the line OB is horizontal and the points P and C are in contact. Take OC as x-axis and the vertical through O as y-axis.

Since there is no slipping, arc IC = arc PI and so, if $\angle COI = \theta$ and $\angle IBP = \phi$

$$a\theta = b\phi \quad . \quad . \quad . \quad . \quad \text{(i)}$$

and BP makes with Ox an angle

$$\theta + \phi \quad . \quad . \quad . \quad . \quad . \quad \text{(ii)}$$

Then if P is the point (x, y)

$$\begin{aligned} x &= \text{sum of projections of } \overline{OB} \text{ and } \overline{BP} \text{ on } Ox \\ &= OB \cos\theta + BP \cos[\pi-(\theta+\phi)], \quad \text{by (ii)} \\ &= (a+b)\cos\theta - b\cos\frac{a+b}{b}\theta, \quad \text{by (i).} \end{aligned}$$

By considering the projections of $\overline{OB}$ and $\overline{BP}$ on Oy we obtain

$$y=(a+b)\sin\theta - b\sin\frac{a+b}{b}\theta.$$

Thus the parametric equations of the epicycloid are

$$x=(a+b)\cos\theta - b\cos\frac{a+b}{b}\theta, \quad y=(a+b)\sin\theta - b\sin\frac{a+b}{b}\theta. \qquad (21.14)$$

As in the case of the cycloid, we may show that if P is joined to L, the other extremity of the diameter IB, PL and PI are the tangent and normal respectively to the epicycloid at P.

21.13. Particular cases of the epicycloid

When $b=a$ we have, from (21.14),

$$x=2a\cos\theta - a\cos 2\theta, \quad y=2a\sin\theta - a\sin 2\theta$$

This is the equation of a cardioid, for if r is the distance from P to C, the point $(a, 0)$

$$r^2=(x-a)^2+y^2$$

$$\begin{aligned} r^2/a^2 &= (2\cos\theta-2\cos^2\theta)^2+(2\sin\theta-2\sin\theta\cos\theta)^2 \\ &= 4(1-\cos\theta)^2 \end{aligned}$$

$$\therefore\ r=2a(1-\cos\theta) \quad . \quad . \quad . \quad . \quad . \qquad \text{(i)}$$

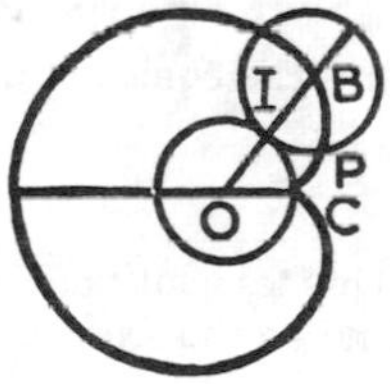

Fig. 100 (b)

Since $a=b$, $\theta=\phi$ and CP is parallel to OB, i.e. θ is the angle between OC and CP. Thus (i) is the polar equation of a cardioid referred to C as pole and OC as initial line.

The curve is shown fig. 100 (b).

21.14. The hypocycloid

The equations of the corresponding hypocycloid are obtained by changing the sign of b in (21.14). They are

$$x=(a-b)\cos\theta + b\cos\frac{a-b}{b}\theta, \quad y=(a-b)\sin\theta - b\sin\frac{a-b}{b}\theta \qquad (21.15)$$

When $a=2b$ these equations give $x=2b\cos\theta$, $y=0$, i.e. the diameter through C of the fixed circle (fig. 101 (*a*)).

When $a=4b$, we have from (21.15)

$$x=\tfrac{1}{4}a(3\cos\theta+\cos 3\theta)=a\cos^3\theta,\quad y=a\sin^3\theta.$$

This curve whose cartesian equation is $x^{2/3}+y^{2/3}=a^{2/3}$ is shown in 101 (*b*). It is called the astroid.

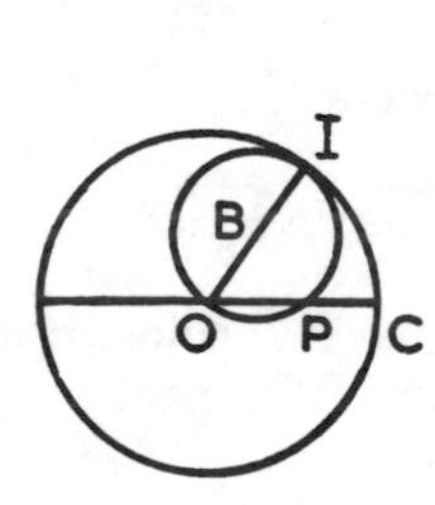

Fig. 101 (*a*)

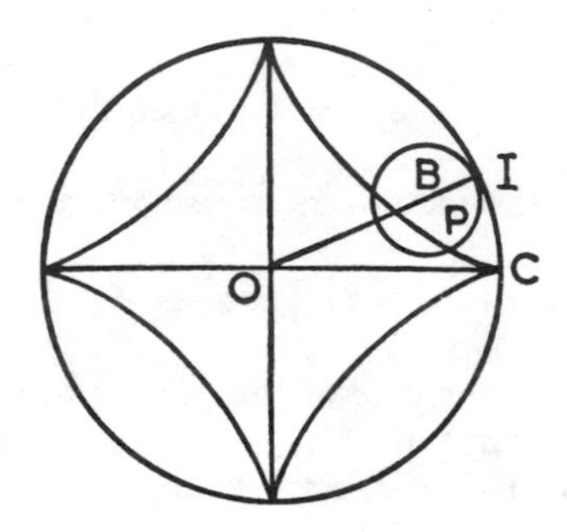

Fig. 101 (*b*)

Example 9

Show that the coordinate axes intercept a constant length on the tangent at any point of the astroid $x=a\cos^3\theta$, $y=a\sin^3\theta$.

Since $x=a\cos^3\theta$, $y=a\sin^3\theta$, $dy/dx=-\tan\theta$,

and the equation of the tangent at any point is

$$y-a\sin^3\theta=-\tan\theta\,(x-a\cos^3\theta)$$

i.e.

$$y+x\tan\theta=a\sin\theta.$$

This tangent makes intercepts $a\cos\theta$ and $a\sin\theta$ on Ox and Oy respectively and so the axes intercept on the tangent a constant length a.

Exercises 21 (*b*)

1. Show that the finite area bounded by the parabola $l/r=1+\cos\theta$ and a chord through the focus which makes an angle $\frac{1}{3}\pi$ with the axis is $16l^2/9\sqrt{3}$.

2. Sketch roughly the curve whose polar equation is $r=a\tanh\frac{1}{2}\theta$.

   Find the length s of the arc and the area A of the sector measured from the radius vector at $\theta=0$ to the radius vector (r, θ) and show that
   $$2A=a(s-r).$$ [L.U.]

3. Sketch the curve whose equation is $r=a(\sqrt{2}\cos\theta-1)$, and show that the area of the smaller loop is $\frac{1}{2}(\pi-3)a^2$. Find also the area of the larger loop. [L.U.]

4. Show that

$$\int_0^\pi \frac{d\theta}{1+e\cos\theta}=\frac{\pi}{\sqrt{(1-e^2)}}\ (0<e<1).$$

If each focal radius vector of an ellipse is produced a constant length c, show that the area between the curve so formed and the ellipse is $\pi c(2b+c)$, b being the semi-minor axis of the ellipse. [L.U.]

5. Prove that the area bounded by the curve
$r^2(a^2\sin^2\theta+b^2\cos^2\theta)=(a^2-b^2)b^2\cos^2\theta$, $(a>b>0)$ is $\pi b(a-b)$. [L.U.]

6. O is the pole and OX the initial line of polar coordinates. The straight line $r\cos\theta=2$ intersects OX at A and the curve $r=3+2\cos\theta$ at B and C. Prove that the area of the triangle OAB is $2\sqrt{3}$.

Find also the area of the smaller of the two portions into which the line BC divides the area enclosed by the curve $r=3+2\cos\theta$. [L.U.]

7. Starting from the point whose polar coordinates are $(a, 0)$, a point P moves in a plane in such a way that the direction of its motion always makes the same angle $\frac{1}{4}\pi$ with the radius vector OP. Prove that the equation of the locus of P is $r=ae^\theta$.

Prove that the difference of the areas described by the radius vector OP as θ increases (i) from $2(n-1)\pi$ to $2n\pi$, (ii) from $2n\pi$ to $2(n+1)\pi$ is $a^2e^{4n\pi}\sinh^2 2\pi$. Shade on a sketch the area so calculated when $n=2$. [L.U.]

8. Prove that the parabola $y^2=x$ divides the circle $x^2+y^2=2$ into two portions whose areas are in the ratio $9\pi-2:3\pi+2$. [L.U.]

9. If the area bounded by the curve $r=f(\theta)$ and the radii vectores $\theta=\theta_1$ and $\theta=\theta_2$ be rotated through four right angles about the initial line, prove that the volume of the resulting solid of revolution is given by

$$\tfrac{2}{3}\pi\int_{\theta_1}^{\theta_2} r^3\sin\theta\,d\theta.$$

(Use the method of Example 7, page 489).

10. Sketch the curve whose polar equation is $r=\cos^2\theta$. Find the area of a loop, and the volume of the solid formed by rotating it about the x-axis. [Durham.]

11. Show that the curve $r=1+2\sin\theta$ consists of an outer and an inner loop. Show that the area enclosed by the inner loop is $\frac{1}{2}(2\pi-3\sqrt{3})$. Find the volume of the solid formed by the rotation of this area through two right angles about the line $\theta=\frac{1}{2}\pi$. [L.U.]

12. Trace roughly the curve $r\cos^3\theta=a\cos 2\theta$ and show that the volume generated by revolving the loop about the line $\theta=\frac{1}{2}\pi$ is $64\pi a^3/105$. [L.U.]

13. A point O bisects a radius of a circle of radius a, and from O perpendiculars are drawn to the tangents to the circle. Taking O as the pole and the diameter of the circle through O as initial line, find the polar equation of the locus of the feet of these perpendiculars. Show that this curve encloses an area $9\pi a^2/8$ and that, if it is rotated about its axis of symmetry through two right angles, the volume generated is $5\pi a^3/3$. [L.U.]

14. The area enclosed by the lemniscate $r^2=a^2 \cos 2\theta$ is rotated through two right angles about the line $\theta=\frac{1}{2}\pi$. Prove that the volume of the solid so formed is $\frac{1}{8}\sqrt{2}\pi^2a^3$. [L.U.]

15. Show that the volume generated by revolving through four right angles about the line $\theta=\frac{1}{2}\pi$ the area bounded by this line and that part of the cardioid $r=a(1+\cos\theta)$ contained between $\theta=\frac{1}{2}\pi$, and $\theta=\pi$ is $\frac{1}{8}\pi(16-5\pi)a^3$. [L.U.]

16. The curve $r=a\ (1+\cos\theta)$ is rotated through an angle of π radians about the initial line. Find the superficial area and the volume of the solid so generated. [L.U.]

17. A solid is formed by rotating the area between the two loops of the curve $r=a(1+2\cos\theta)$ through four right angles about the initial line. Find its volume. [L.U.]

18. Give a rough sketch of the curve $r=a+b\cos\theta$ when a and b are both positive and $a>b$. The area interior to both this curve and the circle $r=a$ is rotated through two right angles about the initial line. Find the volume generated. [L.U.]

19. A curve is given by the parametric equations $x=\sin^2 t$, $y=\sin^3 t\cos t$. Find the slope of the tangent at the point "t", and sketch the curve. Calculate the area enclosed by this curve. [Durham.]

20. Sketch the curve $x=a\sin 2t$, $y=b\cos t$, and find the area of a loop. [Leeds.]

21. Give a sketch of the curve $y^2=x^2(x+1)$. By putting $y=tx$, express the coordinates x, y in terms of t, and find the area of the loop of the curve. [L.U.]

22. Sketch the curve $ay^2=x^2(4a-x)$. Show that the equation of its loop may be written in the form

$$x=4a\sin^2 t,\ y=8a\sin^2 t\cos t,$$

and find the area of the loop. [Durham.]

23. Prove that the curve $x=\cos^3 t$, $y=2\sin^3 t$ is of length $28/3$ and encloses an area $3\pi/4$. [L.U.]

24. The coordinates of a point on a curve are given parametrically by the equations

$$x=3at/(1+t^3),\ y=3at^2/(1+t^3).$$

Show that, for positive values of t, the point describes a loop in the first quadrant inside the square formed by the axes and the lines $x=2^{2/3}a$, $y=2^{2/3}a$. Find the area of the loop. [L.U.]

25. A plane curve is given by equations $x=f(t)$, $y=g(t)$. Show that the volume generated when the sectorial area OPQ is rotated through 2π about the x-axis, O being the origin and P, Q being the points of the curve where t has the values t_1, t_2 respectively, is

$$\tfrac{2}{3}\pi\int_{t_1}^{t_2} y\left(x\frac{dy}{dt}-y\frac{dx}{dt}\right)dt.$$

Prove that the volume generated when the area enclosed by the curve $x=a(1-t^2)$, $y=2a(1+t)$ and the y-axis is rotated through 2π about the x-axis is five times the volume generated when the same area is rotated about the y-axis. [L.U.]

26. By expressing the coordinates of a point on the curve $(x+y)^3=axy$ parametrically in the form $x=at(1-t)^2$, $y=at^2(1-t)$, find the area of the loop of the curve and the volume of the solid formed by rotating this area through four right angles about the axis of x. [L.U.]

27. Sketch roughly the shape of the curve $x^{2/5}+y^{2/5}=a^{2/5}$, any point of which can be put in the form $(a\cos^5 t,\ a\sin^5 t)$.

Prove that the curve lies entirely within the annulus formed by the two circles whose centres are at the origin and whose radii are a and $\frac{1}{4}a$.

Show also that, if any tangent to the curve cuts the coordinate axes in the points A and B, then, O being the origin, $OA^{2/3}+OB^{2/3}=a^{2/3}$. [L.U.]

28. Find the area enclosed by the curve

$$x=3a\cos\theta-a\cos 3\theta,\ y=3a\sin\theta-a\sin 3\theta. \qquad \text{[L.U.]}$$

29. A circle of radius a rolls, without slipping, inside a circle of radius $4a$. Show that, by a suitable choice of axes, the equation of the curve described by a point on the circumference of the rolling circle may be written $x=4a\cos^3\theta$, $y=4a\sin^3\theta$ and find its cartesian equation. If this curve be revolved about the x-axis, show that the surface area of the solid formed is $192\pi a^2/5$. [L.U.]

30. If a circle of radius a rolls outside a circle of radius $3a$, show that the locus traced out by a fixed point on the rolling circle is given by

$$x=4a\cos\theta-a\cos 4\theta,\ y=4a\sin\theta-a\sin 4\theta,$$

the origin being the centre of the fixed circle.

If r be the distance of a point on the locus from the centre of the fixed circle, and p the perpendicular from this centre to the tangent at the point, prove that $r^2=9a^2+16p^2/25$. [L.U.]

31. A circle of radius a rolls externally on a fixed circle of radius $2a$. Show that, referred to axes through the centre of the fixed circle, the equations of the curve described by a point on the circumference of the rolling circle can be written in the form

$$x=3a\cos\theta-a\cos 3\theta,\ y=3a\sin\theta-a\sin 3\theta.$$

Prove also that in this curve, $p=4a\sin\frac{1}{2}\psi$, where p is the perpendicular from the origin to the tangent, and ψ is the angle the tangent makes with the axis OX of the above coordinates. [L.U.]

32. Find $\frac{ds}{d\theta}$ for the epicycloid

$x=(a+b)\cos\theta-b\cos\{(a+b)\theta/b\},\quad y=(a+b)\sin\theta-b\sin\{(a+b)\theta/b\}$
and deduce the length of the curve traced out in one revolution of the rolling circle.

33. A fixed circle, centre O and of radius $4a$, touches a circle of radius a externally at A. The smaller circle rolls round the fixed circle, and, when C is the point of contact, P is the point on the rolling circle which was originally at A. If angle $AOC=\theta$, show that the coordinates of P can be expressed in the form

$$x=5a\cos\theta-a\cos 5\theta,\quad y=5a\sin\theta-a\sin 5\theta.$$

The line OC produced meets the circle of radius a at T. Show that TP is a tangent to the locus of P.

If $OP=r$ and p is the perpendicular from O to the tangent TP, prove that $r^2=16a^2+5p^2/9$. [L.U.]

34. Sketch the cycloid $x=a(t+\sin t)$, $y=a(1-\cos t)$ from $t=0$ to $t=\pi$. If O, A, B are the points on the curve at which $t=0, \frac{1}{2}\pi, \pi$ respectively, show that the lengths of the arcs AB and OA are in the ratio $(\sqrt{2}-1):1$. If the curve is rotated about the x-axis, show that the areas of the surfaces generated by AB and OA are in the ratio $(2\sqrt{2}-1):1$. [L.U.]

Find the centroid of each of the areas in Nos. 35-39:

35. The area in the first quadrant bounded by the curve $r=a\cos 2\theta$ and the line $\theta=0$.

36. The area enclosed by the first quadrant loop of the curve $r=a\sin 2\theta$.

37. The area bounded by the cardioid $r=a(1-\cos\theta)$ and the lines $\theta=0$ and $\theta=\frac{1}{2}\pi$.

38. The area bounded by one loop of the curve $r=a\cos 3\theta$.

39. The area enclosed by a loop of the lemniscate $r^2=a^2\cos 2\theta$.

CHAPTER 22

CURVATURE

22.1. Curvature

Let P and Q (fig. 102) be points on the arc of the curve $y=f(x)$ at distances s and $s+\delta s$ measured along the curve from a fixed point A. Let the tangents at P and Q make angles ψ and $\psi+\delta\psi$ respectively with Ox where $-\frac{1}{2}\pi<\psi<\frac{1}{2}\pi$. Then $\delta\psi$ measures the change of direction along the curve between P and Q.

$\dfrac{\delta\psi}{\delta s}$ is defined as the *average curvature* of the arc PQ, and the *curvature* at P (denoted by κ) is defined as the limit of $\dfrac{\delta\psi}{\delta s}$ as Q tends to P along the curve, i.e. as $\delta s \to 0$. Hence

$$\kappa=\frac{d\psi}{ds}. \qquad (22.1)$$

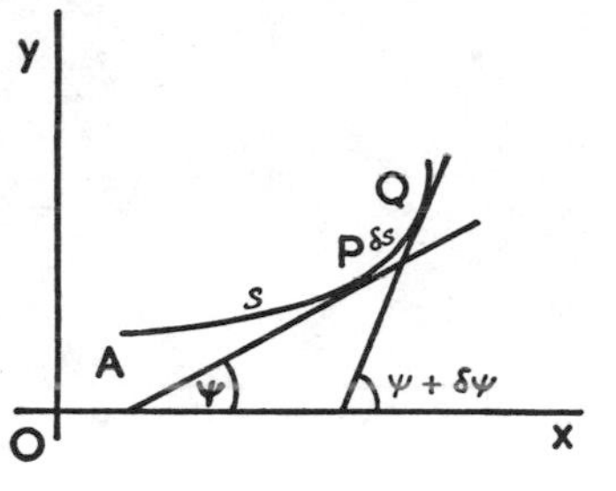

Fig. 102

If ψ and s increase together, κ is positive, otherwise κ is negative.

22.2. Curvature of a circle

In the case of a circle of radius ρ (fig. 103), the angle $\delta\psi$ between the tangents at the extremities of an arc of length δs is equal to the angle subtended by the arc at the centre of the circle. Hence

$$\delta s=\rho\delta\psi,$$

$$\frac{\delta s}{\delta\psi}=\rho=\text{constant}$$

and so $$\frac{d\psi}{ds}=1/\rho.$$

It follows that the curvature of a circle is constant and equal to the reciprocal of its radius.

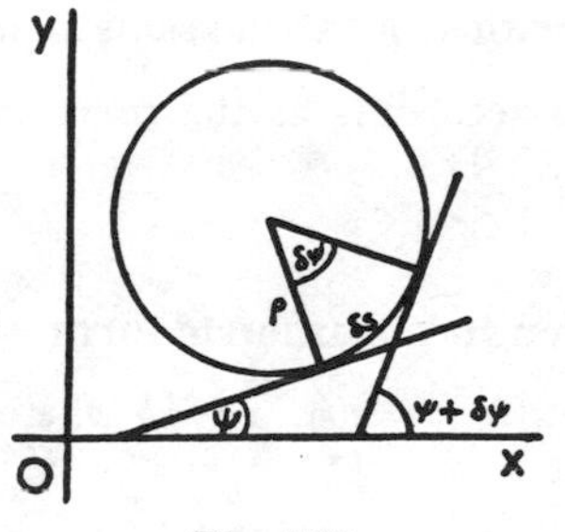

Fig. 103

22.3. Radius of curvature

The reciprocal of the curvature at any point P on a curve is called the *radius of curvature* at P and is denoted by ρ. Thus

$$\rho = 1/\kappa = \frac{ds}{d\psi}.$$

The term radius of curvature is used because ρ is the radius of the circle whose curvature is the same as that of the given curve at P.

22.4. Formula for ρ in terms of cartesian coordinates

By definition,
$$\rho = \frac{ds}{d\psi} = \frac{ds}{dx}\bigg/\frac{d\psi}{dx} \quad . \quad . \quad . \quad . \qquad \text{(i)}$$

and since
$$\frac{dy}{dx} = \tan\psi \ (-\tfrac{1}{2}\pi < \psi < \tfrac{1}{2}\pi),$$

$$\frac{d^2y}{dx^2} = \sec^2\psi \frac{d\psi}{dx}$$
$$= (1+\tan^2\psi)\frac{d\psi}{dx}$$

i.e.
$$\frac{d\psi}{dx} = \frac{\dfrac{d^2y}{dx^2}}{1+\left(\dfrac{dy}{dx}\right)^2}.$$

Also, adopting the convention that ds/dx is positive, we have by (20.10), p. 454,

$$\frac{ds}{dx} = \sqrt{\left\{1+\left(\frac{dy}{dx}\right)^2\right\}}$$

and so, by (i),
$$\rho = \frac{\left\{1+\left(\dfrac{dy}{dx}\right)^2\right\}^{3/2}}{\dfrac{d^2y}{dx^2}}. \qquad (22.2)$$

Thus, when s is measured so as to increase with x, ρ has the same sign as $\dfrac{d^2y}{dx^2}$ and is therefore positive or negative according as the curve is concave upwards or downwards.

22.5. Formula for ρ when a curve is given in parametric form

If a curve is given by the equations $x=x(t)$, $y=y(t)$, $\rho=\dot{s}/\dot{\psi}$ where dots denote differentiation with respect to t.

Now
$$\psi = \tan^{-1}(dy/dx) = \tan^{-1}(\dot{y}/\dot{x}).$$

Hence $$\frac{d\psi}{dt}=\frac{1}{1+(\dot{y}/\dot{x})^2}\left(\frac{\dot{x}\ddot{y}-\ddot{x}\dot{y}}{\dot{x}^2}\right)=\frac{\dot{x}\ddot{y}-\ddot{x}\dot{y}}{\dot{x}^2+\dot{y}^2}.$$

Also, $\dfrac{ds}{dt}=\sqrt{(\dot{x}^2+\dot{y}^2)}$, s being measured so as to increase with t,

$$\therefore \quad \rho=\frac{(\dot{x}^2+\dot{y}^2)^{3/2}}{\dot{x}\ddot{y}-\ddot{x}\dot{y}} \tag{22.3}$$

22.6. Formula for ρ in polar coordinates

If a curve is given by the equation $r=f(\theta)$, we write $\rho=\dfrac{ds}{d\theta}\Big/\dfrac{d\psi}{d\theta}$ and use the relation

$$\psi=\theta+\phi, \quad \text{see (21.3), page 477,}$$

which gives $$\frac{d\psi}{d\theta}=1+\frac{d\phi}{d\theta} \quad . \quad . \quad . \quad . \tag{i}$$

Now $$\tan\phi=r\frac{d\theta}{dr}=\frac{r}{r'},$$

where dashes denote differentiation with respect to θ.

Hence $$\sec^2\phi\,\frac{d\phi}{d\theta}=\frac{r'^2-rr''}{r'^2}$$

i.e. $$\left(1+\frac{r^2}{r'^2}\right)\frac{d\phi}{d\theta}=\frac{r'^2-rr''}{r'^2},$$

$$\frac{d\phi}{d\theta}=\frac{r'^2-rr''}{r^2+r'^2}.$$

Hence, by (i),

$$\frac{d\psi}{d\theta}=\frac{r^2-rr''+2r'^2}{r^2+r'^2} \text{ and by (21.2), p. 477, } \frac{ds}{d\theta}=\sqrt{\{r^2+r'^2\}},$$

s being measured so as to increase with θ.

Thus $$\rho=\frac{ds}{d\theta}\Big/\frac{d\psi}{d\theta}=\frac{(r^2+r'^2)^{3/2}}{r^2-rr''+2r'^2}. \tag{22.4}$$

Formula (22.4) may be expressed in another form by the substitution $r=1/u$. Then $r'=-\dfrac{1}{u^2}u'$, $r''=\dfrac{2}{u^3}u'^2-\dfrac{1}{u^2}u''$,

and so $$\rho=\frac{(u^2+u'^2)^{3/2}}{u^3(u+u'')}. \tag{22.5}$$

22.7. Formula for ρ when the p, r equation of a curve is given

If a curve is given by the equation $r=f(p)$, we again use the relation $\psi=\theta+\phi$ to obtain

$$\kappa=\frac{d\psi}{ds}=\frac{d\theta}{ds}+\frac{d\phi}{ds}.$$

But since, by § 21.3, $\sin\phi=r\dfrac{d\theta}{ds}$ and $\cos\phi=\dfrac{dr}{ds}$

$$\kappa=\frac{1}{r}\sin\phi+\cos\phi\frac{d\phi}{dr}$$

$$=\frac{1}{r}\frac{d}{dr}(r\sin\phi)$$

$$=\frac{1}{r}\frac{dp}{dr} \quad \text{by (21.8), page 478.}$$

Hence
$$\rho=r\frac{dr}{dp}. \tag{22.6}$$

It is often convenient to find the p, r equation of a polar curve and to use (22.6) to find the radius of curvature.

Example 1

Find ρ at any point of the cycloid $x=a(t-\sin t)$, $y=a(1-\cos t)$.

$$\dot{x}=a(1-\cos t)=2a\sin^2\tfrac{1}{2}t, \quad \dot{y}=a\sin t,$$

$$\ddot{x}=a\sin t, \qquad \ddot{y}=a\cos t,$$

$$\dot{s}^2=\dot{x}^2+\dot{y}^2=2a^2(1-\cos t)=4a^2\sin^2\tfrac{1}{2}t,$$

and
$$\dot{x}\ddot{y}-\ddot{x}\dot{y}=a^2(\cos t-1)=-2a^2\sin^2\tfrac{1}{2}t.$$

When $0<t<2\pi$, we must take $\dot{s}=2a\sin\frac{1}{2}t$ since s is conventionally assumed to increase with t.

Then, by (22.3),
$$\rho=\frac{8a^3\sin^3\frac{1}{2}t}{-2a^2\sin^2\frac{1}{2}t}=-4a\sin\tfrac{1}{2}t.$$

When $2\pi<t<4\pi$, we must take $\dot{s}=-2a\sin\frac{1}{2}t$, and in this case

$$\rho=+4a\sin\tfrac{1}{2}t.$$

Continuing in this way we see that for each arch of the cycloid ρ has a negative value.

Example 2

P is the point (r, θ) *on a curve given in polar coordinates, and p is the length of the perpendicular from O, the pole, to the tangent at P. Prove that*

$$\frac{1}{p^2}=u^2+\left(\frac{du}{d\theta}\right)^2, \text{ where } u=1/r.$$

Find the p, r equation of the conic $a=r(1+2\cos\theta)$ *and prove that* ρ, *the radius of curvature at P, is given by* $\rho p^3=ar^3$. [L.U.]

The result $\dfrac{1}{p^2}=u^2+\left(\dfrac{du}{d\theta}\right)^2$ is proved in § 21.4, and applying it to the conic

$$a=r(1+2\cos\theta) \quad . \quad . \quad . \quad . \quad . \quad \text{(i)}$$

which may be written $u=\dfrac{1}{a}(1+2\cos\theta)$

we have
$$\frac{1}{p^2}=\frac{1}{a^2}\{(1+2\cos\theta)^2+4\sin^2\theta\}$$
$$=\frac{1}{a^2}(5+4\cos\theta).$$

Substituting for $\cos\theta$ from (i) we obtain the p, r equation of the conic

$$\frac{1}{p^2}=\frac{1}{a}\left(\frac{2}{r}+\frac{3}{a}\right).$$

Differentiating this equation with respect to p, we have

$$\frac{1}{p^3}=\frac{1}{ar^2}\frac{dr}{dp}$$

i.e.
$$r\frac{dr}{dp}=\frac{ar^3}{p^3}$$

Hence by (22.6)
$$\rho p^3=ar^3.$$

22.8. The circle of curvature

The line which makes an angle $+\frac{1}{2}\pi$ with the positive tangent to a curve at P is called the positive normal at P, and if from P a length PC equal to the value of ρ at P is measured along this normal, C is called the *centre of curvature* at P, and the circle with centre C and radius ρ is called the *circle of curvature* at P.

The definition of the centre of curvature at P implies that C always lies at a distance equal to $|\rho|$ from P along the normal drawn on the concave side of the given curve at P.

If C (ξ, η) is the centre of curvature at $P(x, y)$ on a curve $y=f(x)$, we see from fig. 104 that

$$\xi=x-\rho \sin \psi, \ \eta=y+\rho \cos \psi. \tag{22.7}$$

In fig. 104, ψ is a positive acute angle and ρ is positive, but the

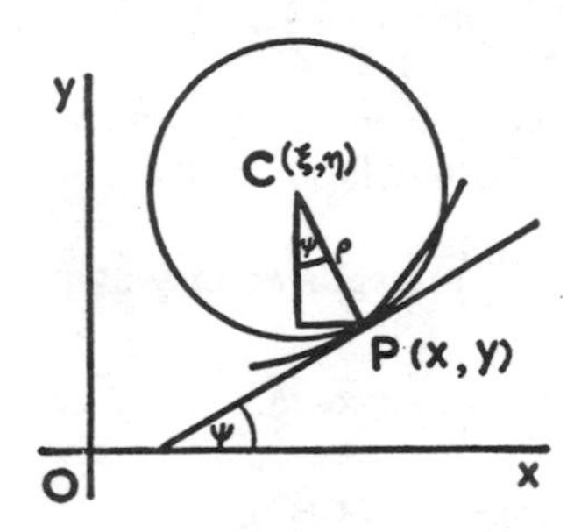

Fig. 104

above formulae are true in all cases if ρ is given its proper sign and if $-\frac{1}{2}\pi<\psi<\frac{1}{2}\pi$. When ψ is measured in this range

$$\sin \psi=\frac{\dfrac{dy}{dx}}{\sqrt{\left\{1+\left(\dfrac{dy}{dx}\right)^2\right\}}} \text{ and } \cos \psi=\frac{1}{\sqrt{\left\{1+\left(\dfrac{dy}{dx}\right)^2\right\}}},$$

the positive square root being taken in all cases. Substituting for ρ from (22.2) we get

$$\xi=x-\frac{\dfrac{dy}{dx}\left\{1+\left(\dfrac{dy}{dx}\right)^2\right\}}{\dfrac{d^2y}{dx^2}}, \quad \eta=y+\frac{1+\left(\dfrac{dy}{dx}\right)^2}{\dfrac{d^2y}{dx^2}}. \tag{22.8}$$

22.9. The evolute

The locus of the centres of curvature of a curve is called the evolute of the curve. The equation of the evolute may be found by eliminating x and y between equations (22.8) and the equation of the given curve. Another definition of the evolute of a curve is given in Example 27, page 427.

22.10. Contact of two curves

Suppose that the two curves $y=f(x)$, $y=g(x)$ intersect at a point P where $x=a$. Then, if we write $F(x)=f(x)-g(x)$, a is a root of the equation

$$F(x)=0 \quad . \quad . \quad . \quad . \quad \text{(i)}$$

and so, $F(a)=0$.

If $F(a)=0$, $F'(a)=0$ and $F''(a)\neq 0$, a is a double root of (i) (see § 2.7) and so the given curves intersect in two coincident points at P, i.e. they touch at P.

If $F(a)=0$, $F'(a)=0$, $F''(a)=0$ and $F'''(a)\neq 0$, a is a triple root of (i) and the given curves intersect in three coincident points at P. They are said to have three-point contact at P.

The conditions for three-point contact at P are $f(a)=g(a)$, $f'(a)=g'(a)$, $f''(a)=g''(a)$ and $f'''(a)\neq g'''(a)$, i.e. the two curves have the same values of y, $\frac{dy}{dx}$ and $\frac{d^2y}{dx^2}$ at P. It follows from (22.2) that they have the same curvature at P.

Conversely, if two curves which intersect at P have the same slope and the same curvature at P they have three-point contact there, and, in particular, a curve has three-point contact with its circle of curvature.

22.11. Miscellaneous examples

Example 3

A plane curve touches the x-axis at O, the origin. The tangent at P, a point of the curve, meets Ox at T making an angle ψ with Ox. If the ordinate of P is $\sin^4\psi$, show that the length of the arc OP is equal to $\frac{4}{3}PT$.

Show also that, if the perpendicular to the x-axis at T meets the normal at P in G, the radius of curvature at P is equal to 4PG. [T. U.]

Let s be the length of arc of the curve measured from O where $\psi=0$ up to the point $P(x, y)$ on the curve (fig. 105).

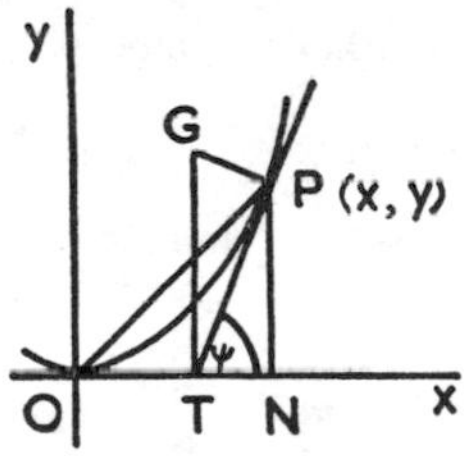

Then

$$y=\sin^4\psi \quad . \quad . \quad . \quad . \quad \text{(i)}$$

$$\frac{dy}{d\psi}=4\sin^3\psi\cos\psi\text{; and since } \frac{dy}{ds}=\sin\psi,$$

$$\frac{ds}{d\psi}=\frac{ds}{dy}\cdot\frac{dy}{d\psi}=4\sin^2\psi\cos\psi \quad . \quad . \quad \text{(ii)}$$

Integrating, we obtain $\quad s=\frac{4}{3}\sin^3\psi+C.$

But when $\psi=0$, $s=0$ $\quad\therefore C=0.$

Fig. 105

Hence $\quad$ arc $OP=\frac{4}{3}\sin^3\psi=\frac{4}{3}PT,$

since, by (i), $\quad PT=y\operatorname{cosec}\psi=\sin^3\psi \quad . \quad . \quad . \quad \text{(iii)}$

If ρ is the radius of curvature at P,

$$\rho=4\sin^2\psi\cos\psi \text{ by (ii)}$$

and

$$PG=PT\cot\psi=\sin^2\psi\cos\psi \text{ by (iii)}$$

$$\therefore \rho=4PG.$$

Example 4

Show that the centre of curvature corresponding to the point $(at^2, 2at)$ on the parabola $y^2=4ax$ is the point (X, Y) where $X=a(2+3t^2)$, $Y=-2at^3$.

Deduce the cartesian equation of the evolute of the parabola and find the points of intersection of the two curves. [L.U. Anc.]

The coordinates of C, the centre of curvature may be found by using (22.8). Alternatively, from (22.7) the coordinates of C are given by

$$X=x-\rho\sin\psi=x-\frac{ds}{d\psi}\cdot\frac{dy}{ds}=x-\frac{dy}{d\psi}=x-\dot{y}/\dot{\psi},$$

and, similarly, $Y=y+\dot{x}/\dot{\psi}$,
where dots denote differentiation with respect to t.

On the parabola $\qquad x=at^2,\ y=2at$

$$\dot{x}=2at,\ \dot{y}=2a$$

so that $\qquad dy/dx=\tan\psi=1/t,$

$$\psi=\tan^{-1}(1/t),$$

and $\qquad \dot{\psi}=-\dfrac{1}{1+t^2}.$

Hence $\qquad X=at^2+2a(1+t^2)=a(2+3t^2) \quad . \quad . \quad . \quad$ (i)

$$Y=2at-2at(1+t^2)=-2at^3 \quad . \quad . \quad . \quad \text{(ii)}$$

and eliminating t between (i) and (ii) we have

$$t^6=Y^2/4a^2=(X-2a)^3/27a^3$$

i.e. (X, Y) lies on the curve

$$27ay^2=4(x-2a)^3.$$

This is the equation of the evolute of the parabola. It meets the parabola where $(x-2a)^3=27a^2x$, whence $x=-a$ (repeated) and $x=8a$.

The parabola and its evolute meet in two real points $(8a, \pm 4a\sqrt{2})$; the other common points are imaginary.

Example 5

Find $\dfrac{d^2y}{dx^2}$ and the radius of curvature at the point $\theta(0<\theta<\pi)$ on the tractrix

$$x=a\ (\log\cot\tfrac{1}{2}\theta-\cos\theta),\ y=a\sin\theta,$$

and show that $\dfrac{d^2y}{dx^2}$ is positive.

Also show that if (ξ, η) are the coordinates of the centre of curvature at the point (x, y), then $\xi=x+a\cos\theta$, $\eta y=a^2$. [L.U.]

$$x=a(\log\cot\tfrac{1}{2}\theta-\cos\theta),\ y=a\sin\theta \quad . \quad . \quad . \quad \text{(i)}$$

$$\frac{dx}{d\theta}=a(-\operatorname{cosec}\theta+\sin\theta),\ \frac{dy}{d\theta}=a\cos\theta.$$

Hence $\qquad \dfrac{dy}{dx}=-\tan\theta \quad . \quad . \quad . \quad . \quad . \quad . \quad . \quad$ (ii)

and $\qquad \dfrac{d^2y}{dx^2}=-\sec^2\theta/a(\sin\theta-\operatorname{cosec}\theta)=\sin\theta/a\cos^4\theta.$

In the range $0<\theta<\pi$, $\dfrac{d^2y}{dx^2}$ is positive, and so ρ is positive if we measure s so as to increase with x.

From (22.2), $$\rho=\frac{(1+\tan^2\theta)^{3/2}}{\sin\theta}.a\cos^4\theta,$$

and since $\dfrac{ds}{dx}$ is positive, $(1+\tan^2\theta)^{1/2}$ must be positive.

Hence $\rho=\pm\dfrac{\sec^3\theta}{\sin\theta}\cdot a\cos^4\theta=\pm a\cot\theta$, the positive sign being chosen when $0<\theta\leqslant\frac{1}{2}\pi$, the negative sign when $\frac{1}{2}\pi\leqslant\theta\leqslant\pi$. Calculating ξ and η from (22.8), we have

$$\xi=x+a\cos\theta,$$
$$\eta=y+a\cos^2\theta/\sin\theta=a^2/y \text{ by (i).}$$

i.e. $$\eta y=a^2.$$

Exercises 22

1. Sketch the curve given by $x=a\cos^3 t$, $y=a\sin^3 t$.
   Find the radius of curvature at the point P whose parameter is t.
   If C is the centre of curvature at P, prove that the tangent at P divides OC (where O is the origin) in the ratio $1:3$. [Sheffield.]

2. For the catenary $y=c\cosh(x/c)$ prove that $\rho=y^2/c$, and show that if K is the centre of curvature for the point P, and the normal at P meets Ox at G, then P is the mid-point of KG. [L.U.]

3. Sketch the curve $x=a\sin 2t$, $y=b\cos^3 t$ when $a>0$, $b>0$.
   (i) Prove that the radius of curvature at the point $t=0$ is $-4a^2/3b$.
   (ii) Find the area that the curve encloses. [Sheffield.]

4. Find the centre of curvature corresponding to a point t on the cycloid
   $$x=a(t+\sin t),\quad y=a(1+\cos t).$$ [Durham.]

5. Determine the points of the hyperbola $xy=c^2$ at which the radius of curvature takes its numerically least value. Determine the centres of curvature for these points. [Durham.]

6. A curve is given parametrically by the equations
   $$x=a(2\cos\theta-\cos 2\theta),\ y=a(2\sin\theta-\sin 2\theta).$$
   Prove that, as θ increases from 0 to 2π, the tangent rotates through an angle 3π. Find, in terms of θ, the radius of curvature at any point, and determine the coordinates of the centre of curvature when the radius of curvature has its maximum value. [Sheffield.]

7. At the point $P(x, y)$ on a plane curve, the centre of curvature is $C(X, Y)$ and the inclination of the tangent to the x-axis is ψ. Prove that
   $$X=x-dy/d\psi,\ Y=y+dx/d\psi.$$
   P is the point (x, y) on the curve $x=a\cos^3\frac{1}{3}\phi$, $y=a\sin^3\frac{1}{3}\phi$ and C is the centre of curvature corresponding to P. A point N divides CP internally so that $CN=2NP$. Find the locus of N. [L.U.]

8. If $x=\tan\theta-\theta$, $y=\log\sec\theta$, $-\frac{1}{2}\pi<\theta<\frac{1}{2}\pi$, and s is measured from the origin, show that $s=\sec\theta-1$. Determine the radius of curvature in terms of θ. [L.U.]

9. Make a careful sketch of the curve whose equation is $y(1+x^2)=x-x^2$. Find the radius of curvature of this curve at the origin, and the co-ordinates of the corresponding centre of curvature. [L.U. Anc.]

10. Prove that, for the curve $axy=x^3-2a^3$, $(a>0)$ the ordinate, y, has a minimum value at the point $(-a, 3a)$, and find the equation of the circle of curvature at this point. [L.U.]

11. Find the coordinates of the centre of curvature of the ellipse
$$x^2/a^2+y^2/b^2=1$$
at the point $(a\cos\theta, b\sin\theta)$, and show that at this point
$$a^2b^2\rho^2=(a^2\sin^2\theta+b^2\cos^2\theta)^3,$$
where ρ is the radius of curvature.

Show that the locus of the centre of curvature has the equation
$$(ax)^{2/3}+(by)^{2/3}=(a^2-b^2)^{2/3}$$ [L.U.]

12. A, A'; B, B' are the ends of the major and minor axes of an ellipse of centre O and C is the fourth vertex of the rectangle whose other vertices are A, O, B. Prove that the line through C perpendicular to AB cuts AA' in the centre of curvature at A and BB' in the centre of curvature at B. [L.U.]

13. Find the equation of the circle of curvature at the point $(a, 2a)$ on the parabola $y^2=4ax$. Determine the other point at which this circle meets the parabola. [L.U. Anc.]

14. Prove that the centre of curvature at the point $P(at^2, 2at)$ on the parabola $y^2=4ax$ has coordinates $x=2a+3at^2$, $y=-2at^3$.

A segment PQ of length a is measured along the tangent to the parabola at the point P in the direction of t increasing. Prove that the equation of the locus of Q is
$$x=a\{t^2+t/\sqrt{(1+t^2)}\},\quad y=a\{2t+1/\sqrt{(1+t^2)}\},$$
and prove that the normal at Q to this locus passes through the centre of curvature at the point P on the parabola. [L.U.]

15. Prove that the line $lx+my+n=0$ touches the curve $x^4+y^4=a^4$ if $l^{4/3}+m^{4/3}=(n/a)^{4/3}$.

Prove also that the radius of curvature of the curve at its points of intersection with the lines $y=\pm x$ is $2^{1/4}a/3$. [L.U.]

16. The coordinates of a point P on a curve are given by the equations
$$x=a\cos^3 t,\ y=a\sin^3 t,$$
where t is a parameter; prove that the radius of curvature at P is $3a\sin t\cos t$.

If the normal to the above curve at the point P cuts the x-axis at G and the y-axis at g, and if ρ is the radius of curvature at P, prove that $\rho^2=9\,PG.Pg$. [L.U.]

17. Sketch the graph of the plane curve $x=\sin^3\theta$, $y=3\cos^3\theta$ for values of θ lying in the range 0 to 2π.

P is the point on the curve for which $\theta=\pi/3$ and C is the centre of curvature at P. Show that the equation of the line CP is $\sqrt{3}y=x$ and find the length of CP. [L.U.]

18. If (x, y) are the coordinates of a point P on the curve $y=\log(\cos x)$, find the radius of curvature at P, and show that the projection of the radius of curvature on the y-axis is a constant.

Find the coordinates of Q, the centre of curvature at P, and show that the radius of curvature at Q, of the locus of Q, is $\tan x \sec x$, where x is the abscissa of P. [L.U.]

19. Prove that the curve represented by the equation $y^3=x(x+2y)$ possesses a minimum ordinate when $x=1$, and show that the radius of curvature of the curve at this minimum point is $\frac{1}{2}$.

Prove also that the locus of the mid-points of chords parallel to the x-axis is the straight line $x+y=0$, and sketch clearly on the (x, y)-plane the general shape of the curve. [L.U.]

20. Prove that the locus of the centres of curvature of the rectangular hyperbola $x=ct$, $y=c/t$ is $(x+y)^{2/3}-(x-y)^{2/3}=(4c)^{2/3}$. [L.U.]

21. With the usual notation, show that the radius of curvature of a curve is given by $r\,dr/dp$.

A line is drawn through the origin meeting the cardioid

$$r=a(1+\cos\theta)$$

in the points P and Q and the normals at P and Q meet in C. Show that the radii of curvature at P and Q are proportional to PC and QC respectively. [L.U.]

22. Obtain the (p, r) equation of the curve $r=a \operatorname{sech}\theta$, and find its radius of curvature. [Leeds.]

23. Sketch the curve $r=1+2\cos\theta$ which consists of two loops. Show that its (p, r) equation referred to the origin O is $r^4=p^2(3+2r)$, and find an expression for ρ. If the initial line cuts the loops at A and B, show that the radii of curvature at A and B are in the ratio 27 : 5. [L.U.]

24. The normal at the point P on the curve $r^2=a^2\cos 2\theta$ intersects the initial line at G. Find the radius of curvature at the point P and the length of PG in terms of the radius vector to P.

Show also that at the point P, where PG is a maximum, the radius of curvature is $\frac{4}{3}PG$. [L.U.]

25. If P is any point on the curve $r^2=a^2\cos 2\theta$ and Q is the intersection of the normal at P with the line through the pole O at right angles to the radius vector OP, prove that the centre of curvature corresponding to P is a point of trisection of PQ. [L.U.]

26. Find the p, r equation of the conic $a=r(1+e\cos\theta)$ and prove that ρ, the radius of curvature at P, is given by $\rho p^3=ar^3$. [L.U.]

27. A point moves so that the product of its distances from two fixed points, A and B, is constant and equal to $3c^2$, where $c=\frac{1}{2}AB$. By taking the mid-point of AB as pole, show that the polar equation of the locus of the point can be expressed in the form

$$r^4-2c^2r^2\cos 2\theta-8c^4=0.$$

Find the (p, r) equation of the curve, and hence show that its curvature at any point is $(3r^4-8c^4)/6c^2r^3$. [L.U.]

28. Find the radius of curvature at any point of the curve $r^n=a^n\cos n\theta$, and show that the length of the intercept made by the circle of curvature on the radius vector is proportional to the length of the radius vector. [L.U.]

29. Draw a rough sketch of the curve $r^3=a^3\sin 3\theta$, showing that it consists of three loops. If P is the point (r, θ) on the curve, show that the tangent at P makes an angle 4θ with the initial line, and that $\rho=a^3/4r^2$. If the circle of curvature for the point P intersects OP at Q, where O is the pole, find the ratio of OQ to OP. [L.U.]

30. Show that the locus of Y, the foot of the perpendicular from the pole on to the tangent at P to the curve $r=a(1+\cos\theta)$ is the curve $R=2a\cos^3\frac{1}{3}\psi$ where $OY=R$, and ψ is the angle between the initial line and OY.

Show that the radius of curvature ρ of the locus of Y at Y is $\frac{3}{4}OP$, and that $OY.\rho$ is proportional to $OP^{5/2}$. [L.U.]

31. P is the point (r, θ) on the curve whose equation is $r=ae^{\theta\cot\alpha}$ and I is the centre of curvature at the point P. Find the length of PI and show that it subtends a right angle at the origin, O.

Find also the locus of I and the radius of curvature of this locus in terms of OI. [L.U.]

32. A parabola with focus at the origin and latus rectum $4a$ has equation $r=a\sec^2\frac{1}{2}\theta$ in polar coordinates. Prove that the angle ϕ between the tangent to the parabola at any point P and the radius vector at P is equal to $\frac{1}{2}(\pi-\theta)$, and obtain the relation between r and p for this parabola.

Prove that the radii of curvature at the vertex and at one end of the latus rectum are in the ratio $1:2\sqrt{2}$.

33. A circle of radius a rolls externally on a fixed circle of radius $2a$. Show that, referred to axes through the centre of the fixed circle, the parametric equations to the curve described by a point P on the circumference of the rolling circle may be expressed in the form

$$x=3a\cos\theta-a\cos 3\theta,\quad y=3a\sin\theta-a\sin 3\theta.$$

Show that the radius of curvature at the point θ is $3a\sin\theta$.

A is the point of contact of the two circles and K is the other point of intersection of PA with a circle of radius $a/2$ touching the fixed circle internally at A. Show that PA is the normal at P to the locus of P and deduce that K is the centre of curvature at P of the locus of P. [L.U.]

34. The rectangular coordinates of a point on a curve are given by

$$x=a \sin t-b \sin (at/b), \; y=a \cos t-b \cos (at/b).$$

Find the maximum distance of a point on the curve from the origin, and show that the curvature at such a point of maximum distance is $(a+b)/4ab$. [L.U.]

35. The parametric equations of a curve are

$$x=a(\cos \theta+\theta \sin \theta), \; y=a(\sin \theta-\theta \cos \theta),$$

where θ is the parameter and a is a constant.

Find the radius of curvature, ρ, in terms of θ, and the coordinates of the centre of curvature. Show that the centre of curvature lies on a circle of radius a. [L.U.]

CHAPTER 23

DIFFERENTIAL EQUATIONS OF THE FIRST ORDER

23.1. Definitions

Any relation between the variables x, y and the derivatives dy/dx, $d^2y/dx^2, \ldots$ is called an *ordinary differential equation.* The term *ordinary* distinguishes it from a *partial* differential equation which involves partial derivatives.

The *order* of a differential equation is that of the highest derivative occurring in it; the *degree* of a differential equation is that to which the derivative of the highest order is raised when the equation is expressed in a rational integral form.

Throughout this chapter we shall frequently write y' and y'' for dy/dx and d^2y/dx^2 respectively.

23.2. Formation and solution of differential equations

The equation $$x^2+y^2=c^2 \qquad \text{(i)}$$
where c is an arbitrary constant or parameter, represents a family of concentric circles with centres at the origin; c is constant for each circle but varies from circle to circle. Differentiating (i) with respect to x, we have

$$2x+2y\,y'=0,$$
$$y'=-x/y \qquad \text{(ii)}$$

This is the differential equation of the family of circles given by (i). It expresses a property common to all the members of the family, viz. that the tangent at the point $P(x, y)$ on any one of them is perpendicular to the radius which passes through P. The differential equation is of the first order.

If we start with a relation such as

$$y=ax+b/x \qquad \text{(iii)}$$

which contains two arbitrary constants a and b, it is necessary to differentiate twice before we can eliminate the constants.

We have $$y'=a-b/x^2 \qquad \text{(iv)}$$
$$y''=2b/x^3 \qquad \text{(v)}$$

and eliminating a and b between (iii), (iv) and (v) we obtain the second order differential equation

$$x^2y''+x\,y'=y.$$

The process of finding a relation from which a given differential equation is derived is known as solving (or integrating) the differential equation. The above examples suggest that the differential equation obtained from a relation involving n arbitrary constants will be of the nth order. Conversely, it may be expected that if we can solve a differential equation of the nth order, the most general solution (or integral) will contain n arbitrary constants. Such a solution is termed the *general solution* or the *complete primitive*. There may, however, be other solutions in addition to the complete primitive (cf. § 23.12).

If particular values are assigned to the n arbitrary constants, a particular solution, called a *particular integral*, is obtained. For example, (i) is the general solution of the differential equation (ii); the circle $x^2+y^2=4$ is a particular integral of the differential equation.

Below are given methods of solution for some of the simpler, commonly occurring types of ordinary differential equation.

23.3. Equations of the first order and first degree

This chapter is mainly devoted to the methods of solution of differential equations of the first order and first degree, i.e. to equations which may be written in the alternative forms

$$M(x, y)+N(x, y)y'=0,$$

$$M(x, y)dx+N(x, y)dy=0,$$

where M and N do not involve derivatives of y. The general solution contains one arbitrary constant which we shall denote by C.

23.4. Variables separable

If a differential equation of the first order and first degree can be written in the form

$$X\,dx+Y\,dy=0,$$

where X is a function of x alone, and Y is a function of y alone, the variables are said to be *separable* and the solution is obtained by direct integration:

$$\int X\,dx+\int Y\,dy=C.$$

Example 1

Solve the differential equation $(1-x^2)^{\frac{1}{2}}(dy/dx)+1+y^2=0.$ [Durham.]

This equation may be written in the form

$$\frac{dy}{1+y^2}+\frac{dx}{(1-x^2)^{\frac{1}{2}}}=0,$$

whence, on integration, $\tan^{-1} y+\sin^{-1} x=C$. This is the general solution.

Example 2

Find the general solution of the differential equation

$$xy(1+x^2)(dy/dx)-(1+y^2)=0.$$ [Sheffield.]

The given equation may be written in the form

$$\frac{y\,dy}{1+y^2}-\frac{dx}{x(1+x^2)}=0$$

i.e.
$$\frac{y\,dy}{1+y^2}-\left(\frac{1}{x}-\frac{x}{1+x^2}\right)dx=0,$$

whence
$$\tfrac{1}{2}\log(1+y^2)-\log x+\tfrac{1}{2}\log(1+x^2)=\log C,$$
$$\tfrac{1}{2}\log\{(1+x^2)(1+y^2)\}=\log Cx,$$

and so the general solution is
$$\sqrt{\{(1+x^2)(1+y^2)\}}=Cx.$$

Note.—If logarithmic functions occur in the integration of a differential equation, the constant of integration is frequently written in the form $\log C$ in order to simplify the form of the general solution.

Exercises 23 (*a*)

For brevity y' is written for (dy/dx).

1. Obtain a differential equation of the second order by eliminating the constants A and B from the equation $\phi=A/r+B$. [Sheffield.]
2. If a and b are arbitrary constants, find the second order differential equation whose solution is $y=a\sec x+b\tan x$. [L.U.]

Solve the differential equations in Nos. 3-9:

3. $y'=x^2/y$.
4. $y'=e^x\tan y$.
5. $y'=y/(x^2-4)$.
6. $(x^2+1)y'=y^2+4$.
7. $xy\,y'=1+x^2+y^2+x^2y^2$.
8. $y\,y'=e^{x+2y}\sin x$. [L.U.]
9. $2x^3\,y'=y^2+3xy^2$, given that $y=1$ when $x=1$. [L.U.]

23.5. The homogeneous equation

The differential equation

$$M\,dx+N\,dy=0 \quad . \quad . \quad . \quad . \quad \text{(i)}$$

is said to be *homogeneous* when M and N are homogeneous functions *of the same degree* in x and y. In this case M/N is a function of y/x and (i) may be written in the form

$$dy/dx=f(y/x) \quad . \quad . \quad . \quad . \quad \text{(ii)}$$

If $y/x=v$, so that $y=vx$ and $dy/dx=v+x\,dv/dx$, (ii) takes the form

$$v+x(dv/dx)=f(v),$$

and, separating the variables, we obtain the equation

$$\frac{dv}{f(v)-v}=\frac{dx}{x},$$

which may be integrated as in § 23.4, the original variable being restored by substituting y/x for v in the result.

Example 3

Solve the equation $dy/dx=(x^2+2y^2)/xy$, and find the particular solution for which $y=0$ when $x=1$.

If $y=vx$, the equation becomes

$$v+x\frac{dv}{dx}=\frac{1+2v^2}{v},$$

$$x\frac{dv}{dx}=\frac{1+v^2}{v},$$

$$\frac{v}{1+v^2}\,dv=\frac{dx}{x}.$$

Integrating, we obtain

$$\tfrac{1}{2}\log(1+v^2)=\log x+\log C,$$

which may be written $\sqrt{\{1+v^2\}}=Cx,$

i.e. $\sqrt{\{x^2+y^2\}}=Cx^2.$

If $y=0$ when $x=1$, $C=1$. Hence the particular solution is

$$\sqrt{(x^2+y^2)}=x^2.$$

23.6. Equations reducible to homogeneous form

The equation

$$\frac{dy}{dx}=\frac{ax+by+c}{a'x+b'y+c'} \quad . \quad . \quad . \quad . \quad \text{(i)}$$

is not homogeneous but may be reduced to homogeneous form when $a/a'\neq b/b'$ by writing $Y=ax+by+c$ and $X=a'x+b'y+c'$. Then

$$\frac{dy}{dx}=\frac{Y}{X}$$

and

$$\frac{dY}{dX}=\frac{dY}{dx}\Big/\frac{dX}{dx}=\left(a+b\frac{dy}{dx}\right)\Big/\left(a'+b'\frac{dy}{dx}\right)$$

$$=\left(a+b\frac{Y}{X}\right)\Big/\left(a'+b'\frac{Y}{X}\right)$$

i.e.

$$\frac{dY}{dX}=\frac{aX+bY}{a'X+b'Y}.$$

This equation is homogeneous in X and Y and may be solved as in § 23.5.

If $a/a'=b/b'$ the substitution $z=ax+by$ will reduce (i) to an equation in which the variables can be separated.

Example 4

Solve the equations

$$\text{(i)}\quad \frac{dy}{dx}=\frac{4x-2y+4}{2x+y-2},$$

$$\text{(ii)}\quad \frac{dy}{dx}=\frac{2x+3y+2}{4x+6y-3}.$$

(i) Let $Y=4x-2y+4$ and $X=2x+y-2$; then the given equation becomes

$$\frac{dY}{dX}=\frac{4X-2Y}{2X+Y}$$

and the substitution $Y=vX$ gives

$$v+X\frac{dv}{dX}=\frac{4-2v}{2+v},$$

$$X\frac{dv}{dX}=\frac{4-4v-v^2}{2+v},$$

i.e.
$$\frac{(2+v)}{v^2+4v-4}\,dv+\frac{dX}{X}=0.$$

Integrating, we have

$$\tfrac{1}{2}\log(v^2+4v-4)+\log X=\log C,$$

$$X\sqrt{\{v^2+4v-4\}}=C,$$

or
$$Y^2+4XY-4X^2=C^2.$$

Restoring the original variables we obtain

$$4(2x-y+2)^2+8(2x+y-2)(2x-y+2)-4(2x+y-2)^2=C^2$$

and simplifying,

$$4x^2-4xy-y^2+8x+4y=C', \text{ where } C'=\tfrac{1}{8}C^2+4.$$

(ii)
$$\frac{dy}{dx}=\frac{2x+3y+2}{4x+6y-3}.$$

Here, since the coefficients of x, y in the numerator and denominator are in the same ratio, we let $z=2x+3y$; then the given equation becomes

$$\frac{1}{3}\left(\frac{dz}{dx}-2\right)=\frac{z+2}{2z-3},$$

which leads to

$$\frac{dz}{dx}=\frac{7z}{2z-3},$$

i.e.
$$(2-3/z)dz=7dx.$$

Integrating, we have

$$2z-3\log z=7x+C,$$

i.e.
$$\log(2x+3y)=2y-x+C'.\quad (C'=-C/3).$$

Exercises 23 (*b*)

Solve the differential equations:

1. $x(y-3x)y'=2y^2-9xy+8x^2$. [Liverpool.]
2. $(x^2+y^2)y'=2xy$. [L.U.]
3. $xy^2 y'=x^3+y^3$. [Durham.]
4. $x(y+4x)y'+y(x+4y)=0$. [Sheffield.]
5. $(2x+y)\, y'=6y-4x$. [Durham.]
6. $(x+2y-3)\, y'=2x-y+1$. [L.U.]
7. $(x+y-2)\, y'=x+y+2$. [Durham.]
8. $(2x-y)^2 y'+(x-2y)^2=0$. [Sheffield.]
9. $(3x+y+3)\, y'+2(x+3)=0$. [Sheffield.]
10. $(2x-4y-8)\, y'=3x-5y-9$. [Sheffield.]
11. $y'=y/x+x \sin (y/x)$. [Durham.]
12. $4x^2 y'=4xy-x^2+y^2$, given that when $x=1$, $y=2$. [Liverpool.]

23.7. Exact equations

The equation
$$\frac{dy}{dx}=\frac{4x-2y+4}{2x+y-2}$$
which is solved in § 23.6, Example 4 (i), by the standard method may be written in the form
$$(4x-2y+4)dx-(2x+y-2)dy=0,$$
$$(4x+4)dx-2(y\,dx+x\,dy)-(y-2)dy=0,$$
i.e.
$$\frac{d}{dx}\left\{(2x^2+4x)-2xy-(\tfrac{1}{2}y^2-2y)\right\}=0.$$
Integrating, we have, $2x^2+4x-2xy-\frac{1}{2}y^2+2y=C$.

The equation $\dfrac{dy}{dx}=\dfrac{ax+by+c}{a'x+b'y+c'}$ may be solved in this way if $b=-a'$.

The equation is then said to be *exact*.

An exact equation is of the form
$$\frac{\partial f}{\partial x}dx+\frac{\partial f}{\partial y}dy=0.$$
This can be written $df(x,y)=0$ and so the solution is $f(x,y)=\text{constant}$.

If the equation

$$M\,dx+N\,dy=0 \quad . \quad . \quad . \quad \text{(i)}$$

is exact, there is a function $f(x, y)$ such that

$$M=\frac{\partial f}{\partial x} \text{ and } N=\frac{\partial f}{\partial y}$$

$$\therefore\ \frac{\partial M}{\partial y}=\frac{\partial^2 f}{\partial x\,\partial y}=\frac{\partial N}{\partial x} \quad \text{by (19.1), page 405.}$$

Hence, if (i) is exact,

$$\frac{\partial M}{\partial y}=\frac{\partial N}{\partial x} \quad . \quad . \quad . \quad . \quad \text{(ii)}$$

The converse result is true, but is harder to prove ; (ii) is the condition for (i) to be exact.

The method of solving exact equations is illustrated in the following example :

Example 5

Solve the equation $(3x^2+y+1)dx+(3y^2+x+1)dy=0$.

In the above notation $M=3x^2+y+1$, $N=3y^2+x+1$, and since

$$\frac{\partial M}{\partial y}=\frac{\partial N}{\partial x}=1,$$

the given equation is exact.

Now if $f(x, y)=C$ is its solution,

$$M=f_x=3x^2+y+1 \quad . \quad . \quad . \quad . \quad \text{(i)}$$

and

$$N=f_y=3y^2+x+1 \quad . \quad . \quad . \quad . \quad \text{(ii)}$$

Integrating (i) with respect to x, we have, *for each value of* y,

$$f=x^3+xy+x+\text{constant}.$$

But the value of this constant may depend on the value of y and so

$$f=x^3+xy+x+\phi(y) \quad . \quad . \quad . \quad . \quad \text{(iii)}$$

where $\phi(y)$ is a function to be determined.

Differentiating (iii) partially with respect to y, we obtain

$$f_y=x+\phi'(y).$$

But from (ii)

$$f_y=3y^2+x+1.$$

Hence

$$\phi'(y)=3y^2+1$$

and so

$$\phi(y)=y^3+y,$$

the addition of an arbitrary constant being immaterial.

Substituting this value in (iii), we have

$$f(x, y)=x^3+xy+x+y^3+y;$$

and so the solution of the given equation is

$$x^3+y^3+xy+x+y=C.$$

Exercises 23 (c)

Show that the following equations are exact and integrate them:

1. $xy'+y=e^x$.
2. $(1/x)\,y'-y/x^2=2x$.
3. $(ax+hy+g)\,dx+(hx+by+f)\,dy=0$.
4. $(1-\cos 2x)\,dy+2y\sin 2x\,dx=0$.
5. $(3x^2+2y+1)\,dx+(2x+6y^2+2)\,dy=0$.

23.8. Integrating factors

If a differential equation $M\,dx+N\,dy=0$ becomes exact when multiplied throughout by a suitable factor, this factor is known as an *integrating factor*.

For example, if we solve the equation

$$3y\,dx+2x\,dy=0 \quad . \quad . \quad . \quad . \qquad \text{(i)}$$

by separating the variables, we obtain

$$x^3y^2=C,$$

and, taking the total differential of this result,

$$3x^2y^2\,dx+2x^3y\,dy=0.$$

Comparison of this equation with (i) shows that x^2y is an integrating factor of (i).

Example 6

Show that the equation $(3xy-2ay^2)dx+(x^2-2axy)dy=0$ has an integrating factor which is a function of x alone. Solve the equation.

Suppose that the integrating factor is $f(x)$; then the equation

$$(3xy-2ay^2)f(x)dx+(x^2-2axy)f(x)dy=0 \quad . \quad . \qquad \text{(i)}$$

is exact, and so

$$\frac{\partial}{\partial y}\left\{(3xy-2ay^2)f(x)\right\}=\frac{\partial}{\partial x}\left\{(x^2-2axy)f(x)\right\}$$

i.e.

$$(3x-4ay)f(x)=(x^2-2axy)f'(x)+2(x-ay)f(x),$$

and hence

$$\frac{f'(x)}{f(x)}=\frac{1}{x}.$$

This equation is satisfied if $f(x)=x$.

Substitution of this value in (i) leads to the exact equation

$$(3x^2y-2axy^2)dx+(x^3-2ax^2y)dy=0 \quad . \quad . \qquad \text{(ii)}$$

If $F(x, y)=$constant is the solution, (ii) is identical with

$$F_x\,dx+F_y\,dy=0$$

whence

$$F_x=3x^2y-2axy^2 \quad . \quad . \quad . \quad . \qquad \text{(iii)}$$

and

$$F_y=x^3-2ax^2y \quad . \quad . \quad . \quad . \qquad \text{(iv)}$$

From (iii) and (iv) $F(x, y)$ may be determined as in Example 5 and the solution of the given equation shown to be $x^3y-ax^2y^2=C$.

Example 7

Show that the equation $dx+\{1+(x+y)\tan y\}dy=0$ *has an integrating factor of the form* $(x+y)^n$, *where n is a constant. Solve the equation.*

If $(x+y)^n$ is an integrating factor, the equation

$$(x+y)^n\,dx+\{1+(x+y)\tan y\}(x+y)^n\,dy=0$$

is exact.

Hence
$$\frac{\partial}{\partial y}(x+y)^n=\frac{\partial}{\partial x}\Big\{\{1+(x+y)\tan y\}(x+y)^n\Big\}$$

i.e.
$$\begin{aligned} n(x+y)^{n-1}&=n(x+y)^{n-1}\{1+(x+y)\tan y\}+(x+y)^n\tan y\\ &=(x+y)^{n-1}\{n+(n+1)(x+y)\tan y\},\\ n&=n+(n+1)(x+y)\tan y\\ \therefore\ n&=-1. \end{aligned}$$

Hence $1/(x+y)$ is an integrating factor of the given equation and the equation

$$\frac{dx}{x+y}+\left(\frac{1}{x+y}+\tan y\right)dy=0$$

is exact.

Comparison with $f_x\,dx+f_y\,dy=0$ gives

$$f_x=1/(x+y) \qquad \text{(i)}$$

and
$$f_y=1/(x+y)+\tan y \qquad \text{(ii)}$$

From (i) and (ii), $f(x, y)=\log(x+y)+\log(\sec y)$ and the solution of the given equation may be written in the form $x+y=C\cos y$.

23.9. The linear equation of the first order

The equation

$$\frac{dy}{dx}+Py=Q \qquad \text{(i)}$$

where P and Q are functions of x alone, involves y and dy/dx to the first degree only and is known as the first-order *linear* equation.

To find a method of solution, we consider first the particular case when $Q=0$. The equation is then

$$\frac{dy}{dx}+Py=0 \qquad \text{(ii)}$$

which may be written

$$(1/y)dy+Pdx=0$$

and integrated to give

$$\log y+\int Pdx=\log C$$

or
$$ye^{\int Pdx}=C.$$

If we verify this solution by differentiating with respect to x we get

$$\frac{d}{dx}(ye^{\int Pdx})=0. \quad . \quad . \quad . \quad . \qquad \text{(iii)}$$

Now $$\frac{d}{dx}(e^{\int Pdx})=\frac{d}{du}(e^u)\frac{du}{dx}, \text{ where } u=\int Pdx$$

$$=e^u P$$

$$=Pe^{\int Pdx},$$

$$\therefore \frac{d}{dx}(ye^{\int Pdx})=e^{\int Pdx}\left(\frac{dy}{dx}+Py\right) \quad . \quad . \quad . \quad . \qquad \text{(iv)}$$

Hence the left-hand side of (iii) is that of (ii) multiplied by $e^{\int Pdx}$. It follows that $e^{\int Pdx}$ is an integrating factor of (ii) and also of (i).

To solve (i) we multiply each side by $e^{\int Pdx}$ and obtain

$$e^{\int Pdx}\left(\frac{dy}{dx}+Py\right)=Qe^{\int Pdx}$$

i.e. $$\frac{d}{dx}(ye^{\int Pdx})=Qe^{\int Pdx}, \text{ by (iv)}$$

$$\therefore ye^{\int Pdx}=\int Qe^{\int Pdx}dx+C. \qquad (23.1)$$

The student should verify that it is unnecessary to include an arbitrary constant in $\int Pdx$ when determining the integrating factor.

The following results are useful for expressing the integrating factor in its simplest form:

$$e^{\log x}=x\ ;\quad e^{n\log x}=(e^{\log x})^n=x^n\ ;\quad e^{-n\log x}=1/e^{n\log x}=1/x^n.$$

Example 8

Solve the equation $dy/dx+y\cot x=2\cos x$.

Here $$P=\cot x, \quad \int Pdx=\log\sin x$$

$$\therefore e^{\int Pdx}=\sin x.$$

Hence, from (23.1), $y\sin x=\int \sin 2x\, dx+C$

i.e. $$\sin x=C-\tfrac{1}{2}\cos 2x.$$

Example 9

Solve the equation $y+x(x+1)(dy/dx)=x(x+1)^2e^{-x^2}$ [L.U.]

This equation may be written in the form

$$\frac{dy}{dx}+\frac{1}{x(x+1)}y=(x+1)e^{-x^2}.$$

Here $$P=1/x-1/(x+1),$$

$$\int P dx=\log x-\log (x+1)=\log x/(x+1)$$

and so the integrating factor is $x/(x+1)$.

Hence $$yx/(x+1)=\int xe^{-x^2}\,dx+C \quad \text{from (23.1)}$$

$$=C-\tfrac{1}{2}e^{-x^2}$$

i.e. $$xy=(x+1)(C-\tfrac{1}{2}e^{-x^2}).$$

Exercises 23 (*d*)

Solve the differential equations given in Nos. 1-15:

1. $y'\sin x-2y\cos x=e^x\sin^3 x$. [L.U.]
2. $y'\cos x-4y\sin x=6\cos^2 x\sin x$. [Liverpool.]
3. $x^2y'+xy=\log x$. [Durham.]
4. $y'\tan x+2y=x\operatorname{cosec} x$. [Sheffield.]
5. $x(1+x)y'-y=3x^4$. [Sheffield.]
6. $(1-x^2)y'+xy=(1-x^2)^{3/2}e^{\cos x}\sin x$. [Durham.]
7. $y'+xy-x^3=0$, where $y=0$ when $x=0$. [Sheffield.]
8. $(x-2)(x-3)y'+2y=(x-1)(x-2)$. [Liverpool.]
9. $$x(1-x^2)y'+(3x^2+1)y=(1+x)^3.$$
   Find the solution to the equation which remains finite as x tends to zero. [L.U. Anc.]
10. $(x-1)y'-2y=(x-1)^4\cos^2 x$. [Durham.]
11. $x^2y'+x(3+2x)y=e^{-2x}+e^{-3x}$. [Durham.]
12. $xy'+(1+x)y=x\sin x$. [Sheffield.]
13. $(1+x^2)y'-xy=(1+x^2)x^2$. [L.U.]
14. $(1+3x)y'+(3-9x)y=3$. [L.U.]

15. $xy'-y=x^3\cos x$, with $y=0$ when $x=\pi$. [L.U.]

16. (i) Reduce the differential equation $(x+1)yy'-y^2=x$ to a linear form by writing $z=y^2$, and solve it, given that when $x=0$, $y=1$.

(ii) Solve the equation $(2x+y)y'=x+2y$, by writing $y=vx$, given that when $x=1$, $y=0$. [L.U. Anc.]

23.10. Bernoulli's equation

The equation

$$\frac{dy}{dx}+Py=Qy^n$$

in which P and Q are functions of x alone and n is constant is known as *Bernoulli's equation* and is reducible to linear form.

Suppose first that $n\neq 1$ and divide throughout by y^n so that the given equation becomes

$$y^{-n}\frac{dy}{dx}+Py^{1-n}=Q.$$

The form of this equation suggests the substitution

$$y^{1-n}=v,\quad (1-n)y^{-n}\,dy/dx=dv/dx,$$

which reduces the given equation after multiplication by $(1-n)$ to the linear equation

$$\frac{dv}{dx}+(1-n)Pv=(1-n)Q.$$

This equation may be solved by the method of § 23.9.

If $n=1$, the given equation may be solved by separating the variables.

Example 10

Solve the equation $x\,dy/dx+y=y^2x^2\log x$. [L.U.]

When this equation is divided throughout by x it is seen to be Bernoulli's equation with $n=2$. Proceeding as above, we then divide throughout the equation by y^2 and put $v=y^{-1}$ so that $dv/dx=-y^{-2}dy/dx$.

Then

$$\frac{dv}{dx}-\frac{v}{x}=-x\log x \quad . \quad . \quad . \quad . \quad \text{(i)}$$

The integrating factor is $e^{-\int dx/x}=e^{-\log x}=1/x$; hence from (i),

$$\frac{d}{dx}(v/x)=-\log x,$$

$$v/x=C-(x\log x-x),$$

i.e.

$$1/xy=C+x(1-\log x).$$

Exercises 23 (*e*)

Solve the following differential equations:

1. $y'+y\cot x=y^2\sin^2 x$. [L.U.]
2. $y'=y\tan x+y^3\tan^3 x$. [L.U.]
3. $y'=x/y+y$. [L.U.]
4. $2y'\sin x-y\cos x=y^3\sin x\cos x$.
   Also find the particular solution for which $y=-1$ when $x=\frac{1}{2}\pi$. [Sheffield.]
5. $y'+y=xy^3$. [L.U.]
6. $y'=2y\tan x+y^2\tan^2 x$. [L.U.]
7. $2y'-y(2x+1)/(x^2+x+1)=xy^3/(1-x)$. [L.U.]
8. $(x^2-1)y'+xy=e^x y^{-2}\sqrt{(x^2-1)}$. [Durham.]

23.11. Change of variable

Differential equations of the first order and first degree which are not of the foregoing types may sometimes be solved by a suitable change of variable.

Thus, for example, in an equation of the form

$$dy/dx=f(ax+by+c)$$

the substitution $u=ax+by+c$ is indicated.

An equation of the form

$$\{xf(y)+F(y)\}dy/dx=\phi(y)$$

becomes a linear equation if written in the form

$$\phi(y)(dx/dy)-xf(y)=F(y),$$

where x is the dependent and y the independent variable.

A few substitutions which are frequently useful are listed below, but in most cases an appropriate substitution is suggested by the functions which occur in the equations under consideration. Thus, for example, the Bernoulli equation solved in § 23.10 may also be solved by the substitution $u=xy$. This substitution is suggested by the presence of the function x^2y^2 and by the expression $y+x(dy/dx)$ which is equivalent to $\dfrac{d}{dx}(xy)$.

The substitution $u=x^2+y^2$ is suggested by $x\,dx+y\,dy$ and the substitution $u=y/x$ by $x\,dy-y\,dx$.

If both $(x\,dx+y\,dy)$ and $(x\,dy-y\,dx)$ occur, simplification may be obtained y a change to polar coordinates, for we have

$$x^2+y^2=r^2 \text{ and } y/x=\tan\theta$$

so that

$$x\,dx+y\,dy=r\,dr \text{ and } x\,dy-y\,dx=r^2\,d\theta.$$

Example 11

Solve the equation $dy/dx=(x+y)/(x-y)$.

This equation is homogeneous and may be solved by the method of § 23.5. Alternatively it may be written in the form

$$x\,dx+y\,dy=x\,dy-y\,dx,$$

which, expressed in terms of polar coordinates, reduces to $dr/r=d\theta$.

The general solution, $r=Ce^{\theta}$, is the equation of a family of equiangular spirals.

Example 12

If $dy/dx=(x+y)^2$ *and* $y=\frac{1}{2}$ *when* $x=\frac{1}{2}$, *calculate to three significant figures the value of* y *when* $x=0{\cdot}7$. [L.U.]

The substitution $x+y=u$, $dy/dx=du/dx-1$ leads to the equation

$$\frac{du}{dx}=u^2+1.$$

The general solution is

$$\tan^{-1} u=x+C,$$

i.e.

$$\tan^{-1}(x+y)=x+C.$$

But when $x=\frac{1}{2}$, $y=\frac{1}{2}$ and so $C=\frac{1}{4}\pi-\frac{1}{2}$.

Hence

$$\tan^{-1}(x+y)=x+\tfrac{1}{4}\pi-\tfrac{1}{2}$$

i.e.

$$y=\tan(\tfrac{1}{4}\pi+x-\tfrac{1}{2})-x.$$

When $x=0{\cdot}7$,

$$y=\tan(\tfrac{1}{4}\pi+0{\cdot}2)-0{\cdot}7$$

$$=0{\cdot}808 \text{ from tables.}$$

Example 13

Transform the equation $(2xyy'+x^2-y^2)(x^2+y^2)^{\frac{1}{2}}=xyy'-y^2$, *where* $y'=dy/dx$, *into one involving* θ, r *and* $dr/d\theta$. *Hence, or otherwise, solve the equation.* [Sheffield.]

The given equation may be written in the form

$$(x^2+y^2)^{\frac{1}{2}}\{2xy\,dy+(x^2-y^2)dx\}=xy\,dy-y^2\,dx,$$

i.e.

$$(x^2+y^2)^{\frac{1}{2}}\{x(y\,dy+x\,dx)+y(x\,dy-y\,dx)\}=y(x\,dy-y\,dx),$$

and on changing to polar coordinates we obtain

$$r\{r\cos\theta(r\,dr)+r\sin\theta(r^2\,d\theta)\}=r\sin\theta(r^2\,d\theta),$$

which reduces to

$$dr/d\theta+r\tan\theta=\tan\theta.$$

Separating the variables we have

$$\frac{dr}{r-1}+\tan\theta\,d\theta=0,$$

which leads to

$$(r-1)\sec\theta=C$$

or

$$r=1+C\cos\theta.$$

The solution is generally left in polar form.

Miscellaneous Exercises 23

Solve the differential equations given in Nos. 1-24 :

1. (i) $(4y+3x)y'+y-3x=0$;
 (ii) $xy'-2y=x+1$,
 given that when $x=\frac{1}{2}$, $y=1$. [Sheffield.]

2. (i) $(5y+x)y'=5x+y$;
 (ii) $y'\sin x-y\cos x=\sin^3 x$. [Sheffield.]

3. (i) $\dfrac{1}{t}\dfrac{dx}{dt}=1-\dfrac{x}{1-t^2}$;
 (ii) $4(y-x)y'=3(3x+4y)$. [Leeds.]

4. (i) $y'=y\tan x-2\sin x$;
 (ii) $(x^2-3y^2)x\,dx=(y^2-3x^2)y\,dy$. [Leeds.]

5. (i) $y'+y\tan x=\sin 2x$;
 (ii) $x(x+y)y'=x^2+y^2$;
 (iii) $(\cos x-x\cos y)dy-(\sin y+y\sin x)dx=0$. [Leeds.]

6. (i) $(x^2+1)y'+4xy=5(x^2-1)$;
 (ii) $x(x^2+3y^2)y'=y(3x^2+y^2)$. [Leeds.]

7. (i) $y'\sin x-y\cos x=\sin x-(1+x)\cos x$
 (ii) $xy\,y'=2y^2-3xy+2x^2$. [Sheffield.]

8. (i) $y'+y\tan x=\cos 2x$;
 (ii) $(4x-3y)y'=3x+4y$. [Durham.]

9. (i) $4x^2y'=4xy-x^2+y^2$;
 (ii) $(1+x^2)y'=2x(1+y+x^2)$;
 (iii) $y'=(x+4y)^2$. [Sheffield.]

10. (i) $xy'=2y+x^{n+1}y-x^n y^{3/2}$;
 (ii) $(x+3y+8)y'=3x+y$. [L.U.]

11. (i) $y'+(2x+1)y=2x^2+x+1$;
 (ii) $(2x-2y-4)y'=2x+7y+5$. [L.U.]

12. (i) $(4x+2y+1)y'=2x+y-3$;
 (ii) $y'+y\cot x=y^2\cos^2 x$. [L.U.]

13. (i) $x(1-x^2)y'+(2x^2-1)y=x^3y^3$;
 (ii) $xy'-y=(x^2+y^2)^{\frac{1}{2}}$. [L.U.]

14. (i) $(3x+2y-1)y'=x+2y-3$;
 (ii) $4y'\sin^2 x-y\sin 2x=y^3\cos x$. [L.U.]

15. (i) $(x^2+y^2)y'=x^2+xy$;
(ii) $(x-y)y'=2x+y-3$;
(iii) $2(1+x)y'-(1+2x)y=x^2(1+x)^{\frac{1}{2}}$. [L.U.]

16. (i) $(3x+2y-4)y'=3y-2x+7$;
(ii) $y'+y\tan x=y^3\sec^6 x$. [L.U.]

17. (i) $x(1-x^2)y'+(1-2x-x^2)y=1$;
(ii) $y^2y'=x^2+xy-y^2$. [L.U.]

18. (i) $y'=2x-y$;
(ii) $(x+2y-5)y'=2x-y$. [L.U.]

19. (i) $y'\cos x+y\sin x=x\sin 2x+x^2$;
(ii) $(x-8y+7)y'=x-y$. [L.U.]

20. (i) $(3y-2x)y'=2x+3y$;
(ii) $y'+y=xy^3$. [L.U.]

21. (i) $(x^2+xy)y'=x^2+xy-y^2$;
(ii) $y'+y\cot x=(y\sin x)^{\frac{1}{2}}$.

Show also that the only solution of the latter equation which remains finite as $x\to 0$ is $y\sin x=\sin^4\frac{1}{2}x$. [L.U.]

22. (i) $(x^2-x)y'+y=(x^2-x)\log x$;
(ii) $xy'-y=(x^2+y^2)^{\frac{1}{2}}$;
(iii) $xy'+3y=x^2y^2$. [L.U.]

23. (i) $x(y^3+x^3)y'=y^4+2x^3y-x^4$;
(ii) $y'\sin x+2y\cos x=\cos x$. [L.U.]

24. (i) $xyy'=y^2+x^2e^{y/x}$;
(ii) $x(1+x)y'+(2+x)y+3x+2x^2=0$. [L.U.]

25. (i) Solve the equation $(x^2+y^2)y'+2x(x+y)=0$.
(ii) Find the integral curves of the equation $(y'-y)e^x+1=0$. Show that, in general, every curve of the system has EITHER a real point of inflexion, on the line $y=0$, OR a real point for which y is a minimum, on the curve $y=e^{-x}$. [L.U.]

26. (i) Solve the differential equation $x^2y'+1+(1-2x)y=0$.
(ii) Prove that the differential equation $M\,dx+N\,dy=0$, where M and N are functions of x and y, is exact if

$$\frac{\partial M}{\partial y}=\frac{\partial N}{\partial x}.$$

Show that a constant α can be found so that $(x+y)^\alpha$ is an integrating factor of $(4x^2+2xy+6y)dx+(2x^2+9y+3x)dy=0$, and hence integrate the equation. [L.U.]

27. Show how to solve the differential equation $y'+Py=Qy^n$, where P and Q are functions of x only.

Find the solution of the equation $xy^2 y'-2y^3=2x^3$, which is such that $y=1$ when $x=1$. [L.U.]

28. (i) It is known that, when multiplied by a certain power of x, the equation $(5x^2+12xy-3y^2)dx+(3x^2-2xy)dy=0$ becomes exact. Find this integrating factor and solve the equation.

(ii) Obtain the solution of the equation $(x-1)y'+xy=(x-1)e^x$ for which y and x vanish together. [L.U. Anc.]

29. (i) Solve the equation $xyy'=(x+y)^2$.

(ii) Show that the equation $x^2 y'=1-2x^2y^2$ may be reduced to a linear differential equation of the first order by the substitution $y=1/x+1/z$.

Hence, or otherwise, solve the equation. [L.U.]

30. (i) Solve the equation $x(x+1)y'+y=2x$.

(ii) By means of the substitution $y^2=u-x$, reduce the equation $y^3 y'+x+y^2=0$ to homogeneous form and hence, or otherwise, solve it. [L.U.]

23.12. Clairaut's equation

This equation has the form

$$y=px+f(p) \quad \quad \text{(i)}$$

where $p=dy/dx$.

Differentiating (i) with respect to x, we have

$$p=\left(x\frac{dp}{dx}+p\right)+f'(p)\frac{dp}{dx},$$

$$\frac{dp}{dx}\{x+f'(p)\}=0.$$

Hence either

$$\frac{dp}{dx}=0$$

or

$$x+f'(p)=0 \quad \quad \text{(ii)}$$

If $\dfrac{dp}{dx}=0$, $p=C$ (constant) and, substituting this value in (i) we have

$$y=Cx+f(C) \quad \quad \text{(iii)}$$

This solution which contains one arbitrary constant C is the complete primitive or general integral of (i). It represents a family of straight lines with parameter C (cf. § 23.2).

Another solution is obtainable by eliminating p between (i) and (ii). This solution contains no arbitrary constant and is called a *singular solution* of (i).

Differentiating (iii) partially with respect to C, we obtain the relation

$$x+f'(C)=0 \quad . \quad . \quad . \quad . \quad \text{(iv)}$$

and elimination of C between (iii) and (iv) gives the equation of the envelope of the family of straight lines represented by (iii) (cf. § 19.13). But this equation is the same as that obtained by eliminating p between (i) and (ii). Hence the singular solution represents the envelope of the family of straight lines given by the general integral.

Example 14

Find the general integral and singular solution of the equation $y=px-\log p$, where p denotes dy/dx. [L.U.]

$$y=px-\log p \quad . \quad . \quad . \quad . \quad . \quad \text{(i)}$$

Differentiating with respect to x, we have

$$p=p+(x-1/p)\frac{dp}{dx}.$$

Hence either $\dfrac{dp}{dx}=0$ or $p=1/x$.

If $\dfrac{dp}{dx}=0$, $p=C$ (constant) and from (i) $y=Cx-\log C$.

This is the general integral of (i).

Substituting $p=1/x$ in (i), we obtain the singular solution $y=1+\log x$.

Exercises 23 (f)

Obtain the complete primitive and the singular solution of the differential equations in Nos. 1-5 :

1. $y=px+a/p$, $(p=dy/dx)$.
2. $y=px+2\sqrt{(ap)}$.
3. $y=px+p^3$.
4. $y=px+a\sqrt{(1+p^2)}$.
5. $y=px+p-p^2$.
6. Show that Clairaut's equation $y=px+f(p)$, $(p=dy/dx)$ has a family of straight lines as its complete primitive and their envelope as its singular solution.

   Solve the differential equation $e^{y+p}=(1+p)e^{px}$. [Durham.]
7. Obtain the complete primitive and the singular solution of the differential equation $2y=2px-\log \sec^2 p$, where $p=dy/dx$. [Sheffield.]
8. By means of the substitution $x^2=X$, $y^2=Y$ (or otherwise), reduce the equation $x^2+y^2-xy(p+1/p)=c^2$ to Clairaut's form and find the complete primitive and singular solution. [L.U.]
9. Show that the equation $(px-y)(px-2y)+x^3=0$, where p denotes dy/dx, may be reduced to Clairaut's form by means of the substitution $y=vx$. Hence find its complete primitive and singular solution. [L.U.]

10. The feet of the perpendiculars from the point $(c, 0)$ to the tangents to a certain curve lie on the circle $x^2+y^2=a^2$. Obtain the differential equation of the curve in the form $y^2-2xyp+(x^2-a^2)p^2+c^2-a^2=0$ where $p=dy/dx$, and show that $y=mx\pm\sqrt{\{a^2(1+m^2)-c^2\}}$ is a solution. [L.U.]

23.13. Geometrical applications

Geometrical properties of a curve are sometimes expressible in terms of a differential equation of the first order. The solution of this equation represents the family of curves which possesses the given property.

23.14. Tangents and normals in cartesian coordinates

Let P (fig. 106), be any point on a plane curve and let the tangent and normal to the curve at P meet the axis of x at T and G respectively. Then, if PN is the ordinate of P, TN and NG are called the subtangent

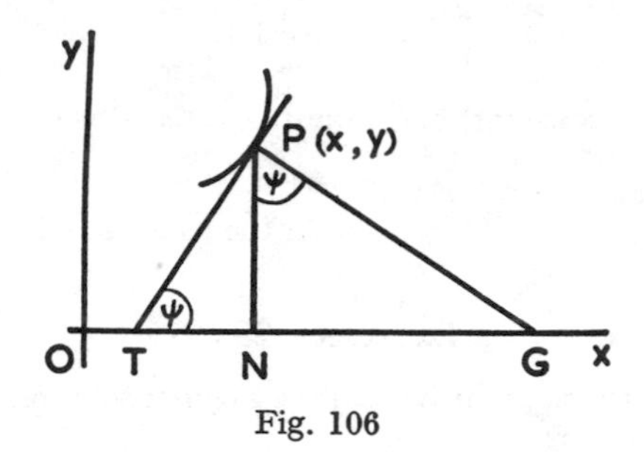

Fig. 106

and subnormal at P; the lengths PT and PG are known respectively as the tangent and normal at P.

If P is the point (x, y), and (X, Y) are the coordinates of any other point on the tangent PT to the curve at P, then the equation of PT is

$$Y-y=y'(X-x).$$

In this equation (x, y) are the coordinates of the point of contact and $y'=dy/dx$ is obtained by differentiating the equation of the curve and substituting in the result the values of x and y.

The equation of PG, the normal at P, is $(X-x)+y'(Y-y)=0$.

Now $y'=\tan\angle xTP=\tan\angle NPG=\tan\psi$ $(-\frac{1}{2}\pi<\psi<\frac{1}{2}\pi)$, and $NP=y$.

Hence for the subtangent, subnormal, tangent and normal the following expressions are obtained:

$$TN=y\cot\psi=y/y',$$
$$NG=y\tan\psi=yy',$$
$$PT=|y\operatorname{cosec}\psi|=|y\sqrt{\{1+(1/y')^2\}}|,$$
$$PG=|y\sec\psi|=|y\sqrt{\{1+y'^2\}}|.$$

23.15. Tangents and normals in polar coordinates

In fig. 107, $P(r, \theta)$ is any point on a plane curve, O is the pole and Ox the initial line. The perpendicular to OP drawn through O meets the tangent and normal at P in H and K respectively. OH and OK are known as the polar subtangent and the polar subnormal respectively; the lengths PH and PK are called the polar tangent and the polar normal at P.

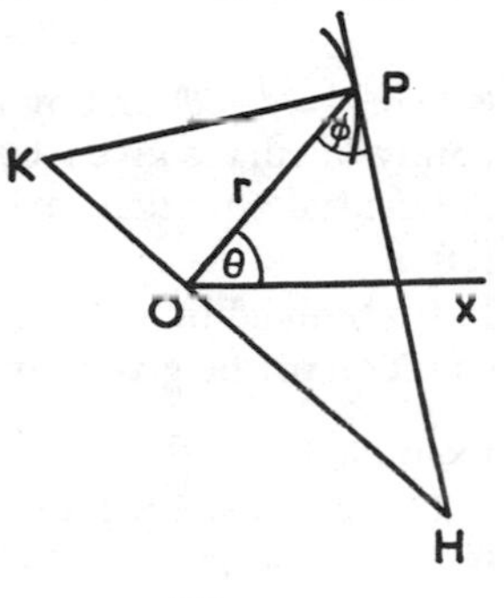

Fig. 107

Now, in the usual notation

$$\angle OPH = \angle PKH = \phi$$

and $$\tan \phi = r/r',$$

where $$r' = dr/d\theta \text{ (see § 21.3).}$$

Hence

$$OH = r \tan \phi = r^2/r',$$
$$OK = r \cot \phi = r',$$
$$PH = | r \sec \phi | = | r\sqrt{\{1 + (r/r')^2\}} |,$$
$$PK = | r \operatorname{cosec} \phi | = | r\sqrt{\{1 + (r'/r)^2\}} |.$$

23.16. Orthogonal trajectories

Two families of curves, such that each member of one family cuts every member of the other family at right angles, are called *orthogonal trajectories* of one another.

From the equation $f(x, y, c) = 0$ representing a one-parameter family of curves we can form a differential equation of the first order

$$F(x, y, y') = 0 \quad . \quad . \quad . \quad . \quad \text{(i)}$$

which is the differential equation of the family.

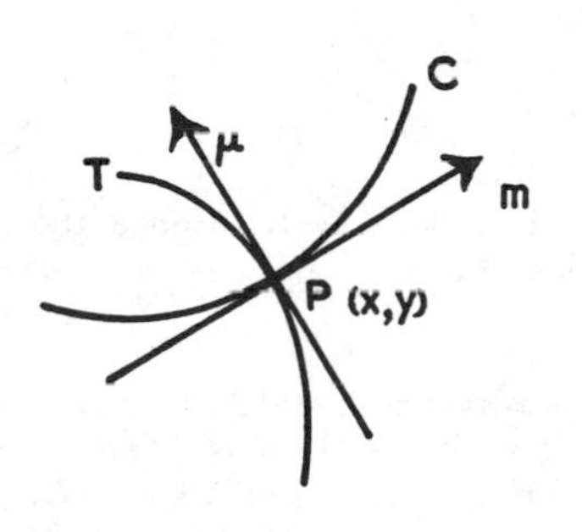

Fig. 108

Now through any point $P(x, y)$ there passes a curve C of the given family and a curve T which is a member of the family of orthogonal trajectories. Let m and μ be the gradients of the tangents at P to the curves C and T respectively (fig. 108).

Since C is a curve of the given family, by (i) $F(x, y, m) = 0$. But C and T cut orthogonally and so $m = -1/\mu$.

$$\therefore F(x, y, -1/\mu) = 0. \quad . \quad \text{(ii)}$$

Now, by definition of μ, $\mu = y'$ for the curve T and at P, x and y are the same for both curves. Hence if we write y' for μ in (ii) we shall obtain the differential equation of the family of orthogonal trajectories. This equation is

$$F(x, y, -1/y') = 0.$$

It is obtained by writing $-1/y'$ for y' in the differential equation of the given family.

For a family of curves given by the polar equation

$$f(r, \theta, c)=0 \quad . \quad . \quad . \quad . \quad \text{(iii)}$$

we may establish the differential equation of the family

$$F(r, \theta, r/r')=0,$$

(where $r'=dr/d\theta$) and by reasoning similar to that used in the case of a family of curves given by a cartesian equation, we may show that the differential equation satisfied by the orthogonal trajectories to family (iii) is

$$F(r, \theta, -r'/r)=0.$$

It is obtained by writing $rd\theta$ for dr and $-dr$ for $rd\theta$ in the differential equation of the given family.

Example 15

The normal at any point $P(x, y)$ on a certain curve meets the axes of x and y respectively at points Q and R on opposite sides of P and such that

$$RP/RQ=x^2/y^2.$$

Find the equation of the curve, given that it passes through the point (1, 1). [L.U.]

The equation of the normal at P is

$$X-x+y'(Y-y)=0.$$

The normal meets the x-axis at Q, where $X=x+yy'$, $Y=0$. If O is the origin and P' is the projection of P on the x-axis

$$OP'/OQ=RP/RQ.$$

Hence
$$x/(x+yy')=x^2/y^2$$
i.e.
$$y'=(y^2-x^2)/xy.$$

The substitution $y=vx$ reduces this equation to

$$v\,dv+dx/x=0$$

which yields
$$\tfrac{1}{2}v^2=\log\,(C/x)$$
or
$$y^2/x^2=\log\,(C^2/x^2).$$

This curve passes through the point (1, 1) if $\log C^2=1$. Hence the equation of the required curve is $y^2=x^2(1-\log x^2)$.

Example 16

The normal at a point P of a curve meets the x-axis at G, and N, the foot of the ordinate of P, lies between G and the origin O. If $OG=OP$, find the differential equation of the system of curves for which this condition holds, and integrate it. Find also the equation of the orthogonal trajectories of the system and show that they are parabolas. [L.U.]

Let $P(x, y)$ be any point on a curve of the system. Then, as in Example 15

$$OG=y\frac{dy}{dx}+x$$

and
$$OP=\sqrt{(x^2+y^2)}.$$

If
$$OG = OP$$
$$y\,dy/dx + x = \sqrt{(x^2+y^2)} \quad . \quad . \quad . \quad . \quad . \qquad \text{(i)}$$
This is the differential equation of the system. To integrate it let
$$x^2+y^2=r^2$$
so that
$$x + y\,dy/dx = r\,dr/dx.$$
The equation then becomes
$$dr/dx = 1$$
$$r = x + C$$
i.e.
$$x^2+y^2=(x+C)^2$$
$$y^2 = C(C+2x) \quad . \quad . \quad . \quad . \qquad \text{(ii)}$$
This is the equation of a family of parabolas with a common focus at the origin and common axis Ox.

By § 23.16 the differential equation of the orthogonal trajectories of the system is
$$x - y\,dx/dy = \sqrt{(x^2+y^2)}$$
$$dy/dx = y/\{x - \sqrt{(x^2+y^2)}\} = -\{x + \sqrt{(x^2+y^2)}\}/y$$
$$\therefore\ y\,dy/dx + x = -\sqrt{(x^2+y^2)}$$
and, comparing this equation with (i), we obtain the solution
$$y^2 = A(A-2x) \quad . \quad . \quad . \quad . \qquad \text{(iii)}$$
This equation represents a family of parabolas with a common focus at the origin and common axis Ox; also, if we write $A = -C$ we see that (iii) represents the same system as (ii). On account of this property the given system is said to be self-orthogonal.

Curves (ii) and (iii) intersect in real points only when A and C have the same signs and so each member of the system intersects orthogonally an infinite number (but not all) of the members of the system.

Example 17

Find the equation of the system of orthogonal trajectories of the family of curves whose equation in polar coordinates is $r = a(1-\cos\theta)$, where a is a parameter.

Sketch clearly a typical curve of each system. [L.U.]

If
$$r = a(1-\cos\theta)$$
$$\log r = \log a + \log(1-\cos\theta)$$
and hence
$$\frac{1}{r}\frac{dr}{d\theta} = \frac{\sin\theta}{1-\cos\theta} = \cot\tfrac{1}{2}\theta.$$

This is the differential equation of the given system. The differential equation of the system of orthogonal trajectories is
$$-r\frac{d\theta}{dr} = \cot\tfrac{1}{2}\theta$$
i.e.
$$\frac{dr}{r} + \tan\tfrac{1}{2}\theta\,d\theta = 0.$$

Hence $$\log r+2 \log \sec \tfrac{1}{2}\theta=\log C,$$
$$r=C \cos^2 \tfrac{1}{2}\theta$$
or $$r=C'(1+\cos \theta), \text{ where } C'=\tfrac{1}{2}C.$$

A typical curve of each system of cardioids is shown in fig. 109.

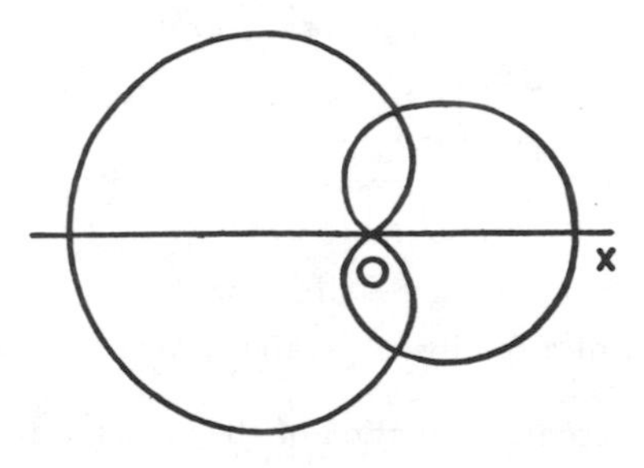

Fig. 109

Exercises 23 (*g*)

1. The tangent at a point of a curve meets the axis of y at M, the parallel through the point to the axis of y meets the axis of x at N, and O is the origin. If the area of the triangle OMN is constant, show that the curve is a hyperbola. [L.U.]

2. Find the orthogonal trajectories of the family of curves defined by the equation $3x^2+y^2+\lambda x=0$, where λ is a variable parameter. [L.U.]

3. The tangent at any point P of a curve meets the y-axis at Q and O is the origin. If $OP=OQ$, show that the curves having this property are parabolas, whose common axis is the y-axis. [L.U.]

4. P is a point on a plane curve and the ordinate from P intersects the line $y=x$ at Q, where Q is between P and the x-axis. The tangent at P to the curve intersects the line through Q parallel to the x-axis at R, where R is between Q and the y-axis. If $RQ=2$, find the equation of the family of curves.

   Show that the member of the family which passes through the point (0, 3) is $y=x+2+e^{x/2}$. [L.U.]

5. The tangent at any variable point P (in the positive quadrant) of a plane curve meets the positive part of the x-axis in T; O is the origin and M the foot of the perpendicular from P on the y-axis. If the area of the trapezium $OTPM$ is constant and equal to $3a^2$, prove that the curve is a cubic of the family $xy=Ay^3+2a^2$, where A is an arbitrary constant. [L.U.]

6. The tangent at a point P on one of a family of curves meets the y-axis in Y and the fixed straight line $y=mx$ in T. Find the differential equation of the family such that $YT=TP$, and show that it consists of parabolas touching the fixed line at the origin. [L.U.]

7. Find the orthogonal trajectories of the system of ellipses

$$x^2 - xy + y^2 = c.$$ [Sheffield.]

8. The tangent at any point $P(x, y)$ of a plane curve meets the x-axis in Q. If the area of the triangle OPQ, where O is the origin, is constant and equal to a^2, find the differential equation of the curve.

Show that the curves having this property form a family of hyperbolas with a common asymptote and sketch the curve that passes through the point $(0, a)$. [L.U.]

9. Show how to solve the equation $dy/dx + Py + Qy^n = 0$ when P and Q are functions of x only.

The tangent at the point (x, y) on a plane curve meets YOY' at T. O is the origin. Find the equation of the curve if $OT = y^{n+1}/a^n$, where a is a constant. [L.U.]

10. P is a point on a curve and the tangent at P meets the x-axis at T. The normal at P meets the y-axis at M, and the line through P parallel to XOX' meets the y-axis in L. L lies between the origin and M. If $OT = LM$, find the differential equation of the curve. Integrate this and find the curve which passes through $(1, 1)$. [L.U.]

11. The normal to a plane curve at a variable point P meets the axes of x and y at Q and R respectively. The orthogonal projection of PQ on the x-axis is equal to that of PR on the y-axis. Find the equation of the curve, given that it passes through $(1, 1)$ but not through $(0, 0)$. [L.U.]

12. $P(x, y)$ is a point on a curve C. The tangent at P meets the axis OX at T and has gradient p. The normal at P meets the axis OY at G. Prove that the gradient of TG is $(x+py)/(y-px)$.

If, for all positions of P for which T and G are determinate, TG is perpendicular to OP, show that C is a circle of the system of circles touching OX at O.

Write down the differential equation of the system of orthogonal trajectories of this system of circles. [L.U.]

13. The tangent at a point P on a plane curve meets the x-axis in the point T and PN is the ordinate at P. If $AO.AN = AT^2$, where A is the point $(-1, 0)$ find the differential equation satisfied by the curve and hence show that the curve is a member of the family $(x-ky)^2 = 4ky$. [L.U.]

14. The tangent at a point P of a plane curve meets the axis of y at the point Q. O is the origin. If $OP = 2OQ$, find the differential equation of the locus of P and solve it, showing that the locus is a curve of the third degree. [L.U.]

15. The tangent and normal to a curve at the point P intersect the x-axis at A and B, and the foot of the perpendicular from P to the x-axis is M. Find and solve the differential equation of the family of curves for which $2MP^2 = OM.AB$, where O is the origin.

Find also the differential equation of the orthogonal trajectories to these curves. [L.U.]

16. The tangent to a plane curve at a point $P(x, y)$ meets the axes of x and y in A and B respectively. If $PA : PB = kx : y$, where k is a constant, find the differential equation of the family to which the curve belongs.
    Integrate the equation, and show that the curves of the family are rectangular hyperbolas with their centres on a fixed straight line. [L.U.]

17. A plane curve has the property that the tangents from any point on the y-axis to the curve are of constant length, a. Find the differential equation of the family to which the curve belongs and integrate it.
    Show that the orthogonal trajectories of the curves are circles. [L.U.]

18. The ordinate at a point P on any one of a family of curves meets the x-axis in Q, and R is the foot of the perpendicular from Q to the normal at P to this curve. If for all positions of P the length of PR is k, find the differential equation of the family of curves.
    Find also the equation of the particular curve of the family which passes through the point $(0, k)$. [L.U.]

19. N is the foot of the ordinate at a point P on a plane curve, and T, U are the respective points at which the tangent and normal at P meet the axis of x. If $TN - NU = 2ON$, where O is the origin, find the differential equation of the curve, and hence show that the curve belongs to a family of parabolas.
    Find the two members of the family which pass through the point $(-\frac{3}{4}, 1)$. [L.U.]

20. P is any point on a plane curve, T is the foot of the perpendicular from P to the x-axis, and O is the origin of coordinates. The perpendicular at P to the radius vector OP is met at Q by a line through O parallel to the tangent to the curve at P. Given that $PQ = OT$, find the equation of the curve if it passes through the point $(1, 1)$ [L.U.]

21. Prove that a differential equation of the form

$$p^2 + p\phi(x, y) - 1 = 0, \quad (p = dy/dx)$$

    represents a system of plane curves such that two pass through every point and intersect at right angles.
    Find the system of curves for which $\phi(x, y) = -2y/x$. [L.U.]

22. The tangent at a point $P(x, y)$ on a plane curve meets the x-axis at T. If $PT = ay^2$ where a is a constant, find the differential equation of the family of curves to which this curve belongs and integrate the equation.
    Show that the orthogonal trajectories of the family are the curves $ay = \cosh(ax + b)$. Find the equations of the two curves, one from each family, which pass through the point $(0, 1/a)$. [L.U.]

CHAPTER 24

LINEAR DIFFERENTIAL EQUATIONS WITH CONSTANT COEFFICIENTS

24.1. The operator D

In this chapter we shall write Dy to denote dy/dx and $D^r y$ to denote $d^r y/dx^r$ so that the symbol D represents the differential operator d/dx.

With this notation the rules of differentiation give

$$D(u+v) = Du + Dv,$$
$$D^m(D^n y) = D^n(D^m y) = D^{m+n} y$$

when m and n are positive integers,

and $$D(ay) = aDy \text{ when } a \text{ is a constant.}$$

Using these results we write, when a and b are constants,

$$\frac{dy}{dx} - ay = (D-a)y,$$

$$\frac{d^2y}{dx^2} - (a+b)\frac{dy}{dx} + aby = \{D^2 - (a+b)D + ab\}y,$$

where any function of D is interpreted as operating on the function which follows it.

Now
$$\begin{aligned}(D-a)(D-b)y &= (D-a)(Dy-by)\\ &= D^2y - (a+b)Dy + aby\\ &= \{D^2-(a+b)D+ab\}y;\end{aligned}$$

and, similarly, $(D-b)(D-a)y = \{D^2-(a+b)D+ab\}y.$

Hence a second-order operator with constant coefficients may be factorised in the algebraic sense into two linear factors whose order of arrangement is immaterial. For example

$$\begin{aligned}(3+2D-D^2)(3\cos x + 3\sin x) &= 3(3-D)(1+D)(\cos x + \sin x)\\ &= 3(3-D)(2\cos x)\\ &= 6(3\cos x + \sin x).\end{aligned}$$

The same result is obtained from $3(1+D)(3-D)(\cos x+\sin x)$.

In the same way any function which contains only positive integral powers of D may be factorised and the factors arranged in any order.

24.2. Applications of the operator D

(a) *Operators applied to the function e^{kx} where k is constant.*

By differentiation, $De^{kx}=ke^{kx}$,

$D^n e^{kx}=k^n e^{kx}$, when n is a positive integer,

and it easily follows that if $\phi(D)$ is any polynomial in D

$$\phi(D)e^{kx}=\phi(k)e^{kx}. \qquad (24.1)$$

Thus, for example $(D^3-D^2+2)e^{3x}=(27-9+2)e^{3x}=20e^{3x}$.

(b) *Operators applied to products of the form $e^{kx}V$, where k is a constant and V is a function of x.*

By the product rule for differentiation

$$D(e^{kx}V)=e^{kx}(D+k)V$$

and, more generally, when n is a positive integer, Leibniz's theorem gives

$$D^n(e^{kx}V)=e^{kx}(D+k)^n V.$$

Hence, if $\phi(D)$ is any polynomial in D,

$$\phi(D)(e^{kx}V)=e^{kx}\phi(D+k)V. \qquad (24.2)$$

For example, $(D-1)(e^{3x}\sin x)=e^{3x}(D+2)\sin x$

$=e^{3x}(\cos x+2\sin x)$,

and $(D^2-4)\{e^{2x}(x^2+1)\}=e^{2x}(D^2+4D)(x^2+1)$

$=e^{2x}(8x+2)$.

24.3. The linear equation with constant coefficients

In a linear differential equation, y and its derivatives occur separately in the first degree only, and the linear equation of order n with constant coefficients is of the form

$$(D^n+a_1D^{n-1}+a_2D^{n-2}+\ldots+a_{n-1}D+a_n)y=f(x) \quad . \qquad \text{(i)}$$

where the a's are given constants and $f(x)$ is a given function of x.

Throughout this chapter we shall write $\phi(D)$ to denote the function

$$D^n+a_1D^{n-1}+a_2D^{n-2}+\ldots+a_{n-1}D+a_n.$$

With this notation we may write (i) in the form

$$\phi(D)y=f(x) \quad . \quad . \quad . \quad . \qquad \text{(ii)}$$

and, in the case when $f(x)=0$, we obtain the equation $\phi(D)y=0$, which is known as the *reduced equation* of (ii).

24.4. An important property of the reduced equation $\phi(D)y=0$

If $y_1, y_2,\ldots, y_n$ are n solutions of the equation $\phi(D)y=0$ and $A_1, A_2,\ldots, A_n$ are arbitrary constants, then $A_1y_1+A_2y_2+\ldots+A_ny_n$ is also a solution; for

$$\phi(D)\{A_1y_1+A_2y_2+\ldots+A_ny_n\}$$
$$=\phi(D)A_1y_1+\phi(D)A_2y_2+\ldots+\phi(D)A_ny_n$$
$$=A_1\phi(D)y_1+A_2\phi(D)y_2+\ldots+A_n\phi(D)y_n=0.$$

24.5. The solution of the reduced equation $\phi(D)y=0$

We first consider the equation

$$(D-a)y=0$$

i.e.

$$\frac{dy}{dx}-ay=0$$

for which e^{-ax} is an integrating factor.

We have

$$\frac{d}{dx}(e^{-ax}y)=0$$

and so

$$y=Ae^{ax},$$

where A is an arbitrary constant.

When $a\neq b$, the second order equation

$$\{D^2-(a+b)D+ab\}y=0. \quad . \quad . \quad . \quad \text{(i)}$$

which may be written in the form $(D-b)\{(D-a)y\}=0$ is clearly satisfied when

$$(D-a)y=0$$

i.e. when

$$y=Ae^{ax}.$$

(i) may also be written in the form

$$(D-a)\{(D-b)y\}=0$$

and so is satisfied by

$$y=Be^{bx}.$$

Hence by § 24.4, (i) is satisfied by $y=Ae^{ax}+Be^{bx}$. This function which contains two arbitrary constants if $a\neq b$ is the general solution of (i).

The nth order equation

$$\phi(D)y=0 \qquad (24.3)$$

may be treated in the same way if it is assumed that $\phi(D)$ has n distinct linear factors $(D-k_1)$, $(D-k_2),\ldots,$ $(D-k_n)$, where the k's are real or complex. These factors may be arranged in any order and hence (24.3) will be satisfied when

$$(D-k_1)y=0,\ (D-k_2)y=0,\ldots,\ (D-k_n)y=0.$$

Solving each of these equations, we have

$$y=A_1e^{k_1x},\ y=A_2e^{k_2x},\ \ldots,\ y=A_ne^{k_nx}.$$

It follows from § 24.4 that

$$y=A_1e^{k_1x}+A_2e^{k_2x}+\ldots+A_ne^{k_nx} \qquad (24.4)$$

is the general solution of (24.3) when the k's are all distinct, i.e. when $\phi(D)$ has no repeated factors.

If, however, $k_1=k_2=k_3=\ldots=k_r$, while k_{r+1}, $k_{r+2},\ldots,$ k_n are distinct, the first r terms in (24.4) can be combined into a single term of the form Ae^{k_1x}, the number of arbitrary constants in (24.4) is reduced to $n-r+1$ and (24.4) does not represent the general solution of (24.3).

In this case $(D-k_1)$ is an $r-$fold factor of $\phi(D)$.

24.6. The solution of $\phi(D)y=0$ when $\phi(D)$ has a repeated factor $(D-k_1)^r$

If $\qquad \phi(D)\equiv(D-k_1)^r(D-k_{r+1})(D-k_{r+2})\dots(D-k_n),$
the equation $\phi(D)y=0$ is satisfied when

$$(D-k_1)^r y=0 \quad . \quad . \quad . \quad . \quad \text{(i)}$$

Since one solution of equation (i) is $A_1e^{k_1x}$, to find its general solution we put $y=e^{k_1x}V$, where V is a variable dependent on x.

Then $\qquad (D-k_1)^r\{e^{k_1x}V\}=0$

i.e. $\qquad e^{k_1x}\,D^rV=0$ by (24.2)

and so $\qquad D^rV=0.$

By r successive integrations we obtain

$$V=A_1+A_2x+A_3x^2+\dots+A_rx^{r-1},$$

where the A's are arbitrary constants.

Hence, corresponding to the factor $(D-k_1)^r$ in $\phi(D)$, there is the term $(A_1+A_2x+A_3x^2+\dots+A_rx^{r-1})e^{k_1x}$, containing r arbitrary constants, in the solution of (24.3). The other terms are as found in § 24.5, and the general solution is

$$y=(A_1+A_2x+A_3x^2+\dots+A_rx^{r-1})e^{k_1x}+A_{r+1}e^{k_{r+1}x}+\dots+A_ne^{k_nx} \quad (24.5)$$

In practice the values of $k_1, k_2,\dots$ in any particular case are found by taking $y=e^{kx}$ as a trial solution of the given equation. The method is illustrated in the following examples.

Example 1

Solve the equation $\dfrac{d^2y}{dx^2}-3\dfrac{dy}{dx}-4y=0.$

This equation is $\qquad (D^2-3D-4)y=0 \quad . \quad . \quad . \quad . \quad$ (i)

If $y=e^{kx}$ is a solution

$$(D^2-3D-4)e^{kx}=0$$

i.e. $\qquad (k^2-3k-4)e^{kx}=0$ by (24.1)

and hence $\qquad k^2-3k-4=0 \quad . \quad . \quad . \quad . \quad$ (ii)

(ii) is known as the *auxiliary equation* of (i); its roots are -1 and 4. Hence, by (24.4), the general solution of (i) is

$$y=Ae^{-x}+Be^{4x},$$

where A and B are arbitrary constants.

Example 2

Solve the equation $(D^2-4)y=0.$

The auxiliary equation is $k^2-4=0$; its roots are $k=\pm2$. Hence the general solution is $y=Ae^{2x}+Be^{-2x}$.

There is an alternative form of solution, for $e^{2x}=\cosh 2x+\sinh 2x$ and $e^{-2x}=\cosh 2x-\sinh 2x$.

Hence $$y=(A+B)\cosh 2x+(A-B)\sinh 2x$$
$$=a\cosh 2x+b\sinh 2x,$$
where a and b are arbitrary constants.

Example 3

Solve the equation $(D^2+9)y=0$.

The auxiliary equation is $k^2+9=0$; its roots are $k=\pm 3i$.

Hence the general solution is

$$y=Ae^{3ix}+Be^{-3ix} \quad . \quad . \quad . \quad . \quad \text{(i)}$$

But $e^{3ix}=\cos 3x+i\sin 3x$ and $e^{-3ix}=\cos 3x-i\sin 3x$.

Hence $$y=(A+B)\cos 3x+i(A-B)\sin 3x$$

i.e. $$y=a\cos 3x+b\sin 3x \quad . \quad . \quad . \quad . \quad . \quad \text{(ii)}$$

When the roots of the auxiliary equation are purely imaginary, the solution should be given in trigonometrical form as in (ii) and not in exponential form as in (i).

Useful alternative forms to (ii) are $y=R\cos(3x+\alpha)$, $y=R\sin(3x+\alpha)$, where R and α are arbitrary (see Example 4 of § 1.5).

Example 4

Solve the equation $(D^2+4D+13)y=0$.

The auxiliary equation is $k^2+4k+13=0$; its roots are $k=-2\pm 3i$.
Hence the general solution is

$$y=Ae^{(-2+3i)x}+Be^{(-2-3i)x}$$
$$=e^{-2x}(Ae^{3ix}+Be^{-3ix})$$
$$=e^{-2x}(a\cos 3x+b\sin 3x), \text{ as in Example 3,}$$

or $$y=Re^{-2x}\sin(3x+\alpha).$$

Example 5

Solve the equation $(D^3+D^2-D-1)y=0$.

The auxiliary equation is $k^3+k^2-k-1=0$, or $(k-1)(k+1)^2=0$; its roots are $k=-1, -1, 1$.

Hence by (24.5) the general solution is $y=(A+Bx)e^{-x}+Ce^{x}$.

Example 6

Solve the equation $(D^4+8D^2+16)y=0$.

The auxiliary equation is $(k^2+4)^2=0$; its roots are

$$k=+2i, +2i, -2i, -2i.$$

Hence, by an extension of (24.5) the general solution is

$$y=(A_1+B_1x)e^{2ix}+(A_2+B_2x)e^{-2ix},$$

which may be expressed in the form

$$y=(a_1+b_1x)\cos 2x+(a_2+b_2x)\sin 2x.$$

24.7. Solution of the equation $\phi(D)y=f(x)$ when $f(x)\neq 0$

Let $y=u$ be the general solution of the reduced equation $\phi(D)y=0$, i.e. the solution containing n arbitrary constants; and let us adopt as a trial solution of the equation

$$\phi(D)y=f(x) \quad . \quad . \quad . \quad . \qquad \text{(i)}$$

$y=u+v$ where v, like u, is a function of x.

Then $$\phi(D)(u+v)=f(x)$$

i.e. $$\phi(D)u+\phi(D)v=f(x).$$

But $$\phi(D)u=0, \therefore \phi(D)v=f(x) \quad . \quad . \quad . \quad . \qquad \text{(ii)}$$

Thus v is any function of x which can be found to satisfy (i), and since the n arbitrary constants required in the general solution of (i) are contained in u, no arbitrary constant need appear in v.

In other words the general solution (G.S.) of (i) is the sum of two terms u and v: u, the general solution of the reduced equation $\phi(D)y=0$, which contains n arbitrary constants and is known as the *complementary function* (C.F.); v, a *particular integral* (P.I.), i.e. a function which satisfies (i) and which contains no arbitrary constant.

A particular integral may sometimes be found by inspection, but since, in general, this is not the case, we go on to consider methods of finding one for equation (i).

From (ii), $$\phi(D)v=f(x)$$

and we write symbolically $$v=\frac{1}{\phi(D)}f(x).$$

Our problem is to find a meaning for the operator $\dfrac{1}{\phi(D)}$.

24.8. Inverse operators

We define $\dfrac{1}{\phi(D)}y$ as a function z (if one exists) such that $\phi(D)z=y$. According to this definition the operator $\dfrac{1}{\phi(D)}$ is the inverse of the operator $\phi(D)$ and it can be shown that $\dfrac{1}{\phi(D)}$ can be broken up into factors (which may be taken in any order) or into partial fractions. It can also be shown that if u and v are functions of x,

$$\frac{1}{\phi(D)}(u+v)=\frac{1}{\phi(D)}u+\frac{1}{\phi(D)}v \qquad (24.6)$$

In the particular case where $\phi(D)=D$, we have by definition

$$D\left\{\frac{1}{D}y\right\}=y \quad . \quad . \quad . \quad . \quad \text{(i)}$$

But

$$D\left\{\int y\,dx\right\}=y$$

and so the operator $\frac{1}{D}$ or D^{-1} indicates the operation of integration.

In the same way $\frac{1}{D^r}$ or D^{-r}, where r is a positive integer, indicates the operation of r-fold integration.

Again, provided the constant of integration is correctly chosen,

$$\int \{Dy\}\,dx=y$$

i.e.

$$\frac{1}{D}(Dy)=y \quad . \quad . \quad . \quad . \quad \text{(ii)}$$

and so from (i) and (ii)

$$D\left(\frac{1}{D}\right)y=\frac{1}{D}(Dy)=y.$$

When we are using the inverse operator $\frac{1}{D^r}$ to find a particular integral, the constants of integration are omitted.

24.9. Applications of inverse operators

(a) *Inverse operators applied to the function e^{kx}.*

By (24.1),

$$\phi(D)e^{kx}=\phi(k)e^{kx},$$

and by repeated integration, we obtain

$$\frac{1}{D^n}e^{kx}=\frac{1}{k^n}e^{kx}$$

when n is a positive integer. This suggests that

$$\frac{1}{\phi(D)}e^{kx}=\frac{1}{\phi(k)}e^{kx} \text{ if } \phi(k)\neq 0. \qquad (24.7)$$

To prove this, we have to show that

$$\phi(D)\frac{1}{\phi(k)}e^{kx}=e^{kx},$$

and this follows from (24.1).

Example 7

Solve the equation $(D^2-8D+16)y=64+e^{3x}$.

The C.F., being the solution of the reduced equation $(D-4)^2y=0$, is $(A+Bx)e^{4x}$.

A P.I. is given by

$$y=\frac{1}{(D-4)^2}(64+e^{3x}).$$

$$=\frac{1}{(D-4)^2}64+\frac{1}{(D-4)^2}e^{3x} \text{ by (24.6).}$$

Treating 64 as $64e^{0x}$ and applying (24.7), we have

$$\frac{1}{(D-4)^2}64e^{0x}=\frac{1}{(-4)^2}64e^{0x}=4.$$

Also
$$\frac{1}{(D-4)^2}e^{3x}=\frac{1}{(-1)^2}e^{3x}=e^{3x} \text{ by (24.7).}$$

Hence a P.I. is $4+e^{3x}$ and the G.S. is $y=(A+Bx)e^{4x}+4+e^{3x}$.

Example 8

Solve the equation $(D^2+2D+2)y=5\cos x$.

The roots of the auxiliary equation $k^2+2k+2=0$ are $k=-1\pm i$.
Hence the C.F. is $e^{-x}(a\cos x+b\sin x)$.
To find a P.I. we assume that (24.7) is valid when k is complex so that

$$\frac{1}{D^2+2D+2}5e^{ix}=\frac{5e^{ix}}{1+2i}$$
$$=(1-2i)e^{ix}$$
$$=(1-2i)(\cos x+i\sin x) \quad . \quad . \quad \text{(i)}$$

Then since $5\cos x$ is the real part of $5e^{ix}$, the required P.I. is the real part of (i), i.e. $\cos x+2\sin x$.

The G.S. is $y=e^{-x}(a\cos x+b\sin x)+\cos x+2\sin x$.

Example 9

Solve the equation

$$\frac{d^3x}{dt^3}+\frac{2d^2x}{dt^2}-\frac{dx}{dt}-2x=sin$$

when x *satisfies the conditions* (i) $x=0$, $t=0$, (ii) $dx/dt=0$, $t=0$, (iii) x *remains finite as* $t\to\infty$. [L.U. Anc.]

The equation is $(D^3+2D^2-D-2)x=\sin t$, where $D\equiv d/dt$

i.e.
$$(D+2)(D-1)(D+1)x=\sin t.$$

The C.F. is $x=Ae^{-2t}+Be^{t}+Ce^{-t}$.

Since $\sin t$ is the imaginary part of e^{it}, we evaluate

$$\frac{1}{(D+2)(D^2-1)}e^{it}=-\frac{e^{it}}{2(2+i)} \text{ by (24.7)}$$
$$=\tfrac{1}{10}(i-2)(\cos t+i\sin t).$$

The required P.I. is $\frac{1}{10}(\cos t-2\sin t)$. Hence the G.S. is

$$x=Ae^{-2t}+Be^{t}+Ce^{-t}+\tfrac{1}{10}(\cos t-2\sin t).$$

But when $t\to\infty$, x remains finite, $\therefore\ B=0$.

When $t=0$, $x=0$, $\therefore\ 0=A+C+\frac{1}{10}$ . . . . (i)

When $t=0$, $dx/dt=0$ $\therefore\ 0=-2A-C-\frac{1}{5}$ . . . . . . (ii)

From (i) and (ii) $A=-\frac{1}{10}$, $C=0$

$\therefore$ the solution satisfying the given conditions is $x=\frac{1}{10}(\cos t-2\sin t-e^{-2t})$.

(b) *Inverse operators applied to products of the form $e^{kx}V$, where k is a constant and V is a function of x.*

By (24.2), if V is a function of x,

$$\phi(D)e^{kx}V=e^{kx}\phi(D+k)V \quad . \quad . \quad . \quad \text{(i)}$$

This result suggests that

$$\frac{1}{\phi(D)}e^{kx}V=e^{kx}\frac{1}{\phi(D+k)}V \qquad (24.8)$$

To verify (24.8) we have to show that

$$\phi(D)\left\{e^{kx}\frac{1}{\phi(D+k)}V\right\}=e^{kx}V.$$

Now, by (i), the left-hand side of this equation is equal to

$$e^{kx}\phi(D+k)\frac{1}{\phi(D+k)}V$$

i.e. to $e^{kx}V$; hence (24.8) is proved.

Example 10

Solve the equation $(D^2+D-2)y=2\cosh 2x$.

The C.F. is $Ae^{x}+Be^{-2x}$.

A P.I. is given by $\dfrac{1}{(D-1)(D+2)}(e^{2x}+e^{-2x})$,

and by (24.7) $\dfrac{1}{(D-1)(D+2)}e^{2x}=\frac{1}{4}e^{2x}$.

If we apply this method to find $\dfrac{1}{(D-1)(D+2)}e^{-2x}$ the denominator of the operating function vanishes because the factor $(D+2)$ becomes zero. We therefore use (24.7) to find $\dfrac{1}{(D-1)}e^{-2x}$ and then use (24.8).

We write

$$\frac{1}{(D+2)(D-1)}e^{-2x}=\frac{1}{(D+2)}\cdot\frac{1}{(D-1)}e^{-2x}$$

$$=\frac{1}{(D+2)}(-\tfrac{1}{3}e^{-2x}) \text{ by (24.7)}$$

$$=-\frac{1}{3(D+2)}(e^{-2x}.1)$$

$$=-\tfrac{1}{3}e^{-2x}\frac{1}{D}(1) \text{ by (24.8), taking } V=1,$$

$$=-\tfrac{1}{3}xe^{-2x}.$$

Hence a P.I. is $\frac{1}{4}e^{2x}-\frac{1}{3}xe^{-2x}$ and the G.S. is

$$y=Ae^{x}+Be^{-2x}+\tfrac{1}{4}e^{2x}-\tfrac{1}{3}xe^{-2x}.$$

Example 11

Solve the equation $(D^2+2D+5)y=4e^{-x}\sin 2x$.

The roots of the auxiliary equation $k^2+2k+5=0$ are $k=-1\pm 2i$.
Hence the C.F. is $e^{-x}(a\cos 2x+b\sin 2x)$.
A P.I. is the imaginary part of

$$\frac{1}{(D+1-2i)}\cdot\frac{1}{(D+1+2i)}4e^{(-1+2i)x}=\frac{1}{i}\cdot\frac{1}{(D+1-2i)}e^{(-1+2i)x} \text{ by (24.7)}$$

$$=-ie^{(-1+2i)x}\frac{1}{D}(1) \text{ by (24.8), taking } V=1,$$

$$=-ixe^{-x}(\cos 2x+i\sin 2x).$$

The required P.I. is $-xe^{-x}\cos 2x$. Hence the G.S. is

$$y=e^{-x}(a\cos 2x+b\sin 2x-x\cos 2x).$$

(c) *Inverse operators applied to polynomials*

If $f(x)$ is a polynomial in x and $a\neq 0$,

$$\frac{1}{D+a}\cdot f(x)=\frac{1}{a(1+D/a)}\cdot f(x)=\frac{1}{a}(1+D/a)^{-1}.f(x)$$

$$=\frac{1}{a}\{1-D/a+D^2/a^2-D^3/a^3+\ldots\}.f(x)$$

This may be shown by operating on the right-hand side with $D+a$.

Example 12

Solve the equation $(D^2+6D+8)y=x^3e^{-2x}$.

The C.F. is $Ae^{-2x}+Be^{-4x}$.
A P.I. is given by

$$\frac{1}{(D+2)(D+4)}x^3e^{-2x}=e^{-2x}\frac{1}{D(D+2)}x^3, \text{ by (24.8)}$$

$$=\frac{1}{2}e^{-2x}\frac{1}{D}\cdot\left(1-\frac{D}{2}+\frac{D^2}{4}-\frac{D^3}{8}+\ldots\right)x^3$$

$$=\frac{1}{2}e^{-2x}\frac{1}{D}(x^3-\tfrac{3}{2}x^2+\tfrac{3}{2}x-\tfrac{3}{4})$$

$$=\frac{1}{2}e^{-2x}\left(\frac{x^4}{4}-\frac{x^3}{2}+\frac{3x^2}{4}-\frac{3x}{4}\right).$$

The general solution is $Ae^{-2x}+Be^{-4x}+\frac{1}{8}e^{-2x}(x^4-2x^3+3x^2-3x)$.
It is usually best to perform the operation $1/D$ last.

Example 13

Solve the equation $(D^2+2D+2)y=50x \cos 2x$.

The C.F. is $e^{-x}(A \cos x+B \sin x)$.
Also, $x \cos 2x$ is the real part of xe^{2ix} and so to find a P.I. we evaluate

$$\frac{1}{D^2+2D+2}50xe^{2ix}=\frac{1}{(D+1+i)}\cdot\frac{1}{(D+1-i)}50xe^{2ix},$$

$$=50e^{2ix}\frac{1}{(D+1+3i)}\cdot\frac{1}{(D+1+i)}\cdot x,$$

$$=\frac{50e^{2ix}}{(1+3i)(1+i)}\left(1-\frac{D}{1+3i}+\ldots\right)\left(1-\frac{D}{1+i}+\ldots\right)x,$$

$$=\frac{50e^{2ix}}{4i-2}\left(x-\frac{1}{1+3i}-\frac{1}{1+i}\right),$$

$$=-(2i+1)e^{2ix}\{5x+(4i-3)\},$$

$$=e^{2ix}\{(11-5x)+i(2-10x)\}.$$

The real part of this expression, i.e. $(11-5x)\cos 2x-(2-10x)\sin 2x$, is a P.I. of the given equation. Hence the G.S. is

$$y=e^{-x}(A \cos x+B \sin x)+\{(11-5x)\cos 2x-(2-10x)\sin 2x\}.$$

24.10. Use of operators in the evaluation of integrals

The theory of operators may be used to evaluate certain types of integrals.

Example 14

$$\int x^4 e^{-2x} dx = \frac{1}{D} x^4 e^{-2x}, \text{ where } D \equiv d/dx$$

$$= e^{-2x} \frac{1}{D-2} x^4 \text{ by (24.8)}$$

$$= -\tfrac{1}{2} e^{-2x} (1 + D/2 + D^2/4 + D^3/8 + D^4/16 + \ldots) x^4$$

$$= -\tfrac{1}{2} e^{-2x} (x^4 + 2x^3 + 3x^2 + 3x + \tfrac{3}{2})$$

the constant of integration being omitted.

24.11. Simple harmonic motion ; damped harmonic motion

If a particle moves along the x-axis in such a way that its displacement from O at time t is x, the velocity of the particle at time t is $\dot{x}$ and its acceleration is $\ddot{x}$, where dots denote differentiation with respect to t.

If the acceleration of the particle is proportional to its displacement from O and is directed towards O, the differential equation of the motion of the particle is $\ddot{x} = -n^2 x$.

This is the differential equation of *simple harmonic motion.* Its solution may be written in either of the forms

$$x = a \cos nt + b \sin nt$$

or

$$x = R \cos (nt + \alpha),$$

where a, b, R and α are arbitrary constants.

If, in addition, there is a resistance proportional to the velocity, the motion of the particle is referred to as *damped harmonic motion* and the differential equation of the motion is of the form

$$\ddot{x} + 2p\dot{x} + n^2 x = 0,$$

where the constant p is positive. There are three cases depending on the nature of the roots of the auxiliary equation.

Example 15

Solve the equation $\ddot{x} + 4\dot{x} + 29x = 0$.

The roots of the auxiliary equation $k^2 + 4k + 29 = 0$ are $k = -2 \pm 5i$, and the G.S. is $\quad x = e^{-2t} (a \cos 5t + b \sin 5t)$,

or $\quad x = Re^{-2t} \cos (5t + \alpha)$.

Example 16

Solve the equation $\ddot{x} + 4\dot{x} + 3x = 0$.

The roots of the auxiliary equation $k^2 + 4k + 3 = 0$ are $k = -1, -3$, and the G.S. is $\quad x = Ae^{-t} + Be^{-3t}$.

Example 17

Solve the equation $\quad \ddot{x} + 4\dot{x} + 4x = 0$.

The roots of the auxiliary equation $k^2 + 4k + 4 = 0$ are $k = -2, -2$, and the G.S. is $\quad x = e^{-2t}(A + Bt)$.

In all three cases, as $t \to \infty$, $x \to 0$, i.e. the motion ultimately dies out and the particle tends to a position of rest at the origin.

In Example 15 the motion may be roughly described as being oscillatory with constant period $2\pi/5$ and with decreasing amplitude Re^{-2t}. It can be shown that successive amplitudes form a decreasing geometrical progression and the motion is said to be *slightly damped.*

In Example 16 the motion is said to be *heavily damped* for both x and $\dot{x}$ tend to zero as $t \to \infty$ and so the particle ultimately comes to rest at the origin. If A and B are of opposite sign, the particle may pass through the origin once before coming to rest, but the motion is not oscillatory and is known as "*dead-beat*".

In Example 17 the motion is said to be *critically damped.* If A and B are of opposite sign, the particle may pass through the origin once before coming to rest there, but the motion is not oscillatory. It is very similar to that of Example 16.

24.12. Forced oscillations

If, in addition to the force which causes simple harmonic motion and the force which causes damping, the particle is acted on by any other force depending only upon the time, the differential equation of the motion takes the form

$$\ddot{x} + 2k\dot{x} + n^2x = f(t)$$

and the motion is said to be *forced.*

In the solution of this equation the complementary function u represents the general solution for free oscillations, i.e. when there is no applied force $f(t)$. The particular integral v represents the effect of the applied force $f(t)$ on the displacement x.

Since $x = u + v$ and we have already shown that $u \to 0$ as $t \to \infty$, it follows that $x \to v$ as $t \to \infty$. For this reason v is sometimes called the *steady state* and the part u which dies away is called the *transient.*

Example 18

If $\ddot{x} + 4\dot{x} + 29x = \cos 5t$, find x in terms of t and deduce that when t is large the motion of the particle is approximately simple harmonic and of period $2\pi/5$.

We have shown in Example 16 that the C.F. is $e^{-2t}(a \cos 5t + b \sin 5t)$.

A P.I. is given by the real part of

$$\frac{1}{(D+2-5i)(D+2+5i)}e^{5it}, \text{ where } D \equiv d/dt.$$

Now $$\frac{1}{(D+2-5i)(D+2+5i)}e^{5it} = \frac{e^{5it}}{4(1+5i)} = \frac{1}{104}(1-5i)(\cos 5t + i \sin 5t)$$

and the real part of this expression is $\frac{1}{104}(\cos 5t + 5 \sin 5t)$.

Hence the G.S. is

$$x = e^{-2t}(a \cos 5t + b \sin 5t) + \tfrac{1}{104}(\cos 5t + 5 \sin 5t).$$

When t is large, $x = \frac{1}{104}(\cos 5t + 5 \sin 5t)$ approximately, i.e. the motion of the particle is approximately simple harmonic and of period $2\pi/5$.

Example 19

Solve the equation $\ddot{x}+n^2x=a\cos pt$, given that $x=\dot{x}=0$ when $t=0$.

The C.F. is $A\cos nt+B\sin nt$.

(i) When $p\neq n$, a P.I. is the real part of $\dfrac{1}{D^2+n^2}ae^{ipt}$, where $D\equiv d/dt$.

Now $\dfrac{1}{D^2+n^2}ae^{ipt}=\dfrac{a}{-p^2+n^2}(\cos pt+i\sin pt)$ by (24.7).

Hence the required P.I. is $(a\cos pt)/(n^2-p^2)$ and the G.S. is

$$x=A\cos nt+B\sin nt+(a\cos pt)/(n^2-p^2).$$

Substituting $x=\dot{x}=0$ when $t=0$ we find that

$$B=0 \text{ and } A=-a/(n^2-p^2),$$

whence $$x=a(\cos pt-\cos nt)/(n^2-p^2).$$

(ii) When $p=n$, a P.I. is the real part of

$$\frac{1}{(D+in)}\cdot\frac{1}{(D-in)}ae^{int}=\frac{a}{2in}\,\frac{1}{(D-in)}e^{int} \text{ by (24.7)}$$

$$=-\frac{ia}{2n}e^{int}\frac{1}{D}(1) \text{ by (24.8), taking } V=1,$$

$$=-\frac{iat}{2n}(\cos nt+i\sin nt).$$

The required P.I. is $\dfrac{at}{2n}\sin nt$ and the G.S. is

$$x=A\cos nt+B\sin nt+\frac{at}{2n}\sin nt.$$

Substituting $x=\dot{x}=0$ when $t=0$, we find that $A=B=0$

and so $$x=\frac{at}{2n}\sin nt.$$

The result in (ii) may be deduced as the limit as $p\to n$ of the result in (i)
It will be noted that in case (ii), where the period of the forced oscillations is the same as the period of the free vibrations, $x\to\infty$ as $t\to\infty$. The condition in which the frequencies of the forced and free vibrations are equal is known as the state of *resonance*.

Exercises 24 (*a*)

For brevity, y'', y' are written for d^2y/dx^2, dy/dx, and $\ddot{x}$, $\dot{x}$ for d^2x/dt^2, dx/dt respectively. Unless otherwise stated, $D\equiv d/dx$.

Solve the differential equations in Nos. 1-36 :

1. $\ddot{x}+9x=e^{-t}$, given that $x=0$ and $\dot{x}=1$ when $t=0$. [Durham.]
2. $y''+4y=x-\sin 3x$. [Liverpool.]
3. $y''+y=e^{-x}\sin x$. [L.U.]

4. $\ddot{x}-4\dot{x}+3x=t+e^{2t}$. [Durham.]

5. $y''+y'=1+x^2$. [L.U.]

6. $y''+4y'+4y=18xe^{-2x}$. [L.U.]

7. $y''-2y'+y=e^x \cos x$. [L.U.]

8. $y''+4y'+3y=\sin x$. [Durham.]

9. $y''-5y'+4y=xe^x$. [Durham.]

10. $y''+3y'+2y=2xe^{-x}$. [Leeds.]

11. $y''+4y'+13y=e^{-2x} \cos x$. [L.U.]

12. $y''-3y'+2y=e^x(1+x)$. [L.U.]

13. $y''-4y'-5y=\cos x$. [L.U.]

14. $4\ddot{y}-4\dot{y}+5y=17 \cos t$, given that $y=2$, $\dot{y}=-7\frac{1}{2}$ when $t=0$. [Leeds.]

15. $y''-5y'+4y=xe^x$. [Durham.]

16. $y''-2y'+2y=\sin 2x$. [Durham.]

17. $a^2y''-2aby'+4b^2y=4b^3x^2$, where a and b are non-zero constants. [L.U.]

18. $y''+4y'+4y=\cosh 2x$. [L.U.]

19. $y''-2y'+y=(x+1)^2e^{2x}$. [L.U.]

20. $y''+2y'+3y=e^{-x}+\cos 2x$. [Durham.]

21. $y''+2y'+2y=x+\sin x$. [Durham.]

22. $y''-9y=\cosh 3x+x^2$. [L.U.]

23. $\ddot{s}+6\dot{s}+25s=24 \cos 4t$. [Durham.]

24. $y''-4y'+5y=xe^{2x} \sin x$. [Sheffield.]

25. $y''-2y'+2y=e^x \sin x$, given that $y=0$ when $x=0$, and $y'=0$ when $x=2\pi$. [Durham.]

26. $y''+4y'+5y=(1+e^{-2x}) \cos x$. [Liverpool.]

27. $2y''+3y'-2y=x^2+\sin 2x$. [L.U.]

28. $(D^3+4D)y=\sin x-x$. [Sheffield.]

29. $(D^3+2D^2+4D+8)y=\cos x+2 \sin x+5+6x$. [Sheffield.]

30. $(D^5-3D^3-4D)y=2x^3+3+\sin 2x$. [Sheffield.]

31. (i) $(D^2+4)(D-1)^2y=16e^{2x}+6 \cos x$;
(ii) $(D^3+D^2+D+1)y=e^{-x}+x^4$. [Sheffield.]

32. $(D^4+3D^2-4)y=170e^x \sin^2 x$. [Sheffield.]

33. $(D^3+D^2+D+1)x=\cos t$, where $D\equiv d/dt$, with the initial conditions $x=Dx=D^2x=0$ at $t=0$. [L.U. Anc.]

34. $(D^3-3D^2+7D-5)y=5x^2-14x+6$, where $y=Dy=D^2y=0$ when $x=0$. [Sheffield.]

35. (i) $(D^4-1)y=\sinh x+\sin 2x$;
(ii) $D^2(D+1)y=x+\sin x$. [Sheffield.]

36. (i) $(D^2-2D+5)y=\cos 2t$, where $D\equiv d/dt$;
(ii) $(D^3-D^2+4D-4)y=5e^t$, where $D\equiv d/dt$. [L.U.]

37. (i) Find the general solution of the equation

$$(D^4+3D^2-4)y=x^2+6 \sin x+24e^{2x}.$$

(ii) The abscissa of a point moving along the x-axis satisfies the equation $\dot{x}+2x=4ae^{-4t}$, where $a>0$.

When $t=0$, $x=a$. Find the maximum value of x for $t>0$. [Sheffield.]

38. (i) Find the general solution of the equation

$$(D^5-D)y=1+6e^{2x}+\sin 2x.$$

(ii) The abscissa of a point moving along the x-axis satisfies the equation $\ddot{x}+9n^2x=-6an^2 \cos nt$, where a, n are positive constants.
It is given that $x=-2a/3$, $\dot{x}=0$ when $t=0$. Find the maximum speed of the point. [Sheffield.]

39. Find the particular integral of $\ddot{x}+2h\dot{x}+(h^2+p^2)x=ke^{-ht} \cos pt$, where x is the distance, t the time, and h, p, k are constants. Show that it represents a vibration of variable amplitude, and that this amplitude is a maximum when $ht=1$. [L.U.]

40. The differential equation of a vibrating system is

$$\ddot{x}+2\dot{x}+5x=4e^{-t} \cos 2t,$$

where x represents distance and t time.

Solve the equation completely and find the particular solution satisfying the initial conditions $x=\dot{x}=0$ when $t=0$.

Show that the particular solution represents an oscillation of varying amplitude, the maximum value of the amplitude being $1/e$. [L.U.]

24.13. The homogeneous linear equation

This name is applied to an equation of the form

$$a_0x^n\frac{d^ny}{dx^n}+a_1x^{n-1}\frac{d^{n-1}y}{dx^{n-1}}+a_2x^{n-2}\frac{d^{n-2}y}{dx^{n-2}}+\ldots+a_{n-1}x\frac{dy}{dx}+a_ny=f(x),$$

where the a's are constants.

Such an equation may be reduced to one with constant coefficients by the substitution $x=e^t$ if $x>0$ (or $x=-e^t$ if $x<0$).

Then $$\frac{dy}{dx}=\frac{dy}{dt}\cdot\frac{dt}{dx},$$

i.e. $$\frac{dy}{dx}=e^{-t}\frac{dy}{dt} \quad . \quad . \quad . \quad \text{(i)}$$

Thus the operators d/dx and $e^{-t}d/dt$ are equivalent ; and if we write $d/dx\equiv D$ and $d/dt\equiv \mathscr{D}$

$$Dy=e^{-t}\mathscr{D}y$$

$$D^2y=e^{-t}\mathscr{D}(e^{-t}\mathscr{D}y)=e^{-2t}(\mathscr{D}-1)\mathscr{D}y \text{ by (24.2)}$$

and $$D^3y=e^{-t}\mathscr{D}\{e^{-2t}(\mathscr{D}-1)\mathscr{D}y\}=e^{-3t}(\mathscr{D}-2)(\mathscr{D}-1)\mathscr{D}y.$$

These results give

$$xD\equiv\mathscr{D},\ x^2D^2\equiv\mathscr{D}(\mathscr{D}-1),\ x^3D^3\equiv\mathscr{D}(\mathscr{D}-1)(\mathscr{D}-2), \text{ and so on.}$$

Example 20

Solve the differential equation

$$x^2\frac{d^2y}{dx^2}+2x\frac{dy}{dx}-2y=x^3+\log x. \quad \text{[Durham.]}$$

By the standard substitution $x=e^t$,

$$\{\mathscr{D}(\mathscr{D}-1)+2\mathscr{D}-2\}y=e^{3t}+t, \text{ where } \mathscr{D}\equiv d/dt.$$

$$\therefore (\mathscr{D}^2+\mathscr{D}-2)y=e^{3t}+t.$$

The C.F. is Ae^t+Be^{-2t}.

A P.I. is given by

$$\frac{1}{(\mathscr{D}^2+\mathscr{D}-2)}(e^{3t}+t)=\tfrac{1}{10}e^{3t}-\tfrac{1}{2}\{1-\tfrac{1}{2}\mathscr{D}(1+\mathscr{D})\}^{-1}t$$

$$=\tfrac{1}{10}e^{3t}-\tfrac{1}{2}\{1+\tfrac{1}{2}\mathscr{D}(1+\mathscr{D})+\ldots\}t$$

$$=\tfrac{1}{10}e^{3t}-\tfrac{1}{2}(t+\tfrac{1}{2}).$$

Thus the G.S. is

$$y=Ae^t+Be^{-2t}+\tfrac{1}{10}e^{3t}-\tfrac{1}{4}(2t+1)$$

$$=Ax+B/x^2+\tfrac{1}{10}x^3-\tfrac{1}{4}(2\log x+1).$$

Exercises 24 (*b*)

For brevity, y'', y' are written for d^2y/dx^2, dy/dx respectively.

1. By the substitution $x=e^t$ transform the equation

$$x^2y''-3xy'+4y=6x^2\log x+6/x$$

and hence solve it.

Find also the particular solution such that $y=0$ and $y'=1$, when $x=1$.

2. Find the complete solution of the equation

$$x^2y''-4xy'+6y=x^3. \quad \text{[Liverpool.]}$$

3. If y is a function of x, $x=e^t$ and $D\equiv d/dx$, $\Delta\equiv d/dt$, prove that

$$x^2D^2y=\Delta(\Delta-1)y.$$

Find the solution of $x^2y''-xy'+2y=2x$, which makes $y=1$ and $y'=0$ at $x=1$. [L.U.]

4. (i) Solve the equation $y''-2y'+y=3\sinh x$.

(ii) If $x=e^z$, show that $xy'=dy/dz$ and $x^2y''=d^2y/dz^2-dy/dz$.

Hence, or otherwise, solve the equation

$$x^2y''-3xy'+5y=3x+\log x.$$

[L.U.]

5. (i) Solve the differential equation $y''-4y'+4y=e^{2x}$.

(ii) If $x=e^\theta$ and D and $\mathscr{D}$ denote the operators d/dx and $d/d\theta$ respectively, prove that $x^2D^2=\mathscr{D}(\mathscr{D}-1)$.

Solve the differential equation $x^2y''+y=x^2$. [L.U.]

Solve the differential equations given in Nos. 6-24 :

6. (i) $2y''=x^3+y+y'$, given that $y=0$ and $y'=1$ when $x=0$;

(ii) $x^2y''+3xy'+y=x^2$. [L.U.]

7. (i) $3y''-5y'+2y=e^x(1+x)$;

(ii) $x^2y''+5xy'-5y=\{1+\cos(\log x)\}/x^2$. [L.U.]

8. (i) $y''-2y'+2y=e^x(1+\cos x)$;

(ii) $x^2y''-xy'+y=x^2+\log x$. [L.U.]

9. (i) $y''-5y'+6y=x^3e^x$;

(ii) $x^2y''+4xy'+2y=(x-1/x)^2$. [L.U.]

10. (i) $y''-y=\cosh x$;

(ii) $x^2y''-xy'-3y=x^2\log x$. [L.U.]

11. (i) $y''+y'+y=e^x\sin x$;

(ii) $x^2y''-3xy'+4y=x^2$. [Leeds.]

12. (i) $y''-2y'+2y=e^x\cos x$;

(ii) $x^2y''-xy'+y=4x^3$. [L.U.]

13. (i) $y''+4y'+5y=4x+6e^{-3x}+\sin x$;

(ii) $x^2y''-xy'+y=2x\log x$, given that $y=1$, $y'=0$ when $x=1$. [L.U.]

14. (i) $y''-6y'+9y=e^{3x}(1+x)$;

(ii) $x^2y''-2xy'-4y=x^2+2\log x$. [L.U.]

15. (i) $y''-2y'+5y=e^x\sin 2x$;

(ii) $x^2y''-xy'+y=x^3$. [L.U.]

16. (i) $y''+6y'+9y=x^4+x+e^{-3x}$;

(ii) $x^2y''-2xy'+2y=(x+1)^2$, given that $y(1)=\frac{3}{2}$, $y(e)=\frac{1}{2}$. [Liverpool.]

17. (i) $y''+y'+y=e^x(x+\sin x)$;
(ii) $4x^2y''+xy'-y=x+\log x$. [L.U.]

18. (i) $y''+y=e^x\cos x+x^2$;
(ii) $x^2y''+4xy'+2y=(\log x)/x^2$. [L.U.]

19. (i) $y''-y'-2y=e^{-x}(1+\sin x)$;
(ii) $x^2y''+5xy'+3y=(1+1/x)^2\log x$. [L.U.]

20. (i) $y''-4y'+13y=e^{2x}\sin 3x$;
(ii) $9x^2y''+3xy'+y=x\log x$. [L.U.]

21. (i) $y''-4y'+5y=e^{2x}\sin x$;
(ii) $x^2y''-3xy'+4y=x^2\log x$. [L.U.]

22. (i) $y''-4y'+4y=\cosh 2x+x\sinh 2x$;
(ii) $x^2y''+4xy'+2y=x-1/x$. [L.U.]

23. $2x^3y'''+4x^2y''+xy'-y=3x$.

24. $x^3y'''-3x^2y''+6xy'-6y=12x^4-x^2$.

24.14. Simultaneous linear equations with constant coefficients

The application of the theory of operators to the solution of simultaneous linear equations with constant coefficients is demonstrated in the following examples.

Example 21

Solve the equations

$$\frac{dx}{dt}+y=\sin t+1,$$

$$\frac{dy}{dt}+x=\cos t,$$

subject to the conditions $x=2$, $y=1$, *when* $t=0$.

The equations may be written in the form

$$Dx+y=\sin t+1 \quad \text{(i)}$$

$$x+Dy=\cos t \quad \text{(ii)}$$

where $D\equiv d/dt$.

Operating on (i) with D, we have

$$D^2x+Dy=\cos t \quad \text{(iii)}$$

and from (ii) and (iii)

$$(D^2-1)x=0$$

which yields $\quad x=Ae^t+Be^{-t} \quad$ (iv)

But, from (i) $\quad y=\sin t+1-Dx$

i.e. $\quad y=\sin t+1-Ae^t+Be^{-t}$.

The given conditions, $x=2$ and $y=1$ when $t=0$ lead to the equations $2=A+B$ and $0=B-A$ so that $A=B=1$. Hence the particular solutions are

$$x=e^t+e^{-t}$$
$$y=\sin t+1-e^t+e^{-t}.$$

It should be observed that if y is found from (ii),

$$Dy=\cos t-Ae^t-Be^{-t}$$
$$\therefore y=\sin t-Ae^t+Be^{-t}+C \quad . \quad . \quad . \quad . \quad \text{(v)}$$

and substituting from (iv) and (v) in (i) we find that $C=1$.

Again, if y is found by eliminating x between (i) and (ii),

$$(D^2-1)y=-2\sin t-1$$
$$\therefore y=ae^t+be^{-t}+\sin t+1 \quad . \quad . \quad . \quad \text{(vi)}$$

Substitution from (iv) and (vi) into (i) or (ii) gives $a=-A$ and $b=B$, so that there are in fact only two independent arbitrary constants in the solution of (i) and (ii).

It is useful to note that the number of independent arbitrary constants required in the solution of the equations

$$f_1(D)x+F_1(D)y=\phi_1(t)$$
$$f_2(D)x+F_2(D)y=\phi_2(t),$$

where f_1, f_2, F_1 and F_2 are polynomials with constant coefficients and $D\equiv d/dt$, corresponds to the degree in D of the determinant

$$\begin{vmatrix} f_1(D) & F_1(D) \\ f_2(D) & F_2(D) \end{vmatrix}.$$

To avoid introducing more arbitrary constants than are required, when one of the unknowns (in this case x) has been determined, the other (in this case y) should if possible be found by means of a relation which does not involve the derivatives of y.

Example 22

Solve the simultaneous differential equations

$$2\frac{di_1}{dt}+i_1+\frac{di_2}{dt}=\cos t,$$

$$\frac{di_1}{dt}+2\frac{di_2}{dt}+i_2=0,$$

subject to the conditions that $i_2=i_1=0$ when $t=0$. [L.U.]

The equations are

$$(2D+1)i_1+Di_2=\cos t \quad . \quad . \quad . \quad . \quad \text{(i)}$$
$$Di_1+(2D+1)i_2=0 \quad . \quad . \quad . \quad . \quad \text{(ii)}$$

where $D\equiv d/dt$.

We eliminate i_1 by operating with D on (i), and with $(2D+1)$ on (ii)

$$D(2D+1)i_1+D^2i_2=-\sin t,$$

$$D(2D+1)i_1+(2D+1)^2i_2=0,$$

whence

$$(3D^2+4D+1)i_2=\sin t.$$

The C.F. is $Ae^{-t}+Be^{-\frac{1}{3}t}$.

A P.I. is the imaginary part of

$$\frac{1}{3D^2+4D+1}e^{it}=\frac{1}{4i-2}e^{it}$$

$$=-\tfrac{1}{10}(1+2i)(\cos t+i\sin t).$$

Hence a P.I. is $-\frac{1}{10}(\sin t+2\cos t)$ and the G.S. is

$$i_2=Ae^{-t}+Be^{-\frac{1}{3}t}-\tfrac{1}{10}(\sin t+2\cos t).$$

The given equations are each of the first order so that the expressions for i_1 and i_2 should contain only two arbitrary constants. To avoid introducing further constants we eliminate Di_1 between (i) and (ii) and obtain i_1 in terms of i_2 and its derivatives.

We multiply (ii) by 2 and subtract from (i) obtaining

$$i_1-(3D+2)i_2=\cos t$$

i.e.

$$i_1=\cos t+(3D+2)\{Ae^{-t}+Be^{-\frac{1}{3}t}-\tfrac{1}{10}(\sin t+2\cos t)\}$$

$$=Be^{-\frac{1}{3}t}-Ae^{-t}+\tfrac{1}{10}(3\cos t+4\sin t).$$

Applying the given conditions, we find that

$$0=B-A+\tfrac{3}{10},\ 0=A+B-\tfrac{1}{5}.$$

Hence $A=\frac{1}{4}$, $B=-\frac{1}{20}$,

and

$$i_1=\tfrac{1}{20}(6\cos t+8\sin t-e^{-\frac{1}{3}t}-5e^{-t}),$$

$$i_2=\tfrac{1}{20}(5e^{-t}-e^{-\frac{1}{3}t}-4\cos t-2\sin t).$$

Exercises 24 (*c*)

For brevity, $\dot{x}$, $\dot{y}$ are written for dx/dt, dy/dt respectively.

1. Given that x and y are functions of t such that $\dot{x}=3x-y$, $\dot{y}=x+y$, and $x=1$, $y=0$ at $t=0$, show that $x-y=e^{2t}$. [Liverpool.]

2. Solve the simultaneous equations $\dot{y}+ay=x$, $\dot{x}+ax=y$, given that $x=0$ and $y=1$ when $t=0$. [L.U.]

3. A point (x, y) moves in accordance with the equations

$$\dot{x}+2y=5e^t,\ \dot{y}-2x=5e^t.$$

It is given that $x=-1$ and $y=3$ when $t=0$. Show that the point moves in a straight line. [Sheffield.]

4. Find x and y in terms of t, given that $\dot{y}+y=3x$, $\dot{x}+2x=2y$, and that $x=0$ and $\dot{y}=\frac{1}{2}$ when $t=0$. [L.U.]

Solve the pairs of simultaneous differential equations given in Nos. 5-18:

5. $\dot{x}+2x+y=0$, $\dot{y}+x+2y=0$, subject to the conditions that $x=1$ and $y=0$ when $t=0$. [L.U.]

6. $\dot{x}=4x-2y+e^t$, $\dot{y}=6x-3y$. [Leeds.]

7. $3\dot{x}+3x+2y=e^t$, $4x-3\dot{y}+3y=0$. [L.U.]

8. $\dot{y}-2x=\cos 2t$, $\dot{x}+2y=-\sin 2t$. [L.U.]

9. $\dot{x}+3x-2y=1$, $\dot{y}-2x+3y=e^t$, given that, when $t=0$, $x=y=0$. [L.U.]

10. $\dot{x}+\dot{y}+2x+y=e^{-3t}$, $\dot{y}+5x+3y=5e^{-2t}$, given that, when $t=0$, $x=-1$ and $y=4$. [L.U.]

11. $\dot{x}+x-y=te^t$, $2y-\dot{x}+\dot{y}=e^t$. Find the particular solution for which $x=y=0$ when $t=0$. [L.U.]

12. $\dot{x}+5x-2y=t$, $\dot{y}+2x+y=0$, given that $x=0$ and $y=0$ when $t=0$. [L.U.]

13. $2\dot{x}+\dot{y}=2t-x-2y$, $\dot{x}+\dot{y}=3t+x-3y$.
   If $x=0$ and $y=-\frac{1}{5}$ at the initial instant $t=0$, show that the point (x, y) describes the curve $5x+1=e^{x-y-1/5}$.

14. $\dot{x}+\dot{y}+x=0$, $2\dot{x}+\dot{y}-y=1$, subject to the conditions $x=y=0$ at $t=0$. [L.U.]

15. $2\dot{x}-\dot{y}+3x=2t$, $\dot{x}+2\dot{y}-2x-y=t^2-t$, given that $x=1$ and $y=1$ when $t=0$. [L.U.]

16. $\dot{x}+x-2\dot{y}=\cos t$, $\dot{x}-\dot{y}+6y=0$. [L.U.]

17. $3x+\dot{y}-5y=e^t$, $\dot{x}-5x+3y=t$, given that when $t=0$, $x=-3/7$ and $y=-4/7$. [L.U.]

18. The coordinates of a point moving in the x-y plane satisfy the differential equations $\dot{x}+y=\sin 2t$, $\dot{y}-x=2\cos 2t$.
   It is given that $x=1$, $y=0$ when $t=0$. Prove that the path of the point is given by $y^2=(1-x^2)(2x+1)^2$. [Sheffield.]

19. The coordinates x, y of a point P moving in the x-y plane satisfy the equations
$$\dot{x}-y=2\cos t, \quad \dot{x}+\dot{y}+2x+2y=3\sin t-\cos t.$$
   When $t=0$, $x=1$ and $y=-2$. Find the coordinates of P when $t=\frac{1}{2}\pi$. [Sheffield.]

20. By using $\zeta=u+iv$, or otherwise, solve the simultaneous equations
$$m\frac{du}{dt}=eE-evH,$$
$$m\frac{dv}{dt}=euH,$$
   where m, e, E, and H are constants. If $u=dx/dt$ and $v=dy/dt$, find x and y as a function of the time t from your solution for ζ, and show further that if $x=y=u=v=0$, when $t=0$, $x=(E/\omega H)(1-\cos\omega t)$, $y=(E/\omega H)(\omega t-\sin\omega t)$ where $\omega=eH/m$. [L.U.]

24.15. Change of variable

We have shown how the homogeneous linear equation may be integrated by means of a change of independent variable. Other types of change of variable are illustrated in the following examples.

Example 23

By means of the substitution $u=\cos x$, solve the equation

$$\sin x\frac{d^2y}{dx^2}-\cos x\frac{dy}{dx}=y\sin^3 x.$$

If $u=\cos x$, $\frac{du}{dx}=-\sin x$

Hence
$$\frac{dy}{dx}=-\sin x\frac{dy}{du} \quad . \quad . \quad . \quad . \quad . \quad . \quad \text{(i)}$$

and
$$\frac{d^2y}{dx^2}=-\cos x\frac{dy}{du}-\sin x\frac{d^2y}{du^2}\cdot\frac{du}{dx}$$

$$=-\cos x\frac{dy}{du}+\sin^2 x\frac{d^2y}{du^2} \quad . \quad . \quad . \quad . \quad . \quad \text{(ii)}$$

Substituting in the given equation from (i) and (ii), we obtain the equation

$$\frac{d^2y}{du^2}-y=0$$

or
$$(D^2-1)y=0, \text{ where } D\equiv d/du.$$

$$\therefore y=Ae^u+Be^{-u}$$

i.e.
$$y=Ae^{\cos x}+Be^{-\cos x}.$$

Example 24

Find n such that the substitution $y=zx^n$ transforms the differential equation

$$x^2\frac{d^2y}{dx^2}+2x(x+2)\frac{dy}{dx}+2(x+1)^2y=e^{-x}\cos x$$

into one with constant coefficients. Hence solve the original equation, and show that in all solutions, y is small when x is large and positive. [L.U.]

If
$$y=zx^n, \ \frac{dy}{dx}=x^n\frac{dz}{dx}+nx^{n-1}z$$

and
$$\frac{d^2y}{dx^2}=x^n\frac{d^2z}{dx^2}+2nx^{n-1}\frac{dz}{dx}+n(n-1)x^{n-2}z.$$

With these values, the expression

$$x^2\frac{d^2y}{dx^2}+2x(x+2)\frac{dy}{dx}+2(x+1)^2y$$

becomes on simplification

$$x^{n+2}\frac{d^2z}{dx^2}+2\{x^{n+2}+(n+2)x^{n+1}\}\frac{dz}{dx}+\{2x^{n+2}+2(n+2)x^{n+1}+(n+1)(n+2)x^n\}z.$$

Thus, if we give n the value -2, the given differential equation reduces to

$$(D^2+2D+2)z=e^{-x}\cos x \quad . \quad . \quad . \quad \text{(i)}$$

where $z=x^2y$ and $D\equiv d/dx$.

The C.F. is $z=e^{-x}(A\cos x+B\sin x)$.

Now $e^{-x}\cos x$ is the real part of $e^{(i-1)x}$,

and

$$\frac{1}{(D+1-i)(D+1+i)}e^{(i-1)x}$$

$$=\frac{1}{D+1-i}\cdot\frac{e^{(i-1)x}}{2i} \quad \text{by (24.7)}$$

$$=-\tfrac{1}{2}ie^{(i-1)x}\frac{1}{D}\cdot 1 \quad \text{by (24.8)}$$

$$=-\tfrac{1}{2}ix\,e^{-x}(\cos x+i\sin x).$$

The real part of this function is $\frac{1}{2}xe^{-x}\sin x$ and this is a P.I. of (i).

Hence
$$z=e^{-x}\{A\cos x+B\sin x+\tfrac{1}{2}x\sin x\}$$

i.e.
$$y=\frac{e^{-x}}{x^2}\{A\cos x+B\sin x\}+\frac{e^{-x}}{2x}\sin x.$$

Since, as $x\to+\infty$, $(e^{-x}\cos x)/x^2\to 0$, $(e^{-x}\sin x)/x^2\to 0$ and $(e^{-x}\sin x)/x\to 0$, then, in all solutions, y is small when x is large and positive.

Exercises 24 (*d*)

For brevity, y'', y' are written for dy^2/dx^2, dy/dx respectively.

1. By changing the independent variable by means of the transformation $x=z^2$, or otherwise, solve the differential equation
$$2xy''+y'+2y=x^2.$$
[L.U.]

2. Transform the differential equation
$$x^2y''+(3x^2+4x)y'+(2x^2+6x+2)y=0$$
by the substitution $x^2y=z$.

   Hence or otherwise solve the equation, subject to the conditions $y=e^{-2}$ when $x=1$, and $y=e^2$ when $x=-1$. [L.U.]

3. By means of the substitution $y=zx^{-1/2}$, transform the differential equation $4x^2y''+4xy'+(4x^2-1)y=0$, and obtain the solution for which $y=0$ when $x=\frac{1}{2}\pi$ and $y=1$ when $x=\pi$. [L.U.]

4. Transform the equation $x^2y''+(4x^2+6x)y'+(3x^2+12x+6)y=0$ by the substitution $y=z/x^3$.

   Hence, or otherwise, solve the equation, given that $y=e^{-1}$ and $y'=-4e^{-1}$ when $x=1$. [L.U.]

5. By using the substitution $z=x-y$, or otherwise, find a function y of x which satisfies the equation
$$y'^2-2y'-\frac{x^2}{1-x^2}=0$$
and for which $y=0$ when $x=0$. [Sheffield.]

6. If $t = \sin x$ and if y is a function of x, prove that

$$\frac{d^2y}{dx^2} = \frac{d^2y}{dt^2}\cos^2 x - \frac{dy}{dt}\sin x.$$

Transform the differential equation

$$y'' + y' \tan x + y \cos^2 x = 2e^{\sin x}\cos^2 x$$

into an equation connecting y and t, where $t = \sin x$. Solve the resulting equation, and hence find the solution of the given equation which satisfies the conditions $y = 1$ and $y' = 0$ when $x = 0$.

7. Show that the constant n may be chosen so that, by the substitution $y = x^n z$, the differential equation $x^2y'' + 4x(x+1)y' + (8x+2)y = \cos x$ reduces to the form $z'' + az' + bz = \cos x$, where a and b are constant.
Hence, or otherwise, solve the given equation. [L.U.]

8. Using the substitution $z = \sqrt{x}$, or otherwise, solve the equation

$$4xy'' + 2(1 - \sqrt{x})y' - 6y = e^{-2\sqrt{x}}.$$ [L.U.]

9. Prove that, in general, at every point on any curve in the (x, y) plane

$$\frac{d^2x}{dy^2} + \left(\frac{dx}{dy}\right)^3 \frac{d^2y}{dx^2} = 0.$$

Transform the differential equation

$$\frac{d^2y}{dx^2} + 3\left(\frac{dy}{dx}\right)^2 = (2x - y)\left(\frac{dy}{dx}\right)^3,$$

so that y is the independent variable and x is the unknown function of y. Hence obtain the general solution of the given differential equation. [L.U.]

10. If $x = \cosh z$, prove that $(x^2 - 1)y'' + xy' = d^2y/dz^2$.
Solve the equation $(x^2 - 1)\, y'' + xy' - y = x$. [L.U.]

11. If y is a function of x and $x = \tan u$, prove that

$$(1 + x^2)\frac{dy}{dx} = \frac{dy}{du}$$

and calculate $\dfrac{d^2y}{du^2}$ in terms of x, $\dfrac{dy}{dx}$, $\dfrac{d^2y}{dx^2}$.

Find a solution involving two arbitrary constants of the equation

$$(1 + x^2)^2y'' + 2x(1 + x^2)y' = \tan^{-1} x.$$

What is the solution for which $y = 0$ and $y' = 1$ when $x = 0$? [Sheffield.]

12. Transform the differential equation

$$y'' \cos x + y' \sin x + 4y \cos^3 x = 8 \cos^5 x$$

into one having t as independent variable, where $t = \sin x$, and hence solve the equation. [L.U.]

13. By the substitution $x = \sinh t$ transform the equation

$$(1 + x^2)y'' + xy' + y = 1 + x^2$$

and hence solve it.
Find the solution for which $y = 0$ and $y' = 0$ when $x = 0$. [L.U.]

CHAPTER 25

SPHERICAL TRIGONOMETRY

25.1. Spherical triangle

A plane cuts a sphere in a circle ; the circle is called a great circle if the plane passes through the centre of the sphere, otherwise it is called a small circle.

The figure formed on the surface of a sphere by the minor arcs of three great circles is known as a spherical triangle. The three arcs are the sides, the angles between the arcs are the angles, and the points of intersection of the arcs are the vertices of the spherical triangle.

As in the case of a plane triangle, we denote the angles of a spherical triangle ABC (fig. 110) by A, B, C. The sides BC, CA, AB are measured

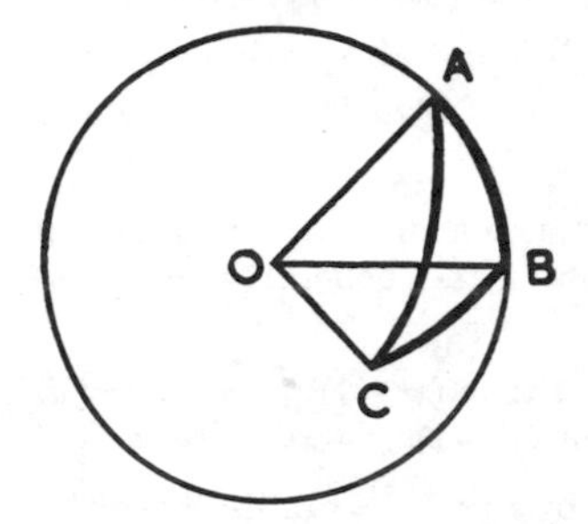

Fig. 110

by the angles a, b, c which they respectively subtend at O, the centre of the sphere. If the radius of the sphere is taken as the unit of length, $BC=a$, $CA=b$ and $AB=c$; when the radius is r, the sides are ra, rb, rc respectively.

The angle A of triangle ABC is the angle between the tangents at A to the arcs AB and AC. These tangents are both perpendicular to OA and hence angle A is the angle between the planes OAB and OAC. We shall refer to the angle between the arcs AB and AC as the angle BAC unless confusion with the angle between the straight lines AB and AC is likely to arise.

25.2. Area of a lune

Planes of two great circles intersect along a diameter of a sphere and cut off on the surface of the sphere two pairs of equal areas called lunes. For example, in fig. 111 the great circles $ABA'B'$ and $ACA'C'$ divide the sphere into four lunes :— $ABA'CA$, $ACA'B'A$, $AB'A'C'A$ and $AC'A'BA$. The angle θ between the planes of the circles is called

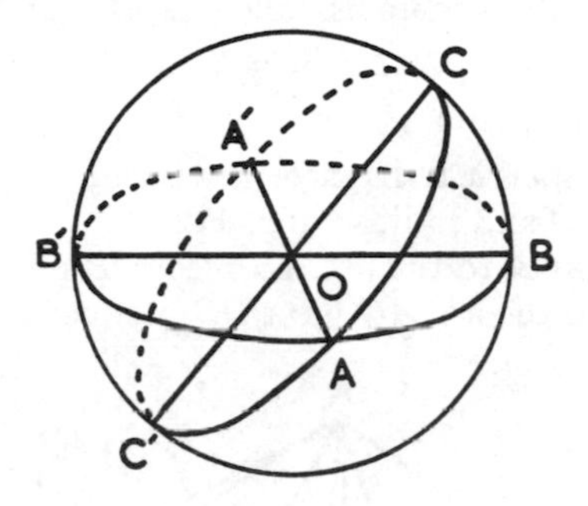

Fig. 111

the angle of the lune and the area of a lune is proportional to its angle. Hence if θ is measured in radians and r is the radius of the sphere

$$\frac{\text{area of lune}}{\text{area of surface of sphere}}=\frac{\theta}{2\pi},$$

$$\therefore \text{ area of lune}=\frac{\theta}{2\pi}\times 4\pi r^2=2r^2\theta.$$

25.3. Area of a spherical triangle

Let the great circles of which AB, BC, CA are arcs intersect again at the points A', B', C' (fig. 111). Then A', B', C' are the opposite extremities of the diameters through A, B, C. We shall use ΔABC to denote the area of the spherical triangle ABC.

$$\text{Area of lune } ABA'CA=2r^2A.$$

$$\text{i.e.} \quad \Delta ABC+\Delta A'BC=2r^2A \quad . \quad . \quad . \quad \text{(i)}$$

$$\text{Area of lune } BCB'AB=2r^2B.$$

$$\text{i.e.} \quad \Delta ABC+\Delta B'CA=2r^2B \quad . \quad . \quad . \quad \text{(ii)}$$

$$\text{Area of lune } CAC'BC=2r^2C.$$

$$\text{i.e.} \quad \Delta ABC+\Delta C'AB=2r^2C \quad . \quad . \quad . \quad \text{(iii)}$$

From (i), (ii) and (iii) we have by addition

$$2\,\Delta ABC+\{\Delta ABC+\Delta A'BC+\Delta B'CA+\Delta C'AB\}=2r^2(A+B+C).$$

But by symmetry $\Delta C'AB=\Delta CA'B'$ and triangles ABC, $A'BC$, $B'CA$, $CA'B'$ make up a hemisphere

$$\therefore\ 2\,\Delta ABC+2\pi r^2=2r^2(A+B+C),$$

$$\Delta ABC=r^2(A+B+C-\pi).$$

$A+B+C-\pi$ is called the spherical excess of the triangle and is generally denoted by E. Thus

$$\Delta ABC = r^2 E. \tag{25.1}$$

If the radius of the sphere is taken as the unit of measurement

$$\Delta ABC = E.$$

25.4. The cosine formula

Let ABC be a spherical triangle described on the surface of a sphere with centre O and radius r (fig. 112).

Let the tangent at A to the arc AB meet OB produced at D and let the tangent at A to the arc AC meet OC produced at E. Join DE.

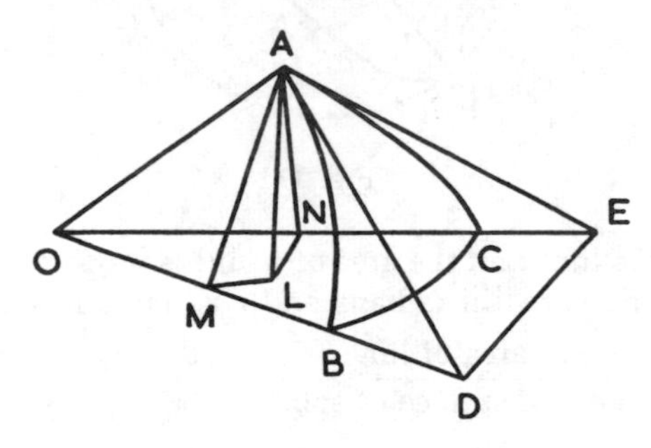

Fig. 112

Then $\angle DAE$ is angle A of the spherical triangle, and from the plane triangle DAE

$$\cos A = \frac{AE^2+AD^2-ED^2}{2AE.AD} \quad . \quad . \quad . \quad \text{(i)}$$

But $AE = r \tan b$, $AD = r \tan c$ and from ΔDOE

$$DE^2 = OE^2+OD^2-2OE.OD \cos a$$
$$= r^2(\sec^2 b + \sec^2 c - 2 \sec b \sec c \cos a).$$

Hence, from (i), $$\cos A = \frac{\sec b \sec c \cos a - 1}{\tan b \tan c},$$

or, multiplying numerator and denominator by $\cos b \cos c$,

$$\cos A = \frac{\cos a - \cos b \cos c}{\sin b \sin c}. \tag{25.2}$$

By cyclic interchange of letters corresponding formulae are obtained for $\cos B$ and $\cos C$.

(25.2) is known as the *cosine formula* and it enables us to find the angles of a spherical triangle when the three sides are known. Written in the form

$$\cos a = \cos b \cos c + \sin b \sin c \cos A \tag{25.3}$$

it enables us to find one side of a spherical triangle when we know the other two sides and the angle between them.

NOTE.—In fig. 112, b and c are both less than $\frac{1}{2}\pi$ but the cosine formula can be shown to be true in all cases.

25.5. The sine formula

If in fig. 112 we draw AL perpendicular to plane BOC and LM, LN perpendicular to OB and OC respectively in plane BOC, then AM and AN are perpendicular to OB, OC respectively and so $\angle AML$ is the angle between the planes AOB, BOC.

Hence $\angle AML = B$ and similarly $\angle ANL = C$.

$$\therefore AL = AM \sin B = AN \sin C$$

$$\therefore r \sin c \sin B = r \sin b \sin C$$

i.e.
$$\frac{\sin b}{\sin B} = \frac{\sin c}{\sin C},$$

and similarly each ratio can be shown to be equal to $\dfrac{\sin a}{\sin A}$.

Hence
$$\frac{\sin a}{\sin A} = \frac{\sin b}{\sin B} = \frac{\sin c}{\sin C}. \tag{25.4}$$

This is the *sine formula.*

25.6. The cotangent formula

If we substitute for $\cos a$ from formula (25.3) in the formula

$$\cos b = \cos c \cos a + \sin c \sin a \cos B,$$

we obtain $\cos b = \cos^2 c \cos b + \sin c(\sin a \cos B + \sin b \cos c \cos A)$.

Writing $\cos^2 c = 1 - \sin^2 c$ and dividing throughout by $\sin b \sin c$ we have

$$\cos c \cos A = \sin c \cot b - \frac{\sin a}{\sin b} \cos B$$

i.e.
$$\cos c \cos A = \sin c \cot b - \sin A \cot B. \tag{25.5}$$

By interchanging the letters, five other formulae of the same type are obtained.

Of the two angles and two sides which occur in (25.5), one of the angles is contained between the two sides and may be called the inner angle, while one of the sides lies between the two angles and may be called the inner side. With this notation the formula may be stated in the form

cos (inner side) cos (inner angle)
= sin (inner side) cot (other side) − sin (inner angle) cot (other angle).

25.7. The polar triangle

The normal to the plane of a circle through its centre is called the axis of the circle. The axis of a circle drawn on a sphere is a diameter of the sphere and its extremities are called the poles of the circle. The poles of a great circle are equidistant from the plane of the circle, but in the case of a small circle, the nearer pole is usually called *the* pole of the circle.

If ABC and $A'B'C'$, two triangles on the surface of a sphere, are so related that A' is the pole of the great circle BC on the same side of this circle as A, B' is the pole of the great circle CA on the same side of this circle as B, and C' is the pole of the great circle AB on the same side of this circle as C, $A'B'C'$ is called the polar triangle of triangle ABC (see fig. 113).

If $A'B'C'$ is the polar triangle of ABC, ABC is the polar triangle of $A'B'C'$. For if O is the centre of the sphere, $\angle B'OA = \frac{1}{2}\pi$ since B' is the pole of CA and $\angle C'OA = \frac{1}{2}\pi$ since C' is the pole of AB. Hence OA is perpendicular to plane $B'OC'$ and so A is the pole of $B'C'$. Similarly B is a pole of $C'A'$ and C is a pole of $A'B'$.

Also, $\angle AOA' < \frac{1}{2}\pi$ since A and A' are on the same side of plane BOC, and since AO is perpendicular to plane $B'OC'$, it follows that A and A' are on the same side of $B'OC'$. Similarly, B and B' are on the same side of $C'OA'$, and C and C' are on the same side of $A'OB'$. Hence ABC is the polar triangle of $A'B'C'$.

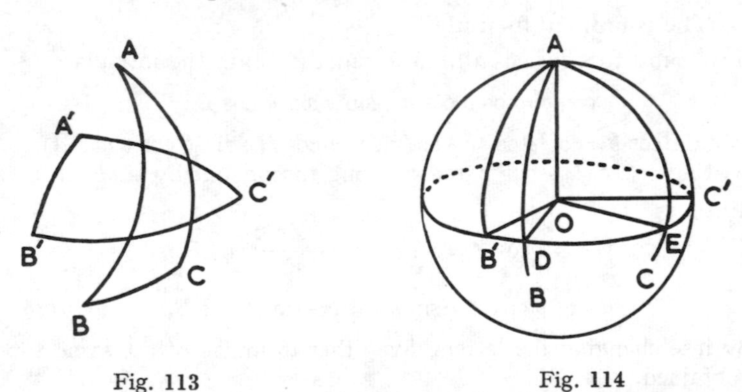

Fig. 113 Fig. 114

25.8. Sides and angles of polar triangles

Let arc $B'C'$, produced if necessary, meet the arcs AB and AC, produced if necessary, at the points D and E respectively (fig. 114). Then since B' is the pole of CA

$$\angle B'OE = \tfrac{1}{2}\pi.$$

Similarly $\angle C'OD = \frac{1}{2}\pi$

$$\therefore \angle B'OE + \angle C'OD = \pi$$

i.e. $\angle B'OC' + \angle DOE = \pi$

$$\therefore \angle B'OC' = \pi - A.$$

If we denote the angles and sides of the polar triangle $A'B'C'$ by A', B', C' a', b', c' respectively this result gives

$$a' = \pi - A$$

and, similarly, $b' = \pi - B$, $c' = \pi - C$.

Since ABC is the polar triangle of $A'B'C'$, it follows that

$$a = \pi - A',\ b = \pi - B',\ c = \pi - C',$$

i.e. $A' = \pi - a,\ B' = \pi - b,\ C' = \pi - c.$

Because of these relations, a triangle and its polar triangle are called supplemental triangles.

25.9. Supplemental formulae

By applying to the polar triangle $A'B'C'$ any formula connecting sides and angles, a supplemental formula may be obtained for triangle ABC involving the sides and angles opposite to the angles and sides which appear in the original formula.

For example, applying (25.3) to triangle $A'B'C'$ we have

$$\cos a' = \cos b' \cos c' + \sin b' \sin c' \cos A'$$

which since $a' = \pi - A$, etc., gives

$$\cos A = -\cos B \cos C + \sin B \sin C \cos a$$

and so $$\cos a = \frac{\cos A + \cos B \cos C}{\sin B \sin C}. \tag{25.6}$$

25.10. Right-angled triangles

From the formulae already established we may deduce the following results in the case when $A = \frac{1}{2}\pi$.

From (25.4) $$\sin B = \frac{\sin b}{\sin a}, \quad \sin C = \frac{\sin c}{\sin a}. \tag{25.7}$$

From (25.3) $$\cos a = \cos b \cos c. \tag{25.8}$$

Substituting from this equation in

$$\cos b = \cos c \cos a + \sin c \sin a \cos B$$

we obtain $$\cos b = \cos b \cos^2 c + \sin c \sin a \cos B$$

$$\therefore \cos B = \cos b \frac{\sin c}{\sin a} = \frac{\cos a}{\cos c} \cdot \frac{\sin c}{\sin a} \quad \text{by (25.8)}$$

i.e. $$\cos B = \frac{\tan c}{\tan a} \text{ and, similarly, } \cos C = \frac{\tan b}{\tan a}. \tag{25.9}$$

From (25.6) $\cos a = \cot B \cot C,$ (25.10)

and the corresponding formulae for cos b, cos c give

$$\cos b = \frac{\cos B}{\sin C}, \quad \cos c = \frac{\cos C}{\sin B}. \tag{25.11}$$

From (25.7) and (25.9) $\tan B = \dfrac{\sin b}{\tan c \cos a}$, $\tan C = \dfrac{\sin c}{\tan b \cos a}$

and so by (25.8) $$\tan B = \frac{\tan b}{\sin c}, \tan C = \frac{\tan c}{\sin b}. \tag{25.12}$$

Also $$\tan B \tan C = \frac{1}{\cos b \cos c} = \frac{1}{\cos a} \quad \text{by (25.8)}$$

i.e. $$\cos a = \cot B \cot C. \tag{25.13}$$

The above results may be conveniently listed as shown below:

$\sin b = \sin B \sin a$	$\sin c = \sin C \sin a$	(25.7)
$= \tan c \cot C$	$= \tan b \cot B$	(25.12)
$\cos B = \cos b \sin C$	$\cos C = \cos c \sin B$	(25.11)
$= \tan c \cot a$	$= \tan b \cot a$	(25.9)

$$\cos a = \cos b \cos c \tag{25.8}$$
$$= \cot B \cot C. \tag{25.10}$$

25.11. Napier's rules

The results of § 25.10 are embodied in two simple rules given by Napier. We define the five *circular parts* of a right-angled spherical triangle (fig. 115 (a)) as the two sides which include the right angle

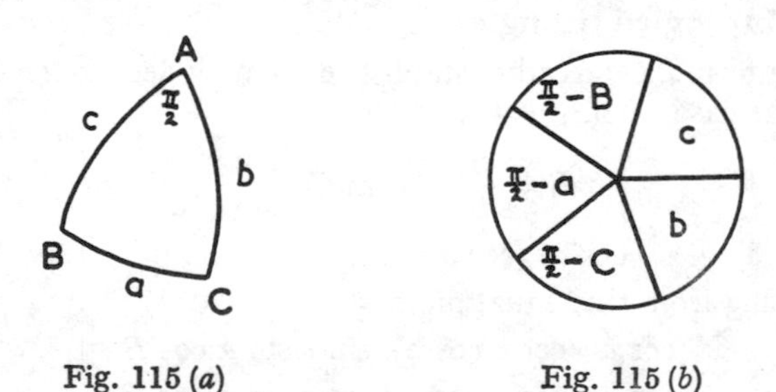

Fig. 115 (a) Fig. 115 (b)

and the complements of the other three parts, the right angle being omitted.

We arrange these five parts round a circle (fig. 115 (b)) in the order in which they naturally occur in the triangle. We select any one of these parts and call it the *middle part*, the two parts next to it being termed the *adjacent parts* and the two remaining parts the *opposite parts*. Then Napier's rules state:

(i) sine of the middle part = product of cosines of opposite parts,

(ii) sine of the middle part = product of tangents of adjacent parts.

Taking each part in turn as middle part, we obtain the formulae given in § 25.10.

Each of the ten formulae of § 25.10 involves a different group of three elements formed from B, C, a, b, c; but only ten such groups can be formed, so the above formulae enable the three remaining elements to be calculated in every case when two of the five are given.

25.12. Quadrantal triangles

A quadrantal triangle is one in which one side is a right angle. A quadrantal triangle may be solved by applying Napier's rules to the polar triangle.

25.13. Measurements on the earth's surface

The earth is not a true sphere, but is very nearly one. For many purposes it is a sufficiently good approximation to assume it to be a sphere of radius 3900 miles.

In fig. 116, NS is the axis of the earth and O is its centre. The section of the earth's surface by any plane through the axis NS is called a *meridian* and the meridian NGS through Greenwich is called the *prime meridian.* The great circle whose plane is at right angles to NS is called the equator. Suppose that the equator cuts the prime meridian at A.

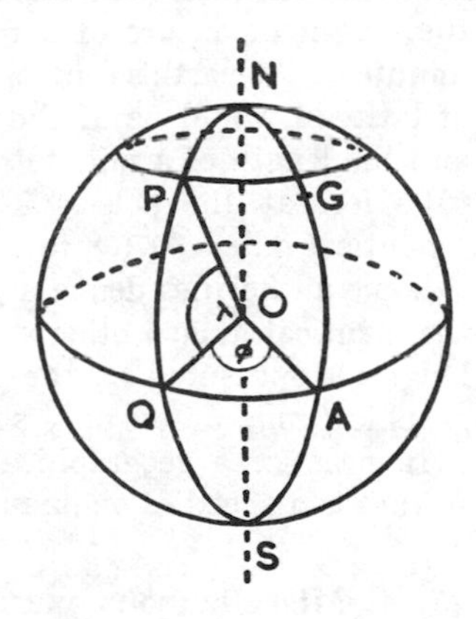

Fig. 116

Let P be any point on the earth's surface and let NPQ, the meridian through P, cut the equator at Q. We fix the position of P by the arcs AQ and QP which, expressed in angular measure, are called respectively the *longitude* (ϕ) and the *latitude* (λ) of P.

ϕ is measured east or west from the prime meridian up to 180° in either direction; λ is measured as north or south from the equator. Planes perpendicular to NS cut the surface of the earth in small circles called *parallels of latitude.* On such a circle all points have the same latitude; on a meridian all points have the same longitude.

The arc NP (or, if P lies south of the equator, SP) is called the *co-latitude* of P. It is $90° - \lambda$.

The *bearing* of a place K from a place H is the angle (measured clockwise) between the northward drawn meridian through H and the arc HK of the great circle drawn through H and K.

In figs. 117 (*a*) and (*b*) the bearings of K from H are 50° and 220° respectively. The direction of K from H may also be described as

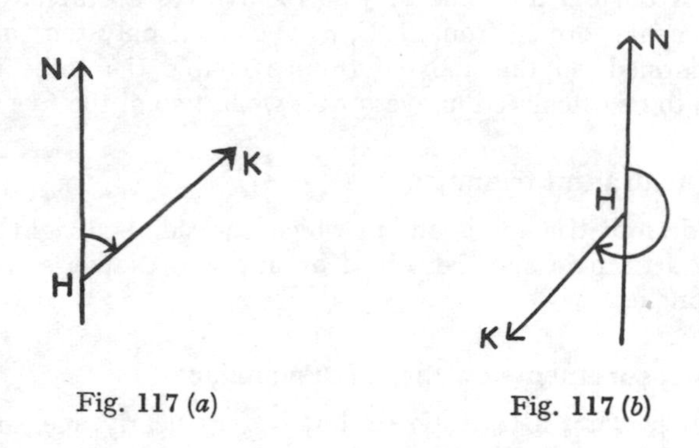

Fig. 117 (*a*) Fig. 117 (*b*)

N. 50° E. or 50° E. of N. in fig. 117 (*a*) and S. 40° W. or 40° W. of S. in fig. 117 (*b*).

A ***nautical mile*** is defined as the mean length of a minute of latitude between the equator and one of the earth's poles. A minute of latitude is the length of an arc of a meridian which subtends an angle of one minute at the earth's centre. If the earth were a sphere, every minute of latitude would be of the same length, but the earth is a spheroid and the length of a minute increases from 6046 feet at the equator to 6108 feet at the poles. The mean length is 6076·8 feet. Hence 1 nautical mile = 6076·8 feet.

Some authorities define a ***geographical mile*** as being identical with the nautical mile; others define it as the length of a minute of longitude measured at the equator. Defined in this way, the geographical mile differs only slightly from the nautical mile.

In practice a geographical or nautical mile is taken as 6080 feet. A knot is a speed of one nautical mile per hour.

25.14. Miscellaneous examples

Example 1

If D is any point on the side BC of a spherical triangle ABC prove that

(a) $\cos AD \sin a = \cos b \sin BD + \cos c \sin DC$,

(b) $\cot AD \sin A = \cot b \sin \angle BAD + \cot c \sin \angle DAC$.

(*a*) In fig. 118, $\angle ADB + \angle ADC = \pi$

$$\therefore \cos \angle ADB + \cos \angle ADC = 0 \quad . \quad . \quad . \quad . \quad \text{(i)}$$

and $$\cot \angle ADB + \cot \angle ADC = 0 \quad . \quad . \quad . \quad . \quad \text{(ii)}$$

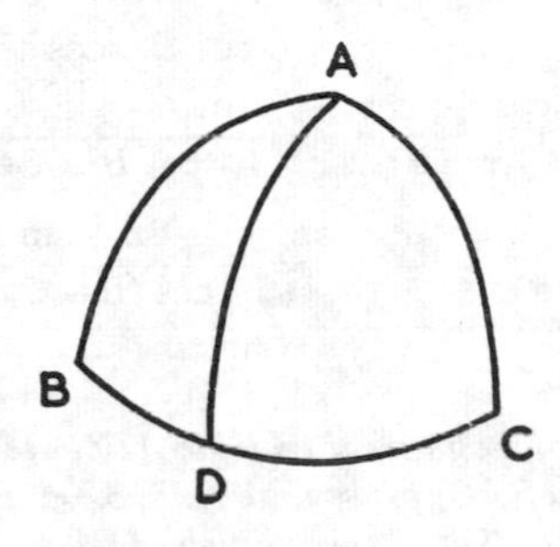

Fig. 118

We apply the cosine formula to Δ^s ADB and ADC

$$\cos \angle ADB = \frac{\cos c - \cos AD \cos BD}{\sin AD \sin BD}$$

$$\cos \angle ADC = \frac{\cos b - \cos AD \cos DC}{\sin AD \sin DC}.$$

Hence by (i)

$$\sin DC (\cos c - \cos AD \cos BD) + \sin BD (\cos b - \cos AD \cos DC) = 0$$

$$\therefore \cos b \sin BD + \cos c \sin DC = \cos AD \sin (BD + DC)$$

$$= \cos AD \sin a.$$

(*b*) Formula (25.5) gives

$$\cot B = \frac{\sin c \cot b - \cos c \cos A}{\sin A}$$

and applying this result to triangles ADB and ADC we have

$$\cot \angle ADB = \frac{\sin AD \cot c - \cos AD \cos \angle BAD}{\sin \angle BAD}$$

and $$\cot \angle ADC = \frac{\sin AD \cot b - \cos AD \cos \angle DAC}{\sin \angle DAC}.$$

Hence by (ii)

$$\sin \angle DAC (\sin AD \cot c - \cos AD \cos \angle BAD)$$
$$+ \sin \angle BAD (\sin AD \cot b - \cos AD \cos \angle DAC) = 0$$

$$\therefore \cot b \sin \angle BAD + \cot c \sin \angle DAC = \cot AD \sin (\angle BAD + \angle DAC)$$

$$= \cot AD \sin A.$$

Example 2

If ABC is a spherical triangle in which $A=\frac{1}{2}\pi$, prove that

$$\sin(B+C)=\frac{\cos b+\cos c}{1+\cos b\cos c}.$$

By (25.8), (25.10) and (25.11)

$$\frac{\cos b+\cos c}{1+\cos b\cos c}=\frac{\dfrac{\cos B}{\sin C}+\dfrac{\cos C}{\sin B}}{1+\cot B\cot C}$$

$$=\frac{\frac{1}{2}(\sin 2B+\sin 2C)}{\cos(B-C)}$$

$$=\sin(B+C).$$

Example 3

If the sides AB, AC of a spherical triangle ABC are each 90° and BC=135° and E is the mid-point of AC, prove that ∠AEB=∠EBC and that the area of the triangle EBC is twice that of the triangle ABE. [L.U.]

In fig. 119, O is the centre of the sphere, $\angle AOB=\angle AOC=90°$, and $\angle BOC=\angle BAE=135°$; also $\angle ABC=\angle ACB=90°$.

Applying the cotangent rule to triangles AEB and EBC in turn we have

$$\cos AE\cos\angle BAE=\sin AE\cot AB-\sin\angle BAE\cot\angle AEB$$

$$\therefore \cot\angle AEB=1/\sqrt{2}.$$

Also, $$\cos BC\cos\angle BCE=\sin BC\cot EC-\sin\angle BCE\cot\angle EBC$$

$$\therefore \cot\angle EBC=1/\sqrt{2}.$$

But $\angle AEB<180°$ and $\angle EBC<90°$.

$$\therefore \angle EBC=\angle AEB=\alpha \text{ radians (say).}$$

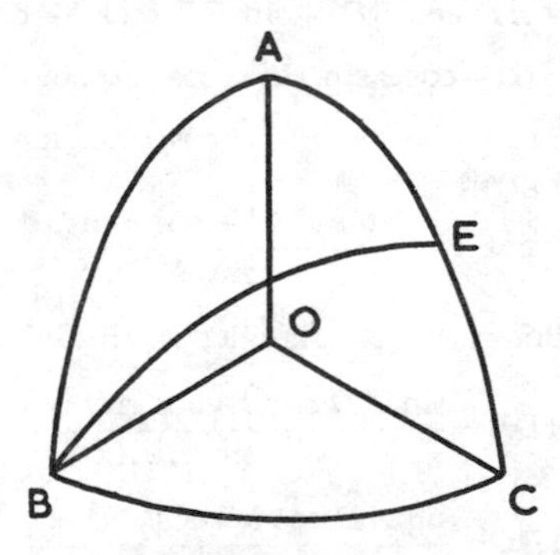

Fig. 119

Hence $\angle ABE=\frac{1}{2}\pi-\alpha$ and $\angle BEC=\pi-\alpha$.

The spherical excess of $\triangle EBC=\frac{1}{2}\pi+(\pi-\alpha)+\alpha-\pi=\frac{1}{2}\pi$ radians, and the spherical excess of $\triangle ABE=\frac{3}{4}\pi+\alpha+\left(\frac{\pi}{2}-\alpha\right)-\pi=\frac{1}{4}\pi$ radians.

Hence by (25.1) $\triangle EBC=2\triangle ABE.$

Example 4

Assuming the earth to be a sphere of radius 4000 miles find the great circle distance between two points A and B whose latitude and longitude are: (A) 40° 16′ N., 18° 33′ E., and (B) 0°, 58° 12′ E. Find also in degrees east of south the angle of departure at A of the great circle route. [L.U.]

If C represents the north pole, then in the usual notation

$$b=90°-40°\ 16'=49°\ 44',\ a=90° \text{ and } \angle ACB=39°\ 39'.$$

Also
$$\cos c=\cos a \cos b+\sin a \sin b \cos C$$
$$=\sin 49°\ 44' \cos 39°\ 39'$$
$$\therefore\ c=54°\ 1'=0{\cdot}9428 \text{ radians},$$

and
$$AB=4000\times 0{\cdot}9428=3771 \text{ miles}.$$

By the sine rule

$$\sin A=\frac{\sin a \sin C}{\sin c}=\frac{\sin 39°\ 39'}{\sin 54°\ 1'}=0{\cdot}7885.$$

But since B lies south of A, $\angle CAB$ is obtuse $\therefore\ A=127°\ 57'$

i.e. the angle of departure at A of the great circle route is 52° 3′ E. of S.

Example 5

The most southerly latitude reached by the great circle joining a place P on the equator to a place Q in north latitude λ is φ. Prove that the difference of longitude between P and Q is $\sin^{-1}$ (tan λ cot φ), and find the angle between the meridian through Q and the great circle PQ. [L.U.]

Let A be the north pole and PR be the equator (fig. 120). Then ϕ is the angle between the equatorial plane and the plane of the great circle PQ so that $\angle APQ=90°-\phi$.

$$AP=q=90°,\ AQ=p=90°-\lambda.$$

Then by the cotangent formula,

$$\cos q \cos \angle PAQ=\sin q \cot p-\sin \angle PAQ \cot \angle APQ,$$
$$\therefore\ \sin \angle PAQ=\cot (90°-\lambda) \tan (90°-\phi),$$
$$\angle PAQ=\sin^{-1} (\tan \lambda \cot \phi).$$

This is the difference in longitude between P and Q.

From the sine formula,

$$\frac{\sin \angle AQP}{\sin q}=\frac{\sin \angle APQ}{\sin p},$$

$$\sin \angle AQP=\frac{\sin (90°-\phi)}{\sin (90°-\lambda)}$$

$$\therefore\ \angle AQP=\sin^{-1}\left(\frac{\cos \phi}{\cos \lambda}\right).$$

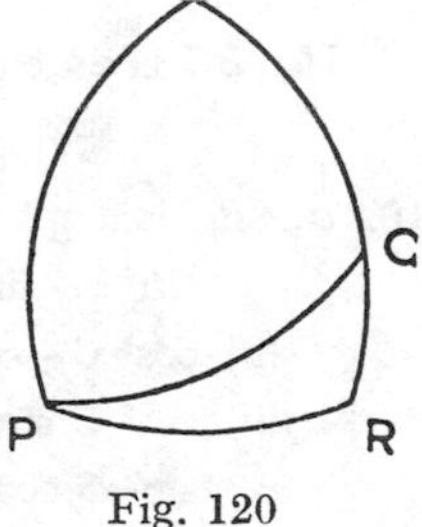

Fig. 120

This is the angle between the meridian through Q and the great circle PQ

Exercises 25

In Nos. 1-8 prove the following results for any spherical triangle ABC:

1. $$\frac{\sin a}{\sin A}=\frac{\sin b}{\sin B}=\frac{\sin c}{\sin C}=\frac{\sin a \sin b \sin c}{\sqrt{(1-\cos^2 a-\cos^2 b-\cos^2 c+2\cos a\cos b\cos c)}}.$$

2. If A' is mid-point of BC, $\angle BAA'=\theta$ and $\angle CAA'=\phi$,

(i) $$\cos AA'=\frac{\cos b+\cos c}{2\cos\frac{1}{2}a},$$

(ii) $$\cot\theta=\cot A+\frac{\sin C}{\sin A\sin B},$$

(iii) $$\cot\phi-\cot\theta=\frac{\sin(b+c)\sin(b-c)}{\sin b\sin c\sin A},$$

(iv) $$\cot AA'B=\frac{\cot C-\cot B}{2\cos\frac{1}{2}a}.$$

3. $$\frac{\sin(A+B)}{\sin C}=\frac{\cos a+\cos b}{1+\cos c}.$$

4. If AD bisects $\angle CAB$ internally,
$$\cot AD=\frac{\cot b+\cot c}{2\cos\frac{1}{2}A}=\frac{\sin(b+c)}{2\sin b\sin c\cos\frac{1}{2}A}.$$

5. (i) $$\frac{\sin(B-C)}{\sin A}=\frac{\cos c-\cos b}{1-\cos a},$$

(ii) $$\frac{\sin(b+c)}{\sin a}=\frac{\cos C+\cos B}{1-\cos A}.$$

6. $\sin b\sin c+\cos b\cos c\cos A=\sin B\sin C-\cos B\cos C\cos a$.

7. $\sin(b+c)\sin(b-c)\sin^2 A=\sin(B+C)\sin(B-C)\sin^2 a$.

8. $\sin a\cos B=\cos b\sin c-\sin b\cos c\cos A$.

9. If ABC is an equilateral spherical triangle, prove that
$$\cos A=\tfrac{1}{2}(1-\tan^2\tfrac{1}{2}a) \text{ and } \sin\tfrac{1}{2}A=\tfrac{1}{2}\sec\tfrac{1}{2}a.$$

10. If ABC is a spherical triangle in which $A=\frac{1}{2}\pi$, prove that

(i) $\sin^2 b+\sin^2 c-\sin^2 a=\sin^2 b\sin^2 c$,

(ii) $\cos^2 C+\cos^2 a-\cos^2 c=\cos^2 C\cos^2 a$,

(iii) $\cos^2 C\sin^2 a=\sin(a-c)\sin(a+c)$,

(iv) $\sin^2 C\cos^2 a=\sin(C-c)\sin(C+c)$,

(v) $$\cos(B+C)=\frac{-\sin b\sin c}{1+\cos a}.$$

11. Two ports are in the same latitude l, their difference of longitude being 2λ. Show that the distance saved in sailing from one port to the other along a great circle, instead of due east or west, is

$$2r\{\lambda \cos l - \sin^{-1} (\sin \lambda \cos l)\},$$

where r is the radius of the earth.

Calculate the distance thus saved if the latitude is 60° and the difference of longitude 90°, taking the radius of the earth as 3960 miles. [L.U.]

12. Prove that in any spherical triangle $\cos a = \cos b \cos c + \sin b \sin c \cos A$. A ship is proceeding uniformly along a great circle. At a given moment the latitude is observed to be l_1; after the ship has travelled distances s and $2s$ the latitudes are observed to be l_2 and l_3 respectively. Show that

$$\cos s = \sin \tfrac{1}{2}(l_1 + l_3) \cos \tfrac{1}{2}(l_1 - l_3)/\sin l_2.$$

Also express the total change of longitude in terms of the three latitudes. [L.U.]

13. In a spherical triangle XYZ, the angle XZY is a right angle. Prove that

$$\cos YXZ = \cot XY \tan XZ.$$

From the vertex A of the spherical triangle ABC a great circle arc is drawn to meet the side BC at right angles at the point P.

Prove that $$\frac{\cos BAP}{\cos CAP} = \frac{\cot BA}{\cot CA}.$$

Find an expression for $\sin AP$ in terms of the sides of triangle ABC. [L.U.]

14. (i) If in a spherical triangle ABC the great circle arc AN cuts BC at right angles at N, establish the results

(a) $\dfrac{\cos BN}{\cos CN} = \dfrac{\cos AB}{\cos AC}$, (b) $\tan NC \cot AC = \cos C$.

(ii) Evaluate NA, NB, NC when $BC = 54°$, $AC = 65°$ and $C = 78°$. [L.U.]

15. If ABC is a spherical triangle with a right angle at C, prove the formulae

(i) $\sin a = \tan b \cot B$, (ii) $\cos c = \cos a \cos b$.

An aeroplane flies in a great circle course from a point A (lat. 30° N., long. 10° E.) to a point B on the equator, the initial direction of departure being 20° E. of S. Find the longitude of B, and the length of the journey, taking the earth to be a sphere of radius 4000 miles. [L.U.]

16. Prove ONE of the following formulae for a spherical triangle ABC:

(i) $\cos a = \cos b \cos c + \sin b \sin c \cos A$,

(ii) $\dfrac{\sin a}{\sin A} = \dfrac{\sin b}{\sin B} = \dfrac{\sin c}{\sin C}$.

An aeroplane flies along a great circle route from a point A (lat. 0°, long. 30° W.) to a point B (lat. 45° N., long. 120° E.). Find (i) the distance travelled, (ii) the direction in which the aeroplane is heading when it reaches B, (iii) the latitude of the most northerly point on the route. (Take the radius of the earth as 3960 miles.) [L.U.]

17. With the usual notation, prove the formula

$$\cos a = \cos b \cos c + \sin b \sin c \cos A$$

for a spherical triangle.

Find in square miles (correct to three significant figures) the area of an equilateral triangle whose sides are 1500 miles long, drawn on the surface of a sphere of radius 4000 miles. [L.U.]

18. A spherical triangle ABC has sides a, b, c. State the formula for $\cos A$ in terms of the trigonometrical functions of a, b and c. If the angle C is a right angle prove that $\cos A = \tan b/\tan c$ and $\tan A = \tan a/\sin b$.

CA is the Greenwich meridian and A is in latitude 52° N.; CB is the equator and the longitude of B is 30° W. Find the area of the triangle ABC to four significant figures, taking the radius of the earth as 3960 miles. [L.U.]

19. A ship sails on a great circle arc from a point A on the equator to a point B in latitude 45° N. and longitude 30° W. of the meridian through A. Taking the radius of the earth to be 3960 miles, find the distance sailed to the nearest mile.

Find, also, the headings of the ship in degrees and minutes West of North at the beginning and at the end of the run. [L.U.]

20. Prove the cosine formula for a spherical triangle ABC,

$$\cos a = \cos b \cos c + \sin b \sin c \cos A,$$

and show that when a, b, c are very small this formula is equivalent to the cosine formula for a plane triangle.

In a spherical triangle ABC, $A = \frac{1}{4}\pi$, $b = c = \frac{1}{2}\pi$, and D is the mid-point of the side AC. If $\angle BDC = \alpha$, show that $\cos \alpha = 1/\sqrt{3}$ and show that the areas of the triangles ABD and ABC are in the ratio

$$(3\pi - 8\alpha) : \pi.$$

[L.U.]

21. Show that the area of a spherical triangle ABC is given by

$$R^2(A + B + C - \pi),$$

where R is the radius of the sphere and the angles are measured in radians.

In the spherical triangle ABC the angle C is 90° and $a = 30°$, $b = 90°$. Find the side c and the area of the triangle. [L.U.]

22. A ship sails from a place A on the equator along a great circle which makes an angle of 60° with the equator. Find the difference in the longitudes of A and the place B on the path of the ship where it reaches the latitude 30° for the first time. Find also the area included by the path of the ship, the meridian through B, and the equator, giving your answer to three significant figures. (Take the radius of the earth to be 3960 miles.) [L.U.]

23. The angle C of the spherical triangle ABC is a right angle. Prove that

$$\tan a = \cos B \tan c = \tan A \sin b.$$

The perpendicular drawn from C meets AB at D. Prove that

$$\sin^2 CD = \tan AD \tan BD.$$

[Leeds.]

HINTS AND ANSWERS

Exercises 1 (*a*), p. 5

1. $A=6$, $B=9$, $C=2$.

2. $A=-3$, $B=3$, $C=6$, $D=1$.

3. (i) $(2x+y-4)(x-2y+3)$; (ii) $(3x+2y-4)(2x-3y+1)$; (iii) $(3x-y)(2x+y-1)$.

4. (i) $R=\sqrt{2}$, $\alpha=45°$; (ii) $R=2$, $\alpha=60°$; (iii) $R=5$, $\alpha=53° 8'$.

5. $a=\frac{1}{8}$, $b=-\frac{1}{6}$, $c=\frac{1}{12}$; $\Sigma n^7=\frac{1}{24}n^2(n+1)^2(3n \; +6n^3-n^2-4n+2)$.

Exercises 1 (*b*), p. 7

1. $-2(a+b+c)(b-c)(c-a)(a-b)$.

2. $-(b-c)(c-a)(a-b)(bc+ca+ab)$.

3. $(a+b+c)(a^2+b^2+c^2-bc-ca-ab)$.

4. $-(b-c)(c-a)(a-b)(a^2+b^2+c^2+bc+ca+ab)$.

5. $-(b-c)(c-a)(a-b)(bc+ca+ab)$.

6. $-(b-c)(c-a)(a-b)\{3(a^2+b^2+c^2)+5(bc+ca+ab)\}$.

7. $(b-c)(c-a)(a-b)(a^2+b^2+c^2+bc+ca+ab)$.

8. $5(b-c)(c-a)(a-b)(a^2+b^2+c^2-bc-ca-ab)$.

9. $3abc(b+c)(c+a)(a+b)$.

10. $-(b-c)(c-a)(a-b)\{a^3+b^3+c^3+bc(b+c)+ca(c+a)+ab(a+b)+abc\}$.

Exercises 1 (*c*), p. 13

1. $3/(x+4)-2/(x+3)$.

2. $2/(x-4)+1/(x+2)$.

3. $\frac{1}{2}\{1/x+1/(x-2)-2/(x+1)\}$.

4. $2/(2-x)+3/(3-x)-2/(1-x)$.

5. $8/(4-x)-8/(4+x)-x$.

6. $1/(x-1)+5/(2x+1)-7/(2x+3)$.

7. $1/(x-2)+4/(x-2)^2+4/(x-2)^3$.

8. $1/(x-1)+2/(x+1)+3/(x+1)^2$.

9. $\frac{1}{2}\{(1/x+1/(x+2)-2/x^2\}$.

10. $1/(x-2)-2/(x+1)-4/(x+1)^2$.

11. $1/(2x-1)-2/(x+2)^2$.

12. $2/(2x-7)+1/(x+4)^2-1/(x+4)$.

13. $1/(1+x)-x/(1+x^2)$.

14. $1/x-(x+1)/(x^2+9)$.

15. $1/(x+1)+(x-1)/(x^2-x+1)$.

16. $2/x+1/(2-x)+1/(2-x)^2-1/(2+x)-1/(2+x)^2$.

17. $1/x-x/(9+x^2)-9x/(9+x^2)^2$.

18. $1-1/x^2-\frac{1}{3}\{2/(x+1)+(1-2x)/(x^2-x+1)\}$.

19. $\frac{1}{18}\{(13x-15)/(x^2-2x+3)+5/(x+1)-6/(x+1)^2\}$.

20. $1/(x-1)+2/(x-1)^2+2/(x-1)^3$.

21. $2x/(x^2+1)+(3-x)/(x^2+1)^2$.

22. $(1+x)/(1+x+x^2)^2-1/(1+x+x^2)+1/(1-x)^2-1/(1-x)$.

23. $x/(x^2+2)-4x/(x^2+2)^3$.

24. $1/(1-x)^2-1/(1+x)^2+2/(1-x)^3-2/(1+x)^3$.

Exercises 2, p. 22

1. $y^4-13y^2+36=0$; $x=-1, -2, 3, 4$.

2. $4y^4-13y^2+9=0$; $x=-1, -3, -\frac{1}{2}, -\frac{7}{2}$.

3. $x=2, 3, \frac{1}{2}, \frac{1}{3}$. **4.** $x=-1, -1, \frac{1}{2}, 2$. **5.** $x=-1, 2\pm\sqrt{3}, 3\pm2\sqrt{2}$.

6. (i) $x=2, 5, -3, -6$;
(ii) $bx+cy+az=0$; $x=\pm(5\sqrt{2})/6$, $y=\mp(\sqrt{2})/6$, $z=\mp(7\sqrt{2})/6$.

7. $r^2x^3+(2pr-q^2)x^2+(p^2-2q)x-1=0$; $(2pr-q^2)/r$.

8. $4x^3-13x^2+43x-49=0$. **9**. $p(3q-p^2)-3r$. **10.** $q^3=p^3r$; $x=\frac{2}{3}, 2, 6$.

11. $(4p^4-12p^2q+2q^2)/(2p^2+q)^2$. **12.** Write $\alpha+\beta+\gamma=a-\delta$, etc.

14. $\alpha^2+\beta^2+\gamma^2=p^2$; $\beta^2\gamma^2+\gamma^2\alpha^2+\alpha^2\beta^2=-2pr$;
$x^3-2p^2x^2+p(p^3-2r)x+r(2p^3+r)=0$.

15. $x^3(pq-r)+x^2(2pq-p^3-3r)+x(pq-3r)-r=0$.

16. $2, \frac{1}{2}, 2\pm\sqrt{3}$. **17.** (i) $2p^2$, (ii) $\frac{1}{2}, 2, \frac{1}{6}(5\pm i\sqrt{11})$ where $i^2=-1$.

19. $b^2x^4-2(2h^2-ab)x^2+a^2=0$; $y^4-12y^2+4=0$; $x=\pm\sqrt{2}, -4\pm\sqrt{2}$.

20. $a+b+c=-p$, $bc+ca+ab=q$, $abc=-r$.

21. (i) $y^4-5y^2+6=0$; $x=-3\pm\sqrt{2}, -3\pm\sqrt{3}$.

22. $x=y=z=0$; $x=k/(a+1)$, $y=k/(b+1)$, $z=k/(c+1)$ where
$k^2=(b+1)(c+1)+(c+1)(a+1)+(a+1)(b+1)-(a+1)(b+1)(c+1)$.

23. (i) Calculate $1-a^2$ using $a=\frac{1}{2}(x^2-y^2-z^2)/yz$;
(ii) last part: let $x=b+c-a$, $y=c+a-b$, $z=a+b-c$ and use previous result for $(y+z)(z+x)(x+y)$.

24. (i) Given product$=(a^2x^2+b^2y^2+c^2z^2+d^2u^2)+(a^2y^2+b^2x^2)+\ldots+\ldots$;
and $a^2y^2+b^2x^2>2abxy$, etc.
(ii) If $Z\geqslant x$, $x^3y\geqslant ax^2+bx^4$ $\therefore$ $y\geqslant\{\sqrt{(a/x)}\}^2+\{\sqrt{(bx)}\}^2\geqslant2\sqrt{(ab)}$.

25. $a^2+b^2>2ab$ $\therefore$ $a^2-ab+b^2>ab$ and since $a+b>0$, $a^3+b^3>ab(a+b)$ hence $2(a^3+b^3+c^3)>ab(a+b)+bc(b+c)+ca(c+a)$ and so $3(a^3+b^3+c^3)>(a+b+c)(a^2+b^2+c^2)$. Now put $a=y-x$, $b=y+x$ $c=2\sqrt{(y^2-x^2)}$.

28. (i) $x/y+y/x\geqslant2\sqrt{(x/y)}\sqrt{(y/x)}=2$.

Exercises 3 (*a*), p. 35

3. $x=\pm1, -2$. **4.** (i) 1800; (ii) 24. **5.** $x=a, b, -(a+b)$. **6.** $x=2, 3, 6$.

9. (i) $(a+2)(a-1)^2$; (ii) $(b-c)(c-a)(a-b)$;
(iii) $-(a+b+c)(b-c)(c-a)(a-b)$.

16. (i) $p=q=-2$; (ii) $-(a+b+c)^2(b-c)(c-a)(a-b)$.

17. (i) Add $C_1+C_2+C_3$; $-2(a+b)(a-b)^2$.
(ii) Add $C_1+C_2+C_3$; $(1-a)(1-b)(1+a+b)$.

18. Add (C_2+C_1), (C_3+C_1); $8abc$. **19.** $x=-\frac{1}{2}a$ (thrice).

20. $(a+b+c)(b-c)(c-a)(a-b)(a^2+b^2+c^2+bc+ca+ab)$.

21. Add $R_1+R_2+R_3$; $2(a+b+c)^3$.

23. Multiply R_1 by abc and take factors out of the resulting columns; then cf. 9 (iii); $-abc(a+b+c)(b-c)(c-a)(a-b)$.

24. (i) $(\lambda\mu\nu+1)D$; (ii) $x=2, -5/2, 1/2$.

25. (i) Add (R_1+R_2), (R_3+R_4); 0;
(ii) Add $(R_1\times a^2)+(R_2\times ba)+(R_3\times ca)+(R_4\times da)$.

26. 24. **27.** 505. **29.** Add $aR_2+bR_3+cR_4$; $-(ax+by+cz)^2$.

Exercises 3 (*b*), p. 44

1. $x=3, y=5, z=6$.

2. $x=1, y=2, z=2$.

3. $abc+2fgh-af^2-bg^2-ch^2=0$.

4. $\lambda=3, x=4, y=-2$.

5. $\lambda=1, x=-5, y=1$; $\lambda=-1, x=-1/11, y=-15/11$; $\lambda=12, x=\frac{1}{2}, y=1$.

6. $t=0, x=y=z$; $t=3$, the three given equations are identical.

7. (i) $x=-0{\cdot}6, y=1{\cdot}1, z=2{\cdot}7$; (ii) $x=2, y=3, z=4$.

8. $x=12, y=-60, z=60$.

9. $x=1, y=1, z=1$.

Miscellaneous Exercises 3, p. 45

1. Add $(b-c)R_1+(c-a)R_2+(a-b)R_3$; (ii) $2(b-c)(c-a)(a-b)(a+b+c)$.

2. (i) $(c-a)$, $(a-b)$, $(a^2+b^2+c^2+bc+ca+ab)$;
(ii) $\cos 2\alpha-\cos 2\beta=2(\sin^2\beta-\sin^2\alpha)$.

3. (i) $-2(b-c)(c-a)(a-b)(a+b+c)$; (ii) from R_2 take R_3.

4. (ii) Multiply C_1 by $2\sin(B+C)$, C_2 by $2\sin(C+A)$, C_3 by $2\sin(A+B)$ and express resulting elements as sines; then add three resulting columns using $\sin(2A+2B)=-\sin 2C$, etc.

5. (ii) $x=1, 2, -3$.

7. (i) $\sin 2\alpha+3\sin^2\alpha-2$;
(ii) $\lambda=3$; $x=1/6, y=3/2$; $\lambda=14, x=-1/5, y=2/5$.

8. (ii) $4a^2b^2c^2$.

9. (i) $(b-c)(c-a)(a-b)(a+b+c)$.

10. (i) $x=-a$ (thrice); (ii) $-(a^2+b^2+c^2)(a+b+c)(b-c)(c-a)(a-b)$.

12. (i) $x=-3, \pm\sqrt{3}$.

13. (i) $x=3, (1\pm\sqrt{561})/10$.

14. (i) $-3z(x-y)(x+y+z)$; (ii) $x=5, -\frac{1}{2}$.

15. (i) $x=5/3, -1/6$.

16. From R_2 take $R_1\cos\theta$.

17. (i) $(y-z)(z-x)(x-y)(x+y+z)$; (ii) $a+b+c=0$; $5:3:-8$.

18. (i) $n(x+y+z)^3$.

19. (i) $2(b-c)(c-a)(a-b)(a+b+c)$; (ii) $x=1$ (twice), $x=2$.

20. (ii) (a) $x^4(x-1)^3(x+1)$, (b) $-(\alpha-\beta)(\alpha-\gamma)(\alpha-\delta)(\beta-\gamma)(\beta-\delta)(\gamma-\delta)$.

21. $(b-c)(c-a)(a-b)$; $-(a-b)^2(b-c)(c-a)$.

22. $\lambda=1$; $1:-1:0$. $\lambda=-2$; $1:1:\frac{1}{2}\sqrt{2}$. $\lambda=3$; $1:1:-2\sqrt{2}$.

23. (i) $(x+y+z)^2(y-z)(z-x)(x-y)$;
(ii) take $2R_1$ from R_2; $x=0, \frac{1}{2}(-7\pm\sqrt{37})$.

24. (i) From C_1+C_3 take $2C_2 \cos x$; or expand from R_1; (ii) $x=b, c, a^3/bc$.

25. (i) (a) Inconsistent, (b) consistent; $x=4, y=1, z=0$; (ii) $2(x+y+z)^3$.

26. $2xyz(x+y+z)^3$.

27. (i) $x=-5, y=7, z=2$; (ii) To aC_1 add bC_2+cC_3; $(a+b+c)(a^2+b^2+c^2)$.

28. (i) $2(b-c)(c-a)(a-b)(y-z)(z-x)(x-y)$; (ii) $a+b+c=0$.

Exercises 4 (*a*), p. 68

1. (i) Divergent; (ii) convergent; (iii) convergent; (iv) convergent; (v) divergent; (vi) convergent; (vii) divergent; (viii) convergent; (ix) convergent; (x) divergent; (xi) divergent; (xii) convergent.

2. (i) A.C. when $|x|<1$, C.C. when $x=-1$; (ii) A.C. for all values of x; (iii) A.C. when $|x|<1$; C.C. when $x=-1$; (iv) A.C. for all values of x; (v) A.C. when $|x|<1$; C.C. when $|x|=1$.

5. (iv) $u_n \to 0$ since $u_n < \frac{1}{3}(\frac{1}{2})^{n-1}$; (v) $u_n < \frac{1}{3}(\frac{2}{3})^{n-1}$.

8. Convergent if $|x| \leqslant 1$; divergent if $|x|>1$.

Exercises 4 (*b*), p. 73

1. 1. **2.** $\frac{1}{2}$. **3.** 1. **4.** 2. **5.** 0. **6.** $1/2\sqrt{x}$. **7.** $x=2$. **8.** $x=0$.

9. $x=2, x=-1$. **10.** $(2k+1)\pi/4$, k any integer or zero.

11. $2k\pi$, k any integer or zero. **12.** $k\pi/3$, k any integer or zero.

Exercises 5 (*a*), p. 78

1. 2^{2r}; $2^r(2^{r+1}-1)$; $|x| < \frac{1}{2}$.

2. $(x-2)/(x^2+1)+4/(x-2)^2-1/(x-2)$; $(4n+5)2^{-2n-2}+(-1)^n$; $|x|<1$.

3. $(n+1)(a-b)(a^{n+2}+b^{n+2})+2ab(b^{n+1}-a^{n+1})$; $|x|<1/|a|$ or $1/|b|$ whichever is smaller.

4. $a=1, b=4, c=1, d=0$.

5. $a=48, \quad b=3, \quad c=8$; $48/(1-4x)^2-32/(1-4x)+3/(1-x)^2+8/(1-x)$;
$\sum_{n=0}^{\infty} \{4^{n+2}(3n+1)+3n+11\}x^n$; $|x|<\frac{1}{4}$.

6. $b^2/(1-bx)-ab/(1-ax)+a(a-b)/(1-ax)^2$;
$(n+1)a^{n+2}-(n+2)a^{n+1}b+b^{n+2}$; $|x|<1/|a|$ or $1/|b|$ whichever is smaller.

7. $\frac{1}{5}\{1/(x-2)-(x+2)/(x^2+1)\}$; $|x|<1$;
$((x^{2n}))=-\frac{1}{10}\{2^{-2n}+4(-1)^n\}$, $((x^{2n+1}))=-\frac{1}{10}\{2^{-2n-1}+2(-1)^n\}$.

8. Write $\Sigma_1=2(1+3.4x^2/2!+3.4.5.6x^4/4!+\ldots)$,
and $\Sigma_2=2(3x+3.4.5x^3/3!+3.4.5.6.7x^5/5!+\ldots)/x$
then consider the binomial expansions of $(1+x)^{-3}$, $(1-x)^{-3}$.
$\Sigma_1=2(1+3x^2)/(1-x^2)^3$, $\Sigma_2=2(x^2+3)/(1-x^2)^3$.

Exercises 5 (*b*), p. 88

1. (i) $2\left(\frac{1}{1!}+\frac{1}{3!}+\frac{1}{5!}+\frac{1}{7!}+\ldots\right)$; (ii) $2\left(1+\frac{4}{2!}+\frac{16}{4!}+\frac{64}{6!}+\ldots\right)$;

(iii) $1+\frac{3}{1!}+\frac{9}{2!}+\frac{27}{3!}+\ldots$.

2. (i) $1+\frac{2x}{1!}+\frac{4x^2}{2!}+\frac{8x^3}{3!}+\ldots$; $2^r x^r/r!$;

(ii) $1+2x+\frac{3x^2}{2!}+\frac{4x^3}{3!}+\ldots$; $(r+1)x^r/r!$;

(iii) $1-x-\frac{x^2}{2!}+\frac{5x^3}{3!}+\ldots$; $(-1)^{r-1}(r^2-r-1)x^r/r!$.

3. $1+x+\frac{3x^2}{2}+\frac{7x^3}{6}+\frac{25}{24}x^4+\ldots$.

4. $a=1$, $b=3$, $c=1$.

7. $r=4$, $a=\log 2$; $x=-2\log 2$.

8. $x=\log 2$ or $-\log 3$.

9. (ii) $x=1, 6$.

11. $u=\log 0.4$.

13. (ii) When x is large, $\coth x \simeq 1$; $x \simeq k\pi \pm \frac{1}{4}\pi$, where k is any integer.

16. $\tanh 2x = 2t/(1+t^2)$, $\tanh 3x = t(3+t^2)/(1+3t^2)$, $t \equiv \tanh x$. Put $u=\tanh x$ and solve the given equation for k.

Exercises 5 (*c*), p. 93

1. (i) $2x-2x^2+\frac{8}{3}x^3-4x^4+\ldots$; $(-1)^{r-1}(2^r)/r$; $-\frac{1}{2}<x\leqslant\frac{1}{2}$;

(ii) $-\left(\frac{x}{4}+\frac{x^2}{32}+\frac{x^3}{192}+\frac{x^4}{1024}+\ldots\right)$; $-(\frac{1}{4})^r/r$; $-4\leqslant x<4$;

(iii) $\log 3+\left(\frac{x}{3}-\frac{x^2}{18}+\frac{x^3}{81}-\frac{x^4}{324}+\ldots\right)$; $(-1)^{r-1}(\frac{1}{3})^r/r$; $-3<x\leqslant 3$;

(iv) $\log 2-\left(\frac{3x}{2}+\frac{9x^2}{8}+\frac{9x^3}{8}+\frac{81x^4}{64}+\ldots\right)$; $-(\frac{3}{2})^r/r$; $-\frac{2}{3}\leqslant x<\frac{2}{3}$;

(v) $2\log 4+2\left(\frac{x}{4}-\frac{x^2}{32}+\frac{x^3}{192}-\frac{x^4}{1024}+\ldots\right)$; $-2(-\frac{1}{4})^r/r$;

$-4<x\leqslant 4$;

(vi) $-\frac{1}{2}\left(x+\frac{5x^2}{2}+\frac{7x^3}{3}+\frac{17x^4}{4}+\ldots\right)$; $-\frac{1}{2}\{2^r+(-1)^r\}/r$; $-\frac{1}{2}\leqslant x<\frac{1}{2}$;

(vii) $\log\{(1-x^3)/(1-x)\}=x+\frac{x^2}{2}-\frac{2x^3}{3}+\frac{x^4}{4}+\ldots$; $1/r$ unless r is a multiple of 3; $-2/r$ if r is a multiple of 3; $-1\leqslant x<1$;

(viii) $3x-\frac{3x^2}{2}+3x^3-\frac{15x^4}{4}+\ldots$; $\{1-(-2)^r\}/r$; $-\frac{1}{2}<x\leqslant\frac{1}{2}$.

2. Write $1+x+x^2+x^3=(1-x^4)/(1-x)$.

3. $-(3x^2+\frac{7}{6}x^4+\frac{11}{15}x^6+\frac{15}{28}x^8+\ldots)$.

4. (i) $(-3)^r/r(r-1)$, (ii) $(-2)^{r+1}/r(r-1)(r-2)$.

6. $\log\frac{3}{2}$; $\frac{1}{2}\log 3$. **7.** $|x|<1$. **12.** Last part: put $m/n=e^{2\theta}$.

13. $(1-\sin^2\theta)^{\frac{1}{2}}=\cos\theta=(1-\tan^2\frac{1}{2}\theta)/(1+\tan^2\frac{1}{2}\theta)$.
Valid when $-\frac{1}{2}\pi<\theta<\frac{1}{2}\pi$.

14. $|\log\cos\theta-(-2t^2)|=2(\frac{1}{3}t^6+\frac{1}{5}t^{10}+\frac{1}{7}t^{14}+\ldots)<\frac{2}{3}(t^6+t^{10}+t^{14}+\ldots)$,
$t\equiv\tan\frac{1}{2}\theta$. **15.** 2·9444.

16. Last part: consider $\log\{(p+x)/(q+x)\}^{p+q+2x}-\log(p/q)^{p+q}$, $x>0$.

17. Substitute $p=1$, $r=3$; $p=2$, $r=4$ in given result and eliminate log 3 from the resulting equations.

Exercises 5 (*d*), p. 98

1. (i) $-\frac{1}{3}$; (ii) Put $x=1+y$; 1; (iii) $-1/48$; (iv) 1/18; (v) 1.

2. $x\log(1+y)-y\log(1+x)=-\frac{1}{2}xy(y-x)+\ldots$,
$(1+y)^x/(1+x)^y=e^{-\frac{1}{2}xy(y-x)}$, etc.

6. $\log y>x^2$ $\therefore y>e^{x^2}>1+x^2$;
$\log\{1/(1-x^2)\}=(x^2+\frac{1}{2}x^4+\frac{1}{3}x^6+\ldots)>\log y$.

7. $(-1)^{r+1}(a^r+b^r+c^r)/r$.

8. $(a+b)x-\frac{1}{2}(a^2+b^2)x^2+\frac{1}{3}(a^3+b^3)x^3-\ldots$; $-1/a<x\leqslant 1/a$, $a>b$.

9. $3q(3p^3-q^2)$.

10. (i) $x-\frac{5}{2}x^2+\frac{7}{3}x^3-\frac{17}{4}x^4+\ldots$; $((x^n))=-\{1+(-2)^n\}/n$; $-\frac{1}{2}<x\leqslant\frac{1}{2}$.

12. $1+x^2+\frac{1}{2}x^3+\frac{5}{6}x^4+\frac{3}{4}x^5+\frac{33}{40}x^6+\ldots$.

13. To prove $\log(1+1/n)<\frac{1}{2}\{1/n+1/(n+1)\}$, compare given series with
$2\sum_{r=1}^{\infty}\{1/(2n+1)^{2r-1}\}$.

14. From (2), $\sum_{r=2}^{n}\{\log r-\log(r-1)\}>\sum_{r=2}^{n}(1/r)$.

15. When $p>1$, $\log\frac{p+1}{p-1}<2\left\{\frac{1}{p}+\frac{1}{3p^3}+\frac{1}{5p^3}\left(\frac{1}{p^2}+\frac{1}{p^4}+\frac{1}{p^6}+\ldots\right)\right\}$; 1·609.

16. Last part: $\sum_{r=n}^{n+k}\{\log r-\log(r-1)\}>\sum_{r=n}^{n+k}(1/r)>\sum_{r=n}^{n+k}\{\log(r+1)-\log r\}$.

17. (i) Write $n\log\{n/(n+1)\}=\{(n+1)-1\}\log\{1-1/(n+1)\}$.

18. Consider $\log(u_n/u_{n+1})$.

19. Expand $e^{(e^x)}$ first as $1+e^x+e^{2x}/2!+\ldots$ then as $e(e^{x+x^2/2!+x^3/3!+\ldots})$ and compare $((x^3))$, $((x^4))$ in these expansions; $5e$, $15e$.

20. (i) $-2+4\sum_{r=1}^{\infty}x^{2r}/(2r-1)(2r+1)$, $|x|<1$; (ii) See 19; e.

Exercises 6, p. 116

1. $1+i$; modulus $\sqrt{2}$, argument $\frac{1}{4}\pi$.

2. (i) modulus$=1$, argument$=-\theta$; (ii) modulus$=\sec\theta$, argument$=\theta$.

3. (a) $\sqrt{2}$, $\frac{1}{4}\pi$; $\sqrt{2}$, $\frac{3}{4}\pi$; $\sqrt{2}$, $-\frac{1}{4}\pi$.

4. $1, \frac{1}{2}\pi$; $1, \frac{1}{4}\pi$. **5.** $0{\cdot}7+0{\cdot}9i$. **6.** $x-y=1$. **7.** $4x+8y=3$.

10. Centre $(3R, -4R)$, radius $5R$.

11. $3x^2+3y^2-34x+96=0$. **13** (i) $x^2+y^2-x-2=0$.

15. $P_1\equiv r(-\cos\theta+i\sin\theta)$; the circle is $c(u^2+v^2)+4u=0$.

16. (ii) $|z_1|=|z_2|$ and $\arg z_1-\arg z_2=\frac{1}{3}\pi$. Hence $\arg z_1^3-\arg z_2^3=\pi$ $\therefore z_1^3=-z_2^3$, etc.

18. $\rho^2+r^2-2\rho r\cos(\theta-\alpha)$.

19. (i) Divide AB at C so that $AC:CB=\mu:\lambda$. Then $\lambda\overline{OA}+\mu\overline{OB}=(\lambda+\mu)\overline{OC}$ where O is the origin. Thus $\lambda z_1+\mu z_2=z=k\overline{OC}$ so that P is the point on OC such that $OP=kOC$. Hence P lies on a line parallel to AB.

20. $B\equiv(1+\frac{5}{3}\sqrt{2})-i(1+\frac{1}{3}\sqrt{2})$; $D\equiv(1-\frac{5}{3}\sqrt{2})-i(1-\frac{1}{3}\sqrt{2})$.

Exercises 7 (a), p. 121

1. (i) $\sqrt[6]{(2)}\operatorname{cis}\{(8k+1)\pi/12\}$, $k=0, 1, 2$;
(ii) $\sqrt[6]{(2)}\operatorname{cis}\{(8k\pm 1)\pi/12\}$, $k=0, 1, 2$.

2. (i) Modulus$=2\cos\frac{1}{2}\theta$, argument$=-\frac{1}{2}\theta$;
(ii) $\sqrt{2}\operatorname{cis}\{(8k+3)\pi/12\}$, $k=0, 1, 2$.

3. $2(1+i)$, $1+3i$.

4. $\frac{1}{2}\{(3\sqrt{2}-4)+i(6+\sqrt{2})\}$, $\frac{1}{2}\{-(3\sqrt{2}+4)+i(6-\sqrt{2})\}$.

5. $w=\pm 2, \pm 2i$; $z=3, \frac{1}{3}, \frac{1}{5}(3\pm 4i)$.

6. (i) $\pm i$, $\pm\frac{1}{2}(1\pm i)\sqrt{2}$; (ii) $-\frac{1}{2}$, $\frac{1}{2}(-1\pm i)$, one infinite root.

7. $\frac{1}{2}\{1-i\cot(4k+1)\pi/12\}$, $k=1, 2, 3$.

8. $z=\frac{1}{2}\{1\pm 3i(\sqrt{2}+1)\}$, $\frac{1}{2}\{1\pm 3i(\sqrt{2}-1)\}$.

9. 1; $\operatorname{cis}(\pm 2k\pi/5)$, $k=1, 2$.

10. (i) $1+3i$, $-1+i$;
(ii) $\pm(1+i)$, $\pm\frac{1}{2}\{(\sqrt{3}+1)-i(\sqrt{3}-1)\}$, $\pm\frac{1}{2}\{(\sqrt{3}-1)-i(\sqrt{3}+1)\}$.

11. 1; $\operatorname{cis}(\pm 2k\pi/5)$, $k=1, 2$.

13. $\operatorname{cis}(\pm 2k\pi/9)$, $k=1, 2$; $-\operatorname{cis}(\pm\pi/9)$.

16. $\pm(0{\cdot}9808+0{\cdot}1951i)$, $\pm(0{\cdot}1951-0{\cdot}9808i)$. **17.** $\sqrt{2}\operatorname{cis}(\pi/4)$.

19. (i) $|z|=1$; (ii) $\arg z=2p\pi$, where p is any rational number.

20. $-\frac{1}{2}[1+i\cot\{(2k+1)\pi/16\}]$, $k=0, 1, 2, \ldots, 7$.

Exercises 7 (*b*), p. 129

8. $z-a \text{ cis } (k\pi/3)$, $k=0, 1, 2, 3, 4, 5$.

9. $\{x^2-2x\cos(\pi/9)+1\}\{x^2-2x\cos(5\pi/9)+1\}\{x^2-2x\cos(7\pi/9)+1\}$.

10. (ii) $(x+1)\prod_{k=1}^{4}[x^2-2x\cos\{(2k-1)\pi/9\}+1]$.

Exercises 7 (*c*), p. 133

3. $z^5+3z^3+2z-2z^{-1}-3z^{-3}-z^{-5}$.

4. $\theta=2k\pi\pm\frac{1}{3}\pi$, where k is any integer or zero.

5. $32\cos^5\theta-32\cos^3\theta+6\cos\theta$.

7. $256c^8-448c^6+240c^4-40c^2+1$ where $c=\cos\theta$.

8. $32\cos^6\phi-36\cos^3\phi+4$; $\phi=n360°, n360°\pm 60°$, where n is any integer or zero.

10. $8(1-10s^2+24s^4-16s^6)$ where $s=\sin\theta$; let $x=2\sin\theta$, then $\theta=k\pi/8$, $k=1, 2, 3, 5, 6, 7$.

Exercises 7 (*d*), p. 138

1. $u=\{(x-1)\cos\alpha+y\sin\alpha\}\{(x-1)^2+y^2+1\}/\{(x-1)^2+y^2\}$;
$v=\{(x-1)\sin\alpha-y\cos\alpha\}\{1-(x-1)^2-y^2\}/\{(x-1)^2+y^2\}$.

4. (i) (*a*) The major arc PQ of a circle through the points P, Q representing z_1, z_2; (*b*) a hyperbola with the points representing z_1 and z_2 as foci.

(ii) The positive u-axis and the negative u-axis respectively.

7. $z_2+e^{-i\pi/3}(z_3-z_2)=\frac{1}{2}\{(z_2+z_3)+i(z_2-z_3)\sqrt{3}\}$.

9. $\overline{OC}=z_2+k(z_1-z_2)(\cos\alpha-i\sin\alpha)$; $\overline{OD}=z_1+k(z_1-z_2)(\cos\alpha-i\sin\alpha)$.

10. (i) $\overline{OC}=ikz$, $\overline{OB}=(1+ik)z$, $\overline{CA}=(1-ik)z$.

12. $1+i=\sqrt{2}e^{i\pi/4}$; the eight distinct values are ± 1, $\pm i$, $\pm(1+i)/\sqrt{2}$ $\pm(1-i)/\sqrt{2}$.

14. $\frac{1}{2}(\pm\sqrt{3}+i)\sqrt[3]{4}$, $-i\sqrt[3]{4}$.

17. $z=\cot(\theta+r\pi/n)$, $r=1, 2, 3, \ldots, n$.

18. $\cos(4\pi/7)$, $\cos(6\pi/7)$.

19. Modulus$=2^n\cos^n\frac{1}{2}\theta$; amplitude$=\frac{1}{2}n\theta$.

20. (i) $\{b(a^2-1)/(1-at)-a(b^2-1)/(1-bt)-(a-b)\}/ab(a-b)$;
(ii) $\{2/(1-at)+(a^2-1)/(1-at)^2-1\}/a^2$; last part: put $t=\tan\frac{1}{2}\phi$.

Exercises 7 (*e*), p. 146

1. $x=\tanh 2u$, $y=\operatorname{sech} 2u$.

2. $x=\sin u\cosh v$, $y=\cos u\sinh v$.

7. (ii) When $x=\log 3$, $y=\frac{1}{2}\log 2$; when $x=-\log 3$, $y=-\frac{1}{2}\log 2$.

8. (i) $13/3\sqrt{3}$; (ii) $\sin^{-1}3=(4k+1)\pi/2\pm i\log(3+2\sqrt{2})$;
$\tan^{-1}(1+i)=k\pi-\frac{1}{2}\tan^{-1}2+\frac{1}{4}i\log 5$, k any integer or zero.

10. e^z is real if $z=x+ik\pi$; $\cos z$ is real if $z=k\pi+iy$, where y is real and k is any integer or zero.

11. If $z=r \operatorname{cis} \theta$, $|e^z|=e^{r\cos\theta}$, $\arg(e^z)=r\sin\theta$; $r=(k\pi-\theta)/\sin\theta$, where k is any integer or zero; asymptotes are $y=\pm\pi$.

12. 2π apart.

13. (i) $2(i-\sqrt{3})$; $\pm\{(\cosh\theta+\sinh\theta)\sqrt{3}+i(3\sinh\theta-\cosh\theta)\}/4\sqrt{2}$.

14. $y \operatorname{cosec} y=\cosh x$.

16. $\cos z=\cos x\cosh y-i\sin x\sinh y$; $x=(8k+1)\pi/4$, $y=1$, where k is any integer or zero.

20. $-(a^n+b^n)/n$; $(-1)^{n-1}(\cos n\theta)/n$.

Exercises 8 (*a*), p. 151

1. $\frac{1}{6}n(4n^2+21n+35)$.
2. $\frac{1}{2}n(6n^2-3n-1)$.
3. $n(n+1)(2n^2+2n-1)$.
4. $\frac{1}{6}n(n+1)(9n^2+19n+8)$.
5. $\frac{1}{6}n(n+1)(3n^2+5n+1)$.
6. $\{1-(n+1)x^n+nx^{n+1}\}/(1-x)^2$.
7. $\{1+2x-(3n+1)x^n+(3n-2)x^{n+1}\}/(1-x)^2$.

Exercises 8 (*b*), p. 154

1. $(10/7)^{\frac{1}{2}}$. **2.** $(2/3)^{3/2}$. **3.** $2e$. **4.** $5e$. **5.** $\log\frac{3}{2}$. **6.** $\frac{1}{2}\log 3$. **7.** $\frac{3}{2}e$. **8.** 2. **9.** $1-\log 2$. **10.** $\frac{1}{2}(5)^{\frac{1}{2}}$. **11.** 1. **12.** $1/e$.

Exercises 8 (*c*), p. 157

1. (i) $\frac{1}{4}-1/2(n+1)(n+2)$; $\frac{1}{4}$; (ii) $\frac{1}{4}\{1-(4n+3)/(2n+1)(2n+3)\}$; $\frac{1}{4}$.

2. (i) $n(2n^3+8n^2+7n-2)$; (ii) $\frac{1}{2}\{n\cos A+\cos(n+2)A\sin nA\operatorname{cosec} A\}$.

4. 7/12.

Miscellaneous Exercises 8, p. 162

1. $\frac{1}{4}n(n^3+6n^2+5n+4)$.
2. (*a*) $1-1/(n+1)$, (*b*) 348,450.

3. (i) $n/(2n+1)$; (ii) 18,360; (iii) $x\{1-(1+n)x^n+nx^{n+1}\}/(1-x)^2$.

4. $3(e-1)$.
5. $A=-1$, $B=1$, $C=1$.
7. (*a*) $\frac{1}{2}$; (*b*) $5e-1$.

8. $e^x(1+x)$; $e^x(1+3x+x^2)$. **9.** (i) $\frac{2}{3}\sqrt{3}$; (ii) $\{(1+x)\log(1+x)-x\}/x^2$.

11. (i) $4e-2$; (ii) $(1/2^n)\cot(\theta/2^n)-2\cot 2\theta$.

12. (i) $e+1$; (ii) $\tan^{-1}(n+1)-\frac{1}{4}\pi$.

13. $$\frac{\sin^2 nx}{\sin^2 x}=\frac{1}{\sin x}\left(\frac{\sin^2 nx}{\sin x}\right)=\frac{\sin x}{\sin x}+\frac{\sin 3x}{\sin x}+\dots+\frac{\sin(2n-1)x}{\sin x}.$$

Use first series to evaluate $\displaystyle\int_0^{\pi/2}\frac{\sin(2n-1)x}{\sin x}\,dx$, etc.

14. (i) (*a*) 5/4; (*b*) $6e^2$; (ii) last part: equate $((x^{3n}))$ on each side of the identity $(1+x^3)^{3n}\equiv(1+x)^{3n}(1-x+x^2)^{3n}$.

15. $\{(x-5)\log(1-x)-5x\}/x^2$.

16. (i) $\{\tan(n+1)\phi-\tan\phi\}/\sin\phi$;

(ii) Series is $\dfrac{1}{2!}+\dfrac{2x^2}{3!}+\dfrac{3x^4}{4!}+\ldots=\{e^{x^2}(x^2-1)+1\}/x^4$.

17. (i) $e^x(x^2-x+1)-1$; (ii) $2\frac{1}{4}$;

18. (ii) $(2\cos\theta-1)/(5-4\cos\theta)$.

19. $\frac{1}{2}$.

21. (i) $A=-3$, $B=3$, $C=6$, $D=1$.

22. (i) $(1-x)^{-\frac{1}{2}}$, valid if $|x|<1$;

(ii) $e^x(x-1)+1-\dfrac{x^2}{2}-\dfrac{x^3}{3}$, valid for all values of x;

(iii) $-\{(3x^3+6)\log(1-x)+6x+3x^2+2x^3\}/9x^2$, valid when $-1\leqslant x<1$.

24. (i) $\frac{1}{4}$; (ii) $e^{\cos\theta}(2\cos\theta-1)\cos(\sin\theta)-\cos\theta$.
Last part: $n(n+2)<(n+1)^2$, $\therefore n(n+1)(n+2)<(n+1)^3$, etc.

25. (i) $S_n=\frac{1}{3}n(n+2)/(2n+1)(2n+3)$, $S_\infty=1/12$; (ii) $12e-1$;
$e^{\cos\theta}\sin(\sin\theta)$.

26. (i) $A=-1$, $B=4$, $C=4$; (ii) $\lambda=\frac{2}{3}$.

27. (i) $A=6$, $B=9$, $C=2$.

28. (ii) 11/96.

29. $S_n=\{1+3x-(4n+1)x^n+(4n-3)x^{n+1}\}/(1-x)^2$.

30. $\cos(\sin\theta)\cosh(\cos\theta)$.

31. Express u_n in the form

$$\{(An+B)/(n+1)(n+4)\}+\{(Cn+D)/(2n+3)(2n+5)\}$$

and express each of these fractions in terms of partial fractions; $A=C=0$, $B=-1$, $D=5$.

32. (i) $\frac{1}{2}n(n+1)/(2n+1)(2n+3)$.

33. (i) $\cos\{x+\frac{1}{2}(n+1)y\}\sin\frac{1}{2}ny\operatorname{cosec}\frac{1}{2}y$; (iii) $\coth x-1$, $x>0$; $\coth x+1$, $x<0$.

34. $S_n=n(3n+5)/8(3n+1)(3n+4)$, $S_\infty=1/24$;
(i) $5e-1$; (ii) $\frac{1}{3}(2\cos\theta-1)/(5-4\cos\theta)$.

35. (i) Write $(p-2)/p(p+1)(p+3)=(p^2-4)/p(p+1)(p+2)(p+3)$ and express p^2-4 in the form $Ap(p+1)+Bp+C$;
$S_n=(1/36)-(6n^2+15n+1)/6(n+1)(n+2)(n+3)$;

(ii) $\{(n+1)\sin n\theta-n\sin(n+1)\theta\}/2(1-\cos\theta)$.

37. (i) $1-(1-1/x)\log(1-x)$; (ii) $x(\cos x-x)/(1-2x\cos x+x^2)$.

38. (i) $\frac{1}{4}\pi$; (ii) $\frac{1}{10}n(n+1)/(n+5)(n+6)$.

39. (i) $S_n=\frac{3}{4}\{\sin\theta-3^{-n}\sin(3^n\theta)\}$, $S_\infty=\frac{3}{4}\sin\theta$;

(ii) $(\cos\frac{1}{2}\theta)/\sqrt{(2\cos\theta)}$ when $-\frac{1}{2}\pi<\theta<\frac{1}{2}\pi$; 0 when $\theta=\pm\frac{1}{2}\pi$.

40. 1; $\cos\theta+e^{\cos\theta}\{\cos\theta(\sin\theta)-\cos(\theta-\sin\theta)\}$;
$\cos\{\alpha+\frac{1}{2}(n-1)\beta\}\sin\frac{1}{2}n\beta\operatorname{cosec}\frac{1}{2}\beta$.

41. When $n>0$, $(n+1)^2>n^2+n+1$, and so $\cot^{-1}(n+1)^2<\cot^{-1}(n^2+n+1)$.

42. $(25/48)-\frac{1}{4}(4n^3+30n^2+70n+50)/(n+1)(n+2)(n+3)(n+4)$;
$S_n=\tanh n\theta/\sinh\theta$; when $\theta>0$, $S_\infty=\operatorname{cosech}\theta$,
when $\theta<0$, $S_\infty=-\operatorname{cosech}\theta$; $\sin\frac{1}{2}(n+1)\theta\sin\frac{1}{2}n\theta\operatorname{cosec}\frac{1}{2}\theta$.

43. (*a*) (i) $n(n^2+6n+11)/18(n+1)(n+2)(n+3)$;

(ii) prove that cosec $2A = \cot A - \cot 2A$ and deduce that cosec $2^r\theta = \cot 2^{r-1}\theta - \cot 2^r\theta$; $S_n = \cot\theta - \cot 2^n\theta$.

(*b*) $\frac{1}{2}\{e - e^{\cos 2\theta}\cos(\sin 2\theta)\}$.

45. (i) 7/36; (ii) $3(e-1)$; (iii) $e^{\cos\theta}\{\cos(\sin\theta - \theta)\} - \cos\theta - 1$. **46.** $\frac{1}{4}$

47. (i) $(29/36) - (6n^2+27n+29)/6(n+1)(n+2)(n+3)$; (ii) $3(e^{-1}-\frac{1}{2})$.

48. $1/x - 1/(x+1) + 2/x^2 - 2/(x+1)^2$; $S_n = 3 - 1/(n+1) - 2/(n+1)^2$, $S_\infty = 3$; $n = 11$.

49. $\sin\theta(\cos\theta - \sin\theta)/(1 - 2\sin\theta\cos\theta + \sin^2\theta)$.

50. (i) $(1-x^{n+1})/(1-x)$; $\{1 - x^n - nx^n(1-x)\}/(1-x)^2$; 9/4.

51. (i) $\sqrt{3} - \frac{2}{3}$;

(ii) $u_r = \{(r+1)+(2r+1)\}/r(r+1)(2r+1) = 2/2r(2r+1) + 1/r(r+1)$; $S_\infty = 3 - 2\log 2$.

Exercises 9 (*a*), p. 183

8. $xy_{n+1} + (n-m)y_n = 0$, $n > m$.

9. (i) $-b^4/a^2y^3$; (ii) $\frac{1}{3}(a^2/x^4y)^{\frac{1}{3}}$; (iii) $-2xa^3/y^5$. **10.** $-\cot^2\theta \operatorname{cosec}^5\theta$

11. $(\sec^4\theta \operatorname{cosec}\theta)/3a$. **12.** -4480. **14.** $2^n n!$.

Miscellaneous Exercises 9, p. 184

1. (i) $\frac{1}{2}(1-2x)/\{x(1-x)\}^{\frac{1}{2}}$, $\{2x/(x^2+1)^2\}\operatorname{cosec}^2\{1/(x^2+1)\}$.

2. (i) $\frac{1}{2}(9x^2+2x-25)/(3x-1)^{3/2}$, $-3x^2\operatorname{cosec}(x^3)\cot(x^3)$.

3. (i) $\frac{1}{2}(x^5+4)/x^3(x^5-1)^{\frac{1}{2}}$, $(1-\sin x - \cos x)/(1-\cos x)^2$.

4. (*a*) $(2/x^3)e^{-1/x^2}$; (*b*) $(1/2x^2)\operatorname{cosec}^2(1/x)\sqrt{\{\tan(1/x)\}}$;
(*c*) $-e^{-x}(21x^2+32x+33)/(7x-1)^2$.

5. $2a/\sqrt{(1-a^2x^2)}$, $-2e^{\cos x}\sin x/(e^{\cos x}+1)^2$.

6. (*a*) $\frac{1}{2}\cos x(2+2\sin x - \sin^2 x)/\sqrt{\{(2+\sin^2 x)(1-\sin x)^3\}}$;
(*b*) $2x/(2-2x^2+x^4)$.

7. (i) $2x\sin^{-1}(\frac{1}{2}x)$; (ii) $(2/a)\tan(x/a)$; (iii) $4e^{2x}/(1-e^{2x})^2$.

8. (i) $(x+\sin x)/(1+\cos x)$; $ab/(a^2+b^2x^2)$; $\sec x$.

9. (i) $-\tan\frac{1}{2}t$; (ii) $-2\cot 2\theta$.

10. (i) $(\log x)^x\{\log(\log x) + (\log x)^{-1}\}$; (iii) $\tan\frac{5}{2}\theta$.

12. (i) $\frac{1}{2}e$; (iii) $x(n+1+x)^n/(n+x)^{n+1}$.
$\{1+x/n\}^n/\{1+x/(n+1)\}^{n+1} = (n+1)^{n+1}/n^n y = y_0/y$ where y_0 is the value of y when $x=0$.

13. (i) $\frac{3}{4}x^{-\frac{1}{4}}\sec(x^{3/4})\tan(x^{3/4})$, $1/(2\sqrt{x})\sin e^{\sqrt{x}}\cos e^{\sqrt{x}}$.

14. (i) (*a*) $1/(x\log x)$, (*b*) $2/(1-x^2)\sqrt{x}$.

16. (i) $-\frac{1}{2}(x^3-3x^2+2)/(1-x^3)^{3/2}$; (ii) $m\sec^2 x/(1+m^2\tan^2 x)$;
(iii) $\{x/(a^2+x^2)\}\sec\{\frac{1}{2}\log(a^2+x^2)\}\tan\{\frac{1}{2}\log(a^2+x^2)\}$.

17. 0. The given function is constant; its value is log 2.

18. $\sin x + x\cos x$; (i) $1/(1+x)\sqrt{x}$; (ii) $x^2/(x^2-1)$.

19. (i) $\cos(x+\frac{1}{2}n\pi)$; (ii) $-(n-1)!\{(1-x)^{-n}+(-1)^{n-1}(1+x)^{-n}\}$; -2536.

20. 0. The given function is constant; its value is zero.

21. (a) (i) $-1/\{2(1+x)\sqrt{x}\}$; (ii) $x^x(1+\log x)$.

23. (a) $5(3x+8)(3x+1)^4/(2-x)^{11}$; (b) $n\{x+\sqrt{(1+x^2)}\}^n/\sqrt{(1+x^2)}$;
(c) $2/\sqrt{(1-x^2)}$.

24. $2x/\{(x^2+2)\sqrt{(1+x^2)}\}$.

25. (i) $\operatorname{cosech} x$; $1-x^2-3x(1-x^2)^{\frac{1}{2}}\sin^{-1}x$; (ii) $-\frac{1}{2}x^{-3/2}(\sqrt{x}-1)^2$.

26. $\cos x - x\sin x$; (i) $3/\{2(1+x)\sqrt{x}\}$; (ii) $x^{x-1}\{1+x\log x(1+\log x)\}$.

27. $b=-5/4$, $c=3/8$, $d=-1/64$.

28. $2x\cos(x^2)$; (i) $2(x+y)\cos(x+y)^2/\{1-2(x+y)\cos(x+y)^2\}$;
(ii) $\frac{1}{3}\sec^3 2\theta \operatorname{cosec}\theta$.

29. (i) $2ax/\{(a^2+x^2)\sqrt{(a^2+2x^2)}\}$.

31. $\sqrt{(a^2-b^2)}/(a+b\cos x)$; $\frac{1}{2}\sqrt{(a^2-b^2)}/(a+b\cos x)$.

33. (i) $-2/\sqrt{(1-x^2)}$; (ii) -1. **34.** $p(x)=-(1-x^2)$, $q(x)=(2n+1)x$.

35. (i) $\frac{1}{2}/\sqrt{(3x-2-x^2)}$, $-\frac{1}{2}/(x-1)\sqrt{(2-x)}$; (ii) $y_3(1+y_1^2)=3y_1y_2^2$.

36. (i) (a) $-1/\sqrt{(1-x^2)}$; (b) 0;
(ii) $(-1)^n n!\{(x-2)^{-n-1}-(x-1)^{-n-1}-(n+1)(x-1)^{-n-2}\}$.

37. (i) (a) $2/(1+4x)\sqrt{x}$; (b) $(x^2-1)/(x^2-4)$;
(ii) $y_1=\cot\frac{1}{2}(1+a/b)t$;
$y_2=\frac{1}{4}(1/a+1/b)\operatorname{cosec}^3\frac{1}{2}(1+a/b)t\operatorname{cosec}\frac{1}{2}(1-a/b)t$.

38. (a) $\cot x$; $1/x\{1+(\log x)^2\}$;
$\{\log(x+1/x)+(x^2-1)/(x^2+1)\}(x+1/x)^x$.

39. (i) -1; (ii) $-\frac{1}{2}(1+2\cos 2\theta)\operatorname{cosec}2\theta$; (iii) 10.

Exercises 9 (*b*), p. 194

1. $dx/d\theta=-2(\sin\theta+\sin 2\theta)$, $dy/d\theta=2(\cos\theta-\cos 2\theta)$.

2. Tangent is $y=x+1$; max. $=4/27$, min. $=0$.

3. $20x+17y+12a=0$. **4.** $Q\equiv(\frac{1}{4}at^2, -\frac{1}{8}at^3)$.

5. Normal is $125x-20y=363$; tangent is $y=2x$;
max. $(1, 1)$, min. $(-1, -1)$.

7. $y=3x-2$. **8.** $\sinh\log(4/e^2)$.

10. $(-1)^{n+1}x^n/(1+x)$; $(-1)^n x^n/n(1+x)^2$.

11. $x=\frac{1}{3}\pi$ gives minimum value.

13. Max. $(1, -6\log 7)$; min. $(3, 6\log\frac{3}{7}-8)$; $x=7$.

14. $y=6x+8$ at $(3, 26)$; $y=6x-8$ at $(-3, -26)$.

15. max. $(-\frac{1}{3}, \frac{6}{7})$, min. $(-5, \frac{10}{7})$.

16. (i) min. $(-\frac{1}{2}, -\frac{11}{16})$; inflexion $(1, 1)$. **17.** Gradients are -10, $8\log 2-16$.

18. $\frac{3}{8}a^2\sqrt{3}$. **21.** $t(3+t^2)x+2y=3t$.

22. (i) $x=3/5$ (max.), $x=1$ (min.), $x=0$ inflexion.

23. Max.$=3$, Min.$=0$. The greatest and least values of y are 3 and -1 respectively.

24. $x=(27+3\sqrt{78})^{\frac{1}{3}}$ gives max.; $x=(27-3\sqrt{78})^{\frac{1}{3}}$ gives min.

26. Time$=(a^2+x^2)^{\frac{1}{2}}/u+\{(c-x)^2+b^2\}^{\frac{1}{2}}/v$.

31. $V=\frac{1}{3}\pi a^3 \tan^2 \alpha(1+\operatorname{cosec} \alpha)^3$; $25\pi a^3/9$. **32.** h.

34. For all values of α and β, $x=\pi$ gives min. value; when $\alpha \geqslant \beta$, $x=0$ or 2π gives max. value; when $\alpha<\beta$, $x=0$ or 2π gives min. value, and $x=\cos^{-1}(\alpha/\beta)$ gives max. value.

35. $x=a^{2/3}b^{2/3}/\sqrt{(a^{2/3}+b^{2/3})}$.

36. The circles are orthogonal and cut at $(0, 0)$, $\{-2\lambda/(1+\lambda^4), -2\lambda^3/(1+\lambda^4)\}$; area of quadrilateral$=1$; length of common chord $=2\lambda/\sqrt{(1+\lambda^4)}$.

37. $4a^3/3\sqrt{3}=$max. value, $0=$min. value.

For max. value $r_1=a(1+1/\sqrt{3})$; but r_1 must lie between the values a and $a(1+e)$, where e is the eccentricity of the ellipse. Hence if $e>1/\sqrt{3}$, $a(1+1/\sqrt{3})$ is a possible value of r_1 and it gives $r_1 r_2(r_1-r_2)$ its max. value. If $e<1/\sqrt{3}$, $a(1+1/\sqrt{3})$ is not a possible value of r_1, and the max. value of $r_1 r_2(r_1-r_2)$ occurs when r_1 takes its highest possible value $a(1+e)$.

38. $OM=a(\sqrt{17}-3)$.

39. (i) min. $(0, 8)$; max. $(2, 60/e^2)$; (ii) min. value is $a+b+c-3(abc)^{\frac{1}{3}}$.

41. $(a+b)$. **42.** (i) $\sqrt{2}$, 3; (ii) $1/e$.

Exercises 10 (*a*), p. 207

1. $(2x-3)^3/6$. **2.** $\frac{1}{3}\log(3x-4)$. **3.** $1/6(5-3x)^2$. **4.** $\frac{1}{6}\tan^{-1}\frac{1}{3}(5+2x)$.

5. $\frac{1}{4}\sin^{-1}\frac{1}{3}(3+4x)$. **6.** $\frac{1}{3}\sinh^{-1}\frac{1}{2}(4+3x)$. **7.** $\frac{1}{6}\cosh^{-1}\frac{1}{2}(6x-1)$.

8. $-1/7(7x-2)$.

9. $\frac{2}{5}(x-2)(3+x)^{3/2}$ **10.** $\sqrt{(x^2-5)}$. **11.** $\frac{2}{3}\sqrt{(5+x^3)}$. **12.** $\frac{1}{3}\log(5+x^3)$.

13. $\frac{1}{6}\tan^{-1}(\frac{1}{3}x^2)$. **14.** $\frac{1}{3}\sqrt{(9+x^2)^3}$. **15.** $\frac{1}{8}\sin^8 x$. **16.** $-\frac{1}{7}\cos^7 x$.

17. $(5\sin^3 x-3\sin^5 x)/15$. **18.** $-\frac{1}{2}\operatorname{cosec}^2 x$. **19.** $-\frac{1}{4}\log(3-4\tan x)$.

20. $2e^{\sqrt{x}}$. **21.** $\frac{1}{3}(\sin^{-1}x)^3$. **22.** $\frac{1}{4}(\log x)^4$. **23.** $\frac{1}{2}\log(1+x^2)$. **24.** $\sqrt{(4+x^2)}$.

25. $\frac{1}{2}\log(x^2+4x-5)$. **26.** $\sqrt{(x^2+4x-5)}$. **27.** $\log\sin x$. **28.** $\frac{1}{3}\log\sin 3x$.

29. $-\frac{1}{4}\log\cos 4x$. **30.** $-\frac{1}{3}\log(2+3\cos x)$. **31.** $\frac{1}{2}\log(1+2\tan x)$.

32. $\frac{1}{2}\log(1+e^{2x})$.

Exercises 10 (*b*), p. 210

1. $\{3\log(2x-3)+2\log(x+1)\}/10$. **2.** $\{7\log(3x+5)-3\log\ x-1)\}/12$.

3. $-\frac{1}{2}\{\log(3x-1)+\log(x-1)\}$. **4.** $\{3\log x-10\log(x-2)+7\log(x-4)\}/8$.

5. $3\log x+\frac{1}{2}\log(x^2+4)-\frac{1}{2}\tan^{-1}(\frac{1}{2}x)$.

6. $-\frac{3}{2}\{\log(3-x)+\log(3+x)+4/(3+x)\}$.

7. $x-4\log(x+2)-5/(x+2)$. **8.** $\log(x-1)-\log x+2/x-2/(x-1)$.

9. $(1/2\sqrt{13}) \log \{(x+3-\sqrt{13})/(x+3+\sqrt{13})\}$. **10.** $\frac{1}{4} \tan^{-1} \frac{1}{4}(x+3)$.

11. $-2 \log(12-6x-x^2)-(7/\sqrt{21}) \log \{(\sqrt{21}+x+3)/(\sqrt{21}-x-3)\}$.

12. $\log (x^2+2x+26)-\frac{1}{5} \tan^{-1} \frac{1}{5}(x+1)$.

13. $\frac{1}{2} \log (4x^2+16x+25)-(7/6) \tan^{-1} \frac{1}{3}(2x+4)$.

14. $(5/6) \log (3x^2+6x-2)-(2/\sqrt{15}) \log \{(3x+3-\sqrt{15})/(3x+3+\sqrt{15})\}$.

15. $-\frac{3}{2} \log (1+6x-2x^2)+(4/\sqrt{11}) \log \{\sqrt{11}+2x-3)/(\sqrt{11}-2x+3)\}$.

Exercises 10 (*c*), p. 212

1. $-\frac{1}{3}\sqrt{(9-2x)^3}$. **2.** $\frac{1}{2} \sin^{-1} (2x/3)$. **3.** $\sinh^{-1} \frac{1}{5}(x+1)$. **4.** $\cosh^{-1} \frac{1}{5}(x-2)$.

5. $\sin^{-1} \frac{1}{4}(x+3)$. **6.** $\sqrt{(x^2+2x+10)}+2 \sinh^{-1} \frac{1}{3}(x+1)$. **7.** $\sin^{-1} \frac{1}{2}(x-2)$.

8. $\sqrt{(x^2-7x+12)}+(13/2) \cosh^{-1} (2x-7)$.

9. $-\frac{1}{2} \cosh^{-1} \{(8-x)/7x\}$. **10.** $-(1/\sqrt{3}) \cosh^{-1} \{(x+2)/(2x+1)\}$.

11. $(2/105)(15x^2+24x+32)\sqrt{(x-2)^3}$. **12.** $\sqrt{(4+x^2)}$.

13. $\frac{2}{3}(x-8)\sqrt{(4+x)}$. **14.** $\frac{1}{4}\sqrt{(4x^2+16x+25)}-\frac{1}{2} \sinh^{-1} \frac{2}{3}(x+2)$.

15. $\frac{1}{9}\{7\sqrt{3} \cosh^{-1} \frac{1}{8}(3x+2)-6\sqrt{(3x^2+4x-20)}\}$.

Exercises 10 (*d*), p. 218

1. $\frac{1}{3} \cos^3 x-\cos x$. **2.** $\sin x\,(15-10\sin^2 x+3\sin^4 x)/15$. **3.** $\frac{1}{4}(\sinh 2x-2x)$.

4. $\log \cosh x-\frac{1}{2} \tanh^2 x$. **5.** $\frac{1}{2} \log (\operatorname{cosec} 2x-\cot 2x)$.

6. $\frac{1}{3} \log (\sec 3x+\tan 3x)$. **7.** $2 \tan \frac{1}{2}x-x$. **8.** $(7 \sin^5 x-5 \sin^7 x)/35$.

9. $\frac{1}{3} \sec^3 x-\sec x$. **10.** $\tan x+\frac{1}{3} \tan^3 x$. **11.** $-\frac{1}{3} \cot^3 x$.

12. $\frac{1}{3} \tan^3 x+2 \tan x-\cot x$. **13.** $\tan \frac{1}{2}x$.

14. $\tan (\frac{1}{2}x-\frac{1}{4}\pi)$ or $-2/(1+\tan \frac{1}{2}x)$.

15. $(3 \cos 2x-\cos 6x)/12$. **16.** $(5 \sin 2x-\sin 10x)/20$.

17. $(4 \sin 2x+\sin 8x)/16$. **18.** $\frac{1}{5} \log \{(5+\tan \frac{1}{2}x)/(5-\tan \frac{1}{2}x)\}$.

19. $(1/60) \log \{(11 \tan \frac{1}{2}x+1)/(\tan \frac{1}{2}x+11)\}$. **20.** $\frac{1}{12} \tan^{-1} (\frac{3}{4} \tan x)$.

21. $\log (\sin x-\cos x)$. **22.** $4x+\log (\sin x+2 \cos x)$.

23. $x+\log (2+2 \cos x-\sin x)+\log (2-\tan \frac{1}{2}x)$.

24. $\{3 \log (1+t^2)+7 \log (3+t)-13 \log (3-t)\}/15,\ t\equiv\tan \frac{1}{2}x$.

25. $\frac{1}{2}\{x\sqrt{(4+x^2)}+4 \sinh^{-1} \frac{1}{2}x\}$. **26.** $\frac{1}{2}\{x\sqrt{(x^2-9)}-9 \cosh^{-1} \frac{1}{3}x\}$.

Exercises 10 (*e*), p. 221

1. $x^4(4 \log x-1)/16$. **2.** $-(2 \log x+1)/4x^2$. **3.** $\frac{1}{4}(\sin 2x-2x \cos 2x)$.

4. $e^{3x}(3x-1)/9$. **5.** $\{(9x^2-2) \sin 3x+6x \cos 3x\}/27$.

6. $\{(16x^2+1) \tan^{-1} (4x)-4x\}/32$. **7.** $e^{2x}(2 \sin 3x-3 \cos 3x)/13$.

8. $(x+2) \log (x+2)-x$. **9.** $x \tan x+\log \cos x$. **10.** $x \sinh x-\cosh x$.

11. $x \sinh^{-1} x-\sqrt{(1+x^2)}$. **12.** $x \cosh^{-1} 2x-\frac{1}{2}\sqrt{(4x^2-1)}$.

13. $\frac{1}{2}\{x\sqrt{(9+x^2)}+9 \sinh^{-1} \frac{1}{3}x\}$. **14.** $\frac{1}{2}\{x\sqrt{(x^2-9)}+9 \cosh^{-1} \frac{1}{3}x\}$.

15. $x \log (x^2+16)-2x+8 \tan^{-1} \frac{1}{4}x$.

Exercises 10 (*f*), p. 223

1. $\frac{1}{3}$. **2.** Does not exist. **3.** $\frac{1}{8}\pi$. **4.** $\frac{1}{2}\pi$. **5.** 1. **6.** $\frac{1}{2}$. **7.** π.
8. Does not exist. **9.** 2. **10.** $\pi/20$. **11.** 0. **12.** $(2+\pi)/8$.

Exercises 10 (*g*), p. 230

1. 16/35. **2.** $5\pi/32$. **3.** 2/15. **4.** $\pi/32$. **5.** $35\pi/8$. **6.** $3\pi/16$.
7. $\frac{1}{4}\tan^4 x - \frac{1}{2}\tan^2 x - \log\cos x$.
8. $\tan x\,(15+10\tan^2 x+3\tan^4 x)/15$.

Exercises 10 (*h*), p. 233

1. $(16e-38)/e^2$. **2.** (i) $(n-1)!$; (ii) $35\pi/256$.
3. $\pi(2p)!\,(2q)!\,/2^{2p+2q+1}p\,!\,q\,!\,(p+q)!$. **4.** $2^{m+n+1}m\,!\,n\,!\,/(m+n+1)!$.
5. $5\pi/256$. **7.** $(1+n)I_n = x(a^2+x^2)^{n/2} + a^2 n I_{n-2}$.
8. $\pi(2n+1)!/2^{2n+2}n\,!\,(n+2)!$. **9.** 23/15.
10. $\pi(2m)!/2^{2m+2}m\,!\,(m+1)!$. **11.** $\frac{3}{5}\cosh\frac{1}{2}\pi$.
12. (i) $I_n - nI_{n-1} = -e^{-x}x^n$; (ii) $5(\frac{1}{2}\pi)^4 - 60(\frac{1}{2}\pi)^2 + 120$.

Miscellaneous Exercises 10, p. 234

1. (i) $\frac{1}{2}\{\sinh^{-1} x + x\sqrt{(1+x^2)}\}$, $x\sinh x - \cosh x$; (ii) $\frac{1}{2}$; (iii) $\frac{1}{2}$.
2. $\sqrt{(x^2+4x+8)} - 2\log\{x+2+\sqrt{(x^2+4x+8)}\}$;
$\log(x+1) + 2/(x+1) - 1/(x+1)^2$; $-\frac{1}{2}e^{-x}(\sin x+\cos x)$; 10/3.
3. (i) $(1/b)\sin^{-1}(bx/a)$, $\log\{e^x/(1+e^x)\}$, $\frac{1}{4}(2-\log 3)$; (ii) $2\log(2+\sqrt{3})$.
4. (i) Express $\sin 2x+\sqrt{3}\cos 2x$ in the form $R\sin(2x+\alpha)$;
$\frac{1}{4}\log\tan(x+\frac{1}{6}\pi)$; (ii) $\frac{1}{2}(1+e^{-\pi})$.
5. $\log\{(x-2)^3/(x-1)^2\}$, $\{3x^3\cosh^{-1}x - (x^2+2)\sqrt{(x^2-1)}\}/9$;
$(x-1)/\sqrt{(x^2+1)}$.
6. (i) $\pi/\sqrt{3}$; (ii) $2\pi(\frac{2}{3}\sqrt{3}-1)$. **7.** (i) $2-\frac{1}{2}\pi$; (ii) $\frac{1}{2}\sqrt{2}\log(1+\sqrt{2})$.
8. (i) Express denominator as $5\cos\{\theta-\tan^{-1}(4/3)\}$; $\frac{1}{5}\log 3$;
(ii) $-\{\sin(\log x)+\cos(\log x)\}/2x$.
9. (i) $(5\pi-6\sqrt{3}+6)/12$; (ii) $(\pi-2\log 2)/8$.
11. (i) $(3\sqrt{2}-4)/10$; (ii) $\frac{1}{2}(\sqrt{26}-\sqrt{10})+\log\{(3+\sqrt{10})/(5+\sqrt{26})\}$;
(iii) $(2e+\pi)/(4+\pi^2)$.
12. $\frac{1}{4}\log(x-1) - \frac{1}{4}\log(x+1) + \frac{1}{2}\tan^{-1}x + 1/x$;
let $x=e^t$, $x^3\{9(\log x)^3 - 9(\log x)^2 + 6\log x - 2\}/27$; $7\pi/3$.
13. (i) $4\sin^{-1}\frac{1}{3}(2x-1) - 2\sqrt{(2+x-x^2)}$; (ii) $x^4\{8(\log x)^2 - 4\log x + 1\}/32$.
14. (iii) $\cos x + \log\tan\frac{1}{2}x - \cos x\log(\sin x)$.
15. (i) $(3\pi-4)/18$; (ii) $\frac{1}{2}\{a^2\sin^{-1}(x/a) + x\sqrt{(a^2-x^2)}\}$; (iii) $\frac{1}{4}\pi$.
16. (i) $\frac{2}{3}(x-5)\sqrt{(1+x)}$, $e^{ax}(a\sin bx - b\cos bx)/(a^2+b^2)$,
$(4/\sqrt{5})\tan^{-1}\{(1/\sqrt{5})\tan\frac{1}{2}x\}$; (ii) $\frac{1}{2}(a+b)\pi$.

17. (i) $\frac{1}{4}\pi-\frac{1}{2}\log 2$; (ii) $\sinh^{-1}\{x/(x+1)\}-\sqrt{(2x^2+2x+1)}/(x+1)$.

18. (i) $2\tan^{-1}\sqrt{x}$; (ii) $9\log 10-4-\frac{2}{3}\tan^{-1}3$;
(iii) write integrand as $2\cos x/\cos 2x-1/\cos x$;
$(1/\sqrt{2})\log\{(2+\sqrt{2})/(2-\sqrt{2})\}-\frac{1}{2}\log 3$.

19. $\tan^{-1}(\sinh x)$ or $\tan^{-1}e^x$; $\frac{1}{4}+2\log 2$; $\frac{1}{3}\log\frac{1}{4}$.

20. (i) $2\log 2+7/288$;
(ii) $(\frac{1}{4}\sqrt{2})(\tan^{-1}\sqrt{2}-\tan^{-1}\frac{1}{4}\sqrt{2})=(\frac{1}{4}\sqrt{2})\tan^{-1}\frac{1}{2}\sqrt{2}$.

21. (i) $\frac{1}{2}\log\tan\frac{1}{2}x-\frac{1}{4}\operatorname{cosec}^2\frac{1}{2}x$; (ii) $10+9\pi/2$.

22. (i) 0·61; (ii) 22·67; (iii) 3·77.

23. (i) $\frac{1}{4}\pi^2$; (ii) $40/9$; (iii) $\frac{1}{2}a(\pi+2)$.

24. $\log\{(1-x)/(1+x)\}+1/(1-x)$;
$\{2/\sqrt{(a^2-b^2)}\}\tan^{-1}(\{(a-b)/(a+b)\}^{\frac{1}{2}}\tan\frac{1}{2}x)$; $a/(a^2+b^2)$.

25. $\frac{1}{2}\{x^2-x\sqrt{(x^2-1)}+\cosh^{-1}x\}$; $x\tan^{-1}x-\frac{1}{2}\log(1+x^2)$; π.

26. $2(\sqrt{x}-1)e^{\sqrt{x}}$; $2e^{\sin^{-1}\sqrt{x}}(\sqrt{\{x(1-x)\}}-1+2x)/5$;
$\operatorname{cosec}\alpha\log\tan(\frac{1}{4}\pi+\frac{1}{2}\alpha)$.

27. (i) $\frac{1}{2}\pi-1$; (ii) $(\pi-32+18\sqrt{3})/144$;
(iii) $(2/\sqrt{3})\log\{(\tan\frac{1}{2}\theta-2-\sqrt{3})/(\tan\frac{1}{2}\theta-2+\sqrt{3})\}+\log\tan\frac{1}{2}\theta$.

28. (i) $\{2(x^4-1)\log(1+x^2)+x^2(2-x^2)\}/8$;

(ii) $$\int_0^{\pi/2}\frac{\sin\theta\,d\theta}{\sin\theta+\cos\theta}=\int_0^{\pi/2}\frac{\cos\theta\,d\theta}{\sin\theta+\cos\theta}=\frac{1}{2}\int_0^{\pi/2}\frac{\sin\theta+\cos\theta}{\sin\theta+\cos\theta}d\theta=\frac{1}{4}\pi;$$

(iii) 3π.

29. (i) $3\log(x-2)-2\log(x+1)$; $\pi/6$; (ii) $\frac{1}{4}(2\log 2-1)$.

30. (i) $-\sqrt{\{(1-x)/(1+x)\}}$; (ii) $x\tan\frac{1}{2}x$; (iii) $\frac{1}{2}(b-a)\pi$.

31. (i) $\sin^5x(63-90\sin^2x+35\sin^4x)/315$; (ii) $1+\log(8/5^{4/3})$.

32. $\frac{1}{3}\log\{x(x^2+4x+6)\}-(8/3\sqrt{2})\tan^{-1}\{(x+2)/\sqrt{2}\}$.

33. $\{4\log(2+x)-\log(1-x)-3\log(3+x)\}/12$;
$\frac{1}{2}c\sin^{-1}\{(2x-c)/c\}-\sqrt{\{x(c-x)\}}$.

34. (i) $\log\{(x+1)^2/(2x-1)(x-1)\}$; (ii) $\log(\sin\theta+\cos\theta)=\frac{1}{2}\log(1+\sin 2\theta)$.

35. (i) $2x+\log\cos^2\frac{1}{2}x-\log(2+\tan\frac{1}{2}x)^3$.

36. (i) (a) Write integrand as $x\sec x\tan x$; $x\sec x-\log(\sec x+\tan x)$;
(b) $\sqrt{(x^2-1)}+\log\{x+\sqrt{(x^2-1)}\}$.

37. $\pi/32$; $x-\sqrt{(1-x^2)}\sin^{-1}x$;
$(1/\sqrt{3})\log\{(\sqrt{3}\tan\frac{1}{2}\theta-1)/(\sqrt{3}\tan\frac{1}{2}\theta+1)\}$.

38. (i) $-\frac{1}{2}e^{\cos^2x}$; (ii) $\frac{1}{2}x\sqrt{(x^2-1)}-\frac{1}{2}\log\{x+\sqrt{(x^2-1)}\}$.
Express integrand as $2+\cos\theta-3/(2-\cos\theta)$.

39. (i) $-e^{-x}(\sin 2x+2\cos 2x)/5$; $-x\cot\frac{1}{2}x$; (ii) $\frac{1}{4}a^2(\pi-2)$.

40. $x/\sqrt{(1+x^2)}$; $\{3(x^4-1)\tan^{-1}x-x^3+3x\}/12$; $a/(1-a^2)$.

41. (i) $2\{\sqrt{x}-\sqrt{3}\tan^{-1}\sqrt{(x/3)}\}$; (ii) $\frac{1}{2}\{\log\tan(\frac{1}{4}\pi+\frac{1}{2}\theta)+\tan^{-1}(\sin\theta)\}$;
(iii) $\sqrt{(x^2-1)}-\log\{x+\sqrt{(x^2-1)}\}$.

42. (i) $\frac{1}{6}\sin^6 x - \frac{1}{8}\sin^8 x$; (ii) $\tan^{-1}(1+\tan x)$.

43. (i) $\frac{1}{2}\{a^2\sin^{-1}(x/a)+x\sqrt{(a^2-x^2)}\}$;
(ii) $\frac{1}{2}\{(x-4a)\sqrt{(a^2-x^2)}-a^2\sin^{-1}(x/a)\}$.

44. (i) $\sqrt{(x^2+x+1)}-\frac{1}{2}\log\{x+\frac{1}{2}+\sqrt{(x^2+x+1)}\}$;
(ii) $\frac{1}{2}\log 3$; $\{10\tan^{-1}(\frac{1}{3}\tan\frac{1}{2}x)-12\sin x/(5+4\cos x)\}/27$.

45. (i) $\log\tan(\frac{1}{2}x+\frac{1}{4}\pi)$; (ii) $(1/\sqrt{3})\sin^{-1}\frac{1}{4}(3x+2)$;
(iii) $(1/\sqrt{6})\tan^{-1}(\frac{1}{3}x\sqrt{6})$;
(iv) $\log\{(x+1)^4/(x^2-x+1)\}-(8/\sqrt{3})\tan^{-1}\{(2x-1)/\sqrt{3}\}$.

46. (i) $-(5\sin 4x+4\cos 4x)e^{-5x}/41$;
(ii) $\frac{1}{4}\{6\log(x+1)-3\log(x^2+1)-2\tan^{-1}x\}$; (iii) $\tan^{-1}\frac{1}{2}\sqrt{(x-1)}$;
(iv) $\frac{1}{2}\{(x-3)\sqrt{(x^2+2x-8)}+15\log(x+1+\sqrt{\{x^2+2x-8\}})\}$.

47. (i) $\frac{1}{2}\{x^2\tan^{-1}(x+1)-x+\log(x^2+2x+2)\}$; $-(1+2x^3)/6(x^3+1)^2$;
(ii) $\pi/4\sqrt{2}$.

48. (i) $\cosh^{-1}(2x-3)$; (ii) $x/\sqrt{(1+x^2)}$;
(iii) $\frac{1}{2}\sec x\tan x+\frac{1}{2}\log\tan(\frac{1}{4}\pi+\frac{1}{2}x)$.

49. (i) $\frac{1}{4}\{2\log(x-1)-\log(x^2+1)+2\tan^{-1}x\}$;
(ii) $\log(1+\tan\frac{1}{2}x)$; $\frac{1}{8}\{\tan^{-1}x+x(2x^3+x^2-1)/(1+x^2)^2\}$.

50. $\frac{1}{2}\{\tan^{-1}\frac{1}{2}x+\log(x^2+4)-2\log(x-1)-2/(x-1)\}$;
$\{(x+a)^{3/2}+(x-a)^{3/2}\}/3a$; $\pi/(a^2-b^2)$.

51. $x-\log(x^2+2)-(1/\sqrt{2})\tan^{-1}(x/\sqrt{2})+\frac{3}{2}\log(2x+1)$;
$-\cosh^{-1}\{(1+x)/x\sqrt{2}\}$.

52. (i) $-\frac{1}{2}e^{-x^2}(1+x^2)$; $3\pi/16$; (ii) $(8\log 4-\pi)/12$.

53. (i) $3\sin^{-1}\frac{1}{2}(x-1)+\sqrt{(3+2x-x^2)}$; (ii) $\log x-(1+1/x)\log(1+x)$.

54. (i) $3/(1+2x)-2/(1+x)$; $\frac{3}{2}\log(1+2x)-2\log(1+x)$;
(ii) $1+\frac{1}{5}\{8/(x-2)+(2x-1)/(x^2+1)\}$;
$x+\frac{1}{5}\{8\log(x-2)+\log(x^2+1)-\tan^{-1}x\}$.

55. (i) $\frac{2}{3}$; (ii) 1 ; (iii) $\frac{4}{3}$.

56. (i) $(\pi^2-4)/16$; (ii) $\log\{\sqrt{(x+2)}-1\}-\log\{\sqrt{(x+2)}+1\}$;
(iii) $\cosh^{-1}(2x-3)$.

57. $m^n(n)!/\{(m+1)(2m+1)\ldots(nm+1)\}$.

58. Write $u_n=\int_0^{\pi/2}(x\cos^{n-1}x)(\cos x)dx$; $u_4=(3\pi^2-16)/64$;
$u_5=(60\pi-149)/225$.

59. Consider $\int\{\cos n\theta+\cos(n-2)\theta\}\sec\theta d\theta$; $\frac{1}{2}\pi$.

61. $\{3\tan^{-1}(x/a)+3ax/(x^2+a^2)-6a^3x/(x^2+a^2)^2-8a^3x^3/(x^2+a^2)^3\}/48a^3$.

62. 16/315. **64.** $3(16-5\pi)/32$. **65.** $(\pi^4-48\pi^2+384)/16$.

66. If I_n is the given integral, $(2n+3)I_n=2nI_{n-1}$.

67. Substitute for I_{n+1}, I_n and I_{n-1} in L.H.S. of given reduction formula.

68. $\{(2n-1)(2n-3)\dots 5.3.1\}a^n\pi/n\,!$. **69.** $16/15$.

70. $I_{\frac{1}{2}n}=\{x(x^2+a^2)^{n/2}+na^2 I_{\frac{1}{2}n-1}\}/(n+1)$.

72. $\{14+3\sqrt{2}\log(1+\sqrt{2})\}/64a^5$. **73.** (i) $32a^{13/2}/3003$; (ii) $u_{2n+1}=2nu_{2n-1}$.

74. $\pi/32$.

Exercises 11 (*a*), p. 250

4. $2^{-1/3}\left\{1-\dfrac{x^2}{36}+\dfrac{x^4}{6480}+\dots\right\}$

8. $1+\frac{1}{2}x^2-\frac{7}{24}x^4+\dots$. **9.** $\dfrac{d}{dx}[\log\{x+\sqrt{(1+x^2)}\}]=(1+x^2)^{-\frac{1}{2}}$.

13. $1+x-\dfrac{2}{3\,!}x^3-\dfrac{2^2}{4\,!}x^4-\dots$; $\cos x\cosh x=\frac{1}{2}\{e^x\cos x+e^{-x}\cos(-x)\}$.

14. $\log(1+e^{ix})=\log 2+\frac{1}{2}ix-\frac{1}{8}x^2-\frac{1}{192}x^4+\dots$;
$\log(1+e^{-ix})=\log 2-\frac{1}{2}ix-\frac{1}{8}x^2-\frac{1}{192}x^4+\dots$.
$1+\cos x=\frac{1}{2}(1+e^{ix})(1+e^{-ix})$; $\log(1+\cos x)=\log 2-\frac{1}{4}x^2-\frac{1}{96}x^4-\dots$.

15. $\log\{1-\log(1-x)\}=x+\frac{1}{6}x^3+\dots$; $\log\{1+\log(1+t)\}=t-t^2+\frac{7}{6}t^3+\dots$

16. $1+x\cos\alpha+(x^2\cos 2\alpha)/2\,!+(x^3\cos 3\alpha)/3\,!+\dots$; $e^{x\cos\alpha}\sin(x\sin\alpha)$.

17. $1+\dfrac{x^2}{3\,!}+\dfrac{14}{6\,!}x^4+\dots$.

18. Expand $\dfrac{dy}{dx}=(1-x)(1-x^3)^{-1}$ and integrate the result.

19. (i) $\log 2-\frac{1}{2}x^2$; (ii) $-\frac{1}{2}$.

Exercises 11 (*b*), p. 254

1. $y=x-\dfrac{1}{2}\dfrac{x^3}{3}+\dfrac{1.3}{2.4}\dfrac{x^5}{5}-\dfrac{1.3.5}{2.4.6}\dfrac{x^7}{7}+\dots$

2. $(1+x^2)y_{n+2}+(2n+1)xy_{n+1}+(n^2-1)y_n=0$.

$$e^{\sinh^{-1}x}=1+x+\frac{x^2}{2}-\frac{x^4}{2.4}+\frac{1.3}{2.4.6}x^6-\frac{1.3.5}{2.4.6.8}x^8+\dots.$$

$$((x^{2n}))=(-1)^{n+1}\frac{1.3.5.7\dots(2n-3)}{2.4.6.8\dots 2n}.$$

3. $y=1+x+\frac{1}{2}x^2-\frac{1}{6}x^3-\frac{7}{24}x^4+\frac{1}{24}x^5+\dots$.

4. $y=ax+x^2+ax^3/3\,!+2^2.2x^4/4\,!+3^2ax^5/5\,!+4^2.2^2.2x^6/6\,!+\dots$
$((x^{2p}))=2\{(2p-2)^2(2p-4)^2\dots 4^2.2^2\}/(2p)!$ if $p>1$; $p_2=2$;
$((x^{2p+1}))=a\{(2p-1)^2(2p-3)^2\dots 5^2.3^2\}/(2p+1)!$ if $p>1$; $p_1=a$.

7. $-1/45$.

9. $4(1+x)y_{n+2}+2(2n+1)y_{n+1}+\pi^2 y_n=0$.
$y=-1+\pi^2x^2/8-\pi^2x^3/16+\pi^2(15-\pi^2)x^4/384+\dots$.

10. $y=5x+20x^3+16x^5$.

11. $a_{2n-1}=(-1)^{n-1}\{(2n-3)^2(2n-5)^2\ldots5^2.3^2.1^2.2\}/(2n-1)!$ if $n>1$; $a_1=2$;
$a_{2n}=(-1)^{n-1}\{(2n-2)^2(2n-4)^2\ldots6^2.4^2.2^2.2\}/(2n)!$ if $n>1$; $a_2=2$.

12. $m\{(2n-1)^2-m^2\}\{(2n-3)^2-m^2\}\ldots\{9-m^2\}\{1-m^2\}x^{2n+1}/(2n+1)!$

13. $$y=(\log a)^2+\log a^2\left\{\frac{x}{a}-\frac{1^2x^3}{3!\,a^3}+\frac{3^2.1^2x^5}{5!\,a^5}-\ldots\right\}$$
$$+2\left\{\frac{x^2}{2!\,a^2}-\frac{2^2x^4}{4!\,a^4}+\frac{4^2.2^2x^6}{6!\,a^6}-\ldots\right\}$$
$((x^{2n}))=(-1)^{n-1}2\{(2n-2)^2(2n-4)^2\ldots6^2.4^2.2^2\}/(2n)!\,a^{2n}$, if $n>1$
$((x^{2n-1}))=(-1)^{n-1}(\log a^2)\{(2n-3)^2(2n-5)^2\ldots5^2.3^2.1^2\}/(2n-1)!\,a^{2n-1}$ if $n>1$.

14. $$y=1-\frac{1}{2}x^2-\frac{1}{2.4}x^4-\frac{1.3}{2.4.6}x^6-\frac{1.3.5}{2.4.6.8}x^8-\frac{1.3.5.7.}{2.4.6.8.10}x^{10}-\ldots$$

15. $y=7x+56x^3+112x^5+64x^7$. **16.** $(1-x^2)y_{n+2}-(2n+1)xy_{n+1}-n^2y_n=0$.

17. $x+\frac{1}{8}x^4+\frac{3}{56}x^7+\frac{1}{32}x^{10}+\ldots$.

18. $\{1/(2n)!\}(\{2n-2\}^2+1)(\{2n-4\}^2+1)\ldots17.5.1.$

Exercises 11 (*c*), p. 263

1. (i) 1 ; (ii) $\frac{1}{4}$; (iii) 1 ; (iv) $\frac{1}{4}$; (v) $\frac{1}{2}$; (vi) 1 ; (vii) $1/e$; (viii) 1; (ix) -1 ; (x) e ; (vi) 0 ; (xii) $\log a$.

2. $e^{\sin x}=1+x+\frac{1}{2}x^2-\frac{1}{8}x^4+\ldots$; $\frac{1}{3}$. **3.** 1/6. **4.** 1/15.

5. $x^2-\frac{4}{3}x^4+\frac{32}{45}x^6+\ldots$; 22/45. **6.** 1/2560. **7.** p ; $\frac{1}{2}a$.

8. $\log a$. **9.** (i) (*a*) 1, (*b*) $-1/6$; (ii) 1.

10. (i) $\{\log (a/b)\}/\{\log (c/d)\}$; (ii) -2 ;
(iii) put $x=\tan(\frac{1}{2}\pi-v)$ and find $\lim_{v\to0}(v\cot v)$; 1.

11. $-\sin x$; $-1/x^2$. **12.** (ii) $2\log 2$. **13.** $1+x+\frac{1}{2}x^2-\frac{1}{8}x^4$; $\frac{1}{2}$.

14. (ii) $-8/3$.

Exercises 11 (*d*), p. 267

1. 0·4. **2.** 1·205. **3.** 1·06. **4.** 1·468. **5.** 1·18. **6.** 3·82. **7.** 1·41.
8. 2·512. **9.** 0·595. **10.** 4·275. **11.** 1·86. **12.** 1·547. **13.** 0·747.
14. 1·47.

Exercises 12 (*a*), p. 282

1. Take equation of $X'Y'$ as $x/b-y/a=k$; then XY' is $x-y/k=a$ . . . (i) and $X'Y$ is $x/k+y=b$. . . (ii). Multiply (i) by x and (ii) by y and add.

2. Write $\triangle OPQ=\frac{1}{2}x_1x_2(y_1/x_1-y_2/x_2)$; then y_1/x_1, y_2/x_2 are roots of the equation $b(y/x)^2+2h(y/x)+a=0$, and x_1, x_2 are roots of the equation $ax^2+2hx(1-lx)/m+b(1-lx)^2m^2=0$.

4. $(-2, 1)$; 60°.

6. $x^2(ar^2-2gpr+cp^2)+2xy\{hr^2-r(gq+fp)+cpq\}+y^2(br^2-2fqr+cq^2)=0$; $r^2(a+b)-2r(gp+fq)+c(p^2+q^2)=0$.

7. Eliminate y between the equations $ax^2+2hxy-ay^2=0$ and $px+qy=r$; the sum of the roots of the resulting equation in x must be zero.

8. Let the line $lx+my+n=0$ meet the given line pair at $P(x_1, m_1x_1)$ and $Q(x_2, m_2x_2)$. Then $x_1=-n/(l+mm_1)$, $x_2=-n/(l+mm_2)$ where $m_1+m_2=-2h/b$, $m_1m_2=a/b$.

9. The line drawn through the origin perpendicular to $qx+py=pq$ is one of the bisectors of the angles between the given line-pair.

10. The line-pair through $A\ (\alpha, \beta)$ is $b(x-\alpha)^2-2h(x-\alpha)(y-\beta)+a(y-\beta)^2=0$.

11. Let (α, β) be the vertex. Eliminate y between the equations $lx+my+n=0$ and $ax^2+2hxy+by^2=0$; then α is the sum of the roots of the resultant equation;

$$\{2n(hm-bl)/p,\ 2n(hl-am)/p\},\ p\equiv am^2-2hlm+bl^2.$$

12. $2x^2+7xy-2y^2-10x+15y-20=0$.

14. $2gx+2fy+c=0$; $(fh-bg)y=(gh-af)x$.

15. $3x^2-8xy-3y^2=0$; $x^2+y^2-2x-4y=0$.

16. $\lambda=0,\ -585/14$; $3x-y=1$, $3x+y=14$.

17. $\{\tfrac{1}{2}(1-m^2)/m^2,\ \tfrac{1}{2}(1+m^2)/m\}$.

18. The rotation of the given lines is equivalent to a rotation of the axes through $-60°$;

$$x^2(a-2\sqrt{3}h+3b)+2xy(\sqrt{3}a-2h-\sqrt{3}b)+y^2(3a+2\sqrt{3}h+b)=0;$$
$$y^2-\sqrt{3}xy=0.$$

19. $3\tfrac{1}{2}$ sq. units. **20.** $\cos^{-1}\lambda$. **21.** $c=20$.

Exercises 12 (*b*), p. 294

1. $x^2+y^2-2x-1=0$, $x^2+y^2+6x+7=0$. **2.** $x^2+y^2-x-y=0$.

3. $x^2+y^2-3x-7y+14=0$, $x^2+y^2+2x-4y-3=0$.

4. $5x^2+5y^2-8x-8\sqrt{3}y+12=0$. **5.** $k=5$; $(3, 1)$; $3x-4y=5$.

6. $(1-n^2)(x^2+y^2)+4n^2rx=r^2$. **7.** $(1, 1)$; $x^2+y^2-2x-2y-5=0$.

8. $x^2+y^2-2y-1=0$; $(4, -6)$; $x^2+y^2-8x+12y+16=0$.

10. c is the square of the length of the tangent drawn from the origin to the circle; $x^2+y^2=3$.

11. $(-2, 3)$. **12.** $2fy+c=g^2$; $\{-g,\ (g^2-c)/f\}$.

13. $x^2+y^2-15x+4=0$, $3x^2+3y^2+22y=12$. **14.** $x^2+y^2-2x-3y=0$.

15. $(1, 1)$, $(3, 3)$; $x^2+y^2-2x-6y+6=0$, $x^2+y^2-6x-2y+6=0$.

17. If $x^2+y^2+2gx+c=0$, $x^2+y^2+2g'x+c=0$ are the circles, the limiting points are $\{0,\ \pm\sqrt{(-c)}\}$.

19. If $P\equiv(0,\ y_1)$, the circle is $g(x^2+y^2)+f(f+y_1)x+fgy=0$ and the common points are $(0, 0)$ and $(0, -f)$.

21. AB is the diameter of circle $ABCD$.

23. $(x^2+y^2)(1-h)-h(x+y)=0$. **24.** $(x+g)\cos\theta+(y+f)\sin\theta=r$.

25. $2g(g-g_1)+2f(f-f_1)=c-c_1$; $x+2y+2=0$.

27. $(a\cos^2\alpha-2h\cos\alpha\sin\alpha+b\sin^2\alpha)(x^2+y^2)-2x(b\sin\alpha-h\cos\alpha)$ $-2y(a\cos\alpha-h\sin\alpha)+a+b=1$.

28. $x^2+y^2-2p(x\cos\alpha+y\sin\alpha)+\frac{3}{4}p^2=0$; $x^2+y^2-\frac{2}{3}p(x\cos\alpha+y\sin\alpha)=\frac{7}{12}p^2$. The latter equation may be written in the form

$$(x^2+y^2-\tfrac{1}{4}p^2)-\tfrac{2}{3}p(x\cos\alpha+y\sin\alpha+\tfrac{1}{2}p)=0.$$

29. (i) If the fixed point is (a, b), the fixed circle $x^2+y^2=r^2$ and the variable circle $x^2+y^2+2gx+2fy=r^2$, the locus is $2ax+2by=a^2+b^2-r^2$.

30. $\lambda=-5$, $\mu=5$; $(1, \frac{4}{3})$, $\frac{5}{3}$; $(1, -\frac{3}{4})$.

Exercises 13, p. 307

2. $y^2=a(x-3a)$. **3.** $y^2=2a(x-a)$.

4. $x+y=3a$; $Q\equiv(-6a, 9a)$; $\tan^{-1}(-8)$.

6. $m=\pm\{\frac{1}{2}(\sqrt{5}-1)\}^{\frac{1}{2}}$. **7.** $2y^2=9ax$.

13. The equation of the circumcircle of the triangle formed by the y-axis and the tangents at $[t_1]$ and $[t_2]$ is

$$x^2+y^2-ax(1+t_1t_2)-ay(t_1+t_2)+a^2t_1t_2=0.$$

16. $ln=am^2$; $y^2(4b-a)^3=4a^2(x-2b)^3$.

18. Point is $(4a, 0)$; locus is $y^2=2a(x-4a)$.

21. $y^2=2a(x+2a)$. **22.** $2y^2+ax=0$.

23. Parabola is $3y^2=a(x+a)$; fixed point is $(-2a, 0)$.

25. Line is $x+a=0$.

28. If m is the gradient of the given line, the locus is $2x-my=4a(1+2/m^2)$. This is the normal at $(4a/m^2, -4a/m)$.

30. $x_Q=x_P+PQ\cos\psi$, $y_Q=y_P+PQ\sin\psi$, where $\psi=\tan^{-1}(1/t)$, the angle between Ox and the tangent at P; locus is $y^2=4a(x+a)$.

32. If $P\equiv(\alpha, \beta)$, $\alpha=2a-X+Y^2/a$, $\beta=-XY/a$.

33. $x^2+y^2+ax(t_1{}^2+t_2{}^2)+a^2t_1{}^2t_2{}^2=0$, $x^2+y^2-ay(t_1+t_2)+a^2t_1t_2=0$.

36. $y^2=16ax/(8+n^2)$.

37. Find only the ordinate of the orthocentre: $a(t_1+t_2+t_3+t_1t_2t_3)$ where $[t_1]$, $[t_2]$, $[t_3]$ are the points at which the tangents and normals are drawn.

38. $\{\frac{1}{2}a(1+3t^2), \frac{1}{2}at(3-t^2)\}$. **40.** $(-2bt, bt^2)$.

43. The required circle will pass through the origin (see § 13.10). If its equation is $x^2+y^2+2gx+2fy=0$, it cuts the parabola at $[t]$ where $at^3+t(4a+2g)+4f=0$. This equation is identical with (i) of § 13.9 since its roots are the parameters of three conormal points. Hence g and f are determined.

44. $\{2(h-a), 4k\}$. **45.** $2a(x-h)=k(y-k)$; $y^2=2a(x-a)$; $y^2=a(x-3a)$.

46. $7y\pm2(x+6a)=0$. **47.** $y^2=a(x+a)$; P is at A. **48.** $(4t^2+1)=a^2(t^2-1)^2$.

Exercises 14 (*a*), p. 320

1. $13x \cos\phi + 5y \sin\phi = 65$; $5x \sin\phi - 13y \cos\phi + 144 \sin\phi \cos\phi = 0$; $OT.ON = -144$.

3. $bx \cos\frac{1}{2}\theta + ay \sin\frac{1}{2}\theta = ab \cos\frac{1}{2}\theta$; $ay \cos\frac{1}{2}\theta - bx \sin\frac{1}{2}\theta = ab \sin\frac{1}{2}\theta$; $KK' = 2b^2/a$.

5. $x^2/a^2 + y^2/b^2 = 1$. **8.** $x^2/36 + y^2/20 = 1$.

15. From (14.2), P is the point (a^2l, b^2m); locus is $x^2/a^4 + y^2/b^4 = 1/a^2 + 1/b^2$

23. $x^2/a^2 + y^2/b^2 = 4$. **26.** $G \equiv \{(a - b^2/a) \cos\theta, 0\}$; $T \equiv (0, b \operatorname{cosec}\theta)$.

28. (i) $x^2 + y^2 = a^2 + b^2$.

33. $b^2x^2(1 - h^2/a^2) + a^2y^2(1 - k^2/b^2) = 2hkxy$; (i) $x^2/a^4 + y^2/b^4 = 1/a^2 + 1/b^2$; (ii) $x^2/a^2 + y^2/b^2 = 2$.

35. Point of contact is
$\{(a^2 \cos\alpha)/p, (b^2 \sin\alpha)/p\}$, where $p = \sqrt{(a^2 \cos^2\alpha + b^2 \sin^2\alpha)}$;
normal is $x \sin\alpha - y \cos\alpha = \{(a^2 - b^2) \sin\alpha \cos\alpha\}/p$.

36. $\{\sqrt{2}ab^2/(a^2 + b^2), \sqrt{2}ab^2/(a^2 + b^2)\}$.

39. $\{a^3 \sec\theta/(a^2 - b^2), -b^3 \operatorname{cosec}\theta/(a^2 - b^2)\}$.

41. $b^4x^2 + a^4y^2 = a^4b^2$; $(a^2/b\sqrt{2}, -b/\sqrt{2})$, $(-a^2/b\sqrt{2}, b/\sqrt{2})$.

44. $\{a^2b^2h/(b^2h^2 + a^2k^2), a^2b^2k/(b^2h^2 + a^2k^2)\}$; locus is $x^2/a^2 + y^2/b^2 = 1/c^2$.

45. $x^2/a^2 + y^2/b^2 = k(x + y)$ where $a^2/OP + b^2/OQ = 1/k$.

46. Let P be (x_1, y_1); then $PM = y_1 \cot\angle APQ$. PQ is the normal at P. Find $\tan\angle APQ$ using the gradients of AP and PQ.

48. $\cos\theta = (1 - t^2)/(1 + t^2)$, $\sin\theta = 2t/(1 + t^2)$. **50.** $x^2 - xy + y^2 = (lx + my)^2$.

Exercises 14 (*b*), p. 334

1. Foci $(6, 2)$, $(-4, 2)$; points $(0, \frac{11}{4})$, $(0, \frac{5}{4})$.

4. Produce PO to P' so that $OP' = OQ$; produce QO to Q' so that $OQ' = OP$. Then $OP.OQ = OP.OP' = OQ.OQ' = a^2 + b^2 = OS.OS'$ where S, S' are the foci.

5. $bx \cosh u - ay \sinh u = ab$; area $= ab$; locus is $x^2/a^2 - y^2/b^2 = 1$.

7. Points of contact $(6, 1)$, $(-9, 2)$; pole, $(36, -1)$.

17. Asymptotes are $x = a^2h/(a^2 + b^2)$, $y = b^2k/(a^2 + b^2)$.

18. $(y^2 - x^2)^2 = a^2(2x^2 - y^2)$.

23. $\{2c \tan\theta \tan(\theta + \alpha)\}/\{\tan\theta + \tan(\theta + \alpha)\}$, $2c/\{\tan\theta + \tan(\theta + \alpha)\}$.

Exercises 15, p. 344

1. The harmonic mean is l. **14.** $(\frac{1}{2}l, 0)$, $(\frac{1}{2}l \sec^2\alpha, 2\alpha)$.

Exercises 16 (*a*), p. 353

1. 7 ; $(2, -2, -9)$; $(x+2)/(-2)=(y-4)/3=(z-3)/6$.

2. $6x-2y-2z+15=0$; $3x^2+3y^2+3z^2+18x-20y-26z+102=0$.

3. $(-2, 2, -3)$. **4.** $(x-4)/(-5)=(y-2)/8=(z-3)/3$.

5. $(3, -6, 0)$; $2:1$ externally. **6.** 7.

7. $[6/7, -2/7, 3/7]$; $(4, -1, 7)$; $(-8, 3, 1)$. **8.** $\cos^{-1}(\pm 1/\sqrt{2})$.

9. $3, 3, 2\sqrt{3}$; $\cos^{-1}(1/\sqrt{3})$, $\cos^{-1}(1/\sqrt{3})$, $\cos^{-1}(1/3)$. **10.** 2.

11. $(\frac{14}{3}, \frac{4}{3}, -8)$.

Exercises 16 (*b*), p. 360

1. $(3, -4, 4)$. **2.** 14. **3.** $2x+y-3z=4$.

4. $3y+z=5$, $6x-z=16$, $2x+y=7$. **5.** $x/2+y/3+z/4=1$; $p=12/\sqrt{61}$.

6. $(-7, 11, -2)$. **7.** $2x+3y+6z=38$; $\sin^{-1} 3/7$. **8.** $4y-z=0$.

9. $x-3y-2z+11=0$. **10.** $\left[\dfrac{1, 2, -3}{\sqrt{14}}\right]$. **11.** $x-2=y-3=z-4$.

12. $x+3y-2z+17=0$; $[1:1:2]$. **13.** $2x-2y-z+3=0$; 2.

14. $4x-5y-5z=0$. **15.** $4x-10y+z+13=0$. **16.** $(5, 0, 6)$; $(3514)^{\frac{1}{2}}/7$.

17. $\frac{1}{2}(x-1)=y-2=z+1$.

18. $x/4=y/(-5)=z$; the shortest distance is equal to the perpendicular distance of the origin from the first plane ; $5/\sqrt{14}$.

19. $P\equiv(2, 3, -1)$; $Q\equiv(-6, 1, 3)$; $(x+2)/1=(y-2)/(-2)=(z-1)/1$; $(x-1)/(-3)=(y-1)/1=(z+4)/5$.

20. $3x+4y+1=0$; $3x+2y-6z+5=0$.

Miscellaneous Exercises 16, p. 367

1. $\frac{1}{2}\sqrt{35}$. **2.** $x+5y-6z+19=0$. **3.** $x+y+z=11$; $7/\sqrt{3}$.

4. P is $(1/ap, 1/bp, 1/cp)$ where $p=1/a^2+1/b^2+1/c^2$.

5. $x+2y+2z=11$; $(1, 2, 3)$.

6. $\left[\dfrac{5, -2, -3}{\sqrt{38}}\right]$. **7.** $\left[\dfrac{-a, b, 0}{\sqrt{(a^2+b^2)}}\right]$.

8. N divides AB externally in the ratio $18:5$.

9. $x+y+2z=1$, $5x+5y+2z=5$. **10.** $2\sqrt{17}$; $(-2, 3, -5)$.

11. $(3, 1, 1)$; $5x-17y+19z=17$. **12.** $Q\equiv\{a, \frac{1}{2}(y_1+z_1), \frac{1}{2}(y_1+z_1)\}$; $R\equiv\{a+\frac{1}{3}(2x_1-y_1-z_1), a+\frac{1}{3}(2y_1-z_1-x_1), a+\frac{1}{3}(2z_1-x_1-y_1)\}$; $6(a-x)^2+3(z-y)^2=2\{3a-(x+y+z)\}^2$.

13. $(d-a)x+(e-b)y+(f-c)z=\frac{1}{2}(d^2+e^2+f^2-a^2-b^2-c^2)$.

14. $(ax+by+cz+d)/\sqrt{(a^2+b^2+c^2)}=\pm(a'x+b'y+c'z+d')/\sqrt{(a'^2+b'^2+c'^2)}$.

15. $Q\equiv(-1, 1, 5)$, $R\equiv(3, 5, 3)$. **17.** $69x-35y-8z=0$; $(18, 30, 24)$.

20. The two positions of A are $(0, 2, 1)$, $(0, -\frac{2}{5}, -\frac{1}{5})$.

21. The line through $(1, \lambda, 0)$ is $(x-1)/\lambda=(y-\lambda)/(1+\lambda+\lambda^2)=-z/(1+\lambda)$.

22. $x/5=y/(-10)=z/9$; $(10/7, -20/7, 18/7)$.

23. The straight line $x/7=y/(-13)=z/5$. **24.** $7x+y-5z+15=0$.

25. $(4, 3, 2)$; $5x+2y-3z=20$. **27.** $\left[\dfrac{2, 4, 5}{3\sqrt{5}}\right]$.

28. $(4, 2, -1)$; $x-y-z=3$; $4x+5y-z=27$.

29. $7x-8y+3z=0$; $20y+23z=0$, $13x=122$.

30. $(1, 2, 3)$; L' is $x=1$, $y+z=5$; $30°$; If L' is the line of greatest slope in π, the plane through L' perpendicular to π contains the vertical. $[1/\sqrt{2}, -1/\sqrt{2}, 0]$ or $[-1/\sqrt{2}, 0, 1/\sqrt{2}]$.

31. $\left[\dfrac{1, 1, 6}{\sqrt{38}}\right]$.

32. $(la'+mb'+nc')(ax+by+cz+d)=(la+mb+nc)(a'x+b'y+c'z+d')$.

Exercises 16 (*c*), p. 372

1. $2\sqrt{29}$; $(x-5)/2=(y-4)/3=(z-4)/4$. **2.** $\sqrt{3}$; $(2, 2, 2)$, $(1, 1, 1)$.

4. $\sqrt{59}$; $(x-8)/5=(y-1)/(-3)=(z-3)/5$.

5. 13 ; $(x-1)/3=(y-1)/4=(z-2)/12$. **6.** 13 ; $[3:4:12]$.

7. 3 ; $(x-2)/2=(y-1)/(-1)=(z+1)/2$.

9. $4/\sqrt{3}$; $(2, \frac{11}{3}, \frac{8}{3})$, $(\frac{2}{3}, \frac{7}{3}, 4)$. **10.** 22 ; $(x-1)/6=(y+1)/6=(z-4)/7$.

11. $P\equiv(0, 0, 0)$, $Q\equiv(1, 1, -1)$.

13. $(x-r\cos\alpha)/\sin\alpha=(y-r\sin\alpha)/(-\cos\alpha)=z/\cot\gamma$; $\{2r\sin\frac{1}{2}(\alpha-\beta)\}/\sqrt{\{1+\tan^2\gamma\cos^2\frac{1}{2}(\alpha-\beta)\}}$.

14. $(-\frac{46}{5}, 0, -1)$. **15.** $x/a+y/b+z/c=1$.

Exercises 16 (*d*), p. 376

6. Q lies in the plane $z=c(n-m)/(n+m)$ which is normal to AA'. Its locus in this plane is $(m+n)^2(x^2-y^2\cot^2\alpha)=4c^2mn\cos^2\alpha$ when the equations of AP, $A'P'$ are chosen as in § 16.21.

10. $mxz=cy$.

Exercises 17, p. 383

1. Centre $(29/9, 2/9, -35/9)$, radius 12.

2. $(3, 7, 4)$; $3y+4z=27$, $3y+4z+3=0$.

3. $(1, 0, 0)$, $(0, 1, 0)$, $(0, 0, 1)$; centre $(\frac{1}{3}, \frac{1}{3}, \frac{1}{3})$, radius $\sqrt{6}/3$.

4. The equation of the given sphere is of the form $S+\lambda P=0$, where $S=0$ is the equation of the sphere with centre $(1, 2, -1)$, radius 2, and $P=0$ is the equation of the plane. Distance $\frac{1}{2}\sqrt{3}$; plane is $x+3y-3z=0$.

5. Centre $(-4, 8, 9)$, radius 4.

6. $(u^2+v^2+w^2-d)(l^2+m^2+n^2)=(p-lu-mv-nw)^2$;
$2x+2y+z=50$, $3x+4y=86$.

7. $7x-4y-4z+16=0$; $(-4/3, 10/3, -5/3)$.

8. $x^2+y^2+z^2+2x-4y-6z+5=0$; $x^2+y^2+z^2-2x-6y-8z+10=0$.

9. $x^2+y^2+z^2-4a(x+y+z)+8a^2=0$; $x^2+y^2+z^2+6a(x+y+z)-22a^2=0$.

10. $r^2(l^2+m^2+n^2)=(la+mb+nc+p)^2$; $2x+2y\pm z=3$.

11. Locus is $x^2+y^2-8x+6=0$, $z=0$; centre $(4, 0, 0)$, radius $\sqrt{10}$.

12. $x+2y+3z+15=0$, point of contact is $(1, -5, -2)$; $3x-y+2z+3=0$, point of contact is $(-1, -2, -1)$.

14. $x^2+y^2+z^2-4x-6y+2z+5=0$; $x+2y+2z=15$.

15. $9(x^2+y^2+z^2)-18a(x+2y+4z)+169a^2=0$; radius$=(26\sqrt{105})/189$, centre $(169a/189, 338a/189, 676a/189)$.

16. The parallel lines are $x=3$, $y+z=4$; $x=3$, $y+z=-4$.

17. Distance$=\sqrt{p}-\sqrt{q}$. **18.** $\lambda^2\mu^2=4c^4$.

19. $g(l^2+m^2+n^2)=(al+bm+cn)^2$; point of contact $(-lg/k, -mg/k, -ng/k)$ where $k=al+bm+cn$. **20.** $(-al/n, -am/n, 0)$.

22. (ii) $21(x^2+y^2+z^2)+58x-10y+20z+26=0$.

23. Centre $(2, 2, 3)$; sphere $5(x^2+y^2+z^2)-14x-14y-18z=0$.

25. $(3, 3, -2)$; $2x-2y-z+11=0$, $2x-y+2z=9$.

26. $x^2+y^2+z^2+2x-18y-4z-58=0$.

27. $x^2+y^2+z^2-2(ax+by+cz)=0$, centre (a, b, c), radius $\sqrt{(a^2+b^2+c^2)}$
$x^2+y^2+z^2-2x-2y+2(4\pm 3\sqrt{2})z=0$.

30. $x-4y-2z+1=0$; $Q\equiv(11/7, -2/7, 13/7)$.

31. $x^2-3y^2-23z^2-3xy-21yz-7zx-4x-9y-19z+31=0$.

35. $2(x^2+y^2+z^2)-2\sqrt{2}a(x+y+z)+a^2=0$;
$x^2+y^2+z^2-2a(x+y+z)+2a^2=0$;
$\{\frac{1}{2}a(1+1/\sqrt{2}), \frac{1}{2}a(1+1/\sqrt{2}), \frac{1}{2}a(1+1/\sqrt{2})\}$; $\frac{1}{4}(12\sqrt{2}-2)^{\frac{1}{2}}a$.

36. $(1, 2, -3)$, 4; $(2, 3, -2)$, $\sqrt{13}$. **37.** $q^2=2p$; $x\pm 2\sqrt{2}y+4z=20$.

Exercises 18 (*a*), p. 394

2. $x+4y+8z=25$; $3x+8z=25$. **3.** $2(x^2+y^2)=3$, $z=0$.

8. $3x^2+6xy+6y^2=1$, $z=0$. **10.** $\{\alpha(1-c^2/a^2), \beta(1-c^2/b^2), 0\}$.

13. $x+y+7z=10$, $x+y-7z=10$.

14. $x-y-3z=9$, $25x+41y+57z=225$.

15. $36x-9y-64z+49=0$; $12x-3y+8z=13$.

17. $a(\alpha x+\beta-y)=\pm z\sqrt{\{2\beta^2-2a^2(\alpha^2+1)\}}$. **18.** $2x/a-2y/b=\pm(z/c-3)$.

19. $x+y=3$, $x+2y+3z=6$; $(2, 1, 0)$, $(1, 1, 1)$. **20.** $apl+bqm+crn=0$.

21. $x/a+y/a+z/c=1$. **22.** $(18/11, 18/11, 18/11)$.

24. $2x+2y+3z=2$; $(9x-4)/2=(9y-2)/2=(9z-2)/3$; $x/2=y/2=z/3$.

25. $x+y-z+3=0$, $x+y+z=3$. **26.** 6; $\cos^{-1}\{(5\sqrt{14})/21\}$.

Exercises 18 (*b*), p. 400

4. $(x^2+2y^2+3z^2)^2=x^2+4y^2+9z^2$.

5. Conic defined by $4(x-1)^2+3y^2+z^2=4$, $4x+3y+z=0$;
centre $(\frac{1}{2}, -\frac{1}{2}, -\frac{1}{2})$.

6. $(\alpha/\lambda, \beta/\lambda, \gamma/\lambda)$ where $\lambda=\alpha^2/a^2+\beta^2/b^2+\gamma^2/c^2$. **7.** $\sin^{-1}(167/3\sqrt{3385})$.

10. The line defined by the equations $afx+bgy+chz=1$, $alx+bmy+cnz=0$.

12. $(x-4)/10=(y+1)/(-7)=z/2$.

13. (a^2k, b^2k, c^2k), $(-a^2k, -b^2k, -c^2k)$ where $k=1/\sqrt{(a^2+b^2+c^2)}$.

14. $x/a-z/c=\mu(1+y/b)$; $\mu(x/a+z/c)=1-y/b$.

15. $(x-a\cos\theta)/a\sin\theta=(y-a\sin\theta)/(-a\cos\theta)=\pm z/c$; $a=\pm c$.

Exercises 19 (*a*), p. 407

1. $2x/y+y^2/x^2$, $-(x^2/y^2+2y/x)$. **2.** $2y/(x+y)^2$, $-2x/(x+y)^2$.

3. $(1+2x)e^{2x+3y}$, $3xe^{2x+3y}$. **4.** $xy/\sqrt{(x^2-y^2)}$, $(x^2-2y^2)/\sqrt{(x^2-y^2)}$.

5. $y/(x^2+y^2)$, $-x/(x^2+y^2)$.

6. $e^{x-y}\{\log(y-x)-1/(y-x)\}$, $e^{x-y}\{1/(y-x)-\log(y-x)\}$.

7. $2\sin y-y^2\cos x$, $2\cos x-x^2\sin y$.

8. $y(2\log x-3)/x^3$, 0. **9.** $y(2x^2-y^2)/x^2(x^2-y^2)^{3/2}$, $y/(x^2-y^2)^{3/2}$.

Exercises 19 (*b*), p. 410

1. $\delta a \simeq \cos C\,\delta b+\cos B\,\delta c$; the triangle should be right-angled at C.

3. $\triangle ABC=\frac{1}{2}a^2\sin B\sin C\operatorname{cosec}(B+C)$.

4. $D=a^2/\sqrt{(a^2-b^2)}$; $\delta D \simeq \{a(a^2-2b^2)\delta a+a^2b\,\delta b\}/(a^2-b^2)^{3/2}$. **6.** (i) $2xy$.

7. (*a*) 0; (*b*) $(4-\sqrt{3})\pi a/1080\sqrt{2}$. **8.** (ii) $\delta b/b+\delta c/c+\cot A\,\delta A$; $0{\cdot}015$.

Exercises 19 (*c*), p. 427

4. (ii) $\sin(\theta+\alpha)$, where α is constant. (See § 24.11.)

5. (i) $w=xy(x+y)e^{x-y}$.

9. Last part: put $f'(t)=V$ and solve the resulting equation as in § 23.4;
$f(t)=A+B/t^2$.

11. $\phi(t)=A\cos(ct+\alpha)$, where A and α are constants. (See § 24.11.)

15. (i) Last part: put $f'(r)=V$ and solve the resulting equation as in § 23.4; $u=\log ar^n$ where a and n are constants; (ii) $\alpha=\pm 1$.

19. $\partial u/\partial x=g_v/(f_ug_v-f_vg_u)$. **25.** $c=a^2b/7$, $d=8b/7$. **27.** $\sin\theta\cos\theta/r^2$.

28. (i) $V=\log ar^n$ where a and n are constants ;
(ii) $V=r^n(A\cos n\theta+B\sin n\theta)$, where A and B are constants.

38. (i) $27y=x^3$; (ii) $2y=\pm x$; (iii) $y^2=4x+4$; (iv) $x^4+y^4=a^4$.

Exercises 19 (*d*), p. 437

1. (*a*) Volume$=9a^3/2$. (*b*) $PG_1 : PG_2 : PG_3=a^2 : b^2 : c^2$.

2. $x_1^3x+y_1^3y+z_1^3z=3a^4$.

3. $x/\alpha+y/\beta+z/\gamma=3$; $\alpha(x-\alpha)=\beta(y-\beta)=\gamma(z-\gamma)$.

4. $x_1^2x+y_1^2y+z_1^2z=3a^3$.

5. $(y_1+z_1)x+(z_1+x_1)y+(x_1+y_1)z=0$; $\dfrac{x-x_1}{y_1+z_1}=\dfrac{y-y_1}{z_1+x_1}=\dfrac{z-z_1}{x_1+y_1}$;
$Q\equiv(x_1-2y_1-2z_1,\ y_1-2z_1-2x_1,\ z_1-2x_1-2y_1)$; $R\equiv(9x_1,\ 9y_1,\ 9z_1)$.

6. $2(x_1z_1-ky_1)x-2(y_1z_1+kx_1)y+(x_1^2-y_1^2)(z-z_1)=0$;
$\frac{1}{2}(x-x_1)/(x_1z_1-ky_1)=-\frac{1}{2}(y-y_1)/(y_1z_1+kx_1)=(z-z_1)/(x_1^2-y_1^2)$.

7. $x+2y-z=2$; $(x-1)=\frac{1}{2}(y-1)=-(z-1)$.

8. $2\alpha\beta x+(\alpha^2-2\beta\gamma)y-(\beta^2+2\gamma)z=\gamma^2-3$;
$(x-\alpha)/2\alpha\beta=(y-\beta)/(\alpha^2-2\beta\gamma)=-(z-\gamma)/(\beta^2+2\gamma)$.

9. $(1+\lambda)^2x+(1-\lambda)^2y+16z=4a$.

10. $(x-\alpha)/\beta=(y-\beta)/\alpha=-(z-\gamma)/c$.

11. $x+3y=0$ touches at $(18,\ -6,\ 0)$, $x-3y=0$ touches at $(-18,\ -6,\ 0)$;
$y=2/3$, $z=\pm 1/\sqrt{2}$.

12. $2\alpha\beta x+\alpha^2y-\alpha^2z=2\alpha^2\beta$.

13. $(x-4a)=(y-\frac{1}{2}a)/8=(z-\frac{1}{2}a)/8$;
$\{(31+8\sqrt{15})a/16,\ (-16+4\sqrt{15})a,\ (-16+4\sqrt{15})a\}$,
$\{(31-8\sqrt{15})a/16,\ (-16-4\sqrt{15})a,\ (-16-4\sqrt{15})a\}$.

15. $a(b^2-c^2)x+b(c^2+a^2)y-c(a^2-b^2+2c^2-1)z+c^2=0$.

17. $(2x_1z_1-ay_1)x-(2y_1z_1+ax_1)y+(x_1^2-y_1^2)(z-z_1)=0$;
$(x-x_1)/(2x_1z_1-ay_1)=-(y-y_1)/(2y_1z_1+ax_1)=(z-z_1)/(x_1^2-y_1^2)$.

19. Use inequality $a^2+2c^2\geqslant 2^{3/2}ac$.

20. $xx_1+yz_1+zy_1=2$; $(x-x_1)/x_1=(y-y_1)/z_1=(z-z_1)/y_1$; $4x+y-7z=\pm 2$;
$(-10,\ 7,\ -7)$.

21. $x-2z+3a=0$, $x+a=-(z-a)/2$, $y=-a$;
$x-2y+3a=0$, $x+a=-(y-a)/2$, $z=-a$; $2x+y+z+2a=0$.

Exercises 20 (*a*), p. 450

1. Area$=8a^2/15$; volume$=\pi a^3/12$. **2.** $8a^2/15$, $\pi a^3/4$.

3. $y-4a^2x+8a^3=0$. **8.** (i) $a^2(\log 2-\frac{2}{3})$.

11. Area$=3\pi a^2/4$; volume$=\pi^2a^3/4$. **13.** πa^3 $(\log 4-1)$.

14. $\lim\limits_{y\to 1-}(1-y)\log(1-y)=\lim\limits_{z\to 0+}z\log z=0$. (See § 11.12, Example 12.)

15. $\pi(\frac{8}{3}-\sqrt{2})$.

16. Find condition that the tangent at (x_1, y_1) passes through the origin; $y=\pm 2x/3\sqrt{3}$.

17. $\pi(4\pi+3\sqrt{3})/72$.

19. $(-1)^k ae^{-k\pi/a}(1+e^{-\pi/a})/(1+a^2)$; $a/(1+a^2)$.

Exercises 20 (*b*), p. 458

3. Area $=\frac{1}{3}\pi a^2\{67\sqrt{5}-8-\frac{3}{2}\log(2+\sqrt{5})\}$; volume $=96\pi a^3/5$.

4. (i) $8\pi a^2(2\sqrt{2}-1)/3$; (ii) $\frac{1}{2}\pi a^2\{\sqrt{2}+5\log(\sqrt{2}+1)\}$.

5. $2\pi\{\log(\sqrt{2}+1)+\sqrt{2}\}$.

6. $(0, \frac{2}{3}a)$, $(0, -\frac{2}{3}a)$; two tangents at $(-2a, 0)$; $3\pi a^2$.

8. Length $=\log\{(1+x_1)/(1-x_1)\}-x_1$;
area $=(1-x_1)\log(1-x_1)-(1+x_1)\log(1+x_1)+2x_1$;
see Exercises 20(*a*), No. 14.

10. $\frac{1}{3}\sqrt{2}$.

Exercises 20 (*c*), p. 467

1. $\{1/(e-1), \frac{1}{4}(e+1)\}$. **2.** $(1, \frac{1}{8}\pi)$. **3.** $(3/2, 3/10)$. **4.** $(9/5, 9/5)$.

5. $\bar{x}=\bar{y}=9/(15-16\log 2)$.

6. Area $=2(ab^3)^{\frac{1}{2}}/3$; $\bar{x}=3b/5$, $\bar{y}=3(ab)^{\frac{1}{2}}/8$; volume $=4\pi(ab^5)^{\frac{1}{2}}/5$.

7. $256a^2/15$; $\bar{x}=19a/7$. **8.** $\bar{x}=8(\log 2)/3$. **9.** $\bar{x}=3a/8$. **10.** $\bar{x}=8/3$.

11. Volume $=6\pi\sqrt{3}$; surface area $=24\pi$.

12. $V=\frac{1}{2}\pi a^3\sin\alpha$; $S=2\pi\sqrt{3}a^2\sin\alpha$. V and S are greatest when $\alpha=\frac{1}{2}\pi$.

13. Centroid $(r, 2r/\pi)$; area $=2\pi r^2(2+m\pi)/\sqrt{(1+m^2)}$.

17. $\pi(4\pi+3\sqrt{3})/72$. **18.** On the axis 34/7 in. from the upper face.

Exercises 20 (*d*), p. 475

1. $k_x=\frac{1}{3}\sqrt{2}$, $k_y=(\pi-2)^{\frac{1}{2}}$. **2.** $k_x=\sqrt{(1/7)}$, $k_y=\sqrt{(3/5)}$.

3. $k_x=\frac{1}{3}\sqrt{(e^2+e+1)}$, $k_y=\sqrt{\{(e-2)/(e-1)\}}$.

4. $k_x=(1/\log 4)^{\frac{1}{2}}$, $k_y=(6/\log 2)^{\frac{1}{2}}$. **5.** $k_x=4\sqrt{5}$, $k_y=4\sqrt{(3/7)}$.

6. $k_x=\frac{1}{2}\sqrt{(e^2+1)}$, $k_y=\frac{1}{4}\sqrt{(2e^2+10)}$. **7.** $k_x=\sqrt{(14/3)}$, $k_y=\sqrt{(13/3)}$.

8. $k_x=\sqrt{(3/8)}$, $k_y=\sqrt{\{(16\pi^2-15)/48\}}$.

Exercises 21 (*a*), p. 484

1. $8a$. **2.** $3\pi a/2$. **3.** $(r_2-r_1)\sec\alpha$.

4. (i) $\phi=\frac{1}{2}(\pi-\theta)$; $2p^2=ar$; (ii) $\phi=\frac{1}{2}\pi-2\theta$, $pr=a^2$;
(iii) $\phi=2\theta$; $r^3=a^2p$; (iv) $\phi=\theta/n$; $r^{n+1}=ap^n$; (v) $\phi=n\theta$; $r^{n+1}=a^np$;
(vi) $\phi=\frac{1}{2}\pi+3\theta$; $p=r^4$.

5. $-\frac{1}{4}\pi\leqslant\theta\leqslant\frac{1}{4}\pi$; $\frac{3}{4}\pi\leqslant\theta\leqslant\frac{5}{4}\pi$. Greatest width $=\frac{1}{2}\sqrt{2}$.

6. P, Q, R are given by $\theta=\alpha$, $\theta=\alpha+\frac{2}{3}\pi$, $\theta=\alpha+\frac{4}{3}\pi$ respectively.

7. (i) $4a(1-\sin\frac{1}{2}\theta)$; (ii) $\frac{1}{2}(\theta+\pi)$. **8.** $\frac{1}{4}a\{\log(2+\sqrt{5})+2\sqrt{5}\}$.

9. $r=ae^{\frac{1}{4}\theta^2}$. **10.** Draw OM, OM' perpendicular to PT, QT respectively.

11. To find OT^2 draw OM, OM' perpendicular to PT, $P'T$ respectively.

12. $x=f(\theta)$, i.e. $r\cos\theta=f(\theta)$ and so $r=f(\theta)\sec\theta$.

(i) $x=a\cot\theta$ is a straight line parallel to Ox ; $s=a(\cot\theta_1-\cot\theta_2)$;

(ii) $x=a\cos\theta$ is a circle with centre O and radius a ; $s=2\pi a$.

Exercises 21 (b), p. 494

2. $s=a(\theta-\tanh\frac{1}{2}\theta)$; $A=\frac{1}{2}a^2(\theta-2\tanh\frac{1}{2}\theta)$. **3.** $3a^2(\pi+1)/2$.

6. $(22\pi+15\sqrt{3})/6$. **10.** Area$=3\pi/16$; volume$=2\pi/21$.

11. $\pi/12$. **13.** $r=\frac{1}{2}a(2-\cos\theta)$.

16. Surface area$=32\pi a^2/5$, volume$=8\pi a^3/3$. **17.** $20\pi a^3/3$.

18. $\pi(8a^3-6a^2b+4ab^2-b^3)/6$.

19. Slope$=\frac{1}{2}\tan t\,(4\cos^2 t-1)$; area$=\pi/8$. **20.** Area$=4ab/3$.

21. $x=t^2-1$, $y=t(t^2-1)$; $8/15$.

22. $256a^2/15$. **24.** $3a^2/2$. **26.** Area$=a^2/60$; volume$=\pi a^3/420$.

28. $12\pi a^2$. **29.** $x^{2/3}+y^{2/3}=(4a)^{2/3}$.

32. $ds/d\theta=2(a+b)\sin(a\theta/2b)$; $8(a+b)b/a$.

35. $\{128a\sqrt{2}/105\pi,\ 16(8\sqrt{2}-9)/105\pi\}$. **36.** $\bar{x}=\bar{y}=128a/105\pi$.

37. $\{(16-5\pi)a/2(3\pi-8),\ 2a/3(3\pi-8)\}$.

38. Express $\cos^3 3\theta$ in terms of $\cos 9\theta$; $81a\sqrt{3}/80\pi$ from O.

39. $\bar{x}=\pi a/4\sqrt{2}$.

Exercises 22, p. 507

1. $\rho=3a\sin t\cos t$. **3.** (ii) $16ab/15$.

4. $x=a(t-\sin t)$, $y=-a(1+\cos t)$.

5. (c, c) and $(-c, -c)$; centres of curvature $(2c, 2c)$, $(-2c, -2c)$.

6. $\rho=(8/3)a\sin\frac{1}{2}\theta$; $(-\frac{1}{3}a, 0)$. **7.** $x^2+y^2=a^2$. **8.** $\rho=\tan\theta\sec\theta$.

9. $-\sqrt{2}$; $(1, -1)$. **10.** $3x^2+3y^2+6ax-19ay+33a^2=0$.

11. $\{(a-b^2/a)\cos^3\theta,\ (b^2/a-a)\sin^3\theta\}$.

13. $x^2+y^2-10ax+4ay-3a^2=0$; $(9a, -6a)$. **17.** $2{\cdot}25$.

18. $\rho=-\sec x$; centre of curvature is $(x-\tan x,\ \log\cos x-1)$.

22. $2/r^2=1/p^2+1/a^2$, $\rho=\frac{1}{2}r^4/p^3$. **23.** $\rho=pr(3+2r)/(2r^3-p^2)$.

24. $\rho=a^2/3r$, $PG=a^2r/(a^2+2r^2)$. **26.** $a^2/p^2=e^2-1+2a/r$.

27. $p=(r^4+8c^4)/6c^2r$. **28.** $\rho=r\sec n\theta/(1+n)$. **29.** $1:2$.

31. $PI=r\operatorname{cosec}\alpha$; locus of I is $r=a(\cot\alpha)e^{(\theta-\frac{1}{2}\pi)\cot\alpha}$; $\rho=OI\operatorname{cosec}\alpha$.

32. $p^2=ar$. **34.** $a+b$. **35.** $\rho=a\theta$; $(a\cos\theta,\ a\sin\theta)$.

Exercises 23 (*a*), p. 514

1. $r(d^2\phi/dr^2)+2(d\phi/dr)=0.$
2. $(d^2y/dx^2)-\tan x(dy/dx)-y\sec^2 x=0.$
3. $3y^2=2x^3+C.$
4. $\log\sin y=C+e^x.$
5. $Cy^4=(x-2)/(x+2).$
6. $2\tan^{-1}x-\tan^{-1}(\frac{1}{2}y)=C.$
7. $1+y^2=Cx^2e^{x^2}.$
8. $2e^x(\sin x-\cos x)+e^{-2y}(2y+1)=C.$
9. $4x^2=y(1+6x-3x^2).$

Exercises 23 (*b*), p. 517

1. $Cx^4=y^2-6xy+8x^2.$
2. $y=C(x^2-y^2).$
3. $y^3=x^3\log Cx^3.$
4. $x^4y^4=C(x+y)^3.$
5. $4x/(y-2x)=\log\{C(y-2x)\}.$
6. $x^2-xy-y^2+x+3y=C.$
7. $y-x=\log C(x+y)^2.$
8. $(x+y)^3=C(x^2-xy+y^2).$
9. $2x+y=C(x+y-3)^2.$
10. $3x-4y-6=C(x-y-1)^2.$
11. $y=2x\tan^{-1}(Ce^x).$
12. $x(x+y)^2=9(x-y)^2.$

Exercises 23 (*c*), p. 519

1. $xy-e^x=C.$
2. $y=x(x^2+C).$
3. $ax^2+2hxy+by^2+2gx+2fy+C=0.$
4. $y(1-\cos 2x)=C.$
5. $x^3+2y^3+2xy+x+2y=C.$

Exercises 23 (*d*), p. 522

1. $y=(C+e^x)\sin^2 x.$
2. $y=C\sec^4 x-\cos^2 x.$
3. $xy=\frac{1}{2}(\log x)^2+C.$
4. $y\sin^2 x=x\sin x+\cos x+C.$
5. $y(1+x)=x(C+x^3).$
6. $y=(C-e^{\cos x})\sqrt{(1-x^2)}.$
7. $y=x^2-2+2e^{-\frac{1}{2}x^2}$
8. $y(x-3)^2=(x-2)\{(x-1)^2+C(x-2)\}.$
9. $2xy=(1+x)^2+C(1-x^2)^2$; $2y=(1+x)^2(2-x).$
10. $y=\frac{1}{4}(x-1)^2\{(x-1)\sin 2x+\frac{1}{2}\cos 2x+x^2-2x+C\}.$
11. $2x^3y=e^{-2x}\{x^2-2e^{-x}(x+1)+C\}.$
12. $2xy=(1-x)\cos x+x\sin x+Ce^{-x}.$
13. $2y=x(1+x^2)+(C-\sinh^{-1}x)\sqrt{(1+x^2)}.$
14. $y(1+3x)^2=Ce^{3x}-(2+3x).$
15. $y/x=1+x\sin x+\cos x.$
16. (i) $y^2=2x^2+2x+1$; (ii) $x+y=(x-y)^3.$

Exercises 23 (*e*), p. 524

1. $y\sin x=1/(C+\cos x).$
2. $2\sec^2 x=y^2(C-\tan^4 x).$
3. $y^2=Ce^{2x}-(x+\frac{1}{2}).$
4. $y^2(C-\sin^2 x)=2\sin x$; $y+\{(2\sin x)/(3-\sin^2 x)\}^{1/2}=0.$

5. $2/y^2=2x+1+Ce^{2x}$. **6.** $y(\sin^3 x+C\cos^3 x)+3\cos x=0$.

7. $y^2\{x^3+3x^2+9x+9\log(x-1)+C\}=3(x^2+x+1)$.

8. $y^3(x^2-1)^{3/2}=3e^x(x-1)^2+C$.

Miscellaneous Exercises 23, p. 526

1. (i) $(x-2y)^5(3x+2y)^3=C$ (ii) $y=8x^2-x-\frac{1}{2}$.

2. (i) $(x+y)^2(x-y)^3=C$; (ii) $y=\sin x(C-\cos x)$.

3. (i) $x=C\sqrt{(1-t^2)}+t^2-1$; (ii) $(9x-2y)^7(x+2y)^3=C$.

4. (i) $y\cos x=\frac{1}{2}\cos 2x+C$; (ii) $(x^2+y^2)^2=C(x^2-y^2)$.

5. (i) $y=\cos x(C-2\cos x)$; (ii) $(x-y)^2e^{y/x}=Cx$;
(iii) $y\cos x-x\sin y=C$.

6. (i) $y(x^2+1)^2=x^5-5x+C$; (ii) $(x^2-y^2)^2=Cxy$.

7. (i) $y=1+x+C\sin x$; (ii) $(2x-y)^2=Cx^2(x-y)$.

8. (i) $y=\sin 2x+\cos x\log\{C(\sec x-\tan x)\}$;
(ii) $8\tan^{-1}(y/x)=3\log\{C(x^2+y^2)\}$.

9. (i) $(y-x)^2/(y+x)^2=Cx$; (ii) $y=(1+x^2)\log\{C(1+x^2)\}$;
(iii) $2(x+4y)=\tan(2x+C)$.

10. (i) $\log(xy^{-1/2}-1)+\frac{1}{2}x^{n+1}/(n+1)=C$; (ii) $(y-x+4)^2(x+y+2)=C$.

11. (i) $y-x=Ce^{-(x^2+x)}$; (ii) $x+2y+1=C(2x+y-1)^2$.

12. (i) $\log(10x+5y-1)^7=5x-10y+C$;
(ii) $1/(y\sin x)=\log(\operatorname{cosec} x+\cot x)-\cos x+C$.

13. (i) $y^2(C-2x^5)=5x^2(1-x^2)$; (ii) $y+(x^2+y^2)^{\frac{1}{2}}=Cx^2$.

14. (i) $(x-2y+5)^4=C(x+y-1)$; (ii) $y^2\log(C\operatorname{cosec} x)=2\sin x$.

15. (i) $2\sqrt{3}\tan^{-1}\{(2y+x)/x\sqrt{3}\}=\log\{C(x-y)^3(x^3-y^3)\}$;
(ii) $\sqrt{2}\tan^{-1}\{(y-1)/(x-1)\sqrt{2}\}=\log C+\log\{2(x-1)^2+(y-1)^2\}$;
(iii) $2y\sqrt{(1+x)}=Ce^x-(x^2+2x+2)$.

16. (i) $3\tan^{-1}\{(y+1)/(x-2)\}+\log\{(x-2)^2+(y+1)^2\}=C$;
(ii) $(\cos^2 x)/y^2+2\tan x+\frac{2}{3}\tan^3 x=C$.

17. (i) $x(1-x)y+1=C(1+x)$; (ii) $2x/(x+y)+\log\{(x+y)^3(x-y)\}=C$.

18. (i) $y=Ce^{-x}+2(x-1)$; (ii) $y^2+xy-x^2-5y+5=C$.

19. (i) $y=x^2\sin x+C\cos x$; (ii) $(x-1)^2-2(x-1)(y-1)+8(y-1)^2=C$.

20. (i) $(x+3y)^3(2x-y)^4=C$; (ii) $1/y^2=Ce^{2x}+x+\frac{1}{2}$.

21. (i) $(\sqrt{2}-1)\log(x+y\sqrt{2})-(\sqrt{2}+1)\log(x-y\sqrt{2})-2\log x=C$;
(ii) $y=\frac{1}{4}(C-\cos x)^2\operatorname{cosec} x$.
If y is finite when $x\to 0$, $C=1$ and $y=\frac{1}{2}\sin^2\frac{1}{2}x\tan\frac{1}{2}x$.

22. (i) $(x-1)y=x\{x\log x-x-\frac{1}{2}(\log x)^2+C\}$; (ii) $y+(x^2+y^2)^{\frac{1}{2}}=Cx^2$;
(iii) $x^2y(1+Cx)=1$.

23. (i) $Cx^3+6x^2y+3xy^2+2y^3+6x^3\log\{(y-x)^2/x^3\}=0$;
(ii) $y\sin^2 x=C-\frac{1}{4}\cos 2x$.

24. (i) $(x+y)e^{-y/x}=x\log(C/x)$; (ii) $x^2y=x(1-x^2)+1+C(1+x)$.

25. (i) $2x^3+3x^2y+y^3=C$; (ii) $y=\frac{1}{2}e^{-x}+Ce^x$.

26. (i) $y=Cx^2e^{1/x}-(1+2x+2x^2)$;
(ii) $x^4+2x^3y+3x^2y+x^2y^2+6xy^2+3y^3=C$.

27. $y^3=3x^6-2x^3$.

28. (i) I.F.$=x^2$; $x^3(x^2+3xy-y^2)=C$; (ii) $4(x-1)y=(2x-3)e^x+3e^{-x}$.

29. (i) $e^{2y/x}=Cx^3(x+2y)$; (ii) $2xy+1=Cx^3(xy-1)$.

30. (i) $xy=2+(x+1)\{2\log(x+1)+C\}$;
(ii) $\tan^{-1}\{(y^2+x)/x\}=\log\sqrt{\{C(2x^2+2xy^2+y^4)\}}$.

Exercises 23 (*f*), p. 529

1. C.P. $y=Cx+a/C$; S.S. $y^2=4ax$.
2. C.P. $y=Cx+2\sqrt{(aC)}$; S.S. $xy+a=0$.
3. C.P. $y=Cx+C^3$; S.S. $27y^2+4x^3=0$.
4. C.P. $y=Cx+a\sqrt{(1+C^2)}$; S.S. $x^2+y^2=a^2$.
5. C.P. $y=Cx+C-C^2$; S.S. $4y=(x+1)^2$.
6. C.P. $y=C(x-1)+\log(1+C)$; S.S. $y+x+\log(1-x)=0$.
7. C.P. $2y=2Cx-\log\sec^2 C$; S.S. $2y=2x\tan^{-1}x-\log(1+x^2)$.
8. $Y=PX+c^2P/(P-1)$ where $P=dY/dX$; C.P. $y^2=ax^2+c^2a/(a-1)$ where a is an arbitrary constant ;
S.S. $(y+x+c)(y+x-c)(y-x+c)(y-x-c)=0$.
9. $v=Px+1/P$ where $P=dv/dx$; C.P. $y=Cx^2+x/C$; S.S. $y^2=4x^3$.
10. The given equation may be written in the form $(y-px)^2=a^2(1+p^2)-c^2$.

Exercises 23 (*g*), p. 534

For brevity y' is written for dy/dx.

2. $y^3=C(y^2-x^2)$. **3.** $x^2=C(C-2y)$. **4.** $2y'=y-x$; $y=x+2+Ce^{x/2}$.

6. $xy'-2y=-mx$; $y=Cx^2+mx$. **7.** $x-y=C(x+y)^3$.

8. $dy/dx=y^2/(xy-2a^2)$; find dx/dy and solve for x ; x-axis is the common asymptote.

9. $x^n(1/y^n-1/a^n)=C$. **10.** $xy'=x+y$; $y=x\log Cx$; $y-x=x\log x$.

11. $x^{3/2}+y^{3/2}=2$. **12.** $y'=(y^2-x^2)/2xy$. **13.** $y=y'\{x+1\pm\sqrt{(x+1)}\}$.

14. $2(y-xy')=\sqrt{(x^2+y^2)}$; $x^3=C(4y^2+2x^2-Cx)$.

15. $y=\frac{1}{2}x(p+1/p)$ where $p=y'$; $x^2=C(2y-C)$; $y=-\frac{1}{2}x(p+1/p)$.

16. $y'=y^2/kx^2$; curves are $Cxy+kx=y$; line is $y+kx=0$.

17. $y'^2=a^2/x^2-1$; $\pm(y+C)=\sqrt{(a^2-x^2)}-a\log\{a+\sqrt{(a^2-x^2)}\}+a\log x$.

18. $(ky')^2=(y^2-k^2)$; $y=k\cosh(x/k)$.

19. $yy'^2+2xy'-y=0$; $y^2=4(x+1)$; $4y^2=1-4x$.

20. $r(1+1/\sqrt{2})=\sec\theta+\tan\theta$.

21. $y+\sqrt{(x^2+y^2)}=C$; $y+\sqrt{(x^2+y^2)}=C'x^2$.

22. $y'=\pm 1/\sqrt{(a^2y^2-1)}$; $\pm 2(x+C)=y\sqrt{(a^2y^2-1)}-(1/a)\cosh^{-1} ay$; $2x=y\sqrt{(a^2y^2-1)}-(1/a)\cosh^{-1} ay$; $ay=\cosh ax$.

Exercises 24 (*a*), p. 550

1. $30x=11\sin 3t-3\cos 3t+3e^{-t}$. 2. $y=A\cos 2x+B\sin 2x+\frac{1}{4}x+\frac{1}{5}\sin 3x$.

3. $y=A\sin x+B\cos x+\frac{1}{5}e^{-x}(\sin x+2\cos x)$.

4. $x=Ae^t+Be^{3t}+\frac{1}{9}(3t+4)-e^{2t}$.

5. $y=A+Be^{-x}+\frac{1}{3}x^3-x^2+3x$. 6. $y=e^{-2x}(A+Bx+3x^3)$.

7. $y=e^x(A+Bx-\cos x)$. 8. $y=Ae^{-x}+Be^{-3x}+(\sin x-2\cos x)/10$.

9. $y=Ae^{4x}+Be^x-e^x(3x^2+2x)/18$. 10. $y=Ae^{-x}+Be^{-2x}+xe^{-x}(x-2)$.

11. $y=e^{-2x}(A\sin 3x+B\cos 3x+\frac{1}{8}\cos x)$. 12. $y=Ae^x+Be^{2x}-e^x(\frac{1}{2}x^2+2x)$.

13. $y=Ae^{5x}+Be^{-x}-(2\sin x+3\cos x)/26$. 14. $y=(e^{\frac{1}{2}t}+1)(\cos t-4\sin t)$.

15. $y=Ae^{4x}-e^x(B+3x^2+2x)/18$.

16. $y=e^x(A\cos x+B\sin x)+(2\cos 2x-\sin 2x)/10$.

17. $y=e^{bx/a}\{A\cos(bx\sqrt{3}/a)+B\sin(bx\sqrt{3}/a)\}+x(bx+a)$.

18. $y=e^{-2x}(A+Bx)+(e^{2x}+8x^2e^{-2x})/32$.

19. $y=e^x(A+Bx)+e^{2x}(x^2-2x+3)$.

20. $y=e^{-x}\{A\cos(x\sqrt{2})+B\sin(x\sqrt{2})\}+\frac{1}{2}e^{-x}+(4\sin 2x-\cos 2x)/17$.

21. $y=e^{-x}(A\cos x+B\sin x)+\frac{1}{2}(x-1)+\frac{1}{5}(\sin x-2\cos x)$.

22. $y=Ae^{3x}+Be^{-3x}+(27x\sinh 3x-18x^2-4)/162$.

23. $s=e^{-3t}(A\cos 4t+B\sin 4t)+(24\cos 4t+64\sin 4t)/73$.

24. $y=\frac{1}{4}e^{2x}(A\cos x+B\sin x-x^2\cos x+x\sin x)$.

25. $y=\frac{1}{2}e^x\{(2\pi+1)\sin x-x\cos x\}$.

26. $y=e^{-2x}(A\cos x+B\sin x)+\frac{1}{8}(\cos x+\sin x+4xe^{-2x}\sin x)$.

27. $y=Ae^{\frac{1}{2}x}+Be^{-2x}-(5\sin 2x+3\cos 2x)/68-\frac{1}{4}(2x^2+6x+13)$.

28. $y=A+B\cos 2x+C\sin 2x-\frac{1}{3}\cos x-\frac{1}{8}x^2$.

29. $y=Ae^{-2x}+B\cos 2x+C\sin 2x+\frac{1}{5}\sin x+\frac{1}{4}+\frac{3}{4}x$.

30. $y=A+Be^{2x}+Ce^{-2x}+E\cos x+F\sin x-(36x+\cos 2x+6x^4-54x^2)/48$.

31. (i) $y=e^x(Ax+B)+C\cos 2x+E\sin 2x+2e^{2x}-\sin x$;
(ii) $y=Ae^{-x}+B\cos x+C\sin x+\frac{1}{2}xe^{-x}+x^4-4x^3+24$.

32. $y=Ae^x+Be^{-x}+C\cos 2x+E\sin 2x+\frac{17}{2}xe^x+\frac{5}{8}e^x(5\cos 2x+3\sin 2x)$.

33. $x=\frac{1}{4}\{\cos t-e^{-t}+t(\sin t-\cos t)\}$. 34. $y=e^x\sin^2 x-x^2$.

35. (i) $y=Ae^x+Be^{-x}+C\cos x+E\sin x+\frac{1}{4}x\cosh x+\frac{1}{15}\sin 2x$;
(ii) $y=Ax+B+Ce^{-x}+(x^3-3x^2+3\cos x-3\sin x)/6$.

36. (i) $y=e^t(A\cos 2t+B\sin 2t)+(\cos 2t-4\sin 2t)/17$;
(ii) $y=Ae^t+B\cos 2t+C\sin 2t+te^t$.

37. (i) $y=Ae^x+Be^{-x}+C\cos 2x+E\sin 2x-\frac{3}{8}-\frac{1}{4}x^2-\sin x+e^{2x}$;
(ii) $x=a(3e^{-2t}-2e^{-4t})$; $x_{max}=9a/8$.

38. (i) $y=A+Be^{x}+Ce^{-x}+E\cos x+F\sin x-x+(6e^{2x}-\cos 2x)/30$;
(ii) $x=(\cos 3nt-9\cos nt)a/12$; max. speed $=an$.

39. $x=(kt/2p)e^{-ht}\sin pt$.

40. $x=e^{-t}(A\sin 2t+B\cos 2t)+te^{-t}\sin 2t$; the particular solution is $x=te^{-t}\sin 2t$.

Exercises 24 (*b*), p. 553

1. $y=x^2(A+B\log x)+x^2(\log x)^3+2/3x$;
$y=\frac{1}{3}x^2\{3(\log x)^3+9(\log x)-2+2/x^3\}$.

2. $y=Ax^2+Bx^3+x^3\log x$. **3.** $y=x\{2-\sin(\log x)-\cos(\log x)\}$.

4. (i) $y=e^{x}(Ax+B)+\frac{3}{8}(2x^2e^{x}-e^{-x})$;
(ii) $y=x^2\{A\sin(\log x)+B\cos(\log x)\}+(75x+10\log x+8)/50$.

5. (i) $y=\frac{1}{2}e^{2x}(A+Bx+x^2)$;
(ii) $y=\frac{1}{3}x^2+\sqrt{x}\{A\cos(\frac{1}{2}\sqrt{3}\log x)+B\sin(\frac{1}{2}\sqrt{3}\log x)\}$.

6. (i) $y=\frac{2}{3}(4e^{x}-49e^{-\frac{1}{2}x})-x^3+3x^2-18x+30$;
(ii) $y=(A\log x+B)/x+\frac{1}{9}x^2$.

7. (i) $y=Ae^{x}+Be^{2x/3}+\frac{1}{2}e^{x}(x^2-4x)$;
(ii) $y=Ax+B/x^5-\{10+9\cos(\log x)\}/90x^2$.

8. (i) $y=\frac{1}{2}e^{x}(A\cos x+B\sin x+x\sin x+2)$;
(ii) $y=x(A+x)+(1+Bx)\log x+2$.

9. (i) $y=Ae^{2x}+Be^{3x}+\frac{1}{8}e^{x}(4x^3+18x^2+42x+45)$;
(ii) $y=A/x+(B-\log x)/x^2+(x^2-12)/12$.

10. (i) $y=Ae^{x}+Be^{-x}+\frac{1}{2}x\sinh x$;
(ii) $y=Ax^3+B/x-x^2(3\log x+2)/9$.

11. (i) $y=e^{-\frac{1}{2}x}\{A\cos(\frac{1}{2}x\sqrt{3})+B\sin(\frac{1}{2}x\sqrt{3})\}-e^{x}(3\cos x-2\sin x)/13$;
(ii) $y=x^2(A\log x+B)+\frac{1}{2}x^2(\log x)^2$.

12. (i) $y=e^{x}(A\cos x+B\sin x+\frac{1}{2}x\sin x)$; (ii) $y=x^3+x(A\log x+B)$.

13. (i) $y=e^{-2x}(A\cos x+B\sin x)+4(5x-4)/25+3e^{-3x}+(\sin x-\cos x)/8$;
(ii) $y=x(1-\log x)+\frac{1}{3}x(\log x)^3$.

14. (i) $y=\frac{1}{6}e^{3x}(A+Bx+3x^2+x^3)$;
(ii) $y=Ax^4+B/x+(9-12\log x-4x^2)/24$.

15. (i) $y=e^{x}(A\cos 2x+B\sin 2x-\frac{1}{4}x\cos 2x)$;
(ii) $y=x(A+B\log x)+\frac{1}{4}x^3$.

16. (i) $y=e^{-3x}(\frac{1}{2}x^2+Ax+B)+(27x^4-72x^3+108x^2-69x+22)/243$;
(ii) $y=\frac{1}{2}-x(x-2)(1-\log x)$.

17. (i) $y=e^{-\frac{1}{2}x}\{A\cos(\frac{1}{2}x\sqrt{3})+B\sin(\frac{1}{2}x\sqrt{3})\}$
$+e^{x}\{13(x-1)+3(2\sin x-3\cos x)\}/39$;
(ii) $y=Ax+Bx^{-\frac{1}{2}}+3+\frac{1}{6}(x-5)\log x$.

18. (i) $y=A\sin x+B\cos x+\frac{1}{5}e^{x}(2\sin x+\cos x)+x^2-2$;
(ii) $2x^2y=Ax+B-(\log x)(2+\log x)$.

19. (i) $y=Ae^{2x}+e^{-x}(B+9\cos x-10x-3\sin x)/30$;
(ii) $y=A/x+B/x^3-\frac{4}{9}+\frac{1}{6}(2-3/x-6/x^2)\log x+\frac{1}{2}(\log x)^2/x$.

20. (i) $y=e^{2x}(A\cos 3x+B\sin 3x)-\frac{1}{6}xe^{2x}\cos 3x$;
(ii) $y=x^{\frac{1}{2}}(A+B\log x)+\frac{1}{4}x(\log x-3)$.

21. (i) $y=e^{2x}(A\cos x+B\sin x)-\frac{1}{2}xe^{2x}\cos x$;
(ii) $y=x^2(A+B\log x)+\frac{1}{6}x^2(\log x)^3$.

22. (i) $y=e^{2x}(Ax+B+3x^2+x^3)/12+e^{-2x}(1-2x)/64$;
(ii) $y=A/x+B/x^2+\frac{1}{6}x-(\log x)/x$.

23. $y=Ax+B\cos(\frac{1}{2}\sqrt{2}\log x)+C\sin(\frac{1}{2}\sqrt{2}\log x)+x\log x$.

24. $y=Ax+Bx^2+Cx^3+2x^4+x^2\log x$.

Exercises 24 (*c*), p. 557

1. $x=e^{2t}(t+1)$; $y=te^{2t}$. **2.** $x=e^{-at}\sinh t$; $y=e^{-at}\cosh t$.

3. The line is $3x+y=0$. **4.** $x=(e^{-4t}-e^t)/5$; $y=-(3e^t+2e^{-4t})/10$.

5. $x=\frac{1}{2}(e^{-3t}+e^{-t})$; $y=\frac{1}{2}(e^{-3t}-e^{-t})$.

6. $x=e^t(A+4t)+B$; $y=\frac{3}{2}e^t(4t+A-1)+2B$.

7. $x=Ae^{\frac{1}{2}t}+Be^{-\frac{1}{2}t}$; $y=\frac{1}{2}e^t-2Ae^{\frac{1}{2}t}-Be^{-\frac{1}{2}t}$.

8. $x=A\cos 2t-B\sin 2t-t\sin 2t$; $y=A\sin 2t+B\cos 2t+t\cos 2t$.

9. $x=(36+10e^t-45e^{-t}-e^{-5t})/60$; $y=(24+20e^t-45e^{-t}+e^{-5t})/60$.

10. $x=3\sin t-2\cos t+e^{-2t}$; $y=\frac{1}{2}(9\cos t-7\sin t-e^{-3t})$.

11. $x=e^{-t}(A\cos t+B\sin t)+e^t(15t-2)/25$;
$y=e^{-t}(B\cos t-A\sin t)+e^t(5t+11)/25$.
For the particular solution $A=2/25$, $B=-11/25$.

12. $x=\{(3t+1)-e^{-3t}(6t+1)\}/27$; $y=2\{(2-3t)-e^{-3t}(3t+2)\}/27$.

13. $x=Ae^{-5t}+Be^{-t}-\frac{1}{5}$; $y=t-3Ae^{-5t}+Be^{-t}-\frac{2}{5}$.

14. $x=\sin t$; $y=\cos t-\sin t-1$.

15. $x=\frac{1}{8}(7e^{3t/5}+9e^{-t}-8)$; $y=\frac{1}{8}(49e^{3t/5}-9e^{-t}-8t^2-24t-32)$.

16. $x=4Ae^{-2t}+3Be^{-3t}+(5\cos t+7\sin t)/10$;
$y=Ae^{-2t}+Be^{-3t}+(\sin t-\cos t)/10$.

17. $x=(16e^{2t}+e^{8t}-40t-17)/128-\frac{3}{7}e^t$;
$y=(16e^{2t}-e^{8t}-24t-15)/128-\frac{4}{7}e^t$.

19. $x=1+e^{-\pi/2}$; $y=-e^{-\pi/2}$.

20. $u=A\cos\omega t+B\sin\omega t$; $v=(E/H)+A\sin\omega t-B\cos\omega t$;
$x=(A/\omega)\sin\omega t-(B/\omega)\cos\omega t+C$;
$y=(E/H)t-(A/\omega)\cos\omega t-(B/\omega)\sin\omega t+C'$.

Exercises 24 (*d*), p. 560

For brevity z'', z' are written for d^2z/dx^2, dz/dx respectively.

1. $y=A\sin 2\sqrt{x}+B\cos 2\sqrt{x}+\frac{1}{4}(2x^2-6x+3)$.

2. $z''+3z'+2z=0$; $x^2y=e^{-2x}$. **3.** $z''+z=0$; $y\sec x=-\sqrt{(\pi/x)}$.

4. $z''+4z'+3z=0$; $y=e^{-x}/x^3$. **5.** $y=x-\sin^{-1}x$.

6. $(d^2y/dt^2)+y=2e^t$; $y=e^{\sin x}-\sin(\sin x)$.

7. $n=-2$; $x^2y=A+Be^{-4x}+(4\sin x-\cos x)/17$.

8. $y=Ae^{3\sqrt{x}}+Be^{-2\sqrt{x}}-\frac{1}{5}\sqrt{x}e^{-2\sqrt{x}}$.

9. The given equation is found from the relation $\dfrac{d^2x}{dy^2}=\dfrac{d}{dx}\left(1\Big/\dfrac{dy}{dx}\right)\dfrac{dx}{dy}$;

$(d^2x/dy^2)-3(dx/dy)+2x=y$; $x=Ae^y+Be^{2y}+\frac{1}{4}(2y+3)$.

10. $y=Ax+B\sqrt{(x^2-1)}+\frac{1}{2}(x^2-1)^{\frac{1}{2}}\log\{x+\sqrt{(x^2-1)}\}$.

11. $(d^2y/du^2)=(1+x^2)^2(d^2y/dx^2)+2x(1+x^2)(dy/dx)$;
$y=\frac{1}{6}(\tan^{-1}x)^3+A\tan^{-1}x+B$; $y=\frac{1}{6}(\tan^{-1}x)^3+\tan^{-1}x$.

12. $(d^2y/dt^2)+4y=8(1-t^2)$; $y=A\sin(2\sin x)+B\cos(2\sin x)+3-2\sin^2 x$.

13. $(d^2y/dt^2)+y=\cosh^2 t$; $y=A\sin(\sinh^{-1}x)+B\cos(\sinh^{-1}x)+\frac{1}{5}(3+x^2)$;
$y=\frac{1}{5}\{3+x^2-3\cos(\sinh^{-1}x)\}$.

Exercises 25, p. 574

11. 248·6 miles.

12. If ϕ is the total change of longitude,
$\cos\phi=\frac{1}{2}\{2\cos^2 l_2(\sin l_1\sin l_3+1)-\cos^2 l_1-\cos^2 l_3\}\sec l_1\operatorname{cosec}^2 l_2\sec l_3$.

13. $\operatorname{cosec} a\sqrt{(1-\cos^2 a-\cos^2 b-\cos^2 c+2\cos a\cos b\cos c)}$.

14. $NA=62°\ 26'$, $NB=29°\ 58'$, $NC=24°\ 2'$.

15. Longitude $=20°\ 19'$; 2205 miles.

16. (i) 8831 miles; (ii) 39° 14′ E. of S.; (iii) latitude 63° 26′.

17. 991000 sq. miles approx. **18.** $4{\cdot}080\times10^6$ sq. miles.

19. 3610 miles; 26° 34′ W. of N. at start; 39° 14′ W. of N. at end.

21. $c=90°$; $\Delta ABC=\pi R^2/6$. **22.** 19° 28′; $1{\cdot}44\times10^6$ sq. miles.

INDEX

Absolute convergence, 65, 134
Addition of complex numbers, 102, 107
Algebraic equation, 14
Alternating function, 6
series, 64
Amplitude of complex number, 104
Angle between tangent and radius vector, 478
between two lines, 269, 276, 349, 352
between two planes, 360
Approximation to root of an equation, 265
Arc, centroid of, 463
length, 452, 453, 455
in polar coordinates, 477
Archimedes, spiral of, 480
Area bounded by a curve, 441, 486
centroid of, 460
of lune, 562
of spherical triangle, 563
of surface of revolution, 454, 455, 488
of triangle, 269
within closed curve, 486
Argand diagram, 103
Argument of complex number, 104
Arithmetical progression, 53, 149
Arithmetico-geometrical series, 149
Astroid, 458, 494
Asymptote, 446
Asymptotes of hyperbola, 326, 329, 330
Auxiliary circle, 314
equation, 540
Axis, conjugate of hyperbola, 326
major, minor of ellipse, 313
radical of two circles, 288
transverse of hyperbola, 326
Axes, change of, 272
rotation of, 274
translation of, 273

Bernoulli's equation, 523
Binomial series, 67, 74, 152, 247
theorem, 75
Bisectors of the angles between two lines, 271, 277

Cardioid, 480, 493, 534
Central quadric, 389
Centre of curvature, 503
of gravity, mass, 460
radical, 289
Centroid of plane area, 460, 463, 488
of semi-circular area, 464
of solid of revolution, 462
of triangle, 269
Change of axes, 272
of origin, 273
of variable in differential equations, 524, 559
of variable in integration, 204, 216
Chord, focal, 301
of conic bisected at given point, 306, 318, 328
of contact of tangents to circle, 287
of contact of tangents to conic, 301, 314, 328
of polar conic, 340
Circle, auxiliary, 314
director, 316, 328
equation of, 285
of curvature, 503
of intersection of plane and sphere, 379
polar equation of, 338
Circles, coaxal, 289
orthogonal, 285
radical axis of, 288
radical centre of, 289
Circular and hyperbolic functions, connection between, 142
differentiation of, 175
generalised, 141
functions, exponential values of, 135
Clairaut's equation, 528
Closed curve, area within, 486
Coaxal circles, 289
Coefficients, undetermined, 3
Common root of two equations, 19
Comparison tests for series, 59
Complementary function, 542
Complete primitive, 513, 528
Complex number, argument of, 104
definition of, 103
fractional powers of, 120
geometrical representation of, 103
logarithms of, 142
modulus of, 104
vectorial representation of, 106
numbers, addition, subtraction, 102, 107
conjugate, 102
division, 102, 108
geometrical applications, 111
multiplication, 102, 108, 109
roots of equation, 126
terms, series of, 134
variable, circular functions of, 141
exponential functions of, 135
hyperbolic functions of, 141
logarithmic functions of, 142

Concavity and convexity, 192
Concyclic points on conic, 303, 332
Condition for plane to touch quadric, 392
for plane to touch sphere, 379
necessary for convergence, 58
Conditionally convergent series, 65
Conic, 300
polar equation of, 339
Conicoid, 389
Conjugate complex numbers, 102
diameters of conic, 317, 328
Conormal points on conic, 303, 331
Consistency of equations, 42
Contact of plane curves, 504
Continuous function, 72, 403
fundamental property of, 73
variable, function of, 70
Convergence, absolute, 65
tests for absolute, 65
comparison tests for, 59
conditional, 65
general theorems on, 58
interval of, 66
necessary condition for, 58
of alternating series, 64
of geometric progression, 56
of series of complex terms, 134
of series of positive terms, 59
radius of, 66
ratio test for, 62, 65
Convergent series, 55
Coordinates, relation between cartesian and polar, 337
Coplanar lines, 362
Cos $n\theta$ expressed in powers of cos θ and sin θ, 132
Cos $^n\theta$ expressed in multiple angles, 131
Cosine formula for spherical triangle, 564
Cotangent formula for spherical triangle, 565
Cramer's rule, 41
Curvature, 499
centre of, 503
circle of, 503
radius of, 500
Curve tracing (cartesian), 445
(polar), 479
Cyclic function, 6
Cycloid, 491

Damped harmonic motion, 548
Decreasing function, 191
Definite integral, 201, 442
De l'Hospital's rule, 259
Demoivre's theorem, 119
Derivative, definition of, 170
of standard functions, 174-176
partial, 403
second and higher order partial, 405
sign of first, 191
sign of second, 192
total, 414
Determinants, definition of, 26
expansion of, 27
of fourth and higher orders, 33
properties of, 28
solution of equations by, 40
Diameter of parabola, 303
Diametral plane, 397
Differences, method of, 154
Differential coefficient (*see* Derivative)
equation, Bernoulli's, 523
Clairaut's, 528
exact, 517
formation of, 512
homogeneous, first order, 514
homogeneous, linear, 552
linear, first order, 520
linear, with constant coefficients, 538
reducible to homogeneous form, 515
simultaneous, 555
solution of, by change of variable 524, 559
variables separable, 513
relations, 454, 477
total, 414
Differentials, 412
applications of, 416
Differentiation, general rules, 170
logarithmic, 177
of implicit functions, 182
of power series, 68
standard results, 174-176
successive, 179
Directed line, 347
Direction cosines, 347
ratios, 350
Director circle, 316, 328
Directrix, 300, 313, 326, 340
Discontinuous function, 72
Distance between two points, 269, 349
of point from line, 270, 364
of point from plane, 360
shortest between two lines, 370
Divergent series, 55
Double root of equation, 14

e, the number, 80
e^x, properties of, 82
Eccentric angle, 314
Eccentricity of conic, 300, 313, 326, 329, 339
Ellipse, 313
Ellipsoid, 389
Envelope, 425
Epicycloid, 492
Equation, approximation to root of, 265
complex roots of, 126
general of second degree, 278

homogeneous of second degree, 275
intrinsic, 456
multiple roots of, 14, 20
of tangent, normal to plane curve, 190
pedal (p, r), 478
reciprocal, 19
symmetrical functions of roots of, 15
transformation of, 16, 18
Equations, consistency of, 42
use of determinants in solution of, 40
Equiangular spiral, 480
Equiconjugate diameters, 318
Errors, small, 409
Euler's theorems on homogeneous functions, 424
Evaluation of limits, 95, 257
Evolute, 427, 504
Exact differential equations, 517
Expansion in power series, 244, 247, 252
Exponential limits, 87, 262
series, 66, 81, 135
theorem, 81
values of circular functions, 135

Factor, integrating, 519
theorem, 1
Factorisation of cyclic homogeneous polynomials, 6
Factors of polynomial, 1
First moment, 460
Focal chord of parabola, 301
distance, 313, 327
Focus of conic, 300, 313, 326, 339
Forced oscillations, 549
Formula, Wallis's, 228
Fractions, partial, 8
Function, alternating, 6
continuous, 72, 403
cyclic, 6
decreasing, 191
discontinuous, 72
even, 249
expansion of in power series, 244
homogeneous, 6
hyperbolic, 83
implicit, 182
increasing, 191
odd, 249
of a continuous variable, 70
rational, 7
single-valued, 70
symmetrical, 6
Fundamental property of continuous function, 73
theorem of algebra, 14

General equation of second degree, 278
solution of differential equation, 513, 542
Generalised circular, hyperbolic functions, 141
Generator, 398
Geographical mile, 570
Geometrical progression, 56, 149
representation of complex number 103
significance of derivative, 189
Gradient of curve, 189
of straight line, 269
Graphs, cartesian, 445
of hyperbolic functions, 85
polar, 479
Gravity, centre of, 460
Gregory's series, 248
Gyration, radius of, 468

Harmonic motion, damped, 548
simple, 548
series, 58
Homogeneous differential equation, first order, 514
differential equation, linear, 552
equation of the second degree, 275
equations, solution of, 41
function, 6
functions, Euler's theorem on, 424
Hyperbola, 326
conjugate, 329
rectangular, 329, 330
Hyperbolic functions, 83
generalised, 141
graphs of, 85
integration of, 215
inverse, 86
relations with circular functions, 142
Hyperboloid, 390, 397
Hypocycloid, 492

Identity, 1
of polynomials, 2
Imaginary part of complex number, 102
Implicit function, 182
Improper fraction, 7
integral, 221
Increasing function, 191
Indefinite integral, 200
Indeterminate forms, 72, 96, 257
Induction, method of, 151 181
Inequality, 21
Inertia, moment of, 468
Infinite integral, 221
series, 53
for a^x, 83
for e^x, 82
for $\cos x$, $\sin x$, 246
for $\cosh x$, $\sinh x$, 85
for $\log(1+x)$, 91, 247
for $(1+x)^n$, 76, 247

Infinity, 53
Inflexion, point of, 192
Integral, definite, 201, 442
as limit of sum, 441
improper, 221
indefinite, 200
infinite, 221
limits of, 201
Integrals standard, 201-203
Integrating factor, 519
Integration, 200
by parts, 218
by substitution, 204
by successive reduction, 227
miscellaneous substitutions, 216
of hyperbolic functions, 215
of irrational functions, 210
of power series, 68
of powers of sin x, cos x, 214
of powers of tan x, cot x, 215
of products of sines and cosines of multiple angles, 214
of rational functions, 208
of trigonometrical functions, 212
reduction formulae, 227
standard forms, 202
Intersection of line with quadric, 391
of plane with sphere, 379
Interval of convergence, 66
Intrinsic equation, 456
Inverse circular functions, differentiation of, 173, 176
hyperbolic functions, 86
differentiation of, 176
operators, 542

Latitude, 569
Latus rectum, 300, 313, 326, 339, 448
Leibniz's theorem, 180
use in expansions, 252
Lemniscate, 480
Length of arc (cartesian coordinates), 452, 453
(polar coordinates), 477
Limaçon, 480
Limit, idea of, 54
of function of continuous variable, 71
Limiting points, 290
Limits, evaluation of, 95, 257
special; lim x^n as $n \to \infty$, 56
lim nx^n, $|x| < 1$, as $n \to \infty$, 69
lim $(x^n - a^n)/(x-a)$ as $x \to a$, 171
lim $\{(x+h)^n - x^n\}/h$ as $h \to 0$, 171
lim $(\sin \theta)/\theta$ as $\theta \to 0$, 172
lim $(e^h - 1)/h$ as $h \to 0$, 87
lim $t^n \log t$ as $t \to 0+$, 260
lim $x^n e^{-x}$ as $x \to \infty$, 263
lim $(\log x)/x^n$ as $x \to \infty$, 260
lim $(1 + a/n)^n$ as $n \to \infty$, 262
theorems on, 57

Linear first order differential equation, 520
differential equation with constant coefficients, 537
homogeneous differential equation, 552
simultaneous differential equations, 555
Line of intersection of two planes, 356
Logarithm Napierian (natural), 82
Logarithms of a complex number, 142
Logarithmic differentiation, 177
functions, differentiation of, 174 177
series, 67, 90, 143, 245, 247
spiral, 480
Longitude, 569
Lune, area of, 562

Maclaurin's series, 245
Major axis, 313
Maxima, minima, tests for, 192
Maximum, minimum points, 191
Mean value, 444
theorem, 190
Measurements on earth's surface, 569
Meridian, 569
Method of differences, 154
of induction, 151, 181
Methods of summing series, 150
Mid-points of parallel chords of conic, 302, 317, 328
Minor axis, 313
Modulus of complex number, 104
Moment first, 460
of inertia, 468
about perpendicular axes, 469
circular lamina, 473
plane area, 472
rectangular lamina, 471
right rectangular prism, 472
solid of revolution, 473
solid sphere, 474
thin rod, 471
second, 468
Multiple roots of equation 14, 20
Multiplication of complex numbers, 102, 108, 109

n-leaved rose, 480
Napierian logarithm, 82
Napier's rules for solving right-angled spherical triangles, 568
Natural logarithm, 82
Nautical mile, 570
Newton's approximation to root of an equation, 265
Normal to plane curve, 190
to quadric, 391
to surface, 434
Numerical value, 21

Odd function, 249
Operator D, 537
  use in integration, 547
Order of differential equation, 512
Origin, change of, 273
Orthogonal circles, 285
  spheres, 380
  trajectories, 531
Osborn's rule, 84, 142

P, r equation, 478
Pair of straight lines, 275, 278, 281
Pappus, theorems of, 463
Parabola, 300, 339
Parallel axes theorem, 469
Parametric equations, area, 486
  ellipse, 314
  dy/dx, d^2y/dx^2, 182
  hyperbola, 327
  length of arc, 455
  parabola, 300
  radius of curvature, 500
  rectangular hyperbola, 330
  straight line, 271, 350
  surface area, 455
Partial differentiation, 403
  fractions, 8
Particular integral, 513, 542
  solution of differential equation, 513
Parts, integration by, 218
Pedal (p, r) equation, 478
  radius of curvature, 502
Pencils of spheres, 380
Percentage error, 409
Perpendicular axes theorem, 469
  distance of point from line, 270, 364
  of point from plane, 360
  of pole from tangent, 478
Plane bisecting parallel chords of quadric, 397
Plane, equation of, 354, 355, 357, 358, 359
Point of inflexion, 192
Points, limiting, 290
Polar coordinates, 337
  area, 486
  centroid of plane area, 488
  length of arc, 477
  radius of curvature, 501
  surface area, 488
  curves, well-known, 480
  equation of chord of conic, 340
    of circle, 338
    of conic, 339
    of straight line, 338
    of tangent to conic, 341
  form of complex number, 104
  lines with respect to quadric, 397
    with respect to sphere, 381
  of point with respect to circle, 287
    with respect to conic, 301, 314, 328
  plane with respect to quadric, 397
    with respect to sphere, 381
  subtangent, subnormal, 531
  trajectories, 532
  triangle, 566
Pole of line with respect to circle, 287
  with respect to conic, 301, 314, 328
Polynomial, 1, 3
Positive normal, 477
  tangent, 189, 455, 477
Power of point with respect to circle, 287
  with respect to sphere, 379
Power series, 66, 244
  for a^x, 83
  for e^x, 82
  for $\cos x$, $\sin x$, 246
  for $\cosh x$, $\sinh x$, 85
  for $\log (1+x)$, 91, 247
  for $(1+x)^n$, 76, 247
  properties of, 68
Primitive, complete, 513
Principal axes of ellipse, 313
  value, 86, 104, 173
Principle of parallel axes, 469
Progression arithmetical, 53, 149
  geometrical, 56, 149
Projection, 351
  of circle, 314
  of ellipse, 365, 399
Proper fraction, 7
Properties of determinants, 28
  of power series, 68
Proportional error, 409

Quadrantal triangle, 569
Quadratic factors, real, 126
Quadric surface, 388

Radical axis, 288
  centre, 289
  plane, 380
Radius of convergence, 66
  of curvature, 500
  of gyration, 468
  vector, 337
Ratio test, 62, 65
Rational function, 7
  number, 76
Real part of complex number, 102
  quadratic factors, 126
Reciprocal equation, 19
Rectangle, moment of inertia, 471
Rectangular hyperbola, 329
  referred to asymptotes as axes, 330
Reduced equation, 538
Reduction formula for
  $\int \sin^n x\, dx$, $\int \cos^n x\, dx$, 227
  $\int \sin^m x \cos^n x\, dx$, 227
  $\int \tan^n x\, dx$, 229
  $\int \sec^n x\, dx$, 230

Relative error, 409
Remainder theorem, 1
Revolution, surface area of solid of, 452, 454, 455, 488
  volume of solid of, 444
Right-angled spherical triangle, 567
Root common to two quadratic equations, 19
  of an equation, approximation to, 265
Roots and coefficients, relation between, 15
  complex, of an equation, 126
  multiple, 14, 20
  of complex number, 120
  repeated, 14, 20
Rose, n-leaved, 480
Rotation of axes, 274
Roulettes, 491
Rule, Cramer's, 41
  De l'Hospital's, 259
  Osborn's, 84, 142
Ruled surface, 397

Second moment, 468
Section formula, 269, 346
Separable variables, 513
Series, absolutely convergent, 65, 134
  alternating, 64
  arithmetico-geometrical, 149
  binomial, 67, 74, 152, 247
  conditionally convergent, 65
  convergent, 55
  differentiation of power, 68
  divergent, 55
  expansion of function in power, 244, 247, 252
  exponential, 66, 81, 135
  for a^x, 83
  for e^x, 82
  for $\cos x$, $\sin x$, 246
  for $\cosh x$, $\sinh x$, 85
  for $\log(1+x)$, 91, 247
  for $(1+x)^n$, 76, 247
  geometric, 56, 149
  harmonic, 58
  integration of power, 68
  logarithmic, 67, 90, 143, 245, 247
  Maclaurin's, 245
  of complex terms, 134
  of moduli, 134
  of positive and negative terms, 64
  of positive terms, 59
  operations with power, 68
  power, 66
  properties of power, 68
  reducible to exponential, binomial or logarithmic, 152
  related to logarithmic series, 92
  Taylor's, 246
  tests for convergence of, 59, 62, 64, 65
  trigonometrical, 156
Shortest distance between two skew lines, 370
Sides and angles of polar triangle, 566
Sign convention for s, 454, 455, 477
  of dy/dx, 191
  of d^2y/dx^2, 192
Simple harmonic motion, 548
Simultaneous equations, solution by determinants, 40
  linear differential equations, 555
Sine formula for spherical triangle, 565
Sin $n\theta$ expressed in powers of $\cos\theta$, $\sin\theta$, 132
Sin $^n\theta$ expressed in multiple angles, 131
Single-valued function, 70
Singular solution, 528
Skew function, 6
  lines, definition of, 370
    simplest form of equations of, 374
Solution of equation, Newton's method, 265
Sphere, equation of, 378
Spheres, coaxal, 380
  orthogonal, 380
  pencils of, 380
Spherical excess, 564
  triangle, 562
    area of, 563
Spiral equiangular, 480
  of Archimedes, 480
Standard forms (integration), 202-203
  results (differentiation), 174-176
Stationary points, 192
Straight line, equation of, 270, 271
    in space, length, direction and equations of, 349
Subnormal, cartesian, 530
  polar, 531
Substitution integration by, 204
Subtangent, cartesian, 530
  polar, 531
Summation of series by induction, 151
    by method of differences, 154
    using binomial, logarithmic and exponential series, 152
    using complex numbers, 156
    when rth term is polynomial in r, 150
Supplemental formulae, 567
Surface area, cartesian coordinates, 454
    Pappus' theorem for, 463
    polar coordinates, 488
Surface, normal to, 434
    to quadric, 391
  tangent line to, 434
    plane to, 434
    plane to quadric, 391
Symmetrical functions, 6
    of roots of equation, 15

Tangent cartesian, 190, 530
line to surface, 434
plane to quadric, 391
to sphere, 378
to surface, 434
polar, 477, 531
positive direction of, 189, 455, 477
to circle, 286
to conic, 301, 313, 315, 328, 331
to plane curve, 190
to polar conic, 341
to space curve, 433
with gradient m to circle, 286
with gradient m to conic, 302, 316, 328
Taylor's series, 246
applied to evaluation of limits, 257
Test comparison, 59
for maxima and minima, 192
ratio, 62, 65
Theorem, binomial, 75
Demoivre's, 119
exponential, 81
factor, 1
fundamental, of algebra, 14
Leibniz's, 180
Maclaurin's, 245
mean value, 190
of parallel axes, 469
of perpendicular axes, 469
Pappus', 463
remainder, 1
Taylor's, 246
Total derivative, 414
differential, 414
variation, 407
Trajectories orthogonal, 531
Translation of axes, 273
Transverse axis, 326
Trigonometrical series, summation by complex numbers, 156
summation by method of differences, 156
Trochoid, 492
Turning points, 192

Undetermined coefficients, 3
Unity, nth roots of, 121

Vandermonde's theorem, 74
Variables separable, 513
Variation, total, 407
Vectorial angle, 337
representation of complex number, 106
Volume, cartesian coordinates, 444
Pappus' theorem for, 463

Wallis's formulae, 228

Zeros of polynomial, 2